IMPLEMENTATION OF SAFETY AND HEALTH ON CONSTRUCTION SITES

PROCEEDINGS OF THE SECOND INTERNATIONAL CONFERENCE OF CIB WORKING COMMISSION W99/HONOLULU/HAWAII/24-27 MARCH 1999

Implementation of Safety and Health on Construction Sites

Edited by

Amarjit Singh
University of Hawaii at Manoa, Honolulu, Hawaii, USA
Jimmie Hinze & Richard J. Coble
University of Florida, Gainesville, Fla., USA

A.A. BALKEMA/ROTTERDAM/BROOKFIELD/1999

The texts of the various papers in this volume were set individually by typists under the supervision of each of the authors concerned.

Published by
A.A. Balkema, P.O. Box 1675, 3000 BR Rotterdam, Netherlands
Fax: +31.10.413.5947; E-mail: balkema@balkema.nl; Internet site: www.balkema.nl
A.A. Balkema Publishers, Old Post Road, Brookfield, VT 05036-9704, USA
Fax: 802.276.3837; E-mail: info@ashgate.com

ISBN 90 5809 036 1

Printed in the Netherlands

Implementation of Safety and Health on Construction Sites, Singh, Hinze & Coble (eds)
© 1999 Balkema, Rotterdam, ISBN 90 5809 036 1

Table of contents

11 *Ergonomics*

12 *Safety assurance*

13 *Law and regulation*

Implementation of Safety and Health on Construction Sites, Singh, Hinze & Coble (eds)
© 1999 Balkema, Rotterdam, ISBN 90 5809 036 1

Preface

Safety is no accident. *Immense* planning, *persistent* following of standards and specifications, and a *dedicated* cultural orientation are required to ensure excellent safety standings.

Injuries and accidents to workers help no community in any nation. Poor safety hurts not only the worker, but also the company that hired him, the worker's family, the community at large, and the country that may now have one less taxpayer.

The construction industry has the most dismal record of safety among all industrial segments. Perhaps, this is partially due to the rough and tough nature of their work. More likely, it is due to inadequate task planning, poor safety orientation and training, insufficient drug testing, lack of safety incentives, and insufficient incident investigation. All this translates to poor quality of construction safety management.

The aim of construction safety management is not only to have 'zero accidents', but also to doubly confirm safety assurance by having 'zero incidents'. A few construction companies have already achieved zero accidents, but numerous others have a long way to go.

The pressure on production and lower costs should *never* be allowed to endanger safety. Playing with health and lives is simply unethical, not to mention that it is always illegal. Workers and construction personnel have a right to not be harmed. *Everybody* has a right to happiness *and safety*.

The most important need of the hour is to increase our safety efforts across a broad front, and build up a new organizational culture for safety. Research has revealed that the improvement of safety benefits not only the safety record per se, but also time and schedule management, worker happiness, and the overall success of the project. Therefore, there is no reason for us not to focus strongly on the safe work ethic.

Civil engineers are traditionally not taught safety in their academic courses. As a result, they frequently grow up to think that safety is not their responsibility. They end up learning the hard way. Needless to say that a core ethic of all engineering societies is to 'uphold the safety and health of the public'. This quintessential message must be emphasized during undergraduate education and professional practice to reduce safety incidents on construction sites.

The CIB Working Commission 99 for Safety and Health on Construction Sites is dedicated to the improvement of construction safety worldwide. In the short time since its inception in 1995, it has done much to spread the important message through meetings, conferences, and published reports. The set of proceedings here follows the proceedings of the 1st CIB W99 International Conference for the Implementation of Safety and Health on Construction Sites, held in Lisbon, Portugal, from September 4-7, 1996, under the able direction of Luís Alves Dias.

We have a wide variety of topics for this 2nd set of proceedings. We have research and practice papers on medical science pertaining to construction diseases and illnesses. Construction managers must be aware of these medical matters in order to help their site personnel better. Papers are also included on legislation and regulations, owner involvement and insurance, safety during design, education and training, international practices, quality of safety, ergonomics, fatigue and human error, hazard management and indoor pollution, computerized automation to improve safety, work ability, subsurface safety engineering, fall protection, innovative safety technologies, accident statistics, risk analysis, safety culture, safety planning, costs and productivity, program implementation, standards and benchmarking, and auditing for safety. One session has also been devoted to structural safety in the design of engineered structures.

It is apparent that the emerging specialty area of construction safety has many sub-specialty areas, indicating that the war on injury, accident, and illness has to be fought across a wide front.

Our special acknowledgements go to the National Science Foundation for partially funding this conference. Their participation has elevated this conference to a prestigious, scientific level. We are also grateful to the other financial sponsors and endorsers of this conference.

The editors express their gratitude to all authors, members of the scientific and review committee, and members of the international organizing committee. It is their participation that makes this conference a success.

Dr Amarjit Singh, Principal Editor
Dr Jimmie Hinze, Editor
Dr Richard J.Coble, Editor

Implementation of Safety and Health on Construction Sites, Singh, Hinze & Coble (eds)
© 1999 Balkema, Rotterdam, ISBN 90 5809 036 1

Acknowledgements

This conference was held under the aegis of the International Council for Building Research Studies and Documentation (CIB), Working Commission W99. The organizing committee gratefully acknowledges the sponsorship and endorsement of the following organizations:

Sponsors:
- Carter & Burgess, Corporate Headquarters: Fort Worth, Texas, USA
- National Science Foundation;
- Hawaii Occupational Safety and Health Division;
- Hawaii Department of Transportation (Highways);
- University of Hawaii at Manoa;
- Construction Education Concepts.

Endorsers:
- National Institute of Occupational Safety and Health;
- International Commission of Occupational Safety and Health;
- South Africa Institute of Building;
- Bechtel Environmental, Safety, and Health Services (part of Bechtel Construction Operations, Inc.);
- The Association for the Advancement of Cost Engineering International, Hawaii Chapter.

Implementation of Safety and Health on Construction Sites, Singh, Hinze & Coble (eds)
© 1999 Balkema, Rotterdam, ISBN 90 5809 036 1

Organization

Amarjit Singh, *Conference Coordinator,* University of Hawaii at Manoa, Honolulu, USA
Jimmie Hinze, University of Florida, Fla., USA
Richard J.Coble, *CIB W99 Commission Coordinator,* University of Florida, Fla., USA
Chimay Anumba, University of Teeside, UK
Howard Bashford, Arizona State University, Ariz., USA
Michael Baxley, M.D., Florida, Fla., USA
Hazel Beh, University of Hawaii at Manoa, Honolulu, USA
Luís Alves Dias, Instituto Superior Tecnico, Portugal
Roy Duff, University of Manchester, UK
Ian Eilenberg, RMIT University, Vic., Australia
Daniel Halpin, Purdue University, Ind., USA
Awad Hanna, University of Wisconsin-Madison, Wis., USA
Tom Iseley, Iseley Enterprises, Ind., USA
Edward Jazelskis, Iowa State University, Iowa, USA
Charles Kelley, M.D., Straub Clinic, Hawaii, USA
Enno 'Ed' Koehn, Lamar University, Tex., USA
Antti Lakka, VTT Building Technology, Finland
James Langworthy, M.D., Hawaii, USA
Richard Larew, Ohio State University, Ohio, USA
Jeffrey Lew, Purdue University, Ind., USA
Issam Minkareh, University of Cincinnati, Ohio, USA
Boyd Paulson, Stanford University, Calif., USA
Pantaleo Rwelamila, University of Cape Town, South Africa
Steve Rowlinson, University of Hong Kong, People's Republic of China
John Smallwood, University of Port Elizabeth, South Africa
Gary Smith, Iowa State University, Iowa, USA
Marie Sweeney, NIOSH, Ohio, USA
Ali Touran, Northeastern University, Mass., USA

1 Safety management and planning

Implementation of Safety and Health on Construction Sites, Singh, Hinze & Coble (eds)
© 1999 Balkema, Rotterdam, ISBN 90 5809 036 1

Development and implementation of an occupational safety and health management system on construction sites – Experiences from twelve small and medium-sized contractors

U. Helledi
Department of Productivity, Danish Technological Institute, Building Technology, Denmark

ABSTRACT: Twelve Danish contractors have recognised the necessity of improving occupational safety and health management on the construction sites, particularly to reduce the number of occupational accidents. In the period covering 1995-1997 the contractors participated in a network collaboration on a joint project to develop an occupational safety and health management system for small and medium-sized contractors.

The project resulted in the development of a simple, non-bureaucratic management system for each contractor, a system which integrates quality, environment and occupational safety and health as well as guiding their employees' attitude to a planned occupational safety and health management system in a positive direction.

A review of the occupational accidents in the contractors in mid-1998 showed that the serious accidents which occurred before have been completely avoided in the period 1995-1997.

A statistical examination of the occupational accidents occurred have created a good basis for planning a targeted, coordinated and preventive effort in the years to come.

1 INTRODUCTION

It is generally recognised that safety on Danish construction sites is not satisfactory as the level of occupational accidents is high when compared to other industries. The building and construction sector is among the most exposed sectors when it comes to occupational accidents, including serious accidents which are fatal and disabling.

In comparison with the other industrial production sectors the Danish building and construction sector is characterised by a very fragmented structure in the production phase with a large number of independent companies. This type of organisation often makes management assignments in the building sector difficult and complex. It is not unusual that 10 to 20 sub-contractors are involved in a single construction project, often more than one at a time, with varying influence on their own and other sub-contractors' working conditions. A division of decisions and authority between different, legally separate companies have often made placing responsibility diffuse and difficult.

The client, the consulting engineers, architects, general contractor and a number of sub-contractors carry out the building jointly. The sub-contractor who participates in the building is most often the one among 5 to 10 sub-contractors who has submitted a tender with the lowest price to the client. Previous efforts initiated by the public authorities through legislation, education etc. have not provided the results desired.

2 THE PROJECT

This is a description of the participants in the project, the process, the results and effects of the project in addition to the experiences from the process.

2.1 *Project Participants*

The twelve Danish sub-contractors have from 12 to 90 employees and carry out decorative work, masonry, plumbing, concrete work, scaffolding and soil and foundation works.

In the period covering 1995-1997 the contractors participated in a network collaboration on a joint project to develop an occupational safety and health management system for small and medium-sized sub-contractors.

The work done by the contractors was primarily carried out by the statutory safety committee of each contractor. This committee consists of the top manager, selected building site managers and representatives elected by the employees.

The Federation of Co-operative Building Contracting Societies, FCBCS, and the Danish Technological Institute, DTI, functioning as management consultants for FCBCS and the contractors, participated in the project.

2.2 *The Process*

At the first network meeting where the contractors, FCBCS and DTI participated a number of alternative levels of ambition and processes of a development project were presented and discussed. It was agreed to use the network model for mutual inspiration, both to achieve a thorough understanding of the requirements for occupational safety and health management and to develop a simple, yet effective management system. This solution also provided the financial advantage of being able to use the same external management consultant.

It was generally agreed that the system to be developed had to be a simple and non-bureaucratic management system. The external environment was integrated in the project due to the close relation to occupational safety and health, eg. through handling contaminated soil, chemical waste, dust and noise.

As the contractors had already implemented a simple and successful quality management system, they did not wish to have several, parallel systems, but only one integrated management system. The outcome of the project was for each contractor an integrated system for the management of quality, environment and occupational safety and health.

Furthermore, there was general agreement on the fact that the time schedule had to be flexible so that the contractors did not have to deliver results in periods with many business activities, but that they had to present results in less busy periods. It was decided that the project would run for 1½ year and would also contain implementation tasks.

The network collaboration consisted of a number of joint meetings with discussions of selected topics. These were: How does the contractor maintain an overview of the comprehensive legislation, concrete objectives and targets for occupational safety and health management, risk identification, eg. in connection with the erection of scaffolding and the use of ladders, the content of a pro-active management system, experiences encountered in the dialogue with clients, architects, etc. and employees on the construction site, the content of check lists etc. In addition, cases were examined and discussed to obtain a thorough understanding of the problems and possible solutions.

Between the meetings each contractor developed their own management system. Parts of the system were implemented when each contractor was ready. Particularly, the check lists and risk identification were quickly accepted by the management and the employees.

In most of the contractors the work with risk identification, the description of the management system, etc. was done by the safety committee with the participation of the employee representatives and the managing director. Employee influence was then secured through the safety committee.

Each contractor appointed one person who was responsible for the work being done and the managing director approved the results.

During the work the contractors maintained close contact to the management consultant from DTI by telephone and telefax for comments on the results and discussion of new ideas.

In the course of the project two visits were made to the contractor by the management consultant to discuss the results achieved and the impending work as well as matters of internal interest to the contractors. The fact that the employees participated in both the design, the control of the systems and in the management work itself was given high priority even though occupational safety and health management systems are management systems. This ensured that the formalities and documentation was kept on a reasonable and practical level.

FCBCS arranged the network meetings and established a "legal service" for the members as it turned out to be a considerable task for the contractors to obtain an overview of the extensive legislation in the occupational safety and health area. This service consists of an overview of the legislation and is updated continuously.

2.3 *The Results*

Each contractor has a management system which is formalised and documented in a manual. The principle of the manual is illustrated in Figure 1. The system describes the organisation of the current construction site and a number of activities which the employees must follow to achieve the objectives and targets of the contractor. The activities can be divided into four groups: Planning, implementation, checking and corrective actions.

In the planning phase the project material is reviewed and a risk identification is made. The contractors maintain a dialogue with, among others, the consulting engineer and architect to avoid or reduce significant risks. Necessary points of instruction are identified, check points are planned and registered in a check schedule. The emergency plan accounts for the activities to be initiated by the employees if an occupational accident occurs.

The implementation phase consists of an instruction of the employees on the construction site concerning the course of critical activities identified during the planning phase.

The control phase starts just after a critical activity is initiated. Normally, the control activity is most intensive just after a critical activity is initiated. When the site manager is sure that the employees on the site have started their work, eg. the use of mobile lifts, with the safety precautions necessary, the random sampling is reduced.

Following up on these phases is a natural and integrated part of the daily management on the construction sites. An unsatisfactory activity can lead to a new instruction of the employees, a change of material, a change of work process, a new dialogue with eg. the architect or other sub-

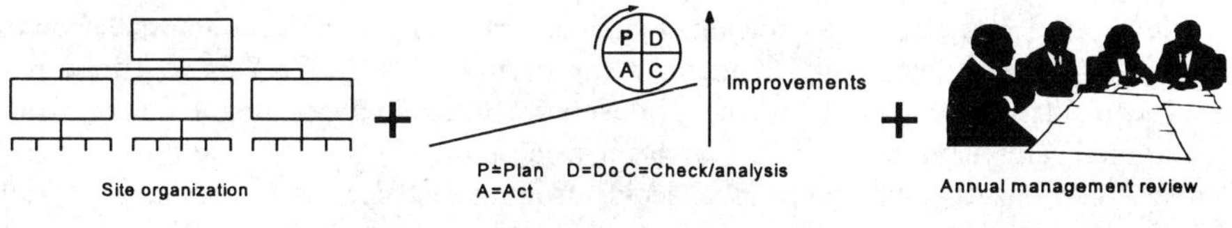

Figure 1. The principle of the management system which includes the Deming-circle

contractors on the site. At the same time the control function is continuously adjusted to new conditions to maintain a satisfactory result.

In addition, the management of the contractor carries out a review once a year to ensure satisfactory results. A management review includes a review of any accidents which have occurred, the wishes and ideas of the employees and client reactions to the management and results of the contractor.

2.4 *The Effect of the Project*

To identify the effect of the management system introduced a follow-up review was made which included an assessment of the number of occupational accidents, types of accident and injuries three years after the project started.

DTI visited the twelve contractors in mid-1998. All contractors had a system to register occupational accidents and documentation existed for a number of years. In order to compare the accidents of the contractors with the national statistics, only registered accidents which entail one absent day from work in addition to the day of the accident are included.

Table 1. The total amount of reported occupational accidents in the 12 contractors, and the frequency.

	1994	1995	1996	1997
Occupational accidents	7	9	14	12
Number of working hours	732,321	746,684	804,646	768,994
Occupational accidents/1000 empl.	20	27	29	26

From Table 1 it appears that the increase in the number of reported occupational accidents from 1994 to 1996 does not reflect the fact that occupational safety and health has deteriorated. This is due to the fact that both management and employees have become more careful in reporting the accidents. Studies have shown (Arbejdstilsynet, 1996a) that the number of registered accidents is 27.1/1000 employees in the building and construction sector and that the number is 18.4/1000 employees for all sectors. The corresponding figure for decorative contractors is 9.7. This must be compared with the fact that studies have shown (Arbejdstilsynet, 1996b) that there is a general underreporting in the construction sector of 50%. The reduction of the number of occupational accidents from 1996 to 1997 is positive but does not currently provide a statistically unambiguous basis on which a positive conclusion may be drawn. Of the twelve contractors studied nine are decorative contractors. The occupational accidents of these contractors have been examined more closely:

Table 2. Degree of seriousness of the occupational accidents. Reported occupational accidents for all Danish decorating contractors for the period 1993-1995 (three years),(Arbejdstilsynet, 1996a), and for the 9 decorating contractors for the period 1995-1997 (three years).

Type of injury	Fatal accidents	Very serious accidents	Other accidents
Decorating contractors (1993-1995)	<1%	20%	80%
9 decorating contractors (1995-1997)	0%	0%	100%

From Table 2 it is clear that serious accidents which occurred in several contractors before the project started have been completely avoided in the period 1995-1997. This registration of occupational accidents supports the statements of the contractors which claimed that their efforts in the occupational safety and health field have been successful.

From Table 3 it appears that the occupational accidents in the nine decorative contractors mainly belong to two categories: Sprains, twists, strains (55%) and open wounds (34%). The main point of difference between the accidents of the contractors and those reported in the national statistics is that the contractor does not have any occurrence of fractures.

Table 3. Types of reported occupational accidents for the 9 decorating contractors and for Denmark's total number of decorating contractors (Arbejdstilsynet, 1996a).

Type of injury	9 decorating contractors (1995-1997) Number	%	All Danish decorating contractors (1993-1995) %
Death	0	0	<1
Amputation	0	0	0
Fractures	0	0	18
Sprains, twists, strain	16	55	37
Open wounds	10	34	19
Burns, scolds, frost bites	1	3	1
Contusions	2	7	11
Cauterization	0	0	2
Poisoning	0	0	<1
Not stated and misc.	0	0	12
Total	29	~ 100	~ 100

From Table 4 it appears that the types of injury registered in the nine decorative contractors were caused by a number of different types of accidents. The distribution of the accidents is more or less equivalent of the distribution in the national statistics. However, the occurrence of the type of accident 'wrong or straining movement' is relatively higher than the occurrence in the national statistics. The type 'slipping, falling items, etc.' has not occurred in the contractors in the three-year period in question.

Table 4. Cause of reported occupational accidents for the 9 decorating contractors and for Denmark's total of registered decorating contractors (Arbejdstilsynet, 1996a).

Injury cause	9 decorating contractors (1995-1997) Number	%	All Danish decorating contractors (1993-1995) %
Fall to lower level	6	21	21
Fall or slip on same level	5	17	18
Wrong or straining movement	8	28	12
Slipping, falling items, etc.	0	0	11
Hand tools	5	17	7
Single elements, etc.	4	14	5
Vehicles, equipment, etc.	0	0	5
Machinery, apparatus	1	3	2
Traffic	0	0	1
Leaks and evaporation	0	0	1
Electric shock	0	0	1
Explosion, fire	0	0	<1
Violence	0	0	<1
Not stated and misc.	0	0	16
Total	29	~100	~100

These data provide a good basis for a prioritised effort aimed at future construction sites, eg. how wrong or straining movements are avoided so that the risk of sprains, twists and strains are reduced.

The efforts to prevent accidents caused by slipping or falling items which often lead to serious accidents must naturally be maintained.

2.5 *Experiences*

The experiences from the process have been that the network collaboration has contributed significantly to the understanding by the contractors of occupational safety and health management and of maintaining the commitment to creating an occupational safety and health management system. Agreement on the project aim of a simple and non-bureaucratic management system, the flexible time schedule and the close contact to the management consultant during the process have ensured that all contractors completed the project. Employee influence through the safety committee which has functioned as the working group of the contractor has contributed significantly to the employees accepting the occupational safety and health management system.

The "legislation service" of FCBCS was a crucial factor in providing the contractors with an overview of the extensive legislation in the occupational safety and health field.

Experiences in implementing the system have shown that the employees follow the guidelines of the management system and use the check lists, carry out informal risk identification and instruct employees on the site, but that they do not necessarily document the planning and control of the sites. The large and complex construction tasks are documented in the form of control schedules and emergency plans, etc. However, a serious piece of work still needs to be done in the twelve contractors before the medium-sized construction tasks are documented. Typically, this has not been done due to work pressure among the building site management of the contractor, the attitude that it is most important to ensure safe and healthy conditions and less important to document this in addition to the lack of external requirements to documentation. The motivation to document planning, control and any corrections to be made has primarily been present in the case of large complex construction tasks. These tasks also benefit the most financially when the ad hoc planning is minimised. Experience showed that the employees who had prepared the documentation found the work relatively simple and fast to carry out. The contractor managers have indicated that they are working towards documentation on an increasing number of construction tasks as it is an excellent tool to carry out construction tasks in a safe, secure and profitable manner.

The experiences of the project have also shown satisfaction with an integrated system including quality, environment and occupational safety and health as it provides an overview of activities and fields of responsibility etc.

The influence of the project on the attitudes of the employees has also led to the registration of occupational accidents being taken more seriously by management and employees.
Improved planning and quick follow-up actions contribute to creating improved communication processes and cooperation between the different organisations on each construction site. The project has emphasized the fact that if safety and health on the site are to be significantly improved, then it requires all the parties involved to make a serious commitment in the field of occupational safety and health. This requires that the clients set requirements for themselves, the consultant engineers, architects and all the contractors and the manufacturers.

3 PROJECT FOLLOW-UP

The study of the occupational accidents, the implementation of the management system and the level of contractor satisfaction with their management system have made it possible to carry out a prioritised follow-up on the project in 1998-1999.

The follow-up will be carried out in each contractor in cooperation with FCBCS and DTI. To support the documentation process a number of good examples will be reviewed with relevant employees. All employees of each contractor will participate in a course on the prevention of falls, proper use of tools and how to prevent injuries due to wrong or straining movements. For the last part mentioned special experts will be used as instructors.

Finally, one or more meetings in the network will ensure experience exchange, attitude formation and inspiration for further development.

The accident statistics will be monitored and coordinated by FCBCS in future years to follow developments and carry out further prioritisation of efforts in this field.

4 CASE STORY

One of the network contractors has carried out a contract on decorating renovation of wooden window frames etc. in a very large building complex. The contract consisted of the first of five stages.

The contract was carried out with the full use of the integrated management system for quality, environment and occupational safety and health.

Before work started, thorough preparations were made. A comprehensive risk identification was made including both quality, environment and occupational safety and health. After this identification was made, top priority was given to the use of a mobile lift, cf. Figure 2. The contractor's plans for using the lift was presented to the client's architect who accepted the solution. Then the control schedule and emergency plan for safety on the site were finished.

In connection with the start of the work the contractor arranged for a safety expert from the enterprise which hired the lift to visit the site. The expert gave on-site instructions on how to use the lift in a safe and efficient manner. Ten of the employees on the site, their safety representative and two building site managers participated in the instruction.

All of these found it easy to check later whether work was progressing in a safe, secure and efficient manner. Small corrections were made as the employees became more familiar with the lift. This led to a positive development in productivity on the site and the contract was carried out efficiently and without any accidents.

This was a concrete effort which contributed to preventing accidents. However, the signals of awareness sent by management in doing this was of equal importance. The evident commitment and focus on safety and health affected employee attitudes and therefore the daily work on the site.

Experiences showed that the formalised management system for quality, environment and occupational safety and health functioned very satisfactorily and that the documentation of the management system being adhered to could be handled even at very busy times. Furthermore, the check lists contributed to making the planning phase considerably more systematic and efficient. Ad hoc planning was reduced significantly and there was time to find profitable solutions.

The client was so satisfied with the work carried out that the decorative contractor was offered to carry out the remaining four stages without a new round of tenders. The decorating contractor will finish the last of the work at the end of 1998.

5 SUMMARY AND CONCLUSIONS

The experiences from the network collaboration of the twelve contractors have been that the work process with joint meetings with topics of mutual interest coordinated with the work carried out in each of the contractors have contributed in a very positive manner to the understanding of occupational safety and health management in the contractors. At the same time the network has had a positive influence on maintaining the contractors' commitment to creating an integrated management system for quality, environment and occupational safety and health on construction sites.

In the small and medium-sized contractors work has been carried out by the safety committee with the participation of the top management and representatives elected by the employees. Employee influence has contributed to support the creation of a simple, non-bureaucratic management system and the implementation of the system on the construction sites.

Presently, mid-1998, the implementation appears successful with regard to carrying out the activities agreed on - planning, implementation, control and corrective actions. The implementation of the documentation of the activities carried out appears on the large and complex construction tasks but is currently lacking on medium-sized contracts.

The effect of the project has been the complete absence of the serious accidents during the period 1995-1997 on the sites which occurred in earlier years. The registration and assessment of the occupational accidents have supported the contractors' clear indications that their efforts in the occupational safety and health field have been successful. The registration and assessment of the effect on the occupational safety and health have also provided a good basis for a prioritised effort in each of the contractors in the future preventive work to ensure safety and health on construction sites.

Experience has shown that the integrated management system contributes to improving cooperation on the construction sites through a pro-active dialogue with the other players. Thus, the project has emphasized that if safety and health on the construction sites are to be seriously improved, it requires all the parties involved to make a serious commitment to in the field of occupational safety and health. This requires that the clients set requirements for themselves, their consultant engineers, architects, contractors and manufacturers.

6 REFERENCES

Arbejdstilsynet (The Danish Working Environment Service, The Secretariat for Analysis and Documentation), 1996a. *Reported occupational injuries in the construction sector in the period 1993-1995.* Copenhagen.

Arbejdstilsynet (The Danish Working Environment Service, The Secretariat for Analysis and Documentation), 1996b. *Under report on occupational injuries 1990-1995.*

Implementation of Safety and Health on Construction Sites, Singh, Hinze & Coble (eds)

Partnering: Incorporating safety management

S.M.Rowlinson & J.Matthews
Department of Real Estate and Construction, The University of Hong Kong, People's Republic of China

ABSTRACT: Implementing the partnering concept in the construction project environment provides an opportunity for the continuous improvement of safety performance. This paper addresses partnering as a concept and draws from examples of partnering in the United Kingdom between a main contractor and its sub-contractors. A number of characteristics of partnering agreements have been identified which can all assist in promoting safety. These characteristics are: continuous evaluation; the project charter; mutual objectives and team building. The context in which partnering in safety can be undertaken is reviewed and a discussion takes place of how the global trend to move away from prescriptive legislation towards performance based legislation in the regulation of safety provides an ideal opportunity to adopt partnering as a methodology for safety improvement.

KEYWORDS

Partnering, site safety, safety management systems, project charter, accident reduction.

1 INTRODUCTION

In the United Kingdom the study undertaken by Sir Michael Latham entitled '*Constructing the Team*' gave the UK industry the focus that it needed in order to improve the relationships between all parties involved in the procurement of buildings. Partnering was seen by many as the key to achieving this improvement.

Within the United Kingdom partnering has become commonplace with many major contractors. Walter (1998) reported that many firms were expecting partnering contracts, either formal or informal, to account for over two thirds of their workloads in the financial year 1998 to 1999. Such approaches have been commonplace in Japan for many years (see, for example, Walker *et al*, 1998)

This paper addresses the issue of adapting partnering arrangements in order to promote construction site safety. Such an approach will go a long way towards restructuring the current iniquitous subcontractor/main contractor relationship that is in part responsible for poor quality and low safety levels on construction sites. By extending the partnering arrangement there is opportunity to not only improve quality and productivity but also safety. Poor safety performance is an issue world-wide and although this paper looks at partnering and safety from a Hong Kong and United Kingdom perspective the methodology devised is generally applicable world-wide.

2 DEFINING PARTNERING

There are no fixed definitions used when defining partnering (Matthews, 1996). Crowley & Karim (1995) and Matthews *et al* (1996) state that partnering is typically defined in one of two ways. Firstly, by its attributes such as trust, shared vision, and long term commitment or secondly by the its process where partnering is seen as a verb and includes developing mission statements, agreeing goals and conducting partnering workshops.

The most commonly cited definition for partnering is that advocated by the Construction Industry Institute (CII, 1991) in their report '*In Search of Partnering Excellence.*' This definition of partnering was formulated after studying 27 partnering case studies in the USA. CII wrote that:

> *".... a long-term commitment between two or more organisations for the purpose of achieving specific business objectives by maximising the effectiveness of each participant's resources. This requires changing traditional relationships to a shared culture without regard to organisational boundaries. The relationship is based on trust, dedication to common goals, and an understanding of each other's individual expectations and values."*

It is this definition which has been used in developing a methodology for introducing partnering into site safety management as it emphasises maximisation of effectiveness or shared resources, a shared culture and trust and common goals. It is these issues which are at the crux of the site safety problem. Subcontractors often find that they are unable to get ready access to safety resources on site, leading to dangerous working practices and accidents. Development of a culture of safety has been a key objective of the Hong Kong government for its construction sites and partnering has the potential to bring about these synergies in a harmonious manner.

3 SAFETY ISSUES

Partnering has the potential to address a number of current issues in the safety process. Some of these issues are raised below and two solutions, brought about by industry level partnering [Third Level Partnering RCF], are described below.

In Hong Kong the Trades Unions are quite weak and so unable to act as gatekeepers in terms of safety performance and legislation. Hence, managers can be seen to favour a performance-based approach as this frees them from regulations and allows them to manage. However, workers tend to favour the use of more prescriptive approaches as a means of self-protection. Thus, an opportunity is provided to ameliorate change these dichotomous views through a partnering approach.

As a further impetus to change, research in the UK has indicated that, since the passing of the Health & Safety at Work Act in 1974, the construction industry has not responded well to the performance-based approach. Dawson *et al* (1988) found that, of 30 firms surveyed, only the large, well-known firms had adopted extensive health & safety management systems and established safety committees and other forums for consultation between employers & employees on health & safety matters. British construction firms of a medium and small size had failed to implement these measures and had thus not embraced the self-regulation approach. Given the fact that the overwhelming majority of Hong Kong construction firms are small to medium in size and they are currently poor safety performers these problems apply at least equally to the Hong Kong industry. Holmes (1996) suggests that in the UK small businesses cannot understand and comply with legal obligations relating to occupational health and safety, the same is true in Hong Kong. Hence, partnering can assist in raising safety standards industry wide by a process of technology transfer from larger to smaller companies, in terms of both hard and soft technologies.

3.1 Partnering in Safety

Safety can be compromised through many aspects of the subcontracting system. For instance, Rowlinson & Lingard (1998) reported:

> "Another factor which may affect the goal setting/expectancy mechanism described above is the reward structure within which site operatives are working. It is common for Hong Kong construction tradesmen to be paid on a piece-rate basis. Where this is not the case, a main contractor may agree a fixed price that is to be paid to a sub-contractor on completion of a given task. In some instances bonuses may be paid for work completed before schedule. In any of these scenarios, it will be rewarding for work to be carried out as quickly as possible. If operatives perceive the value associated with the outcome of completing work quickly to be higher than the value associated with the outcome of attaining a safety performance goal then, under the model, it is likely that goal rejection will occur."

If such issues are raised and addressed in the charter then a significant improvement can be made in safety performance by relieving this pressure through a partnering agreement.

Another example, at the industry rather than company level, is the Hong Kong Government Works Bureau Pay for Safety Scheme. In 1996 the Hong Kong Government conducted discussions with the Hong Kong Construction Association which eventually resulted in the implementation of a pay for safety scheme for Hong Kong Government Works Branch projects. The aim of this scheme was to remove site safety from the pressure of competitive bidding and to attempt to reward those contractors that comply with safety requirements whilst punishing those which do not. Principal Assistant Secretary, Berry stated at an October 1996 meeting of the Lighthouse Club:

> "Hong Kong Construction Association gave its support to the removal of site safety from competitive tendering by means of paying for safety properly carried out, based on an agreed schedule of items and prices. Their view was underlined by a report they had commissioned from Hong Kong University which had noted -
>
> 'research has shown that where safety costs are included in a tender and are accepted by the client the frequency rate of time-lost accidents for major construction projects is decreased considerably'."

This was to be done by including a fixed sum in the bill of quantities to be priced for safety related elements such as personal protective equipment, temporary works, site meetings and safety committees. These items would then be paid for in interim valuations when the project surveyor had checked these items off and certified them. Thus, the aim was to set a fixed sum aside in every contract payment for the provision of a minimum level of site safety infrastructure on each Works Branch project. This scheme was piloted successfully in 1994 and, as a means of regulating the scheme, the Occupational Safety and Health Council introduced an Independent Safety Audit Scheme. Although this is an innovative idea it does not meet wholly with the philosophy of self-regulation and there is a possibility that some contractors will view the provisions required under the pay for safety scheme as the sum total of their commitment to safety.

4 PARTNERING APPROACH TO SAFETY MANAGEMENT

The approach to partnering and safety management discussed below involves all stakeholders of a partnering relationship. In this approach to partnering the main emphasis is between the main contractor and subcontractors. However, in order for this approach to achieve maximum benefit it does necessitate involvement of many other parties outside of this relationship. The advantage of this is not only project wide commitment to partnering and safety, but also an increase in the level of understanding between all project participants.

The approach is divided into three main phases as shown in Fig. 1. below. These are:

Procurement Establishment:
  Project Team
  Package and company identification.

Pre Selection and Notification:
  1st subcontractor interview; and
  2nd subcontractor interview.

Selection and appointment of subcontractor:
  3rd Interview tender clarification;
  Subcontractor selection; and
  Project day.

A 'Project Team' is established for the duration of the project. Although it is understood that the personnel on the team would change during the life of the project, it is proposed that the members of the team during the partnering stages are: the main contractors commercial representation (estimator); design representation (architect and project design manager); engineering representation (structural and services); site management representation (project and contracts manager); and safety management representation (contractor's safety officer and/or site safety officers). The client representation should include where possible: the client; the principal quantity surveyor; architect; and engineering representation (structural and services). The project team should meet regularly in order that project progress and problems can be discussed. Historically, safety representation and site safety personnel have been overlooked when assembling a project team. Including safety personnel at an early stage of establishing a project team is a simple and potentially rewarding task which has been addressed within this paper.

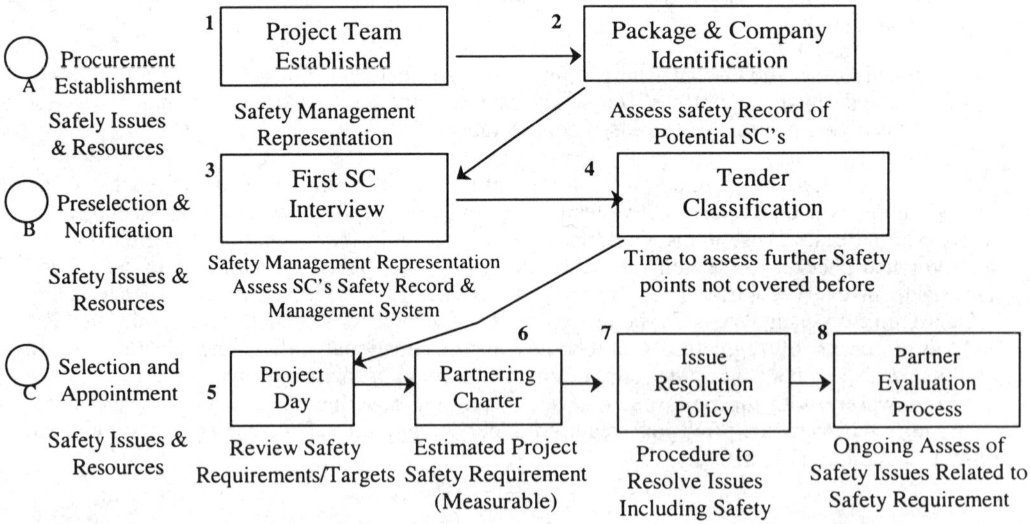

Fig. 1 : Partnering Approach to Safety Management

4.1 Stage 1: Package and Company Identification:
The objective of this stage is to identify all major packages that will accrue benefit from using a partnering approach. Packages are examined under the following headings:

design content, buildability and safety of construction methods and temporary works; complexity of construction;
high subcontract value;
long periods of construction;
early commencement of construction;
high level of aesthetics;
hazard analysis and [safety management systems] Steve, would a 'package' be assessed for a man' sys'?] don't understand this question
long procurement times; and
those trades that could 'add value' with their early input.

Having identified all packages, names of subcontractors are suggested by all project team members during a team meeting. Each team member, including the main contractors site staff, should have an opportunity to comment on the names. The team assesses each company in order to shorten the list. SC's should be assessed primarily on their past performance. Other criteria to be employed include:

ability to undertake this quantity of work;
ability to produce the required standard of work;
ability to undertake the work (not wanting to over 'stretch' the subcontractors);
positive attitude (past experiences);
firm financial background;
where applicable, good in-house design service;
good standards of management (site and head office);
would the main contractor (or client) wish to develop a long term relationship with the subcontractor;
adequacy of safety management system (manuals and procedures);
safety audits;
recent accident rates.

On conclusion of the meeting no more than 5 names should remain for each package. A member of the main contractors project team, probably the estimator, will then contacts each of the subcontractors to invite them to an interview. The team member informs the subcontractors of the type of project, its location and programme time etc. The invitation is followed up by a confirmation letter.

4.2 Stage 2: 1st Subcontractor Interview
The 1st subcontractor interview has three aims. Firstly, to assess the subcontractors ability in terms of attitude, pro-activeness, design capability, honesty, safety awareness/culture, background and work load. Secondly, to introduce the project and the philosophy of 'the team' to the subcontractors and thirdly, if available, to hand over pricing details and other relevant information. The subcontractors are informed of the criteria that their submissions are to be evaluated. These criteria include:

(1) Understanding of the partnership concept;
(2) Response to partnering;
(3) Alternative proposals put forward for their package. This includes design innovation, alternative product specification, safety management system and value engineering;
(4) Indicative price;
(5) Technical ability;
(6) Enthusiasm for the project; and
(7) Past experience of similar work;

During this interview all subcontractors are supplied with: a maximum cost plan price; general arrangements' sections, specifications and approximate quantities. At the end of the meeting each project member assesses the subcontractors using a standard proforma. The proformas is kept by the project estimator. Those subcontractors who do not reach the requirements of the project team are informed by the project estimator that they are not be required to tender on the project.

4.3 Stage 2: Tender Clarification

This interview is used when the first interview acts as an initial subcontractor evaluation whilst the second was used to hand over pricing documentation. The aim of this interview is to give all parties the opportunity to discuss the tender and check for compliance and accuracy following the return of the tender documentation. This meeting is attended by the clients project manager and quantity surveyor, project estimator, project engineer and the project design manager

4.4 Stage 3: Subcontractor Selection

The final decision on which subcontractor is to be used is left primarily to the main contractor, although the clients quantity surveyor can have an input into the decision.

The selection is made based on the criteria communicated to the subcontractor during the first interview. The selections are made at a meeting which is attended by the project design manager, project estimator, main contractors project manager, project buyer, clients quantity surveyor and the clients project manager.

4.5 Stage 3: Project Day

A project workshop is then undertaken. The day is held at a neutral location away from normal meeting places. The aim of the day is to bring together all 'key' members of the project team, to review the progress made to date, where progress needs to be made in the future, and the signing of the partnering charter.

All parties that had worked on the project are represented at the project day. These included the structural and services engineers, architects, tenant, safety officer, main contractor, project managers and all partnering subcontractors.

The day should commence with the stakeholders reviewing the progress made to date on the project. Each partnering party then presented their requirements/objectives for the project and what they could bring to the project. Once the presentations were completed the charter was agreed upon and signed. Once all stakeholders have made a presentation the safety officer reviews the safety performance of the project to date. Issues covered include: safety legislation to be met; current accident rates and trends and establishment of new safety targets; ways of developing and maintaining a safety culture, newly presented method statements and their safety implications, dangerous incident reports, audit follow up procedures, safety manuals, etc. In this manner safety starts to take a prominent position in the hierarchy of project objectives.

4.6 The Partnering Charter

the actual process should be planned out;
the charter needs to be prepared by all, not just by a committee;
charter size should be limited to 1 page;
the charter should where possible be multi coloured;
all key logo's of organisations should be incorporated where possible;
there should be no individual titles or signature blocks; the charter is signed by all who participated in its development;

In order to facilitate safety management the charter should include a safety statement and refer to the projects safety policy/requirements. It is important that any safety targets are measurable in order that progress can be measured over time.

Each party involved in the project is asked to write down 10 objectives that they wanted to achieve during the project before the commencement of the project day. It is suggested that each party establish its own individual meeting where different members of staff, including safety officers, can have their input. Once all objectives had been returned a facilitator identifies the most prevalent objectives and includes them into a provisional project charter. Typical safety objectives that could be included have to be measurable and would include: accident rates, audit results, audit results; incident reports; training courses conducted; worker feedback, induction courses, registered safety staff, etc. Typical safety targets might be zero fatalities, zero injury inducing accidents, etc.

REFERENCES

Construction Industry Institute (1991), *In Search of Partnering Excellence*, Special Publication 17-1.

Crowley, L.G., Karim, M.A. (1995), 'Conceptual model of partnering', *Journal of Management in Engineering*, 11, **5**, Sept/Oct, 33-39.

Dawson, S., Willman P., Clinton, A. and Bamford, M. (1988), *Safety at Work: The Limits of Self-Regulation*, Cambridge University Press, Cambridge, UK.

Holmes, N. (1996), 'Narratives of risk in occupational health and safety: Why the 'good' boss blames his tradesman and the 'good' tradesman blames his tools', *Australia & New Zealand Journal of Public Health.*

Matthews, J. (1996), *A Project Partnering Approach to the Main Contractors – Subcontractor Relationshi*p, unpublished Ph.D., Loughborough University.

Matthews, J., Tyler, A., Thorpe (1996), 'Preconstruction project partnering: Developing the process', *Engineering Construction and Architectural Management*, 13, **1 & 2**, 117-131.

Rowlinson, S. and Lingard, H. (1998), 'Behaviour-based safety management in Hong Kong's construction industry – The results of a field study', *Construction Management and Economics*, E & FN Spon, 16, **4**.

Walker, A., Levett, D., Flanagan, R. (1998), *China; Building for Joint Ventures*, Hong Kong University Press.

Walter, M. (1988), 'The essential accessory', *Construction Manager*, February, 4, **1**, 16-17.

Implementation of Safety and Health on Construction Sites, Singh, Hinze & Coble (eds)
© 1999 Balkema, Rotterdam, ISBN 90 5809 036 1

Factors affecting the safety performance of contractors and construction sites

K.W.Wong, P.C.Chan & K.K.Lo
Department of Building and Real Estate, The Hong Kong Polytechnic University, People's Republic of China

ABSTRACT: This study is part of on-going research concerning the factors affecting the safety performance of the construction industry in Hong Kong. A questionnaire survey was conducted amongst a group of safety practitioners, and the data analysed by the use of SPSS software package. This paper examines the perception of various concerned parties relating to the importance of safety factors affecting safety performance of both contractors and construction sites. The input of the concerned parties is significant because they will shape the construction safety environment, for the local construction industry, for the foreseeable future. The study has shown that there is substantial mismatch between the perception of contractors and clients' employees with that of government officials, in terms of the appropriate amount of people, time, and finance resources for safety, which may cause safety management conflicts amongst the parties.

1 INTRODUCTION

Irrespective of all the efforts made by the Labour Department and other relevant government departments to improve the construction records in Hong Kong, there are still a very high level of accident rate. The numbers of accidents for the construction industry was 15,268 out of a working population of 65,000 in 1995. There was an average of 220 per 1,000 construction workers injured in 1996, including 51 fatalities. Since the consultation paper for the review of industrial safety in Hong Kong was published by the Education and Manpower Branch in 1995, there has been a shift of emphasis from a prescriptive legislative framework, to a new strategy. The new strategy places emphases on encouraging both contractors and workers to manage safety in a self-regulatory manner. Therefore it is important to see how the parties concerned, including contractors, clients and government officials, perceive safety resources for contractors, as well as to seek their views on the major important safety factors affecting the safety performance level of individual companies and construction sites.

2 AIM AND OBJECTIVES

The aim of this study was to prioritise the factors affecting the safety performance of contractors and construction sites, from the perspective of various concerned parties, including government officials, clients' employees, and contractors' employees.

The objectives of this study were (1) to identify the amount of appropriate resources to be given to construction safety in terms of human, time, and financial resources, (2) to rank in order of priority factors which are affecting the safety performance level of construction companies, and (3) to rank, in order of priority, factors which are affecting the safety performance level of construction sites.

3 SURVEY

The subjects of the questionnaire survey were the 38 participants of the Postgraduate Scheme in Occupational Safety and Health (POSH) which is being offered in Hong Kong. The scheme is jointly organised by the University of Western Sydney, Australia and The Hong Kong Polytechnic University. The subjects were mostly practitioners working in construction and related fields. The 38 responses (representing a 100% response rate) were divided into three major groupings, which were (1) 5 number of government officials, (2) 20 number of clients and consultants' employees, and (3) 13 number of contractors' employees.

4 SAFETY RESOURCES

All respondents were requested to answer the questions in a realistic and practical way, and to relate all answers to the Hong Kong construction industry.

4.1 *Human Resources*

One question asked respondents to indicate what they think should be an appropriate number of staff in the safety department, as a percentage of the total numbers of staff in the company. Government officials had the highest expectation at a mean of 5.2%, followed by those who work for contractors at 4%, and those who work for clients and consultants at 2.2%. The results indicated that there is a substantial difference between the perception of government officials and clients and consultants' representatives, in terms of human resources that are available in the safety department of construction companies.

In responding to another related question, about the appropriate numbers of staff who should be partly responsible for safety matters (as percentage of the total number of staff in their organisations), clients' representatives indicated 54.3%, followed by contractors' representatives at 38.3%, and the government officials at 34%. The results suggest that government officials prefer to have a lower human resources input on part-time basis for safety, and in exchange, for a higher degree of allocation of human resources on full-time basis. However, for the clients' representatives it is the opposite way round. The appropriate balance of full-time and part-time safety personnel is an interesting question for further study.

4.2 *Time Resources*

The subjects were asked the appropriate regular interval of their company-level safety committee meetings. Responses from government officials indicated a mean of 1.4 months, followed by the clients' representatives at 2 months and the contractors' representatives at 2.3 monthly intervals. This suggests that the expectations for regular safety meetings at company level from government officials are higher than those who work for clients and contractors.

The appropriate time spent by senior management on safety related issues, as a percentage of their overall workload, was also examined. The mean results were 7.0% from government officials, 5.1% from clients' employees, and 4.8% from contractors' employees. The

Table 1: Appropriate percentage of human resources for safety in construction companies.

	Contractors' employees	Clients' employees	Government officials	Overall average
Percentage of staff in the safety department	4.0%	2.2%	5.2%	3.2%
Percentage of staff partly responsible for safety	38.3%	54.3%	34.0%	46.2%

Table 2: Appropriate time resources for safety in construction companies.

	Contractors' employees	Clients' employees	Government officials	Overall average
Appropriate regular interval for safety committee meetings at company level.	2.3 months	2.0 months	1.4 months	2.0 months
Percentage of time to be spent by senior management on safety	4.8%	5.1%	7.0%	5.3%

Table 3: Appropriate financial resources for safety in construction companies.

	Contractors' employees	Clients' employees	Government officials	Overall average
Percentage of money to be spent on the safety system	0.5%	0.8%	2.1%	0.9%
Percentage of money to be spent on safety education and training	0.08%	0.19%	0.28%	0.16%

pattern of expectation for time resources was consistent with the results from the previous question, i.e. higher expectation from government officials and lower expectation from contractors' employees for the allocation of time resources for safety matters. In general, the contractors' employees expect a much lower time resources allocation for safety than government officials.

4.3 *Financial Resources*

"What do you think should be an appropriate amount of money to be spent on the overall operation of the safety system, as a percentage of your company's annual turnover?" was a question asked in relation to the availability of financial resources. There were substantial differences between the lines of thinking from government officials as compared with that of the clients and contractors' employees. Government officials recorded 2.1%, with 0.8% from clients and 0.5% from contractors' employees. The percentages from the government officials are 2.6 times higher than that of clients' employees and 4.2 times higher than that of contractors' employees. That is to say, the officials have a higher expectation of the amount of money that should be spent on the operation of the safety system by contractors.

There was a question about the appropriate amount of money that should be spent on safety education and training, as a percentage of a company's annual turnover. The government officials indicated a mean of 0.28%, followed by the clients' employees at 0.19% and the contractors' employees at 0.08%. Again, the percentages from government officials are much higher compared with the other two groups, at 1.47 times higher and 3.73 times higher respectively.

After studying the appropriate amount of various kinds of resources for safety, the next stage of the study was to review the relative degree of importance of factors, which affect the safety performance level of construction companies and construction sites. The authors recommend further investigation on time resources by senior management in order to have a better understanding about the effect of this factor on safety performance of construction companies.

5 FACTORS AFFECTING THE SAFETY PERFORMANCE OF CONSTRUCTION COMPANIES

The subjects were requested to rank on a scale from 1 (most important) to 6 (least important), the following factors which affect the safety performance level of construction companies. They are:

- Written safety policy (WSP) and measurable safety targets.
- Communication of the WSP to the various concerned parties.
- Safety system, including safety department.
- Safety committee at company level.
- Safety budget.
- Time spent by the senior management and the company on safety matters.

It is interesting to note that, based on the findings of this survey, most respondents believe that the most crucial factor affecting the safety performance level of construction companies is the amount of safety budget available. Safety system was being identified by the clients' employees as the most important factor, and identified by the officials as the second most important factor. However, the contractors' employees gave the second most important ranking to the time spent by the senior management. It would appear that contractors' employees have a higher degree of expectation on their seniors to give more time to safety matters. If they do, this in turn may motivate the contractors' staff at lower level to take construction safety more seriously. Having said that, it is peculiar to note that the contractors' employees have a lower expectation, than the other two groupings, concerning the time spent by their senior management on safety matters (see Table 2).

6 FACTORS AFFECTING THE SAFETY PERFORMANCE OF CONSTRUCTION SITES

Subjects were asked to rank on a scale of 1 (most important) to 7 (least important) the following factors which affect the safety performance level of construction sites. They are:

- Communication of WSP to the various concerned parties on site.
- Safety system and the associated safety items.
- Safety committee at site level.
- Safety training.
- Safety practice and procedures.
- Monitoring the compliance of safety measures and disciplinary actions for non-compliance.
- Degree of participation by the safety officer and the degree of consultation by site staff to the safety officer.

Referring to Table 5, government officials and clients employees had identified the provision of safety system and the associated safety items as the most important factor affecting the safety performance level of construction sites. However, contractors' employees believed that factors such as "the safety practice & procedures", and "the compliance of safety measures & disciplinary action for non-compliance", are more important than safety system and safety items, with regard to their effect on safety performance of construction sites.

It is interesting to note that government officials ranked safety committees at site level as priority number two, although both the contractors' employees and clients' employees gave this a much lower ranking. The results indicate that government officials are more concerned with the safety system (safety committee on site is part of the safety system), and the two other groupings are more concerned with safety procedures and compliance with safety measures, which are more specific and detailed in nature.

Table 4: Ranking of factors affecting the safety performance of construction company.

	Contractors' employees	Clients' employees	Government officials	Overall ranking
Safety budget	1 (2.0)	2 (2.9)	1 (2.0)	1 (2.4)
Safety system, safety dept.	3 (3.3)	1 (2.7)	2 (3.0)	2 (2.9)
Time spend by senior mgt.	2 (2.8)	4 (3.7)	4 (3.8)	3 (3.4)
WSP & measure. safety target	4 (4.2)	3 (3.1)	3 (3.4)	4 (3.5)
Communication of WSP	6 (4.6)	5 (4.0)	4 (3.8)	5 (4.2)
Safety committee at company level	4 (4.2)	6 (4.5)	6 (4.6)	6 (4.4)

Note: Figures in bracket are mean values.

Table 5: Ranking of factors affecting the safety performance of construction sites.

	Contractors' employees	Clients' employees	Government officials	Overall ranking
Safety system & safety items	3 (3.5)	1 (2.7)	1 (1.4)	1 (2.8)
Safety Practice & procedures	1 (3.1)	3 (3.7)	3 (3.8)	2 (3.5)
Safety training	4 (3.7)	2 (3.2)	5 (4.2)	3 (3.5)
Compliance of safety measures	2 (3.2)	4 (3.9)	6 (4.6)	4 (3.7)
Safety committee at site level	5 (4.2)	7 (4.6)	2 (3.0)	5 (4.3)
Communication of WSP	7 (5.4)	5 (4.1)	3 (3.8)	6 (4.5)
Participation & consultation	6 (5.0)	6 (4.6)	7 (6.0)	7 (4.9)

Note: figures in bracket are mean values.

7 CONCLUSION

Although the sample size for this study was small, the research findings provide useful information to indicate the perceptions of the three groupings. It is important to know that the ranking of factors affecting safety performance, of both contractors and construction sites, are perceived quite differently by government officials and contractors' employees. A larger scale survey is recommended in order to produce a more reliable analysis.

The mismatch in the expectation over the availability of safety resources, in terms of people, time, and finance, may cause conflicts between the various parties. Such conflicts may be particularly pronounced between government officials, and employees of contractors, clients & consultants. The gap between the expectations of these three groups should be minimised to ensure that all the concerned parties are satisfied with the prevailing safety system.

8 ACKNOWLEDGEMENT

The on-going research is sponsored by the staff development fund of The Hong Kong Polytechnic University.

9 REFERENCES

Chan, P.C. & Leung, A. 1991. The role of the site manager, *Australian Institute of Building Papers,* Vol. 4, pp. 57-64.

Education and Manpower Branch, 1995. Consultation paper on the review on the industrial safety in Hong Kong, The Hong Kong Government.

Labour Department, 1996. Report of the Commissioner for Labour, The Hong Kong Government.

Wong, K.W. 1997. Implementation of safety policy and safety system for construction companies in Hong Kong – senior management's perspective. *Proceedings of the 13th annual conference for the Association of Researchers in Construction Management,* Cambridge University, U.K. pp.378-385.

Wong, K.W. 1996. Safety education for construction students – Hong Kong experience. *Proceedings of the CIB Beijing International Conference on Construction Modernization and Education,* Beijing, China. (CD-ROM)

Wong, K.W. 1995. Prosecution and employees' compensation for construction accidents in Hong Kong. *Proceedings of the 11th Annual conference for the Association of Researchers in Construction Management,* University of York, U.K. pp. 696-704.

Implementation of Safety and Health on Construction Sites, Singh, Hinze & Coble (eds)
© 1999 Balkema, Rotterdam, ISBN 90 5809 036 1

Time/space chart drawing techniques for the safety management

A.Ciribini
Diset, Politecnico di Torino, Italy
G.Rigamonti
Diset, Politecnico di Milano, Italy

ABSTRACT: Authors are establishing an emended version of road works time-space chart for building site works planning. Such charts are deemed to be very useful on the behalf of safety planning supervisors, because of the clearness of the activity trends to be monitored. Indeed, safe resource usage could be planned and managed more easily when graphically showed on time-space charts: thus analysing a lot of risks and avoiding a deal of accidents. The paper aims to argue about the operability of the new chart drawing techniques, meanwhile keeping into account advanced network planning tools. Some examples will be illustrated.

1 SAFETY MANAGEMENT AND AVAILABLE PLANNING TECHNIQUES

Safety management of the building sites is, of course, chiefly aimed to reduce (and, sometimes, to remove) accidents and diseases. Accidents and injuries, above all, occur, to the building workers (and to the site officers), i.e. to the human resources which are impacted by various agents: the other kinds of resources available for the construction phase. In the practice, risk analysis tries to increase the safety of the activities performed as far as a specific site is envisaged. Nonetheless, whilst some collective and individual preventive and corrective measures and devices are always applied, hazards are not so significantly lessened.

Furthermore, a deal of accidents is due to the interaction between human resources and the other ones, notwithstanding the inner compliance of the safety measures. The overcrowding and the interference are, indeed, included among the main dangerous factors to be evaluated, as well as human resources working simultaneously nearby the dangerous agents.

Planning techniques are very useful in reducing probabilities of occurence of the accidents: particularly, the construction safety planning supervisor (the safety coordinator at the construction phase) is able, at the utmost, to review the safety-oriented plan of the works written early by the design safety planning supervisor. Indeed, he has a deep knowledge of the plans provided by contractors and subcontractors during the bidding and awarding stage.

The construction safety planning supervisor, who is fully acting as a client's representative (according to the EC Directive 92/50), must integrate and complete the original master plan of the works and the safety plan taking into account the different fragments of the network pertaining to the supply side together with the risk analysis-related documentation.

Nowadays, the planning supervisor is able (Figure 1):

1. to analyse the overlaps (as far as the time is concerned) shown in the bar charts;
2. to resort to the various kinds of relationships and lags in order to lead away the overlapping activities;
3. to resort to interfering, free, and total floats to lead away the overlapping activities;

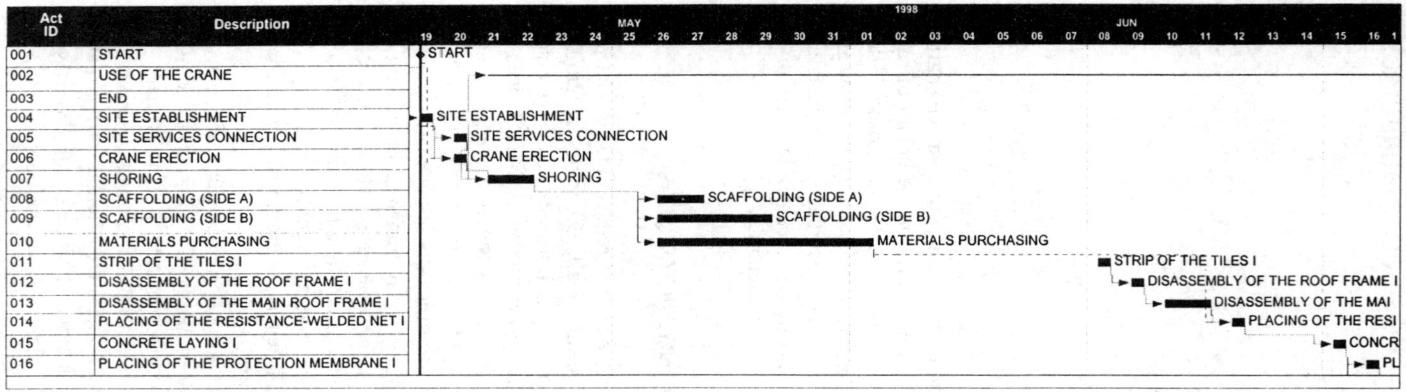

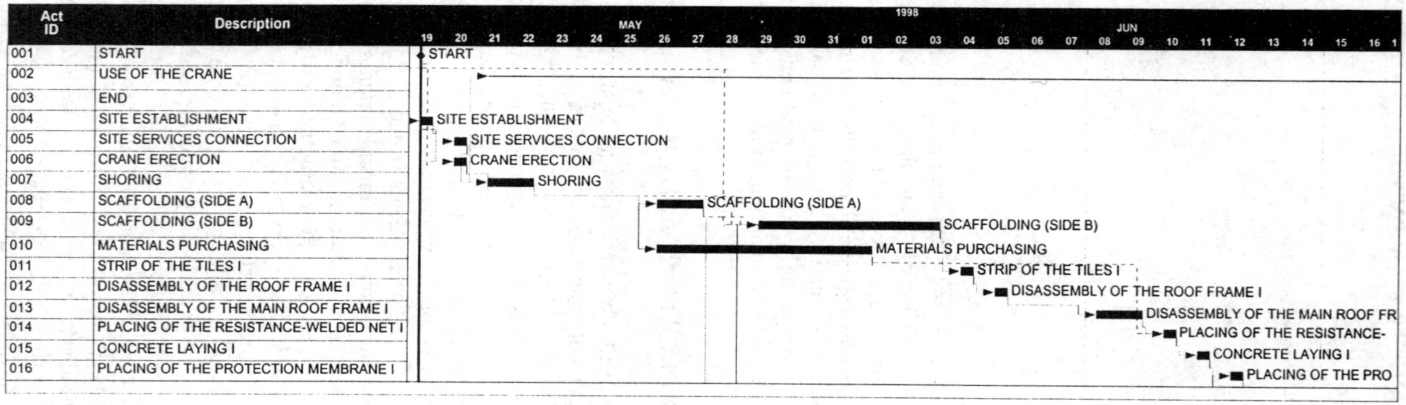

Figure 1.The original and the emended version of the linked bar chart

4. to visualize in the linked bar chart or in the network some specific safety measures to be complied with;
5. to compare simultaneously the histograms concerning each type of human resources involved (engendering an overcrowding index) and of non-human ones (engendering an interference index) in order to smooth and level the peaks identified;
6. to modify the original critical path, quite increasing the duration of the works fixed by the contract.

Unfortunately, clients seldom wish to support such an approach, while planning supervisors neglect the planning tools, owing to widespread amateurism.

On the other hand, the prevention measures can be brought in the linked bar chart both as new activities or as a log for an activity already inserted. Moreover, it is possible to suggest further devices (Figure 2):

1. to insert an activity code related to the displacement of the specific activity: consequently, the main activity has to be split into subactivities (Figure 3);
2. to point out a proximity index according to the amount of overlapping activities and the duration of the overlapping time;
3. to insert an activity code related to the type of contracting party involved or the payment method;
4. to point out a density index according to the amount of building firms simultaneously involved and the overlapping time;
5. to insert an activity code related to the type of dangerous agents and hazard factors previously identified;
6. to insert an activity code related to the degree of inner risk levels early stated;
7. to compare resource histograms and resource distribution curves (linear and non-linear).

By doing so, the stated inner risk level of the different activities considered can be increased through a risk analysis performed upon the bar charts and networks.

Consequently, the construction safety planning supervisor:

1. reviews the detailed plan of works provided by the main contractor, examining links, floats, lags, and criticalities;
2. reviews the fragments of the network provided by the specialty contractors and subcontractors;
3. assesses the overlapping rate amongst activities and resources, according to the different site areas;
4. evaluates the use rate of the single resource, according the the different site areas;
5. evaluates the suggested flowing of the workers in the different site areas;
6. examines the storing measures foreseen by the main contractor;
7. possibly asks to the main contractor some addenda;
8. possibly agrees with the main contractor some planning changes (as links, floats, lags, and criticalities are concerned).

2 SAFETY MANAGEMENT AND NEW PLANNING TECHNIQUES

Such efforts could be furtherly improved through an approach based upon the visualization of the works performed from time to time (Figure 4).

The diagramming techniques hinder a lower level of abstraction than the traditional ones: the authors suggest that it coud be possible to periodically represent the movements of the resources (i.e. workers and dangerous factors/resources) on the site through drawings linked to the bar charts or networks (Figure 5).

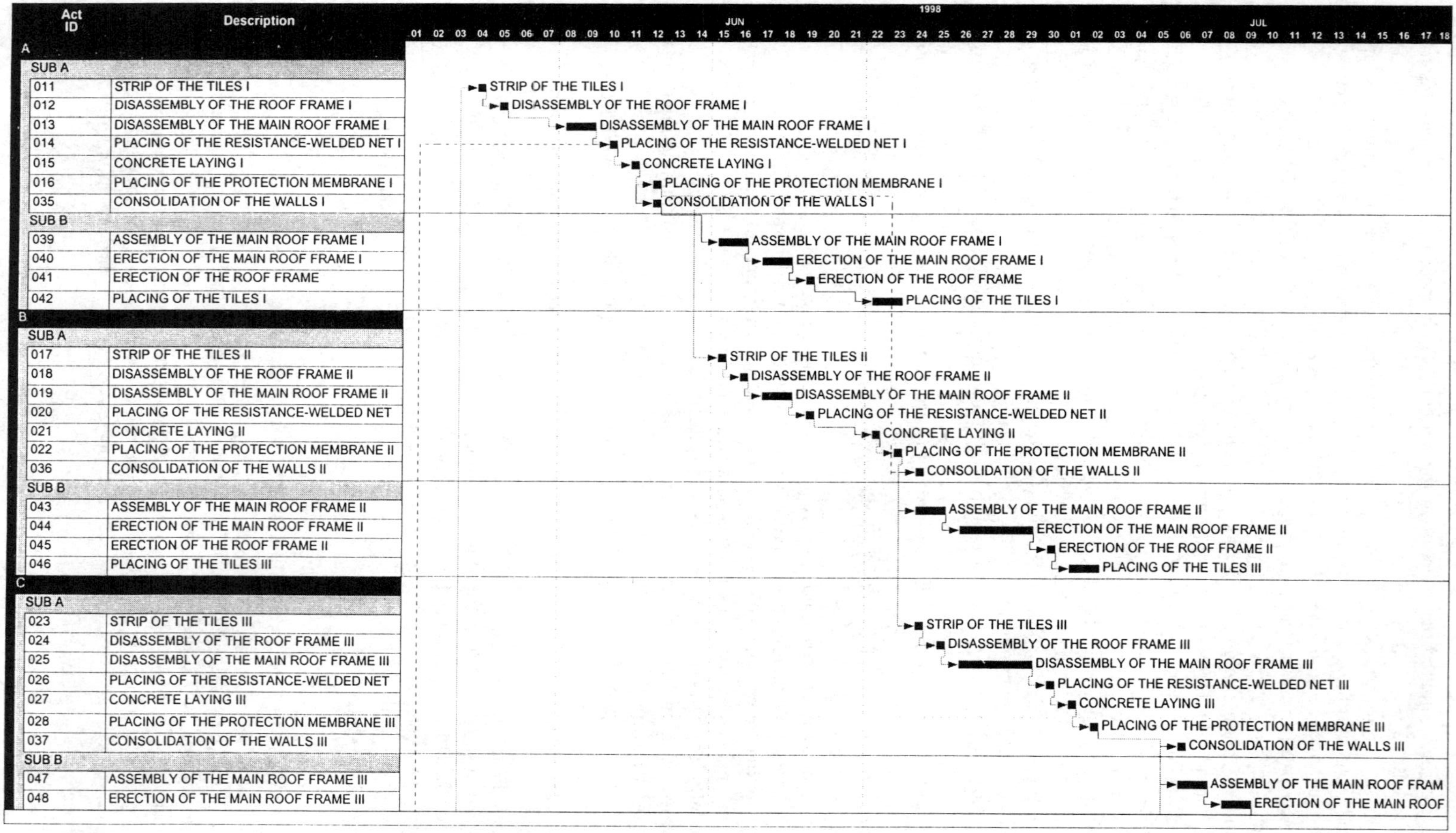

Figure 2. View of the bar chart according to site areas and type of resources.

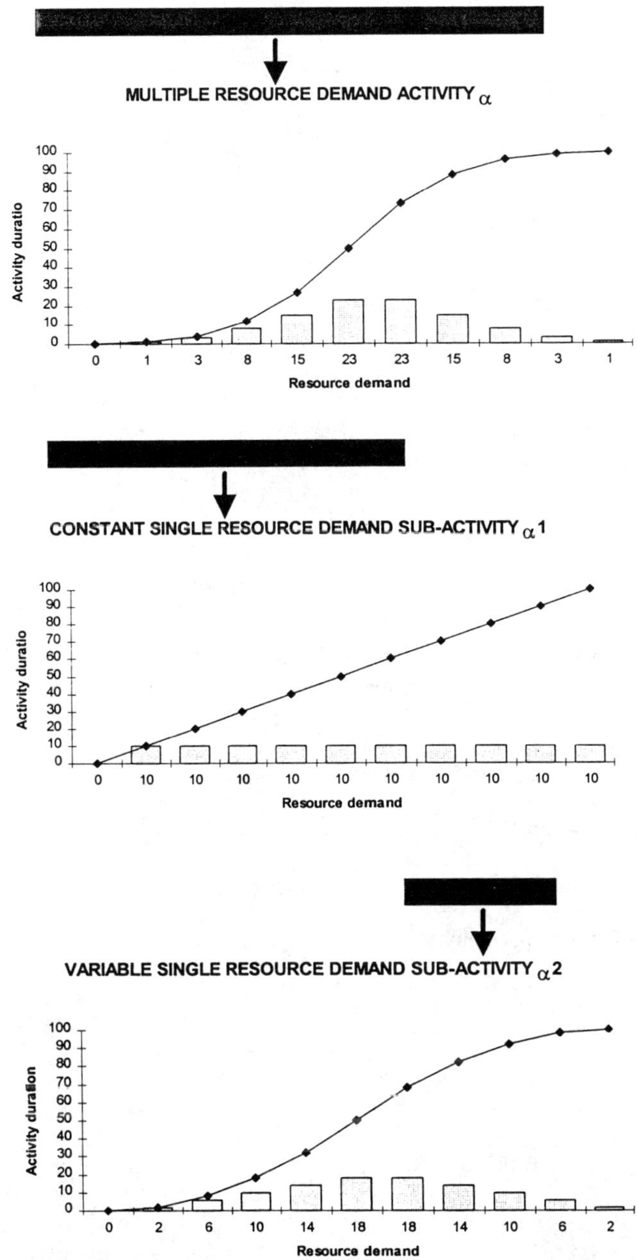

Figure 3.Qualitative splitting of an activity following single resource demand profiles.

Each part of the activity bar should be linked in proportion to a corresponding part of a building element to be erected (Figure 6). In other words, we are trying to draw a fixed percentage of the works scheduled or performed and the forecast interaction between human and non-human resources visiting different activities.

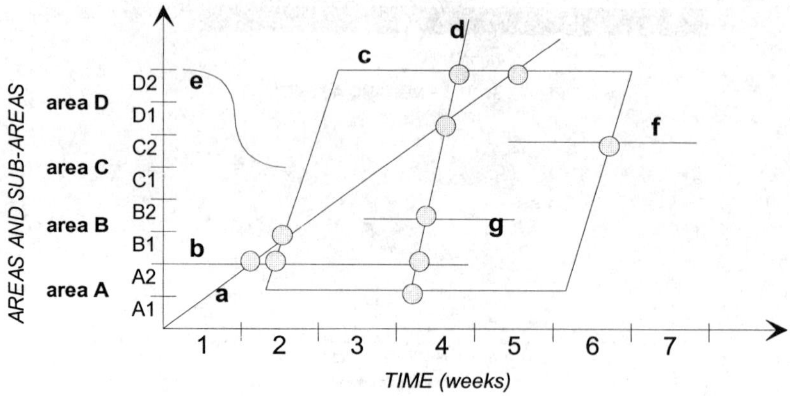

Figure 4. The 2-D time-space chart.

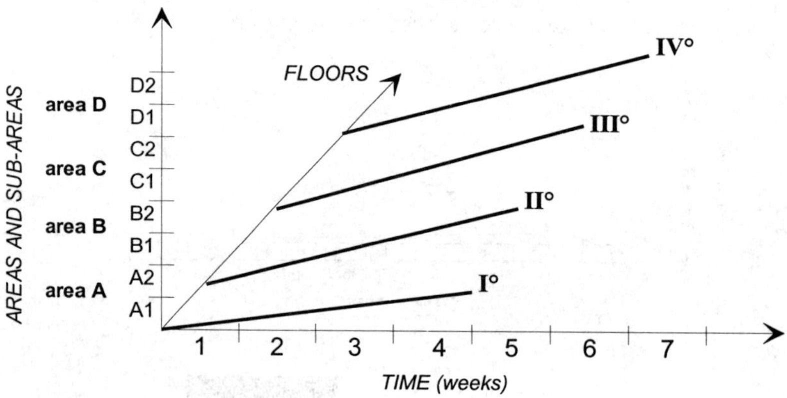

Figure 5. The 3-D time-space chart.

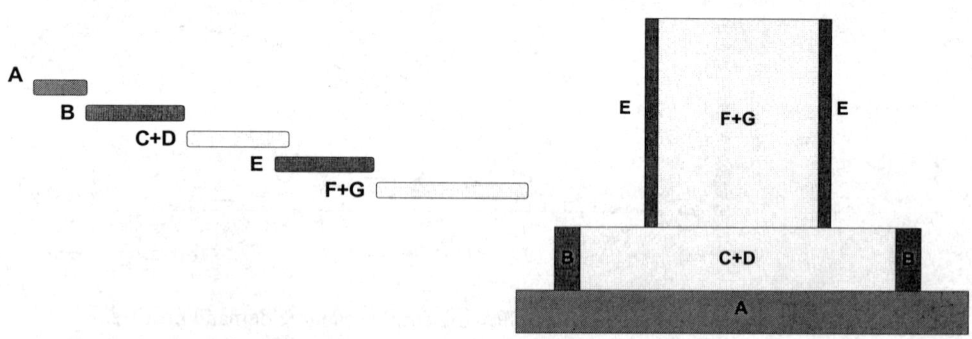

Figure 6.The proportioned schematic drawing of a column foundation interlinked to the bar chart.

Approximately, a filtering is needed following the criterion of the dangerous factors when verifying overlaps and proximities. Such an attempt is really only effective when considering the static dangerous factors, i.e. the shared non-human used by more human resources. The dynamic dangerous factors have to be represented in other ways, because of the importance of the motion.

The space resource is also a critical one: expediting techniques could make easier the best use of the storing areas available during the construction stage on the site.

The effective and timely planning of the material resources demand means, from a safety management viewpoint, a considerable improvement in avoiding critical lack of storing spaces; meanwhile, a shortage of building materials could modify the activity durations, causing unforeseen overlaps and interferences. Clearly, bar charts and networks are not enough, even if simulation techniques and updating techniques for the improvement of the resource planning and the safety management are focused on such tools.

3 CONCLUSIONS

In any case, as we said, there are two separate ways to achieve the goals stated:

1. to develop and adjust some existing planning techniques which link time and space dimensions (time-space charting, line of balance techniques, etc.);
2. to insert drawing techniques closely interlinked to planning tools.

We are proposing two separate approaches for the occurence of the accidents, too:

1. the human resource that is injured is envisaged as allocated to an activity;
2. the same resource is considered as displaced on a stated site area.

Consequently, the dangerous factor (the non human resource) allocated to an activity, moving to other activities could interfere with a human resource that is visiting another activity in a static or in a dynamic way.

Inversely, the dangerous factor that is displaced on the site could interfere with a human resource that is transferring from a site area to another one.

In the first case, the accident happens because of the time overlapping concerning two activities consuming the resources considered.

In the second one, the accident happens because of the space interaction occuring amongst various human and non-human resources, quite apart from the activities.

The first approach is hugely effective when planning the works at the design and the awarding stage, when a whole scenario is needed. From the safety management point of view, the planning supervisor must act modifying resource histograms.

The second approach is preferable when managing and controlling the works in progress, because of the unforecast delays and overlaps. From the safety management point of view, the planning supervisor can examine the resources graphically displaced on the various site areas.

The time-spacing techniques are often used when planning road lincar works or, at least, repetitive building works. We are trying to represent bars (over the time, on the space, at different floors or heights): it could be possible, in such a manner, to periodically identify directly the rate of completion of a fixed work package.

It could be useful to place side by side resource distribution curves and time-space charts.

The diagramming and drawing linked planning techniques could allow:

1. to allocate on the site area resources and safety devices;
2. to easily visualize the resource distribution curves according to the rates of completion of the building element physically drawn;
3. to represent the motion of the site resources;
4. to realize the effects of the change over the time according to the original schedule in terms of interaction and interference;
5. to evaluate the environmental impact of the site;
6. to visualize the dangerous factors.

REFERENCES

Ciribini, A. & G. Rigamonti 1997. Computer Aided Network Techniques and Construction Management : a critical analysis in the italian construction industry. *Proceedings of Worldwide ECCE Symposium, Lahti, Finland.*

Ciribini, A. & G. Rigamonti 1998. Cost control in healthy and safety planning in the italian housebuilding sector. *Proceedings of XV International Cost Engineering Congress, Rotterdam, Netherlands.*

Ciribini, A. & G. Rigamonti 1998. Planning techniques and environmental issues in procurement. *Proceedings of the CIB World Building Congress, Gavle, Sweden.*

Implementation of Safety and Health on Construction Sites, Singh, Hinze & Coble (eds)
© 1999 Balkema, Rotterdam, ISBN 90 5809 036 1

Construction safety management formula for success

John A.Agrilla
Parsons Brinkerhoff Austin, Tex., USA

ABSTRACT: What are the current conditions of our construction work sites? Approximately five percent of the country's industrial work force is employed by the construction industry yet it accounts for fourteen percent of workplace fatalities and nine percent of all disabling injuries. The cost realities of safety issues are dramatic. In 1982 the costs of accidents and injuries were 6.5% of all construction costs. In 1991 the costs was 8% to 15% of all construction costs. We cannot afford not to have a positive proactive approach to safety.

1 INTRODUCTION

A properly designed safety program can produce outstanding results.

- Metropolitan Washington D.C. Airports
  14 million work hours and 0.75 lost time injury rate with no fatalities
  Returned Insurance Premiums estimated at $15 million

- Austin-Bergstom International Airport
  Worked 2 million hours with out a recordable lost time injury, no fatalities
  And a 0.17 lost time injury rate
  Returned Insurance Premiums estimated at $7 million.

2 SAFETY MANAGEMENT

A successful safety management program is based upon the premise that Safety is both a management responsibility and a line function. While Safety Managers help to develop and implement the program, the actual success depends upon management's commitment and the ability of supervisory personnel to insure that the program is carried out during daily operations. The Safety Manager, therefore, fills the unique position of moving between management and line to assure that management commitment remains high and line function needs are met.

The Safety Manager assumes the position of a sports referee, maintaining a level playing field and assuring that both the management and line teams are playing by the same rules. The program objectives must be clearly defined as adding value to the company and/or client. This value can be the reduction of insurance costs thereby returning profit to the company or client. This is accomplished by preventing personal injuries, minimizing property damage, reducing direct and indirect construction cost, increasing project efficiency (maintaining schedule) and providing the client a user friendly quality product.

3 SAFETY TOOLS

The tools necessary to achieve the objectives are referred to as the Three Es of Safety: Safety Engineering, Safety Education and Safety Rule Enforcement. Safety goals must be established early in the program, utilizing the total quality management approach. Using the B.N.A. recordable injury rates rather than insurance loss ratio, or risk experience modification, appears to be the most effective and consistent method for measuring result. Establishing upper and lower control limits with a median that can be changed as needed, provides the flexibility needed that allows line management to remain in the hunt for excellence when a recordable lost time injury occurs.

4 KEY PLAYERS

Key players within the construction framework must be identified as part of the start up procedures for the program. The most common job designations for these people are:

- Project Manager
- Project Safety Manager
- Insurance Risk Manager
- Resident Engineers and Staff
- Contractor's Project Manager
- Contractor's General Superintendent
- Contractor's Safety Superintendent
- Foreman

An organizational chart showing the Safety Chain of Command and Communication is essential to proper program management. In addition, the Safety Management Team (in this case the Construction Management Company) must concentrate on its basic responsibilities which are:

- Review applicable contract documents
- Recommend approval of Contractors Safety Plans
- Review resumes of proposed Contractors Safety Superintendents
- Provide the Program Safety Manager, Resident Engineers and Construction Inspectors to strictly enforce the contract safety provisions, and familiarize bidders and contractors with the Construction Safety Program.

5 HEINDRICK [1]

For management purposes Hendrick's definition of an accident provides the basic understanding of what is considered an accident and what causes contributed to the event.

- The accident is described as the unplanned event that interrupts the normal work flow
- It has causes described as unsafe acts, (88%), unsafe conditions, (10%), that may work in combination
- Approximately 2% of all causes are described as uncontrollable
- Any accident may result in personal injury, property damage or both

Heindrick's 330 theory on accident occurrence also provides an insight for program participants concerning accident occurrence as it relates to "near miss" situations. Heindrick surmised that for every 330 accidents 300 would produce no injury or property damage, of the remaining 30 accidents, 29 would produce minor injury or property damage. The remaining accident would produce a severe injury, high value property damage or both.

1 H. W. HEINDRICKS INDUSTRIAL ACCIDENT PREVENTION

The main focus within the safety program, however, continues to be the role of each construction contractor. Contractor responsibilities include ensuring that all subcontractor's, suppliers, vendors, etc., are informed of their obligations regarding safety, planning and executing all work to comply with stated objectives and safety requirements, preparing and submitting for approval a Contractor's Safety Manager/Superintendent /or Safety Supervisor, and maintaining a safety orientation program.

All of these components are critical to establishing the formula for a successful safety program. The basic procedures that need to be accomplished using these components are to plan the work and work the plan, comply with federal, state, and local laws, ordinances, regulations and requirements, maintain a system of prompt detection and correction of unsafe practices and conditions and establish and conduct an education program to stimulate and maintain the interest and cooperation of all parties involved.

6 RESPONSIBILITY MEANS ACCOUNTABILITY

It seems we can always find the person who is responsible for safety, however holding that person accountable for the results is a different matter. The facts are responsibility without accountability does not work. If safety is a management responsibility and a line function, then line operations must be held accountable for the results good or bad. Accountability measures should be in keeping with company written policies such as retraining, suspension without pay or termination.

The accountability for good results is easy, we hold dinners, give bonuses, and pass out awards. Everyone smiles, we pat each other on the back, let's face it, it's fun.

On the other hand accountability, when things go badly is much more difficult to access. The definition of an accident it self is a problem. It is defined as an unplanned, unforeseen event that interrupts the normal workflow. This works against accountability. How can you hold someone accountable for an unplanned, unforeseen event?

The answer is both difficult and easy at the same time. The military makes it easy by telling an individual if he is in charge, he is responsible and will be held accountable for the safety results. Good safety results will bring excellent efficiency reports and career advancement. Poor safety results will result in poor efficiency reports and limited career advancement.

The individual worker <u>is</u> responsible for his own safety and must be held accountable for poor safety performance, but only after proper Engineering and Education has been accomplished. Should a worker continue to work unsafely after education and correction has been applied, he must be held accountable for his poor performance. This accountability process must be enforced up through the entire chain of command within the organization. If safety performance is poor on a job site the Project Manager and General Superintendent must be held accountable. Should a work crew perform poorly, the Foreman must be held accountable. There can be no responsibility without accountability. The Three Es of Safety, Engineering, Education, Enforcement are so intertwined that the removal or ignoring of one of the Es will capsize a program. The removal of the enforcement/accountability E will make the Engineering and Education process an exercise in frustration.

Accountability for results is essential to an effective Safety Program. It must be applied with consistency, fairness, and swiftness.

7 CONCLUSION

A successful safety program must focus on preventing catastrophic worker injuries, by assuring proper engineering of critical crane lifts, preventing falls from heights, trench collapse, etc., not on preventing minor incidents. An appropriate project staff must be developed and must be supportive of the goals and means and methods of the safety program. In addition, contractor line staff and management must have appropriate resources and must be held accountable for safety results. When all appropriate components of an organization and a program are in place, a safety program can provide excellent results, well below industry norms.

2 International practices in safety management

Implementation of Safety and Health on Construction Sites, Singh, Hinze & Coble (eds)
© 1999 Balkema, Rotterdam, ISBN 90 5809 036 1

Safety and construction in India

E. Koehn
Department of Civil Engineering, Lamar University, Beaumont, Tex., USA
S. Reddy
US Environmental Protection Agency, Dallas, Tex., USA

ABSTRACT: The mechanization of the construction industry is not uniform throughout the world. Especially developing countries, which use labor intensive construction methods and experience high accident and fatality rates, are vastly different compared to developed regions. This is due, in part, to governmental instability, minimum use of equipment, shortages of adequately trained craftsmen, difficulty in acquiring needed materials, and lack of adequate infrastructure such as roads, bridges, harbors, and power generation facilities. This paper investigates construction safety and productivity in one of the developing countries, India. The data should be of universal interest to any contractor considering projects in India and other developing countries. The information presented should assist international contractors with the management, estimating, scheduling, monitoring, and updating of construction operations.

1 INTRODUCTION

The modern world can be divided roughly into three main categories based upon location, political affiliation, or overall wealth. The first world, or developed countries, include the United States, Western European nations, and other industrialized democracies such as Canada and Japan. The second world consists of developed countries, but those that have not met the standards of the first world, such as the Republics of the old Soviet Union. The third world comprises those nations whose development in all areas is not as extensive as those in the first and second worlds and includes various countries in Latin America, Africa, and Asia. For the purpose of this study, countries from the first and second worlds will be called "developed" nations, and those in the third world will be considered "developing" nations. In general, developing countries share a common or almost universal set of construction characteristics that are related to their overall level of development (Merriam 1988).

Today, international projects in many developing countries are constructed by various local firms as well as joint ventures between national and international companies. Construction projects, however, are often complex undertakings. Facilities must be designed in accordance with applicable local codes and standards, culminating in drawings and specifications that describe the work in sufficient detail so that it may be constructed in the field (Levin 1975). The building of a structure involves many skills, materials, equipment, and literally hundreds of other operations. The assembly process must follow a more or less natural order of events that, in total combination, constitutes a pattern of sequential relationships among the various segments of a facility.

Generally, the construction process is subject to the influence of highly variable and sometimes unpredictable factors (Kaming 1997, Schmidt 1970). The construction team, which includes various combinations of architects, engineers, tradesmen, subcontractors, inspectors, owner representatives, material dealers, equipment suppliers, and others, changes from one job to the next. All the complexities inherent to different construction sites, such as subsoil conditions, surface topography, weather conditions, transportation, material supply, utilities and services, local subcontractors, and labor conditions are innate parts of construction. To be precise, the principal costs of a construction project are related to material, labor, and equipment expenses. Since materials and equipment are often imported by larger international firms, their variation generally cannot be controlled in a developing country. Therefore, the most important element associated with the efficient operation of a project is labor productivity, which tends to be an unstudied resource in developing countries.

This paper presents the safety problems and labor requirements for building construction in a typical developing country, India. Recent changes in productivity and the effect of low wage rates are also investigated. These items should be of interest to international contractors in order to achieve the desired cost effectiveness, safety, quality, and duration of construction projects.

2 GEOGRAPHICAL AND ECONOMICAL DATA

India is located in Southern Asia between the Arabian Sea and the Bay of Bengal and is bordered by the countries of China, Bangladesh, Nepal, and Pakistan (Foreign 1997, Information 1995). It is slightly more than one-third the size of the USA with a climate that varies from tropical monsoon in the south to temperate in the northern regions. The topography consists of an upland plain (Deccan Plateau) in the south, flat to rolling terrain along the Ganges River, deserts in West, and the Himalaya Mountains in the north. India's natural resources include coal (fourth largest reserves in the world), iron ore, manganese, mica, bauxite, titanium ore, chromite, natural gas, diamonds, petroleum, and limestone. The country has a population of roughly 936,546,000 with a labor force of approximately 315 million with 94 million involved in the construction industry. As a comparison, it is estimated that the USA has 10 million employed in construction.

India's economy is a mixture of traditional village farming, modern agriculture, handicrafts, a wide range of modern industries, and a multitude of support services. Faster economic growth in the 1980's created a significant increase in per capita private consumption. Nevertheless, a large share of the population, perhaps as much as 40%, is below the poverty line and cannot afford an adequate diet. Poor nourishment causes fatigue which may tend to reduce productivity and increase the accident and fatality rate for workers, especially in the construction industry.

Production, trade, and investment reforms since 1991 have provided new opportunities for Indian businessmen and an estimated 100 million to 200 million middle class consumers now live in the country (Niland 1970). Positive factors for the future include India's strong entrepreneurial class and the government's recognition of the need for market oriented approaches to economic development, involving, for example, upgrading the wholly inadequate communications system. Negative factors include the desperate poverty of hundreds of millions of inhabitants and the impact of the huge and expanding population on an already overloaded environment and infrastructure.

3 CONSTRUCTION INDUSTRY AND BUILDING MATERIALS

The construction industry can be classified into various segments such as the following (Punamia 1987): industrial, housing, commercial, utilities and civil work, and infrastructure. It has been found that construction plays an important role in the development of an economy, especially in a developing country. As an example, the industry has close linkages with growth in the secondary and tertiary sectors of the economy. In India, growing foreign investments and favorable fiscal policies have improved the industrial growth rate to double digits. Increasing capital investment has developed more business for the construction industry which is evident from higher sales and better performance reported by leading construction companies. As an example, recently, construction accounted for 47% of the gross fixed capital formation.

Due to the rapid growth of the industry, new building materials using natural indigenous materials are being developed. The use of earth, brick, thatch, and local stone and gravel, as well as non-conventional building materials made from industrial and agricultural wastes is not uncommon (Kanettkar 1990). Materials such as cement, brick, crushed stone, pre-cast blocks, etc., are also being exported to neighboring countries.

3.1 *Infrastructure and labor organizations*

Despite the availability of financial resources, construction of residential housing is difficult in India and other developing countries due to the lack of basic infrastructure and the availability of adequate space. Random urbanization has taken place at the expense of reducing limited agricultural land. In fact, land speculation is quite high in some parts of India. Research in land economics, utilization of desert or semiarid terrain, and availability of water is being conducted. The social dimension of urbanization, however, has not been taken into adequate consideration resulting in deficiencies in community facilities, especially in low-income areas

In India, roughly 10% of the population is involved with construction. Home building, however, tends to be, specifically, an unorganized and fragmented sector of the industry. Construction workers in India, as well as in other developing countries, are generally uneducated and tend to work for a low wage. Labor unions and trade groups, if they exist, are usually disorganized.

The low wage rate may be the reason for the extensive use of labor intensive methods, and the minimum level of implementation of science and technology in the construction industry, compared with that of the developed nations. Also, in India, as in many countries, engineering education tends to stress theoretical concepts and has a western orientation. Few labor training schools exist which teach skills needed for performing practical construction tasks.

4 LABOR PRODUCTIVITY RATES AND SAFETY

The productivity of labor is generally considered important in construction operations. In fact, in high-cost developed areas, labor expenses may amount to approximately 50% of the total cost of a project. Here, the management of labor productivity is vital to control the projects' cost and duration. In developing regions with low production rates, a large number of laborers tend to be needed to accomplish construction work. The efficient management of this large group involves a considerable amount of organizational skills.

The number of hours required to perform various activities in India is presented in Table 1. The hours were computed using resources available for general estimating calculations (Koehn and Pan 1995, Labor 1996). The data shows that the time required for performing various activities has decreased considerably between 1990 - 1996 except for finishing, masonry, and timber work. Here, the population tends to prefer buildings that reflect their culture and tradition and craftsmen are requested to follow the old artistic methods and skills. Generally, the increase in productivity is due to the following:

- Use of more efficient construction equipment
- Improved methods
- Education and training
- Utilization of modern materials
- Use of proper planning and scheduling techniques such as CPM, PERT, etc.

Nevertheless, construction in India and other developing countries is more labor intensive than that in the developed areas of the globe, involving 2.5 - 10 times as many workers per activity.

Table 1: Comparative Labor Requirement for India

Activity	Unit	1990 Labor Hours*	1996 Labor Hours*	Percent Decrease
Site Preparation				
Clear & Grub	M^2	7.86	5.89	25.1
Excavate drain trench	M^3	38.81	32.98	15.0
Earthwork				
Backfill (light soil)	M^3	292.00	233.60	20.0
Backfill (heavy soil)	M^3	360.87	288.69	20.0
Excavation (heavy soil)	M^3	444.20	399.78	10.0
Excavation (sandy soil)	M^3	398.29	358.46	10.0
Base course (roads)	M^2	0.97	0.90	7.2
Mortar				
Grouting (conc. Block)	M^2	74.16	66.00	11.0
Grouting (cavity wall)	M^2	40.25	36.62	9.0
Masonry				
Brick in c/m	M^3	3792.29	3790.38	0.1
Stone in c/m	M^3	3488.60	3487.90	0.0
Concrete				
Formwork (footing)	M^2	741.25	667.12	10.0
Formwork (foundation)	M^2	1244.20	1132.22	9.0
Formwork (beams)	M^2	598.40	508.64	15.0
Reinforcement (beams)	N	1.12	1.02	8.9
Finishing (floor)	M^2	41.65	36.54	12.3
Timber				
Eaves	M^2	16.79	15.71	6.4
Roof trusses	M^3	11846.50	11628.30	1.8
Plywood partitions	M^2	428.91	399.98	6.8
Flooring				
Wood flooring	M^2	29.71	27.24	8.3
Marble flooring	M^2	263.39	260.91	0.9
Brick flooring	M^2	222.79	222.79	0.0
Parquet	M^2	147.13	147.13	0.0
Timber planks	M^2	525.23	525.23	0.0
Finishing				
Cement stucco	M^2	263.69	250.69	4.9
Acrylic paint, 2 coats	M^2	73.94	70.21	5.0
Bituminous asphalt	M^2	31.11	30.87	0.8
Aluminum paint, 2 coats	M^2	156.90	155.82	0.7

*Note: The labor hour requirements are for 100 units of work.

Typically, workers tend to be uneducated, unskilled, and migrate in a group, with or without their families, throughout the country in search of employment (Koehn and Kothari 1995). Communication problems related to differences in language, religion, and culture tend to inhibit safety on the work site. In addition, high unemployment rates allow contractors to easily replace workers who are not performing at a satisfactory level. Workers, therefore, often take risks which tend to increase the level of accidents and fatalities on the job site.

4.1 *Labor wages and safety*

The labor wage in most developing areas is lower than that in developed nations. For example, the income of a laborer in India is approximately 4000 Rupees per month ($133) which is a weekly wage of about $33. Although the cost of living is minimal, the salary is still much lower than that in a developed country. This may be one of the important factors for the low productivity rates in

Table 2: Labor Rates in India and the USA

Building Construction Trades	Wage Rates* Dollars United States	Wage Rates* Dollars India
Asbestos Workers / Boiler Makers	25.75 / 26.28	1.24 / 1.24
Brick Layer / Brick Layer Helpers	23.57 / 18.58	0.83 / 0.62
Carpenters / Stone Masons	22.91 / 23.79	1.04 / 0.83
Cement Finishers / Rodmen	22.76 / 25.31	0.83 / 0.83
Electricians / Elevator Constructors	26.26 / 27.14	1.04 / 1.04
Equipment Operators - heavy / Pile Drivers	24.43 / 23.02	1.04 / 1.25
Equipment Operators - medium / Equipment Operators - light	23.47 / 22.47	0.93 / 0.83
Marble Setters / Mosaic and Terrazzo Workers	23.62 / 23.10	0.73 / 0.73
Painters (Ordinary) / Painters (Structural Steel)	21.47 / 22.26	0.62 / 0.73
Plasterers / Plasterer Helpers	22.28 / 18.75	0.83 / 0.62
Plumbers / Plumber Helpers	26.79 / 18.89	1.04 / 0.83
Roofers / Roofers Helpers	20.90 / 15.13	0.83 / 0.73
Sheet Metal Workers / Pipe Fitters	25.93 / 26.91	0.83 / 1.04
Structural Steel Workers / Structural Steel Welders	25.55 / 25.55	0.83 / 0.83
Tile Layers / Tile Layers Helpers	22.94 / 18.31	0.73 / 0.62
Trade Average	22.39	0.84

*Note: Wage rates are calculated per hour of work

developing regions. Because of the lower wages, contractors prefer to use labor, if possible, rather than purchasing expensive equipment. In addition, even if equipment is available, laborers, due to minimum training, may be more comfortable with and continue to use the standard tools with which they are accustomed to working.

Table 2 presents a comparison of the wage rates for various activities performed in construction, in India, with those of a highly developed country such as the United States (Labor 1996, Means 1996). As shown, the average salary in India is $.84/hour. This implies that construction workers fall into the category of a low income poverty level group which includes roughly 40% of the population of the country. These workers often do not have enough funds to purchase proper food. Improper nourishment tends to cause fatigue which reduces productivity and increases the probability of accidents and fatalities on the job site. This may be one factor for the high fatality and accident rate in India and other developing countries.

5 CONSTRUCTION PROJECT SAFETY

Construction is, by its very nature, a hazardous undertaking, especially in developing countries. The annual toll of accidents and fatalities in the industry is high, both in terms of individual cost and human suffering. The consequences of construction accidents are not expressible in terms of dollars alone. Money loses much of its significance when personal injury and death are involved. Nevertheless, the financial consequences of accidents should be an important concern to the overall construction industry as well as the individual contractor. In a developed country such as the United States, job site accidents impose a large burden of needless and avoidable expense on the industry. As an example, the direct and indirect cost of accidents in the USA is estimated to increase the cost of construction projects a minimum of 6.5% (Hinze 1991, Usmen 1994). In fact, the Occupational Safety and Health Administration (OSHA) was created in 1970 to establish and maintain an effective and comprehensive safety and health program (Occupational 1970). OSHA regulations were originally based on the voluntary industry standards in existence at the time the act was passed. Today, there is an ongoing effort to develop new standards and review and revise existing standards as new information becomes available. Specifically, OSHA has developed minimum standards and requirements for job-site safety and health for contractors and sub-contractors in the construction industry. These requirements have been tabulated in a comprehensive publication (OSHA 1979).

In developing countries, unsafe conditions exist on many sites, both large and small, and laborers are subjected to numerous hazards (Jaselskis 1993, Koehn 1995). In general, few training programs exist and, therefore, no orientation for new staff or workers is conducted, hazards are not pointed out, and no safety meetings are held. Workers, themselves, consider accidents as due to their negligence and accept that construction is a dangerous occupation. Table 3 illustrates the comparison of fatalities in India with those in the USA (Clough 1995). As shown, the fatality rate in India during 1972 was 8.6 times greater than that in a developed country such as the USA. The data indicate that the rate has been reduced 56% by 1991 in the USA (Hinze 1997). This is a significant decrease that may be due to the effect of OSHA and various other considerations such as the financial aspects of insurance (Hinze 1995). Similar factors have only recently been important in India, and, therefore, the fatality rate, today, is probably only slightly less than that in 1972.

Generally, in developing countries laws to protect laborers may not be strictly enforced. Also, contractors and their employees tend to ignore basic safety rules and regulations. In addition, safety programs and inspection procedures have often not been, or only recently have been, established or utilized to protect workers and reduce on-site hazards. In this regard, companies are also not heavily involved with monetary payments to injured workers. For example, in India the accepted practice for contractors is to pay the family of a victim involved with a fatality the equivalent of roughly 3 month wages. This amounts to approximately 12,000 Rupees or $400.00. This is not a strong monetary incentive for contractors to provide funds to develop programs in order to increase safety on construction work sites.

6 SUMMARY AND CONCLUSIONS

This paper presented data concerning construction productivity and safety in a typical developing country, India. The differences between developed regions such as the USA and developing countries are also noted. As an example, construction tends to be labor intensive in developing areas involving 2 - 10 times as many workers per activity compared with developed regions. High unemployment rates allow contractors to easily replace workers who are not performing at a satisfactory level. Those with jobs, therefore, often take risks which tend to increase accident and fatality rates on the job site.

The labor wage in most developing nations is significantly lower than that in developed countries. For example, the monthly wage in India does not allow laborers to purchase an adequate food supply. Improper nourishment tends to cause fatigue which may slow the rate of wor'· and also increase the number of accidents and fatalities at the job site. In developing countries, unsafe conditions exist on most sites since safety programs, if enacted, are generally not strictly enforced. Workers, themselves, consider accidents as due to their own negligence and accept that construction is a dangerous occupation. In fact, the fatality rate may be approximately 8 - 9 times that of the USA.

Companies are also not heavily involved with monetary payments to injured workers. For example, the accepted practice is to pay the family of a victim involved with a fatality the equivalent of approximately 3 months wages, 12,000 Rupees or roughly $400. Hopefully, new laws designed to protect construction laborers and decrease accidents and fatalities will be increasingly enacted and

Table 3: Construction Fatalities in India and the USA

Country	Year	Fatality Rate per 100,000 Workers
USA	1972	70
India	1972	600
USA	1991	31

enforced in developing countries. Nevertheless, the level of productivity and safety should be of universal interest to any contractor considering projects in developing countries. Specifically, the information presented should assist the international contractors with the management, estimating, scheduling, and monitoring required for efficient construction operations. The results may also assist international consultants during the planning process to achieve the desired quality, safety, cost effectiveness, and duration of construction projects.

7 REFERENCES

Clough, R. H., 1975. *Construction Contracting, 3rd Edition*, New York: John Wiley & Sons.

Foreign Direct Investment and Technology Transfer into India. 1997. *http://www.indiaworld.com.*

Hinze, J. W. 1997. *Construction Safety*. New York: Prentice-Hall, Inc.:4.

Hinze, J. and Applegate, L. 1991. Costs of Construction Injuries. *Journal of Construction Engineering and Management*, ASCE, 177(3): 537 - 550.

Hinze, J. and Russell, D. B. 1995. Analysis of Fatalities Recorded by OSHA. *Journal of Construction Engineering and Management*, ASCE, 121(2): 209 - 214.

Information about India. 1995. New Delhi, India: CIA World Fact Book.

Jaselskis, E. J. and Sauazo, G. A. R. 1993. Comparison of Construction Safety Codes in United States and Honduras. *Journal of Construction Engineering and Management*. ASCE, 119(3): 560 - 573.

Kaming, P. F., Olomolaiye, P. O., Holt, G. D., and Harris, F. C. 1997. Regional Comparison of Indonesian Construction Productivity. *Journal of Management in Engineering*, ASCE, 13(2): 33 - 39.

Kanettkar and Kulkarni. 1990. *Estimation of Buildings, Vol. 2*. New Delhi, India: Lakotia Publishers.

Koehn, E., Kothari, R. K., and Pan, C.-S. (1995). Safety in Developing Countries: Professional and Bureaucratic Problems. *Journal of Construction Engineering and Management*, ASCE, 121(3): 261 - 265.

Koehn, E., Pan, C.-S., Li, J.-Y., Pingili, R. P., and Tanratanawong, S. 1995. Construction Production Factors in International Areas. *1995 AACE Transactions*:C.6.1 - C.6.4.

Labor Productivity and Wage Rates. 1996. Madras, India: Dual Structural Pvt Ltd.

Levin, P. 1975. *Claims and Changes, 4th Edition*. WPL Associates, Inc.

Means Labor Rates for the Construction Industry, 19th Edition. 1996. Kingston, MA: R. S. Means Company, Inc.

Merriam, A. H. 1988. What does Third World Mean? *The Third World States of Mind and Being*, Edited by Norwine, J. and Gonzalez, A., Boston: Unwin Hyman.

Niland, J. R. 1970. *The Asian Engineering Brain Drain*. Boston, MA: Heath Lexington Books.

OSHA Safety and Health Standards (29 CFR 1926/1910). 1979. Washington: DC: U. S. Depart. of Labor, Occupational Safety and Health Administration.

Occupational Safety and Health Act. 1970. Washington, DC: Public Law 91-596, 91st Congress of the United States of America.

Punamia, B. C. 1987. *Building Construction, Vol 3*. New Delhi: Lakshmi Publishers.

Schmid & Testa. 1970. *Systems Building: An International Survey of Methods*. New York, NY: Frederick A. Praeger Publishers.

Usmen, M. A. 1994. Construction Safety and Health for Civil Engineers. *Construction Module*, ASCE.

Implementation of Safety and Health on Construction Sites, Singh, Hinze & Coble (eds)
© 1999 Balkema, Rotterdam, ISBN 90 5809 036 1

Health and safety practices on community projects: The South African experience

Theo Haupt
M.E. Rinker, Senior School of Building Construction, University of Florida, Gainesville, Fla., USA
John Smallwood
Department of Construction Management, University of Port Elizabeth, South Africa

ABSTRACT: Development within the South African context implies human development ranging from enjoying a decent standard of living to participating in community activities, as opposed to solely economic development. It is argued that no development is possible without community participation and that development can only be accelerated if the energies and resources of the people are mobilised and if the people enjoy maximum participation in the decisions which affect their lives and livelihoods. Given that health and safety constitutes an integral part of a decent standard of living and development, and that worker participation is one of the two pillars of health and safety, this paper investigates health and safety practices on community projects.

Findings based on the consultation of workers employed on community projects include: induction is rarely conducted; training is marginal; knowledge is sparse; knowledge of legislation is negligible; workers are never consulted regarding health and safety; reference is never made to health and safety when instruction to execute a task is given; personal protective equipment (PPE) is seldom provided; policy, rules and programme do not exist; goals are not set; Health and Safety Representatives are not appointed; meetings and inspections are not conducted, and workers perceive supervisors not to prioritise health and safety. These findings amplify the need for the implementation of comprehensive health and safety programmes on community projects to ensure that development does not occur at the expense of human life and suffering.

1 INTRODUCTION

"The record of human history is full of macabre experiments on human life, conducted, unwittingly or with forethought, by people in power, who were utterly convinced of the rectitude of their cause." (Schlemmer in Coetzee, 1988)

Policy makers and planners involved with construction projects in a community development context have to guard against contributing to the dilemma so vividly described by Schlemmer. It is necessary to be aware that development must be approached taking full cognisance of the needs of the beneficiaries and must include the elements of decentralisation, community-centredness and an upholding of their freedom and rights. This will include not only a recognition but also an understanding of their values, customs, social structures and political participation.

2 COMMUNITY DEVELOPMENT AND COMMUNITY PARTICIPATION

According to Coetzee (1988) development is about people who experience the implications and practical functioning of the realities in which they constantly and unavoidably find themselves. These people who are the intended beneficiaries of community development projects are the millions in Southern Africa who live in absolute poverty, are either unemployed or underemployed and experience total inequality in respect of access to effective political power. The objective of community development projects should go beyond the mere provision of a physical facility and be to raise the standard and quality of living of community people. They should additionally be given the opportunity to develop their own potential optimally. However before this can become a reality it will be necessary to develop a local understanding with local participation of the social realities within which communities live.

It would seem that there is a movement towards providing opportunities for community participation in community development which could result in social reconstruction and meaningful existence. It is argued that no development is possible without the participation of the community (Coetzee, 1988). In fact the improvement of conditions within a community may only be regarded as such if the community itself has agreed to the changes to be effected. Since all people value respect and desire to be treated as worthy individuals it is necessary that they participate in making fundamental choices especially in respect of issues which will impact on them both directly and indirectly. These choices are made as a result of the way in which they define their own position. This of necessity requires a fundamental paradigm shift against the background of the past political dispensation in South Africa which was characterised by deliberate efforts to dehumanise the majority of the population who are now targeted to benefit from the Reconstruction and Development Programme (RDP) of the present government.

Several international development agencies have as their focus the view that development can only be accelerated if the energies and resources of the people are mobilised and that there should be maximum participation in the making of decisions which affect their lives and livelihoods. Furthermore existing economic and social structures must be utilised to contribute towards full development instead of new ones being created and imposed on them. Economic development should be seen as the activation of idle resources (Cernea, 1985). In this context community development can be defined as being where local people, working in their local environment, not only have the right, but also the responsibility, of choosing their own development objectives and making their own decisions (DBSA, 1993). Through active participation in the entire process of development individuals and communities will be able to improve their quality of life, facilitate growth and contribute significantly to employment creation.

Both Strong (1993) and Gounden (1996) argue for self reliant and people-centred development to ensure a sustainable growth process, the absence of which will result in widening the gap between the advantaged and disadvantaged in South Africa. Not only should this development be self-reliant but it should also be equitable, participatory and sustainable. This implies that communities must be active participants, act in their own interests as they see them, rely primarily on their own resources and exert their own control over development projects.

According to a report of the Economic Development Institute of the World Bank (1988) several factors have contributed to an increasing recognition of the need to address social issues of development. Community involvement in project design and management impacts positively on the efficiency of implementation, cost recovery and project sustainability. Because of their limited capacity government and local authorities are unable to effectively manage the increasing number of development projects and programmes. Furthermore planners need to understand the needs of communities and assess how community life will in fact be affected by development projects.

Szentes (1993) suggests development must meet the criteria of :sustainability; satisfy basic human needs; improve social and economic well-being of people; enable people to reach their potential; build their self-confidence, and lead lives of dignity and fulfilment. This cannot be achieved unless people have opportunities to develop and actively influence the direction and execution of development projects than be mere passive beneficiaries of the benefits of these projects (Paul, 1987). It is also significant to realise that communities themselves have a strong base of skills which is so often ignored (Sapere, 1996 and Galtung, 1993).

It should also be noted that freedom from harm i.e. health and safety, is in fact a lower level need in Maslow's hierarchy of needs (Mitchell, 1978).

3 DESIGNERS AND COMMUNITY PARTICIPATION

Designers have been prone to adopting over-simplistic and paternalistic views of their clients especially in a community development context.

There is a realisation that new emphasis and added responsibilities are being placed on professional consultants to involve local communities in the development process (Taylor and Norval, 1994). The traditional hierarchical "top-down" approach which characterised the way projects were initiated and managed by designers in the past has lost favour with the Department of Public Works which now advocates the implementation of project management principles. It is therefore necessary for them to change their thinking, react and adapt to change.

Sapere (1996) suggests that designers will have to change their role from provider of solutions to enabler and supporter in the development process. Instead of building for the community they need to allocate and direct resources. Rather than be telling people what to do they should be telling them how to find out what to do, and how to find out how to do it. They should support communities in developing their own options by providing guidelines and the tools to do so. They should also develop an understanding of existing people skills and resources and develop people technologies. Finally they should support information transfer for the people by the people to the people. This of necessity will require a major paradigm shift on the part of the designers.

It is also important to realise that designers influence health and safety directly and indirectly. Directly though selection of frame type, design, detail, and specification of materials. Indirectly through selection of procurement system, preparation of contract documentation, and decision regarding contract duration (Smallwood, 1996).

4 RESEARCH

46 Workers employed on a typical rural community low cost housing project in the Western Cape, South Africa were consulted as part of a pilot study of health and safety practices on both rural and urban community projects.

93.5% of the workers consulted were male The mean age was 32.7 years with 50% being in the 20 to 29 year old age group.

65.3% of workers had an educational level of standard 2 to 7 which is equivalent to 4 to 9 years of schooling, provided that they completed each year successfully and consecutively. 19.6% had a further year or two of schooling.

It has been suggested that the educational level of workers would affect their amenability to formal health and safety training and education. This could be overcome by varying the training methods to include visual aids such as video's, role playing and demonstrations which would not require high levels of education to understand.

47.8% of workers had no qualifications whatsoever. However, 39.1% had a recognised building trade. This would suggest that almost half the workers had received no formal training in the correct use of hand tools and equipment. Whatever skills they possessed were self-taught.

Despite the lack of formal training, very few had experienced injuries or illness while working on the current project (average of 2.7 months) or in the constuction industry (average of 7.2 years) (Table 1). This could possibly be attributed to their average experience in the industry. It may also be possible that the workers looked out for each other with regard to safety.

Table 1 Percentage of workers injured or sick.

Injured/Sick	Yes response (%)	
	Current project	Construction
Injured	4.3	19.6
Sick	8.7	26.1

91.3% of workers had not been told anything concerning health and safety upon commencement of work on site. Similarly, 95.7% and 82.6% had not received any health and safety training while working on the current project and in construction respectively. According to Levitt and Samelson (1993), the single most important thing that a job-site manager can do for new workers on a project is to ensure that they are well-oriented. Induction is a critical health and safety intervention given that research indicates the greater percentage (40%) of injured workers to be new employees (OSHA, 1993) i.e. that have been employed for less than a year. Most of the employees on this project have been employed on it for less than a year and could therefore be regarded as new workers. A study undertaken by Hinze (1976) indicated that managers who had direct contact with every new worker had fewer than half as many accidents on their jobs as others who did not.

50% of workers maintained they had no knowledge of health and safety (Table 2).

Table 2 Knowledge of health and safety.

Knowledge of health and safety	Response (%)
A lot	2.2
A fair amount	19.6
Very little	28.2
Nothing	50.0
Total	100.0

However 39.1% of the remaining 50% of the workers knew that they should wear 'safety / protective clothing'. Health and safety training is both a legal requirement and an essential intervention to raise the level of awareness and to empower workers to work in a healthy and safe manner. This is supported by OSHA (1989) in its conclusion that effective management of worker safety and health protection is a decisive factor in reducing the extent and the severity of work-related injuries and illnesses. Effective management would include appropriate health and safety training. Training should be at the core of every health and safety programme (Hinze 1997).

Only 6.5% and 17.4% of workers knew of the Occupational Health and Safety Act (OHS Act) and the Compensation for Occupational Injuries and Diseases Act (COID Act) respectively. The OHS Act requires of employers to inform workers of their responsibilities in terms of the Act.

Table 3 indicates that most workers were never consulted about health and safety.

Table 3 Frequency of consultation of workers about health and safety.

Frequency	Response (%)
Often	2.2
Sometimes	4.3
Seldom	2.2
Never	91.3
Total	100.0

Workers spend more time in and are more familiar with their workplaces than managers. They also have an intimate understanding of their work processes (Hammer, 1981). Consequently they should be consulted regarding health and safety. According to Hinze (1997) safety is a "people thing" and workers are the best resource of every firm.

78.3% of workers stated that reference was never made to the health and safety aspect of a job when instruction to execute a task was given (Table 4).

Table 4 Frequency of reference to health and safety when instruction to execute a task is given.

Frequency	Response (%)
Always	2.2
Often	6.5
Regularly	2.2
Seldom	8.6
Never	78.3
No response	2.2
Total	100.0

Successful completion of a job does not only imply that the desired quality standards be met, costs contained within the agreed budget and completion be achieved within the agreed time, it also includes completion without any fatality, injury and disease. Consequently reference should always be made to the health and safety aspects of a job when instruction to execute a task is given. Every employee who arrives on the site should be treated as a new employee regardless of how long they plan to stay and even if they have worked for the firm or site manager before (Levitt and Samelson, 1993). This orientation process should be a continuous one. Health and safety should be focused upon whenever the nature of the work to be executed changes.

56.5% of the workers maintained that they were never exposed to hazards in their job (Table 5). Although there may have been no hazards, the lack of health and safety induction and training may have precluded the workers from identifying hazards. For example, on one of the days during which the survey was conducted a strong wind was blowing and workers were lifting long lengths of single-span metal roof sheeting onto the open roofs of units without any gloves, hard hats, ladders, scaffolding or harnesses. They were not aware of the potential hazards and risk of injury they were facing. Employees must be trained to be aware of hazards and know safe working practices (Hammer, 1981; Levitt and Samelson, 1993). The firm is responsible for the health and safety of its employees and for the adequacy and health and safety of all construction methods and procedures (OHS Act, 1993).

Table 5 Frequency of exposure to hazards.

Frequency	Response (%)
Often	4.3
Sometimes	21.7
Seldom	17.5
Never	56.5
Total	100.0

Only 4.3% of workers required and received any personal protective equipment (PPE) which by itself in no way eliminates or reduces a hazard, but merely reduces the effects of exposure to a hazard (Marshall, 1994). Not a single worker was wearing a hard hat, 63% were wearing an overall, 8.6% were wearing gloves and 41.3% were wearing safety footwear. Safety footwear and hard hats constitute the minimum requirements in terms of suitable attire. The OHS Act (1993) requires that management should train employees on the proper use and maintenance of PPE. They should in fact be trained to know when and what PPE should be used. From observation of operations on site during the survey it was apparent that many more of the workers should have been making use of PPE. For example glass panes were being carried and inserted into windows without the use of any gloves.

Table 6 indicates the existence of and the extent to which various health and safety related practices occurred. There is a distinct lack of practices whether verbal or written, and formal or informal.

Table 6 Existence and extent of health and safety related practices.

Aspect/Practice	Response (%)			
	Yes	No	Don't know	Total
Policy	2.2	60.8	37.0	100.0
Rules	19.6	47.8	32.6	100.0
Programme	13.0	56.5	30.5	100.0
Goals	6.5	65.2	28.3	100.0
Meetings	8.7	89.1	2.2	100.0
H & S Representatives	13.0	84.8	2.2	100.0
Inspections	8.7	89.1	2.2	100.0

The existence of a formal health and safety programme, policy and rules indicates management commitment and provides the framework within which health and safety can occur. Health and safety goals are important as they motivate, engender teamwork and enable evaluation of performance and improvement. Health and Safety Representatives are not only a legal requirement, but they facilitate health and safety organisation and communication. Meetings and inspections are also both legal requirements and provide a forum for communication and constitute control respectively (Hinze, 1997; Hammer, 1981 & Levitt and Samelson, 1993).

Table 7 Degree of importance of health and safety to supervisors.

Degree of importance	Response (%)
Very	8.7
Important	17.4
Fairly	6.5
Not	67.4
Total	100.0

The majority of workers maintained that health and safety was not important to their Supervisor (Table 7).

Supervisors and foremen manage the workplace and consequently their level of prioritisation of health and safety could be the difference between health and safety or the lack thereof (Hinze, 1997). The influence that they have on the health and safety performance of their crews is not trivial. Safer records i.e. lower incidence of accidents, injuries and illnesses have been noted among foremen who take a personal interest in and keep a watchful eye over particularly their new workers (McMeel, 1979) since the incidence of accidents is highest among new workers.

One comment regarding health and safety was made by 28.3% of workers whereas 52.2% made two comments. This does indicate a level of awareness of health and safety. Generally workers appreciated the need to be careful and to wear PPE.

5 CONCLUSIONS

It is clear from this pilot study that there is a need to prevent accidents from occurring on community projects by improving health and safety at all levels. This particular study did not focus on how the obvious lack of commitment to health and safety by management (supervisors and foremen) impacted on cost, productivity, quality, schedule and the environment. This lack of management commitment alone has the potential for resulting in immense human suffering and being ultimately retrogressive and marginalising development. The study revealed that the community was excluded from becoming involved in health and safety issues prior to and during the actual execution of the project. This is contrary to the objective of community development projects which, as has been stated earlier, should be maximum participation in the making of decisions which affect the lives and livelihoods of members of the community.

6 RECOMMENDATIONS

In order to rectify both the lack of commitment from management and non-involvement of the community a culture of health and safety needs to be developed over time. It is recommended that:

- health and safety issues be included on the agenda during the inception and negotiation stages of all community projects;
- designers become involved on an enabling and supportive basis and design community projects for health and safety in respect of the technology and materials to be used taking into account the skills level of the respective community;
- a comprehensive health and safety plan be designed with input from all stakeholders, and approved which would address all the potential hazards which could be encountered during the execution of the project (Figure 1);
- all workers formally endorse this plan understanding that violation of its provisions could lead to censure;
- a comprehensive health and safety induction and training programme be implemented for all workers prior to the commencement of the project which covers all the provisions of the approved health and safety plan;
- all new workers who arrive on the site undergo the same health and safety induction and training programme before being allowed to work;
- health and safety aspects be focused upon whenever the nature of the work to be executed changes and workers be trained or re-trained accordingly;

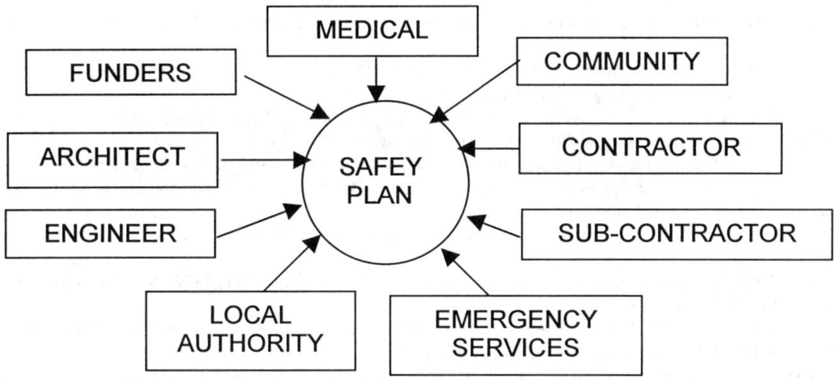

Figure 1. Input from all stakeholders into a health and safety plan for community project

- a competent community person be appointed who will ensure compliance with the approved safety plan and undertake regular safety inspections which will include identification of potential hazards and corrective measures where non-compliance is evident; and
- regular inspections be undertaken to ensure a drug- and alcohol-free worksite.

The implementation of comprehensive health and safety programmes on community projects will ensure that community development does not occur at the expense of human life and suffering.

7 FURTHER RESEARCH

The reason for the uncharacteristic low exposure to injuries and illness by the workers on this project should be investigated further. The possible relationship between levels of formal education or the lack thereof and amenability to formal education and safety training and education should also be researched. Particularly in the rural areas where traditionally workers have lower levels of education than their urban counterparts a model of health and safety training and education needs to be developed.

A study will be extended to include several more sites of other rural and urban community projects.

REFERENCES

Cernea, M. (ed) 1985. *Putting People First*. New York: Oxford University Press.

Coetzee, J.K. (ed) 1988. *Development is for People*. Johannesburg: Southern Book Publishers (Pty) Ltd.

Development Bank of Southern Africa 1993. *Socio-economic enhancement of development projects*. Halfway House: Development Bank of Southern Africa.

Economic Development Institute of the World Bank 1988. *The Role of Community Participation in Development Planning and Project Management*. Washington, DC: The World Bank.

Galtung, J. 1993. People-centred Development through Collective Self-Reliance *In South Centre (ed) Facing the Challenge*. London: Zed Books.

Gounden, S. 1996. Reconstruction and Development Programme : The Construction Industry and Related Procurement Reform *Journal of Construction Procurement*.

Hammer, W. 1981. *Occupational Safety Management and Engineering*. New Jersey: Prentice-Hall.

Hinze, J.W. 1997. *Construction Safety*. New Jersey: Prentice-Hall.

Hinze, J.W. 1976. The Effect of Middle Management on Safety in Construction, *Department of Civil Engineering Technical Report no 29*. Standford University, Stanford.

Levitt, R.E. and Samelson, N.M. 1993. *Construction Safety Management*. New York: John Wiley and Sons, Inc.

Macro Economic Research Group 1993. *Making Democracy work : A Framework for Macro-Economic Policy in South Africa*. Centre for Development Studies.

Marshall, G. 1994. *Safety Engineering, Second Edition*. American Society of Safety Engineers.

McMeel, K. 1979. "The Effect of Foremen on Safety on Super Projects." Unpublished Master's Report, University of Missouri.

Paul, S. 1987. Community Participation in Development Projects : The World Bank Experience. *In Readings in Community Participation.* Washington, D.C: EDI.

Sapere, D. 1996. Back to Basics and Cost-effective solutions *Housing South Africa*. August.

Sawczuk, B 1996. *Risk Avoidance for the Building Team*. London: E&F.N. Spon.

Smallwood, J.J. 1996. The influence of designers on occupational health and safety. *In Proceedings of the 1st International Conference of CIB W99 – Implementation of Safety & Health on Construction Sites, Lisbon, 4-7 September 1996:* 203-213. Rotterdam: AA Balkema.

Strong, M.F. 1993. Achieving Sustainable Global Development. *In South Centre (ed) Facing the Challenge*. London : Zed Books.

Szentes, T. 1993. The Challenge to the South? A Challenge to Humankind as a Whole! *In South Centre (ed) Facing the Challenge*. London: Zed Books.

Taylor, R.G. and Norval, G.H.M. 1994. Developing Appropriate Procurement Systems For Developing Countries *In Proceedings of CIB W92 Simposium*, CIB Publication No 175.

http://www.osha-slc.gov/OshDoc/Fact data/FSNO93-07.html: Improving Workplace Protection for New Workers – New Worker, High Risk! The United States Department of Labour, Occupational Safety and Health Administration.

http://www.osha-slc.gov/FedReg osha data/FED19890126.html: Safety and Health Programme Management Guidelines; Issuance of Voluntary Guidelines The United States Department of Labour, Occupational Safety and Health Administration.

http://www.osha.gov/oshpubs/perpro.html: Personal Protective Equipment The United States Department of Labour, Occupational Safety and Health Administration.

Implementation of Safety and Health on Construction Sites, Singh, Hinze & Coble (eds)
© 1999 Balkema, Rotterdam, ISBN 90 5809 036 1

Philosophy of construction safety management in Japan

T.Watanabe
Kochi University of Technology, Japan

S.Hanayasu
Ministry of Labor, Tokyo, Japan

ABSTRACT: In this paper the history of implementation of safety measures, and law systems related to safety management and their judgments at court in Japan are introduced. Characteristics and philosophy of construction safety management are then summarized.

In practice of the safety management in Japan, achievement of zero accident has been recognized not only ideal but also a political goal. It has been the philosophy of safety management in Japan.

However, some gap actually exists between ideal and reality. Insufficient public owner's consideration on safety is one of the reasons. Much efforts to lessen the gap are made at construction stage; however, important problems occur associated with conducting these efforts.

To construct new philosophy, it becomes more important for each subcontractor, each worker, and a public owner as well as a prime contractor to develop his/her own safety management capability which can flexibly incorporate characteristics of each construction site.

1 INTRODUCTION

Japan has constructed and practiced its own philosophy in safety management. Practice of the philosophy has played an important role in decreasing the number of construction labor accidents. Recently, however, limitations and problems associated with practicing the philosophy are becoming more serious. Necessity of constructing new philosophy is pointed out.

In this paper the history of implementation of safety measures, and law systems related to safety management and their judgments at court in Japan are introduced. Characteristics and philosophy of construction safety management are then summarized. Future issues are finally discussed.

2 LABOR ACCIDENTS IN JAPAN

Figure 1 demonstrates the numbers of labor accident causalities and fatal accidents by year.

In Japan during the high economic growth period, which started in the late 1950's, the heavy industry had rapidly developed, and infrastructure facilities had also been developed at a remarkable speed. This quick restoration of economy had brought many accidents such as large-scale explosions and leakage of chemical substances in the manufacturing industry and rock-fallings in tunneling construction and earth collapse in the construction industry.

The construction industry is also a "dangerous industry" in Japan. Its working population is 10% of the all industries. However, its number of work-related causalities is 30% of all industrial accidents, and fatalities is more than 40% of all industrial deaths. In the construction industry the number of industrial accidents has been steadily decreasing, but its fatalities have

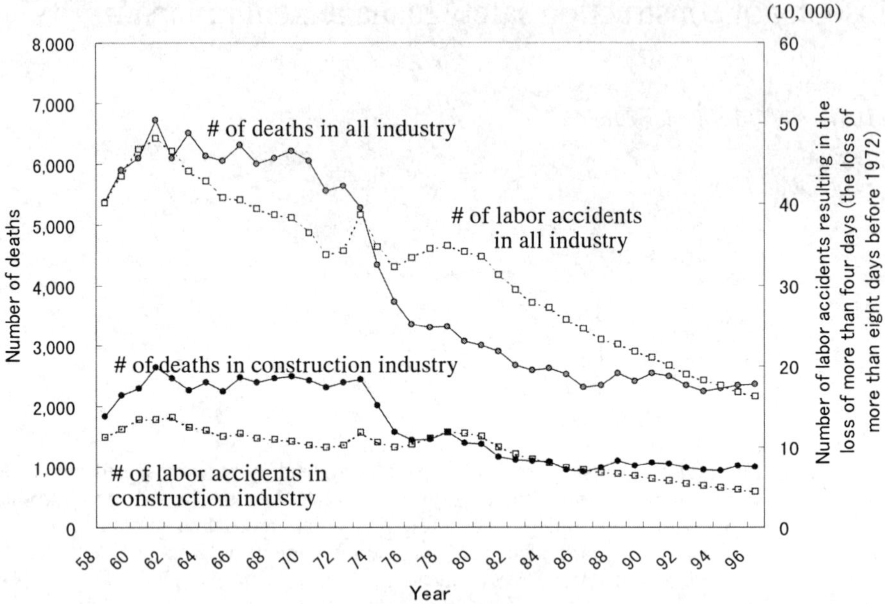

Figure 1: Numbers of labor accident causalities and fatal accidents by year

hovered around 1,000 for more than ten years. New measures to decrease fatalities, in particular, seems necessary to implement.

3 CONSTRUCTION SAFETY REGULATIONS AND SIGNIFICANCE OF INDUSTRIAL SAFETY AND HEALTH LAW

Table 1 summarizes the history of major Construction Safety Regulations in Japan.

Once many (serious) accidents occur with a certain type or under a certain situation, detailed regulations are usually developed and uniformly applied to the corresponding construction type and situation with few exceptions. Development of detailed regulations and their uniform applications with few exceptions is one of the characteristics in safety management in Japan.

Among these regulations, the Industrial Safety and Health Law established in 1972 is one of the most important ones. As demonstrated in Figure 1, the number of causalities and fatalities had been sharply decreasing after its establishment.

Prescription of contractor responsibilities is one of the most significant matters associated with enactment of the Industrial Safety and Health Law.

The Labor Standards Law established in 1947 stipulates that employer's "nonfeasance" obligation: it stipulates that an employer shall not make a worker do a dangerous work. Here, an employer means those who directly give directions to workers such as a site manager. Thus, the representative of a company is not necessarily included as an employer in the Labor Standards Law.

On the other hand, the Industrial Safety and Health Law stipulates contractor's feasance obligations to take measures to protect workers from accidents. Here a contractor means those who run the business in the construction industry and employ workers, and substantially means the representative of a company. That is, in the Industrial Safety and Health Law, the top of a company takes a leadership and responsibilities for performing safety management activities. Its significance and effect was remarkable.

Table 1 : History of Major Construction Safety Regulations in Japan

1945～1955 : Dawn of administration of industrial safety and health post-World War II	
1947	Enactment of the Labor Standards Law [Inclusion of a chapter of industrial safety and health] Enactment of the Ordinance on Industrial Safety and Health [Prescription of safety standards for the structure of scaffoldings, measures to prevent falls, protection from landslides and falling objects, and safety in the use of heavy construction equipment.]
1955～1965 : High economic growth period—increase in the number of labor accidents due to a rapid increase in the amount of construction works	
1959	Amendment of the Ordinance on Industrial Safety and Health [Inclusion of provisions related to scaffolding, pile driving, blasting, and slinging]
1963	Amendment of the Ordinance on Industrial Safety and Health [Inclusion of provisions related to formwork and support structures]
1964	Enactment of the Law Concerning Labor Accident Prevention Associations [Five industries including construction were designated for special concern.] [Encouragement of voluntary safety activities within the industry through formation of the Construction Labor Accident Prevention Association] [Prescription of responsibilities of a prime contractor, overall safety and health controller, and client]
1965～1975 : Frequent occurrence of labor accidents stimulated change in administration of industrial safety and health	
1965	Tightening of safety standards for tunneling construction
1968	Tightening of safety standards for work at high elevation
1970	Tightening of safety standards for high-pressure work
1972	Enactment of the Industrial Safety and Health Law
1975～1985 : Frequent occurrence of conventional types of accidents such as falls, slipping, and stumbling and increase of large-scale accidents	
1980	Amendment of the Industrial Safety and Health Law [For the purpose of providing a greater protection against major accidents, preparation of construction work plans with the participation of qualified personnel and its submission to the Minister of Labor] [Inclusion of rescue measures to be taken by engineering managers in tunneling work]
1985～ : Increase of serious accidents such as the toppling of a mobile crane and increase of incidents at small- and medium-size construction sites	
1992	Amendment of the Industrial Safety and Health Law [Strengthening of requirement for a general safety and health control manager at field office] [Introduction of a new system of head-office safety and health control managers]

Source: Amended Industrial Safety and Health Law and Construction Safety Control and The Principles of Construction Management

4 CONTRACTOR'S LIABILITY FOR DAMAGES RESULTING FROM VIOLATION OF THE LEGAL OBLIGATION TO TAKE WORKER SAFETY MEASURES

4.1 *Development of contractor's liability for damages resulting from violation of the legal obligation to take worker safety measures*

As well as establishment of the Industrial Safety and Health Law, development of contractor's liability for damages resulting from violation of the legal obligation to take worker safety measures has given great influence to Japanese labor safety management. This contractor's liability will be referred to as the contractor's liability for damages in this paper.

This term is not specifically prescribed in any Japanese law. This was first specified in the supreme court's decision in 1975 as employer's obligation of protecting worker's life and health against danger. These had been gradually developed as a clear employer's obligation through many precedents.

This contractor's liability for damages is undertaken to individual worker with whom a contract is made. When a worker has to do a possibly dangerous work, the contractor who makes a contract with a worker is legally obliged to take safety measures to protect a worker from damages to his/her life and health. If contractor's violation of this obligation results in worker's accident and damages, the civil law finds liability for the damages.

The contractor's liability for damages consists of a)prediction obligation, that is to predict a possibility of occurrence of an accident and b)damage avoidance obligation, that is to take measures to avoid damage of accidents which could possibly happen.

Provisions described in the Industrial Safety and Health Law is a part of the contractor's liability for damages.

That is, the Industrial Safety and Health Law consists of a set of regulations, which forms the basis of the contractor's liability for damages, of the most important risk factors considered common to all work places or all types of work. The contractor is obliged to fulfill these regulations on the penal code.

The contractor can first exempt his/her responsibility by observing the Industrial Safety and Health Law and the contractor's liability for damages consisting of the prediction and damage avoidance duties.

After the contractor's liability for damages was developed, the concept of the "safety first" had gradually been prevailing in construction sites in Japan.

4.2 *Characteristics of the contractor's liability demonstrated in precedents*

The contractor's liability for damages forms fundamentals of safety management concept in Japan. This section discusses characteristics of the contractor's liability demonstrated in precedents.

The first characteristic is not to allow the contractor to take balance between securing safety and its cost when taking safety measures. This characteristic is demonstrated through the following precedents. "The defendant's first obligation as the damage avoidance obligation is to make his/her efforts to maintain working environment best achievable at that time." "The idea that defendant's obligation of keeping workers' health is just fulfilled through investment for improving working environment considering its balance with company's profit is hardly accepted."

The second characteristic is to require the contractor to make his/her best efforts in safety education not to let a worker conduct unsafe actions. Contents of education for a worker should be according to his/her ability and experience. One precedent prescribed an example of sufficient education as follows: "Education for a new worker consisted of the followings: first making him understand site's conditions very well, teaching him location of dangerous spots at the cite, safety facilities, and safety plans, prohibiting him from walking alone, making someone well familiar with the cite accompanying him for the first three or four days, giving him instructions specific to each work as well as general instructions at every meeting, and prohibiting him from approaching to holes except for those lifting up and down equipment." Due to sufficiency of the education and installation of necessary safety facilities to protect a worker from falling, contractor's liability was judged fulfilled in the court decision.

Although each worker is supposedly obliged to perform "self-safety protection" duty, the contractor is sometimes required to give education treating each worker as a "child" in reality.

As far as judgments of laws on safety management at court are concerned, the philosophy behind the contractor's liability for damages is not to diminish the number of accidents, but to achieve zero accident. From engineering viewpoints, the safety management might be criticized unreasonable that the balance between securing safety and its cost is not considered and that "self-safety protection" duty is not necessarily obliged to each worker. In practice of the safety management in Japan, however, achievement of zero accident has been recognized not only ideal but also a political goal.

5 OVERVIEW OF CONSTRUCTION SAFETY CONTROL PRESCRIBED IN THE INDUTRIAL SAFETY AND HEALTH LAW

As already mentioned, the Industrial Safety and Health Law, which forms the basis of the contractor's liability for damages, establishes a set of regulations of the most important risk factors. The contractor is obliged to observe these regulations based on the penal code. This section overviews construction safety control prescribed in the Industrial Safety and Health Law.

5.1 *Major construction safety control*

(1) *Contractor responsibilities*

In the Industrial Safety and Health Law each contractor is obliged to take preventive measures to protect his/her worker from labor accidents. Here his/her worker means the one who is directly employed and does actual construction works receiving instructions from the contractor. This obligation is called contractor responsibilities.

Major contractor responsibilities are, for example, a)requesting for consolidating and improving a safety construction plan to a prime contractor, b)implementation of a safety construction plan, c)elimination of unsafe conditions and unsafe actions, and d)provision of education and instructions to workers.

A site representative of a contractor is generally commissioned the contractor responsibilities and rights to fulfill them. The representative of the contractor is obliged to supervise the site representative.

(2) *Responsibilities of an overall safety and health controller*

Since construction works are generally conducted through subcontracting systems with multiple layers in Japan, the whole works are executed under overall control of a prime contractor. The Industrial Safety and Health Law stipulates that one of the prime contractors who directly makes a contract with the client must assign an overall safety and health controller at a site with larger than certain sizes. Responsibilities of the controller are regarding safety and health of each subcontractor and workers. Actual responsibilities are as follows: a)setting up and managing safety and health committees, b)coordinating multiple works, c)inspecting working places, d)providing instructions and assistance in safety and health education for workers, e)preparing a schedule plan and location plans of machine and equipment, f)providing instructions on safety measures in the case of using specific machine or equipment, g)providing a place and information to inform actual conditions of the site, h)deciding rules on signal, sign, warning, and places for collecting materials, i)providing rescuing measures in case of explosion and fire, and j)taking other measures to prevent occurrence of certain types of accidents

(3) *Prime contractor responsibilities*

The Industrial Safety and Health Law stipulates two additional responsibilities of a prime contractor towards a subcontractor: first responsibility is to provide general instructions on observance of the Industrial Safety and Health Law and its related regulations, and second responsibility is to provide technical instructions on taking safety measures at specific sites.

It is not necessarily possible and sometimes completely impossible for a subcontractor to take effective safety measure with his/her sole efforts. This is a reason that these responsibilities of a prime contractor is stipulated.

Even if these two responsibilities are not fulfilled; however, a prime contractor is not punished. These responsibilities are equivalent to instructions. A main objective of the prescription of these responsibilities is to draw more attentions of a prime contractor.

(4) *Client responsibilities*

When a prime contractor conducts construction works, he/she lets each subcontractor use facilities, equipment, and machines such as scaffoldings, temporary passage, formwork and support structures, and a mobile crane. When a prime contractor lets a subcontractor or his/her workers use scaffoldings and specific facilities, a prime contractor is obliged to take sufficient measures to prevent occurrence of a labor accident. In the Industrial Safety and Health Law this responsibility is called the client responsibilities of specific works.

Stipulation of the client responsibilities never means that a subcontractor who uses prime contractor's facility does not have to take any preventive measures against accident. A client and a subcontractor have to take each responsibility. That is, double safety measures are taken by the both players.

5.2 *Characteristics of construction safety control*

In the Industrial Safety and Health Law the responsibilities of the contractor, the overall safety and health controller, the prime contractor, and the client are stipulated.

Although the core measures to prevent occurrence of labor accidents are the contractor responsibilities, these are complemented and consolidated with other responsibilities. This structure of the responsibilities is one of the characteristics of the Japanese systems; however, some drawbacks also exists. In the next section, future issues in construction safety management in Japan are discussed.

6 FUTURE ISSUES IN CONSTRUCTION SAFETY MANAGEMENT IN JAPAN

As already mentioned, achievement of zero accident is ideal in safety management in Japan. However, some gap actually exists between ideal and reality. In this section issues associated with this gap are discussed.

6.1 *A factor to generate some gap between ideal and reality in public works*

A factor to generate some gap between ideal and reality in public works is insufficiency of safety measures taken by a public owner.

In conventional judgments of safety related laws, since the public owner does not employ workers or give work instructions, it has been considered for a long time that the public owner takes neither civil nor penal responsibilities in principle.

In reality, however, the public owner must take responsibilities for preventing accidents from social public viewpoints. During construction, the Japanese public owner has also exercised supervisions and given instructions as a client. For these reasons, it is recently considered that the public owner should carefully consider safety control in ordering works, exercising supervisions, and providing instructions. The Industrial Safety and Health Law also stipulates that those who contract with others for executing works such as an owner of a construction project shall carefully consider in determining construction method, construction period, and etc. not to provide working conditions to impair workers' safety and health.

In reality public owner's consideration on safety is not necessarily sufficient. The following two problems are representative ones.

(1) *Appropriation of safety measurement costs in cost estimation is not necessarily sufficient.*

In most of public works in Japan, materials and a construction method as well as specification of structure is determined in detail. Before bidding, the public owner calculates the price of the construction works based on the detailed determination. This price is called the ceiling price, and it becomes the upper limit for contract price. The more detailed materials and a construction method are determined, the smaller freedom of construction the contractor has. Therefore, the ceiling price has characteristics close to the "absolute price."

The safety measurement costs in the ceiling price is assumed to be included in the Temporary Work Costs, which is analyzed by multiplying the Direct Construction Costs by a certain rate.

That is, in the ceiling price, which has characteristics close to the absolute price, the safety measurement costs are not calculated through summing necessary items but multiplying the Direct Construction Costs by a certain rate. This rate for calculating the Temporary Work Costs increases as the size of construction site becomes small, that is, the Direct Construction Cost decreases. However, the value of this rate represents an average, which does not correspond to all construction site conditions. At some small- and medium-size construction sites, therefore, the safety measurement costs obtained with the rate are often insufficient.

Appropriation of necessary and sufficient safety measurement costs at some small- and medium-size construction sites is an important problem to be solved in future.

(2) *Appropriate design alterations are not always taken by the public owner.*

Contingency fees are said to be insufficient in some small- and medium-scale public work projects in Japan. Furthermore, the deadlines of many public work projects is fixed in March, which is the end month of the fiscal year in Japan. It is often said that, therefore, design alterations to secure safety at work with increase in contract value or delay in construction work are not necessarily accepted. Exercising appropriate design alterations is another important issue in future.

6.2 *Problems associated with reducing the gap between ideal and reality*

Safety control regulations at construction phase are provided in much more detail than at design stage. Detailed safety regulations at construction phase seems considered one of the means to lessen the gap between ideal and reality. However, this means creates some problems.

(1) *Partial fulfillment of the contractor responsibilities*

In the Industrial Safety and Health Law the contractor responsibilities are considered the core measures to prevent labor accidents. The subcontractor is not always a equal partner with the prime contractor, and subcontractor's safety control activities are often conducted at the direction of the prime contractor. Furthermore, responsibilities of the prime contractor become stricter every time the law is amended. In many cases, therefore, the contractor responsibilities of the subcontractor are only partly fulfilled.

(2) *Existence of circumstance in which a worker is "spoiled"*

The Japanese labor contract is a bilateral onerous contract. Furthermore, the Industrial Safety and Health Law stipulates that a worker shall accordingly observe rules which the contractor provides. Summarizing these, the contractor first takes measures to protect workers' life, body, and health, and then a worker has "self-safety protection" obligation accordingly.

One of the contractor's obligations is to make a worker wear the safety belt and the helmet. This type of obligation is often called a "make-obligation." This "make-obligation" is assumed to include not only giving directions and orders to a worker but also supervising and checking a worker whether he/she executes the directions and orders correctly. However, there exists a problem: it is unclear what measures the supervision includes? The concept of supervision is ambiguous.

Recently precedents admitting comparative negligence are increasing. Some precedents demonstrated that, however, the site manager should have intensified supervision obligations in the case of having an imprudent worker at the site. There even exists an opinion that the supervision means going to the site and watching every single movement of all workers incessantly.

In Japan there still exists "spoiled" circumstance in which each worker can be taken care of without observing fundamental of safety: to protect yourself by yourself.

(3) *Problems associated with uniformly applying safety regulations to the corresponding construction type and situation with few exceptions*

In Japan detailed regulations are usually developed for each type of construction or under each situation and uniformly applied to the corresponding construction type and situation with few exceptions. Development of detailed regulations and their uniform applications with few exceptions is one of the characteristics in safety management in Japan.

However, some laws and regulations are criticized for possibly excessive regulations.

Elimination of unsafe conditions is still an important goal in safety control of many small- and medium-scale contractors. At construction sites managed by very large prime contractors, however, unsafe conditions are eliminated very much. The current issue is to eliminate unsafe actions through drawing workers' attentions to safety at such sites.

Under these situations, it becomes more important for each subcontractor and each worker as well as a prime contractor to develop his/her own safety management capability which can flexibly incorporate characteristics of each construction site.

7 CONCLUDING REMARKS

The contractor's liability for damages, the contractor responsibilities, the worker's self-safety protection obligation, and the responsibilities of the public owner are important pillars in construction safety management in Japan. Since the ultimate goal in safety management has been directed not to the performance of safety management but just toward the attainment of results of safety in Japan, the contractor responsibilities, the worker's self-safety protection obligation, and the responsibilities of the public owner have not necessarily been fulfilled.

It will be required to reconstruct the process of safety management to take good balance among the four pillars in future, which could become new philosophy of construction safety management.

ACKNOWLEDGEMENTS

The authors would like to thank Mr. T. Kaku, President of the Japan Crane Association, for his valuable comments.

REFERENCES

Anzai, M. 1993. Prime Contractor Responsibilities and Contractor Responsibilities in Construction Safety Control(in Japanese). Tokyo: Rodo-Kijunchosakai

Kitayama, H. 1993. Amended Industrial Safety and Health Law and Construction Safety Control(in Japanese). Tokyo: Rodo-Shinbunsha

Kunishima, K. & M. Shoji 1995. The Principles of Construction Management. Tokyo: Sankaido

Implementation of Safety and Health on Construction Sites, Singh, Hinze & Coble (eds)
© 1999 Balkema, Rotterdam, ISBN 90 5809 036 1

The attitude of workers toward construction site safety – A study of site safety in Hong Kong

Tony Y.F. Ma
School of Geoinformatics, Planning & Building, University of South Australia, Adelaide, S.A., Australia

Albert P.C.Chan
Department of Building & Real Estate, Hong Kong Polytechnic University, People's Republic of China

ABSTRACT: In Hong Kong, the construction industry has a very poor safety record and the loss of productivity in terms of cost and time is great. Not only are there tremendous direct and indirect cost suffered by government and employers, industrial accidents also bring injuries and fatality to workers. It is said "limited or even negative response of workers to the safety message is one of the problems in implementing safety work". This paper attempts to investigate the workers' attitudes and their awareness of the importance of safety measures in construction site. A survey has been conducted for this purpose. The results show that many workers have limited education and they admitted that it is not their responsibility to comply with safety regulations for the main contractors. Also, due to the multi-levels of sub-contracting, many workers have no sense of belonging to the main contractors and supervisions on safety measures are inadequate.

1 INTRODUCTION

In what way should we portray the unique nature of Hong Kong?. It is only a small city with approximate areas of 402 square miles but with a distinct population of six million of people. According to the Quarterly survey of Construction Output released by the Census & Statistics Department, the gross value of construction work, performed by main contractors amounted to $116 billion in 1996, representing an increase of 16.3% over 1995. However, at the same time, the construction industry of Hong Kong has been notorious for its poor safety record. The reported accidents and injuries have been the highest among other industries and stood at the top from 1982 and onwards. According to the statistics of annual industrial accidents, it reveals that accidents in construction industry alone exceed one-third of the total industrial accidents. Accidents due to stepping on or striking against objects and falling objects represent more than 50% of the total accidents in construction industry (Table 1):

Table 1. Accidents due to stepping on or striking against objects & falling objects

Year	Accidents due to Stepping on or Striking against Objects & Falling Objects (Fatalities)	% of Total Accidents
1993	8673 (15)	52.3%
1994	8023 (11)	48.9%
1995	7815 (15)	51.2%
1996	8967 (9)	54.4%
1997	9582 (9)	51.6%

Source: 5/98 GREEN CROSS, p7 (Hong Kong)

The general perceived factors which contribute to the overall poor performance could be due to the following:-

- Unique nature of construction projects under all sorts of weather conditions;
- High mobility of workforce;
- Generally low level of education of construction workers;
- Dispersal of workforce over wide areas of construction sites;
- Multi-level of sub-contracting within the industry;
- Stringent programme in terms of time for completion and so on.

2 PROBLEMS OF CONSTRUCTION SITE SAFETY IN HONG KONG

Hong Kong is a developed city and majorities of buildings are tall within confined sites. The prosperity of real estate market demands for early completion of construction projects that in turn speeds the return of capital and profits. Because of the topography of the territory many tall buildings are constructed on hillsides and on reclaimed land. These buildings required very deep foundation design (eg. caissons) and using very heavy equipment such as tower cranes for materials transportation and handling. Under the stress of stringent construction programme, this could explain why this industry is the most accident-prone of all industries in Hong Kong. In addition, the working conditions in construction sites are generally poor. Workers are exposed to hot and humid conditions and inclement weather, and are often involved in laborious work; working at height or deep below ground or in dust. Many projects are high-rise buildings. Unfenced floor edges of these tall building under construction and work-platforms without guard-rails are common places from which workers fall easily. Coupled with long working hours to accelerate completion time, accidents occur frequently. Accidents reduce productivity and bring pain and suffering to workers and their families.

In October 1993 edition of the *Lighthouse Club Newsletter* Hong Kong Governor Christopher Patten stated:

"Hong Kong's only real natural resource is its people. We have an excellent record on education, training and retraining, and employment law. But there is one area where an otherwise impressive record is a great deal less impressive – industrial safety. Our record here is unacceptable."

One could still remembered that in June 1993, 12 workers died when a passenger hoist fell down from 20th floor at a gas company construction site in North Point, Hong Kong island. After the lift tragedy, the Governor warned that he would not tolerate Hong Kong's prosperity being built on the blood of its workers and vowed to take tough actions against those who ignore workers' safety!

3 THE SIGNIFICANCE OF COST OF ACCIDENTS

As a result of increasing accident rate in the construction industry, common law claims by employees against their employers for negligence have been increasing rapidly in both number and amount. Because of these increases, premium for Contractor's All Risk insurance and Workmen Compensation has been increased significantly which in turn added to the cost to the clients.

On the other hand , there are many other indirect costs associated with accidents. These include delays to work schedules, time spent with legal proceedings, paperwork and repair of plant as well as damage to contractor's image and harm to workers' morale. A Construction Industry Cost Effectiveness Project Report (The business Roundtable (1982)) in the United States assumed that the indirect costs of accidents were four times greater than the direct costs and stated that this was a conservative estimate! (Rowlinson & Walker 1995)

The Table shows the likely direct and indirect cost as a result of accidents (Table 2):

Table 2. Likely direct and indirect loss due to accidents

Direct Loss to the Victim	Indirect Loss to the Contractor
• Wages	• Loss of production
• Medical cost – pain & suffering	• Damage to equipment
• Promotion prospects	• Increased insurance premium
• Family distress	• Legal penalties
• Psychological damage	• Compensation expenses
• Loss of social intercourse	• Industrial relation problem
• Loss of future working ability	• Loss of goodwill and reputation

4 LEGISLATION AND ITS ENFORCEMENT

The aim of industrial safety legislation is to prevent the occurrence of accidents by laying down proper safety standards and securing compliance with those standards. The legislation has to be comprehensive, specific, and to be kept up-to-date. The legislation concerning construction safety can be dated back to 1955 with the enactment of Factories and Industrial Undertakings Ordinance (F&IU, Chapter 59).

There are three main ordinance which cover construction sites safety in Hong Kong:

1. *The Construction sites (Safety) Regulations (1978)* - lay down the detailed requirements for specific areas of work on construction sites.
2. *The Factories and Industrial Undertakings (Safety Supervisors) Regulations (1986)* – the proprietor of one or more construction sites who is either a principal contractor or a specialist contractor must employ one full-time, registered safety officer if he has 200 or more employees.
3. *The Factories and Industrial Undertakings (Amendment) Ordinance (1990)* – it highlights the general duties of proprietors and person employed. They encourage them to have a wider view of their role and responsibilities with respect to health and safety at work.

The ordinance itself is of course important but the impact it has on the industrial accident rate will depend entirely on how it is to be enforced. Industrial safety work in Hong Kong began in the early 50's with the establishment of the Factory Inspectorate Division in the Labour Department. The Building and Engineering Construction Inspectorate is responsible for taking the enforcement of safety regulations on construction sites. They give advice to workers and employers through site visits and are responsible for the prosecution of those who fail to comply with the site safety regulations. In addition they investigate serious or fatal accidents and make recommendations on the prevention of accidents at sites. As an example of their work, 'failure to fence floor edges with guard rails to prevent fall of person' and failure to provide workers with safety helmets and to ensure wearing helmets' were the two major prosecutions in the construction industry against the Factories and Industrial Undertaking Regulation. These two types of prosecution ranked the highest number of summonses and amount of fines between 1989 to 1991.

4.1 New legislation

Obviously, the Government of Hong Kong has paid much attention to the problems of safety and health problems. *The Occupational Safety and Health Ordinance* which became effective on 23 may 1997 protects the safety and health of all employees at work. According to *News Update* by Labour Department, the new legislation extends the protection under the Factories and Industrial Undertaking Ordinance, which covers 800,000 employees in the industrial sectors, to 1.8 million employees in the non-industrial sectors. The new legislation applies to almost all workplaces, including not only construction sites, factories, financial institutions and banks, retails, commercial premises, offices, but also educational institutions, laboratories, health care centres, departmental stores, hotels and places of public entertainment. Employers and occupiers of premises have the general duties to ensure the safety and health of all their employees at work and others working on their premises. The spirit behind the general duty provisions is to create a suitable and amicable working environment. Whether this new Ordinance will ultimately reduce the number of accidents is still too early to comment. In order to help employers and employees understand the legislation better, the Labour Department and Occupational Safety & Health Council jointly held a conference and workshop on 13th & 14th March 1998. Many speakers pointed out that a "prescriptive law" was not a solution to the problems because duty holders tended to meet only the minimum standards. The onus is for the employers and employees to make a shared responsibility.

5 ATTITUDE OF WORKERS TOWARD SITE SAFETY

> Case Illustration
>
> "A lorry was parked on one side of an access road outside the new site office for unloading materials. At the same time, A 45-tonne dumper loaded with boulders was moving down the road and quickly approaching the lorry from head on direction. The road was not wide enough for the dumper to pass by the lorry safely and therefore the lorry workers waved their hands and shouted to the driver to stop. However, the dumper driver sounded the horn and continued to move on and crushed one of the workers resulting in the fatal accident."
>
> (3/98 GREEN CROSS, p41)

Figure 1. Fatal accident – a case illustration

In the above case (Figure 1), whom should we blame against to bear the responsibility? If the supervisor, the driver and the workers pay more attention to the safety measures beforehand, could the accident be avoided?

There are many research papers on construction safety measures but only few touch on the human aspects. In his paper of "Human Aspects of Construction Safety", Jimmie Hinze, the Associate Professor of Missouri-Columbia indicated the reasons why there have been a large void in construction safety research focusing on the safety performance of individual workers as follows:

- Data collection process at the workers level is a great task.
- Workers are relatively inaccessible because any participation of workers in the study conducted at the site level will cost management a lost of productivity.
- Workers are often so different in so many respects that few conclusive statements could be drawn from the results.

In order to understand the attitude of workers toward construction site safety in Hong Kong, a survey in the form of questionnaire was conducted in 1993. The survey aims to investigate the attitude of workers toward:

- safety regulation and its enforcement,
- safety education and training,
- responsibility and awareness of site safety, and
- change of attitude after a serious accident.

The main objective is to understand their attitude thereby helping government and employers concerned to examine their safety provisions, legislation and investment in safety measures. The attitude of an individual could consist of the general feeling or emotions; the awareness and knowledge of, and the behavioral expectations towards an object. A questionnaire was prepared and there were altogether 42 returns from the workers. The task was not easy at all. It included the interviews in the canteen or next to the site. Even though their supervisors had permitted the survey, many of them were reluctant to answer the questions because they were afraid that something could have happened to their future job security?! On the other hand, conducting interviews on sites is dangerous because of the messy environment. A brief information about the workers was listed below (Table 3):

Table 3. The background of site workers

Information on Workers	(%)
Workers by Trade	Labourer (21%) Carpenter (21%) Plasterer (19%) Steelbenders (10%) Others (29%)
Education	No education (17%) Primary level (40%) Secondary level (43%)
Training	No training (26%) Site apprenticeship (60%) Vocational training (14%)
Experience	More than 10 years (55%) Less than 10 years (45%)

5.1 Summary of the survey results (Table 4)

Table 4. Summary of survey results – attitude toward safety

Attitude toward safety	Summary of Results
• Regulations & its enforcement	• Workers admit that safety regulation can protect them from accidents. • They do not have a general picture on the safety regulations. • They have the idea that the main contractors are responsible for site safety, not the workers by law. • Supervision is not enough.
• Safety education & training	• Workers recognize that there is a direct correlation between unsafe working practice and accidents. • They agree that formal training to workers can reduce accident as knowledge on operation can enforce safety. • In reality, they are reluctant to join the safety training course because earnings will be deducted. • They would join the courses if subsidized by Government in form of cash.
• Responsibility & awareness of site safety	• It is not their responsibility to comply with safety regulations for the main contractors. • Due to the subcontracting system, there is no sense of belonging to the main contractor. • Majority of workers has the consciousness of danger in construction sites. • They complain that safety equipment will slow down their working speed and efficiency. It also brings them feeling uncomfortable.
• Change of attitude after a serious accident	• 86% of the respondents agree that they become aware of safety on site immediately after a serious accident because they worry that the same will happen to them. • Workers take the initiative to examine the safety issues on site before commencing work. • However, they tend to resume their own practice of working after a period of time.

6 CONCLUSION

Based on the study, it appears that the main contractor should increase the safety investment in terms of providing free training, education and incentive (or even penalty) schemes to promote the workers' awareness of safety on sites. A strong feedback from the respondents indicated the inconvenience and discomfort of wearing safety helmet and belt, especially in times of hot and humid summer time.

May be it is time to investigate the suitability of safety tools to suit Hong Kong environments. On the other hand, new workers are usually vulnerable to accidents on job site due mostly to the unfamiliarity with special hazards and problems of a project. Therefore the need for orienting new workers cannot be over-emphasized. As construction sites are scattered all over the places and the long working hours of workers makes them miss easily the enrollment for training courses, it is suggested that Government could use mobile enrollment centre eg. a van to various sites for enrollment and to publicize the safety message and regulations during the appropriate time.

For long term measures, 'license system" is recommended that workers should have completed and passed a safety training course before they can work on construction sites.

One achievement worth mentioning is that starting from 1990, the Housing Authority have committed to organize the site safety campaign every year to promote the message of safety on sites. Since then, the injuries have dropped from 206 to 58 per 1,000 workers for housing projects (Ming Poa 24/8/98). This shows that if we all work together thoughtfully and genuinely, risks of accidents can be reduced gradually.

REFERENCES

Construction output statistics for 1996 4th quarter, Tuesday April 1, 1997, http://www.info.gov.hk.

GREEN CROSS, March 1998, Occupational Safety & Health Council.

GREEN CROSS, May 1998, Occupational Safety & Health Council.

Ming Poa dated 24.8.1998.

News Update 9/7/98, Labour Department, http://www.info.gov.hk/labour/new/nlosh.htm.

Rowlinson & Walker, 1995. *The Construction Industry in Hong Kong*, Longman.

Tang, Lee, Chiu & Ngai, 1997. *Cost of construction site accidents and optimal safety investment*. Asian-Pacific Newsletter on Occupational Health & Safety.

Tse Oi Ping, 1993. *A Study on the Attitude of Workers to construction site Safety in Hong Kong*, unpublished thesis, City Polytechnic of Hong Kong.

Implementation of Safety and Health on Construction Sites, Singh, Hinze & Coble (eds)
© 1999 Balkema, Rotterdam, ISBN 90 5809 036 1

Construction safety in China – The past, present and future

Fang Dongping, Song Hubin & Huang Xinyu
Department of Civil Engineering, Tsinghua University, Beijing, People's Republic of China
Qin Chunfang
The Safety Division of the Ministry of Construction, Beijing, People's Republic of China

ABSTRACT: China, with a population of more than one billion, has maintained the highest economic growth rate in the world for several years now. Meanwhile, standards of living of the citizens are improving, and society is proceeding. In the course of large-scale construction, requirements of better construction safety are increasing. *The Construction Law*, which came into force on Mar. 1st 1998, stipulates specific provisions for construction safety.

In this article, after a brief review of the history of construction safety in China, current regulations and laws on construction safety are introduced, together with the safety management mode of contractors and existing problems. Based on the analysis of current situation at home and abroad, future prospects of construction safety in China are also presented.

1 THE PAST OF CONSTRUCTION SAFETY IN CHINA

As one of the oldest civilizations in the world, China has a history of labor protection in construction for more than a thousand years. For an instance, during the construction of a high tower in the ancient Luoyang City, the carpenters erected protecting curtain to shadow each storey, a method still widely used today (Song 1998).

Early this century, with the development of both traditional industry and modern industry, the number of workers increased and the trade unions struggled in various forms for such rights of workers as greater salaries, less working time, labor protection and better working environment. *The Draft of Labor Protection Law*, enacted on Feb.20th1939, was a specific law on labor protection. *The Draft* stipulated specific organizations to supervise labor protection, measurement to protect women and juveniles, working time and so on.

After the foundation of P.R.China, economic development provided a golden opportunity for the development and prosperity of construction industry. With the principle of *Safety First, Focus on Precaution*, the government took efficient measures to protect the health and safety of laborers.

Construction safety in the 50 years since the foundation of P.R.China can be divided into 3 stages.

The first stage (1949～1957) A stage to set up and develop the system of construction safety. The State Council enacted *Three Regulations*, which greatly stimulated the development of labor protection. In 1957, the national average fatalities per 10,000 capita decreased to only 1.67.

The second stage (1958～1976) A stage of stagnation and declination. Since 1958, construction safety management declined rapidly, the fatalities per 10,000 capita even reached 5.60. Fortunately, this datum decreased to 1.65 in 1965, with the improvement of the economic situation. From 1961 to 1965, 16 design and construction codes were compiled and enacted in total, which were the first state construction standards and codes. After the Cultural Revolution began in 1966, safety performance deteriorated again. Accidents, in which 10 or even more than 100 people were killed, happened continually. In 1970, the fatalities per 10,000 capita was as high as 7.50. The fact that legislation of construction regulations and codes were suspended and many reasonable regulations and experiences were denied since 1966, were the main reasons behind stagnation and declination of construction safety during this period.

The third stage (from 1977 to the present) A stage of recovery and progress. In 1978, fatalities per 10,000 capita was 2.80, in 1980 2.30, and in 1990 1.37. In May 1980, the Construction Ministry enacted *The Safety Operation Technical Code For Construction Workers*. In April 1981, *Ten Measurements To Prevent Accidents Caused By Falling Objects* were presented. Regulations such as *Temporary Regulation To Strengthen Construction Safety Of Construction Companies*, *Construction Safety Regulations For Government-Owned Companies*, *Safety Technical Code Of Electricity On Site* and *Scoring Standards for Construction Safety* were compiled one by one. The fatalities per 10,000 capita has maintained under 1.0 since 1995 (China State Ministry of Construction 1981-1997).

2 CURRENT SITUATION OF CONSTRUCTION SAFETY IN CHINA

2.1 *Strategies and principles*

For years, *Safety First Focus on Precaution* has been the strategy of construction safety management in China, which presents the idea that human life is the most important factor in construction management.

The main principles of construction safety management are *whoever manages construction must manage safety* and *whoever takes charge is responsible for safety*. The former means safety management must be integrated with construction management, while the latter indicates the organizations or persons who take charge of the construction of the project ought to be responsible for the safety thereof.

2.2 *Current system of construction safety management in China*

Since 1983, the system of construction safety management in China has followed the principle of *supervised by the government, administered by the construction industry and checked by the public*, which was pointed out in 1993 in *the Notice On Strengthening Construction Safety Management* issued by the State Council. At present, another principle that workers should obey regulations has been added.

Generally speaking, each contractor should be responsible for its own construction safety management, while safety on site and safety management are assessed by certain authorities of the construction industry, whom are in turn

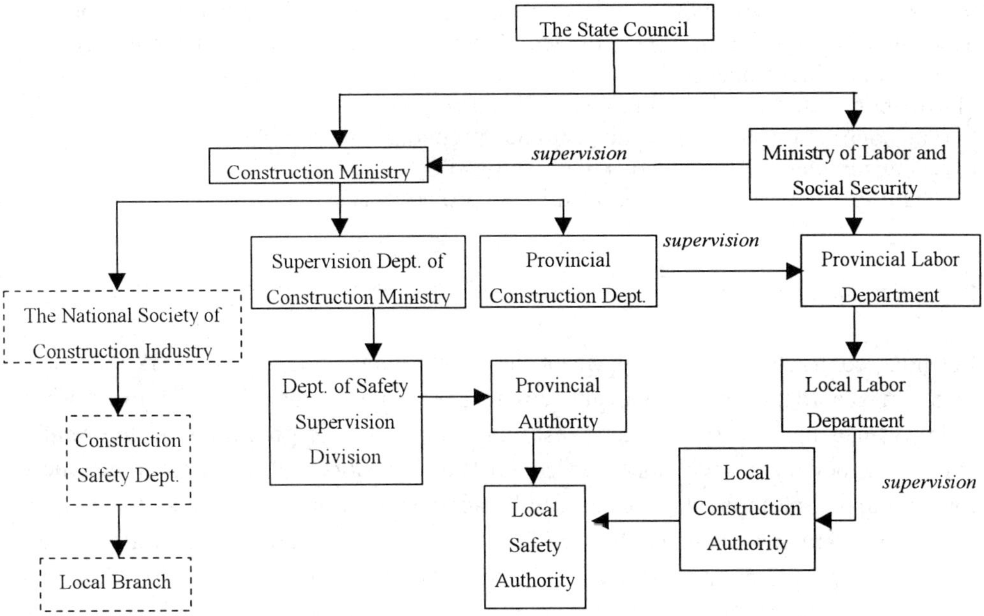

Figure 1. Construction safety management organizations and their relations

supervised by the Labor Administration. Public supervision indicates the rights of trade unions to supervise the implementation of regulations on safety and labor protection.

Compared with the safety management system in the past, the current system has made great progress. Figure1 illustrates the construction safety management organizations in China and their relations.

- Construction Safety Responsibility System

This is the primary system of construction safety management and the core of other safety systems. It includes the clarification of responsibilities taken by such different parties in different levels as the supervision authorities, construction companies and designers etc.

- Accidental Injury Reporting System

In *the Regulation on Reporting Accidental Injury* issued by the State Council, a specific system of reporting accidental injury is stipulated. For instance, if any serious accident happened, report on it should be presented to relative authorities in 24 hours.

- Construction Safety Education and Training System

Safety education and training, through which all the personnel involved can get a clear understanding of construction safety, is a necessary part of safety management.
Safety education and training courses are different for different levels, such as management level, trainers' level and normal workers.

- Accidental Injury Insurance System (Fang 1999)

The workers compensation system in China, beginning in 1950's, mainly deals with provision of economic compensation and other help for those injured in construction accidents. As the Chinese economy develops, drawbacks of the old safety system such as the limited scale of insurance, the lack of reasonable standards to judge the level of the disabled and lack of enforcement hinder the development of construction safety. In

the Construction Law, which came into force on Mar.1st 1998, certain stipulations on accidental injury insurance are provided. The new accidental injury insurance system has the following properties:
(1)Any contractor are obliged to insure for its workers;
(2)The contractors should pay the insurance premium.
(3)The insured are all the persons working on site.
(4)The insurance premium is included in the project cost.

2.3 *Safety Consciousness*

In China, contractors commonly accept that safety management is the work they have to do in accordance with legal and contractual requirements. In order to improve safety management, they have to invest more money - the cost without visible return. Whenever they try to save money, safety costs are usually cut down. Researches show that the direct and indirect losses caused by accident cost much more than the amount spent on safety management. Successful contractors with good returns are always those who paid attention to safety management (Levitt & Samelson 1987). Managers must understand that safety is not a cost, it is an investment with good return. Construction workers should also understand that working in safe environment are their right. The worker's rights to have safe working environment do give pressure to senior management to comply with the law.

2.4 *Safety Management Structure of contractors*

In China, contractors' safety departments are always viewed as a weak department and even it is just part of production (engineering) department. Theoretically, site safety officers have the right to report directly to the safety managers. But in practice, safety officers are supervised by site managers, and site management team will evaluate their achievements and bonus, their position is low and their power is very limited. In this case, though there are detailed safety technical specifications, they are not well executed. To overcome this, contractors must promote the duties of safety officers in the management team by establishing efficient vertical communication channels between safety officers up to general managers and above. Safety officers should report directly to Safety Departments. Their salaries and welfare should also be directed by Safety Departments (Li & Fang 1998).

2.5 *Construction accident records in China*

Figure 2 illustrates the fatalities per 10,000 capita in China since 1981 (China State Ministry of Construction 1981-1997).

The data above have been obtained in certain projects such as building construction, municipal projects, building maintenance and manufacture of construction products.

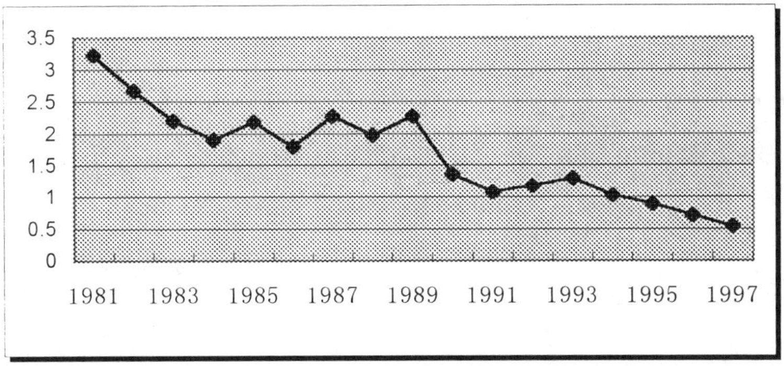

Figure 2. Fatalities per 10,000 capita (1981-1997)

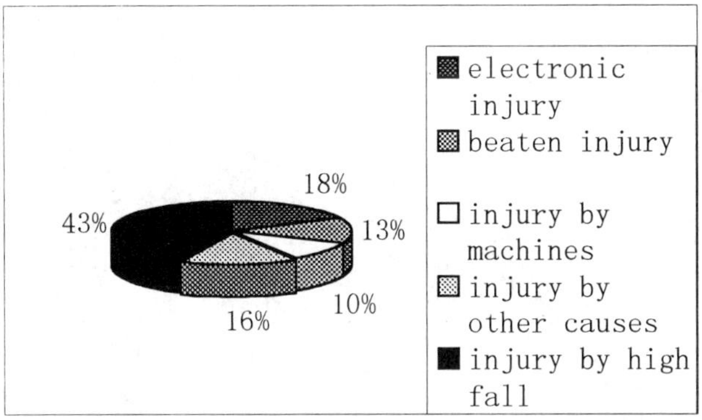

Figure 3. Main categories of construction accidents in China and their percentages

However, the record of accidents is not thorough enough. The Construction Ministry only records accidents in which more than 3 people are killed – other accidents are recorded by local administrations and report to the Construction Ministry monthly. However, only fatalities and serious injuries are reported, no details about slight injuries can be found and concealing of the accidents can not be avoided.

Among reported accidents, high fall, electronic injury, beaten injury and injury by machines are *the Four Main Categories*. Accidents caused by high fall covered 43% of the total accidents (see Figure 3).

3 FUTURE PROSPECTS FOR CONSTRUCTION SAFETY IN CHINA

It is nearly impossible for contractors to form a comprehensive construction safety management system all by themselves. A social and professional environment in which construction companies emphasize construction safety must include efficient government supervision and an attitude of respecting health and safety of labor. Only in such an environment will the companies modify their value standards and strategies to adapt their behavior to social interests. The most important step in creating such environment in China is to strengthen the enforcement. Although *the Construction Law* has come into force, it is not yet fully implemented in this field. Therefore, other

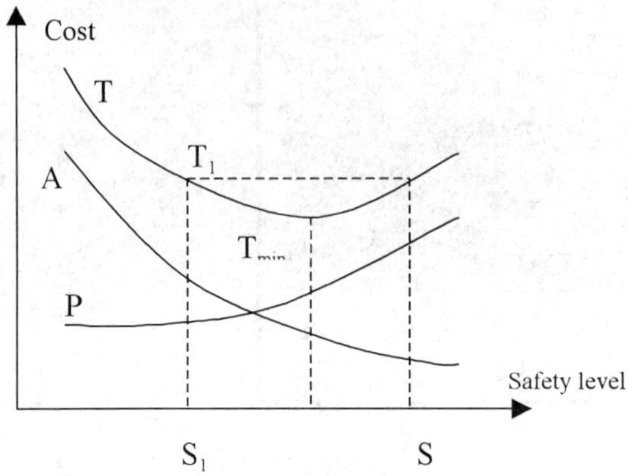

Figure 4. Relation between cost and safety level in normal cases

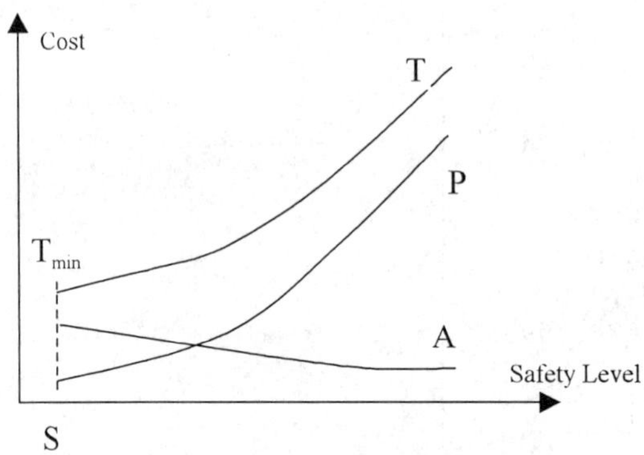

Figure 5. Relation between cost and safety level when the compensation is too low

laws and regulations are necessary to act as supplements to *the Construction Law*.

Secondly, there are problems in China with regards to implementing accidental injury insurance system. These include (1) the determination of premium rating to suit the current situation in China; (2) the determination of relatively precise compensation standards. In normal cases, the cost of safety management (P), accidental losses (A) and the total cost (T=P+A) should satisfy the relationship illustrated in figure 4. However, as the current compensation cost is too low, the actual relationship among P, A and T looks like figure 5. This is caused by the unreasonable assessment as to the value of human life. If we estimate that every fatality could cause the loss of 6,000 working days, the economic losses are 360,000 RMB(43,400 USD), which is 4.5～18 times more than the current compensation, only 20,000～30,000 RMB (2,400～3,600 USD).

It is important to strengthen the construction safety management practices of contractors. The safety roles in each management level must be implemented.

- Company Managers – creating safety culture and safe working place; executing safety policy and supervising the implementation of safety plan; linking up line manager's performance with their safety record; and protecting and encouraging the worker's enthusiasm in safety.
- Project Managers – combining high safety and high productivity standards as the

goal of management and ensuring all relevant bodies and parties, especially sub-contractors, fully understand the goal; maintaining large scale, weekly and daily safety planning throughout the project and ensuring their execution; orienting new hires disregarding their years of experience and no matter how short their involvement on the project will be; keeping a good communication channel with every management level and trying to reward supervisors and craft workers when they do excellent work.

- Foremen – orienting new workers and new crews; and working with crews and orienting them throughout the job.

Education and training to personnel should be classified into different levels according to their abilities. Besides basic safety regulations and laws, courses such as psychology, behavior and pedagogy should also be extended to teachers and trainers. For management personnel, the technology of safety management should also be understood.

Research and development on technology of construction safety are necessary in China. With more investment and the application of modern technology, the situation of construction safety will be improved. Efficient management technology and method will certainly raise the construction safety performance in the future.

4 CONCLUSIONS

Although obvious progress has been achieved in the past years, there are still more to do in the future to improve the safety performance in China. Based on the description and analysis of the past and present of construction safety in China in this paper, following points must be focused on:

- Safety regulations and rules supplement to the *Construction Law* should be made to enable the law to be practical.
- Construction accidental injury insurance system should be implemented to provide economic incentives to encourage the improvement of safety management practices of contractors and to distribute the risk of contractors.
- Efficient management system to improve safety management practices of contractors should be implemented.
- Safety training and education to make the laborers to avoid injuries and to be aware of their legal rights and interests should be strengthened.
- Researches and development of safety technologies should be encouraged.

REFERENCES

China State Ministry of Construction 1989-1997. *China Building Industry Yearbook*. Beijing: China Building Industry Press. (in Chinese)

Fang, D., Cao, X. and Huang, X. 1998. Construction Accidental Injury Insurance in China. *Proceedings of the 2nd International Conference on Implementation of Safety and Health on Construction Sites, Honolulu, Hawaii, USA, 24-27 March 1999.*

Kunishima, M. & Mikio, S. 1996. *The Principles of Construction Management*. Tokyo: Teiji Ishikawa.

Qin, C. & Wei, Z. 1992. *Construction Safety Technical Handbook*. Beijing: China Building Industry Press. (in Chinese)

Levitt, Raymond Elliot & Samelson, Nancy Morse 1987. *Construction Safety Management*. New York: Mcgraw-Hill,

Li, P., Fang, D., Fong, P. and Qin, C. 1988. Managing Construction Safety — A Case Study of The Jiangyin Yangtze River Highway Bridge Project. *Proceedings of International Conference on Safety Science and technology'98*, Beijing, China, 7-10, Oct.1998.

Song, Hubin 1998. *Current Situation of Construction Safety in China and Countermeasure (graduation dissertation)*. (in Chinese)

3 Safety quality

Implementation of Safety and Health on Construction Sites, Singh, Hinze & Coble (eds)
© 1999 Balkema, Rotterdam, ISBN 90 5809 036 1

Barriers to using Environmental, Safety and Health (ESH) record as a competitive advantage in the construction industry

A.B.Ngowi
Department of Building and Quantity Surveying, University of Witwatersrand, South Africa
P.C.Mselle
Department of Civil Engineering, University of Botswana, Botswana

ABSTRACT: Over the recent years, Environmental, Safety and Health (ESH) issues on site have been elevated to the level of company quality concerns which are customer driven and hence subject to market forces. In the process, several proposals have been put forward to enlarge the umbrella of Total Quality Management (TQM) to cover safety and environmental concerns. While ISO 9000 is internationally accepted as the benchmark for quality, safety matters have remained the prerogative of individual countries. This has lead to remarkable differences in safety standards among nations as they are to a large extent influenced by the national cultures.

With a case study in Botswana, this paper argues that, as the construction industry goes global, differences in ESH standards among nations will become a barrier to using an excellent record in these issues as a competitive advantage. The paper concludes by emphasizing the formation of explicit international safety standards that will cut across national cultures and apply to all construction projects.

1 INTRODUCTION

Traditionally, most construction firms have viewed Environmental, Safety and Health (ESH) management as an obligation, driven by regulatory mandates and restrictions. Management has focused on minimizing costs, managing risks, and avoiding unnecessary interference in their primary business activities. Compliance, permitting and other ESH management actions have often been addressed after a firm's broad strategy has been determined, products and services have been defined, and implementation has begun.

However, in response to the converging forces of increased competition, tougher ESH challenges, and more demanding customers, a fundamental shift is taking place. Firms in the construction industry, like those in the manufacturing industry are beginning to actively seek opportunities to gain competitive advantage through ESH excellence. In the manufacturing industry, firms that are integrating ESH factors into business strategy and operations are finding opportunities for ESH management to contribute to business performance, rather than being a hindrance. This does not seem to be the case with the construction industry when viewed globally.

This paper investigates the barriers to using ESH record as a competitive advantage in the construction industry.

2 GLOBALIZATION OF THE CONSTRUCTION INDUSTRY

Competitiveness within the construction industry is increasing as market borders are being expanded through use of widely available telecommunications and increasingly efficient transport systems. Local construction markets, which were previously inaccessible areas, are now open to outside specialists. The barriers that separated economic and vertical market sectors and the firms that operated within them are quickly falling. Competition can arise unexpectedly from anywhere, which means that firms can no longer be overconfident about their market shares. In this competitive environment firms must look to a variety of strategies to assemble the services required to successfully acquire new projects and new revenues. An excellent ESH record in the world that is increasingly becoming aware of these issues is one of the strategies that can enable a firm to enter new markets and sustain its existing clients.

3 COMPETITIVE ADVANTAGE DEFINED

The ultimate aim of strategy is to ensure that the firm creates a competitive advantage and sustain it. Competitive advantage results from a firm's ability to perform the required activities at a collectively lower cost than rivals, or perform some activities in unique ways that create buyer value and hence allow the firm to command a premium price (Porter, 1991). Traditionally, ESH management has not been part of the core activities of a firm, and therefore, it is considered as an additional cost. However, with more demanding clients the situation is gradually changing and ESH issues are becoming part of the core activities of firms.

The main reasons that might cause a firm to achieve favourable positions are initial conditions and pure managerial choices. The influence of initial conditions is that firms may have pre-existing reputations, skills, and in-place activities which integrate ESH issues as a result of history. These initial conditions may reside within an individual firm or, in the environment in which the firm is based. A firm based in the USA, for instance, where ESH issues are strictly enforced, will be more competent in these issues than its counterpart in Mexico although both countries are members of the North American Free Trade Agreement (NAFTA). The managerial choices, or choices independent of initial conditions, define, the firm's concept for competing, its configuration of activities, and the supporting investments in assets and skills. Pure managerial choices lead to the assembly or creation of the particular skills and resources required for carrying out the strategy.

One of the conditions for firms to operate successfully overseas is that they must possess competitive advantages in order to outperform domestic and other foreign contractors (Seymour, 1987). Conversely, for a firm to operate in a global market is an indication that it possesses competitive advantages, which enable it to secure its position. Traditionally, competitive advantages of construction companies have been based on technology, management or finance.

Technological advantage is associated with possession of formidable construction technologies (Strassmann, 1988). For instance, half the world's capacity in ammonia production comes from technologies owned by M.W. Kellogg, while Lummuns Crest's technology is responsible for half of the world's production of ethylene (Murphy, 1983). Management becomes a competitive advantage for contractors who have highly sophisticated computerized project control systems for scheduling, material tracking and organization of sub-contractors. These systems allow them to complete projects as planned and satisfy their clients, ensuring repeat engagement. Project financing becomes a competitive advantage, especially, in countries struggling with their development programmes. Firms that could provide the most attractive financing proposals based on their familiarity with the options available from national export credit agencies, international financial institutions and commercial banks found themselves in a favourable position during the recessionary years of the 1980's (Soubra, 1989).

4 EHS MANAGEMENT AS A COMPETITIVE ADVANTAGE

With the increasing global awareness of ESH issues, an excellent record on these issues may open another avenue of competitive advantage. Unfortunately, ESH management has often been purely reactive, regulatory-driven and focussed on permitting and compliance. For an excellent record on ESH issues to become a competitive advantage as explained above, it has to be fully integrated into a firm's core business practices. Furthermore ESH goals have to evolve from minimizing compliance or remediation costs towards achieving broader goals of cost and risk management while improving ESH performance.

In pursuance of this approach, ESH issues should be considered in the same category as the factors that are necessary for any project to be said to have been successful-namely, cost, time and quality. Ideally, these factors can be looked at as zero cost overruns, zero delays and zero defects. Zero accidents should then be added to this list to complete the conditions necessary for a successful project in the world that is increasingly becoming aware of ESH issues.

The achievement of zero concept can only come about by a process which is characteristically different to that practiced today. The stimulus of achieving zero in all aspects of a project's performance and for a re-engineered process, embodies the ideals of modern management thinking and practice with pragmatism that comes from zero goals. Simply stated, the construction industry cannot evolve by increments to arrive at zero, as this will take too long and will mean that many people are killed or injured. Also, the industry can not look at tools and techniques that address or attack just part of the problem. A holistic approach using modern management tools and techniques to achieve zero targets is needed.

5 EHS STANDARDS

A standard is a procedure, or a product, used as a reference to determine the quality of similar procedures and products. It is established by determining the technical and non technical specifications of the procedure or product (McKechnie, 1972). Many countries have Health and Safety standards that are separate from environmental standards. Similarly authorities that implement these standards are also different. In the USA, for instance, occupational safety and health issues on site are handled by the Occupational Safety and Health Administration (OSHA) while environmental issues are handled by the Environmental Protection Agency (EPA). Accordingly, the employer, that is the construction companies are responsible for occupational safety and health on site, are subject to site inspections and must conform to a set of comprehensive rules and regulations. In Europe, the framework directive for safety and health of workers considers that the employer bears the duty to ensure safety and health of workers in every aspect related to work. In Japan, ESH issues are embraced in quality standards, which form the core of the Japanese business. Managers in construction firms work with the knowledge that the consequences of neglect will be severely punished through such practices as blacklisting by public sector bodies.

What seem to be lacking are internationally accepted standards that can be applied in all countries. Although the International Organization for Standards (ISO) has published a standard containing specifications for environmental management systems (ISO 14001), a standard for safety management systems has not yet been worked out. This has only invited discrepancies in the implementation of these issues in different countries, with the result that some countries have very strict regulations, while others have less strict regulations and in some cases no defined regulations at all. Compliance with regulations differs from country to country as well. Since ESH issues cost front-end money, for a firm to operate in different countries will require to spend different amounts of money too.

6 A STUDY IN BOTSWANA

Since her independence in 1966, Botswana has invested heavily in the development of infrastructure. Given the sophistication of most of the projects, the construction firms owned by citizens have not been able to cope and this has provided an opening for foreign construction firms. The period between 1985 and 1995 witnessed the highest volume of construction work and also the largest number of construction firms from different countries.

6.1Objective of the study

The objectives of the study were to determine:

1. Whether any of the foreign firms has won a tender in Botswana or elsewhere based on its record on ESH issues.

2. The barriers facing the use of an excellent EHS record to win tenders.

6.2Methodology

From the register of construction firms kept at the Central Tender Board (CTB) foreign firms that operated in Botswana between 1985 and now were identified. Some of these firms have left the country, but the majority are still operating in the country. The chief executives of these firms were interviewed using five open-ended questions regarding their experiences on the implementation of ESH in various countries. Where the forwarding addresses of the firms that are no longer operating in Botswana could be obtained efforts was made to contact them by telefax or e-mail.

In selecting the firms for inclusion in the study, care was taken to select from as many different countries as possible. In total nine firms (out of 14 initially contacted) were included in the study; two from the same countries and seven from different countries. The firms have been named A, B, C, D, E, F, G H and I to maintain anonymity as requested by the respondents. The following are the questions that were asked and responses obtained.

Question1: In how many different countries has your firm operated?

The responses to this question are shown in table 1, which also shows the countries of origin of the firms and whether they are still operating in Botswana.

The responses showed that all the responding firms have operated in at least three different countries. Although the question did not specify in which continents these countries are located, firm E responded that it has operated in one continent only - Africa. The rest of the firms have operated in at least two different continents.

Question 2: How does the implementation of ESH regulations in your country of origin differ from the way it is done in Botswana?

The responses to this question established that, in principle the employer bears the duty to ensure safety of workers in every aspect related to work in the countries of origin of the responding firms. The main difference among the countries of the responding firms is the extent to which employees take responsibility of their safety on site. The respondent from U.K. said that in UK safety management is seen in terms of saving lives and saving money. Firms, therefore, employ various methods to motivate their workforce to strive for zero accidents. The

Table 1: Names of firms and the number of countries in which they have been involved.

Name of firm	Country of origin	No. of contract of operation	No. of projects done in Botswana	Status
A	China*	3	5	Present
B	Kuwait	5	3	Present
C	U.K.	5	10	Present
D	Italy	19	1	Gone
E	RSA**	4	4	Present
F	Kuwait	5	4	Present
G	France	22	2	Gone
H	Brazil	8	1	Present
I	Denmark	10	1	Present

* Peoples Republic of China
** Republic of South Africa

responding firm, for instance, had a monthly assessment of its different sites in UK. The site that performed best was awarded a sum of money that was distributed among the workers who displayed safety responsibility as assessed by the various foremen. All respondents indicated that the enforcement of ESH regulations in Botswana is not as strict as in their countries of origin. Specific examples such as use of asbestos, which is allowed in Botswana, were cited as another area in which the countries of origin of the responding firms differ with Botswana.

Question3: Has your firm won/lost any tender as a result of its ESH record?

The responses to this question indicated that firms D, G, and I have won a number of tenders as the result of their excellent ESH record. However, the projects won this way were of specialized nature, which can not be executed by the other responding firms. Firm A, for instance, won two separate tenders for off-shore oil rigs on a Design and Build procurement system in which the clients based their decisions on ESH issues both at the construction and operation stages of the facilities. Similarly, firm I won a tender to construct a bridge connecting an island to the mainland on the basis of the proposal given by the firm to use a combination of a suspended bridge and tunnels situated at the sea bed to minimize disturbance to marine life during construction and operation. The other responding firms indicated that, although ESH issues have been used as one of the criteria for awarding contracts, they have not been high on the list of criteria. Therefore, the ESH record of their firms has not affected their chances of winning contracts.

Question 4: From your experience do any of the countries in which your firm has operated award contracts based on the ESH record of the bidding firm?

This question intended to establish whether the responses given in question 3 were only isolated cases. The responses established that, it is very difficult to compare the safety records of firms originating from different countries because, often, different standards are used in different countries. If a firm enters a market and bids for the first time, it is taken for granted that it has an acceptable record, unless it has been blacklisted in some country of earlier engagement. Incumbent firms are the ones that are thoroughly scrutinized regarding ESH record when they are bidding for work in the same market. However, as responses to question 3 indicated, ESH issues have not been high on the list of criteria in any of the countries as experienced by the respondents.

Exceptions to this observation are projects that involve working under dangerous conditions such as under water, above water or at very high levels. According to respondents from firms D, G and I, often such projects are procured using the Design and Build system whereby the bidding firm has to convince the client that it has an elaborate method to mitigate ESH impacts during the construction and operation of the facility.

Question5: In your opinion, what are the barriers to using an excellent ESH record to win a contract.

The responses to this question established that lack of international standards, culture differences and differences in management styles are the main barriers to using ESH record to win tenders. The existence of different standards in different countries was pointed out by all respondents as being the main barrier because an excellent ESH record in one country may not mean much to another country, which is using different standards.

All respondents pointed out that ISO 9000 series of standards has attempted to force firms to comply with standards but added that they are aware that ISO 9000 is not a standard in itself. Basically, ISO 9000 compels a certificant to do what he/she has promised to do. Therefore, it has not eliminated the differences among different standards. Moreover, only two of the responding firms, i.e. G and I are registered with ISO 9000. The main reason given by the other responding firms for not registering with ISO 9000 is that it is not a requirement in the countries in which they have operated.

The respondents of the firms that originate from Europe, i.e. C, D, G and I said that they were aware of ISO 14001 standard for environmental management as it has been adopted by European Union, but in the countries in which they have operated outside Europe this standard is not used. Two respondents said that they operated using BS 7750 in some countries in Africa and the Middle East.

Responses regarding culture as a barrier to using ESH record to win tenders established that some cultures, such as Japanese and most of South East Asian nations, approach ESH issues differently from other cultures. In the Japanese culture, according to one respondent, quality and safety are linked inextricably because it is believed that where the quality performance of the finished good is low, workers may produce an acceptable result through work practices that are not especially safe or productive in the long run. In such a society, if a firm from another country bids for job, its ESH record will be judged on the basis of holistic performance as it is known in this society, and may stand a little chance to win a tender inspite of a good record in the home market.

Management style, which is heavily influenced by both the organizational and national cultures of the country from which a firm originates may be a barrier to using ESH record to win tenders because it may insist that the standards that are used in a particular project should conform to those in the country where operations are taking place. A record from another country may not be recognized if it does not conform to the one used in the country where the project is based.

7 CONCLUSIONS

This paper has attempted to investigate the barriers to using Environmental, Safety and Health record as a competitive advantage in the construction industry.

The review showed that a competitive advantage is created and sustained when a firm does activities better than its rivals or does activities that are different from (to) what the rivals do. In the case of ESH issues, competitive advantage is created by integrating the issues within the firm's core business practices and by performing the activities better than rivals.

The study in Botswana showed that the foreign firms that operate in the country are also involved in at least two other countries but in none of these countries is ESH record a competitive advantage in the general construction industry. An excellent ESH record is an advantage in projects that involve dangerous working environment such as above and under water. The main barriers given for using ESH as a competitive advantage are lack of international standard, cultural differences and differences in management styles.

REFERENCES

McKechnie, J.L. (ed.) Webster's new universal unabridged dictionary, 2nd Ed., Simon and Schuster, New York, N.Y.

Murphy, K.J. (1983) Macroproject Development in the Third World. Westview Press, Bouilder, CO.

Porter, M.E. (1991) Towards a Dynamic Theory of Strategy, *Strategic Management Journal*, Vol. 12, pp. 95-117

Seymour. H. (1987) The Multinational Construction Industry, Croom Helm Ltd., Beckenham.

Soubra, Y. (1989) The construction and engineering design service sector: Some trade and development aspects, Journal of World Trade, 23(1), 97-124

Strassmann, W.P. (1988) The United States, In Strassmann, W.P. and Wells, J. (eds.) The Global Construction Industry, Unwin Hyman, London, pp. 22-58.

Implementation of Safety and Health on Construction Sites, Singh, Hinze & Coble (eds)
© 1999 Balkema, Rotterdam, ISBN 90 5809 036 1

Appropriate project procurement systems for hybrid TQM

P.D. Rwelamila
Department of Construction Economics and Management, University of Cape Town, South Africa
J.J. Smallwood
Department of Construction Management, University of Port Elizabeth, South Africa

ABSTRACT: Projects will be successful if the objectives of internal and external clients are met. This paper argues that the character of the new era demand an open approach which will allow an inclusive project framework. This framework is possible only if an appropriate procurement system is used. Focusing on one of the project parameters: quality, the paper argues that quality management needs to be integrated with health and safety in order to meet the objectives of both internal and external clients. The integration of Total Quality Management (TQM) with health and safety is defined as hybrid TQM. The paper reports on various surveys which were carried out in South Africa on issues of project quality, health and safety and procurement. The findings suggest that a hybrid TQM system is required for project success and that this is possible through an appropriate procurement system.

1 INTRODUCTION

Project management is about controlling the demands of the project, and making choices, within constraints (Stewart, 1982). The client's objectives (project parameters) are cost, time, quality and utility (Rwelamila and Savile, 1994). Cost generally refers to first cost and in this context, choice means the allocation of budgets and contingencies. Time is better described as timing; choice means the adjustment of timing, timescales and float(s). Quality is the level of specification, and choice therefore means the adjustment of the specification, i.e. to basic, medium or high. Utility includes such things as running costs, maintenance issues, buildability, and flexibility for alterations or other uses. Choice, in this context, usually means making value judgements between higher initial costs and longer-term savings.

For the sake of completeness, it is important to recognise that projects do not occur in isolation, and that consequently they must be considered in relation to the prevailing environment. According to Rwelamila and Savile (1994), this background is specific to the country, year, location, project type, industry etc. There is little choice for the project manager (PM); any changes that occur will probably be due to outside agencies, and will cause a risk of serious disruption to the project. These changes may be in regulations, union policy, markets, technological innovation etc. Thus an appropriate procurement system is necessary for the PM in order to balance the project parameters and allocate risk appropriately. For lack of space and brevity only one project parameter, quality, will be dealt with in relation with the selection of appropriate procurement systems.

This paper endeavours to identify one of the primary reasons, which has contributed to poor quality of construction projects. It will provide evidence which suggest that incorrect choice and use of procurement systems has contributed to neglecting health and safety of project stakeholders and this has consequently contributed to the poor quality of constructed projects in the South African construction industry.

1.1 *Defining terms*

- *construction procurement system:* reporting on an examination of past research and literature, Masterman (1992) refers to phrases such as 'building procurement method', 'procurement form' and procurement path' which have been used by various authorities when referring to this concept.

According to Franks (1984), CPS is 'the amalgam of activities undertaken by a client to obtain a building'. The term construction procurement system (CPS) has therefore been adopted and used throughout this paper. This term is generally used in this paper to describe:

The organisational structure adopted by the client for the management of the design and construction of a building project *(*Rwelamila, 1996).

- *hybrid total quality management (TQM):* as indicated above, quality is defined as 'conformance to established requirements' (Rwelamila, 1995). This definition provides a basis for measurement; i.e. the purpose defined in the brief is either satisfied or not satisfied. Total quality (TQ) is referred to by McCabe *et al* (1996) as a "contemporary organisational initiative" where the "customer is king" – or where the client is king (authors emphasis). The common objective in TQ is for radical improvement in what the client / customer gets by addressing the internal processes. As explained later, there are many clients for a construction project, these are referred to as internal and external clients (ICs and ECs). Client satisfaction is therefore a combination of variables. In order to satisfy all categories of clients it is important to have management which can analyse ways in which it can alter itself. This is often referred to as managing culture (McCabe *et al,,* 1996). This is possible to those who have made the transition from quality assurance (QA) to total quality management (TQM). Using set theorem, TQM could be represented as follows:

$$\mu = TQM = QC + QA \qquad (1)$$

Where μ = a universal quality management set, TQM = total quality management, QC = quality control and QA = quality assurance.

According to Garvin (1988), QC's emphasis is on product uniformity with reduced inspection, QA is about the entire construction site chain, and the contribution of all functional groups to prevent quality failures. TQM, which combines the two variables, brings the client dimension more clearly, the emphasis being on the market and client needs. The client focus qualifies Tyler's (1992) one basic principal 'in search of excellence', namely that the best organization or project stays close to the client / customer and successful projects tend to be safe projects. Based on this argument, the authors have added health and safety in the TQM set and hence the term hybrid total quality management (HTQM). Equation 1 is improved to Equation 2 as follows:

$$\mu = HTQM = QC + QA \qquad (2)$$

Where HTQM = hybrid total quality management. Safe and healthy workplaces compliments quality. According to Crosby (1984), the performance standard for quality is zero defects. The Business Roundtable (1991) maintains that the performance standard of safety is similar, zero injuries being "the only supportable goal since any other leaves the subtle message that injuries will occur and that injuries are acceptable."

- *combination of the terms:* the integration of the above terms is achieved by using equations 3 and 4 below. The equations are focussing on the general performance, in terms of one of the four criteria discussed above: quality. The starting point in looking at performance is to accept the fact that the correct choice and use of the most appropriate CPS is one of the principal requirements for an efficient project management system. But selecting an appropriate CPS needs to be supported by good management, management that is responsible to both internal and external clients, hence hybrid total quality management (HTQM) as discussed above.

Selection of an appropriate CPS $\Rightarrow$ managing (Q) (3)

Where CPS = construction procurement system, $\Rightarrow$ = will lead to and Q = quality.

Managing (Q) as in Equation 3 + applying H & S = +P (4)

Where +P = project success and H&S = health and safety.

*Equation 3:*Equation 3 is based on the argument that once a client is satisfied about real need and feasibility of the construction project within overall budgetary constraints, the instinctive reaction is to retain a consultant to help in the assessment of risk towards devising an appropriate CPS. The client should decide how much risk to accept. No construction project is risk free. Risk can be managed, minimised, shared, transferred or accepted. It can not be ignored. The client who wishes to accept little or no risk should take different CPSs from the client who places on detailed, hands-on control.

The basic decision on the appropriate CPS to take should precede the preparation of the outline (project) brief, since it necessarily affects who shall assist with the design brief as well. That choice of an appropriate CPS must be determined by the nature of the building project and the clients' wishes over acceptance of risk. Such decisions are very difficult. There are a number of publications, for example Masterman (1992), Hughes (1990), and Hughes (1992), which give detailed account of different CPSs, their risk distribution potential and merits and demerits of the same.

Once an appropriate CPS has been determined, the first stage of managing the project parameter (Q) has been reached. You now have a system with its associated project management structure, which will be partly, able to manage the whole spectrum of dynamics associated, with managing quality. Why partly? The fourth equation gives the answer to this question.

E*quation 4:*For the selected CPS to lead to project success (+P), it is important to identify and understand the project client. In fact there are many clients for a building project, e.g. a funding financial institution [external client (EC)], venture capitalist (EC), developer (EC), purchaser on completion (EC), the tenant (EC), contractor [internal client (IC)], and subcontractors (IC) too, have different expectations of the project. Although profit may be a common desire, no one except the end user cares what is built. If the PM does not take it upon him/herself to assess the needs of all the building clients, especially the internal clients (ICs) who ranges from highly skilled to site cleaners, it is unlikely that anyone else will, and project success (+P) may not be achieved.

The project manager should understand that H&S is not only the responsibility of the contractor but designers and all project stakeholders are responsible if the quality of the final product is to be realised. In other words, if the selected appropriate CPS is going to lead to project success, the ICs must feel part and parcel of the whole building project management system and health and safety should be among top priorities at every level of the project system. The quality of the end product will depend on how the PM and his or her subordinates use HTQM to manage the ICs.

2 PROJECT MANAGEMENT IN THE SOUTH AFRICAN CONSTRUCTION INDUSTRY

The dominant CPS in the building sector of the construction industry, is the conventional or traditional construction procurement system (TCPS). It follows a strictly sequential path of four phases: preparation, design, preparing and obtaining tenders and construction.

Briefly, the building project organisation structure of this system is characterised by the appointment of a principal adviser, an architect (the PM), who leads the design team (the team may include structural, building services engineers, a landscape architect etc.) which is assembled at his or her recommendation. The building project is designed and detailed up to a point where the various elements of the design can be taken-off and worked up into a bill of quantities or a schedule of rates. At this stage the building contractor is invited to bid for the construction work and, if successful, is expected to start on site within a few days (quite often) with very little knowledge or understanding of the building he or she is to construct and probably having made no acquaintance with the client for whom the building is to be produced. This project organisation structure is normally formulated parallel with a standard project contract between the client and the building contractor.

The South African building sector purports to use this project organisation structure, and hence the framework of a project quality management system in most building projects, despite the basically transient and unique nature of various building projects. This is contrary to increasingly strong suggestion that, organisational structures should be tailored to meet particular project needs (Barrie and Paulson, 1978; Walker, 1984; Hughes, 1990; Rwelamila and Hall, 1994). As will be demonstrated later, this underlines the nature of project performance problems in the South African building sector.

3 PRACTICE OF SELECTING CPSs & PERCEPTION ON H&S: SURVEYS

This section is devoted to reporting the results of three empirical surveys conducted among construction firms, architects, quantity surveyors, project managers and construction workers. Discussion of these results and their implications and relationship to the theory contained in Equations 1, 2, 3 and 4 will follow after this section. Analysis of the results will be kept very simple, but retaining the logic of the results in relation with the established theme of this paper.

In general, the percentages associated with opinion option are given, but where these results are misleading, this is explicitly stated. Hence, unless otherwise stated, where the term 'respondents' is used, it refers to item respondents as opposed to the overall response rate to the survey.

Questionnaire survey A: Letters and questionnaires to architects, quantity surveyors and main contractors were sent to practices rather than individuals. The sample population of architects was randomly selected from the Institute of South African Architects (Western Cape Chapter) 1996 membership lists. Similarly the sample population of quantity surveyors was drawn from the Association of South African Quantity Surveyors (Western Cape Chapter) 1996 membership lists. The sample population of general contractors was randomly selected from the 1996 Master Builders and Allied Trades Association (Cape Peninsula) membership directory

As stated above, the principle of random sampling was used to select potential respondents where respondents were "randomly" drawn from the potential pool of respondents, thus allowing each member of the representative bodies an equal chance of being selected. The idea behind "random" selection is that it permits the generalization of the results obtained from the survey to be made (Rudestam and Newton, 1992).

Besides the selection process the number of respondents chosen to participate in the survey needs to be considered in order to obtain a representative sample of the target population, as well as to achieve a reasonable level of accuracy. It was concluded that a survey of 30 architects, 30 quantity surveyors and 30 main contractors, thus totaling 90 potential respondents, would be sufficient to attain these objectives.

The number of responses received was extremely rewarding as 17, 15 and 18 completed questionnaires were received from architects, quantity surveyors and contractors respectively. This totals 50 completed questionnaires at a response rate of 56%.

As reported above, the dominant CPS in the South African building sector is the TCPS. This system is discussed in detail elsewhere (Franks, 1984 and Rwelamila, 1996). Because so little exists to confirm the general use of a TCPS, beside the extensive use of TCPS standard contract forms and traditional project management structures, it was necessary to examine the circumstances in which it is used in practice in the South African building sector. The circumstances (statements) where the TCPS is likely to be successful as advocated by Franks (1984) and Hughes (1992) as shown in Table 1, were tabulated, and contractors, architects and

Table 1. Characteristics of the TCPS

A: The client commissions, and takes responsibility for the design of the works
B: The design is complete at the time of selecting the contractor
C: Prime cost sums, including nominated subcontractors do not form the major proportion of the contract sum
D: The architect appointed by the client is adequately experienced to cope with the co-ordination of the design team, to lead the design effort, and to co-ordinate the interface between design and fabrication
E: The client uses the quantity surveyor to plan and control the finance of the project, in conjunction with the architect
F: The client requires the contractor selection process to be based upon the contractor's estimate of price and for the contractor to bear the risk of costs exceeding price
G: The client reserves the right, via nomination, to select the subcontractors for certain parts of the work (but see C above)
H: An acceptable negotiated project contract form is used in order to ensure a fair and familiar distribution of risk
I: The client does not know what else to do, and the consultants do not raise the choice of procurement method as an issue

Source: Franks (1984) and Hughes (1992)

Table 2. Opinions regarding the characteristics of projects (mean %)

Project Characteristics	True	More true than false	Difficult to say	More false than true	False
A	100	-	-	-	-
B	10	-	-	7	83
C	70	-	-	20	10
D	80	-	5	10	5
E	90	10	-	-	-
F	100	-	-	-	-
G	80	20	-	-	-
H	-	-	-	10	90
I	70	30	-	-	-

quantity surveyors were asked to indicate how true those statements were applicable to most of their building projects undertaken within the last 6 years. The results are shown in Table 2.

From Table 2, it is clear that, contractors, architects, quantity surveyors are in agreement that in most of their building projects, TCPS characteristics, **A, C, D, E, F, G and I** apply.

While seven TCPS characteristics apply in the majority of building projects in South Africa, two TCPS principal characteristics do not apply:

- the design is most instances is not complete at the time of selecting the contractor (B);
- a project contract form used, is not normally negotiated in order to ensure a fair and familiar distribution of risk (H).

Each respondent group showed a fairly high general level of agreement with the others on the TCPS characteristics, which are general to most building projects. These are characteristics A, C, D, E, F, G and I.

The majority of respondents, were in agreement that, for the majority of their projects it is false to conclude that "the design is normally complete at the time of selecting the contractor" (B). This suggests that the majority of building project designs are incomplete at the time of selecting the contractor. Furthermore, this suggests that the selection of a contractor is based upon a provisional price, though referred to in the contract as 'a firm price'. Under this situation, the client can not describe with certainty what it is that the contractor is being invited to construct (from the documents available).

Questionnaire survey B: 47 Project management practitioners who were members of the Project Management Institute of South Africa (PMISA) responded to the survey to determine the role of project managers in construction health and safety. The respondents were requested to advise, inter alia, regarding the aspects negatively affected by inadequate health and safety (Table 3).

Table 3. Aspects negatively affected by inadequate health and safety (Response in %)

Aspect	Yes	No	Don't know
Cost	72.3	19.1	8.6
Environment	66.0	23.4	10.6
Productivity	87.2	10.7	2.1
Quality	80.9	17.0	2.1
Schedule	57.4	29.8	12.8
Client perception	68.1	19.1	12.8

In looking at the aspects, which are negatively affected by inadequate safety, Table 3 is helpful. The table shows that productivity and quality predominate over project cost, schedule and client perception.

Questionnaire survey C: A total of 88 skilled, semi and unskilled workers employed by three regional entities of national general contractors were consulted to determine the nature and extent of worker participation. The workers were asked, inter alia, the importance of various aspects related to working life (Tables 4).

Table 4. The importance of various aspects related to working life

Aspect of working life	Rating
Safety	281.0
Health	273.0
Supervisor's attitude	263.0
Site conditions (working)	257.0
Artisan/Assistant's attitude	228.0
Wages	200.0

As clearly indicated in Table 4 above safety and health are considered as priorities in relation to working life.

4 SUMMARY OF FINDINGS AND THEIR IMPLICATIONS

One of the unique features of the TCPS is that it follows a strictly sequential path, where each of these four phases as indicated in the early sections of the paper, can be viewed as separate entities and carried out, to a certain extent, in isolation of the others. The four phases are now used to summarise the findings and their implications to the project, with respect to the characteristics of the TCPS.

4.1 *Summary of findings*

- Preparation: this is the inception phase of the building project when the client establishes his or her needs in principle. This phase is covered under the TCPS characteristic A, which is supported by the majority of respondents as one of the characteristics of public building projects which seems to be adhered to.

- Design: this phase sees the appointment of the design team who develop the project through a series of sub-phases: briefing, feasibility, outline design, scheme design and detailed design with the scheme's configuration and features becoming firmer at each sub-phase. There is no evidence from the questionnaire results to suggest that 'designs are hastily prepared', but there is strong evidence to infer that 'designs are normally incomplete' before going to tender (TCPS characteristic B) – are they incomplete because clients are anxious to see work commence on site or because there is very little understanding of the TCPS?

- preparing and obtaining tenders: tender documentation on the TCPS normally consists of drawings, specification(s) and bills of quantities, with the latter document being prepared by a quantity surveyor (TCPS characteristic E, which seem to apply in Botswana) on the basis of measurements 'taken off' from the designer's drawings in order to provide each tenderer with a common base from which to price the bid. For the TCPS to operate successfully, and to minimise the financial risk to the client, it is imperative that the design is fully developed before the bills of quantities are prepared and tenders invited. If this is not done, as clearly is the case in this study (TCPS characteristic B not being adhered to), excessive variations and disruptions of works which could lead to neglecting H&S are likely to occur.

- Construction: when using the TCPS, an adequate period should be allowed for the contractor to plan the project thoroughly and organise the required resources. Undue haste in making a physical start on site, which seems to characterise the South African situation, may result in managerial and technical errors being made by both the design team and the contractor, which could lead to neglecting appropriate H&S plans.

4.2 *What are the implications of these findings?*

- The characteristics of the majority of building projects in South Africa do not conform to the principal characteristics of the TCPS (characteristics B and H). This strongly suggests that the TCPS is basically used as a default system.

- The extensive use of default TCPSs in the South African building sector suggest that in the majority of projects, PMs do not have appropriate management structures to manage quality. This further suggests that the gap between an appropriate CPS and an inappropriate one (default CPS) remains a hidden challenge to the PM in trying to implement a hybrid TQM.

- The gap between an appropriate CPS and a default CPS leads us to speculate with a strong argument, that any PM's efforts to implement a HTQM under this condition will automatically lead to more pressures, and the possibility of having the synergy between health and safety , and quality are very remote indeed.

A default TCPS is unable to implement a HTQM plan, and the PM's work is reduced to 'fire fighting'. Because the project organisation structure falls short of what is required, the PM's ability to deal with H&S of the IC's as prerequisite in meeting quality requirements is lost. As a last resort, the PM keeps his or her focus on the ECs, and uses the ICs, as cogs in the machine and poor H&S conditions become a common phenomenon across the project system.

5 CONCLUSIONS AND RECOMMENDATIONS

This study has shown that in the majority of building projects, no project characteristics are identified, before selecting a CPS , hence inappropriate CPSs are used. By inference, this is likely to lead to a situation where the ICs are neglected and feel unimportant and hence H&S issues are not priority to project managers. This will finally lead to a situation where all ICs feel that they are not part and parcel of the HTQM – the synergy between H&S and quality get lost and hence the relationships between project stakeholders remain in the peripheral of the project domain.

The extraordinarily use of a TCPS as a 'default system' in most building projects in the South African building sector shows that the sector needs a paradigm shift in its choice of CPSs. Under this culture of default CPSs, there is no chance for synergy between H&S and quality to be part of the building sector; project success in terms of quality will remain a dream until such time when the practice of selecting appropriate CPSs will become a norm.

In order to establish the basic principles of formulating appropriate CPSs, the South African building sector should establish appropriate methods of selecting appropriate CPS. There are a number of ways in which this can be achieved; these are discussed in detail by various authors listed in the reference section of this paper. As a starting point a literature survey should be carried out as an audit of available methods of selecting appropriate CPS. This should be followed by intensive validation and checking exercises in order to establish methods, which are appropriate to the South African building sector.

It is recommended that work be done in establishing a standard flexible contract document, which could be used for any selected CPS. This means that it will be possible to have a contract document, which could go along the above recommendation.

*The financial assistance of the Centre for Science Development (CSD), (HSRC, South Africa) towards part of this research is hereby acknowledged. Opinions expressed and conclusions arrived at, are those of the authors and are not necessarily to be attributed to the CSD.

REFERENCES

Barrie, D.S. & Paulson, B.C. 1978. *Professional Construction Management*. New York: McGraw Hill.

Crosby, P.B. 1984. *Quality without tears the art of hassle – free management*. New York: McGraw-Hill.

Garvin, D.A. 1988. *Managing quality*. New York: Macmillan.

Gordon, C.M. 1994. Choosing Appropriate Construction Contracting Method. *Construction Engineering and Management,* 120, No.1, March: 196-210.

Hughes, W. 1990. Designing Flexible Procurement Systems, *Proceedings of CIB W92: Procurement Systems Symposium*. Zagreb, Yugoslavia.

Hughes, W. P. 1992. An Analysis of Traditional General Contracting. *Construction paper series* No. 12, The Chartered Institute of Building, Ascot.

Masterman, J. W. E. 1992. *An introduction to building procurement systems*. London: E & FN Spon.

McCabe, S., Rooke, J.A. & Seymour, D.E. 1996. The Importance of Leadership in Change Initiatives. In D.A. Langford & A. Retik (eds), *The Organization and Management of Construction: Shaping theory and practice*: 589-598. London: E & FN Spon.

Rwelamila, P. D. & Savile, P. W. 1994. Hybrid value engineering: the challenge of construction project management in the 1990s. *International Journal of Project Management*. 12(3):157-164.

Rwelamila, P.D. 1995. Quality management in the SADC construction industries. *International Journal of Quality & Reliability Management*. 12(8):23-31.

Rwelamila, P.D. 1996. *Quality Management in the Public Building Construction Process*. Unpublished Ph.D. Thesis, University of Cape Town, South Africa.

Rwelamila, P.D. & Hall, K.A. 1994. An inadequate traditional procurement systems? Where do we go from here? *Proceedings of CIB W92: Procurement Systems Symposium*: 107-114. Hong Kong.

Smallwood, J.J. & Rwelamila, P.D. 1996. Total Quality Management of Construction Work: From safety exclusion to a comprehensive management strategy. In D.A. Langford & A. Retik (eds), *The Organization and Management of Construction: Shaping theory and practice*: 859-870. London: E & FN Spon.

The Business Roundtable. 1991. *The Workers' compensation crisis .... safety excellence will make a difference*: p21. New York: The Business Roundtable.

Tyler, W.W. 1992. Total Involvement Safety. *Professional Safety*: 26-29.

Walker, A. 1984. *Construction Project Management*. London: Granada.

Implementation of Safety and Health on Construction Sites, Singh, Hinze & Coble (eds)
© 1999 Balkema, Rotterdam, ISBN 90 5809 036 1

The MASTER project: A partnership with OSHA

Stephen J. Spaulding
Turner Construction Company, Cincinnati, Ohio, USA

ABSTRACT: The number one *purpose* of OSHA is to "encourage employers and employees in their efforts to reduce the number of occupational safety and health hazards at their places of employment and to stimulate employers and employees to institute new and to perfect existing programs for providing safe and healthful working conditions."

In 1993, William Murphy, the Cincinnati Area OSHA Director, had an idea he thought would achieve that goal which resulted in the development of the MASTER Project (Mobilized Alliance for Safety, Teamwork, Education and Results). This program eventually became an OSHA Works 2000 Program to establish certain construction sites as self-compliant through the cooperative efforts of Labor, Management and OSHA, to conserve OSHA resources while achieving a safer workplace.

1 DEVELOPING THE PROGRAM

At that time, my employer, Turner Construction Company, already had an extensive and effective Company Safety Program in place with a well-established "safety first" philosophy. With 100% management backing, a partnership was formed with the OSHA Area Office to develop the specifics of the program — what would be required of contractors and criteria for selection. Under a "gentleman's agreement" between OSHA and Turner, a small strip-mall was selected as an initial project to test the idea and define program parameters (Figure 1). It was intentionally small and located close to the OSHA Area Office. The owner was in agreement, but since the project was already underway, the program had to be grandfathered into construction documents.

To succeed, it was important to ensure there was a clear understanding of a specific and common goal. OSHA sent out specialists, and along with Turner safety professionals, an understanding was reached with the contractors on site of how to make such a program work. The first effort was successful in that it achieved the primary goal of eliminating falls at six feet and over.

Figure 1. Tri-County strip mall construction of 120,000 SF Wal-Mart store, 30,000 SF Marshalls store and 60,000 SF shell building retail space and building pad for future Sam's store, including roads, parking lots and utilities in suburban Cincinnati, OH.

The second project to be designated a test MASTER Project was the Hamilton County Justice Center, a public works project (Figure 2). County officials (owners) saw the merit of the program and contractors (some of whom were already underway) were convinced to participate. In this instance, to achieve that, OSHA walked the job explaining the program concept, helped Turner overcome any contractors' objections and showed them how program criteria could be achieved. [For example, to conform to the MASTER Project six feet and over fall protection rule, tie-off cables were required at the leading edge of the concrete formwork. The contractor's initial reaction was that it had never been done, but "you show us how to do it, and we will." Working together as the program intended, between the Contractor, OSHA and Turner, a workable plan was devised.] It also involved explaining how the MASTER Project program differs from the standard OSHA regulations. [In one case, the bricklayer contractor balked at following MASTER Project tie-off procedures inasmuch as under standard OSHA regulations he could have avoided this by maintaining a "controlled access zone." Under MASTER Project criteria, there are no controlled access zone provisions where fall hazards exist.] This second project helped more concretely to define the program by identifying limitations as well as expectations. And, in reality, practical solutions for some hazards still need to be defined.

Figure 2. Hamilton County Justice Center on a five-acre site in an historic section of Cincinnati, OH. This 160,000 SF, 160-bed facility provides intervention, detention, recreation and schooling as well as support, business, clinical and intake services for children who require the Youth Center's involvement.

2 MASTER PROJECT CRITERIA

After the first two projects were completed, a local OSHA document was created officially identifying the program as the "MASTER Project." [document to be provided at conference] It determined that selection of the applicant and site, in brief, would be based on the following criteria (among others):

- Applicant's Safety & Health Program specifically would include training of workers in the 20 leading causes of injuries and enforcement provisions, and a written safety program would be submitted to the OSHA Area Office and found to be in compliance with published guidelines.
- Management would provide authority for program enforcement for the entire job site including a provision requiring fall protection for ALL employees working at or above six feet from the working surface (with authority to remove from the site anyone in non-compliance).
- Specific signage would be posted at the site designating it a MASTER Project.
- Submission of acceptable 3-year history of Worker's Comp Experience Modification Rate (EMR) and OSHA Form 200, which logs injuries, illnesses and fatalities.
- Controlling contractor (general contractor, construction manager, etc.) and owner agree OSHA would not be required to obtain warrants for any inspection activities at the site.

It should also be noted that to continue as a MASTER Project, the number of recordable injuries at the site must *remain* below the current industry average.

2.1 Site Safety Representatives

A MASTER project is required to have three *types* of on-site safety representatives. First, a job site *Safety Manager* assigned by the general contractor or construction manager to oversee safety for the entire job site who has the appropriate authority to enforce the safety program and correct unsafe conditions. His experience must include, but is not limited to, a minimum of three years experience in overseeing construction safety with a proven track record. He conducts frequent and regular site inspections with the Labor Representative and holds weekly safety meetings with Labor and Contractor Safety Representatives.

Second, a *Labor Representative* (which has to date been someone assigned by a subcontractor from their labor personnel) serves as liaison to the Safety Manager representing workers at the site. He must have completed a (Labor Department) certified construction apprenticeship program (or equivalent) which, at a minimum, includes the OSHA 200 construction course (or equivalent) training. In addition to accompanying the Safety Manager on site inspections, he is to be involved in all accident investigations.

And third, a *Contractor Safety Representative* is appointed by *each* contractor to resolve safety matters and to serve as their liaison to the Safety Manager. He participates in safety inspections involving that contractor's work and attends weekly site safety meetings.

2.2 OSHA Responsibilities and Benefits

All parties must agree OSHA has access to the site to review records and conduct site audits, which might include training if hazards are encountered or expected. [As an example, if an electrical contractor has workers in a plumbing contractor's *unshored* trench, OSHA would review the situation and provide training and information to the electrical contractor on his responsibilities for making the trench safe. In this instance, the plumbing contractor doesn't need to shore the trench in order to safely perform his work, but the electrical contractor does, i.e. only the electricians need to be inside the trench to perform their work.]

On major accidents and fatalities, OSHA conducts enforcement investigations consistent with current OSHA regulations, but on a MASTER Project these investigations are not expanded to cover the entire job site. Any complaints by employees are reported to OSHA, and those classified as "non-formals" are simply presented to the Safety Manager and Labor Representative to investigate and correct. Disputes between parties are resolved by the OSHA Area Director. If the MASTER Project appears on OSHA's regular site inspection list, the inspection is generally limited to a review of compliance with the MASTER Project guidelines. An exception might be where serious or imminently dangerous conditions are expected or exist.

One of the major benefits to OSHA, and ultimately to contractors, is the on-the-job training arrangement provided on a MASTER Project. OSHA staff not only will provide training to workers and representatives of contractors, but the site can be utilized to train OSHA representatives by observing procedures in actual construction site conditions (steel erections, precast wall construction, etc.). This gives them the reality perspective that can be missing from classroom or theoretical training.

2.3 Project Results

Finally, the results of the MASTER Project are measured at completion of the project. It is intended to report successes as well as failures and considers the following factors, particularly in comparison to prior similar projects (size, numbers of employees, risk):

- Number of recordable injuries vs. industry average
- Number of lost time accidents vs. industry average (LWDI)
- Personal interviews/observations of labor and management
- Each contractor's EMR for this site vs. their overall EMR
- OSHA resource expenditures

3 THE MASTER PROJECT AT WORK

The first project selected as a MASTER Project prior to start of work and following the formal program guidelines was in 1995. This Turner project was a research facility for International Paper in Cincinnati, Ohio (Figure 3). Safety documents were submitted and approved and made a part of the formal bidding documents. Both Turner staff and subcontractors were given orientation before start of construction. The project ran as expected and not only was the MASTER Project program implemented as outlined in the guidelines, additional items were added by Turner staff to help ensure its success.

Figure 3. International Paper research facility, general offices and laboratories totaling 163,000 SF on 45 acres in Clermont County, OH (near Cincinnati).

One such item was a special "incentive program." Each month, without a recordable accident, lunch was provided to the entire site with small yet specific recognition of individual efforts. Some of these "luncheons" were attended by both owner and

Turner corporate level management as well as OSHA, even including federal representation by the Asst. Secretary of Labor on one occasion. Remarkably, this project, too, was completed with *no lost time accidents!* In fact, on its merits, the project received the National Liberty Mutual Insurance Company Safety Award and the Liberty Mutual Vice Presidential Award.

At present, Turner has two projects operating on the MASTER Project basis: Dayton Daily News production facility located in Franklin, OH (Figure 4) and the new Paul Brown Stadium, an NFL football stadium for use by the Cincinnati Bengals professional football team and secondarily for professional soccer, concerts and other large outdoor events (Figure 5).

Figure 4. Two-story, 260,000 SF production facility for Dayton Daily News, including 8,000 SF newsprint storage, 18,000 SF press hall, 98,000 mailroom, prepress imaging and platemaking as well as production support areas and office support, breakrooms, and cafeteria.

Figure 5. 66,500 seat, 1,850,000 SF open air sports facility, approximately eleven stories high covering 22 acres and including 7,600 club seats and luxury suites as well as club lounges, team store and administrative offices for the Cincinnati Bengals.

4 IN CONCLUSION

Although each participant may have a slightly different agenda for participation, the end result is of benefit to all — a safer work place. Contractors may initially incur additional monetary costs, however, these are usually more than offset by lowered accident/incident (and subsequently insurance) rates and third party claims. This leads to lowered costs in which owners have a vested interest. Further, OSHA can better utilize increasingly limited personnel who may benefit from training in real-world construction methods.

Figure 6. Six and over fall protection under MASTER Project criteria. Note the use of JLG lift for erection and ironworkers wearing body harnesses at second floor level. (Standard OSHA regulations require use of body harness at 25 feet and above.)

Speaking on behalf of Turner, I know there are measurable positive results from being involved in the MASTER Project program and I look forward to many more of our sites being selected. Not only are we better able to protect our most vital resource — our people, but in Cincinnati we have modified our safety program to include MASTER Project items which exceed minimum standards in OSHA regulations (Figure 6). We have increased the safety knowledge of our people and have given them more heightened awareness through specialized on-site training. Our subcontractors are more actively involved in safety and are providing both better equipment and training for their people as well.

The MASTER Project is still in its infancy. I hope I have given you enough of the basics of the program to want to know more about it. I personally believe in this program and have seen firsthand the difference it has made in the attitude of construction personnel towards safety. If I could leave you with only one thought, it would be that the ultimate goal of this program is to continually advance towards better safety management in the construction industry. As soon as one goal is achieved, another, higher goal is set. And so, the measurement of what constitutes better safety management today will, hopefully, keep moving higher and higher as new and more effective procedures, techniques and relationships are developed.

5 REFERENCES

91st Congress, S.2193. 1970. Occupational Safety and Health Act of 1970, Public Law 91-596. *Congressional Record, Vol. 116 (1970),* U.S. Government Printing Office.

Murphy, William & Spaudling, Stephen J. 1993. The MASTER Project. *U.S. Department of Labor, Occupational Safety and Health Administration, Cincinnati Notice CPL 1.2.,* Cincinnati Area Office.

Implementation of Safety and Health on Construction Sites, Singh, Hinze & Coble (eds)
© 1999 Balkema, Rotterdam, ISBN 90 5809 036 1

Implementing construction safety systems using quality system methodology

Luís M. Alves Dias
Instituto Superior Técnico, Technical University of Lisbon, Portugal

ABSTRACT: The implementation of construction quality systems is now well known by a large number of construction companies world-wide. They are developed based on the ISO 9000 family of standards and can be applied to the organization and the construction site. Using this experience, an approach to implementing construction safety systems based on the elements of ISO 9001 is discussed in this paper. Each of these elements is described and judiciously adapted to construction safety systems, aimed at developing such systems in a systematic and simplified manner and thus contributing to better safety levels on construction sites.

Keywords: Construction Quality Systems; Construction Safety Systems; ISO 9000.

1 INTRODUCTION

The successful experience that the implementation of quality systems in the construction industry and the integration or compatibility of occupational safety and environmental management systems that is expected in the future, explains the methodology presented in this paper. The suitability of each of the elements of ISO 9001 to take occupational safety matters into account is analyzed. In certain cases, some adaptations are suggested according to the nature of the area under consideration (occupational safety).

It is important to emphasize that the reference, in short, to construction safety must be understood as Occupational Safety and Health (OSH). Actually, the implementation of a safety management system must not only be addressed to the prevention of occupational accidents, but also to the prevention of professional diseases.

On the other hand, it must be said that the development of the methodology presented is addressed to the construction industry. The extension to other activity sectors, although possible, may require other adaptations according to the nature of each sector.

Furthermore, the implementation of safety management systems by the main parties involved in the construction process (owners, designers, supervising companies, contractors) must also require adaptations that take the perspective and the contribution of each of these parties to the occupational safety and health into account.

The implementation of this system may be applicable to all organizations that wish to assure employees and show third parties (clients) their commitment, in the minimum, to the laws concerning safety and health.

2 CONSTRUCTION, SAFETY AND HEALTH - EUROPEAN LEGISLATION

The implementation of any safety management system must conform with the existing laws and regulations related with safety and health.

In the European Union, two main Directives are applied to this matter (among others that are also important and must be taken into account): Directive 89/391/CEE on the introduction of measures to encourage improvements in the safety and health of the workers at work; and Directive 92/57/CEE on the implementation of minimum safety and health requirements at temporary or mobile construction sites.

The first one (the so called Framework Directive) is applied to all activity sectors, public and private. It is intended to guarantee a better level of protection for the safety and health of workers through adequate measures which should not be subordinated to purely economic considerations, as it is defined in its foreword.

The second Directive (92/57/CEE and so called Construction Sites Directive) is applied especially to the construction sector, which considers new parties and responsibilities, and new documents on the prevention of professional risks.

The new parties are: the coordinators for safety and health matters at the project preparation stage and at the project execution stage. The duties of the former coordinator include the drawing up, or providing the reason for the drawing up of the safety and health plan and the safety and health file referred to below. Among other things, the second must make any adjustments necessary to the same documents (or cause such adjustments to be made) to take the progress of the work and any changes which have occurred during the construction phase into account.

The new documents are: the prior notice; the safety and health plan; and the safety and health file. The "prior notice" is a document intended to inform the competent authorities that a construction site will start. It is required whenever the construction site is scheduled to last longer than 30 working days and on which more than 20 workers are occupied simultaneously, or the volume of work is scheduled to exceed 500 person-days. The "Safety and Health Plan" (Alves Dias 1996) is a document setting out the rules applicable to the construction site concerned, where necessary taking into account the industrial activities taking place on the site. The "safety and health file" is a document containing relevant safety and health information to be taken into account during any subsequent works.

3 STRUCTURE OF THE PROPOSED SAFETY MANAGEMENT SYSTEM

The proposed safety management system could be structured and documented according to: OSH manual; OSH general procedures; OSH plans; Work instructions or procedures; and Inspection, test and hazard prevention procedures.

The OSH manual is the main document of such a system. It defines the general rules on the matters of occupational safety and health for the whole organization. It contains the OSH policy, the structure of the organization and the responsibilities of all personnel of the organization that influence OSH matters. This manual can be developed to meet the requirements presented in section 4.

The OSH general procedures must complement the information of the manual. They must be detailed with the necessary information needed for the most relevant elements of the manual. For example, the "control of nonconforming product" requirement may refer to a written procedure describing the operational process for this control. It is recommended that all procedures be grouped in one single dossier, usually called the "general procedures manual".

The OSH Plans are documents with particular information concerning each construction site of the organization. They must conform to the OSH manual and to the laws and regulations applicable. They must set up the specific preventive measures to be implemented for a particular construction site, taking into account the working methods that will be used. There will be as many plans in an organization as there are construction sites within the organization.

The OSH system will also include the work instructions or procedures, which are documents describing the work process concerning each construction operation. These working instructions are the base used to identify and assess the hazards involved during the execution of any construction operation and define the appropriate preventive measures that must be implemented to avoid the risk of occupational accidents or diseases. They are essential for the definition of the inspection, test and hazard prevention procedures.

4 REQUIREMENTS OF THE SAFETY MANAGEMENT SYSTEM

This section refers to the requirements of the safety management system following the methodology and structure of ISO 9001 standard (Quality systems – Model for quality assurance in design, development, production, installation and servicing), with the necessary adaptations according to the specifics of the area under consideration (safety and health). These requirements (or elements) are described in 4.1 trough 4.20.

4.1 *Management responsibility*

The organization management should define and document its policy and objectives for OSH and also its commitment to OSH.

This policy may include: recognition of OSH as relevant to the organization performance; commitment to the laws applicable; recognition of the worker's rights who must be involved in OSH matters of the organization; encourage workers to watch over their own safety and to others who may be affected by their actions; periodically review the safety management system; define clearly the responsibilities of each worker in OSH matters; promotion of all necessary actions so that the system is well understood, implemented and maintained at all levels of the organization;

As far as objectives are concerned the management may consider, for instance, a reduction in a given percentage of the occupational index rates compared with past years, or the establishing of maximum values for these indices.

On the other hand, the organization management should include in its structure the person that is appointed to manage OSH system. This person may have the authority and responsibility to maintain, improve and promote the effective implementation of the system in all organization. To a certain extent the success of the system depends upon the skill of this person, who must have the authority and knowledge of OSH matters so that he can identify, assess and control potential risks.

4.2 *Occupational safety management system*

The safety management system relies on the OSH manual and the general procedures that complement it. This manual may be elaborated according to the elements described in this section. A model of the OSH plan referred to in section 3 above must also be included. The structure of this plan should be used in all construction sites introducing the necessary adaptations according to the specific characteristics of the construction under consideration.

A chronogram of the implementation of the system must also be prepared, showing clearly scheduled periods for developing all system documents. Special attention must be paid to the identification and assessment of risks and their appropriate preventive measures for each construction operation. The contractor must prepare this identification and assessment before the construction operation is to be performed, in time to allow approval by the supervisor. The question is to define and document OSH planning.

4.3 *Contract review*

The organization must assure that before any contract is established, it contains specific clauses concerning OSH, which in any case must not be reduced in terms of the applicable laws on safety and health. For this purpose, a list of the laws applicable, as exhaustive as possible, must be prepared and constantly updated.

4.4 *Design control*

This requirement is particularly relevant to organizations that are involved in the design phase as well as the designer companies, supervisors companies when they are involved in design review and also contractors when a design-construction contract is established.

The issue is to assure that safety and health are taken into account during the design and preparation phase of the project, bearing in mind the general principles of prevention referred in the above section 2. This is actually a duty of the designer and the "coordinator for safety and health matters at the project preparation stage" according to the European Directive 92/57/CEE. The question is to identify and assess the risks that may be involved during the execution phase and the future use of the project under construction. For example, an appropriate identification and assessment of the risks involved with a dangerous material may suggest its substitution by another, which is not dangerous or less dangerous. Performing this analysis during the design phase makes it easier to substitute this material with another.

One method of identifying and assessing the risks in a systematic way is to use check lists with the main construction operations and associate risk levels for execution and manipulation of the materials involved to each one. The method used to control the collective protection equipment (safety nets, guard-rails, etc.) and the individual protection equipment (hard hats, safety harness, etc.) must also be defined to ensure they are in good condition during the period they will be used on the construction site under consideration.

4.5 *Document and data control*

The safety management system includes some documents that were mentioned in section 3 above and originate other documents as a result of its implementation. All this documentation must be controlled according to a procedure that must be established and maintained. Examples of these documents include: the prior notice, the safety and health plan, the safety and health file, records of the inspections performed, audit reports, etc..

4.6 *Purchasing*

The purchase of products related to safety and health must be documented and must take into account the existing laws applicable to ensure compliance with the required specifications. Products must be verified against the purchase order on receipt of the same.

The process of evaluation and selection of services related to health and safety must be described according to criteria that should be clearly established (for example, professional qualification of the personnel and their past experience in these kinds of services).

4.7 *Control of customer-supplied product*

Attention must be paid to the products that are supplied by the client in accordance with the contract. Although the client must assume the responsibility not to supply dangerous products, the contractor must perform a risk evaluation of these products in terms of their use and application.

4.8 *Product identification and traceability*

All products, including protection equipment, that are purchased (for instance, hard hats) or produced by the organization (guard-rails for example) including their components, must have unique identification. Data related to purchase, distribution, utilization and maintenance, must be recorded. These records will be used for the organization to purchase or produce new equipment of proven conformity, or to study the causes of non conformity as a result of an accident due to that equipment.

4.9 *Process control*

The risk identification and assessment of each construction operation must be based on the knowledge of the construction methods that will be used. Thus, the preparation of work instruction where these methods are described in detail is an issue, which is important to the correct identification and assessment of the risks involved.

The working instructions must point out the situations that may affect workers safety. Where risks are considered too high the situation must be defined as a "stop point", when verification by a second party must take place to ensure good safety conditions for the work to continue. For instance, before placing concrete in high pile, the stability of workform and the platform must be confirmed to allow concreting in good safety conditions.

4.10 *Inspection and hazard prevention*

The safety and health plan must specify the procedures that must be applied to prevent accidents for each construction site. These procedures must include the identification and assessment of risks associated not only during the execution of each construction, but also with the resources used (labor, materials and auxiliary equipment).

Labor checks to be made may include: they must have the appropriate qualification and training to perform the tasks involved; they must have the appropriate individual protection equipment; and they must be physically and psychologically apt for the work under consideration.

The risks involved in the handling of the materials involved must be identified and assessed and the necessary preventive measures taken. Auxiliary equipment must be checked to see that it is in good working order, i.e., that the recommended periodic servicing has been done. Form sheets to perform these checks can be found in Alves Dias & Pires, 1998. The risks involved in the inspection and prevention process of the construction operation must be also identified and assessed. These inspections must be recorded. A form sheet to perform these checks can be also found in the above publication.

4.11 *Control of inspection, measuring and test equipment*

The equipment used to prove the conformity of products or place that can affect workers' safety and health must be calibrated and maintained in good condition according to the manufacturer instructions. A form sheet to perform these controls can be also found in the publication referred in section 4.10.

Examples of this equipment may include noise and vibration dose meter, equipment to measure the amount of oxygen in confined spaces, equipment to measure toxic gases, etc.

4.12 *Inspection and test status*

The materials involving risks may be clearly labeled. These labels must have information on the care to be taken with handling, storage and application. The equipment (for example, scaffolds, cranes, etc.) that is not in good condition must also be labeled, stating that it must not be used.

4.13 *Control of nonconforming product*

The safety and health plan for each construction site must include an analysis and management of nonconforming materials and equipment to ensure that only those that are not dangerous to health and safety are used.

Safety equipment should be subjected to periodic inspection. Whenever a nonconformity is detected, a record must be made using the appropriate forms, prescribing the necessary corrective and preventive measures. A form sheet to perform these controls can be also found in the publication referred in section 4.10. The creation of delimited areas of the construction site must be considered (usually called "segregated areas") to store dangerous materials and some equipment that can not be removed from of the construction site immediately.

4.14 *Corrective and preventive action*

The identification of non conformities must be followed by appropriate corrective and preventive actions. Examples of non conformities include: inadequate application of legal rules; non consideration of measures established in the safety and heath plan; occurrence of occupational accidents, etc.. A form sheet to record these non conformities and their appropriate corrective and preventive measures can be also found in the publication referred in section 4.10.

4.15 *Handling, storage, packaging, preservation and delivery*

The methods for handling, storage, packaging, preservation and delivery must be established taking into account the risks associated with materials and equipment. Special attention must be paid to the materials involving high risks (explosives and chemical products). The appropriate legal specifications and manufacturers recommendations must be followed.

4.16 *Control of occupational safety records*

In each construction site, the implementation of a safety management system generates records, which must be organized in dossiers and kept during a certain period of time to be established in each case. These records may include: safety and heath plan; safety and health file; control of individual protection equipment distributed to workers; occupational accidents occurring during the construction phase; control of the auxiliary equipment; non conformity records; safety audit reports, etc..

4.17 *Internal occupational safety audits*

As for quality system, the safety management system may be audited to verify whether occupational safety and health activities and related results comply with planned arrangements and to determine the effectiveness of the system. The conformance of the system assures that the organization is committed to comply with regulations but does not assure the compliance with them. To perform a safety audit one can use the same procedures as for quality based on ISO 10011 standard.

4.18 *Training*

Laws on safety and health state that hazard prevention must follow principles, standards and programs for education, training and information to promote safety and health at work. The organization should prepare a training plan for the workers, especially for those who have special responsibilities on safety and health (safety manager, safety and health coordinators, first aid workers, etc.).

4.19 *Servicing*

The organization should prepare a procedure describing the process to be implemented and the responsibilities for all activities related to safety and health after completion of the project according to the contract. Special consideration must be given to the process of updating the safety and health file in the case of any works needed to be executed during the period for the correction non conformities detected.

4.20 *Statistical techniques*

Statistical techniques should be used, whenever considered important, to quantify variables related to safety and health. One example to be considered concerns the statistical analysis of the occupational accidents and accident indices for a particular construction site and for all the construction sites of the organization or other organizations. Examples of indices that can be used are: the incidence index (number of accidents in each thousand workers); the frequency index (number of accidents in each million hours worked); the severity index (number of lost days in each thousand hours worked); and the duration index (number of lost days per accident).

5 CONCLUSIONS

This paper discussed an approach for the implementation of safety management systems in the construction sector. It was based on the well known ISO 9001 standard that is used world wide for quality systems. This paper summarizes a project developed in the Department of Civil Engineering of Instituto Superior Técnico, which was financially supported by the country of Portugal's Institute for the Development of Working Conditions (Alves Dias & Pires, 1998).

The ISO 9001 standard offers a methodology for systematic implementation of these systems, benefiting from the existing experience on the implementation of quality systems.

Knowing that ISO 9001 is currently being reviewed and a new version is expected during year 2000, the approach now presented for safety management systems, should be adapted according to the new methodology. It is the author's belief that all systems applicable to the construction sites (environment, quality and safety) should follow the same methodology. And why not integrated? The discussion is open.

REFERENCES

Alves Dias, L. M. e Fonseca, M. 1996. Safety and Health Plan in Construction (Portuguese version). IST-IDICT. Lisbon, Portugal.

Alves Dias, L. M. & Coble, R. (Ed.) 1996. Implementation of Safety and Health on Construction Sites. Balkema. Rotterdam. The Netherlands.

Hinze, Jimmie W. 1997. Construction Safety. Prentice Hall. New Jersey, USA.

Alves Dias, L. M. and Pires, J. 1998. Quality and Occupational Safety in Construction (Portuguese version). IDICT, Lisbon, Portugal.

Implementation of Safety and Health on Construction Sites, Singh, Hinze & Coble (eds)
© 1999 Balkema, Rotterdam, ISBN 90 5809 036 1

Safety Management Assessment System (SMAS) applied to construction

D.D.Hee
Marine Technology and Management Group, University of California, Berkeley, Calif., USA

R.G.Bea & R.B.Williamson
Department of Civil and Environmental Engineering, University of California, Berkeley, Calif., USA

K.H.Roberts
Center for Risk Mitigation, University of California, Berkeley, Calif., USA

ABSTRACT: The Safety Management Assessment System (SMAS) is a screening method created to help identify and evaluate the roles Human and Organization Factors (HOF) play in system safety. SMAS is founded on a practical and efficient process designed for use by carefully selected and trained system operators. A computer program assists the assessors in capturing uncertainty, evaluating HOF, and archiving and reporting results. A field test of SMAS involving multiple teams of assessors produced practical and consistent results.

1 INTRODUCTION

Construction fatalities for the United States fluctuated around 700 annually from 1985 to 1989 and there appeared to be no trend (OSHA, 1990). The OSHA review of construction accidents during that period shows that in general; age, union affiliation, and construction firm size do not make a difference on the number of accidents. The report also states "Factors other than workplace safety and health conditions may affect the observed trends or the influence of various parameters" (OSHA, 1990). In April 1989 the National Forum on Construction Safety and Health identified six problems: 1. Lack of company programs and rules, 2. Lack of training, 3. Lack of a system of "planning, organizing, motivating, communicating, and controlling", 4. Lack of a set of values, 5. Poor work practices and procedures, and 6. Lack of accountability and responsibility (NIOSH, 1994). Indeed, other industries are also grappling with the fact that accidents continue to occur and there appears to be an accident threshold that cannot be further reduced. Additionally, major "once in a lifetime" accidents occur and cause numerous deaths and extensive damage.

Research has shown that Human and Organization Factors (HOF) have caused or significantly influenced major system failures. Several of these failures are: Union Carbide chemical plant in Bhopal, India (Shrivastava, 1987), Chernobyl Nuclear Power Plant (Meshkati, 1991), Exxon Valdez (Moore, Bea, 1993), and the Piper Alpha Offshore Oil Platform (Moore, Bea, 1993). In early 1998 a U.S. Marine Corps pilot in Italy flew below his authorized altitude, cut a ski gondola cable killing 20 people (Associated Press, 1998). This list of failures will grow as humans and organizations operate systems.

Several industries have begun to evaluate their accidents for human and organization causes. They have found that humans and organizations cause the following percentages of accidents: 60% of U.S. Navy aircraft losses (Ciavarelli, Figlock, 1997), 70% of oil spills in California waters (Gregory, 1997), 80% of high consequence accidents in the marine industry (Moore, Bea, 1993), and 99% of all accidents by the insurance industry (RIMS, 1995). This confirms that humans and organizations do cause a majority of accidents.

2 BACKGROUND

Current system assessment methods focus on equipment failures and rely on historical and test data (Villemeur, 1992). Without failure data these methods rely on expert judgment, which requires truthful system information. Even with accurate information, experts disagree and their assessment results widely vary (Pouchet, 1989). The Relative Risk (Jones, et al, 1995) and the Health Risk Appraisal (MDPH, 1985) methods were used in studies with very little data. Three key findings from these studies are: 1) use system operators as assessors, 2) relative risk can screen and highlight risky areas, and 3) obtain a set of comparison factors. These findings were used in creating the Safety Management Assessment System (SMAS).

Although SMAS was developed to assess an operational system, it can be easily adapted to assess a construction site. The organization is the owner/general contractor, and the operating teams are the subcontractors.

3 SAFETY MANAGEMENT ASSESSMENT SYSTEM (SMAS)

3.1 *Basic Building Blocks*

The basic building blocks of SMAS are:

1) characterization of a construction site with seven modules,
2) development of scenarios to provide context for evaluating HOF, and
3) construction operations observations.

The seven modules used to describe and characterize a construction site are: Structural, Equipment, Procedures, Environment, Operating Team, Organization, and the Interfaces among them. The structural, equipment/hardware, and procedure modules are self-explanatory and are currently assessed by various methods (Villemeur, 1992). The environment module considers both the physical (weather) and the social (political) influences. The majority of SMAS factors reside in the Operating Team and Organization modules. The Operating Team is defined as the worker(s)/ subcontractor who does the work while the Organization is the group having direct and daily influence on the Operating Team, the general contractor/owner.

Scenarios provide a context for evaluating consequence and relative risk. The assessor's first evaluation of HOF is based on documents and briefings. Factors with the worst values are then used to create scenarios. Assessors use these scenarios during the site visit to evaluate the HOF a second time.

SMAS involves observations of construction operations. During the visit, assessors obtain a first-hand look at the construction site and converse with workers to gain insight into the dynamics of the project. These discussions and observations may lead to the creation of additional scenarios.

3.2 *Human and Organizational Factors*

HOF selected for SMAS (Table 1) were based on various industry standards, regulations, textbooks, and results from current research (Hee, Bea, 1997). Research on High Reliability Organizations has defined the following five characteristics as critical to their abilities to operate high technology systems safely: Process Auditing, Appropriate Reward Systems, High Standards of Quality, Appropriate Risk Perception, and Command and Control Functions (Libuser, Roberts, working paper). Based on this background, Table 1 categorizes each factor into one of seven modules. Each factor is further defined by attributes (Hee, Bea, 1997).

3.3 *Unique Features*

The unique features of SMAS are: 1) selecting and training workers as assessors, 2) capturing uncertainty, and 3) the SMAS process.

Table 1. Level 1 (Modules) and Level 2 (Evaluation Factors)

Operating Teams	Organizational	Procedures
Process Auditing Safety Culture Risk Perception Emergency Preparedness Command & Controls Training Communications Requisite Variety	Process Auditing Safety Culture Risk Perception Emergency / Crisis Response Command & Controls Training Communications Resources	Operating Maintenance Safe-work Contractor selection Pre-start up review Emergency response Management of change
Hardware / Equipment	**Structure**	**Environmental**
Pipe capacity Pump capacity Hose capacity Loading arm capacity Fire load capacity Electrical	Berthing capacity Live weight capacity Dead weight capacity Environmental capacity Fire load capacity (passive)	External (Weather) Internal -(Climate control) Social External - (Regulatory, Society)_ Social (Internal) (Within organization and operating team)
	Interfaces	
Operators & other Organizations & other	Procedures & other Environmental & other	Equipment & other Structure & other

Using workers as assessors is the key to a SMAS assessment. Stringent selection criteria include technical proficiency and personal characteristic requirements (e.g. integrity, motivation). The candidate must: be a qualified worker, recognize the importance of HOF in causing accidents, and desire to learn about HOF. Once selected, the candidates are trained on HOF, the SMAS process, and the SMAS computer program. Upon successful completion of training, the candidate is qualified as an assessor.

Uncertainty is inherent in humans and organizations. SMAS captures uncertainty by having assessors select a range of three values. These three values, Most Probable Value (MPV), Best Value (BV) and Worst Value (WV) are used to create a triangular distribution. A triangle with a wider base shows that the assessor has greater uncertainty relative to another with a narrower base. SMAS requires assessors to provide supporting comments since research shows that comments produce more truthful evaluations (Morgan, Henrion, 1990).

3.4 *SMAS Process*

The SMAS process is comprised of five steps:

1. Select a construction site to assess
2. Identify and train an assessment team
3. Conduct Coarse Qualitative Assessment - Identify Factors of Concern
4. Develop scenarios using Factors of Concern
5. Conduct Detailed Qualitative Assessment

The first step, construction site selection, is based on any of the following: accident and near-miss event history, the likelihood and consequences of a major accident, and its periodic assessment schedule.

The second step is selecting and training an assessment team. Selection and training was briefly discussed earlier and is detailed in (Hee, Bea, 1997).

The third step is a coarse qualitative assessment of the seven modules. This assessment reviews the general history of accidents and near-misses at the assessed site and at similar types of construction sites. Site specific information focuses on the structure, equipment, procedures (normal work practices, and emergency / crisis management), operating personnel (all subcontractors), and organizations / management. Results from previous inspections and hazard studies are also evaluated in this step, and briefings are given by representatives of the owner / general contractor organization and working crews.

The result of Step #3 is a list of Factors of Concern (FOC), those factors that could lead to major accidents. As part of the evaluation process, the assessment team records the rationale for identifying each FOC. Experience indicates that these narratives provide the richest source of information about the site. At this step, the assessment process allows for the identification of suggested mitigation actions. Results are reported in standard textural and graphical formats, or in user defined formats (stored in the computer program or produced during an assessment). This information must be confirmed by a site visit. For some sites, this information may be sufficient to exit SMAS, implement mitigation actions, record results, and schedule the next assessment.

If necessary, SMAS proceeds to Step #4, the development of scenario(s) to express and evaluate the FOC. These scenarios are intended to capture the initiating, contributing, and compounding events that could lead to a major accident. They also focus attention on potential high risk construction elements.

Step #5 has the assessors visit the site to conduct a detailed qualitative assessment using FOC and scenarios. During the visit, operations (normal and emergency) and are observed. Multiple visits may be required to observe critical aspects of normal and emergency operations such as crew changes. The visit(s) also confirm the key elements identified during the coarse qualitative evaluation, Step #3. During the visit, discussions occur with workers, and provisions are made for anonymous reporting.

The outcome of Step #5 is a report that confirms or contradicts the coarse qualitative evaluations, and details the measures suggested for FOC mitigation. The rationale for suggested mitigation actions are detailed along with their projected beneficial effects on the FOC. The results of Step #5 are reported in standard and user defined formats.

The assessment team continues SMAS in one of two ways. The first option is to return to the FOC stage and repeat Step #5 using FOC discovered during the site visit. The second option is to terminate SMAS at the end of Step #5, record results, and schedule the next assessment. Those FOC not resolvable with SMAS then become the subject of more detailed quantitative methods, such as Probabilistic Risk Analysis or Quantitative Risk Analysis.

The SMAS computer program guides the assessor through the process, eases data collection, displays the evaluation, and has pre-formatted report forms (Pickrell, Bea, 1997). The assessors capture uncertainty by selecting values with three clicks of the "mouse" and then the computer immediately responds by displaying the resultant triangle. After the third mouse click, the cursor drops down to a text box for the assessor's supporting comments. Further details of the program are included in reference (Pickrell, Bea, 1997).

4 APPLICATION

SMAS was tested on two operational marine terminals in California. The four assessors, two from company X and two from its regulatory agency, were trained and then organized into two teams. Each team independently assessed the same marine terminal. As indicated in Tables 2 and 3, their assessments were very consistent (Hee, Bea, 1997). Assessment comments provided important insights into how the reliability and safety of the systems might best be improved.

One of the most important aspects of SMAS is the training given to workers on Human and Organization Factors. Once the assessment is complete and the report written, the understanding of HOF will remain with the workers and they will use it to overcome communication barriers and continuously improve safety.

SMAS can be easily adapted and applied to any construction site. Since the owner/general contractor usually does not change for the duration of the project, they can be measured and benched marked for continuous improvement. Since there are several subcontractors, each

Table 2. SMAS Module Most Probable Values

Module	Team A	Team B
Structure	3.6	3.3
Procedures	2.4	2.3
Organization	2.6	2.8
Operating Team	2.6	2.4
Interfaces	2.9	3.0
Equipment	2.2	2.0
Environmental	2.7	4.0

Table 3. Ranges of Values for SMAS Modules

Module	Team A	Team B
Structure	3.0 to 4.0	3.0 to 4.0
Procedures	2.0 to 3.0	2.0 to 3.0
Organization	1.0 to 6.0	1.0 to 4.0
Operating Team	1.0 to 7.0	1.0 to 4.0
Interfaces	2.0 to 3.0	3.0 to 3.0
Equipment	1.0 to 4.0	3.0 to 3.0
Environmental	1.0 to 4.0	4.0 to 4.0

would be assessed with SMAS at least once. The training of workers on HOF will be valuable in reducing future accidents.

5 CONCLUSION

SMAS is a method that selects and trains actual workers on HOF and then guides them through a simple process for evaluating their work site. Part of the evaluation involves capturing uncertainty and recording evaluations, both of which are facilitated by the SMAS computer program. A field test of SMAS showed that the process was practical and produced consistent results. SMAS has the potential to overcome communication barriers, empower operators, benchmark, and measure continuous improvement.

6 REFERENCES

Associated Press. 1998. Washington. Marines face disciplinary action. The Oakland Tribune, March 12, 1998.

Ciavarelli T, Figlock R. 1997. Organizational factors in naval aviation accidents. In: Proceedings of The Center for Risk Mitigation, pp 36-46.

Gregory GL. 1997. The California State Lands Commission's marine terminal oil spill prevention program. In: Proceedings of the Center for Risk Mitigation conference, pp 70-78.

Hee DD, Bea RG. 1997. Safety management assessment system (SMAS). Marine Technology and Management Group report, University of California Berkeley.

Jones ED, Banks WW, Altenbach TJ, et al. 1995. Relative risk analysis in regulating the use of radiation-emitting medical devices – a preliminary application, NUREG/CR-6323, UCRL-ID-120051. USNRC Washington DC.

Libuser CB, Roberts KH. (working paper). A conceptual model of risk mitigation. University of California Berkeley.

MDPH (Michigan Department of Public Health). 1985. Health promotion /Risk reduction protocols and screening guidelines for adults. HP-10.

Meshkati N. 1991. Human factors in large-scale technological systems' accidents: Three Mile Island, Bhopal, Chernobyl. Industrial Crisis Quarterly;5:133-154.

Moore WH, Bea RB. 1993. Management of human error in operations of marine systems. Report No. HOE-93-1, Final Joint Industry Project report, Dept. of Naval Architecture and Offshore Engineering, University of California, Berkeley.

Moore WH, Bea RB. 1993. Human and organization errors in operations of marine systems: Occidental Piper Alpha & High Pressure Gas Systems on offshore platforms, Offshore Technology Conference.

Morgan MG, Henrion M. 1990. Uncertainty – a guide to dealing with uncertainty in quantitative risk and policy analysis. Cambridge University Press, Cambridge.

NIOSH (National Institute of Occupational Safety and Health). 1994. 1994 FACT BOOK: National Program for Occupational Safety and Health in Construction. U.S. Department of Health and Human Services, March 1994.

OSHA (Occupational Safety and Health Administration). 1990. Analysis of Construction Fatalities – The OSHA Data Base 1985-1989. U.S. Department of Labor, November 1990.

Pickrell BD, Bea RG. 1997. Assessment of human and organizational factors in operations of marine terminals and offshore platforms. Marine Technology and Management Group report, Department of Civil and Environmental Engineering, University of California, Berkeley.

Pouchet A. 1989. Human factors reliability benchmark exercise – synthesis report, (EUR 12222 EN). CEC Joint Research Centre, Ispra(VA) Italy.

RIMS Conference. 1995. Discussion, 1995, San Francisco.

Shrivastava P. 1987. Managing industrial crises: lessons of Bhopal. Vision Books, New Delhi, India.

Villemeur A. 1992. Reliability, availability, maintainability and safety assessment, vol 1 and 2, John Wiley & Sons, New York.

4 Safety culture

Implementation of Safety and Health on Construction Sites, Singh, Hinze & Coble (eds)
© 1999 Balkema, Rotterdam, ISBN 90 5809 036 1

Nourishing safety culture in the construction industry of Hong Kong

C.M.Tam
Department of Building & Construction, City University of Hong Kong, People's Republic of China
A.P.C.Chan
Department of Building & Real Estate, Hong Kong Polytechnic University, People's Republic of China

ABSTRACT: The construction industry of Hong Kong bears a very poor site safety record. The overall accident rate has gone down a little during the past few years, however; the fatalities have risen dramatically. The annual accident rate per 1000 workers is twice that of the United States, 25 times that of Japan and nearly 30 times that of Singapore. In the past, the Hong Kong government adopted a laissez-fair approach in managing construction safety, hoping that the market force would regulate the safety performance. However, the approach has proven to be ineffective. In combating the poor safety records, the Hong Kong government has lately introduced a series of safety programs, which consist of encouraged and mandatory schemes aiming at nourishing a proper safety culture in the construction industry. The paper describes their effectiveness by referencing to the accident figures of the past ten years in Hong Kong.

1 INTRODUCTION

The number of industrial accidents in Hong Kong remains at a very high level notwithstanding a downward trend in recent years. Despite this trend, the construction industry continues to account for over one third of all industrial accidents and the bulk of the fatalities. The unsatisfactory construction safety record has been a matter of concern to the Hong Kong Government and the public.

All people in Hong Kong know that the construction industry is a problem area. Although it employed less than eight per cent of the total industrial workforce, it consistently accounted for more than one third of all industrial accidents in the last ten years. The average probability of a construction worker getting involved in industrial accidents between 1985 and 1994 is 5.8 times higher than the overall average for all industrial workers. The construction industry is responsible for the bulk of the fatalities in industrial accidents. An average fatality rate of 0.868 per thousand workers in the construction industry between 1985 and 1994, at 10.5 times the overall average for all industrial workers, is clearly unacceptable [Education and Manpower Branch (1995)].

2 GOVERNMENT'S ATTITUDE TOWARDS SAFETY MANAGEMENT IN THE PAST

In the past, Hong Kong Government's attitude towards construction safety was an enforcement approach. This means that safety legislation was enacted to enforce contractors to follow. The Labour Department served as a watchdog. It sent factory inspectors to carry out random checks. This arrangement gives the advantages of being straightforward in implementation and providing some deterrent against those employers who could not be persuaded by other means to observe safety regulations. This has created a culture for contractors in Hong Kong to "avoid" convictions. They window-dressed their safety commitment by posting up posters and slogans on sites; such as "Safety First", etc. However, when there were conflicts between safety and productivity, "productivity" always took the priority.

By then, the Hong Kong government blindly believed in the market-driven mechanism by adopting a laissez-fair approach in managing construction safety, hoping that the market force would regulate the safety performance itself. However, the approach has proven to be ineffective.

Safety legislation and, in particular, the Construction Sites (Safety) Regulations provide a sound basis for good safe building practice. However, the Factory Inspectorate of the Labour Department, whose duty is to enforce the law, does not have adequate resources to scrutinise every site frequently. In 1994, there were only 42 factory inspectors specially assigned to the building, engineering and construction division. They were responsible for approximately 1,300 construction sites. Worse still, most of the experienced staff who were qualified to go out on site were losing due to resignation. Therefore, law breaking was a rife because the risk of prosecution was low. Moreover, the practice of cutting corners for commercial gain was frequently condoned by senior management and ignored by clients and their professional advisors.

3 STEPPING IN OF THE GOVERNMENT TO CHANGE THE SAFETY CULTURE

Under strong social pressure, the Hong Kong government has initiated a series of safety improvement programs starting from 1986. The major aim was to improve construction safety performance by changing the safety culture of the industry.

3.1 *Enactment of the factories and industrial undertakings (safety officers and safety supervisors) regulations in 1986*

The regulations mandated contractors to hire safety officers and safety supervisors in construction sites. Their duties are to assist the contractors in promoting safety and health of workers including:

- Advise the main contractor as to measures to be taken in the interest of safety and health and their implementation.
- Inspect the workplace to identify potential hazards and report the findings with recommendations for correction to main contractors.
- Investigate accidents and dangerous occurrences and report with recommendations for prevention.

- Advise the main contractors of any repair or maintenance in respect of premises, plants and equipment that ought to be carried out (Labour Department 1995).

This was the first initiative by the government to encourage the formation of a self-regulatory framework in safety management. However, this still follows the "watchdog" principle, but shifting the spying responsibility from the government inspectors to the so-called "safety officers" who bear the legal obligations if they do not fully discharge their legal duties. Hence it helps little in changing contractors' attitude in safety prevention.

3.2 *Quality assurance (QA) scheme in 1992*

In 1992, the Housing Department, which provides 50% of the total housing stock as public housing to people in Hong Kong, required that before 31 March 1993, all building contractors must be ISO 9000 certified. Most contractors rushed to obtain the accreditation. The QA scheme encourages a self-regulatory approach in quality management and a comprehensive documentation system. It lays the corner stone for a self-regulatory safety management system. As safety is also considered under the QA system, a comprehensive documentation system forces contractors to take a proactive view in safety management.

3.3 *Punishment by business decisions and encouragement by safety campaigns*

Under the Work Bureau's Technical Circular of No. 22 in 1994, any contractor with six or more convictions on a single government contract within a rolling six month period will be invited to attend a Panel of Enquiry under the Chairmanship of the Secretary for Works. The Panel will evaluate the contractor's safety-related performance and then determine whether he will be suspended from tendering government projects or not (Tse 1998). As a result of this, some contractors were banned temporarily from tendering government projects. This deters contractors from unsafe site practices.

Nearly at the same time, the Housing Authority of the Hong Kong Government has included safety as one of the performance yardsticks of public housing projects in their Performance Assessment Scoring System (PASS). Poorly scored contractors in PASS will be suspended from tendering for a certain period of time, which echoes the practices of the Work Bureau.

Meanwhile, the Housing Authority has administered safety campaigns for their projects, trying to arouse people's safety concern. Presentation of prizes to contractors is openly carried out to cast a safety-concern image to the public and contractors. Further, all building contracts mandated a notice to be posted up on the site hoarding to indicate the up-to date accident figure of the project in order to remind site personnel and the public the importance of site safety.

3.4 *Establishment of a self-regulatory safety management system*

In July 1995, the Manpower and Education Bureau issued a consultation paper on the review of industrial safety in Hong Kong. The paper encouraged a new safety management strategy, which emphasised on self-regulatory approach in safety management. The system basically follows the Quality Assurance System in which it asks construction firms to define, document and implement top management's commitment in safety management, safety management system, safety training, formation of safety committee, implementation of safety audit, setting up of safety benchmark, etc. Since then, a self-regulatory safety management system for the construction industry has been established.

3.5 *Mandatory safety training program*

In June 1996, the Labour Department and the Construction Industry Training Authority (CITA) jointly promoted a safety training program: the "Green Card Scheme". In 1997, the Work's Bureau and the Housing Department both required that all workers of their projects must have completed the Green Card training and acquired a "Green Card". Up to June 1998, there have been more than 20,000 construction workers completed the training. The Government is going to submit a Bill to the Legislative Council in November 1998 to make the "Green Card" training scheme mandatory to all construction personnel; including workers, management and professional site staff, before allowing them entering construction sites. This mandatory safety training scheme aims at educating site personnel the importance of site safety and their rights and duties in site safety.

3.6 *Safety supervision plan*

In 1998, the Buildings Department requires all construction projects to set up a clearly defined site organisation structure, staffed with competent technical supervision personnel. Contractors, engineers and architects must provide method statements, which include all safety measures, for approval before any dangerous construction activities are permitted to carry out. Competent technical staff are classified into five categories according to their qualifications defined by the Buildings Department. These staff should bear legal consequences if their duties are not observed.

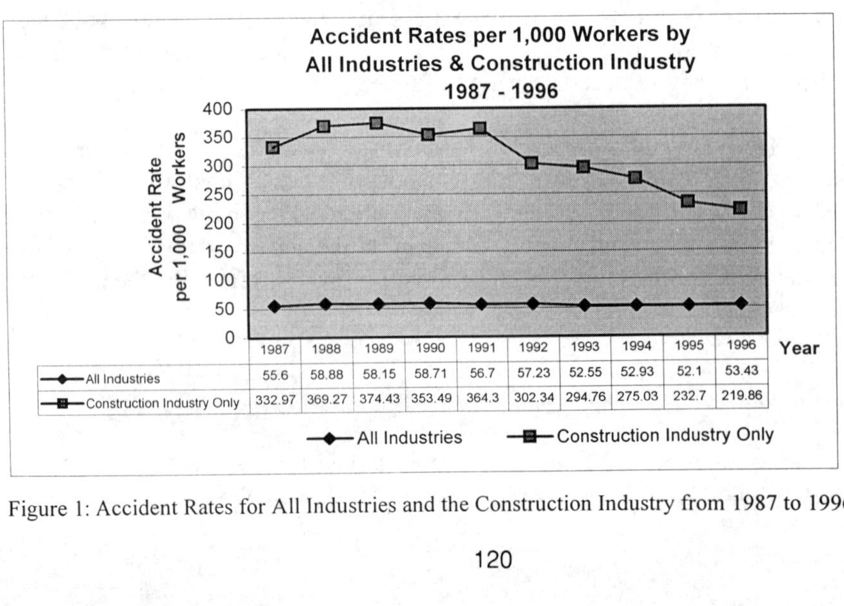

	1987	1988	1989	1990	1991	1992	1993	1994	1995	1996
All Industries	55.6	58.88	58.15	58.71	56.7	57.23	52.55	52.93	52.1	53.43
Construction Industry Only	332.97	369.27	374.43	353.49	364.3	302.34	294.76	275.03	232.7	219.86

Figure 1: Accident Rates for All Industries and the Construction Industry from 1987 to 1996

4 RESULTS

The above effort by the government has resulted in improving accident rates as shown in Figure 1.

From Figure 1, it can be shown that the accident rates of the construction industry have started to drop significantly since 1991 when the QA scheme has been initiated although the overall accident rates for all industries have remained quite constant. Having said that, the accident rate of around 220 per thousand workers is still unacceptable compared to the international norm. The impact of the "Green Card" training and safety supervision plan still needs to be observed. However, it is expected that the accident rates will further drop in the following years.

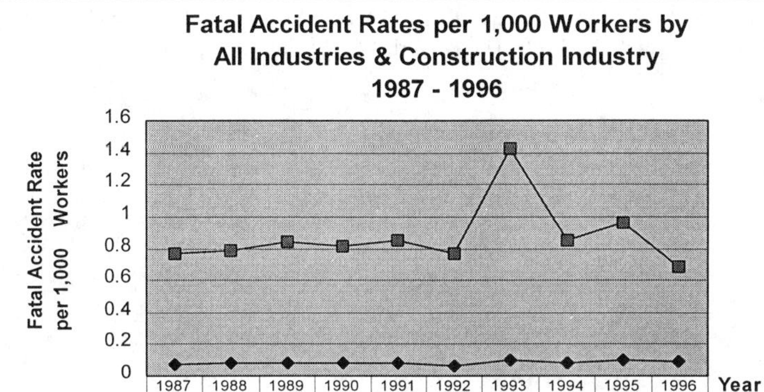

	1987	1988	1989	1990	1991	1992	1993	1994	1995	1996
All Industries	0.077	0.083	0.084	0.086	0.083	0.067	0.103	0.08	0.098	0.088
Construction Industry Only	0.768	0.79	0.837	0.816	0.851	0.771	1.423	0.854	0.96	0.681

Figure 2: Fatal Accident Rates for All Industries and the Construction Industry from 1987 to 1996

From Figure 2, it can be seen that the fatal rate of the construction industry reached a record low figure of 0.681 per thousand workers in 1996. Although the fatal rate reached the peak in 1993, it was solely due to the occurrence of a serious construction accident that killed more than 15 people resulting from the falling down of a workmen's lift in a construction site in North Point. Similarly, the fatal rates are expected to drop in the following years.

5 CONCLUSIONS

After exercising the laissez-fair approach in managing construction safety for decades, the Hong Kong government has realised the poor safety performance in the industry. This triggers off the implementation of a series of safety measures initiated by the Hong Kong Government, intending to cultivate a proper safety culture. These programs have, to certain extent, produced a fruitful result as evidenced by the improving accident rates.

People have started to reckon the importance of site safety. In the following years, with the introduction of more safety awareness programs and accident mitigation measures, it is expected that the site safety performance will be further improved. However, the current safety figures are still not satisfactory. More effort is required from the government to cut the accident rates.

REFERENCES

"Industrial Accidents in the Construction Industry" Hong Kong Labour Department Statistics based on a Building and Construction Site Survey, 1991.

"Consultation Paper on the Review of Industrial Safety in Hong Kong" Education and Manpower Branch July 1995.

Tse, J. 1988. Site Safety and Labour Importation. Construction and Contract News Issue No.1, 1988, pp 5.

Tam C M and Fung I. 1998. Effectiveness of Safety Management Strategies on Safety Performance in Hong Kong. Construction Management and Economics Vol. 16, pp. 49-55, January, 1998.

Labour Department 1995. Guide to the Factories and Industrial Undertakings (Safety Officers and Safety Supervisors) Regulations 1986. The Government Printer, Hong Kong.

Implementation of Safety and Health on Construction Sites, Singh, Hinze & Coble (eds)
© 1999 Balkema, Rotterdam, ISBN 90 5809 036 1

Creating a living community for the elderly – A study on group living for the elderly in Japan

Shinko Sasaki
Tokuyama College of Technology, Japan
Katsuyo Ueno
Kyoto Prefectural University, Japan

ABSTRACT: The purpose of this study has been to look at examples of places where the building has completed and where the plans have been canceled, and by looking at the features of Group Living for the elderly from the positions of individuals and the government administration, to find important factors for these plans to be successful. An analysis of the results shows that for these plans to come together, the importance of the mutual understanding of the values of the way people live and their domestic affairs, the level of cooperation in their livelihood, the length of time from the outset of the project until it is completed, and their age when the project is started are all influencing factors.

1. INTRODUCTION

Health and safety on the construction side is regarded as including the health and safety of occupants after the building's completion. Not only must a safe and healthy construction site be maintained, but after the building's construction, the occupants must be able to live healthy and safe lives while living in the building. This is particularly so in the case of residences for the elderly.

Health is only achieved once the conditions are in place for both physical health and mental/ social health. With regard to the living environment to live healthy lives, physical aspects include barrier-free, ventilation and lighting, which are being adequately understood in research and are being implemented. However, when it comes to mental and social aspects, the complexity of the problem means that initiatives in research and planning are lagging behind.

As mental and social aspects of health in residential environments for the elderly, relationships with neighbors and sense of community with the region are important elements. As if supporting this theory, in the Kobe Earthquake of 1995, it was reported that of the people who were evacuated and lost their relationships with neighbors, 178 elderly people died alone. Without limiting this phenomenon to the period of confusion after disasters, relationships with neighbors have become weaker in city life today, with almost all elderly people living in independent housing. We are far from achieving the level of community required by the elderly. It may be said that in modern society, there has arisen a need for planning of living environments from mental and social aspects, to ensure the health of residents.

Against such a background, the movement of group living construction has started to germinate. This entails the public supply taking into account mutual assistance between occupants and housing construction by groups, with the objective of forming communities for the elderly.

2. CURRENT STATE OF THE ELDERLY IN JAPAN

The percentage of the population accounted for by people aged 65 or more has risen from 4.9% in 1950 to 14.6% in 1995 in Japan, and is estimated to rise to 26.9% by the year 2020, an aged society unprecedented in world terms. Japan is being forced to make a rapid response to this situation (Fig.

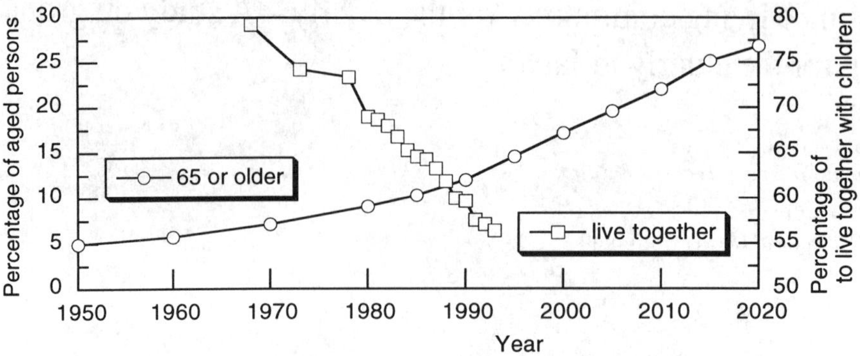

Fig. 1 Percentage of aged persons (65 or older) and to live together with their children

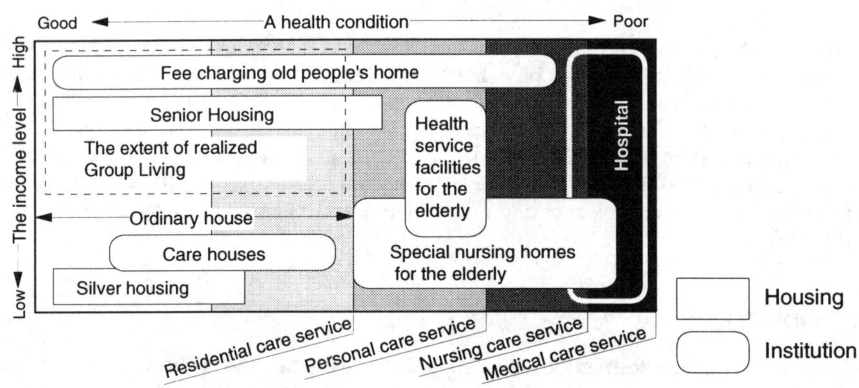

Fig. 2 Organization of housing and an institution for the elderly

1). In the past, it was usual for the elderly in Japan to live together with their children in extended families. However, with the over-estrangement of the population from the countryside and the concentration of the population in cities, the shift towards nuclear families has resulted in an increase in the number of elderly living alone or as couples. The increase in the number of households with only elderly people living in them has made impossible the care by families which has existed in the past, and dramatic increases in the cost of welfare are expected in the future.

3. NEW NEEDS AND HOUSING POLICIES FOR THE ELDERLY

In response to such changes in family structure, as housing policies for the elderly, the government is setting forth the provision of housing which responds to the needs of the elderly, support for families to live together or close-by, and linking with welfare and medical policies. The supply of housing which meets the needs of the elderly started with the provision of housing with living support personnel, such as the Silver Housing Project of 1987 and the Senior Housing of 1990. With regard to welfare, the new Gold Plan results in the strengthening of the care at home and a dramatic increase in the capacity of care facilities, and the introduction of care insurance is enhancing health and medical care. Fig. 2 shows a diagram of the forms of residence of the elderly with the health and income axes.

It is desirable for the elderly that as long as they are healthy, they live in independent housing rather than in rule-laden facilities. From this perspective, Senior Housing and Sliver Housing are being provided to high income earners and low income earners. However, there are still few such housing projects, and the housing places emphasis on the support framework, so adequate consideration is not being given to the formation of communities. For general households which are in the intermediate layer, care houses are provided as facilities, but this group is not provided with independent housing.

There has as yet been no response to the needs of general households who want to live in independent housing within a good community. Group living is a living format which is positioned as one for people with relatively good health and for people with average or above average incomes, and it is hoped that support will be provided for future expansion.

3.1 *Group Living*

Group living entails the housing which comprises small-scale groups. Buildings provide a certain amount of area from each floor space owned by each occupant to create cooperative kitchens, dining rooms and living rooms. This paper uses the term group living to collectively refer to collective housing and co-housing. As construction spaces, such group housing combines housing which maintains the privacy of the individual, and shared spaces which provide opportunities for communication. In the common spaces, the cooperative nature of part of the occupants' lives and mutual assistance between occupants takes place.

3.2 *Review of Literature*

There are many completed case studies of group living in Europe and America. While the characteristics of occupants, form of ownership, construction processes and space structures vary from country to country, they all share the emphasis on the formation of communities. In Sweden there is the series of collective housing research carried out by Vestbro, D. U. (1991 & 1992) and Koyabe, I. (1997), in Denmark Zahle, K. & Mortensen, P. D. (1992) and Ambrose, I. (1993) have carried out research on co-housing, and in the United States Frank, K. A. & Ahrentzen, S. (1991) and Fromm, D. (1991) have brought together research on a variety of forms of such housing. In these series of research, group living is positioned as a new housing format option which has appeared as a result of the pursuit of the authority to make one's own decisions.

3.3 *Objective and Methods*

This paper takes up group living which is housing with the objective of creating communities for the elderly in Japan, as an initiative to ensure the health and safety of people in the latter stages of their lives. The objective of this paper is to clarify the characteristics and conditions for the formation of group living.

This survey targeted the following as typical case studies: four cases of private sector housing in which occupants have participated in construction, three cases of public housing with the objective of community formation after entry, and one case in which construction was suspended. The survey methods used are interview surveys of occupants, representatives or managers from each group.

4. ANALYSIS

In this section, stereotyping was carried out on cases which had been completed or were under construction, and for the case in which construction was suspended, the causes behind the suspension were analyzed. Based on the analytical results, a report was made of a public program which reflects new needs.

4.1 *Stereotyping of Cases Completed and Under Construction*

An overview of the cases surveyed is shown in Table 1. It is possible to stereotype the cases into three types by motivation for construction and characteristics of group formation.

(1) *The False Family Type*

Types which resembled a family-type positional relationship in which the relationship between occupants is close and comprises the head of the family and the family members were classified as the False Family Type. This is the case where friends who have been so for tens of years create housing under the guidance of the leader. The characteristic of occupants is that all members are women who have careers and who have never married or have married but not had children. Examples of points in common between occupants are similar values and experiences such as the same occupation or religion, and occupants have quite a strong cooperative coloring. The content of cooperative activities also entails all members having meals and enjoying hobbies together. This type overlaps with the generation who found it difficult to marry after the war, and it may be regarded as a format unique to Japan which reflects the living requirements of a particular generation.

(2) *The Local Community Type*

These programs are carried out mainly by local municipalities for elderly people living alone, and are managed by cities, towns and villages. Elderly people who live alone in rural agricultural areas have few opportunities for participation in society, and have a tendency to become lonely. The decrease in their physical functions results in greater burden for housework and anxiety in emergencies. To counter these problems, cooperative living facilities were constructed supported by the local community and based on regional circumstances and the intents of the elderly. Living support personnel are regularly dispatched to such facilities. These personnel fulfill the role of coordinator, and give a sense of confidence to housing occupied solely by elderly people. The characteristics of occupants are that they are people who have lived in the same area for a long time, and have the same cultural awareness and the same manners and customs. The case in Prefecture H entails a facility constructed as a winter escape and located in an area with high snowfall coming to be used on a permanent basis. The occupants of the facility were gathered from a region with low population, so there are inter-family relationships which have been fostered over many generations, which makes it easy to form a community. This resulted in a positive rating for the housing. In the case in Prefecture O, however, occupants were drawn from a relatively wide area, and no consideration was given to the formation of groups of occupants. Occupants are also removed from the facility if the many restrictions are not observed, which restricts the independent activities of occupants. It is therefore difficult to say that a good community has been formed.

In the local community type, good communities shall only be created when existing communities in the region are utilized, and the independence of occupants is respected by having relaxed restrictions.

(3) *The Network Type*

In the network-type, housing creation starts with group formation, and the objectives are to respect interaction with neighbors and to achieve a good community. The promoter of such housing invites applications from people with a similar way of thinking, who then form a group and promote construction plans. By doing so it becomes possible for occupants to get to know each other well before moving in, and it becomes possible to utilize this in planning the cooperative activities after moving in, and in designing cooperative spaces and so forth.

The characteristics of occupants are that while the majority are single women, there are also couples and households with children, and there are different living requirements to conventional houses in which the elderly live with their children, and in which couples live alone. Areas in common between occupants include the fact that they "want to live as oneself together with other people who one gets along with," without getting caught up with the existing living customs of the elderly. The largest characteristic is that construction was carried out by healthy senior citizens in their 50s and 60s.

The fact that housing creation starts with group formation has a positive impact on the community after occupants have moved in. In addition, because it is possible to draw group members from a wide range of society, it is predicted that there are many people who can potentially participate in creating this type of housing.

In the cases surveyed above, good communities were formed in housing which respects the self-determination of occupants, in which cooperative lives of occupants are carried out with mutual intent, and independent cooperative activities are formed.

Of the three types, it is hoped that the network-type expands. When the formation of communities is focussed upon, it is suggested that respect for self-determination is an important consideration.

Table 1 Overview of the cases surveyed

Type	The False Family Type		The Local Community Type		The Network Type	
Name	Friend-ship house	House of Beulah	Sunny House	Seiraku-sou	Hot Co-op	LB House
Completed	1976(started in 1975)	1989(started in 1988)	1995	1983	1984(started in 1984)	1999(Planning)
Location	Abiko city (Urban area)	Fukuoka city (Core provincial city)	Pefecture O (Small town)	Prefecture H (A depopulated area)	Sendai city (Core provincial city)	Shimonoseki city (Provincial city)
Convenience	The central part	Suburbia	Country side	Country side	Suburbia	Suburbia
Group composition	A boss & her followers from same workplace	Same religion and work	From same town	From same community	Recruit members from the Women's Lib Group	Advertise members in the newspaper
Description of motive on construction	After the leader's retirment.They definitely needed a house to live together.	The faith motivated the leader to construct the house for her friends who live alone.	A support project for eldery people live alone by prefectural O.	This house was constructed as a winter escape in an area with high snowfall.	They have a general feeling of unrest by living alone. They wish living together.	They want to live as oneself together with other people who one gets along with, without usual style.
Occupants	4 (females)	6 (females)	4 (1male, 3females)	6 (2males, 4females)	3 (females)	9 (regardless of sex)
Age	50-56 (when moved in)	72-84 (when moved in)	70-79 (when moved in)	68-86 (when moved in)	61-70 (when moved in)	48-80(Mainly 50s and 60s)
Households	Singles	Singles	Singles	Singles	Singles	Singles and Couples
Building size	270 sq.m (4units)	249 sq.m (6units)	358 sq.m (6units)	216 sq.m (6units)	136 sq.m (3units)	862 sq.m (8units)
Private facilities*	3bedrooms, Ba, La, DK, Li, E	1or2 bedrooms, Mini-K, E	1bedroom, La, DK	1bedroom, La, K	1bedroom, E	2bedrooms, Ba, La, DK, Li
Common facilities*	Garden	Ba, La, DK, Li, Laundry, Guestroom	Ba, La, Li, D,K, Laundry, Helper's room	Ba, Li, Laundry, Helper's room	Ba, La, K, Li, D, Open space	La,K,Li,D,Free-space,Guestroom,Coordinator's room
Plan of common space and facilities on ground floor* Common	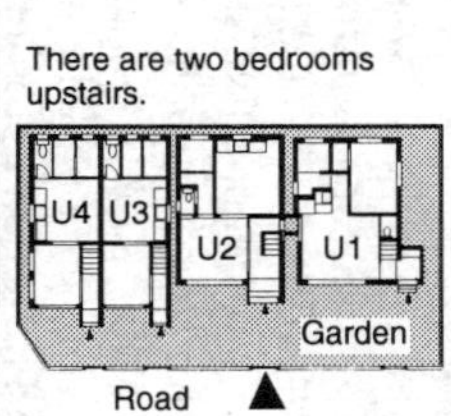	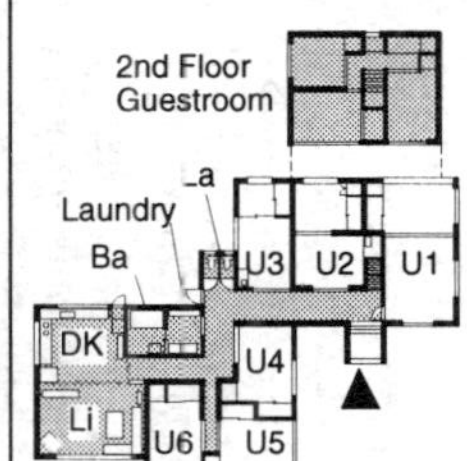	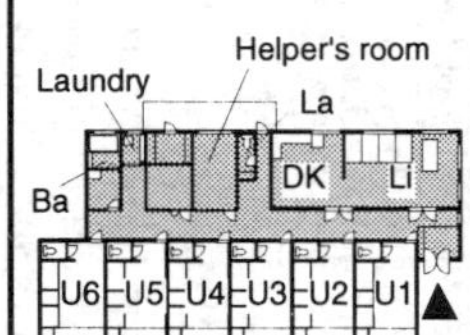	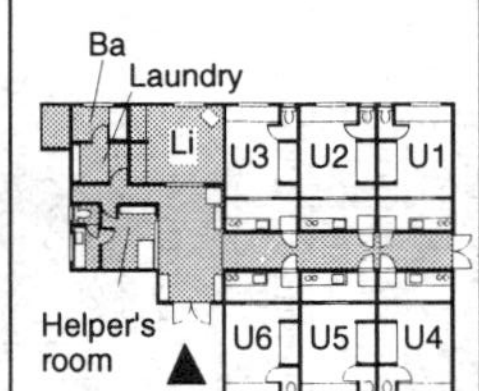	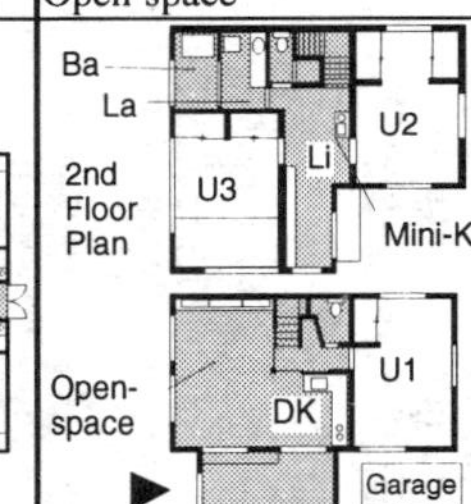	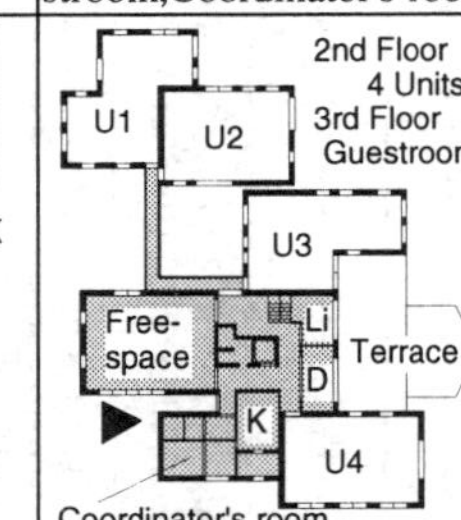
Cooperative activities	Cook together and share every meal. Go for a walk together every morning. Hold silver salon with neighbors once a month.	Cook together and share every meal. Hold morning service. Have a tea break everyday.See a Dr. for a checkup every othermonth.	Cleaning common space take turns at one-week shifts. They are not eager about cooperative activities.	Cleaning common space all together every morning. Hold a tea parties twice a day. Go on an excursion twice a year.	Cooking and cleaning take turns. Hold a handcrafts lesson with a neighborhood at open space twice a month.	People would do the cooperative work to have group dinner. The future plan is to set up a new business for welfare.

* Ba=Bathroom, La=Labatory, D=Dining room, K=Kitchen, DK=Kitchen-dining room, Li=Living room, E=Emergency buzzar, U=Unit

Table2 The process from frormation of the construction group to the suspending of construction

Date	Matters	Notes
1984 April	Set up a propulsion comittee for "Silver Co-op"	Started with 7people & coordinators. The budget for one unit is 20,000,000yen(about $143,000).
1985 October	Organize "The Silver Co-op Society". 1st general meeting. The chairman is Mr.K.	Started with 20people &coordinators. Appeared in the newspaper. The members are increased to 50.
December	Sound HUDC's opinion about the construction project.	-
1986 October	2nd general meeting. Municipality and the housing supply corporation have announced basic plan of K development to public.	The number of housing unit was 50.
1987 October	3rd general meeting.	-
1989 January	Mr.K retires from chairman's post. Dissolve "The Silver Co-op Society". Set up a preparatory committee for construction.	-
1990 Spring	Construction is permitted. Municipality prepare 3,000 sq.m land for K development. And advertise for a group which consists of 45-55 households. The committee advertise for a group which consists of 45-55 households, but applicants are not enough. Start over again.	Recruit 50 households. /May- Jul. Price/ 43,000,000yen (about $286,000) per unit. Each households must include over 60years old.
August	Dissolve the preparatory committee for construction. Organize a construction union of "K Silver Co-op".	-
Autumn	Municipality and the housing supply corporation extend the time limit for the advertise.	-
1991 Spring	A reduction in group members from 50 to 25households.	-
1992 March	2nd advertisement. A reduction in group members from 25 to 15 households, also gross area from 3,000 to 1,000sq.m. 15 households have gathered as the construction union. However, quite a number of them dropped out after all.	Recruit 15 households. /Mar- Apr. Each households must include over 55 years old.
Autumn	The construction union has dissolved. Finally, this project has discontinued.	-

4.2 *Conditions for Formation Viewed From Cases of Suspended Construction*

K Silver Co-op was started from the desire on the part of one citizen to create cooperative housing with care services as people's final housing. Planning was carried out by three parties: occupants, coordinators and housing supply corporations. The main characteristic of this housing was that it placed importance on care and medical care for people to live the latter stages of their lives with peace of mind. Table 2 shows the process from formation of the construction group to the suspending of construction.

<Reasons Why Construction Was Suspended>

The determining factor was the inability to meet the number of people required for a union, but its nature has been laid out below.

(1) The planning period of eight years was too long for elderly people.

(2) Japan entered the bubble economy partway through the planning stage, and the original lot sale price more than doubled.

(3) The qualification for occupants was that they must be at least 60 years old, which meant that people could not pass the inspection for repayment plans for housing loans.

(4) There were great differences in the approaches of occupants towards cooperative spaces, and a lack of understanding of cooperative spirit.

(5) There were many elderly people who were highly reliant, by having peace of mind that the public corporation would supply the land, and leaving planning up to the coordinators.

The first problem in this case was the weakness that among the participants there were many who were very elderly. The second problem was that the planning took a long time, so participants approached physical states in which they were unable to be independent, and the degree of reliance on other increased. As a result, the coordinators went ahead with their planning, without waiting for agreement to be reached. Thirdly, new restrictions were established in the regulations of the public corporation to incorporate groups, which means they failed to change existing systems.

4.3 *New Government Initiatives: Hyogo Revival Collective Program*

The results of the damage survey and revival plan for the Great Hanshin Awaji Earthquake disaster backed up the importance of group living. A report of the process and actual state of living conditions is given below.

According to a survey carried out by Hyogo Prefecture in February 1992, 42% of the people stricken by the earthquake and living in temporary housing were aged 60 or older. It was also difficult for elderly people to recover from the damage of the earthquake with their own strength, so they requested public housing. In the housing revival plan, quality and quantity had to be shifted to old-age policies. In terms of buildings, almost all public housing was converted to Silver Housing. Barrier-free design, 24-hour emergency reporting systems and dispatch of life-support advisors were carried out, which resulted in an improvement in public housing by several stages. However, enhancement in the buildings alone did not solve the problems. The experience of quick-witted support activities in areas where communities had been formed immediately after the disaster, and the lonely deaths of elderly people in temporary housing resulted in calls for a intangible response as to how communities could be formed. Against this background, Hyogo Prefecture placed emphasis on the formation of communities of the elderly, and planned collective housing with the goal of people living together and helping one another, thereby preventing loneliness and enabling people to attain the joy of actively living cooperative lives. There are plans to provide 261 residences by the end of 1998. Here we will cover Katayama Housing, the first example of collective housing, and report the current situation there.

Katayama Housing is a small-scale collective housing project which houses six households. Initially applications were invited for admittance in groups of six people, but the severity of application regulations meant there were no applications who satisfied the conditions, so applications were opened up to individuals. The people who moved into Katayama Housing were two males and four females aged from 63 to 83, and in a relatively good state of health. The area in common between the occupants was the common experience of having experienced the earthquake, and they had the feeling of togetherness of having endured the same hardships. The rules for cooperative living were decided by the six occupants. From the opinion that life becomes restrained if there are too many regulations, the only rules were the handling of common expenses and in time of ceremonial occasions, and a group dinner once a month. However, in practice group dinners are held twice or three times a month, and in the daily lives of the occupants they chat in the cooperative spaces and share food. It is thought that a good community is being formed.

This program is an experiment in building-led community formation in which modifications to the building have been made to enhance cooperative spaces, in order to enable good communities to be formed from occupants who do not know one another initially. However, because little time has passed since the completion of the building, and formation of the community is in a transitional period, so it is impossible to make an adequate evaluation. This paper will only go as far as reporting it, and we hope to continue the study while focusing on future movements.

5. CONCLUSION

The cases surveyed in this study were constructed with repeated trial and error, and as a form of living are still immature. However, these cases included important elements for the formation of communities of elderly people. The results obtained from this study have been brought together below.

(1) Group living for the elderly which has started to appear in Japan started from the need for good communities necessary for elderly people to live safe and healthy lives. These were cases of housing which does not isolate relatively healthy elderly people who have been missed out by housing policies for the elderly.

(2) Current group living was stereotyped into three types. In the false family type, people live as if they would be with real family members, and human relationships are fixed, so if someone leaves it is difficult to add occupants. The local community type entails living in a region in which one is familiar with, and it is only formed if supported by the local community unique to that area, such as the same manners, customs and culture. It is suggested here that emphasis needs to be placed on

the self-determination of occupants, by giving due consideration to group formation and relaxation of living restrictions and so forth. The network-type involves people who want to secure a good community forming a group, with buildings being constructed with participation from occupants. It is thought that there are many people who could potentially participate in this type of housing, and public support is called for in the future.

(3) From the case where construction was suspended the following became clear as problems unique to construction for elderly people: the understanding of new forms of residences; the health to carry out one's plans; and the planning period and the age of people becoming involved with construction.

(4) It became clear in group living that depending on the way groups are formed which foster mutual understanding about housework and the ways of life of occupants, with relaxed setting of regulations and cooperation in daily life, it is possible to form good communities by respecting the self-determination of the individual.

REFERENCES

Ambrose, I. 1993. *Etablering af seniorbofællesskaber, SBI-Meddelelse 97.* Kopenhagen: Statens Byggeforskningsinstitut.

Fromm, D. 1991. *Collaborative Communities.* New York: Van Nostrand Reinhold.

Frank, K. A. & Ahrentzen,S. 1991. *New Households New Housing.* New York: Van Nostrand Reinhold.

Hirayama, Y. 1993. *Community-based Housing.* Japan: Domesu-syuppan.

Hayden, D. 1984. *Redesigning the American Dream.* New York: W. W. Norton & Company.

Housing Bureau. Ministry of Construction 1998. *A date book for housing and welfare of an aged society.* Japan: Foundation for Senior Citizens' Housing.

Koyabe, I. 1997. *Encouragement of Collective Housing.* Japan: Maruzen.

McCamant, K. & Durrett, C. 1988. *Cohousing.* Barkeley: Ten Speed Press.

Sasaki, S. & Ueno, K. 1996. A study on the co-housing for senior citizens in Japan. *Urban Housing Sciences. Vol.15 Autumn: 36-41.*

Ueno, K. 1994. The social system of cooperative in Northern Europe. *The social system of cooperative: 134-156.* Japan: Houritsu Bunka-sya.

Ueno, K. 1996. Development of Co-housing. *Lectures on Contemporary Living. 2 Family and Dwelling:* 181-204. Japan: University of Tokyo's publishing company.

Vestbro, D. U. 1991. Develop forms of dwelling.housing. *Research and Design in Sweden. 201-222.* Stockholm: Swedish Council for Building Research.

Woodward, A., Vestbro, D. U. & Grossman, M. 1989. *Den Nya Generationen Kollektivhus.* Stockholm: Byggforskningsrådet.

Zahle, K. & Mortensen, P. D. 1992. Co-housing in Denmark. *Open House International Vol.17. No.2: 56-65.*

Implementation of Safety and Health on Construction Sites, Singh, Hinze & Coble (eds)
© 1999 Balkema, Rotterdam, ISBN 90 5809 036 1

Fire safety attitudes and management culture in the construction industry

M.C. Puybaraud & J. Hinks
Heriot Watt University, Edinburgh, UK

R. Barham
University of Central Lancashire, Preston, UK

ABSTRACT: Attitude to risk in construction is mainly affected by two factors. The first is the awareness of the possibility of a catastrophic event to which all employees are exposed. The second is the extent to which individuals, both employees and employers, believe that they are capable of controlling the risks that they face. The management team will have an active role throughout the whole project.

In this paper three different examples are presented and a discussion is engendered as to how enhanced fire safety attitudes and management culture can reduce the risk of fire on construction sites.

1 INTRODUCTION:

Fire losses on construction sites in the UK between 1984 and 1995 were estimated to be £152.4 million. During this period two specific large fire incidents should be noted. Both fires occurred during construction and they alone accounted for 90.87% of the total cost of fires in the UK over the whole period (FPA, 1992).

At first sight it might be argued that these two incidents should be considered as exceptional or due to some special circumstances of the period. However, studies revealed a lack of attention to prevention of fire on building sites. Action to promote fire safety in a construction company should not be seen as a legal embarrassment to plague already overburdened managers, but as a normal and necessary part of the company's activities in which workers and management have a common interest. Davies & Tomasin (1996) stated that small contractors have neither the time nor the inclination to keep abreast of legal requirements and technical developments in safety matters. Risk Management can be a significant contributor to the success of a company in identifying the potential sources of risk on a site during the construction of a project and, hence as a consequence, help in reducing potential costs by avoiding fire.

This approach towards fire safety during the construction process has been greatly improved since 1991. First with the introduction in 1992 under the pressure of insurers of a new Code of Practice (BEC,LPC, NCG 1992) and recently with the introduction of a new amendment to the JCT80 Standard Form of Contract regarding Fire Protection during the construction process(JCT, 1980, Amendment 17: Fire Protection, 1997). Following a series of site investigations, those regulatory improvements have been widely recognised and applied on-site by most major contractors in the United Kingdom through the written Fire Safety Plan (FSP) and/or through site practices. Small size companies are still facing a major financial barrier (Davies & Tomasin 1996) and change is therefore slow and cost targeted. But Carey (1992) is arguing that the lack of readily available statistical data makes it difficult accurately to assess what affect, if any, the Code of Practice (BEC, LPC, NCG, 1992) has had on construction fire safety.

The Construction (Design & Management) Regulations 1994 (HMSO, 1994) which came into force on 31 March 1995, require the implementation of a Health & Safety Plan during the construction phase. A set of arrangements should be set out in this Plan for the management and organisation of the project. This can include: Management, Standard Setting, Communication and co-operation, Identification and effective management of activities with risks of health and safety, Emergency procedures, Information and training for people on site, Consultation with people on site etc. Many authors emphasise the behavioural approach to prevention (Mattila, 1988, Sluzer-Azaroff, 1990). The study developed by Sluzer-Azaroff (1990) proved that the behavioural programme is affected by safety even in the difficult setting of building. The results show that behaviour modification may also be an effective tool in companies' own programme. Mattila and his colleagues (1988) obtained "powerful improvement" by determining where major safety "hot-spots" were located and intervening there first. Recently, studies carried out by Marsh (1992) revealed that behavioural goal setting and feedback on construction sites were an effective method to improve safety level. It appears that, generally, intervention works through increasing awareness of safety issues and increasing and improving communication between management and the workforce. Also, Allen & Blackburn (1992) argue that more and more organisations are adopting a proactive approach in order to manage the risk of their business.

In this paper, it is argued that some construction companies are moving towards the development and reinforcement of a Positive Fire Safety Implementation Process which can be generalised in the following Figure 1.

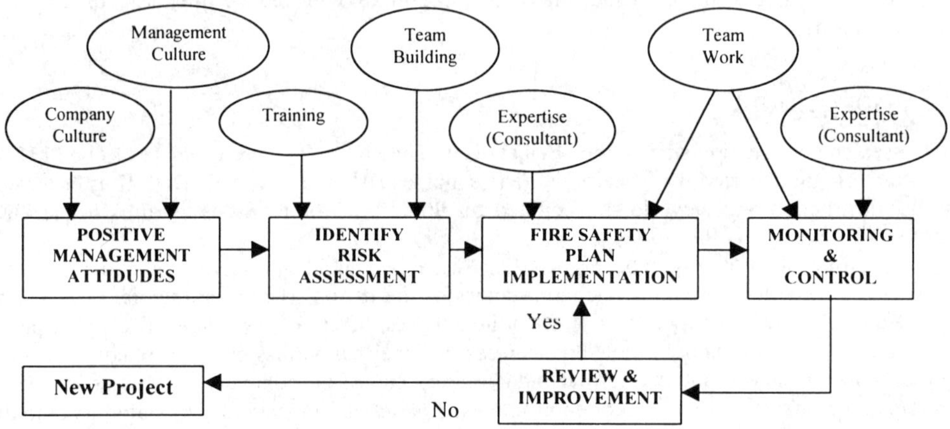

Figure 1: Positive Fire Safety Implementation Process

There is no evidence yet to demonstrate that all companies are following the model proposed in Figure 3. However, the influence of the management culture of the company on employee behaviour and its impact on the control of the risk of fire has been identified as vital. Marsh (1992) identified management attitude as a "central" issue to safety on UK construction sites. Balckburn & Allen (1992) argued that managing the risk to business involves taking a positive approach to the management of safety, starting with the very top of the organisation. Therefore, Figure 3, illustrates a positive fire safety implementation process. The development and implementation of an accurate assessment strategy for identifying performance is now becoming necessary to provide evidences of the effectiveness of such a model.

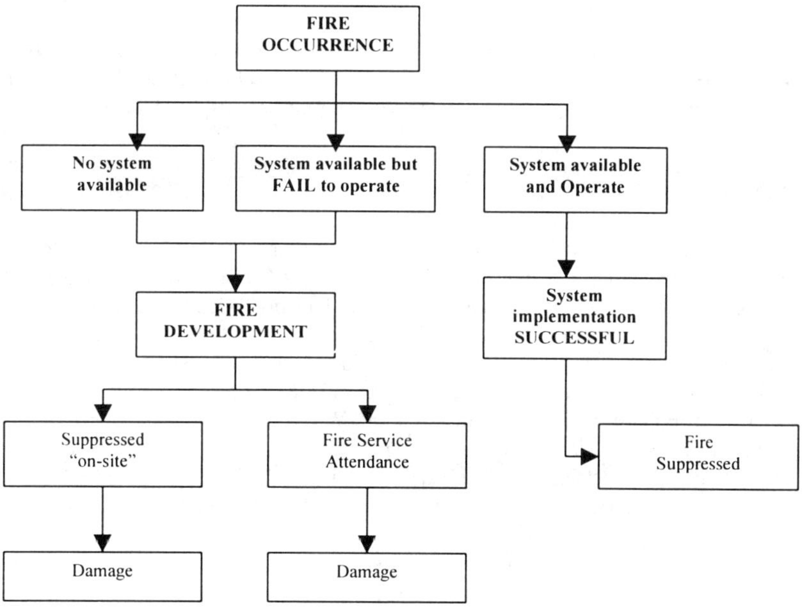

Figure 2: Fault Tree Analysis Approach – fire on construction sites.

2 FIRES ON CONSTRUCTION SITES: A PROCESS OF FIRE OCCURRENCE

There are more than 3000 construction fires annually in the UK. Approximately 100 of them cause more than £50,000 of damage and usually result in complete dislocation of project schedules (HSE 1997a). The HSE estimates that there is one fire per working hour on a construction site somewhere in the UK every working day. This represents, on average 11 construction fires per day (HSE 1997b).

Fires on construction sites is a new area of research and there are no information about the subject. However following a two year investigation and analysis (Puybaraud 1997a, 1997b, 1998) of the available data, it has been concluded that at least 50% of the fires could have been avoided through a better or more effective managerial approach. Attitudes towards Fire Safety on Construction Sites play an important part in the control of Risk during the construction process and lack of attention seems to be the major cause of those fires. Studies conducted by the FPA (1992) on major fires in the UK (loss >£50,000) revealed that more than 22% of major fires on building sites are deliberate or probably deliberate. Moreover, more than 26% involved a source of ignition that could have been controlled through an effective management process. There are still more than 28% of all major fires where the sources of ignition are unknown. The Foreword to the Code of Practice (BEC, LPC, NCG, 1992) says that the majority of fires can be prevented by designing out risks, taking simple precautions and by adopting safe working practices.

Most construction fires have a simple cause and can be dealt with simple precautions. A simple chain of effects that plays a very important role in the break out of a fire. As mentioned earlier, lack of attention can lead to a combination of circumstances that, almost inevitably will result in a break out of fire during the construction process. In the past two decades most research into construction safety has primarily emphasised procedural, behavioural and social aspects (Hadipriono, 1992). Fault-tree models are very often developed to explain the etiology of construction incidents. Hadipriono (1992) defines a fault-tree as a graphic model that shows parallel and sequential causes or events that contribute to a predetermined undesired event. The fault-tree analysis approach for fires on construction sites (Figure 2) is based on a model developed by Barry (1995), the general fault-tree support framework.

Depending when the failure is discovered, it might not be too late to control the situation, but it is never too late to reflect on the process of action to adopt for a future site. This was the philosophy adopted by Bovis after the Minster Court Fire in 1991.

3 THE MINSTER COURT FIRE

This major construction site occurred in London in 1991, and is a prime example of the kind of devastation that such incidents can create. The total losses for this fire were reported by the FPA (1996) to be £105million. The source of ignition has not been clear identified by the Fire Brigade (North Area Fire Brigade, 1991), but from the available evidence it is likely that an introduced ignition source in the form of a naked flame was the most likely source.

Following the fire, the Building Contractor prepared a document to set out the Fire Protection Plan for Base Building Reinstatement Works. The Construction Industry, around that time, had been expressing concern at the particular vulnerability to fire of building in the course of construction or fitting out. The earlier catastrophic fire at Broadgate Phase 8, London, in 1990 had raised specific issues in the respect and Bovis (the Minster Court contractor) had been involved in the full investigation and assessment of the fire damages in partnership with other major organisations. Therefore, Bovis was particularly aware of the devastating effect of fire during the construction process and its risk. But at the time of the Minster Court fire, no Fire Protection Plan (FPP) or Fire Safety Plan (PSP) under the form of a Guidance or Code of Practice has been developed.

The FPP later developed by Bovis Construction Ltd for the Minster Court site draws attention to the importance of a co-ordinated and integrated approach with the Contract of reinstatement and completion of fitting out works.

In 1992 Bovis was engaged in the European Year of Safety, Hygiene and Health Protection at Work. Improving safety and health on European Construction Sites was a vital necessity following the 1987 assessment of accidents on construction sites which revealed a cost to the Community of approximately ECU15,000 million, including indirect costs (Commission of the EU 1993). In January 1992, Bovis presented a Guidance "Fire Prevention during Building Operations" (Bovis, 1992).

4. CONSTRUCTION COMPANIES' RESPONSES

4.1 Large Organisations (e.g. Bovis)

The efforts of Bovis to promote good practice and address new managerial issues have not yet been proven successful. Bovis has not, itself, present any results. However by promoting fire safety within the company at all levels, Bovis seems to have moved towards a new Management Culture by: i/ Developing Fire Awareness, ii/ Making fire safety a Policy statement, iii/ Promoting Fire safety benefits inside and outside the Company, iv/ Implementing appropriate Guidance, v/ Monitoring the implementation and running process of the FPP; but again, there is no evidence provided, as far as the authors know, either by Bovis or the FPA or HSE, to demonstrate any real improvement on Bovis construction sites.

Watson (1994) argues that changing cultures in organisations has become central to management thinking. Promoting job safety in building has been proven to be successful. The results of the study of Mattila & Hyodynmaa (1988) demonstrate that a behavioural safety programme contributed to improving communication in the organisation. They also showed that behaviour modification might also be an effective tool in companies' own safety programme.. However, no one has demonstrated that, by adopting a behavioural safety programme the risk of a fire occurrence would reduce. This is an issue that will need to be address in the near future.

4.2 Medium size companies (e.g. Wimpey and Withey)

Companies like Wimpey Construction UK developed, in 1994, a "Positive Safety Management Concept". At the time this "innovative" measure demonstrated a new approach towards the problem of fire safety, and health and safety in general, in the Construction Industry. Increasing pressure from the Government and the European Commission, to improve the level of safety in the construction industry, reinforced the pressure on contractor through the implementation of new Directive setting new Safety Standard. This new approach should push the Construction Industry to adopt a new vision and attitude to fire safety. First with the introduction of the Construction (Design & Management) Regulations 1994 (HMSO, 1994) and secondly with revisions of Codes of Practice and standard Forms of Contract.

The Safety Plans developed by Wimpey for a site sets out the principles for a positive management of safety in contract. The objectives of this Safety Plan is:

- To plan and implement safety control procedures appropriate to the project.
- To avoid accidents and dangerous occurrences and to promote safe and healthy working environment.

As an example of fire safety concern in a refurbishment project contracted by Withey in London has been reviewed. The work involved was the complete refurbishment of all floors of a four storey building on Leicester Square in London. Withey sought to implement proper standards of fire safety during all aspects of building development. The management have a stated policy of encouraging all those involved in the construction process to achieve a high standard of safety. Locke Carey & Associates Limited were appointed as consultants to develop a Fire Safety Plan (FSP) and carried out on a regular basis site inspections on behalf of the company. The FSP incorporates all those matters specified in the Joint Code of Practice on the Protection from Fire on Construction Sites and Building Undergoing Renovation.

Notwithstanding this, the attitude of the "on-site"personnel is common with a large number of other site management teams in the UK was that:

- Fire safety an obligation imposed under pressure of insurers (or clients).
- "There is nothing to burn on a construction site" is a very common idea.
- The impact of a fire during the construction process is low.

This demonstrating that knowledge of fire safety and of the devastating impact of fire in economic, technical and social terms is often unrecognised or wrong.

5. DISCUSSION

Post 1992, a number of similarities have been identified in the three companies presented above:

- Positive management attitude: a strong positive philosophy has been implemented in company structure. Means for communicating and passing information between the project team and the contractor and its workers on site despite a resistance or lack of appreciation by the latter.
- Responsive Management: Arrangements for securing co-operation between participants for fire safety purposes, arrangements for monitoring systems to achieve compliance with the FPP or FSP, developing training and consultation with all levels.
- Efficiency, Control and Monitoring of on-site standards are key issues in the implementation and development of fire safety culture in construction company. Wimpey members of staff have a responsibility for monitoring safety generally and bringing infringements of safety regulations to the attention of those responsible for enforcement. Transfers of responsibility and awareness are key issues in the positive implementation of a new Management Culture.

In contrast, the fire safety philosophy developed by Withey is user-orientated. By appointing an external organisation like Locke Carey & Associates Ltd, they rely on its expertise to provide the best FSP adapted to both the client's and the contractor needs and requirements. There is a General Awareness of the global problem of fire safety by most site

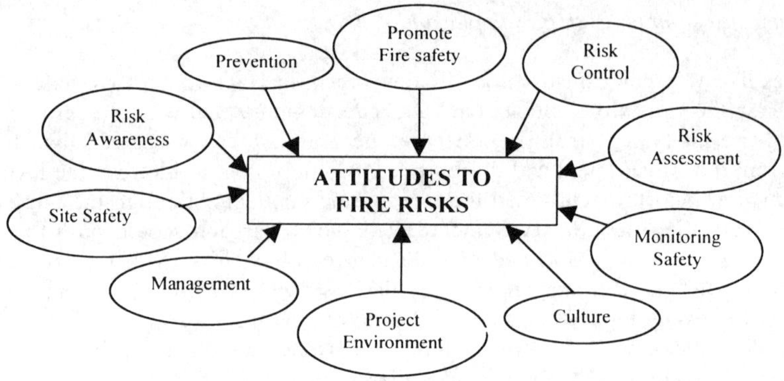

Figure 3: Attitudes to Fire Risk

employees. However they seem to have problem to understanding the need to implement it strictly it. The necessity of a FSP is seen as an obligation rather than a necessity but site managers generally have no objections to its implementation, control and monitoring. The role of the consultant as an advisor is positively welcomed and suggestions for improvement of fire safety standards are respected.

Improvement and modification of Management Culture is usually developed only with time and experience. Preconceived ideas and existing knowledge about fire safety is very often wrong. Davies & Tomasin (1996) argued that the managerial knowledge of many small contractors is based on experience and often lacking in theoretical background. This could be interpreted as a cultural problem. How do you convince a site manager of 20 year's experience that fire is a major problem on his site if he believes fires never occur on site because he never witnessed one? This could be interpreted in different way. It does not prove that fire is not a problem but that there is a difference in the perception and interpretation of fire during the construction process. Further research will try to demonstrate that it is through the implementation of simple measures and a slow modification of attitudes that the perception and awareness of the problem of fire during the construction process is improved.

Attitudes to fire risks involved a large number of issues (Figure 3). Each issue if not taken into account at an early stage of the preparation of the FSP could lead to a major problem. As explained earlier (Figure 1), fires occur on construction sites most often because of "Lack of Attention" – it is what seems to be the conclusion drawn from the low number of data available through the FPA (1992) and the research carried out within the past three years by Puybaraud (1997a, 1997b, 1998). In order to prevent it, it is necessary to consider the issues raised in Figure 3.

Within the last years, the construction sector has been the subject of EU attention. Progress seems to be made in parallel. On one hand the UK Government is concerned about the problem of safety in the construction Industry and improving safety standard (HM Treasury 1996, HMSO 1996). Its actions are reinforced by a strong pressure from the EU (EAHSW 1997). On the other hand no efforts are made, in the UK at least, to improve the existing data and create new ones in order to appreciate the complexity of the problem of fires on construction sites and its real economic, social and technical impact. Basing the approach on assumptions and few evidences does not slow down the production of new standards and regulations to improve fire safety during the construction/refurbishment process or improving the existing documents. So far, this is what happened. Even now it is difficult to identify a clear link between what happened and the series of modifications and/or amendments made to the texts. The Joint Code of Practice for Protection, 1992 (BEC, LPC & NCG, 1992) was a direct consequence of two major fires in London in 1990 and 1991. Puybaraud (1997a) argued that this document was an improved version of the document "Standard Fire Precautions (P5)" published in 1972. The Amendment 17 of the JCT80 Standard Form of Contract does not seem to be justify and add again more pressure on the contractors' shoulders and not especially the clients. The last and not least publication of the HSE is another "Guidance" (HSE, 1997a) which is even more confusing the participants and is again not justify.

In such circumstances, it would be difficult to improve the actual level of protection and fire safety on construction sites. It is time to move towards a harmonised approach in line with the requirements of the contractors – cost benefit and safety based approach, the clients – cost targeted and a satisfactory end product, i.e. a building, the insurance companies carrying the risks, the Government willing to improve the situation and finally the EU concerned by the harmonisation of regulation within the Member States.

Therefore, three major issues will need to be addressed. The first is to provide a detailed range of evidences to assess the causes and consequences of fires on construction sites. Secondly, to address a response through the legal documents either by improving the existing documentation or producing a set of new regulations. Finally, it is of primary importance to measure the impact of any new measures implemented.

6 CONCLUSION

Bovis and Withcy recommend that procedures should be drawn up for any project in accordance with the Joint Code of Practice for Protection, 1992. However, Carey (1997) suggests that there has been some improvement in site fire safety and he believes this to be a direct result of the introduction and application of the Code. This may be pure sophistry. Statistics are still not available to support this suggestion and, therefore, this is an area where further investigation is necessary. Only on the basis of the full investigation will it be possible to assess accurately the impact of the Code and how the participants in the construction industry perceive it.

Much of modern UK health and safety law is goal setting –setting out what must be achieved, but not specifying a single method by which it must be done. This allows the Contractors a lot of flexibility into their managerial approach. Employers must comply with the law and failure to do so is a criminal offence (Section 33 Health and Safety at Work etc. Act 1974).

Bovis used this, in 1992, to develop its own guidance and implement a new Management Culture. However, this work was of little significance for small companies. At least 50% of small companies work for large companies, i.e. as subcontractors, specialist contractors, etc. Small companies have be forced to change and develop a new Corporate approach towards the problem of fire safety on site. Eleven fires a day on construction sites cannot be ignored. This is especially so when more than 96% - 2900 fires out of 3000 in total result in a fire loss <£50,000 – and potentially affect 92% of UK construction companies. Small companies are asking for simplified legislation, better regulation and more guidance as illustrating good practice (EAHSW 1997). It is widely accepted, at least in the UK, that a cost-benefit approach to regulation is appropriate (HM Treasury 1996). Small companies are very concerned about the cost implication of new standards (Davies & Tomasin 1996), but as described in this paper and others (Puybaraud 1997a, 1997b, Carey 1997), simple measures could be taken without a great financial investment. Davies & Tomasin (1996) also argued that the financial resources of small firms are generally insufficient to provide the necessary standard of safety training carried out by many large firms. The Government has been determined, since 1996, to make implementation and enforcement of regulation more business-friendly. The Government has also adopted three principles of good regulation (HM Treasury 1996): i/ Think small first: no regulation without first making sure that firms can cope, ii/Proportionality: no regulation unless the benefits really justify the costs to business and others, iii/ Goal-based: no detailed rules when you can simply specify the objectives.

For the last 7 years we are witnessing a slow change of Management Culture. However, even from the sparse information available it would appear that contrary to popular belief, fires on site are a major problem. Preliminary statistical information points to the continuance of losses. The low number of examples used and discussed in this paper indicates a dearth of research in the evaluation of fire safety on construction sites. Clearly, part of the problem is the lack of published data. The identification of causes, such as "attitudinal negligence" is fraught with problems because of the rarity of major fires and the long periods taken to assess the causes of the incident. The published works reviewed do not provide a solid framework or foundation upon which the current vast problem of fire safety on construction sites can be confidently built. No definite conclusion can presently be drawn about the relative impact of management on the reduction of risk of fire during the construction/refurbishment process of buildings, neither about the effectiveness of different managerial techniques (Fire Safety Plan, Consultant work, developing a Positive Safety Attitude) nor its impact on fire safety attitude. The small number of examples studied largely limits present conclusions.

However, additional avenues of investigations are presenting themselves. These include a review of the process of organisational change and its management for effective fire safety on site incorporating factors such as culture, communication, and adopting new management philosophies. It is now, therefore, important to orientate this research towards the effects of major organisational change on safety culture. The clear need for a study of fire safety during the construction process has now been identified and a thorough investigation of the problem in more depth through a series of field-based studies is now planned.

REFERENCES:

Barry, T. 1995. *An Introduction to Quantitative Risk Assessment in Chemical Process Industries*, SFPE Handbook of Fire Protection Engineering, 2nd Edition, Section 5/Chapter 12. 102-126.

BEC, LPC & NCG 1992. *Fire Prevention on Construction Sites: The Joint Code of Practice on the Protection from Fire of Construction Sites and Building Undergoing Renovation*, ISBN 0 902166720 0.

Blackburn, B. & F. Allen 1992. *The Value of safety Auditing*. Internal Auditing, February 1992: 14-16.

Bovis Construction Limited 1991. *Fire Protection Plan Minster Court*, November 1991, UK.

Bovis Construction Limited 1992. *Fire prevention during building operation*, January 1992, UK.

Bovis Europe 1998. *Portfolio of Work*, http://bovis.com.

Carey, P.W. 1997. *Construction Site Fire Safety: Practical Problems*. 1997 Conference, Kent, UK. 42-46.

CIB UK Kompass 1998. Company Information, ISBN: 0 86268 392 0

Commission of the European Communities (EU) 1993. *Europe for safety and health at work, Safety and health in the Construction Sector*, Luxembourg: Office for Official Publications of the European Communities. ISBN: 92 82660370

Davies, V.J. & K, Tomasin 1996, Construction Safety Handbook, 2nd Edition, ISBN: 0-7277-2519-X.

European Agency for Health and Safety at Work (EAHSW) 1997. *Questionnaire – Priorities and Future Strategies for Occupational Health and Safety Policy in EU Member States*, HSE, 30.4.97.

FPA 1992. Fires during construction, Fire Prevention Magazine, 248: 19.

Hadipriono, F.C. 1992. *Expert System for Construction Safety. I: Fault-Tree Models*. Journal of Performance of Constructed Facilities: 6: 246-260.

Health & Safety Executive (HSE) 1997a. *Fire Safety in Construction Work. Guidance for clients, designers and those managing and carrying out construction work involving significant fire risks*, HSG168 HSE Books 1997, ISBN 0 717613321

Health & Safety Executive (HSE) 1997b. *Construction (Design & Management) Regulations 1994: The health and safety plan during the construction phase*, Construction Sheet No 43.

HM Treasury 1996. *The Setting of safety Standards: A report by an interdepartmental group and external advisers*, UK

HMSO 1994. The Construction (Design & Management) Regulations 1994, No 3140, Health & Safety, ISBN: 0110438450.

HMSO 1996. *Competitiveness, Creating the Enterprise Centre of Europe*, presented to Parliament in June 1996, UK.

Marsh,T 1998. *Behavioural goal-setting and feedback on construction sites*, The Safety & Health Practitioner, January 1995.

Mattila, M. & M. Hyodynmaa 1988. *Promoting Job Safety in Building: An Experiment on the Behavior Analysis Approach*, Journal of Occupational Accidents, 9: 255-267.

Puybaraud, M.C, Barham, R., Hinks, J. 1998. *A Comparison of Fire Safety Legislation and Administration in Europe and Canada*, Cobra '98 Conference, Oxford Brookes University, UK, 2-4 September 1998.

Puybaraud, M.C. & R. Barham 1997a, *Procurement systems and the economic provision of fire safety during the construction process*, CIB W92 Conference, Procurement: The way forward, 20-27 May 1997, Montreal, Canada,

Puybaraud, M.C. & R. Barham 1997b. Addressing the risk of fire during the construction or refurbishment process by better management, Cobra 97 Conference, Portsmouth, UK, 10-12 September 1997.

Sulzer-Azaroff et al. 1990. *Improving Occupational Safety in a Large Industrial Plant: A Systematic replication*. The Journal of Organizational Behaviour Management (USA), 11:99-120.

Watson, T.J. 1994. In search of Management.1st Edition, ISBN: 0 415 09231 0.

Implementation of Safety and Health on Construction Sites, Singh, Hinze & Coble (eds)
© 1999 Balkema, Rotterdam, ISBN 90 5809 036 1

The mechanism of the occurrence of defective housing in Japan

Kazuya Ohnishi
Aichi Konan College, Nagoya, Japan
Katsuyo Ueno
Kyoto Prefectural University, Japan

ABSTRACT: In recent years in Japan, defective housing has been an increasingly serious problem, especially for critical defects such as inferior foundations, unsatisfactory connections between structural members, lack of structural strength, etc., which could pose a danger to occupants. From the perspective of the housing production system in Japan, this paper will clarify the mechanism of the occurrence of defective housing and will offer suggestions for preventive measures.

1. INTRODUCTION

In the Great Hanshin Earthquake of January 1995, roughly 16,000 homes sustained moderate to total destruction, and 5,500 people lost their lives. More than 90% of these victims were crushed to death due to structural collapse resulting from inadequate connection of structural members and/or problems with structural strength due to insufficient bearing walls and eccentric arrangement. The number of such defects that are related to housing safety has not declined since the earthquake, and outside experts are increasingly being consulted. For example, in 1997 the Japan Lawyers' Union held about 1,000 consultations per day for victims of defective housing. Most of these consultations involved critical defects such as inferior foundations, unsatisfactory connections structural members, lack of structural strength, etc., which could pose a danger to occupants[1].

Research in Japan on the mechanism of defective housing dates back about 20 years and includes the work of Ueno[2], Yoshino[3] and Sawada[4]. The purpose of the present study is to explain this mechanism from the perspective of the housing production system in Japan, especially in the period immediately before and after the Great Hanshin Earthquake.

2. MATERIALS & METHODS

To examine the mechanism of defective housing in Japan, the authors collected surveys conducted by specialists such as lawyers and architects, then selected the most typical examples of defective housing and conducted a preliminary survey. From among these surveys, five cases were chosen which presented the best possibility for collaboration and whose content could be understood in general terms, then personal interviews were conducted with residents and with relevant architects, lawyers, etc., on the state of defects and the details from the signing of the contract to the execution of the project. This provided information on direct factors at the site, which in turn helped to clarify the mechanism of the occurrence of defects in housing.

The objects of this study were recently built residences for average citizens in Kobe City (which bore the brunt of the Great Hanshin Earthquake) and surrounding areas such as Osaka City and Kyoto City. The example from Kobe included residences in housing developments on the outskirts of the city, while the example from Ashiya City included relatively high-class homes built by a major housing company. The example from Kyoto was a home built on slopes on the outskirts of the city, and the two examples from Osaka were three-story wooden houses built on narrow lots near the city center.

The personal interviews were conducted for 6 months from January to June 1998 beginning at the offices of lawyers and architects, after which residents of each home were interviewed. These interviews helped to clarify the state of housing defects.

3. RESULTS

The following are descriptions of each surveyed home and defects.

3-1. *Case 1. Poor design, built on a weak foundation (custom-built housing)*

This home, which is located in a major housing development in the northwest part of Kobe, had been custom built a year earlier by a major homebuilder. It was constructed on an embanked slope. Initially, since the foundation was weak, concrete piles were to have been used to support the structure; however, to save 600,000 yen (about $4,300) in construction costs, the site was merely improved and a normal foundation laid. After completion, an enormous sinkhole developed nearby, and the floor began to slant, the doors began to improperly open and close, and wrinkles started to develop in the wallpaper. Although the designer came out to inspect the house, he did not acknowledge these issues as defects. Subsequently, the owner consulted a lawyer and architects conducted a survey to determine that improper construction of the foundation had led to uneven ground subsidence. The builder thus had to take responsibility to install concrete piles to remedy the situation.

The causes of the defects were 1) the designer had little expert knowledge about ground strength, and 2) the builder tried to save money by "cutting too many corners."

3-2. *Case 2. Poor design, built on a weak foundation ("ready-built" home)*

This home, which is located in a corner of a hilly section just northwest of Kyoto which has undergone a lot of residential development in recent years, is a two-story wooden house built 10 years earlier. Immediately after moving in, the owner discovered that the floors were slanted, the doors failed to open and close properly, and that cracks were developing in the walls and foundation. The real estate agent made some simple emergency repairs, but did not take any action when the cracks and slanting became more serious. The owners were ashamed of buying such a defective home, and took it on themselves to make the needed repairs a little at a time. A lawyer was subsequently consulted, and the case is presently being prepared for a hearing in court.

The lot is on an embanked slope which originally had a 10m difference between the highest and lowest sections. However, the retaining wall on the eastern side of the lot was poorly designed and constructed, and improper filling in of earth resulted in insufficient ground strength. All of this led to uneven ground settlement, which was determined to have been directly responsible for the house's defects.

3-3. *Case 3. Lack of skills on the part of the builder and insufficient on-site management system*

This home was constructed by a major builder in the city of Ashiya, the eastern boundary of Kobe, immediately after the owner's previous home had been destroyed by the Great Hanshin Earthquake. The owner, who had lived in emergency prefab housing, noticed several defects in the home immediately before its completion; poor workmanship by the on-site subcontractor caused floors to be slanted, floors to creak, doors to fail to open and close properly, and mold and mildew to develop in the flooring.

Some of the direct causes included poor joining of the structural members, cross-section deficiencies of beams due to cracks of wood, and the use of floorboards whose moisture content caused the adhesive to rot. Although the restoration work in the aftermath of the Great Hanshin Earthquake increased builders' workload by 3-fold or more, the poor workmanship here probably resulted from inadequacies with the major homebuilder's on-site management system and lack of skill on the part of the subcontractor.

3-4. *Case 4. Impractical design for narrow building lot (wooden three-story structure)*

This home is a wooden three-story structure built in an existing urbanized area in southern Osaka. The owner bought the house because of an attractive floor plan. Since the owner signed the contract after construction had commenced, some of the specifications were changed after making arrangements with the real estate agent. However, the real estate agent did not contact the builder about these changes, so they were not incorporated into the actual construction. The completion of the home was postponed by a half-month, and the owner moved in before sewage facilities and gas equipment were still not properly set up. After moving in, the owner discovered such problems as rainwater leaking from windows, shaking, improper opening and closing of doors, and poor interior finishes. The builder fixed the problem about the rainwater leaking from the windows, but the construction company took no action regarding the other defects, even after being contacted by the owner.

There were a number of interacting causes of these defects, including inadequate structural strength arising from a design that was not suited for the narrow building lot, poor construction techniques such as poor use of metal joining materials (nails, etc.) by carpenters who had no detailed knowledge of three-story wooden structures, failure of the architect to oversee construction on-site (he only signed the application at the local city hall), and lack of inspection by local authorities during construction and after completion (they merely conducted a cursory examination of the documents). In addition to the structural defects, the plan that was actually used to construct the residence differed completely from the one submitted to City Hall, and there were several apparent violations of the Building Standards Act building codes on the environmental/sanitation side as well, such as improper ventilation and lighting.

3-5. *Case 5. Improper construction due to construction workers' lack of relevant knowledge (three-story wooden home)*

This home was a three-story wooden structure built in an existing urbanized area of southwestern Osaka. The first floor serves as a garage while the second and third floors serve as living space. It is a typical example of the recent trend in major Japanese cities of homes being built on very small lots. A real estate agent introduced the home to the owner, who bought it because it met his conditions. Immediately upon moving in, however, the owner discovered that the house shook when he walked through the rooms, the floors were slanted, the doors would not open and close properly, and the interior finish was very poorly done. Although the real estate agent was contacted about these problems, little was done about them; in fact, the real estate agent became defiant, claiming that "it is a matter of course that buildings, especially wooden houses, will shake." The construction company was directly notified of the problems, and repairs were made, but they were done only halfheartedly, and the owner was told to "grin and bear it" about the hard-to-repair sections, which were not fixed. An expert was called in to take a survey, which, although it cost 400,000 yen (about $3,000), the homeowner consented to because of worries about his home. As a result, it was determined that causes of the problems were poor workmanship with metal fasteners between structural members and foundation work. The homeowner consulted a lawyer, and the case is being prepared for a court hearing.

4. DISCUSSION

After examining these five case studies, we can raise the following three issues regarding defective housing: (1) The suppliers of housing tend to regard it as a commodity; (2) there are problems with design management and construction techniques; and (3) there are problems associated with the inspection systems, including building codes, of local authorities.

4-1. *Housing as a Commodity*

Before WWII, when a home was built in a community, there was a relationship of trust between the owner of the home and the builders, carpenters, etc., who lived in the community, making it difficult for there to be major mistakes or omissions in the construction process. However, people who moved to the cities after the war had to rely on unknown local homebuilders, etc., and housing became a commodity, something to be sold to an owner. All of the 5 previously cited examples of defective homes were bought from real estate agents or builders as commodities.

Japanese commoditized housing is a collection of products such as building materials and equipment, and this trend has become more pronounced in recent years. Moreover, there is an extremely large amount of money being paid to the manufacturers and distributors of these materials and equipment. Not only does this put pressure on labor costs (for carpenters, etc.), but, as previously mentioned, this is a multi-tiered subcontracting structure, and on-site carpenters' wages are so low that defects may occur through omission. Furthermore, the prefabrication of housing parts means that manufacturers are producing kits in factories which are assembled on site. This assembly does not require any special skills from carpenters so that consistent quality can be maintained. However, this also means that construction workers do not require the same type of skills that they needed before, which, as will be discussed later, can lead to declining technical abilities[5)].

As is well known, land prices in major Japanese cities are very high[6)], so that the average citizen[7)] can only buy land and/or houses that are very small[8)]. Real estate agents are supplying many three-story homes such as seen in Case 5 with a garage on the first floor[9)]. This type of housing is flawed because its structural strength is lower than what the design calls for. If this flaw is not overcome, there is a high possibility that defects will occur. Furthermore, developers are carving out hillsides in suburban areas for large-scale developments, resulting in numerous housing areas that have weak foundations, as in Cases 1 and 2. In such cases, if the design does not provide for a satisfactory foundation, there is a good chance that uneven ground subsidence will occur, which in turn will cause defects in the houses.

4-2. *Design Management and Construction Techniques*

In all five of the case studies, the homeowners did not order their homes directly to the builder or construction company; rather, they signed contracts with real estate agents or housing companies, who then subcontracted the construction these homes to other firms. Thus, in order for Japanese builders to maintain a profit margin based on this multi-tier subcontracting system, the homeowner pays inflated cost for labor and materials, which keeps wages for the people actually doing the work extremely low. At the same time, it is considered part of the job of the designer to change the design during the construction stage, so that designers and construction companies seldom scrutinize the drawings before the actual work begins. This is especially true for ready-built housing. In both Cases 4 and 5, the homeowners were shown only simple sketches of floor plans, etc., and were completely unaware of detailed plans and specifications. In order to conclude contracts based on such incomplete, sketchy plans, it appears that the final subcontractors, who are already cutting labor costs, are reducing the quality of workmanship at the construction stage.

In Japan, there is usually a blanket contract system when ready-built or other such small-scale housing is constructed, and there is almost never a third-party designer/architect at the site to supervise construction. Thus, if omissions are made, they can be overlooked because there is no one to check for them; such was the case with each of the five case studies.

Another problem is that there are few young construction workers nowadays who study their trade for 10 years more, as was done in the past; thus, present-day workers are likely to have poorly developed skills and technical knowledge[5)]. At the same time, there are veteran builders who, as seen in the three-story home in Case 4, do not study the latest techniques but rather rely on their instincts and intuition. At building sites for ready-built housing, etc., it is easy to find and hire such low-cost, semi-skilled labor to keep costs down, thus mistakes will occur during construction[5)].

4-3. *Municipal Inspection System*

According to the Japanese building code, municipal officials conduct only a cursory inspection of floor plans, documents, etc., for small-scale housing projects before construction begins, and usually do not make any inspections on site, either during construction or after completion[10)], meaning that defective workmanship can be overlooked[11)]. Local authorities are obligated to inspect all homes that are being paid with loans at Housing Loan Corporation, and passing the inspection is a precondition for receiving such loans. In Cases 1 and 2 (homes paid with loans at Housing Loan Corporation), inspections were conducted during construction and after completion, and both houses passed. Foundation defects were not apparent at the time; since the homes met the building standards for the loan, they passed inspection, but after the owners moved in, the defects were discovered. In Case 3, an earthquake relief loan, rather than a loan at Housing Loan Corporation, was used to pay

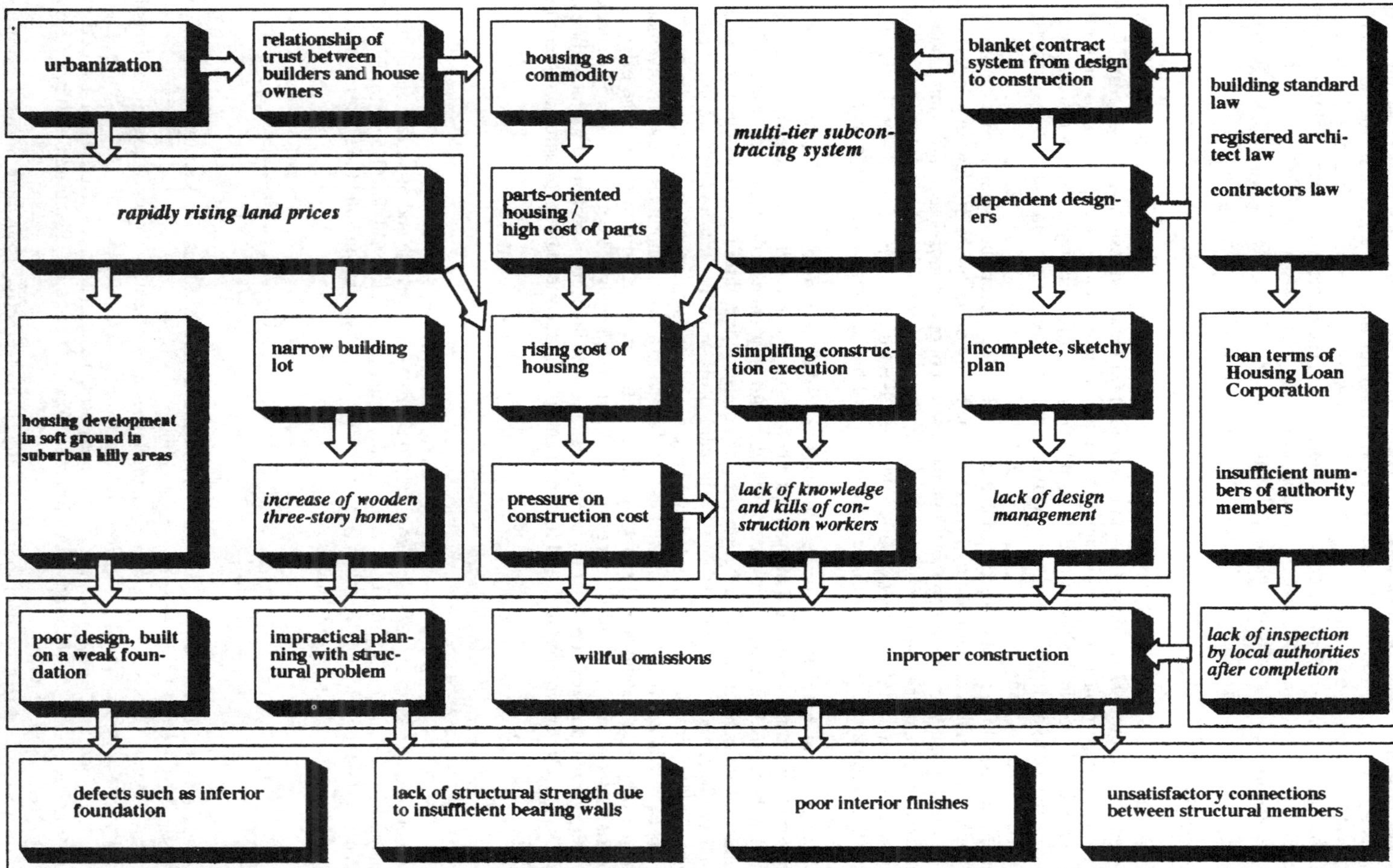

Figure 1. Mechanism of the Occurrence of Defective Housing in Japan

for the home, so no inspection was made. In Cases 4 and 5, the homes did not meet the minimum standards for lot area of 100 m^2 or larger (70 m^2 or larger for floor space), so they did not qualify for loans at Housing Loan Corporation. The structures themselves were in violation of the Building Standards Act but were not inspected by local authorities; the homeowners were not told of this. These are typical examples of narrow, wooden, three-story homes whose defects are overlooked because they are not inspected by municipal building inspectors.

5. CONCLUSION

Ueno[2)] and Yoshino[3)] have already brought to light the state of defective housing in Japan that has come about since the 1970s. Since that time, various problems and issues such as the multi-tier subcontracting system, lack of design management, inadequate municipal inspection systems, etc., have been pointed out, but the basic framework of the construction industry and municipal inspectors has not changed. In addition, a new set of adverse factors have come into play, aggravating the problem of defective housing.

Minor defects, which are the object of Japanese civil codes, occur due to negligence on the part of the builder which has a relationship of trust with the owner of the house. In Ueno et al's research from 20 years ago, it was already apparent that there were not only major defects affecting practical use such as water leakage (which could be found by the occupants of the home), but also critical defects such as structural weakness that could affect the future safety of residents. It was also pointed that companies willfully made omissions, and the problem was treated as a consumer issue.

This study has revealed that in the last 20 years, there has been no change in the framework of the construction industry and local authorities who are supposed to oversee it. The study has also shown that there are also new factors that have come into play, such as the high cost of land which is putting upward pressure on construction costs, the construction of three-story wooden homes on very narrow lots, and the declining knowledge and skills of carpenters and other construction workers. The present survey has revealed that there are several present-day factors involved in the occurrence of housing defects: (1) the trend toward housing as a commodity, (2) the involvement of many levels of contractors and subcontractors in the construction process, (3) the effect of rapidly rising land prices, (4) the blanket contract system from design to construction, (5) the lack of knowledge and skills of construction workers, and (6) the inspection system of municipalities. All of these factors are interrelated, and in combination form the mechanism of defective housing in Japan.

6. ACKNOWLEDGMENTS

The authors would like to express their gratitude to the following people and organization for their assistance with this study: the residents of the homes cited in the study, Toshio Hagio (architect), Suekazu Yuki (surveyor), Kenji Kataoka (Housing Inspection Association), and the National liaison council for Victims of Defective Housing.

NOTES

1) Japan Lawyers' Union, Committee for Consumer Problems: Manual for Victims of Defective Housing. 1997 Civil Law Research Group (Japan)
2) Katsuyo Ueno: Trends in Consumer Housing Issues and Research Topics, Kokumin Seikatsu Kenkyu, vol. 23, no. 4, pp.52-67,1984, Kokumin Seikatsu Center (Japan)
3) Katsuyo Ueno and Shoji Yoshino: The State and Characteristics of Victims of Defective Housing,Report #1. Proceedings of Architectural Institute of Japan, Planning Section, no.359, pp.95-104,1986,(Japan)
4) Kazuya Sawada: Effective Methods for Dealing with Defective Housing Conflicts. Minjiho Kenkyu, 1996, Civil Law Research Group (Japan)
5) Based on an interview with Wooden Housing Industry Association of Japan.

6) In 1995, the average area for housing 30-40 minutes away from central Osaka was 60.5 m^2, with an average land price of 17 million yen (Compiled from data from the Ministry of Construction, Economic Planning Agency, and Japanese Real Estate Research Institute.)
7) In 1995, the average annual income of a working family was 6.85 million yen(Statistics Bureau of Prime Minister's Office, Family Finance Survey)
8) In 1995, the average floor space of a newly-built home was 93.0 m^2, and the average construction cost was 47.5 million yen(Ministry of Construction, Survey on Construction Cost of Private Housing)
9) Since there were no hard statistical data available, the authors used 476 real estate reports from the city of Kyoto (1998). These reports indicated that wooden three-story homes comprised 46% of all houses in the city and that among these, 71% had a garage on the first floor and cost an average of 35 million yen.
10)In June 1998, part of the Building Standards Act was revised. Standards were formulated for structure and fire resistance, information on building approval and inspections was made available to the public, and an interim inspection system was implemented. However, since not all structures are subject to interim inspection, it likely will not be enough to prevent defective housing.
11)Of the applications for building approval that were submitted to city offices in 1995, only 15%of homes were inspected after completion in Kobe; 15% in Kyoto; and 13% in Osaka. The national average was 30%. (Ministry of Construction,Statistics of Building Standards Act Enforcement)

Implementation of Safety and Health on Construction Sites, Singh, Hinze & Coble (eds)
© 1999 Balkema, Rotterdam, ISBN 90 5809 036 1

Promotion of health and safety culture through accident cost awareness

M.A.Al-Mufti
Department of Civil Engineering, The Queen's University of Belfast, UK

ABSTRACT: This paper discusses the idea of developing a computer-based model for costing accidents which have occurred, or potential accidents waiting to happen, and use it as an informative and stimulating means for increasing awareness among staff on construction sites. It is a motivational approach based on the assumption that the fact that accidents are expensive is still the real driving force behind efforts to improve safety in construction. A methodology is suggested and some of the desirable features of such a model are discussed. It needs to be simple, factual, visual, and attractive enough to stimulate a fundamental change in ingrained attitudes and behaviour. The paper also reports on a project carried out to gauge the response of the local construction industry to such an approach based on using a computer template for costing accidents as a first tentative step towards the development of a more comprehensive model.

1 INTRODUCTION

A quick tour of most construction sites around the world can often be a revelation in terms of identifying several accidents waiting to happen at any one time. The severity of the situation will obviously differ from one site to another but it is not unrealistic to say that most if not all construction sites will have potentially hazardous situations which may or may not lead to accidents. Accidents arise from different causes which can generally be classified as physical incidents posing hazardous conditions, and behavioural incidents caused by unsafe acts (Kartam, 1997). In each case there are two elements: the technical cause which may lead to failure, and the procedural errors which allow the faults to occur and go undetected (Jones, 1996). Preventing accidents on site is generally based on satisfying two prerequisites. Firstly establishing and maintaining a safe working environment and secondly the promotion of safe behaviour among all those working on site. This paper focuses on the latter of the two elements and suggests an approach which is aimed at promoting a safety culture on site based on increasing awareness of the consequences and costs of potential accidents on site. Clearly such an approach will be but one component of an effective overall project safety management.

2 COSTING ACCIDENTS

The quest for objective and accurate costing of accidents in construction continues, and to date it still remains the subject of much debate and research. At the heart of this debate there is however a general consensus among engineers that measurement remains the backbone of any objective and scientific approach to defining this problem and finding a solution. In the early 1970s the Confederation of British Industry (CBI) said ' *At the company level, if a readily applied and simple formula could be devised by which the financial loss caused by accidents and diseases could be measured..., it would make a valuable contribution towards reducing industrial accidents and occupational ill health*' (Health & Safety Executive, 1993).

Studies carried out to determine the costs of accidents in construction have produced differing conclusions. Total construction costs investigated at Stanford University were found to be 3% of the total project costs, amounting to 10% of labour costs (Levitt, 1981). Everett (1996) concluded that the total costs of accidents had risen to somewhere between 7.9% and 15% of the total costs of non-residential new construction from the 6.5% proposed by a previous study commissioned by the Business Roundtable in the USA in 1979. In his opinion it appeared that the construction industry was losing the battle against accidents and injuries. More specifically and in terms of costs of individual accidents one out of every six construction workers expect to be injured each year, at an average cost of $18,000 (Kartam, 1997). So estimates differ depending not only on when the studies were carried out but also the methodology and criteria used.

Studies on accident costing have also covered the classification of these costs into the two major components namely direct and indirect costs. In addition the classification has extended to two further related categories namely insured and uninsured costs. Definitions and details of these cost categories are included in various publications on H&S. Figure 1 however, gives a general idea of these categories.

Indirect costs have traditionally been calculated by multiplying the direct costs by an indirect cost multiplier. There have been various estimates of this multiplier ranging from 2 to 20 times direct costs (Everett, 1996). Heinrich, (1959) suggested a multiplier of about 4 and further suggested that in construction the ratio would be higher than 4 to 1. In the well known iceberg analogy, the 1:10 rule of thumb has been suggested to estimate the hidden cost of accidents and a case study on an £8m contract completed in 1992 (Health & Safety Executive, 1993) concluded that the ratio of insured costs to uninsured (hidden) costs incurred by the contractor was 1:11 (Figure 2).

Opinions may differ and estimates may vary regarding the ratio of direct to indirect costs. However, there is a degree of understanding in the construction industry that indirect and impact costs resulting from the adverse effects of accidents on productivity, morale, and on project schedule can be several times the direct costs normally associated with medical

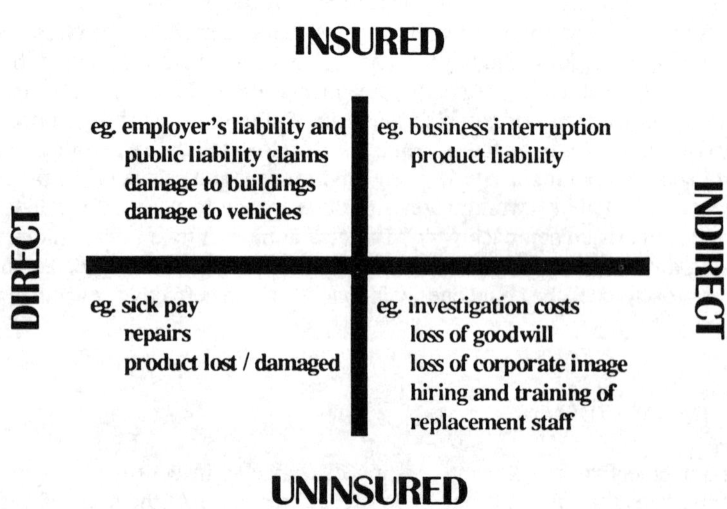

Figure 1. Direct and indirect costs (Health & Safety Executive, 1993)

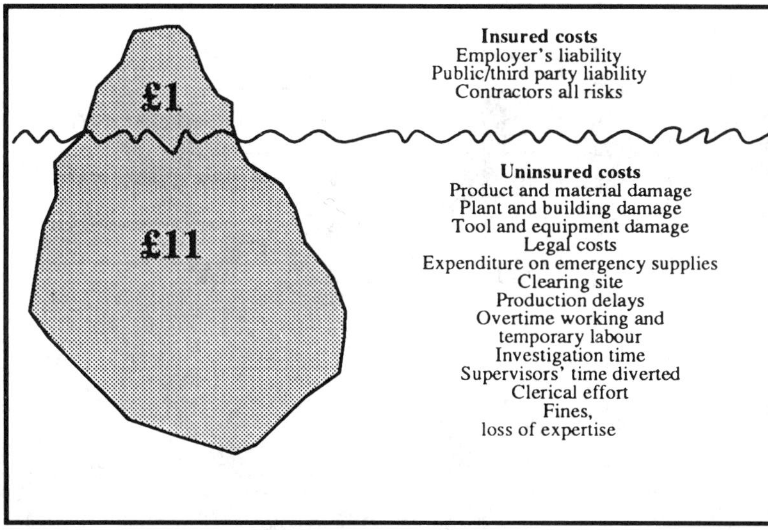

Figure 2. Accident iceberg – the hidden cost of accidents (Health & Safety Executive, 1993)

treatment, hospitalisation, disability pay, spoiled materials, damaged equipment, and remedial work. Moreover, the indirect impact costs must normally be absorbed directly by the project. If this understanding is strengthened particularly at site level then it provides a suitable environment for enhancing H&S awareness and cultivating an overall H&S culture among all those working on construction sites. Such a culture would be based on the well-known principle of prevention rather than cure.

The challenge then is to communicate to personnel and the workforce the adverse effects of potential accidents on site in a manner which is seen to be objective, realistic and easily understood.

3 COMPUTER-BASED ACCIDENT COSTING

In 1989 the Accident Prevention Advisory Unit (APAU) of the Health and Safety Executive (HSE) in the UK began a series of five case studies with organisations from various sectors of industry to develop a methodology to accurately identify the full costs of accidents. The findings were published and a costing methodology was developed by APAU which attempted to identify the cost of all accidental losses that were considered to be preventable and that an organisation committed to loss control would aim to eliminate (Health & Safety Executive, 1993). In 1996 an honours project was assigned to a final year civil engineering student in this department to produce a basic form of a computer-based accident costing model based on HSE's publication. It was envisaged as a first tentative step towards producing a model of practical use in the construction industry. Practising engineers and H&S officials were approached to review the template as it was being developed. Three local construction companies later provided information and data on accidents which had already occurred. The accidents were of diverse nature mainly reflecting the diversity of activities on a typical construction site. Some of the main features of the model are explained below:

(a) It was envisaged that the product is mobile and thus it would be based on the use of a laptop computer.
(b) The intended usage would be analogous to that of an insurance, pension, or an investment scheme salesman. Input the details of the particular case at hand and output

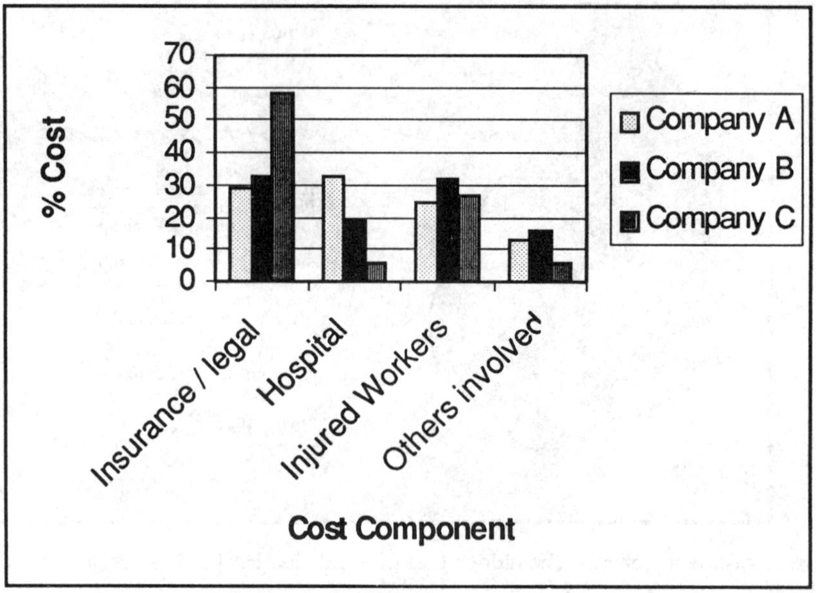

Figure 3. Accident cost components

the required information. In this case input accident or potential accident details and evaluate monetary consequences.

(c) Input and output procedures should not be too lengthy and would need to be completed within a short period of time.

(d) Due to the fact that accident costs cannot be quantified exactly it was decided that the template should be able to illustrate the range in which the expected accident cost will fall. For this Excel's *Scenario Manager* was utilised to produce perceived maximum and minimum accident costs.

(e) The *Note* facility in Excel was used to remind the user of what the accident was or for describing the formulae used to calculate the accident costs.

(f) Input on accident costs was divided into four sections namely human resource, materials, plant, and project delay.

(g) In its present form the presentation of the output costing information can be delivered in tabular or graphical form. The tabular form is modelled on a balance sheet with direct and indirect costs shown separately with a final total cost displayed in bold for the account period. The graphical representation has the expected visual impact and directs attention to areas of vulnerability in H&S on site so that resources can be more effectively targeted to were most needed (Figure 3).

This particular exercise was a modest attempt on the long road of accident costing and further work on extending and improving the model is planned. There was nevertheless a reasonably positive response to the workings of the current template from personnel in the organisations which supplied data and information. It was also significant that there was general consensus on the need for a practical computer-based model for this purpose. The objective being to give a reasonably accurate estimate of the likely costs of accidents or indeed potential accidents. Not many would appreciate a very complex model which confuses and overwhelms the user in a vain attempt for more precise costing.

Accurate and total accident costing is not an easy task nor can it be performed to precise terms. The main difficulties are due to the fact that:

- Accident costs cover a variety of categories, some of which are tangible and can be calculated with relative ease while others, which may be just as significant if not more so, are not.
- Accident costs are incurred at different locations; at the construction site, in the office, at company head office, and in hospitals.
- Accident costs may be borne and handled by different organisations with conflicting interests including the construction company involved, social security, and insurance companies.
- Accident costs are incurred at different time periods and may span a considerable length of time; immediately following the accident, later when replacement workers take over, and on return of the injured persons after recovery. Legal processes to settle compensations or fines may subsequently continue on for a long time.
- It is not just major accidents that cost money. Uninsured losses from day to day accidents can amount to about 10 times the total insurance premium paid in one year (Health & Safety Executive, 1993).

Can a mathematical model be produced to accurately and realistically take all the above into consideration? The answer is probably yes if it is a quick *idea* or a *feel* about the *approximate* costs of accidents that is required. Even with these flexible terms of reference there is still a real challenge in producing a working tool which is seen by the construction industry as being realistic, practical and user friendly. Such a tool does not yet exist or if does then it is still not commonly used by the industry.

The other important function that an accident costing model may perform is to *estimate* or *speculate* on the costs of *potential* accidents or accidents *which are waiting to happen.* This function needs to be seen as an important part of an overall effort aimed at increasing H&S awareness on site and the motivation among site personnel to take a proactive role in maintaining a healthy and safe work environment. The process involved is envisaged as follows:

(a) An H&S assessor is in charge of this task which aims at identifying potential accidents and hazards and providing approximate estimates of the likely costs should they materialise. The assessor can be from an H&S organisation, the planning supervisor (as defined by the CDM regulations) or some independent consultancy. The main requirement being that the assessor has the appropriate level of H&S expertise.

(b) The assessor will carry out the following tasks :

   i. Make a tour of the site to assess the physical conditions and to observe a sample of work practices. At the end of the tour the assessor will have identified potential accidents or undesired events that could occur, assessed their likelihood of occurrence or frequency, and estimated the nature and severity of the resulting consequence. In doing so the assessor would follow formalised methods of observation and measurement for which he would have been trained and qualified.

   ii. Interview personnel at various levels on site in order to assess the degree of knowledge and attitudes.

   iii. Examine H&S documents and confirm that they are adequate and accessible.

   iv. The assessor will then input the information and data into the laptop computer, which contains the accident costing mathematical model. The model is able to calculate a range of financial costs depending on the accident scenarios, which have been input.

   v. A meeting is held with the site manager and foremen during which the H&S assessor will make a presentation to elaborate on his observations and highlighting

Figure 4. Accidents cost money

the potential accidents which were identified earlier. The possible consequences of such accidents in terms of injuries and damage are then discussed. This is then followed by an assessment of the possible range of financial costs, which were determined earlier. These should include worst scenarios where risks / hazards have been identified. For example if a scaffolding is inadequate then the maximum number of workers likely to be on it at any one time can be injured should it collapse. It is at this point that good communication skills are required on the part of the H&S assessor to convince the persons in charge that there are real accidents waiting to happen and drive home the message about the financial consequences.

vi. The above presentation is like getting the site personnel to respond to an alarm. Looking at the output from the computer model is synonymous to visual stimulation in response to a flashing alarm light or a warning siren (Figure 4).

(c) The presentation, which will use modern computer-based visual techniques, should be designed to create maximum impact. However, site personnel present must be convinced of the reliability of criteria and methodology used for costing. In this respect an informative introduction to the workings of the model will be beneficial.

(d) Site personnel may provide the H&S assessor with information about near misses (accidents that did not result in injury) which can then yield information about situations that could potentially cause injury or damage.

(e) The presentation may include examples of previous documented accidents from other sites or projects with some similarities to the situation on the present site. This information may be stored on a database on the computer.

(f) The contents of the presentation should also be communicated to the whole workforce on site. Ideally this should be made when the workforce are assembled together in a comfortable and sheltered part of the site but posters and other printed bulletins may suffice.

(g) It is hoped that this presentation will act as a catalyst for some form of change in behaviour among site personnel, e.g. higher alertness at work, enhanced awareness, deeper and more proactive concern for the well being of oneself and others. The whole idea is that this is an objective and informative exercise for the benefit of the site as a whole and it is not intended to apportion blame or be punitive. In this respect the assessor must try to establish an empathy with site personnel and make sure that there are some common perceptions of hazards and risks.

(h) In the short term it will be expected that the site management will take assertive follow-up preventive action based on the findings of the inspection and on the presentation.

5 CONCLUSIONS

A computer-based mathematical model to assess costs of accidents in construction will undoubtedly be welcome by all parties concerned with H&S issues. The response of some construction professionals to the cost calculating template described above was positive and encouraging. While the product itself was rather simplistic it nevertheless suggested that a more comprehensive model would be acceptable to the industry. Such a model would necessarily incorporate regulatory (required by law) and heuristic (based on experience) criteria. It would have the capability for assessing the costs of accidents based on input information and with reference to a database, which is incorporated in the model. As the model develops it can be used for the purposes of hazard identification, risk assessment and quantification. In its more advanced form in the future it can be used as the basis to set up an H&S expert system for training purposes.

Further however, what is also envisaged here is the use of this model as a motivational tool, which is based on the premise that the driving force for H&S in industry in general is the fact that accidents are expensive. Thus it communicates information in a language which is easily and globally understood by industry, the language of money and numbers. It has to be a tool, which is realistic, reliable and effective enough to stimulate a fundamental change in ingrained attitudes and behaviour. This should be achievable bearing in mind the increasing power of computers and their multimedia capabilities. It is a fascinating challenge to all those whose goal is the creation and maintenance of safe construction sites and who advocate that H&S needs to be managed proactively and not reactively.

REFERENCES

Everett, J. G., 1996, Costs of accidents and injuries to the construction industry, *Journal of construction engineering and management*, Vol 122, No.2, pp 158-164.

Heinrich, H. W., 1959, *Industrial accident prevention*, 4th Ed., McGraw-Hill Book Company.

Jones, Norman (Ed.), 1996, *Safety by design: an engineer's responsibility for safety*, The Hazards Forum, Burlington Press Ltd, Cambridge, UK.

Kartam, Nabil, 1997, Integrating safety and health performance into construction CPM, *Journal of construction engineering and management*, Vol.124, No.2, June, 1997, pp 121-126

Levitt, R.E., Parker, H.W., and Samuelson, N.M., 1981, Improving construction safety performance: the user's role. *Technical report No. 260, Stanford University.*

5 Safety improvement

Implementation of Safety and Health on Construction Sites, Singh, Hinze & Coble (eds)
© 1999 Balkema, Rotterdam, ISBN 90 5809 036 1

Concurrent engineering in construction – An opportunity to improve construction safety

C.J.Anumba
Department of Civil and Building Engineering, Loughborough University, UK

ABSTRACT: The thesis of this paper is that the adoption of concurrent engineering principles and practices in the construction industry offers a unique opportunity to address construction safety issues in a proactive way. The paper argues that the poor safety record of the construction industry can be attributed, in part, to the inadequacies of the conventional construction process and procurement routes. In particular, traditional approaches to project delivery offer little opportunity for the integration of safety considerations and, in some cases, actually engender unsafe working practices. The opportunity to improve construction safety through concurrent engineering is discussed and the benefits of this proactive approach outlined.

1 INTRODUCTION

The construction industry is notorious for its poor safety record when compared with other industries (Bentil 1990, Stanton & Willenbrock 1990, Jaselskis & Suazo 1994, King & Hudson 1985). Every year, a considerable amount of time is lost as a result of work-related health problems, accidents and 'safety incidents'. In addition to the personal injuries and fatalities often sustained, there are huge cost implications which the construction industry can ill-afford. These costs have been classified as direct and indirect costs (Bentil 1990, Hinze and Appelgate 1991). Direct costs are insured costs which are usually covered by the worker's compensation insurance while indirect costs are uninsured costs (such as loss of productivity, transportation costs, management costs in dealing with an accident, and general disruption) which are often difficult to quantify.

1.1 Health and safety statistics

The construction industry's poor safety record is reflected in numerous health and safety statistics. In the UK, the most reliable statistics are those published by the Health and Safety Executive (HSE), an executive agency charged with the responsibility for developing and enforcing the laws, codes and regulations which govern safe working in all industries. It should be noted that many accidents go unreported; thus the HSE figures represent only a tip of the iceberg. The basis for the classification of accidents often varies from country to country thereby allowing room for misinterpretation. However, fatal injury figures are generally regarded as a good indicator of safety levels (King & Hudson 1985). Fatality figures for the UK construction industry (125 in 1986/87 - incidence rate of 15.5 per 100 000; reducing to 74 in 19984/95 - incidence rate of 9.1 per 100 000 - HSE 1995) are considered to be, on average, four to five times that of the manufacturing sector (King & Hudson 1985, Stokdyk 1994, HSE 1989). In comparison, fatality levels in the US construction industry exceed 2000 persons per annum; this corresponds to an incidence rate of about 34 fatalities per 100,000 workers (Hinze & Appelgate 1991). The UK figures can be considered low when compared with developing countries which tend to have very high casualty rates (King & Hudson 1985, Lingard & Rowlinson 1994, Suazo & Jaselskis 1993).

1.2 Factors responsible for site accidents

There are numerous factors responsible for accidents on construction sites. These include the following: falls from heights, falling objects, tripping over objects at ground level, strenuous work, inclement weather, hazardous substances used or generated in the course of construction, 'force majeure' (e.g. earthquakes, etc.), site conditions, collapse or overturning of site structures, site transportation, unsafe practices (methods, equipment, etc.), inadequate training, poor supervision/monitoring of site activities, lack of facilities and structures for anticipating and dealing with health and safety risks, and human error. Of the above, falls are considered to represent the worst case accounting for up to 50% of all accidents (HSE 1995).

The above factors are exacerbated by the unique nature of the construction industry. King & Hudson (1985) identified some of the special problems of the construction industry as: temporary duration of work sites, seasonal employment, extensive use of migrant labour, small size of most construction firms, extensive use of sub-contractors, effects of weather (including spells of long working hours to compensate for bad weather), high labour turnover, welfare problems of construction workers, and competitive tendering. They also discussed details of how these factors contribute to the poor safety record of the construction industry.

This paper argues that, in addition to the above factors, the construction project delivery process is an important factor. It suggests that the limitations of the conventional construction process (discussed in the next section) can be overcome through the adoption of concurrent engineering principles and practices by the construction industry.

2 LIMITATIONS OF THE CONVENTIONAL CONSTRUCTION PROCESS

An often-overlooked factor in the poor safety record of the construction industry is the project delivery process itself. Previous studies have focused far too much on the micro-level causative agents without adequate consideration of the macro-level factors, such as the construction environment and the project delivery process, which also impact upon safety. The traditional approach to project development does not provide a conducive atmosphere for proactive safety management. Some of its features that hinder safety include the following:

2.1 Use of sequential processes

The conventional construction process is basically sequential. For a typical building project, the disciplines involved - architects, structural engineers, building services engineers, quantity surveyors, contractors, sub-contractors, materials suppliers, etc. - often work sequentially with, for example, architects completing their design before the structural design starts, and so on. This 'over the wall' approach to product development was characteristic of the manufacturing industry before the industry's adoption of concurrent engineering. It is much criticised for not providing for 'downstream' input to the design stage (Evbuomwan & Anumba 1996) . This means that contractors, who are responsible for site operations usually have little or no influence at the design stage, during which the most critical decisions are made. Thus, there is little room for collaborative decision-making, resulting in sub-optimal and, in some cases, unsafe solutions being adopted. Furthermore, there is no effective mechanism for true life-cycle analysis of design options and their long-term implications. Under these circumstances, the early assessment of the safety aspects of design and/or the associated construction and maintenance processes is often neglected or inadequately undertaken. This means that safety problems that could have been eliminated by design are often undetected until it is too late. Although improvements in procurement methods (such as Design and Build; Design Build Finance and Operate, DBFO; and the Public Finance Initiative, PFI; etc.) have sought to improve the conventional processes, the basic sequential processes remain and have far-reaching implications for construction safety.

2.2 Lack of cost provisions for safety

Within the traditional construction process, little or no provision is usually made for the cost of safety measures. With the various procurement methods, tenders are normally based either on a lump sump (with or without incentives for cost savings) for the whole contract, or a set of prices for individual work items or packages. Very rarely is 'safety' included as an explicit item to be costed, with prospective contractors expected to have safety costs built into their unit rates for the work items. Clients, who have to pay for construction work, also do not make specific cost provisions for construction safety, except in highly sensitive projects. This results in a situation where the contractor is required to ensure safety on site but has no resources with which to do so.

2.3 Competitive tendering

King and Hudson (1995) have already highlighted the practice of competitive tendering as a factor in construction safety. When contractors are involved in competitive bids for a project, the key to securing the contract is to be the lowest tenderer, who is not necessarily the most efficient or competent contractor. In order to keep the tender value down, contractors tend to base their estimates on the cheapest possible means of undertaking a given task without adequate consideration of the safety implications. In many cases, safety compromises are made on site to compensate for the low rates allowed for in tenders. Reputable contractors who make adequate provisions for the cost of safety measures in their tenders often find that their tenders are higher than their competitors'.

2.4 Adversarial environment

The adversarial environment within which construction projects take place is well known. It is a legacy of the industry's fragmentation and has adverse implications for construction safety. Within such an environment, there is usually a culture of 'buck passing' with none of the parties ready to take responsibility when things go wrong. In fact, potential hazards are sometimes neglected and not flagged by one party on the grounds that it is another party's responsibility. Also, there is limited collaborative decision-making and the will to work together to resolve a problem is often not there.

2.5 Legal framework

The limited legal framework within which construction projects are undertaken exacerbates the above characteristics of the conventional construction process. Until the Construction, Design and Management (CDM) Regulations came into force in the UK in 1995 (HSE 1994), there was no legal requirement for clients and designers to make adequate provisions for construction safety. Safety was primarily the contractor's problem, although individual workers had a responsibility to use personal protective equipment where these are provided. The CDM Regulations require clients to ensure that only 'competent people' are appointed to work on a project, that all information that relate to safety is provided to the contractor and that adequate resources (including time) are available to ensure the safe execution of the project. Designers have to ensure that their designs reduce or eliminate as many risks to health and safety as can reasonably be identified at the design stage. Where risks cannot be avoided, adequate information has to be provided to all relevant parties.

The above catalogue of limitations in the conventional construction process is not intended to be exhaustive but is adequate to demonstrate the need for changes in the project delivery process.

3 CONCURRENT ENGINEERING FUNDAMENTALS

To address the above limitations of the conventional construction process, a radical review of existing procedures is necessary. This has to be undertaken with a view to integrating all the various functional disciplines involved in a construction project - architects, civil/structural engineers, contractors, quantity surveyors, building services engineers and materials suppliers - within a multi-functional matrix team so that all key issues (including construction safety) can be addressed early in the project life-cycle. This can be achieved by means of a concurrent engineering approach to project delivery.

3.1 Definition

Concurrent Engineering, which is also known as Simultaneous Engineering or Parallel Engineering, is a concept that has come into being within the last decade. It has resulted from ever-increasing pressures on manufacturers to be more competitive (in terms of product quality, cost, durability, etc.) and more responsive to change. Despite growing interest in Concurrent Engineering, there is no universally accepted definition. Stephanon & Spiegl (1992) state that 'Concurrent Engineering involves conducting engineering operations in such a way that all functional considerations from design to manufacture are taken into account, and solutions to potential problems developed as early as possible'. Other definitions include those by Winner et al (1988), Kannapan & Marshek (1992), and Cleetus (1992).

3.2 Goals and principles

The definitions may differ in semantics but they are all indicative of the goals, principles and benefits of concurrent engineering. Broughton (1990) summarises the aim of concurrent engineering as being 'to achieve reduced lead times and improved quality and cost by the integration of design and manufacturing activities, and by maximising parallelism in working practices'. Other key goals and principles include:

- Proper analysis and establishment of customer requirements and specifications;
- Development of conceptual solutions that are modular, easy to manufacture and assemble;
- Integration of the manufacturing process and product design that best matches needs and requirements;

- Designing the interface between subsystems within a product to take account of tolerances as well as designing the product to be robust;
- Adopting a systems approach to product development and taking into account the entire product life-cycle;
- Continually focusing on improvement of the product and manufacturing process;
- Location of multi-functional teams together when possible to facilitate better communication;
- Reduction of product lead times and product costs;
- Paralleling the design process.

3.3 Benefits

There are numerous benefits associated with the adoption of the above goals and principles of concurrent engineering. These have been discussed in detail by Clealand and Bursic (1992), Evbuomwan and Sivaloganathan (1994) and Evbuomwan et al (1994) in the context of product development, and can be summarised as follows (with the specific benefits of concurrent engineering to construction safety being presented later on in the paper):

1. Segregation, isolation and the 'over the wall' syndrome are virtually removed from a company and the various divisions can work together in an integrated manner;
2. The overall product development process is shortened as steps along the way are handled in parallel;
3. Increased capacity for global competition and ability to get lower costs and high quality products to customers on time;
4. Fewer design errors and mistakes and a reduced number of engineering changes;
5. Reduction and possible elimination of design reviews and product iterations;
6. Enhanced communication and co-operation between designers, managers, and other professionals involved in the product development process;
7. Greater employee involvement in the organisation;
8. New products with better customer satisfaction, lower costs and higher quality can be released into the market;
9. Companies can be more responsive to customers.

4 IMPROVING CONSTRUCTION SAFETY THROUGH CONCURRENT ENGINEERING

4.1 The opportunity for proactive safety management

It is being increasingly recognised that proper management is vital for reducing hazards and accidents on site (Levitt & Samelson 1987, Seals & Hale 1992, Hislop 1991, Carey 1991, Stokdyk 1994). A proactive safety management approach is required to ensure that potential hazards are identified early and eliminated either through careful design or by the implementation of appropriate control measures. This has obvious economic advantages as it is much cheaper to combat risks at source than to contain problems when they do occur at a much later stage.

As evident from the aforementioned goals, principles and benefits of concurrent engineering, there is an opportunity to address construction safety in a proactive way within a concurrent engineering environment. Concurrent engineering facilitates the early, collaborative evaluation of various aspects of a constructed facility's life-cycle, including safety. Within a concurrent engineering environment, all the principal participants in a construction project are encouraged to have input at the early stages of the design phase, so as to accommodate the full range of requirements, expertise and perspectives. This means that downstream participants, such as contractors, can comment on design proposals that impose unrealistic safety burdens on those that have to construct the facility. Where appropriate, they can also suggest alternative ways that the identified problems can be overcome.

Concurrent engineering's emphasis on the need for continued focus on meeting the needs of clients and end-users (or customers) ensures that consideration is given to how they might be affected by the constructed facility over its life-cycle. Thus, there is a unique opportunity to address early in the design process, the safety requirements of clients and end-users at all stages of the facility's life-cycle, particularly the facility utilisation/operation and facility management and maintenance stages.

4.2 Life-cycle consideration of safety in a concurrent engineering environment

In undertaking an analysis of the life-cycle implications of a constructed facility, there is major scope to assess safety issues at each of the critical stages of a project - design, construction, facility utilisation/operation, facility management and maintenance, and facility disposal or decommissioning. The safety considerations at each of these stages will differ and are discussed below:

Design: The key issue at the design stage is to avoid design features that will pose safety problems at the construction and subsequent stages in a facility's life-cycle. As far as possible, risks should be eliminated through careful design, substitution of hazardous materials, avoiding unsafe construction operations and/or sequences, eliminating the necessity for elaborate temporary works, etc.

Construction: Safety problems at the construction stage usually relate to several factors including the site layout design (Anumba & Bishop 1997), the construction methods and sequences adopted, the adequacy of the temporary works, the site management regime, weather conditions, human error, and the safe handling of hazardous materials used or generated in the course of construction operations. Most of these are issues that can be addressed in a proactive way within a concurrent engineering environment. The contractor has a vital role to play in raising potential problems early so that the project team can resolve these early.

Facility Utilisation: It is important that safety problems likely to affect the client and/or facility end-users are avoided. Concurrent engineering's focus on meeting the client's requirements entails an active, rather than passive, client involvement in the project delivery process, and provides an avenue for potential problems to be raised and resolved early. Design provisions that adversely affect the safe utilisation or operation of a facility can be revised and it is important that the project team has a clear understanding of the use characteristics of the facility.

Facility Management: The effective management of a facility is affected to a large extent by decisions and provisions made at the earlier stages of design and construction. It is important, therefore, that the safety problems posed by poor maintenance access, the requirement for elaborate rigs for routine cleaning, confined spaces, very frequent maintenance, etc. are avoided through design. The involvement of facility managers and end-users in a collaborative concurrent engineering project team is vital in this regard.

Facility Disposal/Decommissioning: The disposal or decommissioning of a facility often involves demolition or dis-assembly of the facility. This has inherent safety and environmental implications that can be more readily identified and resolved within a concurrent engineering environment rather than the traditional sequential construction process. To avoid safety problems, it is important that the facility collapses in the predicted manner; does not endanger the lives of the demolition team or third parties; has no adverse effect on neighbouring structures; does not generate hazardous substances in the course of disposal; and can be safely reused (in part) if necessary.

It is noteworthy, from the foregoing, that the most important benefits of a concurrent engineering environment are as follows:

- the safety assessment for each project stage is undertaken early in the construction process, when hazards can be eliminated or reduced by appropriate design provisions;
- there is collective input from the all the key project participants in the identification and resolution of safety problems, resulting in enhanced cross-disciplinary knowledge of safety issues;
- the early assessment of safety requirements will enable their cost implications to be made explicit and funds set aside to cover these;
- the enhanced communication within a concurrent engineering environment provides for early notification of safety concerns to all parties in the project team.

5 SUMMARY AND CONCLUSIONS

There is no doubt that improving the construction industry's safety record is an important objective that will improve the industry's image, efficiency, financial performance ratios, productivity, and service performance standards (Veltri 1989). A proactive approach is vital in this regard and the key elements of such an approach should include the following: an objective analysis of the requirements; elimination of risks through design and forward planning, development and implementation of procedures for dealing with known risks; institution of structures for managing unforeseen risks and accidents; setting up a clear safety management structure that all workers can easily identify; implementation of a regime for safety training; and the continuous monitoring and review of adopted measures (Anumba & Bishop 1997). Traditional 'over the wall' approaches to project delivery have failed to adequately support proactive safety management, and it is evident that new approaches are required.

Concurrent engineering is an innovative philosophy that is fully supportive of a proactive approach to safety management, and has proved successful in other industries, notably manufacturing. It offers significant scope for project teams to assess, very early in the design process, the safety implications (at all stages in the life-cycle of a facility) of design decisions as they are made. The collaborative input of downstream participants - services engineers, contractors, sub-contractors, material suppliers, clients and/or end-users, facility managers, and demolition contractors - is of the essence in this regard, and is actively encouraged within a concurrent engineering environment.

The adoption of concurrent engineering in construction will have a positive impact on safety, in addition to the other numerous improvements in lead times, cost, quality, and client satisfaction.

REFERENCES

Anumba C. J. and G. Bishop 1997. Importance of safety considerations in site layout and organisation. *Canadian Journal of Civil Engineering*, 24:229-236.

Bentil, K. K. 1990. Construction site safety: a matter of life and costs. *Cost Engineering*, 32(3):7-10.

Broughton, T. 1990. Simultaneous engineering in aero gas turbine design and manufacture. In *Proc. 1st Int. Conf. on Simultaneous Engineering*, 4-5 December, London, pp 25-36.

Carey, J. 1991. Getting business to think about the unthinkable. *Business Week*, Issue 3219, June 24, pp 104-107.

Cleland, D.I. and K.M. Bursic 1992. *Strategic technology management - systems for products and processes*. American Management Association, New York, 1992.

Cleetus, K.J. 1992. Definition of concurrent engineering, *CERC Technical Report Series Research Note CERC-TR-RN-92-003*, Concurrent Engineering Research Centre, West Virginia University, Morgantown, USA, pp 1-5.

Evbuomwan, N.F.O. 1994. Design function deployment - a concurrent engineering design system. *PhD Thesis*, City University, London.

Evbuomwan, N.F.O. and S. Sivaloganathan 1994. The nature, classification and management of tools and resources for concurrent engineering. *Proc. 1st Int. Conf. on Concurrent Engineering: Research and Applications*, Paul, A. J. and M. Sobolewski (Eds), August 29-31, pp 119-126.

Evbuomwan, N.F.O. and C.J. Anumba 1996. Towards a concurrent engineering model for design and build projects. *The Structural Engineer,* 74(5):73-78.

Hinze J. & L. L. Appelgate 1991. Costs of construction injuries. *ASCE Journal of Construction Engineering and Management*, 117(3):537-550.

Hinze, J. & F. Wiegand 1992. Role of designers in construction worker safety. *ASCE Journal of Construction Engineering and Management*, 118(4):677-684.

Hislop, R. D. 1991. A construction safety program. *Professional Safety*, 36(9):14-20.

HSE 1994. *Construction (design and management) regulations*. Health and Safety Executive, UK.

HSE 1995. *Health and safety statistics 1994/95 annual report.* Health and Safety Executive, UK.

Jaselskis, E. J. & G. A. R. Suazo 1994. A survey of construction site safety in Honduras. *Construction Management and Economics*, 12:245-255.

Kannapan, S.M. and K.M. Marshek 1992. A schema for negotiation between intelligent design agents in concurrent engineering. *Intelligent Computer Aided Design*, Brown D.C. et al (Eds), Vol. 4, Chap. 36, Elsevier Science Publishers BV, North Holland, pp 1-25.

King, R. W. & R. Hudson 1985. *Construction hazard and safety handbook*. London: Butterworth.

Levitt, R. E. & N. M. Samelson 1987. *Construction safety management*, New York: McGraw Hill.

Lingard, H. & S. Rowlinson 1994. Construction site safety in Hong Kong. *Construction Management & Economics*, 12:501-510.

Seals, D. & J. Hale 1992. Safety in numbers. *Management Accounting*, 73(11):45-48.

Stanton, W. A. & J. H. Willenbrock 1990. Conceptual framework for computer-based construction safety. *ASCE Journal of Construction Engineering and Management*, 116(3):383-398.

Stephanon, S. E. and F. Spiegl 1992. *The manufacturing challenge*, Van Nostrand Reinhold Publishers, New York.

Stokdyk, J. 1994. No falling back, *Building*, June 3, pp 38-39.

Suazo, G. A. R. & E. J. Jaselskis 1993. Comparison of construction safety codes in United States and Honduras. *ASCE Journal of Construction Engineering & Management*, 119(3):560-572.

Veltri A. 1989. Strategically planning the safety function. *Occupational Hazards*, 51(11):59-62.

Winner, R.I., Pennel, J.P., Bertrand, H.E., and M.M.G. Slusarczuk 1988. The role of concurrent engineering in weapons system acquisition. *IDA Report R-388*, Institute of Defence Analysis, Alexandra, Virginia, USA.

Implementation of Safety and Health on Construction Sites, Singh, Hinze & Coble (eds)
© 1999 Balkema, Rotterdam, ISBN 90 5809 036 1

Psychological research in the development of the new safety 'Sticker Sign' for construction sites

H. Mutoh
Research and Development Institute, Takenaka Corporation, Chiba, Japan

ABSTRACT: On construction sites, safety signs are indispensable to ensure workers' safety. However, the following problems with the old type of signs were clarified through our survey. (1) It takes time and effort to set or remove them. Some materials and tools such as wire and cutter are necessary. (2) Signs cannot be put on finished surfaces because the surfaces may be worn. (3) The old type of sign is rather expensive. To solve these problems, the new Sticker Sign was developed. It has adhesive on the reverse side like a seal. It has four merits. (1) It can be stuck on finished surfaces without damaging the surface. (2) Tools are not required to set or remove it. (3) Sticker Signs are much cheaper than the old type of signs. (4) The design and size are standardized. This study is based on environmental psychology. The location and position of signs were observed at construction sites to find problems, and workers evaluated the attractiveness and intelligibility of designs in order to find the best designs.

1. INTRODUCTION

1.1 *Functions and problems of safety signs at construction sites*

On construction sites, it is essential to ensure workers' safety. Therefore, various measures are taken. Handrails or nets are set up around openings to prevent accidents due to falls. Safety managers regularly patrol the construction site to check workers' behavior and safety.

The environment of construction sites varies as work advances, and workers change quite often. For that reason, safety signs are indispensable for informing workers, and those people who are unfamiliar to the construction site, about such things as danger. The signs can respond rapidly to change in circumstances because they could be set or removed easily.

Safety management requires a lot of time and effort on the part of the manager. It is necessary to cut down these factors.

1.2 *Outline and originality of this study*

This paper reports the details of the development of a new safety sign named 'Sticker Sign'. Like a seal, it has an adhesive on its reverse side making it easy to stick on a wall. It has many advantages over the old type of signs. It can be stuck on finished surfaces without damaging the surface. No tools are required to set or remove it, and it is much cheaper than the old type of signs. Although there are already tiny seals are affixed to a worker's helmet to show credentials, the idea has not spread to use in warning signs, prohibitive signs and information signs, within the limits of the Japanese construction trade.

The characteristic of this study is that it is based on research in environmental psychology. The location and position of signs were observed and analyzed at forty construction sites to find problems. The attractiveness and intelligibility of the signs were evaluated by workers. Past studies and surveys on safety signs or labels have concentrated on various fields especially in traffic, public and products. Most of the studies have related on the design of guide signs. Design guidelines have been suggested (Martin & Shelley 1997) and the influence of various sign design elements on legibility, identification or comprehensibility were examined. The elements are concerned with color (Gurney & others 1977, Braun & others 1995), font (Garvey & others 1997), graphic image (Young 1997), allocation of the components of the message (Young & others 1995) and so on. Other studies include the evaluation of comprehensibility of meanings (Cooper 1989, Easterby & Hakiel 1981) and placement (Hoskovec & others 1974). However, few studies on safety signs at construction sites exist.

This study is different from those past studies in the following respects. The result of the study resulted in the development of a new product. It was important to solve the problems of maintenance and management found through the field survey in addition to legibility and design identification. That is, the time and effort required to set and remove signs were reduced.

2. Exploration of problems with safety signs

To find the problems of using safety signs, the state and maintenance of signs on forty construction sites were investigated based on an environmental psychological approach, that is, the signs' attractiveness and intelligibility. The problems common to almost all construction sites can be summarized in three categories as seen in Table 1.

Table 1. The problems common to almost all construction sites (The numbers in parentheses at the end of each problem correspond to the three targets in 3.1).

Category	Evaluated item	Problems
(1) Design	Conciseness of meaning	Designs are not united between construction site. *(3)
		Different designs are used to communicate the same meaning. *(3)
		Similar signs are used to communicate different meanings. *(3)
		Some signs do not express concretely how to behave. *(3)
	Color	Some signs are different from the standard (Japan Industrial Standard 1998a). *(3)
	Graphic image and letter	Some signs have either only graphic images, or only letters. *(3)
		The letters are too small to read from a distance. *(3)
(2) Location and position	Location	No signs are found in dangerous locations. *(2)
	Position	Some signs are not prominent because they are in the dark. *(2)
		Some signs are not conspicuous because building materials hide them. *(2)
		Some signs do not stand out because they are fixed parallel to workers' eyes. *(2)
(3) Maintenance and management	Maintenance	All signs cannot be affixed to finished surfaces. *(2)
		It takes much time and effort to fix or remove the signs. *(2)
	Management	Many signs remain after they have become unnecessary. *(2)
		There are not enough signs for guides or supplement. *(3)
		A place to keep signs is not near the working area.
		The cost of managing the signs is high. *(1)

(1) Design

The design and size of signs are not standardized. There are a large number of designs, which are used to communicate the same meaning. Designs are not uniform between construction sites, and what is worse, different designs are used on the same construction sites (Photo 1). In the Japanese construction trade, the Japan Construction Safety and Health Association standardize thirteen sign designs (Fig. 1). However, other signs are being used at almost all of the construction sites. In addition, some designs are not intelligible at a glance, because similar signs are used to communicate different meanings (Fig. 2), or the letters are too small to read from a distance.

(2) Location and position

The most serious problem is the lack of signs in dangerous locations. Maintaining the signs falls behind the changing circumstances because it takes time to set or remove them. On the other hand, it is feared that some signs are not recognized because of their inadequate position.

Photo 1. An example of the use of different designs to communicate 'Do not enter'

Fig. 1. Examples of signs standardized by the Japan Construction Safety and Health Association (from the left to the right: Do not enter, No smoking, Fire extinguisher, Watch your step, High voltage)

Fig. 2. An example of similar design conveying different meanings (from the left to the right: Do not enter, Watch your step, Mind your head, Safety passage and Use a safety belt)

(3) Maintenance and management

The old type of signs cannot be used on finished surfaces. Many signs are left after they become unnecessary because it needs much effort to remove them. The cost of managing the signs is high. In the case of an office building with floor area of fifty thousand square meters, about six thousand signs were used throughout the construction period, at a cost of over eighty thousand dollars.

3. DEVELOPMENT OF THE NEW 'STICKER SIGN'

3.1 *Target of the development*

To solve the problems shown in Table 1, three targets were held as follows. In Table 1, the numbers put in parentheses at the end of each problem corresponds to these targets. Many problems can be solved by these measures.

(1) Cost reduction

To reduce the cost of the sign itself, the material was changed from plastic board to vinyl chloride sheet.

(2) Effort saved on maintenance and management

The cause of many problems in maintenance and management, besides location and position, is that it

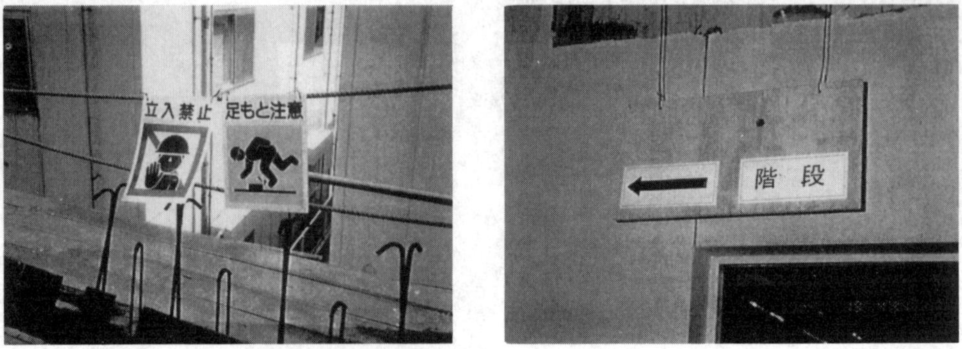

Photo 2. Typical example of setting the old type of signs

takes a lot of time and effort to fix and remove signs. The old type of sign is usually fastened to scaffolds or wire fences with wire or string put through in tiny hole in the corners to the sign. Therefore, some tools like clipper or cutter is required for cutting. The location and position of setting should be restricted (left of Photo 2). Though the old type of sign is often stuck on walls with tape, they cannot be put on the finished surfaces because the strong adhesive wears off the wall surface. Some extra devices are necessary to fix signs to finished surfaces (right of Photo 2). To solve these problems, a sticker formula was adopted and the most appropriate adhesive was chosen. It does not require extra materials or tools, and can be used on finished surfaces.

(3) Uniformity of designs

The main cause of the design problems is that there are no established standards for selecting the best design. To clarify the best design, the attractiveness and intelligibility of signs were evaluated by workers. Furthermore, to show information more concretely, the kinds of guide and supplementary signs were increased.

3.2 *Measures for reduction cost and effort (target (1) and (2))*

To reduce the cost of the sign itself, the material was changed from plastic board to vinyl chloride sheet. The sticker formula was adopted for finished surfaces because little time and effort is required. To choose the most appropriate adhesive, the gluing power of six samples was examined. If the power is too strong, the sign may damage the surface even though it stays adhered to the surface. Conversely, if the power is too weak, signs may fall off without damaging the surface.

In conformity with the prescribed method (Japan Industrial Standard 1998b.), six samples of twenty-five millimeters in width were examined. The samples were pasted on five materials used frequently for buildings, as shown in Table 2. This table shows the test results of the six samples. Gluing power was measured after twenty-four hours at a temperature of twenty-three degrees and a humidity of sixty-five percent. Based on this result, the adopted sample has acrylic adhesive. Moreover, the gluing power of the adopted sample with repetitive peeling and over a long period was measured. It was confirmed that the adhesive can be stuck repeatedly and keeps over a long period (Table 3).

Table 2. The results of gluing power of six samples (gram): 'damaged' means that the sample tore off the surfaces of materials, and 'fell' means that the sample fell by itself.

Materials	Adopted one	Sample A	Sample B	Sample C	Sample D	Sample E
Plaster board	50	damaged	damaged	900	2220	75
Vinyl cloth	80	damaged	damaged	damaged	damaged	100
Plywood board	60	1300	250	100	1200	fell
Painted board	100	damaged	damaged	1350	2000	90
Mortar wall	35	3800	2250	1350	2000	95

Table 3. The results of gluing power in the adopted sample (gram)

	Gluing power with repetitive peeling		Gluing power after a long period	
Materials	After one time	After five times	A week later	Two weeks later
Plaster board	30	20	70	80
Vinyl cloth	70	40	60	60
Plywood board	50	45	70	80
Painted board	50	40	60	80
Mortar wall	40	25	40	40

3.3 *Examination of workers' evaluation of designs (target (3))*

To find the best designs from a large selection, many color samples were evaluated. One hundred and fifty-four workers and fifteen members at the public estimated the attractiveness and intelligibility of forty-four samples. Four samples used frequently at construction sites were chosen from every eleven meaning shown in Table 4. The reason why the members at the public were included as subjects is because workers tend to evaluate familiar signs as attractive and intelligible on preliminary evaluation. The designs evaluated to be attractive and intelligible by both workers and members at the public were chosen. Workers evaluated the samples using a three-grade scale and the members at the public evaluated the samples using a five-grade scale.

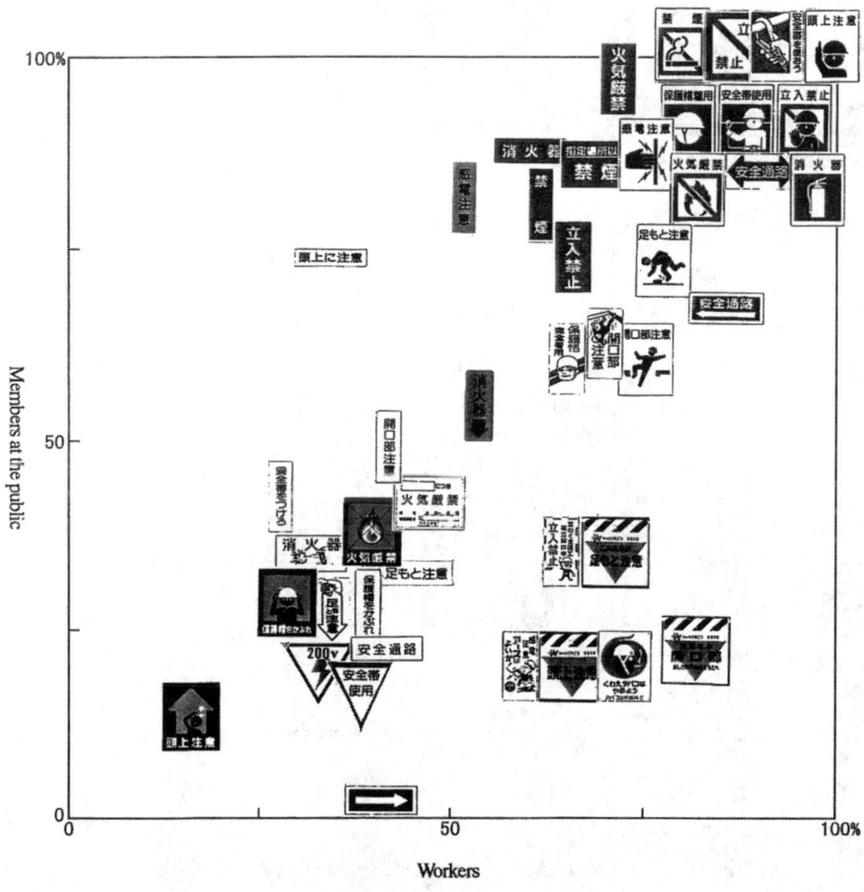

Fig. 3 The results of the examination –the proportion of subjects who evaluated the sample attractive and intelligible

The proportion of subjects who evaluated the sample attractive and intelligible was calculated. The samples plotted in the upper right of Fig.3 were evaluated to be attractive and intelligible by both workers and members at the public. All of the samples had both letters and graphic images, especially pictographs, and expressed their meaning concretely. On the contrary, the samples plotted in the lower left of Fig. 3 were evaluated to be unattractive and unintelligible. Most of the samples had only letters or only graphic images, which did not directly relate to the meanings.

According to the results, the designs were united. Pictographs were designed on twenty signs of waning signs, prohibitive signs and information signs (Fig. 4). In addition, ninety signs for guides and supplementary were added. They show the place or locality, arrows, and period and person responsible for the work. Because module of signs was fixed, it makes a good show when they are combined.

Table 4. Eleven meaning of samples used in the evaluation

Category	Meaning	Category	Meaning	Category	Meaning
Prohibition	Do not enter.	Warning	Watch your step.	Information	Wear a helmet.
	Do not use fire.		Here is an opening.		Use a safety belt.
	No smoking.		Mind your head.	Guidance	Fire extinguisher.
			High voltage.		Safety passage.

Prohibition signs (Left: Do not enter; Center: No smoking; Right: Do not use fire)

Warning signs (Left: Mind your head; Center: Watch your step; Right: High voltage)

Information signs (Left: Wear a helmet; Center: Use a safety belt; Right: Keep tidy)

Guide signs (Left: Office of Joint Venture; Center: Material stockyard; Right: Rest room for women)

Supplementary signs (Left: Finished surface; Center: Under dismantling of support for mold; Right: Period and person responsible for the work)

Fig. 4. Example of Sticker Sign

4. CONCLUSIONS

4.1 *Merits of Sticker Sign*

(1) Sticker Sign must be stuck wherever it is necessary. It must be put in the location required and in the most suitable position. In the case of the old type of signs, the location and position of setting should be restricted, because the signs need to be fastened to scaffolds or wire fences.
(2) Sticker Sign can be pasted on finished surfaces without damaging the surfaces. Although the old type of sign is often stuck on wall with tape, they cannot be put on finished surfaces because they wear off the surface with strong adhesives. Sticker Sign is also suitable for repair work.
(3) Sticker Sign can be easily set without any tools (Photo 3). It drastically reduces the effort required for setting. To fix the old type of signs, it takes a lot of time and effort because materials such as wire or string are needed for fastening and some tools like clipper or cutter is required for cutting.
(4) Sticker Sign can be easily stuck and removed repeatedly unless the surface becomes dirty. Maintaining the signs does not fall behind the changing circumstances of construction site.
(5) Sticker Sign is forty percent cheaper than the old type of signs.
(6) The design of the Sticker Sign was improved by adoption of pictographs popular in public signs.
(7) The kinds of signs for guides and supplementary material were increased. It becomes possible to show information more concretely because supplementary signs can be combined with warning signs, prohibitive signs and information signs (the left of Fig. 5). The right on the Fig. 5 shows an example of guide sign. Because an arrow sign was separated from information signs, it needs only to fit the direction of arrow sign to show a way for the destination. Until now, different signs were prepared in every direction because the both signs were integrated. Because module of signs was fixed, it makes a good show when they are combined.
(8) It is easy to duplicate signs at the construction site.

Photo 3. Setting of Sticker Sign on plaster board

Fig. 5. Examples of combination of signs

4.2 *Present situation of utilization and improvement on Sticker Sign*

Sticker Sign was applied at over one hundred and sixty construction sites until August 1996 and has been evaluated well. The effect of this development was confirmed through the survey at several construction sites using the Sticker Sign. Almost all of the problems shown in Table 1 were solved.

To meet the demands of use outside, a new one with a stronger adhesive was added in October 1997. It is suitable for sticking on outside tile or concrete wall. It can be used in the same way as the one for interior finishing.

ACKNOWLEGEMENT

I would like to express my gratitude for Ms Edit Nagy, a guide researcher at the Building Research Institute at that time who gave me a hint at the beginning of this research. I give my thanks for direction given by the Japan Construction Safety and Health Association.

REFERENCES

Braun, C. C., P. B. Kline & N. C. Silver 1995. The influence of color on warning label perceptions. *International Journal of Industrial Ergonomics*. 15(3): 179-187.

Cooper, B.R. 1989. Comprehension of traffic signs by drivers and non-drivers. *Research Report – Transport and Road Research Laboratory*. 167: 24

Easterby, R. S. & Hakiel, S. R. 1981. Field testing of consumer safety signs: The comprehension of pictorially presented messages. *Applied Ergonomics*. 12(3): 143-152.

Garvey, P. M., Pietrucha, M. T. & Meeker, D. 1997. Effects of font and capitalization on legibility of guide signs. *Transportation Research Record*. 1605: 73-79.

Gurney, G. F., McNaught, E. D. & Bryden, J. E. 1977. Evaluation of Adirondack Park Highway signs. *Research report – New York State Department of Transportation, Engineering Research and Development Bureau*. 56: 44.

Hoskovec, J., Stikar, J. & Raouf, A. 1974. The effectiveness of vertical traffic signs: Experimental approach. *Studia Psychologica*. 16(3): 220-223.

Japan Industrial Standard 1998a.General code of safety colors, Z 9101.

Japan Industrial Standard 1998b. Testing methods of pressure sensitive adhesive tapes and sheets, Z 0237.

Martin, B. J. & Deppa, S. W. 1997. Human factors in the revised ANSI Z535.4 standard for safety labels. *Proceedings of the Human Factors and Ergonomics Society 1997*. 2: 821-825. Human Factors and Ergonomics Society, Inc., Santa Monica, CA, USA.

Young, S. L., Wogalter, M. S., Laughery, K.R., Magurno, A. & Lovvoll, D. 1995. Relative order and space allocation of message components in hazard warning signs. *Proceedings of the Human Factors and Ergonomics Society 1995*. 2: 969-973. Human Factors and Ergonomics Society, Inc., Santa Monica, CA, USA.

Young, S. L. 1997. Role of pictorials in environmental safety signs. *Proceedings of the Human Factors and Ergonomics Society 1997*. 2: 797-800. Human Factors and Ergonomics Society, Inc., Santa Monica, CA, USA.

Implementation of Safety and Health on Construction Sites, Singh, Hinze & Coble (eds)
© 1999 Balkema, Rotterdam, ISBN 90 5809 036 1

Electrical contractor safety management evaluation

G.R.Smith
Iowa State University, Ames, Iowa, USA
A.A.Oloufa & J.W.Kranz
The Pennsylvania State University, Indiana, Pa., USA

ABSTRACT: This paper describes research conducted toward the development of a safety management evaluation system. The three key aspects of the system are a company's management philosophy, written safety plan and field compliance monitoring. Two elements, management philosophy and written safety plans are evaluated in this paper. Six electrical contractors provided data and supporting information. A management philosophy score was developed from principles of safety management identified in the literature. Questions covering the principles were created to evaluate company management philosophy. An extensive checklist was developed to evaluate written safety plans. The management scores and safety plan scores were compared to company experience modification ratings.

1 INTRODUCTION

General business management styles have changed over the last forty years from the 'Classical' chain of command style in the 1950's to today's management systems where more effort is placed on decision making at the lowest level. In this changing management period, safety is often managed in the 'Classical' style, i.e. management makes rules and the workers are supposed to follow them. According to Petersen (1996) "we don't seem to be integrating safety into our changing management philosophies." The safety performance of many contractors can be taken as evidence that current safety management systems are not working as effectively as they could. Rising accident costs and insurance premiums, increasing regulation of safety, potential criminal liability for managers involved in safety violations, and rising litigation costs are further indications that problems continue unabated. A need exists to develop evaluation tools for construction firms enabling them to benchmark their overall safety programs. Improvements in their safety management process might be tracked by such a system of interrelated measurements. This paper is limited to examining the relationship between the company management philosophy, the written safety plans and the company's safety experiences as reflected by their experience modification rating. The third element of the process, field performance measurement and reporting requires further development (Smith and Arnold, 1997).

2 BACKGROUND

A safety management system begins with top management's commitment to establish a company philosophy and atmosphere that demonstrates safety is a top priority. Krause (1994) captured many of these concepts in an article that relates Total Quality Management (TQM) implementation principles with safety management implementation principles. Usually, a detailed written safety plan translates company safety commitment into an operational tool. To be effectively implemented, the written safety plan must be well organized and easy to understand. The plan must involve all aspects of the company's business and allow for employee involvement. Employee involvement in the development of the safety plan increases compliance and participation during implementation. To insure continued compliance and improvement, the safety plan must be periodically monitored at the jobsite to insure compliance and provide data on performance.

2.1 *Management philosophy*

A Company's safety management philosophy sets one important standard for the safety performance of all employees. Once the company managers have established a clear safety philosophy, diligent efforts can then be expended to draft an effective safety program. (Hinze, 1997). In order to have an effective safety management system integrated into all aspects of the company's business, management must make safety a top priority. The management philosophy section of the system should include the activities of policy making, objective setting, and strategy development. The *Safety Management Opinion Survey* was designed to develop a rating of the contractor's safety management philosophy. The measurement is a score against a selected set of standard safety principles. It can be used to understand the company's safety "climate" and identify areas needing attention according to selected set of safety principles.

2.2 *Petersen's principles for management's philosophy*

Although many companies practice good safety management, a comprehensive set of operating principles for safety management was difficult to find. The following ten principles of safety management are the elements of Petersen's Principles (1996).

1. An unsafe act, an unsafe condition, or an accident: all these are symptoms of something wrong in the management system.
2. Certain sets of circumstances can be predicted to produce severe injuries. These circumstances can be identified and controlled.
3. Safety should be managed like any other company function. Management should direct the safety effect by setting achievable goals and by planning, organizing, and controlling to achieve them.
4. The key to effective line safety performance is management procedures that assign accountability.
5. The function of safety is to locate and define the operational errors that allow accidents to occur. This function can be carried out in two ways: (1) by asking why- searching for root causes of accidents, and (2) by asking whether or not certain known effective controls are being utilized.
6. The causes of unsafe behavior can be identified and classified. Some of the classifications are overload (improper matching of the person's capacity with the load), traps, and the worker's decision to err. Each cause is one that can be controlled.
7. In most cases, unsafe behavior is normal human behavior; it is the result of normal people reacting to their environment. Management's job is to change the environment that leads to the unsafe behavior.
8. There are three major subsystems that must be dealt with in building an effective safety system: (1) the physical, (2) the managerial, and (3) the behavioral.
9. The safety system should fit the culture of the organization.
10. There is no one right way to achieve safety in an organization; however, for a safety system to be effective, it must meet certain criteria. The system must: force supervisory performance, involve middle management, have top management commitment, have employee participation, be flexible, and be perceived positively.

These were chosen to represent an 'ideal' company safety philosophy for this research. They represent one possible baseline for evaluation and comparison. Furthermore, the company's management philosophy should be reflected in the written safety plan.

2.3 *Written safety plan*

All construction companies need to have a clear, concise, but comprehensive written safety plan. One of the employer responsibilities required by OSHA is to "Establish or update operating procedures and communicate them so that employees follow safety and health requirements" (U.S. Dept. of Labor, 1992). To meet that requirement, most companies publish and distribute a written safety program or plan. Although each company's plan is different, OSHA regulations and the company's work activities usually determine the items that are included in the written plan. A comprehensive checklist of items that should be included in a subcontracting trade-specific safety plan was developed.

3 RESEARCH APPROACH

Initially 22 electrical contractors, ranging in size from an annual volume of $500,000 to $70,000,000 were asked to participate in this research. They had to be willing to respond to surveys and follow-up questions, provide a copy of their written safety plan for evaluation, and respond to other requests for information. Only eight contractors were willing to complete and return both surveys, and out of those eight, only six contractors would provide a copy of their written safety plan for evaluation.

3.1 *Limitations of the results*

It should be clear to the reader that this work is based on a very small sample and as such it is best considered as a pilot effort. The researchers feel sufficient data was collected for meaningful comparisons, but not for developing industry-wide conclusions. In the following section, the results of the surveys and the comparison of participating contractors are reported as a demonstration of the process. The surveys are not provided as part of this paper.

4 MANAGEMENT SAFETY OPINION SURVEY

Because a company must recognize their own operations philosophy before writing their objectives into a safety plan, they need a way to assess and reflect on this philosophy. To accomplish that objective, a Safety Management Opinion Survey was created. Ten subject areas were developed from Petersen's 10 Principles. A total of 41 questions about management were generated. However, the survey specifically contained 20 questions that related specifically to Petersen's Principles that are the focus of this paper. The respondents were asked to rate their agreement or disagreement to various responses to questions about safety management principles.

4.1 *Survey findings*

A fair amount of variation among the responding firms was noted while classifying responses. Two groups of questions were segregated for consideration. The first was for employee related questions and the second was for management related questions. Recognizing that the data set is severely limited from a survey standpoint, the only evaluation provided for the six participants is a qualitative view of the consistency in their response patterns. The employees are still generally considered part of the problem and not part of the solution as suggested by the group of responses below. Responses were from 5 (strongly agree) to 1 (strongly disagree).

Employee grouped questions

- Unsafe acts cause construction accidents. (4.8)
- Employee safety is the top priority in my company. (4.9)
- Our company management procedures assign accountability for safety to each individual worker. (4.1)
- In evaluating unsafe behavior, employees will make mistakes that cause accidents. (3.2)
- Our safety policy has employee participation in planning. (3.1)

There is an indication in the survey responses that suggests management still considers itself removed from direct influence over safety performance

Management grouped questions

- Failures of the management system cause construction accidents. (2.6)
- Our company management procedures assign accountability for safety to the project manager (3.5)
- Unsafe behavior is a result of the environment constructed by management (2.6)
- Our safety policy involves middle management (3.4)
- Our safety policy has top management showing its commitment. (3.1)

5 SAFETY MANAGEMENT SURVEY

The Safety Management Survey was created to gather data related to the electrical contractors' management practices and field level implementation issues. Table 1 provides responses to key safety management practice information.

Table 1. Safety Management Survey Selected Responses

	QUESTION	CONTRACTOR					
		A	B	D	F	H	I
1	Annual Volume ($1,000,000)	20	3	1	3	50	70
2	Annual Employee Hours (1000s)	247	39	13.5	35	434.7	700
3	Current EMR	0.690	1.160	0.968	1.169	0.958	1.000
9	PM Safety Performance Tracked	N	N	N	N	Y	N
18	Orientation Safety Training	Y	Y	N	N	Y	Y
19	Management Safety Training	N	N	N	Y	Y	Y
29	Accident Reporting	PM	PM	TM	TM	TM	TM
30	Job Hazard Analysis	N	N	N	N	Y	N
32	Incentive Programs	N	N	N	N	N	N

Y – Yes, N – No,
PM – Project Management, TM – Top Management,

6 WRITTEN SAFETY PLAN EVALUATION

A safety plan checklist containing 124 items was developed from safety regulations and suggested plan requirement lists found throughout the literature. Each plan was evaluated against the checklist. If the item on the checklist was identified in the contractor's plan, a yes (Y) response was recorded, if not, a No (N) response was recorded.

Table 2. Safety Plan Checklist Scores

Measure	CONTRACTOR					
	A	B	D	F	H	I
Yes – Response Count	64	57	9	25	105	111
Percentage of Total Checklist	52%	46%	7%	20%	85%	90%

It is notable from a management perspective that none of the written safety plans included the following items included in the checklist:

- Provisions for an annual review and update with any new regulations.
- Plans for "safety climate" surveys among employees.
- Provisions for feedback from employees.
- Accountability of supervisor's through safety performance appraisals.
- Provisions for OSHA 10 hour or 40 hour certifications

7 COMPARISONS

After the safety plans were summarized, the data from the safety plans was compared with the survey results.

7.1 *Petersen scores vs. safety plan score*

Figure 2 shows the plot for the six contractors participating in this work. This plot was prepared to evaluate whether a contractor's level of agreement with Petersen's Principles would be reflected in their detailed safety plans.
Hypothetically, where management would give strong support to Petersen's Principles, their safety plans would be more detailed. Regression data in provided in Table 4 suggests this relationship is weak given only the 6 data points.

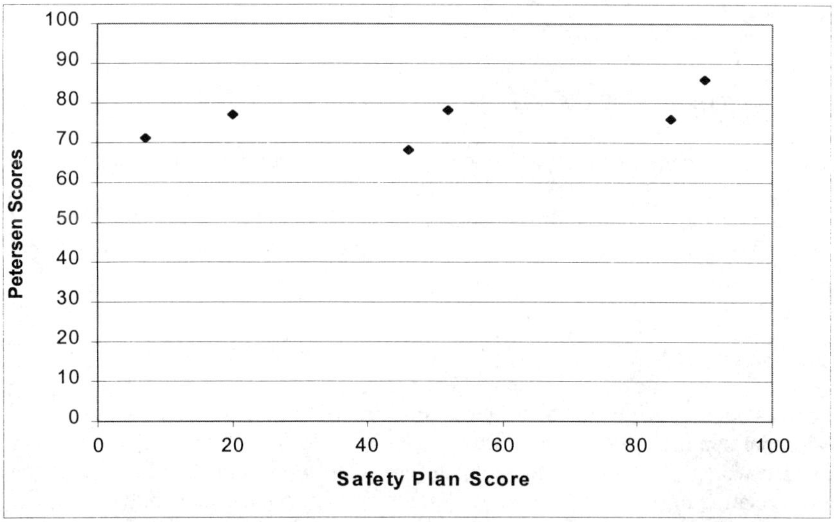

Figure 1. Safety Plan Scores Compared to Petersen Principle Scores

Table 3. Regression Statistics for Figure 1

Regression Statistics					
Multiple R		0.5961142			
R Square		0.35535			
Adj. R Square		0.194190			
Standard Error		5.5915488			
Observations		6			
ANOVA	df	SS	MS	F	Significance F
Regression	1	68.938326	68.938346	2.2049385	0.211743879
Residual	4	125.06167	31.265418		
Total	5	194			

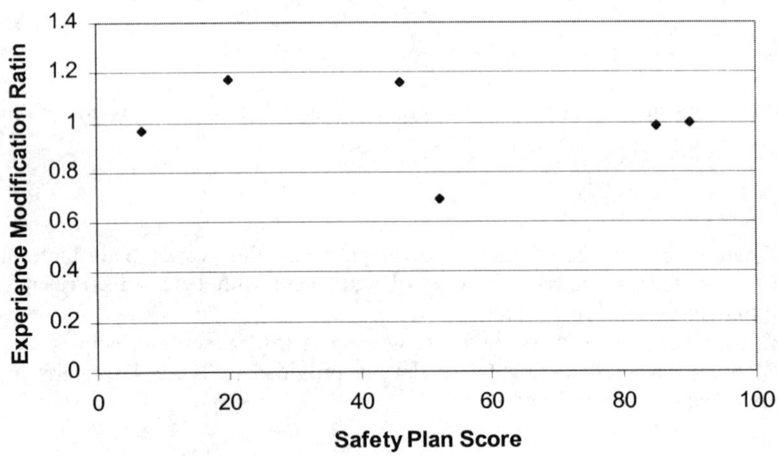

Figure 2. EMR Compared to Safety Plan Scores

7.2 *Petersen scores vs. EMR*

Assuming that an EMR adequately reflects a company's safety practice performance, the anticipation was that higher scoring safety plans would be have lower EMR values. Analysis shows no significant relationships. The only feature of a regression analysis that held promise was the slope coefficient was negative, which is consistent with the concept that EMR improves with increasing Safety Plan Scores.

7. 3 *Ratio comparison*

One last analysis was considered for the overall relationships within the data. The assumptions needed for the comparison are: 1) An EMR reflects a company's safety effort; 2) the number of hours a company works is an important factor to consider along with their EMR; 3) the Petersen Score is a relative measure of the company safety management philosophy; and 4) the Safety Plan Score reflects the implementation of safety at the project site. Since these factors have different scales, a method to compare the data was developed on the basis that the "best practice" or highest score would be chosen. The other scores would be measured relative to this value. Three ratios result:

Table 4. Summary Comparison Data

Contractor	Work Hours (1000's)	Work Hour Ratio	Safety Plan Score	Safety Plan Ratio	Petersen Principle Score	Petersen Principle Ratio	EMR
D	13.5	0.019	7	0.078	71	0.826	0.968
F	35.0	0.050	20	0.222	77	0.895	1.169
B	39.0	0.056	46	0.511	68	0.791	1.160
A	247.0	0.353	52	0.578	78	0.907	0.690
H	434.7	0.621	85	0.944	76	0.884	0.958
I	[700.0]	1.000	[90]	1.00	[86]	1.00	1.00

Note: Values in [ ] are the used as the baseline for factor calculations

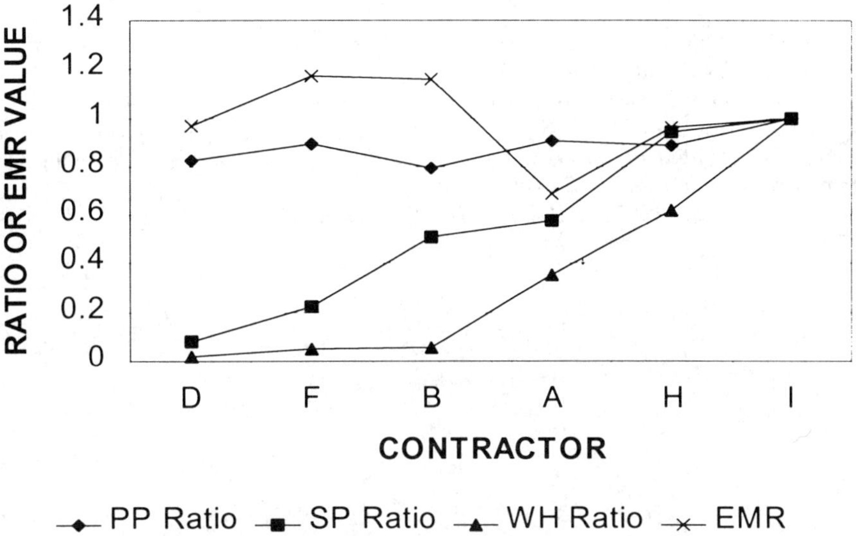

Figure 3. Overall Comparison Using Factor Ratios

Safety Plan Ratio (SP Ratio) = Contractor's SP Score / Maximum SP Score
Petersen's Principles Score Ratio (PP Ratio) = Company PP Score / Maximum PP Score
Work Hour Ratio (WH Ratio) = Contractor WH/Maximum WH
The EMR values were not converted to Ratios. The data is provided in Table 5. Figure 4 shows a plot of the three calculated ratios to the EMR.
The most obvious relationship can be observed with the 'SP Ratio' and the 'WH ratio.' Within the six firms shown, as their work hours increase, their safety plan coverage tends to increase. Managing a larger workforce requires better standard operating procedures for management systems. In this case the safety plans show the relationship. The Petersen Principle (PP) Ratio results in a very flat relationship. It could be concluded that management at all six firms are equally dedicated to safety or that the measurement method does not capture the actual differences in management. The EMR line shows no clear relationship to the factor ratios, other than smaller firms (based on work hours) have higher EMR values. This further supports Hinze, et al.'s (1995) comment that "EMR is clearly not an appropriate sole measure of safety performance for all companies."

8 SUMMARY

The intent of this research was to investigate two parts of a three part Self-Audit Process for Contractor Safety Management that could be implemented by electrical contractors. In developing the system, the safety philosophies, plans, and policies of a diverse group of electrical contractors was studied. The findings from the analysis portion were tentative at best.

As a pilot study on safety management measurements, the approach and comparison technique the project were considered to be successful. Recommendations and findings were statistically limited due to the constraints imposed by the data. Further investigation is needed using a larger pool of data to provide proper statistical evaluations.

9 REFERENCES

Hinze, Jimmie W. 1997. *Construction Safety*, Prentice Hall, Columbus, OH.

Hinze, J., D.C. Bren, and N. Piepho, 1995. "Experience Modification Rating As Measure Of Safety Performance", *Journal of Construction Engineering and Management,* 121(4), December, 455-458.

Krause, T.R., 1994. "Safety and Quality: Two Sides of the Same Coin," *Quality Progress*, American Society for Quality, October 1994, 51-55.

Petersen, D. 1996. *Safety By Objectives, 2nd Edition,* Van Nostrand Reinhold, New York, NY.

Smith, G.R. and T.M. Arnold, 1997. "Safety Performance Measurements for Masonry Construction," CIB *Publication 209, Safety and Health on Construction Sites*, Working Commission W99, October 1997, 270-281.

U.S. Department of Labor, 1992. *All About OSHA*, Publication 2056 OSHA

Implementation of Safety and Health on Construction Sites, Singh, Hinze & Coble (eds)
© 1999 Balkema, Rotterdam, ISBN 90 5809 036 1

Perceptions of occupational health and safety among electricians

Christine K. Miller & John G. Everett
Department of Civil and Environmental Engineering, University of Michigan, Ann Arbor, Mich., USA

ABSTRACT: Most construction workers are exposed to situations where they feel that their personal safety is compromised. A survey on occupational health and safety was distributed to a group of electricians to try to capture their concerns and perceptions about safety on the job. A classification scheme was developed from their responses. The data was used to evaluate possible changes that can be made on job sites as suggested by the workers surveyed. The responses were also compared to reported injuries and fatalities, and citations by the Occupational Safety and Health Administration (OSHA).

1 INTRODUCTION

Construction is one of the most hazardous industries. Construction trade workers account for 5% of the civilian workforce but received a disproportionate 9% of all disabling injuries and 18% of all occupational fatalities (*Accident Facts* 1997). Most construction workers are exposed to situations where they feel that their personal safety is compromised. What events or circumstances embody these feelings of unsafeness? Is it factors outside the work environment, the work environment itself, the type of work being performed or equipment used? What improvements can be implemented within the work environment to improve a worker's feeling of safe or not safe? What defines a safe work environment from a construction workers' perspective?

Traditionally, most of the statistics reported from the construction industry focus on fatality rates and injury rates. Safety standards regulated by the Occupational Health and Safety Administration (OSHA) attempt to reduce fatality and injury rates in the workplace. Every year more safety regulations are imposed and more safety equipment is required on the job. While this is very important, a critical voice in all this advancement seems to be going unheard. This is the voice of construction workers.

The objective of this paper is to present the perspective and ideas that typical construction trade workers have on the many issues surrounding safety on the job. The paper will first discuss the questionnaire that was administered to a group of electricians. The changes proposed by the workers are appraised and evaluated. The responses are then compared to OSHA citations and penalties, and injuries reported by the Bureau of Labor Statistics. While this survey consisted entirely of electricians, the results indicate that it would be beneficial to survey other trades in the construction industry.

2 WORKER SAFETY QUESTIONNAIRE

In order to capture the thoughts and experiences of workers in the construction industry, a questionnaire was developed and administered to a selection of construction trade workers. The questions in the survey asked workers to identify two of the normal work activities that they perform that present the greatest possibility of injury. Workers were asked to explain why it is possible to be injured while performing these activities and to rate the degree of seriousness on a scale of 1 to 5 with a 5 being "very serious" and a 1 being "not serious." Workers were asked to describe any injuries that they had knowledge of occurring while performing the tasks they identified. The survey also asked workers to suggest practical ways to improve the safety of each activity with respect to the construction method, equipment and tools, or training.

The respondents were asked for personal information such as age, years of construction experience, and membership in a trade union. If the worker was a member of a trade union then he or she could indicate their status as either an apprentice or a journeyman. Lastly, the survey asked respondents for any additional suggestions or comments that they might have with regards to construction safety.

3 TRADE WORKER SAMPLE

This paper describes the results of surveys administered to a group of electricians. The *Occupational Outlook Handbook* published by the U. S. Department of Labor describes the working conditions for electricians. It states that electricians "may stand for long periods and frequently work on ladders and scaffolds often working in awkward or cramped positions." In addition, it lists that possible risks of injury could result from electrical shock, falls, and cuts. It also claims that "more electricians are trained through apprenticeship (schools) than workers in other construction trades" (*Occupational* 1997). There were 736,000 electricians employed in the United States in 1995.

4 DISCUSSION OF SURVEY RESPONSES

The International Brotherhood of Electrical Workers (IBEW) Apprenticeship Training School in Ann Arbor, Michigan agreed to distribute the questionnaire. The sample of surveys returned consisted of 54 apprentices and 26 journeyman. The average age of the respondent was 30 years and the average years of construction experience were 7. All 80 electricians were members of the electricians' trade union. The surveys were compiled into a database and a classification scheme was developed to categorize their responses. Although the sample obtained represents only a small portion of this construction trade, the responses are useful in comparing actual OSHA citations with the workers' perception of what is unsafe on the job site.

4.1 Activities and injuries

The first question on the survey asked workers to list the two normal work activities they perform that present the greatest possibility that they might be injured. They were asked to identify specifically the reason why they might be injured from these activities.

Table 1. Occupational injuries among electricians

Cause of Injury [1]	Percentage of Injuries [2]
Fall	20%
Struck by	18%
Overexertion	13%
Electric shock	11%
Other	38%

Source: *Construction Accidents* (1992).

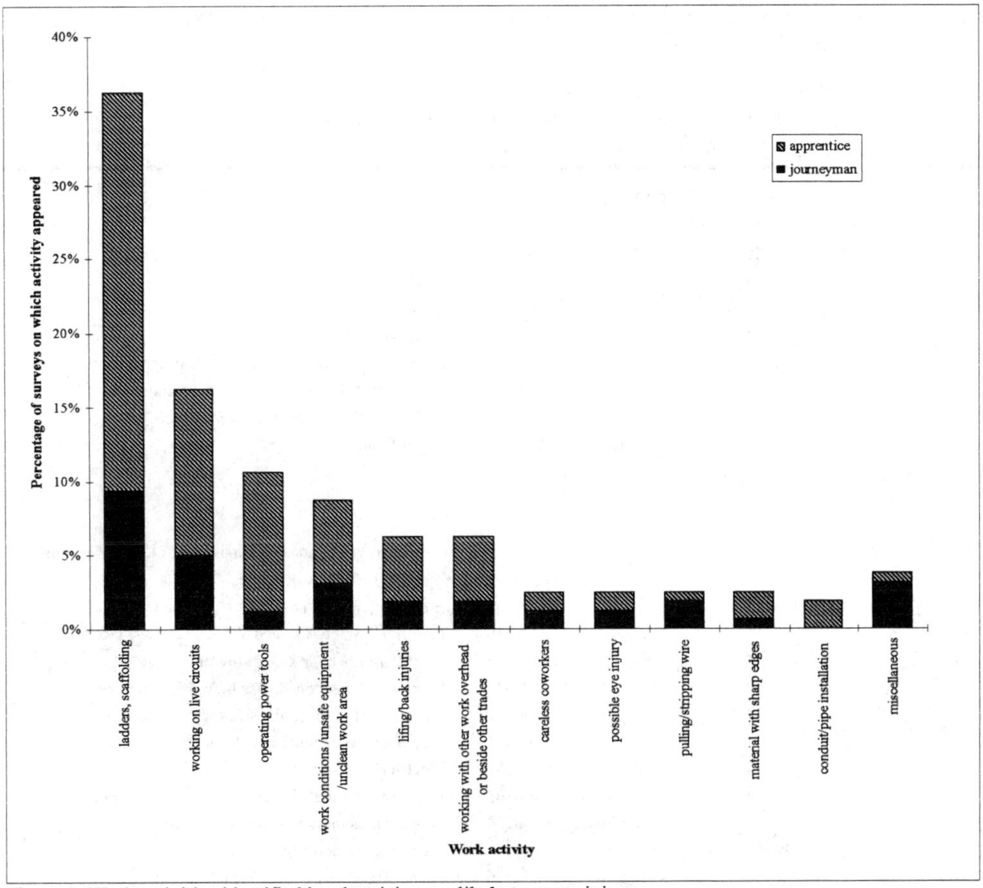

Figure 1. Work activities identified by electricians as likely to cause injury.

Respondents listed a total of 160 activities as presenting the greatest possibility of injury. Figure 1 lists the responses by category as the percentage of surveys on which each activity appeared. The responses given by the sample of journeymen corresponded closely with the responses from the apprentices. The largest number of responses (36%) given by the electricians rated activities requiring work on ladders, scaffolding and lifts as their riskiest activity. Injury from falls was the reason these activities were identified.

Lost-time injuries for electricians are shown in Table 1. The leading cause of injury for electricians was falls from either an elevation or the same level (20%) (*Construction Accidents,* 1992). OSHA violations that involved fall protection, ladders, scaffolding, and stairs comprised 14% of all federal OSHA citations and 13% of all penalties for SIC (Standard Industrial Classification) 1731, Electrical Work, for the period from October 1995 through September 1996. The number of citations and penalty amount in dollars is shown in Table 2.

The electricians gave a number of reasons why they might be injured from working on ladders, scaffolding, or lifts. They included not only the falls themselves, but also leaning or reaching too far off ladders and scaffolding, loss of balance causing a fall, poor footing to place ladders on, not enough room to set ladders up properly, and using old, unstable ladders. The workers themselves could prevent some of the risks they identified by improving their work ethic; moving ladders when necessary and placing ladders on sturdy footing. Management could reduce other risks by providing sturdy, well built ladders and ensuring that there is adequate room for the electricians to work. The electricians were also concerned about working on ladders with other trades working below, using power tools on

Table 2. Standards Cited by Federal OSHA for SIC 1731 Electrical Work, October 1995-September 1996. (Penalties shown reflect current rather than initial amounts.)

	Number of Citations [1]	Penalty ($) [2]	OSHA Standard [3]	Description [4]
Total	1051	$581,491		
	50	42,734	19260501	Fall Protection Scope/Application/Definitions
	28	19,813	19260451	Scaffolding
	25	5,853	19261052	Stairways
	24	5,901	19261053	Ladders
	6	350	19261060	Stairways & Ladders, Training Requirements
	5	1,200	19261051	Stairways & Ladders, General Requirements
	3	645	19260503	Fall Protection Training Requirements
	1	1,375	19100027	Fixed Ladders
Subtotal	142	$77,870		
Percentage	14%	13%		
	235	79,509	19260505	Elec. Wiring Methods, Components & Equip, Gen'l use
	121	68,301	19260403	Electrical, General Requirements
	120	55,779	19260404	Electrical, Wiring Design & Protection
	50	54,965	19260416	Electrical, Safety-Related Work Practices, Gen Rqts
	25	17,010	19260417	Electrical, Lockout & Tagging of Circuits
	8	13,100	19100333	Electrical, Selection & Use of Work Practices
	5	2,500	19100147	The Control of Hazardous Energy, Lockout/Tagout
	4	7,000	19100335	Electrical, Safeguards For Personnel Protection
	4	10,350	19100332	Electrical, Training
	4	1,400	19100305	Electrical, Wiring Methods, Components & Equipment
	3	4,150	19260950	Power Transmission, General Requirements
	2	1,050	19260400	Electrical, General Introduction
	1	100	19260956	Power Transmission, Underground Lines
	1	3,000	19260955	Power Transmission, Overhead Lines
	1	3,000	19260952	Power Transmission, Mechanical Equipment
	1	4,200	19260407	Electrical, Hazardous (Classified) Locations
	1	1,050	19100304	Electrical, Wiring Design & Protection
Subtotal	586	$326,463		
Percentage	56%	56%		

top of ladders, accidentally dropping objects on other workers, and carrying heavy objects up and down ladders.

The second most commonly identified (16%) group of work activities that might lead to injury involved work with or on live electrical circuits. Not surprisingly, electrical shock was identifies as the reason for injury. The Bureau of Labor Statistics reported that electrical shock was the fourth leading cause of injuries to electricians comprising 11% of all injuries. OSHA violations that involved electrical safety comprised 56% of all federal OSHA citations and 56% of all penalties for SIC 1731, Electrical Work, for the period from October 1995 through September 1996. The number of citations and amount of penalty for the violations are shown in Table 2.

Other activities identified as risky by the respondents included operating power tools, using unsafe equipment, working in an unclean work area, lifting heavy material, and working beside or underneath workers from a different trade. The sentiments of the workers closely follow the additional injury causes reported by the Bureau of Labor Statistics. These include being struck by an object (18%), and overexertion (13%) (*Construction Accidents* 1992).

Common reasons for injury given were not using proper safety apparel or equipment, tripping over material left laying around the job site, drowsiness on the job, not being attentive on the job, physical fatigue, accidents with power tools, and eye injuries. The Bureau of Labor Statistics reported eye injuries as the most frequent body part affected by an injury for electricians (*Construction Accidents*, 1992). Workers also cited external factors such as working beside other trades, careless or untrained co-workers, inexperience, and rushing on the job. OSHA citations target some of these areas. For example OSHA issued 25 citations for housekeeping on the job site.

Again the solution for these risks requires a combination of effort from both management and the workers. Management should make the use of safety apparel, equipment, and protective eyewear mandatory at all times. Management can also require workers to keep job sites clean. On the other hand, the workers are responsible for drowsiness, physical fatigue, and carelessness on the job.

4.2 Knowledge of Injuries

Workers were asked to list and describe anyone they knew who was injured performing one of the activities they gave as presenting the greatest possibility of injury. For the 160 activities given by the electricians, 152 responses were given about injury knowledge. 37% of workers said that they had no knowledge of an injury occurring while performing the activity. 16% simply responded "yes" that they knew someone who was injured while performing one of the activities. 44% of the respondents actually described an injury that occurred while performing one of these normal work activities. Table 3 shows the number of responses for each type of injury.

Some of the workers who indicated that an injury occurred from a ladder fall did not indicate the severity of the injury so they were classified as "fall from ladder." The most common injuries involved broken bones, cuts, burns, abrasions, and back injuries. Falls from ladders or scaffolding, lifting heavy material, material with sharp edges, and pulling wire were the causes of the majority of these injuries.

4.3 Construction Method

A number of practical ways to improve safety were suggested by the electricians. Suggestions were made with respect to the construction method, equipment and tools, and training. The improvements suggested with respect to the construction method included clean work areas, always checking the floor for levelness and grates before setting up a ladder, and tying off when working on ladders. It was suggested that workers would be safer if they were alert on the job site, worked together, and observed safe work habits. Table 4 lists the electricians' responses that appeared more than once.

Other responses that appeared only once include designing fewer bends in conduit, slowing down other equipment being driven on the job site, and improving the design of safety goggles. Other improvements for work on ladders suggested by workers were minimizing the workers' time on ladders by preassembling items, and providing a tool rest on ladders.

4.4 Training

The workers gave a number of suggestions for training improvements. The majority of workers indicated that there was a need for increased training in the correct use of ladders, scaffolding, and lifts, and also the safety features of these devices. OSHA cited fall protection training, and stairway and ladder training 9 times, but the penalties only amounted to $995.00. Proper use of equipment and power tools was also given as an area where training should be improved.

The Bureau of Labor Statistics reported that overexertion comprised 13% of all injuries to electricians and workers indicated that training should be given on the proper techniques for lifting and bending. (*Construction Lost-Time* 1992) Workers also wanted to see more training on general safety and legitimate safety meetings once a week where they could

Table 3. Type of Injury Reported by Survey

Type of injury [1]	Percentage of responses [2]
Fall from ladder	8%
Fractures	8%
Cuts, burns, abrasions	7%
Mild electric shock	5%
Back injury	5%
Eye injury	4%
Twisted ankle/tripping	3%
Injury while using ladder	2%

Table 4. Electricians' suggestions to improve safety

Count [1]	Improvement [2]
8	clean work areas
5	check floor where you are going to set up ladder
5	check ladder before using
5	do not work on live circuits
4	safe work habits
4	be alert on the job site
3	use tie off for ladders
3	use larger crews
2	work together as a team
2	use proper lifting techniques
2	use ladders properly
2	safety meetings
2	follow OSHA-MIOSHA regulations
2	wear hard hats
2	improve ladder design
2	make sure equipment is in proper working order
2	manufacture items without sharp edges
2	manufacture plastic cover for live parts of panel
2	protective eyewear

discuss safety issues. OSHA accurately reflected these sentiments with 32 OSHA citations for Construction Safety Training and Education amounting to almost $40,000 in penalties.

4.5 Equipment and Tools

Most of the improvements suggested for equipment and tools dealt with ladders, lifts, and scaffolding. Workers said these devices should always be in good condition and checked prior to use. Others suggested that lifts or scaffolding be used in place of ladders whenever possible. Workers said ladders could be improved by being taller, sturdier, and lighter in weight or made out of fiberglass. One worker suggested that wood ladders should be used and that threaded nuts should be used to attach ladder steps to the ladder instead of rivets.

Insulated tools, newer tools and properly maintained tools were improvements listed by workers. Workers also thought material handling carts would helping with lifting activities. Improvements in worker safety apparel were also listed. The responses included using safety glasses, insulated gloves, back support braces, and tool pouches to free the worker's hands.

5 CONCLUSION

The result of this safety survey indicates that the perceptions that electrical workers have about safety on the job correspond closely with the actual injuries that are occurring on the job site. Workers are most concerned with falls and the safe use of ladders, scaffolding, and lifts. Statistics show that 20% of injuries to electricians are caused by falls from either an elevation or the same level. It seems reasonable that 14% of OSHA citations involved ladders, fall protection, and scaffolding.

Electricians are next most concerned with electrical shock from working on or with live electrical circuits. This is the fourth most common injury (11% of all injuries) but represents 56% of all OSHA citations and penalties. Clearly OSHA has emphasized this hazard in its focused inspections.

The other actual causes of injuries, struck by (18% of all injuries) and overexertion (13%) also correspond with workers' concerns but receive relatively little attention from OSHA. Proper use of personal protective equipment such as head, eye, and face protection may mitigate some of the struck by injuries. For several years, OSHA has been attempting to implement an ergonomics standard aimed at the problem of overexertion injuries, but no such standard has yet been legislated.

A number of the improvements suggested by the workers could be easily implemented, but it would involve cooperation between the workers and management. Some of the suggestions are simple, such as keeping the work site clean, and some are more difficult and costly such as providing new tools. Overall, the responses indicated that the workers take safety very seriously and their concerns are mirrored in the actual injuries occurring on the job site.

REFERENCES

Accident Facts for 1997. (1997). Itasca, Illinois: National Safety Council.

Construction Accidents: The Workers' Compensation DataBase 1985-1989. (1992). Washington, D.C.: Dept. of Labor, Occup. Safety and Health Admin

Construction Lost-Time Injuries: The U.S. Army Corps of Engineers Data Base 1984-1988, (1992). Washington D.C.: Dept. of Labor, Bureau of Labor Statistics.

Fatal Workplace Injuries in 1993: A Collection of Data and Analysis. (1995). Washington D.C.: Dept. of Labor, Bureau of Labor Statistics.

Fatal Workplace Injuries in 1994: A Collection of Data and Analysis. (1996). Washington D.C.: Dept. of Labor, Bureau of Labor Statistics.

Occupational Outlook Handbook, (1997). Washington D.C.: Dept. of Labor, Bureau of Labor Statistics.

Standards Cited for SIC 1731 (1997). Washington D.C.: Dept. of Labor, Occupational Safety and Health Admin.

Implementation of Safety and Health on Construction Sites, Singh, Hinze & Coble (eds)
© 1999 Balkema, Rotterdam, ISBN 90 5809 036 1

Safety in building management and maintenance: The Italian *Fascicolo dell'Opera* (Building Maintenance Safety File)

A. Gottfried
Building Engineering and Territorial Systems Department, Polytechnic of Milan, Italy
M.C. Dejaco
Building Engineering and Territorial Systems Department, Polytechnic of Milan & Building Engineering and Territorial Department, University of Pavia, Italy

ABSTRACT: The entry into effect of the European Directives in the area of design and construction of buildings has led to the definition of new tools, in particular in the sphere of safety and maintenance of building structures. In the complex operating context of Italy, which is not yet ready to adopt in full the new requirements, the Italian legal provisions (which remain partially incomplete) have not contributed to providing clarity. In particular, as far as the Fascicolo dell'Opera (Building Maintenance Safety File) is concerned, a critical analysis and the proposal of a differentiated document of transition in the direction of unified operating procedures that depend on a more consolidated culture of design are called for.

1 INTRODUCTION

The legislative "revolution" currently in progress in Italy in the building industry sector is bringing about important changes from the points of view of organization and assignment of management responsibilities linked to the different phases of the building process.

There would at long last seem to be some initial signs of attention paid to the control of safety in the phases of design and management – and not only in the phase of execution – of building structures, as well as in the "univocal" attribution of the corresponding responsibilities.

In particular, the attention directed to the health and safety of workers, understood as subjects directly involved in the different phases of the building process and use of the structure (those placing the order for the building, those carrying out the work, those using the building), is performing a function of augmenting awareness in regard to problems that up to now have been altogether disregarded or, at best, only partially taken into consideration.

In this connection, the need clearly emerges for a sound knowledge of the building structures as a whole (both from the architectural point of view and from the point of view of the design, construction and installation of plants and systems). The problems linked to the availability of documentation providing information on the building may arise only for pre-existing buildings, in so far as the present legal provisions contemplate, for newly built structures, the filling-in of a set of documents from which may be drawn the information necessary for safe management of buildings.

In particular, in this regard, reference may be made to the following legislative provisions:

1. The so-called "Merloni law" (Act No. 109/94 and subsequent integrations, such as Act No. 216/95), which introduces the Maintenance Plan document;

2. Legislative Decree No. 494/96, which introduces the document Fascicolo dell'Opera (Building Maintenance Safety File), to be drawn up in the design phase and to be updated in the phase of execution of the work.

Whereas the latter legislative provision defines with sufficient clarity the structure and contents, of an information nature, of the Fascicolo dell'Opera, the Merloni Law fails to provide a definition of the actual contents of the Maintenance Plan, postponing any explanation to a subsequent explanatory regulation. Such a situation has led to very few operators taking the trouble to organize and provide documents suitable for the required purpose, with the consequent creation of a serious gap in information both as regards the phases of management of the works and as regards the structuring and understanding of the Fascicolo dell'Opera, which is a tool aimed at ensuring safety in the context of maintenance of building structures.

2 THE FASCICOLO DELL'OPERA ACCORDING TO THE ITALIAN LEGISLATION

The Legislative Decree No. 494/96, which is an implementation of the EEC Directive 92/57/CEE, has introduced different documents to be created in the design phase for the purposes of safety of building-site operators, embracing both the phase of construction and that of management of the building.

During the phase of design of works, the drafting of appropriate plans of safety linked to the context of execution of the works themselves is called for. For the subsequent phase of management of the buildings, it is required to draw up the so-called Fascicolo dell'Opera, which is a record that must be kept up-to-date in the course of service life of the building.

The legal provisions furnish a proposal of structuring of this building maintenance safety file according to the following modalities (see Fig. 1):

1. Part A - Routine and extraordinary maintenance of the structure, in turn subdivided into two parts, the first of which regarding works of overhaul (italian definition usually refers to car and mechanical maintenance), the second works of sanatory (italian definition means: legitimation of an irregularity by the authorities) and repair;
2. Part B - Equipping of the structure; also in this case a double cataloguing according to works of overhaul and works of sanatory and repair is envisaged.

The present division, though perhaps desirable from a formal standpoint, unfortunately has led to certain difficulties of interpretation linked to the headings of the different parts of which the file itself is made up. In fact, in a procedural reality where modalities of management and maintenance of buildings have always received little or no consideration, the introduction of an operating tool that involves differentiated definitions of the types of maintenance has led to confusion, and consequently to an attitude of refusal in regard to the proposed instrument.

2.1 *Fascicolo dell'Opera – Part A*

In Part A of the building maintenance safety file (see Fig. 1), structured as a form consisting of a number of columns, two sections may be distinguished for the heading of the first column, and consequently for the contents thereof: for the section headed "works of overhaul" these works are indicated as being "for the compartment", whereas for the section headed "sanatory and repair", the wording "compartments of the structure with indication of the kinds of work involved" is introduced.

The main difference, as emerges from the examples, is that, whereas for the "overhaul" document only the break-down of the building structure into its structural-architectural components (compartments) is provided, in the section headed "sanatory and repair" the operations of management and maintenance (kinds of work, i.e., sets of trades involved in carrying out given management and maintenance operations) to which the aforesaid compartments may be subject are added.

The subsequent columns are structured as follows:

1. "Indispensable" – In this column it must be stated whether the intervention on the component of the building, be this of a routine or of an extraordinary nature, is indispensable or not;
2. "Frequency" – The indication is to be provided of the time interval between two consecutive interventions for each of the items into which the building under consideration has been broken down;
3. "Firm appointed" – The firm that is to carry out the maintenance operations must be indicated;

PART A - Building maintenance **SANATORY AND REPAIR**

Routine and extraordinary maintenance of the structure

	Compartments of the structure with indication of the kinds of work involved	indispensable		frequency	firm appointed	potential risks	safety equipment provided	auxiliary devices on lease	remarks
		yes	no						
	1	2	3	4	5	6	7	8	9
1.	*Systems in the property*								
a.	sewerage								
	sanatory works								
	.....								

PART B - Equipping of the structure **SANATORY AND REPAIR**

	documentation for	available		N° of design and/or record	site	remarks
		yes	no			
	1	2	3	4	5	6
1.	*Equipments and systems in the property (plans)*					
a.	gas system					
b.	water supply					
c.	sewerage system					

Figure 1. The Building Maintenance Safety File according to italian Legislative Decree No. 494/1996

4. "Potential risks" – Here the risks involved in the operations of inspection and maintenance set down for the different components into which the building is broken down (see Column 1 of the file) must be described;

5. "Safety equipment provided" – In this field, the fixed devices and equipment that are necessary during the carrying-out of inspection and maintenance operations are to be indicated;

6. "Auxiliary devices on lease" – These are the mobile devices and equipment that are necessary for carrying out the inspection and maintenance operations in conditions of safety;

7. "Remarks" – This field must contain the set of information that the various operators of the building process (designers, installers, maintenance staff) deem advisable to point out for the purposes of proper conduct of operations of inspections and maintenance to be carried out on the individual components of the building.

The schematization of Part A of the file presented above is defined as one of the possible structurings. Hence, the indications provided by the Legislative Decree may be interpreted as guidelines of good conduct for the purposes of guaranteeing the drawing-up of a document which is easy to interpret and provides a common basis of interpretation for operators working in the building sector.

2.2 *Fascicolo dell'Opera – Part B*

Part B of the file (see Fig. 1) is also subdivided into a first section regarding "works of overhaul" and a second section regarding "works of sanatory and repair", with a similar subdivision in the heading of the first field whereby to the first subdivision into structural-architectural components (compartments) is added the specification of the kinds of work.

In the first field of this section of the file, there reappear then the contents of the first column of Part A of the file, the subsequent four columns proposing, instead, the following:

1. "Available" – The purpose here is to provide indications on the presence and availability of information of a design and/or descriptive type, rather than technical and management reports regarding the break-down of the building into compartments according to the first column of Part B of the file;

2. "No. of design and/or record" – Here the reference number of the design and/or of the informative document available is to be given;

3. "Site" – This identifies the location of the documents described in the two previous

columns;

4. "Remarks" – If necessary, additional information and comments may be entered on what is set down and pointed out in the previous columns of Part B.

Part B of the file thus assumes a clear connotation of filing and location of the set of information and documents regarding the building in question. This section must not be limited to the indication of the design documents alone, but must be kept updated so that it may provide information relevant to the actual state of individual components or of the building as a whole.

2.3 *Analyses and comments*

If, on the one hand, the Fascicolo dell'Opera, as this is presented by the legal provisions in force, which envisage its drafting right from the design phases of the building operations, appropriately meets the need to study and plan the maintainability of the building structures from their very conception, on the other hand, it clashes with the operating realities consolidated in Italy.

The transposition of the European Directives into Italian legislation has been done somewhat slavishly without taking into due account the particular operating characteristics of the local market: in Italy, a design culture in the sense of a complex and complete set of documents, is not widespread, and likewise it is difficult to define a break-down of the building complex that may be univocally accepted by design and maintenance operators.

The break-down according to "compartments" and correlated "kinds of work" has been the source of considerable puzzlement and conflicting interpretations on the part of design operators, and this has led to personal interpretations such as to render comparison between documents presented by different subjects more than arduous.

In many cases, in the Italian reality it is possible to come across widely varying interpretations as regards modalities and schedules of intervention on components and structures (both in the architectural context and in the context of plants and systems). In fact, if on the one hand there is an availability of information provided by test laboratories and technical sheets on building products, materials and components, on the other hand, in the actual operating context the little know-how that does exist is linked to the skills of individual designers or of specific maintenance operators.

In the context of the fields destined to the definition of the risks connected to execution of the work of management and maintenance, no comments need to be made, except to say that there is a lack of clarity in the definition of the "equipment provided" and the "auxiliary devices provided", which may give rise to some problems of interpretation (in this regard, see Section 2.1, Points 5 and 6); these problems, however, may be fairly easily overcome.

Part B of the file does not present particular problems of interpretation other than those connected with the definition of the first column ("documentation for"), which, in any case, depends upon what is set down in Part A of the file.

The only consideration that may be made once again regards the operating context of Italy, whereby it is difficult to manage the collection and systemization of the sets of documentation regarding a building design, especially if this concerns a complex structure.

It must be remembered that if the Legislative Decree No. 494/96 establishes the subject responsible for filling in the file during the stages of design and execution of the work, it does not do the same for the subsequent period of useful service life of the structure. As regards this latter period, the maintenance and updating of the set of documentation and of the file as a whole must be entrusted to competent subjects in order to maintain the real knowledge of the building even following on significant interventions of maintenance and restructuring.

In short, it is correct to state that the file, as this is proposed by current legislation, is "difficult" to interpret, and consequently difficult to fill in, in particular on account of the use of definitions that are somewhat far removed from the customary language used by people operating in the building sector. Furthermore, there is encountered an extreme compacting of a set of information that may, and indeed must, be supplied at different times by the numerous operators involved in the building process.

The set of linguistic imprecisions and the "compacting" of the information in the present production context is a cause of an attitude of practically total refusal on the part of the operators in regard to this operating instrument.

3 PROPOSAL FOR A "DIFFERENTIATED" FASCICOLO DELL'OPERA

The present proposal for the Fascicolo dell'Opera does not develop an operating instrument altogether different from what has been proposed by the law, but seeks to render it more rational and consistent in its structure and language with the Italian operating reality.

In a general sense, it has not been deemed advisable to change the information fields defined in the notes to the Legislative Decree No. 494/96. Their headings, on the other hand, have been changed to make them more understandable, in the sense of being better correlated to the definitions commonly used in the sector.

At the same time, it has been decided to define three main sections for the file, dividing in a more rational way the information called for, as well as the responsibilities for filling in the file of the different operators involved in the building process:

1. Planning of interventions
2. Contextualization of the interventions
3. Available documentation

This division has been made, whilst maintaining, as has been said, practically unaltered the information fields to be filled in.

In the present context, in order to provide a univocal indication for the break-down of any building structure, it has been decided to refer to UNI standards, and in particular to UNI 8290 "Residential building - Technological system - Classification and Technology", the appendix of which defines a scheme of classification of the technological system that constitutes a building.

In the context of this "classification", which, however questionable and susceptible to improvement, can be univocally recognized by everybody, it is possible to enter the set of the elements making up a building, from the macro-component to the micro-component. In particular, the design documents to which reference may be made are the Work Breakdown Strategies (WBSs) necessary for the analysis and definition of the safety plans for the building site, as well as the specifications and descriptions of the works to be carried out.

The use of these documents makes it possible to consider and enter in the file, in an adequate and univocal way, the work items of the building structure which will (or may) undergo maintenance operations in the course of the useful service life of the building.

3.1 *Fascicolo dell'Opera: planning of interventions*

This first part of the file, entitled "Planning of interventions" has in turn been subdivided into two distinct sections (see Fig. 2):

1. "1.1 - Planning of interventions of inspection during use", referring to Section "A1 - Overhaul" of the Fascicolo dell'Opera as proposed by the Legislative Decree No. 494/96;
2. "1.2 - Planning of maintenance interventions", referring to Section "A2 - Sanatory and repair" of the Fascicolo dell'Opera as proposed by the Legislative Decree No. 494/96.

The aim of the first section (1.1) is to record the inspections and checks of a "visual" type that do not involve substantial intervention on the parts of the building (inspection during use), whilst the second section (1.2) defines and specifies the type of maintenance operations that must be periodically carried out or that may be expected to be necessary for proper maintenance of the structure (routine and extraordinary maintenance).

The only structural difference between the two sections is represented by what is contained in the first column: Section 1.1 contains exclusively the wording "technical element", which derives from the subdivision of the building according to what was mentioned previously, whilst Section 1.2, in addition to the technical element, specifies the "work item" linked to it; i.e., the maintenance operations are given for each of the technical elements of which the building is made up.

The subsequent columns of Sections 1.1 and 1.2 involve the following information:

1. "Indispensable" – This provides indications regarding the advisability or otherwise of carrying out maintenance. In certain cases, it enables a distinction to be made between routine

1.2 - Contextualization of Maintenance Interventions						1.2 (A2 - Sanatory and Repair)
	Technical element / Work items	**Indispensable**		**Frequency**	**Firm appointed**	**Remarks**
		Yes	**No**			
	1	**2**	**3**	**4**	**5**	**6**
1.	**Load-bearing structure: foundations**					
a.	Concrete structures					
a.1	Piling and foundations of building					
	• Inspection and checks on settling					
2.	**Load-bearing structure: above ground**					
a.	Vertical					
a.1	Columns					
	• Inspection and checks					

Figure 2. Examplification of Part 1.2 for the "differentiated" Building Maintenance Safety File.

programmed maintenance interventions and unscheduled maintenance interventions;

2. "Frequency" – This defines the frequency of the interventions. Clearly, in the design phase the aim will be to provide rough indications, which must be updated by the person carrying out maintenance in the course of the effective use and management of the building structure as a whole;

3. "Firm appointed" – Here the name must be given of the subject entrusted with the task of management and maintenance, whether this be an individual person or a specialized firm;

4. "Remarks" – It is possible to indicate a set of information regarding the technical element and the work items linked thereto, rather than indications for other reference documentation.

As may be noted, the contents of the present section of the file, with the exclusion of the column "Technical element/Work items" may easily be entered and filled in during the phase of management of the building structures, whereas filling in the column during the design phase may entail deficiencies.

An attempt, then, has been made to transfer to a special section of the file a set of information which may be pre-defined in the design phase, but which it is preferable and more functional to complete in the phase of assignment and execution of the works of management of the building, also in order to safeguard the operating capacity of the subjects who are responsible for management and maintenance.

The information to be given in the file must also respect the contents of the maintenance plan for the building, as these are defined by Act No. 109/94. Unfortunately, in this case the lack of the specific explanatory regulation is frequently a cause of the failure to provide the documentation on the part of operators and designers, so that the premises for a proper correlation of the various documents fail to materialize

3.2 *Fascicolo dell'Opera: contextualization of interventions*

This second part of the proposed file (see Fig. 3), entitled "Contextualization of interventions", is also divided into two sections:

1. "2.1 - Contextualization of interventions of inspection during use", referring to Section "A1 - Overhaul" of the Fascicolo dell'Opera as proposed by the Legislative Decree No. 494/96;

2. "2.2 - Contextualization of maintenance interventions", referring to Section "A2 - Sanatory and Repair" of the Fascicolo proposed by the Legislative Decree No. 494/96.

These two sections have the function of defining the problems of safety in the context of the interventions of inspection during use and of maintenance. This section of the file must be defined and filled out in the context of the design phases of the building process.

In this case, the first column of the two sections reproduces exactly the contents of the first column of the corresponding sections of "Contextualization of interventions". Comparing this text with the file according to the provisions of the Legislative Decree No. 494/96, it may be

2.2 - Contextualization of Maintenance Interventions — ***2.2 (A2 – Sanatory and Repair)***

	Technical element/Work items	Potential environmental hazards	Working equipment required	Safety devices provided	Remarks
	1	2	3	4	5
1.	**Load-bearing structure: foundations**				
a.	Concrete structures				
a.1	Piling and foundations of building				
	• Inspection and checks on differentiated settling				
2.	**Load-bearing structure: above ground**				
a.	Vertical				
a.1	Columns				
	• Inspection and checks	Falling from a height	Equipment for activities at a height		

Figure 3. Examplification of Part 2.2 for the "differentiated" Building Maintenance Safety File.

seen how the information fields have remained the same. It has been deemed advisable to divide the types of information so as to concentrate in this second section the contents regarding safety according to the following points:

1. "Potential environmental hazards", i.e., the possible conditions of risks with which the operator responsible for carrying out inspections and/or maintenance may be confronted as he tackles the task assigned to him (the risks connected to execution of the operations listed in Column 1 instead form part of the analysis connected to the safety plans of individual operators, according to the Legislative Decree No. 626/94);
2. "Working equipment required" – The equipment necessary for carrying out the operating functions and linked also to the definitions of the potential environmental risks are indicated. Consequently, also the specific individual protection devices are included;
3. "Safety devices and safety measures during use" – All the precautions contemplated in the design phase as being appropriate for favouring the execution of the operations of inspection and maintenance in compliance with worker safety standards are indicated;
4. "Remarks" – As in the previous fields, the contents of this one may be extremely variable, but in all cases aimed at furnishing indications for the purposes of safety in the phases of management and maintenance of the building structures.

It emerges clearly how the contents of this second section may be considered a complementary, but self-sufficient, part with respect to the fields entered in Sections 1.1 and 1.2 of the previous paragraph.

Whereas the first section, on account of its being strictly linked to the definition of the functions of management and maintenance, must be filled in after the stages of construction of the building, the second section must be defined during the stage of design of the work, contemplating the measures necessary for a better and safer use of the structures by the maintenance operators.

3.3 *Fascicolo dell'Opera: available documentation*

This section of the file (see Fig. 4) re-proposes the schematization envisaged by the Legislative Decree No. 494/96, associating, however, the phases of "overhaul" with those of "sanatory and repair", in that there are no substantial modifications in the documentation that is to be made available in the two cases.

In particular, the first column of the sheet gives the list of technical elements corresponding to that of Sections 1.1 e 2.1, hence omitting the subdivision into "work items". This does not mean to say that, for the individual technical element, the presence of technical documentation and/or important documentation fails to be considered and reported in the framework of the execution of the functions of inspection and maintenance (work item).

	3.0 - Available documentation					3.0 (B - Inspection, Sanatory and Repair)
	Documentation	**Available**		**Document references**	**Traceability of documentation**	**Remarks**
		Yes	**No**			
	1	**2**	**3**	**4**	**5**	**6**
1.	**Load-bearing structure: foundations**					
a.	Concrete structures					
a.1	Piling and foundations of building	YES		3026-300 Location of piles on lay-out, May 1998	Design Sector	
2.	**Load-bearing structure: above ground**					
a.	Vertical					
a.1	Columns	YES		3026-301 Foundation plan, May 1998	Design Sector	

Figure 4. Examplification of Part 3.0 for the "differentiated" Building Maintenance Safety File.

The ensuing fields contain the following indications:

1. "Document references" – Document references with indication of the document (drawing, report or technical sheet) regarding the technical element considered;
2. "Traceability of documentation", i.e., indication of place (office, study, premises, etc.) where the originals of the documents referred to in the previous column are deposited;
3. "Remarks" – Various indications regarding the documents are provided; for example, the replacement and/or non-traceability, with the reason or reasons adduced for the change made.

Also this last part of the file must be kept up-to-date.

4 CONCLUSIONS

The version of the Fascicolo dell'Opera (Building Maintenance Safety File) proposed is based on the diffusion of a more consolidated design culture which today is frequently lacking in the context of Italian building production. It is clear how the contents of the file must derive from the definitions and structuring of complex documents and instruments, from which part of the information necessary for filling in the file is to be drawn.

In the current operating context, the management and maintenance problems are not perceived as forming part of the building process, which instead is held to be concluded with the testing and consignment of the works ordered. Also the filling-in of the documents regarding the definition of the procedures for management and maintenance of the building structures depends uniquely upon the reconstruction of the known data subsequent to the phases of construction.

The proposed solution of subdivision of the file into two distinct sections, the first having a clearly "maintenance" connotation, and the second with typically "design" implications, forces the operators of design to investigate in greater depth the problems of management and maintenance, at least in the definition of those measures that are able to guarantee worker safety.

In so doing, it will perhaps be possible to increase the interest of Italian designers also in regard to maintenance problems in such a way as to develop the knowledge of instruments and documents which, if envisaged right from the phase of conception of the structures, will be able to improve the usability thereof in the course of their useful service life.

REFERENCES

Dejaco, M.C. 1997. Building Management and Maintenance Problems: the "Building Management Safety File". In A. Gottfried & M.L. Trani (eds), The Construction Planning and Execution phase Coordinator: 541-557. Rimini: Maggioli, Italy.

Dejaco, M.C. 1996. Real Estate Management and Maintenance: Methods and Tools. Ph.D. Thesis, ISET Dept. Polytechnic of Milan, Italy.

Ferruti, A. & Pais, R. 1996. Health and Safety on Building Construction. Roma: EPC.

6 Work ability on construction sites

Implementation of Safety and Health on Construction Sites, Singh, Hinze & Coble (eds)
© 1999 Balkema, Rotterdam, ISBN 90 5809 036 1

Work ability assessment of construction workers

Juhani Juntunen & Helena Rytkönen
LEL Employment Pension Fund, Helsinki, Finland

ABSTRACT: In addition to medical disability, the concept of work ability also includes several other elements such as working conditions and work tasks, as well as socioeconomic and motivational factors. Futhermore, there are many features inherent in construction work which have to be taken into consideration when individual work ability is assessed. They include physically heavy work, seasonal variation of working conditions, high unemployment rate, constantly changing work sites, short periods of employment and high risk of accidents. This report gives an account of the results of two research projects concerned with the issue of work ability assessment of construction workers.

1 INTRODUCTION

In Finland, the major part of the workers covered by The Temporary Employees' Pension Act (LEL) are employed in the construction industry. As part of the functions of the LEL Employment Pension Fund, research has been carried out over the years on factors affecting perceived health, work ability and disability in construction workers. Work ability is a continuum theoretically ranging from absolute health to death. At a certain stage of this continuum, disability pension has to be considered. Although disability pensions are highly regulated by national legislation, the health-based and work-related reasons for the disability of workers are universal. Surveys and analyses of disability pension statistics provide suitable and relatively uncostly methods for studying the association between health and work ability and the predictors of disability pension among construction workers.

2 SURVEYS

Two consecutive surveys were conducted in 1989 and 1994 in order to monitor the work ability of construction workers in Finland (Rytkönen et al, 1997, Liira et al, 1998, Liira & Leino Arjas, 1998). The questionnaires contained 56 questions or groups of questions covering health issues, employment status, working conditions and living habits. The response rate for the second questionnaire was 75%, and a total of 1189 construction workers answered the questionnaire in 1994.

A deep economic depression prevailed in the early 1990s in Finland, and at the time of the follow-up about 50% of the construction workers were unemployed, a condition which is known to affect the perceived health of workers (Bartley, 1994, Lahelma 1992, Martikainen & Valkonen, 1996). Unemployment was long-standing and 25% of the construction workers were unemployed for more than 24 months between 1989 and 1994. One-fifth of the construction workers changed occupations during the follow-up period. Their new occupations involved tasks that were physically less demanding than those of the former ones.

The perceived strain of the work tasks decreased and social support increased between the years 1989 and 1994 and satisfaction with conditions of employment declined (Table 1). Perceived health status and work ability deteriorated sligthly (Table 2). The mass unemployment witnessed in the construction industry in the 1990s was probably a major contributory factor to these developments. Previous unemployment, advanced age and life-style factors were predictors of unemployment during the follow-up.

Table 1. Work conditions of active construction workers in 1989 and 1994, mean (SD), n=342

	1989	1994	p
Ergonomic strain (4-20)	12.5 (3.3)	11.9 (3.5)	0.01
Physico-chemical hazards (6-30)	19.4 (4.3)	18.6 (4.8)	0.01
Difficult physical work conditions (4-20)	10.4 (2.9)	9.9 (3.0)	0.01
Work interest (3-14)	10.5 (2.0)	10.3 (2.2)	
Dissatisfaction with work conditions(5-25)	11.5 (2.7)	13.0 (3.5)	0.001
Social support at work (6-30)	18.0 (4.0)	18.7 (4.8)	0.05
Support and instruction offered by the superior (4-17)	11.5 (2.3)	11.8 (2.6)	

(t- test: significance of the divergence from 0)

Table 2. Health indicators (rather or very good health) of active construction workers in 1989 and 1994

	1989 %	1994 %	Number of cases
Perceived health	61.21	56.47	379
Work ability	80.74	75.35	369
Physical fitness	49.34	49.61	377

A total of 205 construction workers were granted disability pension between 1990 and 1994. Advanced age, long-term disease, deteriorated health status, decreased physical fitness and subjective work ability strongly predicted retirement on disability pension during the follow-up period (Table 3). Long-term diseases were predictive throughout the whole follow-up period, while subjective work ability declined only during the first years of follow-up.

Table 3. Baseline variables as predictors of disability pensions in construction workers (bivariate models, dichotomous variables)

	Risk ratio		Confidence interval of 95% Lower	Upper
Long-term disease	3.58	***	2.46	5.21
Perceived health	4.57	***	3.06	6.83
Musculoskeletal disorders	2.53	***	1.72	3.71
Stress symtoms	2.12	***	1.45	3.09
Physical fitness	3.17	***	2.04	4.93
Work ability index	2.91	***	1.99	4.24
Subjective work ability	3.17	***	2.11	4.77
Strain of work tasks	2.63	***	1.82	3.80
Physical strain of work	1.63	*	1.06	2.49
Mental stress at work	1.63	**	1.14	2.34
Physico-chemical hazards	1.26		0.86	1.84
Ergonomic strain	1.69	**	1.18	2.43
Work interest	1.07		0.73	1.57
Dissatisfaction with work conditions	1.05		0.74	1.50
Support and instruction offered by the superior	0.94		0.65	1.36
Social support at work	1.54		1.07	2.23
Life control	1.31		0.91	1.88
Inadequate income	1.67	*	1.08	2.57
Smoking	1.72	**	1.19	2.49
Frequent use of alcohol	1.62		0.96	2.73
Exercise	1.12		0.76	1.65

***p<0.001; **p<0.01; p<0.05

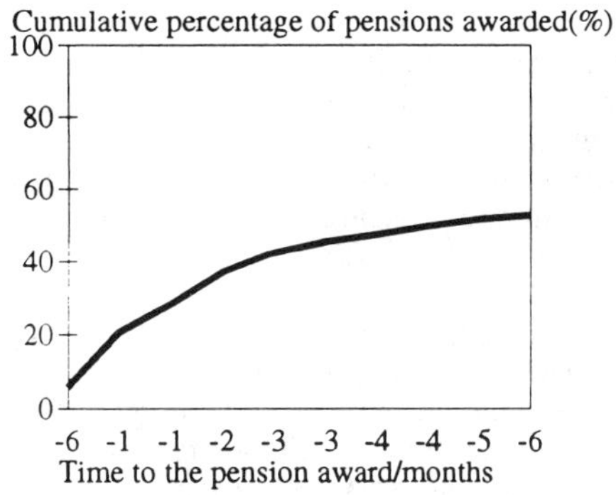

Figure 1. Disability pension awards as a percentage of the number of denials in the construction industry in the years 1980, 1981, 1985 and 1986. Delay from denial.

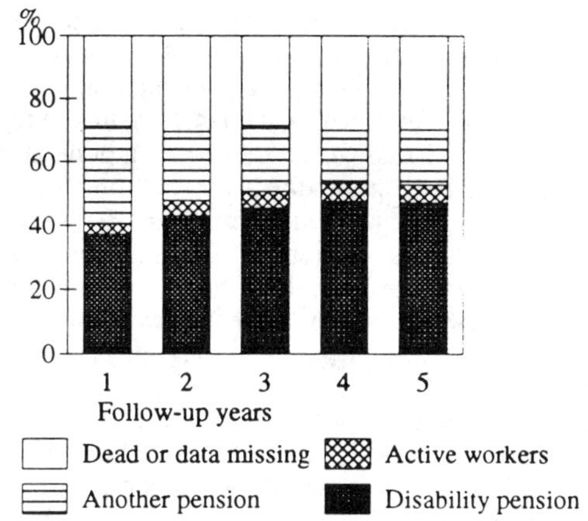

Figure 2. The situation of the applicant during the five years following the year of the denial, 1981, 1985, 1986, disability pension applications

3 DISABILITY PENSION STATISTICS

Disability pension statistics reflect general health status of the employees in the construction industry. Musculoskeletal diseases and mental disorders are the main disease categories underlying disability of workers. In the construction industry, about 80% of the disability pension applicants are awarded a pension.Denial of a disability pension application causes many problems for the occupational health care system. The assumption behind a denial is that the person is capable of continuing his/her work. About half of the applicants having been denied pension start the appeal process, while the other half seemingly accept the denial.

In order to evaluate the economic impact of pension denials to the occupational health care system it is necessary to know what actually happens to the workers after the denial. The sample of our study comprises about 2000 construction workers who were denied a disability pension in the years 1980-81, 1985-86 and 1990-91 and who did not file a pension application within the three preceding years (Juntunen et al, 1996).

An analysis of the events following pension denial showed that within three years of the denial the disability pension and individual early retirement pension awards increased rapidly in number and levelled out thereafter (Fig. 1). Within five years, about 45% of those whose applications had originally been denied received an award and approximately one-fifth were still at the disposal of the labor market (Fig. 2). At the pension institute, disability adjudication practice seems to have become established in that the incidence of pension awards subsequent to denial has remained more or less stable over the period under review.

The factors of advanced age and the diagnosis of musculoskeletal disease related to the denial significantly spoke in favor of an award. Considering the awards subsequent to denial, age dependency was apparent for the other disease categories expect for the category of mental disorders. A comparison of the individuals who were still at the disposal of the labor market with those who had retired showed that the former were younger, more often women than men, more frequently suffering from musculoskeletal diseases and less frequently from mental disorders than those who had retired. They had also accrued a lower pension.

4 SUMMARY AND CONCLUSIONS

Due to the features inherent in construction work, assessment of disability in construction workers differs to some extent from that of the general population of workers. The requirements of physical fitness for the heavy work explain why musculoskeletal disorders seem to dominate as the main cause of disability pensions, as suggested in earlier studies (e.g.

Vingård et al, 1992). An analysis of the appeal process of disability pension denial among construction workers in Finland revealed that within five years of the denial about one half of the workers had been awarded a pension while one-fifth was still at the disposal of the labor market. Due to highly national characteristics of the Finnish social security system the results can only be compared within the working population of Finland. A population -based Finnish study confirmed the pronounced associations between physical and mental workplace factors and the incidence of retirement on disability pension (Krause et al, 1997).

The general health problems of manual workers obviously apply to construction workers as well. Some occupations in the construction industry require rather advanced skills, while many of the workers are less educated. The working conditions and health requirements in different occupations within the construction industry may show considerable variation. Therefore, further epidemiologic studies are important to reveal those occupation-related factors which may affect the work ability of workers. The data accumulating from work ability studies of construction workers should be employed in developing health promotion and early rehabilitation programs within the framework of the occupational health care services in the construction industry.

REFERENCES

Bartley, M. 1994. Unemployment and ill health: understanding the relationship. *J. Epid. Community Health* 48: 333-337.

Juntunen, J., Hänninen, K. & Lundquist, B. 1996. Disability pension denied-what then?. Study of the events following pension denial in the LEL branches over the period 1980-1994, based on records. *LEL Publications 28: 1-31.*

Krause, N., Lynch, J., Kaplan, G., Cohen, R., Goldberg, D. & Salonen, J.. Predictors of disability retirement. *Scand. J. Work Environ. Health* 23: 403-413.

Lahelma, E. 1992. Unemployment and mental well-being: elaboration of the relationship. *Int.J.Health Ser.* 22: 261-274.

Liira, J. & Leino-Arjas, P. 1998. Some predictors and consequences of unemployment in construction and forest work. *Scand. J. Work Environ. Health* (in press).

Liira, J., Matikainen, E., Leino-Arjas, P., Malmivaara, A., Mutanen, P., Rytkönen, H. & Juntunen, J. 1998. Work-ability of middle-aged Finnish construction workers- a follow-up study in 1991-1995. *Int. J. of Ind. Ergonomics* (in press).

Martikainen, P.T. & Valkonen, T. 1996. Excess mortality of unemployed men and women during a period of rapidly increasing unemployment. *Lancet* 348: 909-912.

Rytkönen, H., Liira, J., Hänninen, K. & Leino-Arjas, P. 1997. Working conditions, perceived health and work ability in construction and forest workers in 1989-1994. *LEL Publications* 30: 1-54.

Vingård, E., Lars, A., Evy, F. & Christer, H. 1992. Disability pensions due to musculo-skeletal disorders among men in heavy occupations. *Scand. J. Soc. Med.* 20: 31-36.

Implementation of Safety and Health on Construction Sites, Singh, Hinze & Coble (eds)
© 1999 Balkema, Rotterdam, ISBN 90 5809 036 1

Putting environmental health and safety on the R.O.A.D.

Christopher Campion
Safety Institute of Australia, NSW Division, Sydney, N.S.W., Australia

ABSTRACT: This paper discusses a methodology for analyzing projects for risk and opportunity primarily in the en user and end user maintenance phase of the life cycle of facilities or structures. The process develops via structured think tanks involving consultants, construction personnel and the end users. The hypothesis is that the earlier you identify potential hazards the earlier you can use the "hierarchy of control" to remove reduce or eliminate them. This is beneficial to the Construction Team as hazards are designed out and solutions in e.g. safe anchorage points on structural steel. The paper outlines the method timing and records that need to be kept and has examples included.

1. INTRODUCTION

Environment, Health & Safety issues as management disciplines have been marginalised for many years.

The time to integrate them as a part of normal management practice is overdue and should lead to time, cost and quality gains for our customers and ourselves.

2. THE LEND LEASE EH&S POLICY

The Lend Lease goal is "to achieve leadership in sustainable environmental practices and the promotion of the health and safety of our stakeholders". We believe all damaging occurrences are preventable.

The core of the Lend Lease Property Group EH&S Policy is the progressive elimination of the potential for Class One damage to occur in our work. The best way to achieve this to set out to produce "working and living environments that optimize performance" remembering that "Class One damage is damage which permanently alters the future of the individual or the environment".

Class 1	Permanently alters the future
Class 2	Temporarily alters the future
Class 3	Inconveniences the future

Because of the primacy of Project Management in our business, it is this discipline which can give the greatest contribution to achieving the EH&S policy of the company.

3. WHY IS EH&S SO IMPORTANT

3.1 Creation of Wealth

Fundamentally the company is in business to create wealth. To do this our process of management seeks to identify risks and opportunities in all parts of the delivery process of our product. It seeks to maximize the positive aspects of any element/discipline and to control or reduce the negative (risk elements).

A company does not create wealth in an economic or social sense by damaging people or the environment.

The costs to company reputation, to individuals, to individuals' families and to the environment can be enormous. A few brief examples will clearly illustrate this point.

- Actual loss of profit to Lend Lease incurred by Woodstore Refurbishment collapse, in excess of $5 million.
- Six families extensively damaged in this incident as their breadwinners put permanently out of the work force.
- Clean up of Homebush Bay, Site of the 2000 Olympics.

Current estimates exceed $100M due to years of uncontrolled waste dumping. These are graphic examples, but what about the not so obvious implications.

3.2 The Law

Acts or omissions that occasion harm to people or the environment are criminal offences under EH&S law in Australia and in most jurisdictions around the world. Liability is strict, this means it is no defence to say that we did not intend to damage people or the environment. The company, as an entity itself, can be fined in excess of $1m per incident and individuals (employees) can face fines and potentially up to 25 years imprisonment for criminal manslaughter. Fines and individual prosecutions are being imposed more regularly by all courts in all jurisdictions than ever before.

3.3 Ethics

Lend Lease Property Services belief in human optimization means that acts or omissions that negate performance are not responsible social or business activities. It follows that designing and building environments that damage people e.g. slippery floors leading to broken hips in elderly customers are not ethical activities. The same applies for destroying or educing the utility of the environment.

Consigning people to the pain of economic poverty and hardship via lump sum workers compensation payments is not morally supportable and simply not fair. Lend Lease property Services recognizes this and its employees can be provided with an additional payment of 5 x annual salary if the employee is found to be totally and permanently disabled. Obviously, this is a reactive and remedial measure and it would be much better if the energy exchange that caused the damage was identified and avoided in the first place.

4. DUE DILIGENCE

Running a business using 'Due Diligence' is an important aspect of corporate life and good management. 'Due Diligence' means everyone making sure the company's operations are legal and ethical. The EH&S 'due diligence' objective must be to record the process of continuous improvement in order to monitor EH&S performance and reward the individual company/client! To do this PROOF of what is being achieved is required.

What is "Proof?" It is the body of evidence written, oral, and visual that can be presented to categorically prove something has taken place. In EH&S we need to prove three things:

- The existence of an effective EH&S management system.
- That improvement in living and working conditions has occurred, and
- That a reduction in the potential for damage to occur to people/property/environment has been achieved.

To achieve this the Lend lease Property Services Group uses a dynamic Environment, Health and Safety System based upon a continuous cycle of improvement.

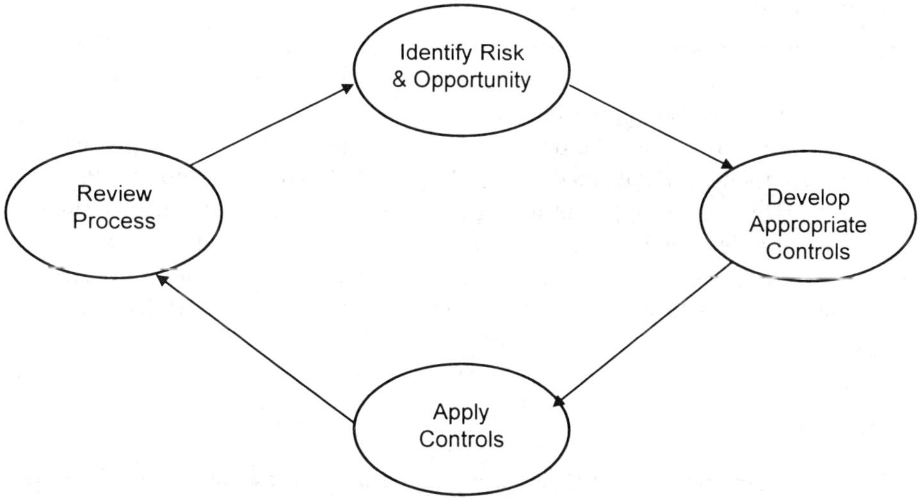

Fig. 1. The Cycle of Continuous Improvement

'Due Diligence' means the continuous management of the hazards and opportunities identified in the delivery of our product.

To do this we need to continuously do the following:

- Identify hazards and opportunities.
- Assess the risk of the hazard and develop the opportunities.
- Develop controls for the hazard.
- Assure compliance to controls.
- Review the process.

5. THE R.O.A.D. MAP or HOW TO GET SAFELY TO YOUR DESTINATION

5.1 What is the R.O.A.D?

It is a means of recording risk and opportunities and what we have or are doing with them throughout the design and construction phases of our projects.

5.2 Assessment to be undertaken

EH&S risk management is a continuous process which begins with hazard identification. Hazard identification means identifying sources of potentially damaging energy or loss in the end user and maintenance and construction of our projects.

To identify hazards we use the 'R.O.A.D' (Risk and Opportunity in Design) and the EH&S Risk/Opportunity Checklists.

5.3 When is it done?

Risk and opportunity assessments must be done again and again during the life of a project. Risk and opportunity assessments done once at the start of a project only identify "big picture" risk and opportunities.

Initially, risk assessment must be done as early as possible in the concept stage of design or alternatively on receipt of the 'Project Risk and Opportunity Profile' (PROP) prepared by the development manager. This assessment must identify major risks and opportunities for construction, end user and end user maintenance, and identify those areas of design needing immediate focus. The risk and opportunity assessment should also allocate risks, which must be dealt with when tender packages are being developed.

5.4 Who does it?

The Project Manager should co-ordinate a team of people. The team should include the Project Manager, consultants, the Facility Manager, Company EH&S Specialist (if existing) and nominated other company construction personnel and subcontractor personnel for the project.

5.5 The Consultants Role

Adequate briefing of consultants is vital. All consultants need to understand the concept of environmental optimization that Lend Lease Property Services seeks. It is necessary to include consultants in the overall risk and opportunity assessment to achieve two things:

- Understanding of the discipline and the tools to use to optimize performance, and to develop
- Enthusiasm for the objective.

5.6 How is it done?

The best way to prepare the assessment is using a think tank format. The team should consider elements of the project from the perspective of the end user, and end user maintenance The team should also look at the construction elements of the project and, using the "Risk and Opportunity in Design" (ROAD) checklist, identify risks and opportunities.

5.7 What is prepared?

The risk and opportunity assessment is recorded on the "ROAD" document, which dedicates actions to eliminate risk, and identifies opportunities to consultant or trade packages. Once an opportunity or risk has been allocated the nominated persons must report progress eliminating the risk or developing the opportunities.

A sample "R.O.A.D" checklist is attached. Essentially the checklist will:

- Identify the risk/opportunity;
- Record the proposed solution:
- Identify the consultant or trade package and (importantly);
- Record the action and status of each risk/opportunity.

5.8 When is the R.O.A.D. checklist updated?

As each package of work developed, the R.O.A.D. document is updated as risk and opportunities identified are either eliminated or highlighted as needing controls.

The R.O.A.D. is a 'living' checklist. The status of the R.O.A.D. must be reviewed at each design meeting. Specific R.O.A.D. meetings should be highlighted on the design programme.

R.O.A.D. Process

Assessment:	Building element assessment at pre-construction using R.O.A.D. hazard/ opportunities checklists;
Recording:	Else R.O.A.D. Checklist
Management:	Show R.O.A.D. meetings on design programme.
Action:	Establish action and status list.

Review:	Update and report status at each design review.
AFT/AFC Process:	Actions from R.O.A.D. issues to be considered prior to AFT/AFC.

6. ACHIEVING ENVIRONMENT, HEALTH AND SAFETY EXCELLENCE

Environment, Health and Safety excellent must be integrated into our normal business activities in order to achieve excellence in the design and delivery and acceptance of our product. EH&S gives us the opportunity to build excellence into our product for the customers benefit.

Excellence is positive and the opposite of mediocrity. Excellence means enhancing the EH&S aspects of the environment or product we supply. Mediocrity means giving little or no consideration to those inputs. Seeking excellence asks all stakeholders to take a longer view of what we do and its potential EH&S impact and requires us to believe that it is better to pay upfront to realise savings down the line.

A continuous and rigorous examination of the 'why, how, when and what' of our projects should theoretically delivery a more effective, efficient, timely, safer and lower cost product.

It follows that delivering excellent rather than mediocre product should give Lend Lease Property Services a competitive advantage in our markets.

The ability to be able to record through continuous risk assessment the maximizing of the opportunity to achieve safer, more efficient, environmentally enhancing product is a distinctive competence that will be accepted by clients that can see the benefits they have personally accrued through the process.

7. SUMMARY

This paper identified the following:

- Potential hazards can be mitigated if the hierarchy of control is used immediately after identifying the hazard using R.O.A.D. (Risk Opportunity in Design).
- Elements of R.O.A.D. include assessment, recording of hazards, management of programme, action, and review.

Implementation of Safety and Health on Construction Sites, Singh, Hinze & Coble (eds)
© 1999 Balkema, Rotterdam, ISBN 90 5809 036 1

Improvement of working conditions in the masonry sector

U. Berg
Bau-BG Frankfurt am Main, Germany

ABSTRACT: The weight of building blocks causes problems in the masonry sector. The association of the German construction BGs („Arbeitsgemeinschaft der Bau-Berufsgenossenschaften“) has initiated discussions on this subject. The main results which lead to a Code of Practice „The handling of building blocks in masonry“ are presented here. The weight of building blocks to be laid by hand has been limited, and proposals have been made for ergonomic improvements in masonry. Nevertheless, some questions still remain open and further developments are still required to improve working conditions in the masonry sector, especially in the following fields: ergonomics, mechanisation, organisation, and industrial medicine.

1 STARTING POINT

Building blocks are mass produced blocks of fired or bound material, in specific sizes, which are destined for use in wall building. Over the past few years building blocks have become increasingly heavy in the German construction industry as a result of their growing dimensions. For instance some big building blocks have a weight of more than 50 kg. The increases in weight are gradually resulting in more and more complaints of the construction workers – especially masons – that the weight of heavy building blocks used in masonry has become excessive. In particular, masons are complaining of trouble with or ailments of the joints and spine. Whether or not the complaints are connected to the weight of the building blocks used in masonry has always been an open question. Consequently, the association of the German construction BGs (Bau-BGs) – responsible for safety and health in construction work as well as industrial accident injuries insurance for the construction industry – decided to look into the problems affecting the masonry sector.

Due to the complexity of masonry, the questions connected with the working conditions – especially the stress and strain of the masons – could not be answered from the viewpoint of any single discipline. Therefore the task was undertaken in an interdisciplinary manner. The problems involved in laying of building blocks have been discussed by a group of scientists from various specialized areas and by active masons on the basis of a comprehensive list of problems and questions.

The conclusions were as follows: Masonry work by hand may constitute a health hazard if the loads are excessive, regardless of the type, size, and weight of the building blocks. This means that the above mentioned complaints by masons are justified and that working conditions in the masonry sector have to be improved. Consequently, the weight of building blocks laid by hand should be limited. The main results of the discussions and recommendations of the experts lead to a new Code of Practice for the German construction industry titled „The handling of building blocks in masonry“.

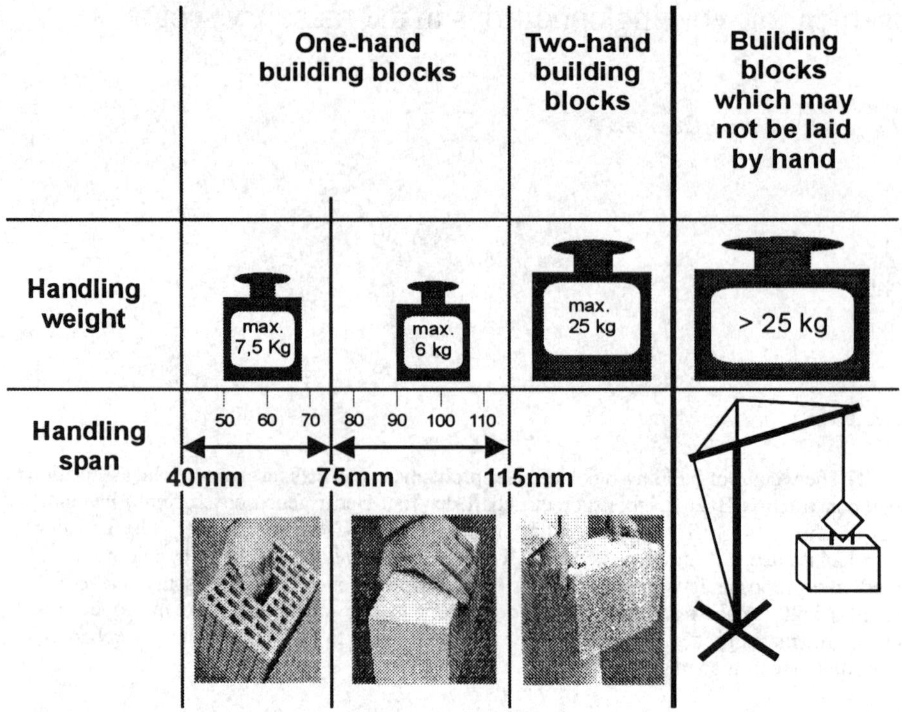

Figure 1 Limit values for the handling weight of the one-hand and two-hand building blocks

Figure 2 Prohibitive symbol

2 CODE OF PRACTICE

The lifting, carrying and positioning of building blocks necessitate postures and movements which cause a high level of physical stress. In the long term this can lead to excessive physical strain of the masons. Particular hazards to health can arise when construction workers handle large building blocks on site. These building blocks are being used to an increasing extent in the effort to achieve economical building methods in construction.

Work-related physiological investigations indicate that a limit value is reached at a block weight of approximately 25 kg when laying blocks by hand. This means that frequent lifting, carrying and positioning of blocks with higher weight factor can lead to health impairment or damage. The Code of Practice „The handling of building blocks in masonry“ based on work-related and proven scientific knowledge. It is a summary of those requirements which contribute to a reduction in the physical strain involved in the handling of building blocks.

This Code of Practice is applicable to the handling of building blocks and should contribute towards an appropriate design of building blocks, whether destined for laying by hand or by machine. It should also contribute towards the use of appropriate brick laying equipment and machinery on construction sites in place of manual block laying operations and in case of large building blocks with more than 25 kg weight.

3 HANDLING OF BUILDING BLOCKS

The recommendations concerning building blocks distinguish between one-hand blocks and two-hand blocks. One-hand blocks are building blocks which can be grasped between the thumb and fingers of one hand. Normally they can be laid in position with one hand without incurring excessive physical strain. Building blocks with a gripping span of at least 40 mm up to a maximum of 115 mm are held to be one-hand blocks.

The gripping span is subject to the requirement of limit values for the handling weight (fig. 1) which is the block weight of the building blocks as they are handled on-site, including any moisture consequent upon the production process and the weather:

- The handling weight may not amount to more than 6 kg in the case of a gripping span of more than 75 mm up to a maximum of 115 mm.
- The handling weight may not amount to more than 7.5 kg in the case of a gripping span of 40 mm to 75 mm.

Two-hand blocks are building blocks in respect of which the limit values for one hand blocks are exceeded and which have a handling weight of not more than 25 kg (fig. 1). This means that bearing in mind the working conditions and environment and the specific working methods and motions involved in masonry the maximum admissible weight of building blocks handled by masons is limited. Building blocks weighing more than 25 kg should no longer be laid by hand, only by machine. This continues to apply even where a second person is available to give assistance. Blocks with a handling weight of more than 25 kg may only be emplaced using block laying equipment or machinery.

Furthermore, the packs of building blocks in which the handling weight of the individual blocks amounts to more than 25 kg should carry a written notice and a prohibitive symbol (fig. 2). This symbol on the packaging or the attached packing note indicates that these blocks must not laid by hand.

4 ERGONOMIC DESIGN OF BUILDING BLOCKS (HANDLING AIDS)

Furthermore, building blocks should be equipped with grasping aids (holes or handles) to facilitate their handling. The ergonomic design and position of these handling aids should be sufficient to enable the building blocks to be grasped, held , moved, and laid easy and safely with the least possible effort. Therefore in the mentioned Code of Practice is described how to design building blocks to be laid by hand in an ergonomically favourable way especially with gripping holes or gripping pockets.

Gripping holes are openings in the upper surface of building blocks, in the form of thumb gripping holes or 4 finger gripping holes. Gripping pockets are slots in the vertical joint sides of two-hand blocks which can be located in the vicinity of the upper side of the block (topside), or alternatively on the underside (in the form of so-called undergrips).

The handling aids must have minimum dimensions which are required in the Code of Practice also. The edges of handling aids must not have any ridges. The edges and the corners of gripping holes should be rounded off.

5 BLOCK LAYING EQUIPMENT AND MACHINERY

Where building blocks are too heavy to be laid by hand suitable block laying equipment must be employed. The machinery for laying of building blocks must be adapted to the work process to be undertaken and the working conditions on site. This should be designed in accordance with ergonomic principles and should be as versatile as possible in terms of the ability to adapt to prevailing block laying techniques. These are for instance the usage of mini-cranes or block laying elevator handling machines.

Building blocks designed for positioning by block laying equipment or machinery must be constructed for this purpose. They must have a suitable format which enables the block laying equipment to grasp one or several blocks simultaneously, both simply and safely. This applies in particular to the arrangement of spur holes and the design of gripping tongues.

6 HEIGHT-ADJUSTABLE WORKSTATIONS

In the process of laying building blocks – especially by hand – measures should be taken to avoid the masons or bricklayer having to undertake unnecessary bending actions. Height-adjustable (if possible infinitely adjustable) and moveable workstations (scaffolding platforms) can be used for this purpose. An improvement in working conditions is achieved if a higher level is provided for the building blocks, packs of blocks and mortar, and a lower level as a standing surface for the bricklayer. In addition the packs of blocks should be supplied in a user-compatible form with blocks stacked at the point of usage, so that the blocks can be laid without additional movement of the hands.

7 SUMMARY

The improvement of the working conditions in masonry is important and necessary for the health of the construction workers especially for masons and bricklayers. The requirements of the German Code of Practice „The handling of building blocks in masonry“ is a first step to lower the risk of ailments by limiting the weight of the blocks to be laid by hand. The recommendations are able to contribute to make the laying of building blocks easier without neglecting economic objectives. Despite the results obtained many questions relating to the masonry sector remain open. Further development to improve the working conditions is still required. An exchange of information and cooperation on an international level would be useful to come up with suggestions.

7 Cost and productivity

Implementation of Safety and Health on Construction Sites, Singh, Hinze & Coble (eds)
© 1999 Balkema, Rotterdam, ISBN 90 5809 036 1

The cost of accidents in the South African construction industry

J.J.Smallwood
Department of Construction Management, University of Port Elizabeth, South Africa

ABSTRACT: Accidents result in injury to persons, loss of life, damage to property and in cases the environment. Consequently, accidents contribute to the cost of construction, directly through increased compensation and insurance, and indirectly through decreased productivity, quality non-conformances and schedule overruns.

The cost of accidents (COA) is easily understood by all stakeholders involved in the construction process and represents a tangible measure which can be related to project financial accounts and both the income statement and balance sheet of a contractor. Measuring the COA enables a cost benefit analysis of a preventative financial investment in health and safety which will indicate health and safety to be a profit centre. It also raises the awareness of the absolute impact of accidents.

In an endeavour to determine the COA and current practice relative to all other aspects of accidents a descriptive survey was conducted among general contractors who are members of the Building Industries Federation of South Africa (BIFSA).

Findings include, inter alia: accidents negatively affect productivity, cost, schedule and quality, and result in administration costs and increased documentation; adequate health and safety has financial benefits; contractors generally do not calculate the cost of accidents; the extent to which contractors record, investigate and follow up injuries increases commensurately with the severity thereof, and there is a need for publicity regarding the total cost of accidents and benefits of prioritising health and safety.

1 INTRODUCTION

Inadequate or the lack of health and safety results in considerable human suffering both to the injured and in many cases their families, particularly the next of kin of fatalities.

To this end a literature and descriptive survey was conducted to determine the:

- Level of injuries;
- Impact of accidents;
- Financial cost of accidents;
- Ratio between the direct and indirect cost of accidents:
- Cost of health and safety effort, and
- Financial benefits of health and safety effort.

Table 1 indicates the nature and extent of the injury problem in South African construction. It should be noted that occupational health and safety statistics are only available six years in arrear in South Africa.

Other salient statistics according to the Compensation Commissioner (CC), Federated Employers Mutual Assurance (FEMA) and the South African Reserve Bank (SARB) are:

- The Disabling Injury Incidence Rate (DIIR) of 2.03 in 1990 means that 2.03 workers per 100 received disabling injuries (CC, 1995).

Table 1. Injuries per working day and per R100m (A$28.09m) construction completed in South Africa for the year 1990 (South African Reserve Bank, 1991 and Compensation Commissioner, 1995).

Class of injury	Total No.	No./Class of injury/	
		Working day	R100m Construction completed
Fatalities	189	0.8	0.8
Permanent disablement	946	4.1	4.0
Temporary disablement	7 645	33.0	32.1
Medical aid cases	10 108	43.6	42.5

- The Severity Rate (SR) of 3.86 in 1990 means 3.86 days were lost per worker (CC, 1995).
- The total of 1 620 046 days lost as a result of fatal and non-fatal accidents in 1990 is the equivalent of 6 983 workdays lost for every workday (FEMA and CC, 1994).
- The Fatality Rate in 1990 was 53.51/100 000 workers (CC, 1995). This does not compare favourably with the rates of selected countries for 1992: Japan (19); United States of America (18.6); Germany (14); Australia (11); Canada (7.4); Sweden (6), and The Netherlands (3.3) (Center to Protect Workers' Rights, 1997).

2 COMPENSATION INSURANCE

In terms of the Compensation for Occupational Injuries and Diseases Act, 1993, contractors are required to insure their workmen against injury and disease incurred in the course of their employment.

Building and civil engineering contractors currently pay R2.19 and R2.63 respectively per R100 wages in compensation insurance which is included as a labour overhead i.e. 2.19% and 2.63% respectively.

Although on average, insurance only contributes R2 600 to the cost of completing R1m of construction, a contractor depending upon their Claims Ratio may receive either a loading or rebate on their assessments.

The Claims Ratio represented by the percentage claims constitute of premiums, is reviewed annually for each contractor for a triennial period (FEMA, 1986). Contractors with a Claims Ratio between 62% and 24% are entitled to merit rebates on their compensation insurance ranging from 2.5% to 50% respectively. Conversely contractors with a Claims Ratio between 76% and 103% qualify to have their assessment rates loaded ranging form 15% to 50% resepectively.

Rising medical costs, among others, are likely to contribute to a constant periodic increase in assessment rates.

3 COST OF HEALTH AND SAFETY

According to The Business Roundtable (1991) data collected from a significant sample of contractors working at various construction sites in the United States of America in 1980 indicated that the cost of administering a construction health and safety programme usually amounts to about 2.5% of direct labour costs. These costs include: salaries for health and safety and certain administrative personnel; health and safety meetings; inspection of tools and plant and equipment; site inspections; personal protective equipment (PPE); health and safety programme, & miscellaneous supplies and equipment.

Based on two South African construction projects direct labour costs constitute 10.64% of project value in which case the cost of administering a health and safety programme of 2.5% of direct labour cost is equivalent to 0.266% (0.025 x 10.64%) of project value i.e. R2 660 per R1m.

More extensive research conducted in South Africa (Smallwood, 1992) determined the actual cost of health and safety to be 0.22% of project value based on 16 projects which either had, or would have qualified for a BIFSA Star Grading status of between 3 and 5 stars.

4 COST OF ACCIDENTS

According to Hinze (1994) the costs of accidents can be categorised as being either direct or indirect.

Direct costs tend to be those associated with the treatment of the injury and any unique compensation offered to workers as a consequence of being injured.

The direct costs are covered by workmen's compensation insurance premiums and historical records can be reviewed to determine the amount of expenditure attributed to each particular injury.

The indirect costs are those which are hidden and for which no historical record is kept. However, indirect costs are identifiable, but only 'hidden' in that they are not attributed to injuries in a bookkeeping sense.

According to Levitt & Samelson (1993) indirect costs include: reduced productivity for both the returned worker(s) and the crew or workforce; clean-up costs; replacement costs; stand-by costs; cost of overtime; administrative costs; replacement worker orientation; costs resulting from delays; supervision costs; costs related to rescheduling; transportation, and wages paid while the injured is idle.

5 RELATIONSHIP BETWEEN DIRECT AND INDIRECT COST OF ACCIDENTS

Research conducted by the University of Washington (Hinze, 1992) determined the indirect costs (excluding claims and material damage costs) to be more than 1.67 times the direct costs of accidents. Thus an injury that causes direct costs of $1 000 results in indirect costs of $1 670 and consequently a total cost of $2 670. However, for minor injury cases (direct costs less than $50), the indirect costs were determined to be more than 5 times the direct costs.

Another study conducted by Hinze (Levitt & Samelson, 1993) determined the ratio to be 2.94 for non lost-time accidents and 2.53 for lost-time accidents. However, when Hinze included the cost of liability claims as a result of litigation the ratios change to 4.2 for non lost-time, and to 20.3 for lost-time accidents respectively.

A further study of 834 injuries (Hinze, 1994) determined the ratio between the indirect and direct costs to be 985 for medical case injuries and 0.23 for lost time injuries (excluding claims). When the claims costs are included the ratios increase to 1.18 and 2.06 respectively.

However, according to Grossman (1991) a study by the Construction Industry Institute Safety Task Force concluded that the indirect costs are 20 times the direct cost.

Research conducted by Levitt & Samelson (1993) determined non lost-time accidents generally to have larger indirect cost ratios than lost-time accidents.

6 FINANCIAL BENEFITS OF HEALTH AND SAFETY

According to Korman (Bentil, 1992), a panel erector who developed a formal safety programme was able to reduce its workers' compensation claims from US$800 000 a year to US$56 000 in 1989. Another contractor who had no formal safety programme in 1986 paid workers' compensation claims in excess of US$500 000 after working nearly 600 000 person hours. In 1987, after hiring a full-time safety director and implementing a formal safety programme their claims were reduced to US$300 000 after working nearly 700 000 person hours. By 1988, the amount of claims had been reduced to less than US$115 000 after working nearly 600 000 person hours and by mid 1989, claim amounts had been reduced to less than US$50 000 after working over 600 000 person hours i.e. over a period of 3½ years they reduced their claims by 90%.

Bentil (1992), also cites findings by Du Pont Safety Services which include that construction firms that make safety a priority can reduce lost workday accidents by an average of 37% in the first year and 10-20 percent each subsequent year.

In 1988 M.B. Khan Construction Company's workers' compensation insurance premium exceeded US$500 000 for the first time. According to Synnett (1992), they introduced a completely refurbished accident prevention and safety programme, whereafter total claims paid averaged US$104 000 over a period of 3 years. He states that when the latter is added to the minimal fixed costs of the self-insured programme, conservative overall savings for the entire programme are likely to exceed US$725 000!

According to Young (1996) the Los Angeles Metro project began with a single health and safety professional in the construction manager's organization. After 5 years of increasingly negative publicity as a result of inadequate health and safety, the project managers increased their investment in health and safety by approximately tenfold. The Washington Metro project also increased staff over time and implemented a financial incentive programme for contractors which has been attributed with saving US$10m per year.

Young (1996) also cites the Denver International Airport which increased its health and safety management staff from 6 to 39 and realised a corresponding four fold decrease in the cost of claims per hour worked. After 3 years the savings amounted to approximately US$35m.

7 RESEARCH

114 General contractors who were either members of the Building Industries Federation of South Africa (BIFSA) or the South African Federation of Civil Engineering Contractors (SAFCEC) responded to a national postal survey using a structured questionnaire.

The mean level of subcontracting was 13.4% for 'labour only' and 21.9% for 'full' subcontractors. This relatively low level of subcontracting was reinforced by the high mean number of workers employed by the contractors (Table 2).

42.6% of contractors worked mostly at ground level, 34.8% between 0 – 2 floors and 12.2% between 0 – 10 floors, the remaining 10.4% worked at other levels.

Table 3 indicates that the frequency to which contractors investigated incidents and injuries generally increased commensurately with the severity thereof.

The frequency to which contractors followed up incidents and injuries with preventative action also increased commensurately with the severity thereof (Table 4).

Table 2. Mean number of workers employed

Category	No.		
	Permanent	Temporary	Total
Skilled	50.9	2.6	53.5
Semi-skilled	127.5	5.1	132.6
Unskilled	50.0	43.0	93.0
Total	228.4	50.7	279.1

Table 3. Frequency of incident and injury investigation

Type	Frequency (%)							
	Always	Often	Regularly	Seldom	Never	Don't know	N/A	Total
Incidents	40.5	9.6	17.5	26.3	6.1	0.0	0.0	100.0
First Aid injuries	40.0	14.8	21.7	14.8	6.1	1.7	0.9	100.0
Medical Aid injuries	58.3	8.7	15.7	8.7	4.3	1.7	2.6	100.0
Disabling injuries	67.5	2.6	9.6	7.0	3.7	2.6	7.0	100.0
Fatalities	71.8	0.0	0.9	3.7	7.3	1.8	14.5	100.0

Table 4. Frequency of follow up of incidents and injuries with preventative action

Type	Frequency (%)							
	Always	Often	Regularly	Seldom	Never	Don't know	N/A	Total
Incidents	45.2	10.4	18.3	17.4	8.7	0.0	0.0	100.0
First Aid injuries	45.6	13.1	21.1	10.5	7.0	1.8	0.9	100.0
Medical Aid injuries	57.1	11.4	14.9	9.6	2.6	1.8	2.6	100.0
Disabling injuries	69.9	4.4	9.7	3.5	2.7	1.8	8.0	100.0
Fatalities	76.4	0.9	1.8	1.8	4.5	0.9	13.7	100.0

Table 5. Aspects negatively affected by accidents

Aspect	Response (%) Yes	No	Don't know	Total
Productivity	89.5	7.9	2.6	100.0
Cost	87.8	10.5	1.7	100.0
Worker morale	73.5	23.9	2.6	100.0
Schedule	72.1	22.5	5.4	100.0
Industrial relations	63.1	30.6	6.3	100.0
Client satisfaction	53.0	36.5	10.5	100.0
Public relations	51.8	36.6	11.6	100.0
Quality	51.8	42.1	6.1	100.0
Environment	32.7	52.2	15.1	100.0
Other	36.4	27.3	36.3	100.0

Productivity and cost predominated among aspects negatively affected by accidents. The next most frequently cited aspects were worker morale and schedule (Table 5).

Table 6 indicates that administration costs, increased documentation, first aid costs, increased compensation insurance and clean up costs are ranked within the top five in terms of the overall impact of accidents, based on an impact index computed as follows:

$$\text{Impact Index} = \frac{4n_1 + 3n_2 + 2n_3 + 1n_4}{4(n_1 + n_2 + n_3 + n_4 + n_5)}$$

Where n_1 = frequency of 'Always'
n_2 = frequency of 'Often'
n_3 = frequency of 'Regularly'
n_4 = frequency of 'Seldom'
n_5 = frequency of 'Never'

Table 6. Impact of accidents

Impact	Impact Index	Frequency (%) Always	Often	Regularly	Seldom	Never	Don't know	Total
Administration costs	78.9	61.7	13.0	6.2	13.0	5.2	0.9	100.0
Increased documentation	73.4	45.1	25.7	10.6	11.5	6.2	0.9	100.0
First-aid costs	68.0	39.5	18.4	18.4	16.7	5.2	1.8	100.0
Increased compensation insurance	53.4	30.1	13.3	8.8	24.8	16.8	6.2	100.0
Clean up costs	52.9	23.2	22.3	9.8	25.0	16.1	3.6	100.0
General complications	49.0	23.9	11.5	18.6	25.7	16.8	3.5	100.0
Reduced productivity after injury	47.4	12.3	24.8	17.7	31.0	12.4	1.8	100.0
Loss of crew efficiency	44.3	15.2	13.3	22.3	27.7	18.8	2.7	100.0
Material damage or loss	40.2	11.5	17.7	8.8	39.8	19.5	2.7	100.0
Transportation costs	39.7	13.4	12.4	11.6	42.0	18.8	1.8	100.0
Idle workers	39.7	15.9	12.4	15.1	28.3	26.5	1.8	100.0
Funeral costs	37.4	22.2	1.9	3.7	35.2	29.6	7.4	100.0
Plant and equipment downtime	36.6	12.3	12.3	10.5	38.6	24.5	1.8	100.0
Worker claims	36.3	13.4	8.9	11.6	41.1	23.2	1.8	100.0
Rescheduling costs	33.2	8.0	10.6	8.0	48.6	23.0	1.8	100.0
Overtime costs	30.6	8.8	10.6	6.2	40.7	31.9	1.8	100.0
Legal costs	29.3	10.0	7.3	6.4	37.3	34.5	4.5	100.0
Retraining costs	27.6	8.1	3.6	4.5	55.0	27.0	1.8	100.0
Extension of time	25.9	8.0	6.2	4.5	43.8	36.6	0.9	100.0
Third party claims	23.1	4.5	3.6	7.2	44.2	36.9	3.6	100.0
Subcontractor claims	20.0	7.2	2.7	2.7	34.2	50.5	2.7	100.0

With the exception of the sustainability of the environment the majority of contractors responded that health and safety had positive results (Table 7).

Table 7. Results of health & safety

Aspect	Response (%)			
	Yes	No	Don't know	Total
Reduced accident costs	86.8	7.9	5.3	100.0
Increased productivity	78.1	15.8	6.1	100.0
Quality	73.7	20.2	6.1	100.0
Compensation insurance rebates	72.0	14.0	14.0	100.0
Sustainability of the environment	49.1	20.9	30.0	100.0

Table 8. Percentage contribution by indirect cost to the total cost of accidents.

Class of injury	Mean cost (%)	Response (%)			
		Know	Don't know	N/A	Total
Fatality	36.7	11.4	12.3	76.3	100.0
Disabling injury	30.7	21.1	24.5	54.4	100.0
Medical Aid	29.6	30.1	46.9	23.0	100.0
First Aid	30.4	34.5	50.5	15.0	100.0

Table 9. Percentage rebate received by contractors.

Range (%)	No	(%)
0	0	0.0
1 – 25	10	37.0
26 – 50	15	55.6
51 – 75	1	3.7
76 – 100	1	3.7
Total	27	100.0

Only 38.9% of contractors calculated the direct cost, and 18.6% the indirect cost of accidents. The mean percentage the indirect cost of accidents constitutes of the total cost of accidents is substantially lower than the percentages cited in the literature (Table 8).

44.2% of contractors did not know whether they had received a rebate on compensation insurance or not. 23.9% had received and 31.9% had not received a rebate. Table 9 quantifies the range of rebates received.

34.5% of contractors did not know whether they had received a loading on compensation insurance or not. 11.5% had received and 54.0% had not received a loading. Table 10 quantifies the range of loadings received.

Table 11 indicates that the extent to which injuries are recorded increases commensurately with the severity of the injuries.

The incidence rates shown in Table 12 represent the number of injuries per 100 workers. The industry rates are for 1990, they being the latest rates available. Compared to industry, the contractors surveyed attained a worse medical aid rate, a better disabling rate and the same fatality rate. A first aid injury rate is not available for the industry.

Table 13 indicates that compensation insurance rebates and publicity regarding cost dominated among aspects relative to accidents requiring more attention.

60.5% of contractors viewed health and safety as very important, 22.8% as important and 14.9% fairly.

Table 10. Percentage loading received by contractors

Range (%)	No	(%)
0	3	23.1
1 – 25	7	53.9
26 – 50	2	15.3
51 – 75	0	0.0
76 – 100	1	7.7
Total	13	100.0

Table 11. Record of injuries

Type of injury	(%)			
	Record	Don't record	Don't know	Total
First Aid	66.7	20.7	12.6	100.0
Medical Aid	77.5	10.8	11.7	100.0
Disabling	86.5	3.6	9.9	100.0
Fatalities	91.0	2.7	6.3	100.0

Table 12. Incidence rates for contractors surveyed (1994) and the industry (1990)

Type of injury	Rate	
	Survey	Industry
First aid	19.27	-
Medical aid	4.18	2.37
Disabling	1.61	2.03
Fatalities	0.05	0.05

Table 13. Degree of attention required relative to accidents in South African construction

Aspect	Degree of attention (%)				
	More	Adequate	Less	Don't know	Total
Accounting (Contractor)	30.4	46.4	2.7	20.5	100.0
Accounting (Industry)	32.4	39.6	2.7	25.3	100.0
Compensation insurance rebates	55.8	31.9	0.9	11.5	100.0
Loading of compensation insurance	19.8	39.6	21.6	18.9	100.0
Publicity regarding cost	52.7	26.8	2.7	17.8	100.0

8 CONCLUSIONS

The conclusions relative to the objectives of the study stated in the introduction are reviewed below.

Given the fatality rate of selected construction industries worldwide, the level of injuries in South African construction is unacceptably high, the cost of which contributes to the cost of construction, directly or indirectly through, inter alia: lower productivity, quality non-conformances and schedule disruptions. Generally, literature indicates the indirect cost of accidents to be greater than the direct cost, but are commensurately less significant in terms of percentage contribution to the total cost as the severity of the injury increases. However, South African research findings indicate the converse.

Although there is a cost associated with providing a healthy and safe workplace the benefits in terms of inter alia, lower compensation insurance premiums, constitute sufficient motivation to prioritise health and safety and allocate resources thereto.

Contractors clearly view health and safety to be an important aspect of construction. However, more attention needs to be paid to publicising the cost of accidents and the calculation thereof.

ACKNOWLEDGEMENTS

The author is grateful to the contractors who responded to the national postal survey and to Danie Venter, Director, Institute for Statistical Consultation and Methodology, University of Port Elizabeth, for the statistical processing of data.

REFERENCES

Bentil, K.K. 1992. Construction Site Safety: The Silent Profit Centre. *Construction Business Review*. September/October: 52-54.

Center to Protect Workers' Rights (CPWR). 1997. The Construction Chart Book. The U.S. Construction Industry and Its Workers. Washington: CPWR.

Compensation Commissioner. 1995. Report on the 1990 Statistics. Pretoria.
Compensation Commissioner. 1994 Personal contact: Mrs Wendels.
Federated Employers Mutual Assurance Company Limited. 1986 *Guide to Workmen's Compensation.* Johannesburg.
Federated Employers Mutual Assurance Company Limited. 1994 Personal contact: Mr D Joss.
Grossman, D. 1991. Construction Industry Builds a Safe Workplace. *Safety & Health*, April: 48-51.
Hinze, J. 1992. Indirect costs are a major portion of jobsite injury costs. *Concrete Construction*. March.: 229.
Hinze, J. 1994 Quantification of the indirect costs of injuries. *In Proceedings of the Fifth Annual Rinker International Conference Focusing on Construction Safety & Loss Control*, 521-534. Gainesville: University of Florida.
Krause, T.R. 1993. Safety & Quality: Two sides of the same coin. *Occupational Hazards.* April: 47-50.
Levitt, R.E. & Samelson, N.M. 1993. *Construction Safety Management.* 2nd Edition. New York: John Wiley & Sons Inc.
Smallwood, J.J. 1992. Does safety and loss control cost money? *SA Builder/Bouer*, May: 26,28 & 30.
South African Reserve Bank. 1991. *Quarterly Bulletin 1990.* Pretoria.
Synnett, R.J. 1992. Construction Safety: A Turnaround Program. *Professional Safety.* October: 33-37.
The Business Roundtable. 1991. *Improving construction safety performance.* New York: The Business Roundtable.

Implementation of Safety and Health on Construction Sites, Singh, Hinze & Coble (eds)
© 1999 Balkema, Rotterdam, ISBN 90 5809 036 1

Safety costs money and can save money

Anthony J.Joseph
Department of Safety Sciences, Indiana University of Pennsylvania, Ind., USA

ABSTRACT : Safety measures, quality requirements and expected profit influence the costing of construction projects. The cost dynamics of these three considerations are interrelated and independent variables. High quality control and assurance of tasks require more inspections or testing. Additional safety features or measures on tasks can mean more personal protective equipment or safety gadgets. These requirements will clearly incur additional cost. Using 52 examples of steel erections, this study showed that actual profit can increase by five to 10% for an investment of one to 3% in safety. In addition, by including safety in every task of a construction project the tendency to compromise safety measures due to additional high costs or inadequate appropriation of funds will be reduced. The study concluded that safety costs money and can save money.

1 INTRODUCTION

The goals of establishing accident prevention programs are to preserve the health and lives of employees through the prevention of injuries, illnesses and accidents; prevent accidents that can result in damage to property, and promote safe and healthful work conditions as the key element in day to day operations. Both management commitment and employee involvement are necessary to accomplish these goals. Unfortunately, these goals are often overshadowed in construction projects by operational and cost goals.

Fall hazards create the greatest exposure to injury in the construction industry with the most common fall hazard from elevations. These elevations include ladder, elevated workplaces, through floor openings, and from edges (OSHA 1990 and BLS 1990). Steel erection projects are notorious for avoiding any type of fall protection, even when working hundreds of feet off the ground, as shown in Figure 1.

Figure 1. An Example of Steel Erection Project

In evaluating the cost effectiveness of safety programs, most injury rate estimates rely on data from surveys such as the Bureau of Labor Statistics, analysis of workers' compensation claims, emergency room surveillance, and proportionate mortality studies to provide the baseline. The data provided by these surveys do not allow an accurate picture to be painted. Most of these surveys suffer from under reporting and imprecise estimates of the number of workers at risk. For example, Glazner and others suggested that the occupational injury rates and workers' compensation payment rates on a major construction project were greater than twice the Bureau of Labor Statistics's rates (Glazner 1997). This is not unique as reported by several other researchers such as Moran (Moran 1995) and Pollack (Pollack and Keimig 1987). By studying selected cost elements of 52 steel erection projects distributed in 10 States and 24 different contractors, a unique opportunity was provided to examine the relationship between safety investment and profit. The final cost of these projects ranged from 1.2 to 3 million dollars. These projects were started and completed between 1989 and 1994.

The cost of accidents is an important financial consideration to most organizations. Yet, most organizations rely on the insurance cost as their only positive data for appraising safety accomplishment or deciding how far to go financially and morally in supporting the company's accident-prevention efforts. Insurance costs are only a fraction of the total accident related costs. The root of the problem is that the cost of accidents is hidden in the accounting records. They are not easily retrievable, and line managers have not become sensitized to ask the accounting systems to produce these accident costs (Joseph 1993). Further discussion of this topic is beyond the scope of this study. The cost of accidents devastated many construction contractors.

Profit is the goal of most businesses. Therefore, in determining a bid for a construction project safety measures and quality requirements are two important elements that influence the price. The cost dynamics of these two elements are interrelated. High quality control and assurance of tasks, such as certification of the soundness of the welds will require more inspections or testing. More safety features or measures on tasks, such as full fall protection instead of body harness onto a tie line will require more personal protective equipment and safety gadgets. Fortunately, most accidents on construction sites are the result of inadequate safety measures or failure to enforce safety programs.

2 PILOT SURVEY OF STEEL ERECTION PROJECTS

2.1 *Definitions and Terms*

The following definitions and terms are used in this paper.

Accident cost: The direct cost as recorded in the accounting ledgers of the contractor. Note that in the literature accident cost is defined by the human capital method which breaks down cost into two categories, namely direct cost such medical expenses, and indirect cost such as lost of productivity.

Contractor: An organized group of workers registered with the Associated Builders and Contractors Inc. for the sole purpose of providing construction services.

Profit: Profit is the difference between contracted price and the cost incurred by the contractor in completing the project before taxation.

Project Cost: The final sum paid by the owner or main contractor to the contractor or subcontractor for doing steel erection services.

Safety Investment: The direct cost of providing safety personnel, personal protective equipment, safety gadgets, training of employees, medical and safety insurance, and any other direct costs related to promote safety approved for purchase before the commencement of the project.

Steel Project: All the activities associated with the erection of steel framing, roofing and flooring on a construction site.

SAFETY COST SURVEY

Dear Contractor,

Thank you for agreeing to participate. As we discussed on the telephone, identify projects in which 95% or more of the tasks performed were related to steel erection. To the best of your knowledge complete the following questions as they relate to a specific project started and completed within the last five years. *Note: Use separate sheet for each project.*

1. What is the final contracted sum of the project? $..
2. Did the scope of your contract included more than steel erection? If yes, briefly identify the type and sum involved....
3. What was your average number of workers each day on the project?
4. Days worked on the project? *Complete the following*
   Starting date:
   Ending date:
   Total number of lost days:
   Number of days per week worked:
5. What is the average number of hours worked per day on the project?
6. What was your direct safety investment or cost? *Complete the following*
   - *Personnel cost (Wages, salaries and benefits)* $
   - *Training* $
   - *PPE* $
   - *Safety gadgets* $
   - *Insurance (WC, medical and safety)* $
   - *Others (briefly state....* $
7. How many recordable non-fatal accidents occurred during the life of the project?
8. What was your total accident costs not paid by insurance including first aid cases? *Note: Do not include fines* $
9. What was your budgeted or planned profit margin of this project ? $... or .....%

Note: % = (contracted price minus total cost of fulfilling contract divided by contracted price) x 100

10. What was your actual profit? $ or .... %

Figure 2. Questionnaire for Pilot Survey on Steel Erection Safety Cost

2.2 *Data Collection and Treatment of Data*

A self administered questionnaire consisting of 10 questions was mailed to 29 construction companies who were listed in the Associated Builders and Contractors Inc. database, 1995. These companies were selected randomly for an initial telephone interview. After the initial telephone interview the questionnaire was mailed to the contractor if an expression of willingness to participate was communicated. The questionnaire was designed to keep the identity of the contractor and project anonymous, see Figure 2. This condition was a prerequisite for most of the companies to participate in the pilot survey. So, the issues of validity, completeness, and accuracy of data provided by these companies were not challenged.

Twenty-four contractors responded with 52 different projects. Forty-eight of these projects were as subcontract. All of the contractors were required to bid for their projects and have effective safety programs. These projects were completed between 1989 and 1994. The number of accidents was standardized using OSHA's base factor of 200, 000 working hours per year, that is 100 full-time workers at 40 hrs for 50 weeks per year. The project was rejected if the cost of the steel work was less than 95% of the final contracted sum of the project. Days worked on the projected if not given by the contractor was calculated using number of weeks multiplied by the number of days worked per week minus the number of lost days. This value of worked days was then multiplied by the average number of hours worked per day to obtained total working hours.

This study did not consider the following factors as they influence profit margin, accident rates and safety investment.

a. The changing work force on the project;
b. Indirect cost of safety investment and accidents;
c. The root causes of accidents;
d. The duration of the project;
e. The year the project was started and completed;
f. The distribution of accidents and
g. The contracted sum of the project.

The study assumed that the direct cost of safety is the safety investment of the project. This does not include fines imposed for safety violations. It also assumes that all accidents cost money, therefore a reduction in profit.

2.3 *Results and Discussion*

The results of the survey are presented in Figures 3 to 6. A scattered plot and a trend line are depicted in each figure. The linear correlation coefficient and slope of the trend lines are given in Table 1. The graphs in Figure 3 suggest that there is a strong linear relationship between safety investment and accident rates. Despite a weak linear relationship of profit and accident rates as shown in Figure 6, accidents clearly reduce profit. Figure 4 clearly shows that profit will increase when safety investment is increased. The scattered plot in Figures 3, 4 and 6 plateaued, therefore suggesting that an optimum point after which the rates of return are diminishing.
Analyzing four projects from one contractor strengthened the above argument that safety investments can increase profit. The data of the four projects are presented in Figures 7 to 10. The contractor experienced decreased accident rates with increase safety investments as shown in Figure 7, and increase profits as the accident rates decrease as shown in Figure 10. It is reported that the profit of this contractor on steel erections projects is now greater than 80% of the estimated value. It is my contention that this contractor is highly profitable in steel erection because of its commitment to the assurance of safety. This is reflected in one goal of this contractor, which is to reduce its accident rate to zero by integrating safety into every task. The executives, managers, and supervisors of the contractor are committed to the continuous improvement of its operations to provide a safe and healthful place to work.

Table 1. Trend Lines Characteristics

Figures	Correlation Coefficient	Slope
3	0.865	-2.67
4	0.740	0.413
5	0.922	0.733
6	0.656	-0.5

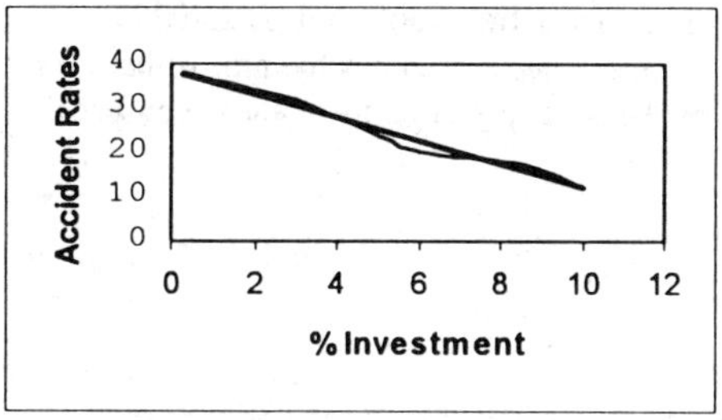

Figure 3 Accident Rates and Investment

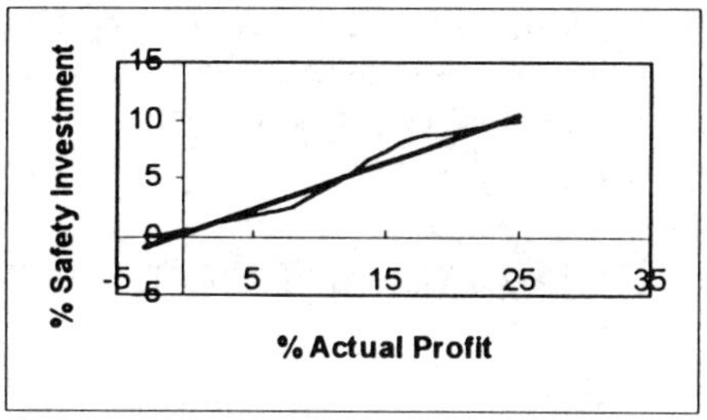

Figure 4 Safety investment and Actual Profit

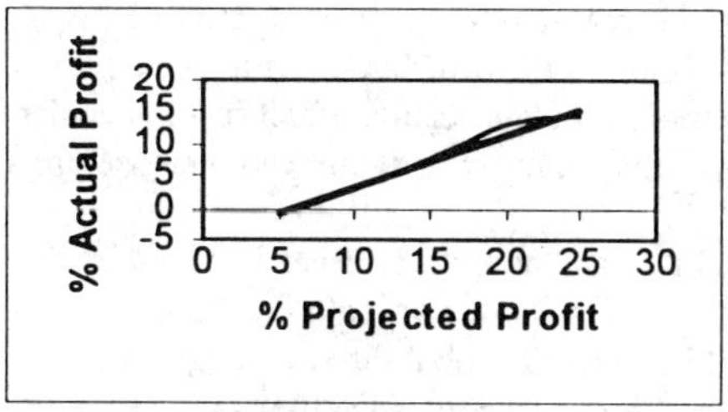

Figure 5 Actual and Projected Profit

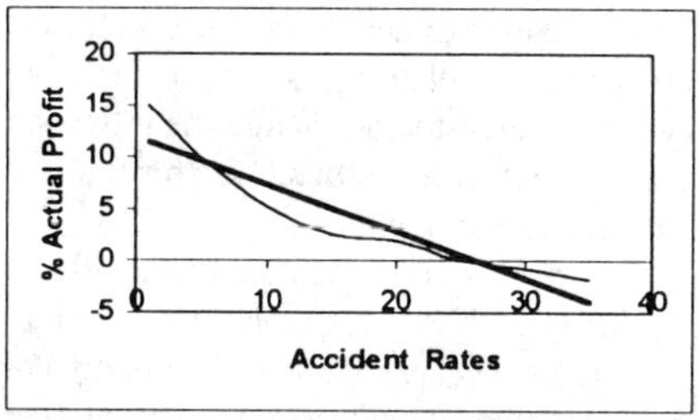

Figure 6 Actual Profit and Accident Rates

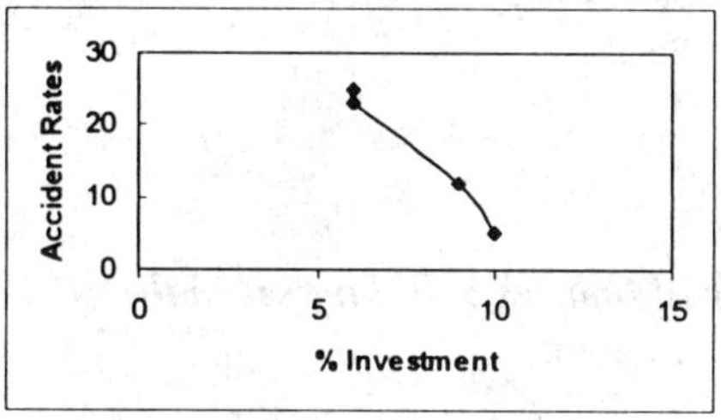

Figure 7 Accident Rates and Investment of a Contractor

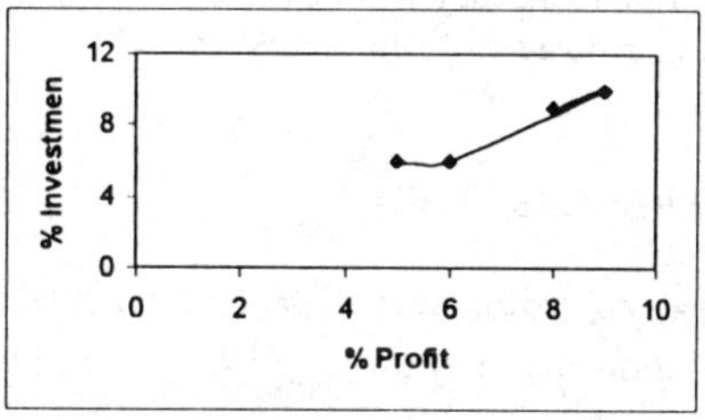

Figure 8 Investment and Profit of a Contractor

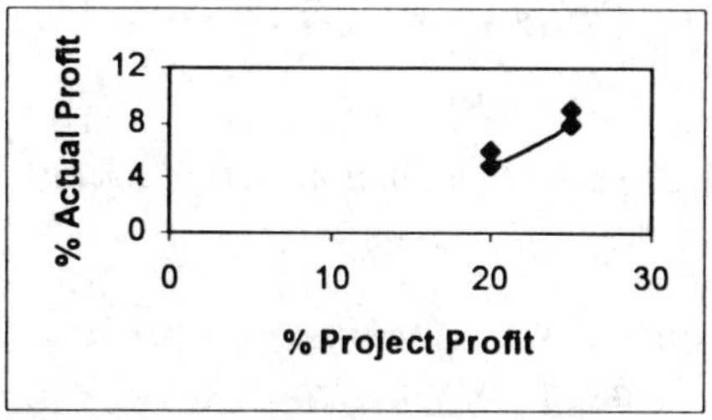

Figure 9 Actual and Projected Profit of a Contractor

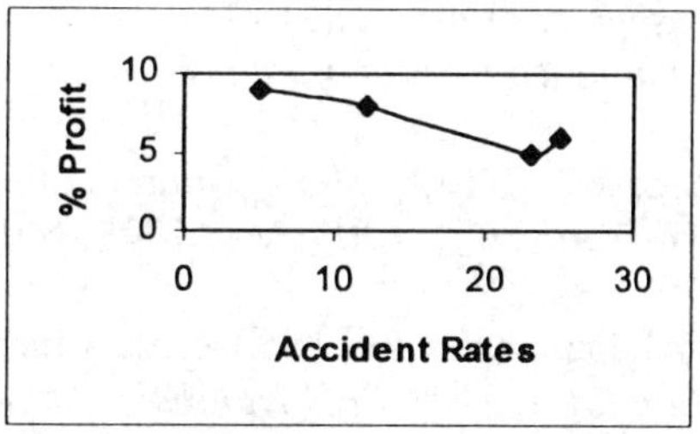

Figure 10 Profit and Accident Rates of a Contractor

The elimination of hazards, and the proper communication, instruction, tools and personal protective equipment to deal with hazards that cannot be eliminated, promotes the atmosphere for accident free job sites. In so doing the contractor has increased its profit and was more competitive in bidding.

3 SUMMARY AND CONCLUSIONS

In this paper, the justification for safety investment was explored. Fifty-two projects were examined to discover relationships between safety investment, profit and accidents. The experience of a contractor was highlighted to justify the argument that safety cost money and can increase profit. The literature postulated that indirect cost of accidents is at least three times more than the direct cost. If this is true, then reduction in accident rates will inevitably exponentially increase profit. The study did not consider the influence of the changing work force; the indirect cost of accidents; the root causes of accidents; the duration of the project; the year the project was started and completed; the distribution of accidents or the contracted sum of the project on profit, accident rates and safety investment. While the influence of these factors may significantly affect profit, accident rates and safety investment, the results of this study show that safety investment can increase profit, and decrease accidents.

Safety investment can also influence the probability and amount of fines imposed by a regulatory authority for violations of safety. This aspect of cost is more speculative than predictable, because there are too many uncertain variables such as probability of an agency inspecting the site. Despite the speculative nature of fines, it can be concluded that safety measures reduce fines. Investments in safety measures are difficult to justify because of fines, or when the safety record or performance of the contractor is good. Therefore, there is the tendency to compromise safety measures. By incorporating safety measures in every task of a construction project this tendency will be reduced. This additional investment in safety will cost the contractor money. However, it can save lives and increase the profit margin.

REFERENCES CITED

Bureau of Labor Statistics (BLS) 1990. *Occupational Injuries and Illnesses in the United States by Industry, 1988.* U.S. Department of Labor.

Glazner JE et al 1997. Construction Injury Rates May Exceed National Estimates: Evidence from the Construction of Denver International Airport. *National Occupational Injury Research Seminar, West Virginia September 1997.* Unpublished.

Joseph A. 1993. The Cost of Accidents. *MS Thesis, Safety Sciences Department, Indiana University of Pennsylvania.* Unpublished.

Moran J. B. (ed.) 1995. Counting Injuries and Illnesses in Construction. *Appl. Occup. Environ. Hgy. 10(5),* May 1995: 449-451.

Occupational Safety and Health Administration (OSHA) 1990. *Analysis of Construction Fatalities: The OSHA Database 1985-1989.* U.S. Department of Labor.

Pollack E.S. and D.G. Keimig (eds.) 1987. *Counting Injuries and Illnesses in the Workplace: Proposals for a Better System.* National Academy Press, Washington, DC.

Implementation of Safety and Health on Construction Sites, Singh, Hinze & Coble (eds)
© 1999 Balkema, Rotterdam, ISBN 90 5809 036 1

Social costs of construction accidents in Hong Kong

K.L.Ngai & S.L.Tang
Civil and Structural Engineering Department, Hong Kong Polytechnic University, Hunghom, Kowloon, People's Republic of China

ABSTRACT: This paper describes a methodology to investigate the social costs that society incurs due to construction accidents in Hong Kong. The social costs, social benefits and safety investment are defined, identified and quantified. A survey is currently being conducted for contractors and a number of government departments which comprise the Labour Department, the Social Welfare Department, the Occupational Safety and Health Council (OSHC), the Court, the Police Force, the Fire Department, and educational institutions including the Hong Kong Polytechnic University and the Construction Industry Training Authority (CITA). Using the statistical data, the social costs incurred by and the safety resources invested by contractors and government departments in 1996 and 1997 can be evaluated. The difference between 1996 and 1997 social costs, expressed in HK$ per fatal accident and HK$ per non-fatal accident, will represent the social benefits (the social costs in 1996 is expected to be higher than that in 1997 due to the increase in safety investment in 1997). As a result, the impact of safety investment on construction projects from society's point of view can be established.

1 INTRODUCTION

In Hong Kong, the number of industrial accidents, especially in the construction industry, remains at a very high level although there has been a downward trend in recent years (Hong Kong Government 1995). The accidents rates of all industries and, in particular, the construction industry for the years 1992-1996 are shown in table 1. The statistical data indicated that although the construction industry employed less than 8% of the total industrial workforce, it consistently accounted for more than one third (about 40% in 1996) of all industrial accidents in the last five years. The risk of a worker involved in construction accidents between 1992 and 1996 was 5 times higher than the overall average for all industrial workers. Most of the fatalities in industrial accidents also occurred in the construction industry; the average fatality rate between 1992 and 1996 was 0.938 per thousand workers and was 10.8 times higher than the overall average for all industrial workers. Such poor accident records in the Hong Kong construction industry is unacceptable.

In the United States, the total costs of accidents were somewhere between 7.9% and 15% of the total costs for nonresidental, new construction (Everett & Frank 1996). It was also shown that the ratio of indirect to direct costs for medical-case injuries was 4.2 and for lost-workday injuries was 20.3 (Hinze & Appelgate 1991). The above two pieces of research were carried out in the United States. There is no doubt that the costs of accidents are substantial, but the authors are doubtful about their high figures and their suitability of application in Hong Kong. This piece of investigation work is to find out what happens in Hong Kong.

Table 1. Industrial accident rates of all accidents in Hong Kong in 1992-96. (Notes: All industries include the manufacturing industry, construction industry, catering industry, transportation, storage, communication and utilities.)

		1992	1993	1994	1995	1996
Construction Industry	No. of Injured Workers	18,815	16,573	16,422	15,268	16,469
	No. of Fatal Injury Accident	48	80	51	63	51
	No. of Workers Employed	62,232	56,226	59,710	65,611	74,907
	No. of Injured Workers per thousand Workers	302.34	294.76	275.03	232.7	219.86
	No. of Fatal Injured Workers per thousand Workers	0.771	1.423	0.854	0.96	0.681
All Industries	No. of Injured Workers	55,343	46,797	44,164	41,001	40,251
	No. of Fatal Injury Accident	65	92	67	77	66
	No. of Workers Employed	966,995	890,535	834,371	787,003	753,302
	No. of Injured Workers per thousand Workers	57.23	52.55	52.93	52.1	53.43
	No. of Fatal Injured Workers per thousand Workers	0.067	0.103	0.08	0.098	0.088

2 FINANCIAL COSTS AND SOCIAL COSTS

The definition of accident must be defined in order to explain what are financial losses and social costs of accidents. The law in Hong Kong has such a definition: an accident is one where the employee is suffering from fatal injury or is absent from work through injury for more than three consecutive days (Employees' Compensation Ordinance, Chapter 282, Hong Kong).

2.1 *Financial costs*

Financial costs represent the losses incurred by the private investors, such as contractors, due to the occurrence of construction site accidents. In financial analysis, market prices are always used to represent benefits and costs (Tang 1996). In the case of a construction site accident, the financial losses incurred by the contractor include loss due to the lost productivity of the injured worker, loss due to the inefficiency of the injured worker after resuming work, loss due to medical expenses, fines and legal expenses, loss of productivity of other employees immediately after the occurrence of the accident, loss due to damaged equipment or plant, loss due to damaged material or finished work, and loss due to idle machinery or equipment.

2.2 *Social costs*

Social costs represent losses incurred by society due to the occurrence of construction site accidents. Social costs are defined as any items that will result in an utilization of national resources. For instance, if a construction site accident happens the social losses incurred by society include all the financial losses mentioned above except fines (which is internal transfers rather than social losses), the costs of the resources utilized by the government departments such as the Labour Department, the Social Welfare Department, the Court, the Police Force, the Fire Department, and the Hospital Authority in tackling the accidents.

The following highlights the differences of financial costs and social costs:

a. Financial costs are based on the private investor's (contractor's) point of view. Social costs are based on the society's point of view.
b. Most financial costs (but not all) are also social costs.
c. Some costs incurred by society are much higher than those incurred by private investors. For instance, a contractor pays for his injured worker HK$68.00 per day for hospitalization because medical treatment in Hong Kong is heavily subsidized by the Government, and this HK$68.00 per day is the financial loss of the contractor. However, the social cost (actual cost incurred in treating the injured worker) is much higher than the figure and is about HK$ 1,770.00 per day (on average) in 1997.

3 SOCIAL BENEFITS AND SAFETY INVESTMENT

Social benefits are defined as any items that will result in a saving of national resources. In this paper, social benefits refer to the resources saved owing to the reduction in the number of fatal and/or injured workers. For instance, the number of fatal and injury workers per thousand workers in the construction industry in 1995 were 0.96 and 232.7 respectively and in 1996 were 0.681 and 219.86 respectively (see Table 1). Therefore, the social benefits gained by the society are the resources saved owing to the reduction of the 12.84 injury workers per thousand workers and 0.279 fatal workers per thousand workers. As the saved resources can be used somewhere else in the society, these can represent benefits to the society.

Safety investment by the society in construction industry is defined as the sum of two parts. The first part is the resources on safety invested by contractors in construction projects and the second part is the resources invested by government departments. These will be explained in detail in section 5 of this paper. Some may wonder why the resources invested by contractors are included in the society's safety investment. It is because if the contractors do not invest the money in safety, they will invest the money somewhere else such as buying more plant and equipment. The increased safety facilities can contribute to the safety aspects of the projects. So the resources invested by contractors on safety will be treated as a part of the safety investment of the society.

4 IDENTIFICATION AND QUANTIFICATION OF FINANCIAL LOSSES AND SOCIAL COSTS

The identification of financial losses have been given briefly in section 2.1 of this paper and in detail in Reference 9 (Tang et al. 1997). The following will only discuss about those of social costs.

The social costs due to occurrence of accidents are as follows:

a. *The productive years of the injured worker.* To evaluate the loss of the productive years of a worker, the method stated in the "Employees' Compensation Ordinance" published by the Hong Kong Government is adopted. The ordinance establishes the compensation of an injured worker for the case of permanent total incapacity and the case of permanent partial incapacity, with reference to earnings, age and the extent of loss of earning capacity of the injured worker.

b. *Families and relative losses.* This refers to the opportunity costs of housewives' work and relatives' work to take care of the injured workers.

c. *Fire Department and rescue services.* Costs are incurred by society to provide rescue services such as the ambulance transportation and first-aid services. Besides, fire-engines services and the wages of the related staff are also social costs. It is found that the average costs per accident, including the human resources and the operation cost of equipment of the Fire Department in 1996 and 1997 were HK$ 9014.00 and HK$ 9,885.00 respectively.

d. *Losses due to the medical expenses and hospitalization.* In Hong Kong, the hospitalization fees of local residents in public hospitals were HK$60.00 and HK$ 68.00 per day per person in 1996 and 1997 respectively. However, the actual expenses that the hospitals incurred were HK$1,570.00 and HK$1,770.00 per day per person in 1996 and 1997 respectively. The losses incurred by the society are the actual expenses.

e. *The Hong Kong Police Force.* When a construction site accident is reported to the police, the latter will tackle the case and carry out immediate actions. The police also maintains discipline on site and assists factory inspectors from the Labour Department in investigating the accident. This is also a cost to the society.

f. *The Social Welfare Department.* This includes the administration/personnel costs of the Social Welfare Department to provide assistance to the injured worker.

g. *The Labour Department.* This includes the costs for regular site inspection for prevention of accidents and the costs for investigation and reporting if accidents occur.

h. *The Court.* When a serious or a fatal accident happens, the Court will carry out an investigation to find out the reasons for the injury or the death of the worker, especially when there is any argument between the employer and the family of the employee. This is another cost to society.

5 IDENTIFICATION AND QUANTIFICATION OF SAFETY INVESTMENT

As discussed in section 3, safety investment on construction projects is divided into two parts, including the safety investment by contractors and the safety investment by the government.

Safety investment by contractors includes the following cost items :

a. *Investment in safety administration personnel.* This includes staff on site and in the head office. Under the current regulation in Hong Kong, the employment of safety officers and safety supervisors are required at construction sites to implement safety programmes and improve site safety. For some big contracting companies, the employment of safety managers and/or senior safety officers at their head offices is required to set up safety programme, carry out internal safety audits and coordinate site safety staff. The salaries of these personnel and their supporting staff (e.g. clerk/typist) are considered as a part of safety investment.

b. *Safety equipment.* This includes the purchasing of safety boots, goggles, helmets, safety fences, first-aid facilities and any other facilities which have to do with the provision of safety on construction sites.

c. *Safety training and promotion.* This includes the costs of safety training courses, such as tool box talks and induction courses for workers organised by contractors; the costs of printing pamphlets and posters and the making of safety advertising boards and banners; the organising of safety campaigns and the rewarding in monetary terms to individual workers or subcontractors who achieve a good safety standard and so on.

Safety investment by the government includes the following cost items:

a. *Investment in safety administration personnel.* This includes the safety inspectors, safety advisors, senior safety advisors and other supporting staff in the Labour Department and the Occupational Safety and Health Council (OSHC) who are responsible for safety in the construction industry.

b. *Safety training and promotion.* This includes the costs of printing of pamphlets and posters, the making of safety advertising boards and banners, the organising of safety campaigns etc. Besides, the costs of safety training courses offered by education institutions such as the Hong Kong Polytechnic University (HKPU) and the Construction Industry Training Authority (CITA) should also be included as safety investment by the government.

6 CALCULATION OF SOCIAL COSTS, SOCIAL BENEFITS AND SAFETY INVESTMENT

To calculate the social costs, the social benefits and the safety investment, a survey is being conducted for contractors and a number of government departments which comprises the Labour Department, the Social Welfare Department, the Occupational Safety and Health

Council (OSHC), the Court, the Police Force, the Fire Department, and education institutions including the Hong Kong Polytechnic University and the Construction Industry Training Authority (CITA).

Using the statistical data, the social costs incurred by and the safety resources invested by contractors and the government in 1996 and 1997 can be evaluated. The difference between 1996 and 1997 social costs, expressed in HK$ per fatal accident and HK$ per non-fatal accident will represent the social benefits (the social costs in 1996 is expected to be higher than that in 1997 due to the increase in safety investment in 1997). As a result, the impact of safety investment on social benefits for construction projects can be established.

7 CONCLUSION

As the rate of accidents in the Hong Kong construction industry remains at a very high level and the costs of accidents, especially the social costs incurred by society, are substantial, there is no doubt that the government, promoters and contractors should increase their safety investment in construction projects to improve site safety.

Recently, the Hong Kong government took a number of measures such as the implementation of the Safety Supervision Plan (starting 1998) (Hong Kong Government 1998) and that of the Pay for Safety Scheme (PFSS) and the Independent Safety Audit Scheme (starting 1997) (Hong Kong Government 1996a, 1996b) to improve site safety.

A large scale survey is now being conducted to obtain accident data and other relevant information. Afterwards, data analysis will be carried out. The impact of safety investment on social benefits for construction projects in Hong Kong will then be established. It is hoped that figures will be available at the time of presentation of this paper at the conference.

REFERENCES

Everett, John G. and Peter B. Frank Jr. 1996. Costs of accidents and injuries to the construction industry. *Journal of Construction Engineering and Management*. 122(2): 158-164.

Hinze, Jimmie and Lisa Lytle Appelgate 1991. Costs of construction injuries. *Journal of Construction Engineering and Management*. 117(3): 527-550.

Hong Kong Government 1993. *Employees' Compensation Ordinance,* Chapter 282.

Hong Kong. Government 1995. Consultation Paper on the Review of Industrial Safety in Hong Kong. *Education and Manpower Branch, Hong Kong Government.*

Hong Kong Government 1996a. Works Branch technical circular number 4/96: Pay for safety scheme (PFSS). *Works Branch, Hong Kong Government Secretariat, 4 March 1996.*

Hong Kong Government 1996b. Works Branch technical circular number 5/96: The independent safety audit scheme. *Works Branch, Hong Kong Government Secretariat, 5 March 1996.*

Hong Kong Government 1998. Technical memorandum for supervision plans. *Building Department, Hong Kong, February 1998.*

Tang, S.L. 1996. *Economic Feasibility of Projects.* Revised Edition, McGraw Hill.

Tang, S.L., H.K. Lee and K. Wong 1997. Safety Cost Optimization of Building Projects in Hong Kong. *Construction Management and Economics*, 15(2): 177-186.

Implementation of Safety and Health on Construction Sites, Singh, Hinze & Coble (eds)
© 1999 Balkema, Rotterdam, ISBN 90 5809 036 1

Economic overview of construction safety

Cristina Reis
Universidade de Trá-os-Montes e Alto Douro, Vila Real, Portugal
Alfredo Soeiro
Faculdade de Engenharia da Universidade do Porto, Portugal
Fernando Santos
ENGIL SA, Porto, Portugal

ABSTRACT: The study addressed the economic analysis of construction safety plans and measures versus the partial or total lack of these. The data collected was obtained from several sources including insurance companies, official statistics, construction companies, hospitals, professional associations and publications. The evaluation of costs of accidents in Construction was made using the examination of real events and the simulation of other type of accidents. The costs of prevention were collected from the several participants especially the construction companies. The final goal was to have an economic comparison of accidents and safety plans leading to eventual motivation. The accident costs definition followed a traditional approach of classification in direct and indirect costs. The cost associated with safety plan implementation and consequent tentative of accident elimination were quantified to allow comparisons and a related cost benefit analysis. These costs are divided between design, construction and supervision. While the prevention costs are fixed and constant the accident costs were affected by the frequency index showing an expected value for this type of accidents. The analysis of these values show that since the insurance costs are not affected by the ratio of occurrence one verifies that these present a lower risk than any type of accident. The economic potential influence of some type of accidents is more important than the total costs involved when the frequency index is considered. Speacial attention should be given to the sfatey plans to these accidents if the economic analyss would be considered.

1. INTRODUCTION

This study addresses the economic analysis of construction safety programs and the impact of the partial or total lack of accident prevention. The data was obtained from several sources, including insurance companies, official statistics, construction companies, hospitals, professional associations and publications. The evaluation of the costs of accidents in construction was made on the basis of real events and simulations of other types of accidents. The costs of prevention were obtained from several of the participants in this study, especially construction companies.

The ultimate object of the study was to obtain an economic comparison of accidents and their prevention that would motivate the implementation of safety programs. Accident costs were defined according to the traditional classification of direct or indirect costs. Costs associated with the implementation of safety programs and their consequent attempts to eliminate all accidents were quantified for purposes of comparison and with a view to performing the respective cost-benefit analysis. These costs were subdivided as design, construction and supervision costs. Whilst prevention costs were fixed and constant, accident costs were calculated according to the frequency index and the estimated value for these types of accidents.

The analysis of these values demonstrated that as insurance costs are not affected by the ratio of occurrence, these present a lower risk than any other type of accident. The results indicate that there is a clear economic advantage to the implementation of safety programs in construction.

2. THE CONSTRUCTION SAFETY PROGRAM AT THE DESIGN STAGE

The cost of a Construction Safety Program (CSP) at the project level was quantified by the AFA, Adao da Fonseca project department, using that for the Ermesinde railway line as an example. Two persons are usually involved in the CSP: 1 engineer who is responsible for developing the program and who does 90% of the work and 1 designer who is responsible for the remaining 10%.

As regards the cost of developing this CSP, the engineer was paid USD 2800 and the designer USD 333 due to the complexity of this task that involved dealing with several specific cases. An average of USD 222 was spent on stationary, copies and revisions. The honoraria are determined by the type of project, the amount of work involved, its nature, the deadlines and the expected number of workers involved in the construction work.

3. THE ENGIL, S.A. PREVENTION, HYGIENE AND SAFETY AT WORK POLICY

3.1 *Introduction*

The policy of safety at work has become an increasingly important concern to the company given that in Europe, 15% of all labor accidents and 30% of all fatal accidents in industry occur in the building trade. According to the latest available data, the costs of accidents in construction represent approximately 3% of the volume of trade in this sector.

A major factor of consideration is the peculiarity of the building industry in that construction sites, by nature, are usually portable, that is, each construction project has its own yard and as such production varies from site to site. This situation aggravates the risk as the type of prevention and action implemented in one yard may not apply to another, even though one project may be similar to another.

In ENGIL's experience, and this is supported by the statistics, implementing a proper safety program contributes to a clear decrease in the frequency and seriousness of accidents, increases workers' motivation, reduces the number of potential accidents and is a most effective way of detecting the most qualified workers.

3.2 *The Structure of the Company*

In 1997, ENGIL's constant concern with improving the safety and health conditions at its construction yards, the increasingly widespread geographic dispersion of its activities and its decentralization of decision-making and responsibilities led the Company to reformulate its corporate structure and to replace the central safety executive body with several executive and consultative bodies.

3.2.1 *Safety Committees*

The following safety committees are responsible for defining the Company's accident prevention, hygiene and safety policy:

- Central Safety Committee (CSC)
- Regional Safety Committees (RSC)
- Site Safety Committees (SSC)

The Central Safety Committee reports directly to the Board of Directors and acts as its consultative and informative body. Its most important remit is as follows:

(a) To contribute to defining the objects of the Company's Accident Prevention, Hygiene, Health and Safety at Work policy;
(b) To set the guidelines that will enable the policy to attain the objects set by the Board of Directors;
(c) To coordinate and integrate the activities of the several RSCs and to promote the standardization of the Company's current procedures and criteria.

The Regional Safety Committees report to the CSC. These bodies are responsible for encouraging and supervising the application, at a regional level, of the Company's Hygiene, Health and Safety at Work policies and the CSC's directives. Their most important remit is as follows:

(a) To make a monthly examination of the incidence of accidents in their region;
(b) To support and encourage the Accident Prevention and Work Safety activities of the executive bodies;
(c) To propose and investigate all accidents and incidents that, due to their nature or consequences, may bring to light markedly abnormal situations;
(d) To keep the CSC informed of their activities through the Minutes of their monthly meetings and other informative documents.

The Site Safety Committees report to the RSC responsible for the area in which their site is located and are responsible for acting on and implementing the Company's safety policy and directives on their respective sites. SSCs, consisting of a minimum of 3 members, are created on all sites where so required by the number of workers (Article 57, CCTJ – Collective Bargaining Agreement). Their duties are as follows:

(a) Make regular inspections of all and installations and materials of interest to Hygiene and Safety at Work;
(b) Ensure compliance with all legislation, collective bargaining agreement clauses, internal regulations and instructions on Hygiene and Safety;
(c) Request and examine suggestions from workers on matters of hygiene and safety and present them at Committee meetings;
(d) In the case of workers who are hired for the first time or who have been transferred from another site, provide training, instructions or advice in matters of hygiene and safety at work, as required, and encourage their use of the safety equipment that is made available to them.

3.2.2 *Prevention and Safety Offices*

The Prevention and Safety Offices, just like the RSCs, are created according to sectors. Given that these are executive bodies, there are currently only two such sectors: Head Office (HAPSO) and Northern Branch (NPSO), The HPSO is responsible for implementing the Company Hygiene, Health and Safety policy everywhere that ENGIL is active (including inter-regional areas) except for those under the direction of the Northern Branch (NPSO) whose volume and diversity of work justify the creation of an independent sector.
Prevention and Safety Offices, as corporate executive bodies, are attached to the Construction Work Support Departments. They are composed of civil engineers and project and site prevention and safety specialists whose principle remit is as follows:

- Develop and implement the Site Safety and Health Program
- Inspect and audit site safety
- Prepare monthly accident incidence reports (accident frequency index)
- Submit annual activity reports

- Issue technical opinions on projects
- Offer training and information to every employee at all levels
- Publish Safety information and technical instructions
- Create check-lists for assessing risks
- Investigate new materials, equipment and technology of value to Prevention and Safety at Work in the building trade.
- Attend meeting of in-house Safety Committees (CSC, RSC)
- Contact and work with external entities and experts in this sector
- Encourage contacts and the exchange of information with schools, universities and other bodies
- Create a Safety document database
- Develop a Prevention and Safety Quality Management System with a view to standardizing procedures
- Keep projects and sites informed of all that is of great value regarding the application of better safety conditions for workers.

3.2.3 *Organization Chart*

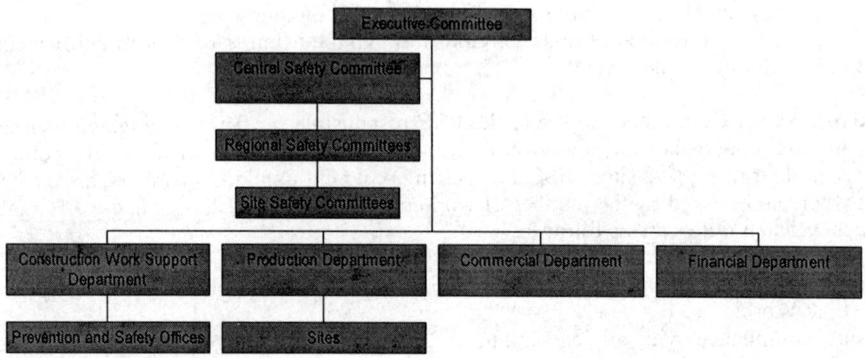

Figure 1 – ENGIL, S.A. Organization

4. COST OF DEVELOPING THE CONSTRUCTION SAFETY PROGRAM

Two ENGIL projects were studied: I – Tapada do Outeiro, a USD 15.9 million power station built over two years and II – Edificio das Antas, a USD 2.2 million housing complex built over ten months. In the case of the Tapada do Outeiro, the cost of the wages paid to the safety team consisting of a safety expert, a foreman (who dedicated 10% of his time to safety) and two unskilled workers, for an average of 160 hours/normal months + 40 hours/month overtime, represented a total of 200 man hours/month on safety, and two carpenters whose cost is entered here in terms of the overtime they worked.

In the case of the Edificio das Antas, although the safety expert was only employed as a member of the team until February 1998 as his services were dispensed with once the project entered the finishings stage, the remainder of the team was as above and the cost of their wages were those of an average of 160 hours/normal months + 40 hours/month overtime, a total of 200 man hours/month on safety, and "half" a carpenter whose cost is entered here in terms of the overtime he worked.

The above labor costs include wages, holiday pay, one month's bonus pay, social security payments, taxes, individual work accident insurance, absences, etc., in other words, all applicable staff charges required under law.

As regards the safety equipment, the cost of both collective and individual protective equipment was calculated. Other costs were those of the civil liability insurance coverage for each site.

The Edificio das Antas project has not yet been completed and as such this study only refers to the period from its beginning in August 1997 to June 1998. Thus, as the value of the civil liability insurance coverage will devalue as the work continues, this was calculated as the total value of the insurance x total value of the project / total cost of the work.

5. FIXED STRUCTURAL COSTS

In addition to the above costs, there also are the structural charges, i.e., those related to labor, training, administrative overheads and fixed assets. The labor charges are those incurred by two engineers, one secretary and a safety expert, for an average of USD 60,000 for 1997. Travel charges are included in this item and represent an average of USD 556 per employee. Thus, these three employees (the engineers and the safety expert) spend an average of USD 16,700 over 11 months.

As regards training in the field of safety, an average of USD 2778 is spent annually which, added to the cost of travelling to the training costs (USD 2778), represents a sum total of USD 5556. Fixed assets (office equipment, tables, chairs, computers, etc.) totaled USD 299730 in 1997. Depreciation of these assets is calculated over 4 years, a value of USD 7440 per annum. Lastly, the administrative overheads (stationary, diskettes, printer ink, postage, etc.) are approximately USD 450/month, i.e., USD 5340 per annum.

We also quantified the cost of the safety and health program at the level of each construction site, although this proved somewhat difficult. The administrative costs (all the support to the project), those of the engineers, designers and safety experts involved in developing the Site Safety and Health Program, expenditure in stationary, copies, revisions, etc. and estimated salaries are entered as structural costs. An average of fifteen (15) days is necessary for developing a Site Safety and Health Program.

Site Safety Meetings are held twice a month for both projects. These meetings at the Tapada do Outeiro site cost USD 260 per meeting for a total of approximately USD 12,440 and, at the Edificio das Antas site, USD 200 per meeting for a total of USD 4,000.

Normally, a site is only inspected every 3 weeks and only by one person. Only in exceptional cases do these inspections occur more frequently and their cost is included under structural costs.

Five days were spent in studying the Tapada do Outeiro project and two days were spent on the Edificio das Antas project.

Table 1 – Tapada do Outeiro: Safety Implementation Costs
(in thousands of PTE)

Activity	Cost	Cost of Prevention	Percentage
Labor	747 118	19 746	2.64%
Equipment	258 367	392	0.15%
Material	1 037 914	5 420	0.52%
Sub-contracts	812 703	400	0.05%
Other	12 903	8 000	62.00%
Total	2 869 005	33 958	1.18%

Table 2 – Edificio das Antas: Safety Implementation Costs
(in thousands of PTE)

Activity	Cost	Cost of Prevention	Percentage
Labor	165 755	3 615	2.18%
Equipment	47 782	965	2.02%
Material	164 248	1 125	0.68%
Sub-contracts	14 875	0	0.00%
Other	5 142	766	14.90%
Total	397 802	6 471	1.63%

6. COMPARISON OF COSTS

Collecting information regarding the economic aspect of construction accidents proved difficult due to the sensitive nature of this information and to the legal implications of its publication. The solution we therefore adopted was to make an exhaustive list of all the types of costs that could be related to the different types of construction accidents. Naturally, our primary source was MAPFRE Portugal, the insurance company that sponsored the initial research.

There were no accidents on either of the two projects we studied. In order to evaluate the benefits of safety plans, we used the accident rates for Portugal during 1996 and simulated accidents for both projects taking into account the size of the project, the number of workers involved, the region and the type of injuries suffered. The injuries were chosen from the list of most frequent types of accidents: death, head injury, eye injury, chest injury, severe hand injury, arm injury, leg injury, foot injury and slight hand injury. Nine cases were selected for each project and the respective economic analysis was performed.

6.1 *Tapada do Outeiro*

Table 3 – Tapada do Outeiro: Costs related to Safety

Cost	PTE	USD
Insurance	8 000	44 450
Development of the Company Safety Program	160	890
Development of the Site Safety Program	2 920	16 225
Implementation of the Site Safety Program	25 958	144 210
Structural Costs	3 963	22 020
Accidents	0	0
Total Cost of Prevention	41 023	227 905

The accident frequency index was used to compare the cost of prevention and the cost of accidents. Whilst prevention costs were fixed and constant, total accident costs (CA) were calculated according to the frequency index (FI) and the estimated value for these types of accidents by means of the following formula:

$$CA = \frac{\text{N° of total accidents} \times \text{N° of workers employed} \times \text{Cost of the Accidents} \times \text{Duration of the Work in Months}}{\text{Total number of workers} \times 12 \text{ months}}$$

Table 4 – Tapada do Outeiro: Estimated Accident Costs without Safety Program

Cost	PTE	USD
Death	382	2 125
Head injury	20 655	114 750
Eye injury	19 820	110 110
Arm injury	16 858	93 655
Severe hand injury	37 375	207 640
Chest injury	32 431	180 170
Leg injury	22 529	125 160
Slight hand injury	24 209	134 495
Foot injury	35 299	196 105
Insurance Premium	8 000	44 450
Total Costs	185 127	1 028 040

6.2 *Edificio das Antas*

Table 5 – Edificio das Antas: Costs related to Safety

Cost	PTE	USD
Insurance	766	4 255
Development of the Company Safety Program	300	1 670
Development of the Site Safety Program	1 271	7 060
Implementation of the Site Safety Program	5 705	31 690
Structural Costs	346	1 925
Accidents	0	0
Total Cost of Prevention	8 388	46 600

Table 6 – Edificio das Antas: Estimated Accident Costs without Safety Program

Cost	PTE	USD
Death	80	445
Head injury	4 303	23 905
Eye injury	4 129	22 940
Arm injury	3 512	19 510
Severe hand injury	7 787	43 260
Chest injury	67 565	37 535
Leg injury	4 693	26 070
Slight hand injury	5 043	28 020
Foot injury	7 354	40 855
Insurance Premium	766	4 255
Total Costs	44 423	246 795

Table 7 – Ratio between Accident Costs and Safety Costs

Project	Estimated Cost of Accidents	Cost of Prevention	Relation between Cost of Accidents and Cost of Prevention
Tapada do Outeiro	PTE 185 127 000 (USD 1 028 485)	PTE 41 023 000 (USD 227 905)	4.5
Edificio das Antas	PTE 44 423 000 (USD 246 795)	PTE 8 388 000 (USD 46 600)	5.3

7. CONCLUSIONS

The analysis of the estimated cost of accidents as compared to the cost of preventing these shows that there are clear economic benefits to implementing prevention and safety programs and that construction companies should have an effective Construction Safety Policy. The relation between the cost of prevention and the cost of accidents that could be expected to occur in the absence of a Site Prevention and Safety Program was of 22% for the Tapada do Outeiro project and 19% for the Edificio das Antas project. (Table 7)

8. BIBLIOGRAPHY

1. *"Health and Safety"*, 1996. The costs of accidents at work, Series Booklet, London, UK.
2. I.D.I.C.T. 1994. "*Do projecto ao estaleiro*", Portugal.
3. I.D.I.C.T. 1997. "*Coordenacao da Seguranca*", Portugal.
4. Inspeccao Geral do Trabalho 1994. "*Estatisticas dos acidentes de trabalho*", Portugal.
5. Miguel, A. 1995. "*Manual de Higiene e Seguranca no Trabalho*, Porto Editora", Portugal.
6. Reis, C. and Soeiro, A. 1997. "*Analise Economica de Seguranca na Construcao*", MAPFRE, Portugal.

Implementation of Safety and Health on Construction Sites, Singh, Hinze & Coble (eds)
© 1999 Balkema, Rotterdam, ISBN 90 5809 036 1

Fraud in construction and building: A safety, management and insurance perspective

Jonathan E.Turner
Wilson & Turner Incorporated, Memphis, Tenn., USA

ABSTRACT: This article discusses the vital issues involved in investigating fraud in the construction process, and its hidden and subtle effects on construction safety. The introduction of deliberate untruth, has reverberating effects throughout the selection of vendor process, employee management, and precautionary actions of on-site and senior management. Using two recent case examples where fraud has dramatically affected the construction process, we can explore the relationship and effects that the fraud had on the safety plan and the resulting unforeseen issues. While still atypical in safety circles, the topic of management fraud is a tremendous factor in the safe resolution of modern construction projects. Like Wells (1997), in the experience of this author the introduction of fraud in the construction process affects the whole project, increasing the broad dangers from increased losses in safety.

1 FRAUD IN CONSTRUCTION

The following case studies are from the author's own experience and involved recent construction projects in the United States. The first involves federally funded low income housing (Turner 1996), and the second deals with two state funded school renovation projects (Turner 1998). Both of these projects became independently mired in fraud when senior managers initiated complex fraud schemes. In both cases the schemes escalated until they finally were discovered and collapsed. As the cases are still involved in a variety of litigation, I am referring to the core facts without naming any of the parties.

1.1 *Case study one*

This case involves an international service company, with substantial operations in the United States. In the late 1980's, a division that dealt with industrial maintenance nationwide was expanding in the Northeast into a new line - small construction projects. As the recession took hold in the early 1990's, the company's main business line was depressed, and construction activities grew to encompass most of this division's business in Massachusetts. It began to bid on, and win, federally funded construction projects of increasing size.

This developed into an entirely new line of business for the company, but one that was closely related to the core line, handling industrial property maintenance. In light of the changing business climate of the Northeast in the early 1990's, it seemed a natural growth area, providing regular revenues with acceptable profit margins.

The branch soon expanded its small construction efforts into a full scale contracting company. Divisions dealing with construction, flooring, roofing and painting were created - all in addition to the traditional main line, maintenance. Within a few years, the branch began bidding successfully on multi-million dollar projects, progressing on to projects over ten million dollars.

Given the company's vast experience in contracting, bidding on state and federally funded projects seemed a natural next step. It was bidding to provide essentially the same services it was providing its industrial customers. More importantly, the effort was successful at each level. The revenues from the new service areas not only made up for the declines in the maintenance division; they exceeded them.

In this case, the company transformed itself at the local level from a mechanical service company into a construction company. In making this transformation, the company began hiring construction managers who had limited knowledge of the company's core business. As the construction division grew, local management, who had limited knowledge of the construction business, became more dependent on the construction managers for information.

The operations manager, hired for his knowledge and expertise in all types of construction, quickly put together, and won, the most ambitious bid the branch had attempted. He was not long after when the general manager left. Now, as general manager, he had total control of branch operations. In this capacity, he greatly expanded the construction business, which was his area of expertise, and let the traditional business decline. This expansion created the first opportunity for safety issues to arise. The manager focused on developing these large construction projects without engaging experienced management or allowing for staff training.

As the projects quickly grew in both size and number, the office staff was overwhelmed with the attendant paperwork. Unfamiliar with the requirements of federally funded construction projects, they failed to properly control the costs or the administration of the projects. Yet, despite these difficulties, the financial statements showed regular revenues; and the company president was relieved to see an increase in profitability at a time of recession in the northeast. This overshadowed the overall lack of depth of the construction management.

Under the new general manager, the company succeeded in landing project after project, and was recognizing revenues with healthy profit margins. Having discovered that he could operate without intrusion as long as his records indicated profitability, he began making sure that his reports always reflected steady profits. As the projects advanced to a stage when they should have neared completion, he "cooked the books" to insure satisfaction at the corporate level. Since no one from the corporate office understood the construction business, he had no trouble settling any issues that arose.

With this protection in place, he began to defraud the company. He initiated a series of frauds, many of which became inter-related, and very efficiently siphoned off cash from the federally funded projects. He hired his best friend as a subcontractor, and utilized his invoices to initiate these frauds. When he was successful at that level, he spread the scope of the fraud to include almost every "shakedown" known to the construction industry. He and his partner threatened workers, lied on federal forms, falsified invoices, created companies, used nominee companies, and stole supplies to further their schemes.

Their most successful venture was in the use and abuse of under educated manual workers. They invented front companies that employed often illiterate and immigrant workers at hard labor. To maximize their thefts, they provided no training, insurance or safety materials to the laborers. As the scheme matured, injuries and accidents occurred. The fraudsters threatened and bullied the workers to minimize discovery, though the incidents would inevitably come to light.

They depended on the delayed reporting and posting of costs, and the automatic posting of accompanying revenue and profits, to hide the malfeasance. When one project would approach budget, they would simply shift costs to another; and, as they booked to different accounts, there was almost no way to determine that the frauds were being committed. Their use of these improper subcontractors, allowed them to control the outgoing costs, while providing a layer of protection against the worker's compensation and accident claims.

They used the funds and supplies from the fraud to live well, purchasing luxury automobiles and boats, and renovate a historic ballroom that the two secretly purchased. Further, they arranged to have supplies that were used on their side-jobs billed to company accounts. But it was the labor on the ballroom renovation that ultimately placed the workers at risk. They required the immigrant workers to put in 8 hours on the housing projects, and then another 8 hours on the ballroom.

Over 18 months numerous workers were required to work these long hours handling dangerous machinery, climbing on scaffolding, and carrying heavy loads of supplies and equipment. Inevitably, this led to workers getting injured and trying to make claims against the front companies. While the fraudsters effectively dissuaded some of the workers, over time these claims came through and were pressed against the company.

Much like a Ponzi scheme, however, this sort of fraud depends on ever increasing sources of revenue to cover the defalcation (McIlvaine 1997). Over the course of two years the fraudsters were able to siphon off several million dollars worth of funds and supplies. When they won a bid for a 12 million-dollar project, they transferred the losses from several other projects, which were winding down, to the new project and began executing the fraud there as well. Because of the increasing claims, they used some of the funds to pay off some of the injured workers in a desperate attempt to hide the fraud from senior management. To coordinate the conspiracy the general manager even went so far as to hire his partner in crime as his second in command.

As they moved through the available funds in the last project, they continued to bid on other, larger projects. But with no awards forthcoming, they frantically started bidding on more and more projects. They needed the continuing revenue stream, one large enough to hide what had grown into a multi-million dollar fraud.

Unable to land another project large enough to hide the mounting losses, they could no longer conceal their plight from corporate management. Management was first led to believe that all of the problems were limited to one smaller project. The fraudsters diverted attention in that direction until it was no longer possible to hide the problems with the large project. At that point, corporate internal auditors and investigators became involved, and the frauds quickly fell apart.

Unfortunately once the frauds had ended, it was no simple task to unravel the records that made up a dozen construction projects and thousands of subcontractor and supplier invoices. Initial fears focused on the staggering size of the losses, but were gradually replaced by concerns of what was concealed by the internal fraud. The fraudsters, who had both left the company independently, had hidden the fraud such that the initial suspicions amounted to only $200,000 in the face of anticipated losses of $2 million.

The company moved quickly once it discovered the internal fraud, filing the appropriate criminal referrals and giving notice to the crime policy carrier. The ensuing criminal investigation resulted in guilty pleas by all of the participants charged. Investigation of the fraud revealed that the losses from the scheme totaled in excess of $6 million.

The loss figure includes additional costs to cover claims from unpaid workers, worker's compensation and medical claims, and other employee claims resulting from the scheme. While no serious injuries occurred on these jobs, the convoluted fraud scheme served to hide many of the real injury statistics until after the scheme was discovered. And because of their methods, it is still not possible to know if all of the accidents and injuries have been accounted for.

1.2 *Case study two*

This next case dealt with a national service company, primarily structured along a franchisee/branch basis. In the early 1990s the company decided to strengthen its disaster recovery services and began implementing the roll out of a national construction services function. The first location came on line by the middle '90s and became a banner success. Subsequent operations came on line in other regions, and the company was well on its was to copying the model nationwide. As part of the aggressive path, the lead branch initiated a transfer from residential remodeling to commercial remodeling, quickly committing the company to two massive school modernization projects, totaling over 15 million dollars.

This lead branch was entering new territory, and developing a balanced progression to transfer to a mix of residential and commercial work. As the country was going through a series of annual natural disasters, the company seemed poised to make a bold move into a new line of business.

The branch expanded its residential remodeling, winning contracts with insurance companies and even other construction firms to provide services. Focusing on the general contracting aspect, the branch kept its personnel limited to project supervisors, managers, and other construction professionals. Quickly the blending into commercial work took off, the Manager convinced the corporate parent to sign a contract for a multi-million dollar school modernization as a test, and followed it up with another in less then 30 days. The second contract was for in excess of 10 million dollars. These contracts required the company to complete multi-phase renovation and new construction around an open school environment, a challenge for even very experienced construction managers.

As with the first case, the company performed well initially. The transformation provided a strong growth projects, as well as regular revenues and exceptional growth, while maintaining respectable, but not excessive, profitability. The company hired a senior construction manager to head the branch, and depended on him to provide reporting and information to the corporate headquarters. The key to convincing management was a clear understanding that all of the subcontractors would be insured and bonded to protect the workers and the company.

Unlike the first case, this manager kept the staff small and mobile, allowing him to maintain tight control over all aspects of the business (Turner & Segers 1997). He effectively had total control over the operations, well removed from corporate proximity. He quickly changed out most of the office staff replacing them with young and inexperienced personnel, who were easily cowed by his aggressive management style. In this manner, he focused his time and attention on the budding commercial business and let the traditional areas decline.

As the two projects progressed, the inexperienced office staff was overwhelmed trying to keep up with the subcontracts, vendors, and other paperwork. Hampered by their lack of experience, they often times simply followed his instructions blindly, setting the stage for the fraud to follow. In spite of this, the financial reporting always indicated steady numbers along the predicted path. It would not be until later that it was learned that the numbers were being created, rather then compiled. During this time also, they quickly dropped the requirement that the subcontractors present proof of insurance and bonding. After setting up and establishing these relationships they carefully hid this important lack of documentation from senior management.

Upon discovering that he could run his "kingdom' as he saw fit, as long as his financial reporting met expectations, he "cooked the books" to insure satisfaction at the corporate level. Since no one from the corporate office understood the construction business, he had no trouble resolving (at least, to their satisfaction) the issues that arose. When they did question various aspects, he relied upon their belief that everything was covered by both insurance and bonding.

Thus protected, he began defrauding the project funds. He used unqualified contractors, illegal workers, uninsured subcontractors, recent immigrants, and other parties who he could coerce into some level of performance, while he milked substantial funds out of the contracts for his personal benefit. When accidents occurred, he threatened, coerced silence or bought a solution, regularly resorting to complex lies to cover the expenditures and hide any reporting requirements. Taking advantage of uneducated and often illiterate workers, he created the appearance of progress through sloppy or unsafe practices and conditions.

Since most of the subcontractors were unqualified, and the project was a modernization, phased around school students, this created a highly dangerous situation for both the workers and the children. The workers were relying on the work done by the other subcontractors. Several times during the construction phase the school board representative found subcontractors building and preparing areas contrary to the specifications. Had the company not discovered the fraud and acted responsibly to correct all of the deficiencies, children could have come to subsequent harm.

As part of this scheme, he used and threatened workers, lied on federal forms, falsified invoices, created companies, used nominee companies, and stole supplies to further his schemes. The key to hiding the fraud was the delayed reporting period in large-scale construction. In this case one of the large projects was scheduled for 12 months, but eventually ran 18; and the second was scheduled for 24 months, and was finally completed in 28. Similarly to the first case study, he depended on the delayed reporting and posting of costs, and the automatic posting of accompanying revenue and profits, to hide the malfeasance. When one project would approach budget, he shifted costs to another; and, as they booked to different accounts, the false entries were extremely effective at masking the fraud.

As with many similar frauds, the scheme was concocted to allow him to maintain an extravagant lifestyle and support a gambling habit. While his coworkers were aware of his acts, they were successfully cowed into submission and senior management would not learn of this important information until after the scheme had collapsed under its own weight.

From a safety perspective, this scheme placed a large number of workers directly in harm's way. Much of the construction involved multi-story masonry, built by his under-qualified workers. When accidents occurred, he threatened to fire the entire team if the matter was pressed.

The company moved quickly once it discovered the internal fraud, filing the appropriate criminal referrals and giving notice to the crime policy carrier. The ensuing criminal investigation resulted in a guilty plea by the former manager. Investigation of the fraud revealed losses in excess of $4 million.

As with the first case, this loss figure includes additional costs to cover claims from unpaid workers, worker's compensation and other related claims resulting from the scheme. Fortunately there were no serious injuries during the school modernizations, but, again, the convoluted fraud scheme hid the injury statistics until after the scheme was discovered. Since these jobs have just recently closed, it is also impossible to know if all of the accidents and injuries have been reported.

2 ANALYSIS

Both stories have striking similarities, beyond the fact that they involve management frauds. They illustrate the inherent danger that can attach to any owner, developer, or construction firm, company or subcontractor that works in today's environment. These managers' actions exposed everyone associated with the projects to a range of safety issues; increased risk of injury (from poor performance), exposure to worker's compensation or environmental liability claims (from ineligible/uninsurable entities), and the often overlooked need for employee dishonesty insurance.

These cases all involved managers who were willing to create or ignore the required insurance and expose all associated with the project to the dangers from improperly trained employees and fabricated or substandard work product. Despite the risk of this type of frauds, both projects were managed along traditional constructions methods, reliance on site managed reporting, blind bidding, lack of pre-qualification requirements, and other practices designed to ensure lowers costs, but at an increased risk to safety and financial stability.

While the trend in these cases is complex contract forms, no contract will protect you from active dishonesty. As these cases illustrate, the use of possibly well intentioned but unqualified elements to hide or obscure fraud exposes the remaining elements to unrealized risks. In these cases, there were higher incidences of worker injury, serious and unreported dangers in both site and environment, and other improper situations that were tolerated. The result of this tolerance was increased severity of injury, and a dramatically higher incidence of injury due to accidents related to poor quality work product.

3 CONCLUSIONS

Fraud presents a unique problem to all business groups. But the construction industry, with its almost unique reliance of disparate and unrelated elements fusing together to function along time specific guidelines, is at a special risk. There is no way to ensure that fraud will not occur, and no cost effective way to insure against the full cost of fraud. As such, each company must begin to pro-actively address fraud risks to identify potential weaknesses, address these areas early, and move quickly in the face of discovery. For only through aggressive management of can a company limit its losses. In both of the above cases, complacency allowed the companies to absorb millions of dollars in losses before they addressed the problems. These are clearly figures that would kill lesser organizations. In both of the included cases they did destroy the construction divisions of these organizations.

REFERENCES

McIlvaine, A., 1997. To Catch a Thief. *Human Resources Executive* Vol. 11 No. 12: 77-80.

Turner, J. E., 1996. Strategies for Preventing Employee Fraud in Times of Transition. *John Liner Review* Vol. 10 No. 2: 40-48.

Turner, J. E. & Segers, P. J., 1997. Employee Accomplices: The Fifth Column in Fraud Cases. *The Examiner* Vol. 3 No. 4: 1-3.

Turner, J. E., 1998. Successfully Resolving Issues of Employee Dishonesty. *John Liner Review* Vol. 12 No. 2: 47-58.

Weinstein, M., 1996. Conventional Wisdom. *Risk & Insurance* Vol. 8 No. 6: 32-38.

Wells, J. T., 1997. *Occupational Fraud and Abuse*. Austin: Obsidian.

8 Education and training

Implementation of Safety and Health on Construction Sites, Singh, Hinze & Coble (eds)
© 1999 Balkema, Rotterdam, ISBN 90 5809 036 1

Teaching health and safety to construction engineers

J.Christian
Department of Civil Engineering, University of New Brunswick, N.B., Canada

ABSTRACT: The paper describes a course in Construction Engineering which has been offered to civil, forest, and geological engineering students at the University of New Brunswick in Canada over the past twenty years. A detailed analysis of the course is made covering topics including site procedures and planning; excavation and trenches; stabilization, piling and cofferdams; concrete formwork and design; falsework and scaffolding; case histories; costs of accidents, workers compensation; liability; health and safety regulations applied to construction and details of hazardous materials information systems, fall arrest systems, hygiene and legal implications. The reasons for changes that have been made in the past are also discussed. Recent more rigid requirements of the Canadian Engineering Accreditation Board are described regarding safety, health and hazardous materials on construction sites. The course described has recently successfully met these more rigid requirements.

1 INTRODUCTION

The course described is concerned with the construction of temporary works, construction methods and safety, the provincial occupational health and safety act and code requirements, site planning, structural falsework, scaffolding, formwork, shoring, bracing, trench excavation and earthworks operations. The course provides the students with a detailed understanding of the Act and Regulations of the provincial Workplace, Health, Safety and Compensation Commission (WHSCC). It also provides a detailed understanding of many national codes of practice and standards regarding safety and describes international best practice procedures.

The course is offered over a 13 week term, with two 50 minute lectures each week.

The required texts for the course include the New Brunswick Occupational Health and Safety Act, the New Brunswick Workplace Hazardous Materials Information System Regulation, the General Regulations of the New Brunswick Occupational Health and Safety Act and a book of class notes which has been built up over the years and is modified every year. The Act and Regulations in other Canadian provinces are not identified but are very similar. This means that a student graduating in one province will be very familiar with the regulations in another province.

2 COURSE OUTLINE

The course is divided into three segments: Temporary Works, Safety Analysis, and Health and Safety. The lecture topics are as follows:

1. Introduction (1)

A. TEMPORARY WORKS

2. Site Procedures, Layout and Planning (1)
3. Excavations and Trenches (3)
4. Stabilization, Piling and Cofferdams (3)
5. Concrete Formwork, Formwork Design (4)
6. Falsework and Scaffolding (2)

B. SAFETY ANALYSIS

7. Case Histories/Case Studies (3)
8. Cost of Accidents, Workers Compensation (1)
9. Liability (1)
10. Loads and Failures (1)

C. HEALTH AND SAFETY

The Workplace, Health, Safety, Compensation, Commission of New Brunswick Act and Regulations applied to Construction,
Workplace Hazardous Materials Information System, Fall
Arrest Systems, First Aid in Construction Hygiene, Legal Implications (6)

The numbers in parenthesis indicate the number of lecture hours for each topic. Civil, forest and geological engineering students attend the course. It is a senior level course because students are required to have already taken the basic rudimentary engineering courses and there is an expectation that most of the students will have been exposed to work on construction sites for several of their summer employment periods. For these reasons, students are required to have completed 110 credit hours out of a total of 180 to 200 credit hours for the programs. The civil engineering students are also required to have taken an introductory course in Construction Engineering as a prerequisite. There are similar requirements for the forest engineering students. The course is a compulsory course for civil and forest engineering students. Until the present, the course was optional for the geological engineering students. However, following a recent Canadian Engineering Accreditation Board visit, the Board has now required that the course must be mandatory for geological engineering students. Other Engineering departments offer other courses concerning health and safety.

Most of the assignments in the course are structured so that the students answer the question from a "safety" point of view, by complying with standards and codes of practice, for example, in the design of formwork and the design of sheet piling assignments. One group assignment gives details of fictitious accidents which have occurred on construction sites and the students are asked to comment on why the accidents happened and how the accidents could have been prevented. Another group assignment asks 60 specific questions concerning information found in the provincial Health and Safety Act and Regulations. This ensures, in theory, that the students thoroughly read and discuss the Act and Regulations which they regard as rather "boring" and feel that for fun-loving engineering students better use could be made of their Friday and Saturday evenings. However, they realize that the knowledge that they gain is very important.

3 TEMPORARY WORKS AND SAFETY ANALYSIS

The following subtopics are dealt with in the course with a special emphasis being placed on "safety" aspects:

Site Procedures

- preparations & preliminaries
- permits (building, demolition, foundations, structural)
- contract documents
- preliminary layout survey
- global positioning system
- materials supply
- demolition (sub-contractor)
- haul roads
- site fabrication
- works layout
- WHMIS & MSDS
- temporary works and services
- planning for the job
- protective gear
- accident prevention checklists
- due diligence
- construction site planning
- construction signs

Excavations and trenches

- unit weight
- coefficient of bulking
- coefficient of compaction
- worked example. excavation
- W.H.S.C.C.N.B. excavations and trenches
- trench failures
- trench supports
- safe slopes
- stabilization of soils
- pumping & dewatering
- trenchless technology
- sheet piling
- cofferdams & caissons
- worked examples
  - pumping
  - sheet piling
- appendices

Concrete Formwork and Formwork Design

- formwork introduction
- concrete pressures
- formwork details
- formwork design
- worked examples
- formwork types
- transporting & pumping concrete
- removal of formwork
- practice & safety
- cold weather
- tests
- appendices

Falsework and Scaffolding

- definitions - falsework - shoring
- types of scaffolding, properties
- loads, effective length
- design of scaffolding
- systems
- props
- inspection - safety
- rolling scaffolds

Case Histories/Case Studies

Stabilization, Piling and Cofferdams

- earthworks
- cuttings, pits and trenches
- stabilization
- groundwater lowering systems
- piling

Cost of Accidents, Workers Compensation

- safety management
- financial incentives
- hidden costs of accidents
- typical costs of accidents
- workers compensation

- cooling tower collapse
- walkway collapse
- trench support
- composite steel & concrete frame
- culvert
- staircase
- Fredericton water tower collapse

Liability

- case study
- legal aspects
- due diligence

Load and Failures

- engineering judgement
- loads
- uncertainties

Several videos are introduced into the lectures to enhance the students' interest and understanding. The videos cover scaffolds, trenching, and safety aspects on two local mega construction projects (C$1b & C$0.85b).

4 HEALTH AND SAFETY APPLIED TO CONSTRUCTION AND FORESTRY

Several professional speakers from the provincial Workplace, Health, Safety, Compensation Commission (W.H.S.C.C.) are invited to teach the segment of the course which covers the provincial Act and Regulations applied to construction and forestry. Health, hygiene and legal matters are dealt with as well as safety issues. In the past the speakers came into the university to speak during the regular 1 hour slots scheduled for the course. If their segments took two or three hours they would return on another date, suitable to them, to complete the presentation. This became somewhat disjointed for the students, not only for the health and safety presentations but also for the regular course topics which were covered. Discontinuities occured.

About six years ago a short course was instituted for the health and safety segment. Including slides and a video, the short course extended from early afternoon to late evening and took nine hours, including two breaks. Even this format proved to be very difficult because several other classrooms, courses and laboratories had to be rescheduled on the particular afternoon the short course was scheduled. For the past few years the short course has been presented for four hours on two evenings, usually the same evening in consecutive weeks. This has proved to be a very time effective format for both the over one hundred students that attend the course some years and the presenters. A good continuity has been created for the first time.

The segment of the short course which covers safety issues regarding the Provincial Act and Regulations emphasizes the following topics:

Act	Regulations
definitions	interpretation
duties	food and rest periods
Joint Health and Safety Committee	first aidroom
right to refuse	sound level dBa v duration
powers & duties of officers	hearing protection
appeals	headwear
enforcement	footwear

Regulations (continued)

	guardrails
gloves	allowable unit stresses
fall-arresting systems	roofs
safety nets	cat walks
stockpiled material	fixed ladders
hazardous substances	portable ladders
storage batteries	scaffolds
general duties of employers	excavations and trenches
traffic safety	pits and quarries
buildings and structures	diver-medical requirements

The segment which covers health issues is given by a professional hygienist and deals with hygiene, air quality, physical agents, respiratory and hearing protection, the Provincial Workplace Hazardous Materials System Regulations (WHMIS) which includes Material Safety Data Sheets (MSDS)

The segment of the short course which covers the legal issues is given by a lawyer who devotes all her time to health and safety legal issues for the provincial government. It covers several case studies in occupational health and safety. The case studies cover lockout procedures causing an accident, a shot blasting accident, a mining accident and safety procedures, a drowning accident on a bridge construction site, and a punch press accident and procedures. The students are asked to comment and discuss each case study during the class. The discussions are very spirited and educational; some of the engineering students are often surprised by the stringent interpretation of the law regarding the responsibilities of the managers and engineers. The term "due diligence" is once again brought into the discussions.

5 CANADIAN ENGINEERING ACCREDITATION BOARD REQUIREMENTS

In 1965, the Canadian Council of Professional Engineers established the Canadian Engineering Accreditation Board to accredit Canadian undergraduate engineering programs that meet or exceed educational standards acceptable for professional engineering registration in Canada. A detailed questionnaire is completed by the engineering faculty prior to an accreditation visit by a team of senior engineers. A qualitative and quantitative analysis of the curriculum content is performed to ensure that the programs meet the fairly stringent criteria.

One area in which the criteria have become more stringent is in the area of professional aspects and in particular how the students are made aware of, and exposed to, public worker safety and health considerations. This means that each engineering department, when completing the questionnaire, must explicitly state in which course and how public safety and health considerations are covered. As has been previously mentioned the Accreditation Board now require that some courses which were optional in some programs are made mandatory, so that the health and safety considerations are covered.

6 CONCLUSIONS

Over the last twenty years enormous changes have taken place in the teaching of safety and health on construction sites in civil and construction engineering departments at some universities. These changes have been brought about by the need for change, the far sightedness of some engineering departments, the necessity of including such an important topic in the curriculum, and the more stringent requirements of Accreditation Boards.

REFERENCES

J. Christian. 1981. Management, Machines & Methods in Civil Engineering. New York. John Wiley.

Occupational Health and Safety Act. 1997. Ch. 0-0.2. Workplace, Health, Safety, Compensation Commission of New Brunswick.

Regulation 91-191. OHSA. 1997. Workplace, Health, Safety, Compensation Commission of New Brunswick.

WHMIS 88-221. Workplace, Health, Safety, Compensation Commission of New Brunswick.

Canadian Engineering Accreditation Board. 1997. Accreditation Criteria and Procedures. Canadian Council of Professional Engineers.

Implementation of Safety and Health on Construction Sites, Singh, Hinze & Coble (eds)
© 1999 Balkema, Rotterdam, ISBN 90 5809 036 1

College's emphasis on construction safety [1]

R.J.Coble, J.H.Hinze, M.J.McDermott & B.R.Elliott
M.E.Rinker, Sr. School of Building Construction, University of Florida, Gainesville, Fla., USA

ABSTRACT: The research contained in this paper is focused on construction safety education, with respect to the extent that safety is taught at the college and/or university level. Construction safety is a critical subject since it is basic to all aspects of the construction process, but construction education programs at the college/university level have varied approaches and emphasis on this subject. Construction graduates will face multiple issues related to safety throughout all phases of construction projects in which they are involved. Because construction injuries are such a large part of a construction company's exposure to liability, especially in regard to workers' compensation exposure, building construction graduates are often placed in positions where knowledge of construction safety is essential to their job performance. It is therefore critical that the institutions of higher learning provide a thorough and practical education which introduces students to the identification of job site hazards, the proper use of protective safety devices, and a reasonable understanding of the OSHA regulations as they pertain to the construction industry. The focus of this research was to inquire about safety education at the colleges/universities which offer programs in construction science. This study was designed to provide a general idea of how, and to what extent, construction safety is promoted within their curriculum. This paper presents the results of the study, which was conducted during the Spring 1997 semester.

Key Words: Construction education, construction safety, college engineering curriculum, OSHA training.

1 INTRODUCTION

Construction worker injuries can seriously compromise the completion of a project and have a significant impact on the profitability of a firm. Even in the absence of any litigation, the high costs of medical treatment can severely impact corporate profits. This realization in the construction industry has become more widely recognized in recent years. In light of this new awareness, it is prudent that graduates of building science programs have training in the area of safety. But to what extent are the various building science programs responding to this need? The results of a survey of a selection of building science programs provide an answer to this question, conducted during the Spring of 1997.

[1] This paper first appeared in the American Professional Constructor, Dec. 1998, published by The American Institute of Constructors.

The need for construction safety education is not an issue of consensus among construction educators at the college level. This study was designed to evaluate the extent that construction safety education was being emphasized in construction programs at the college level throughout the United States. In addition to the obvious importance placed on the study of OSHA construction regulations, by legislative mandate, there are also cost considerations that affect bottom-line profits. Because most graduates from a construction education program can expect to be in a management position at some point during their careers, it is imperative that these students are well prepared in the area of construction safety management, from preliminary estimating to project close out.

2 RESEARCH METHODOLOGY

To obtain information regarding the extent that safety is integrated in construction programs, a survey was developed that asked for some basic information. The questionnaire used for this report was designed to gather information from colleges and universities offering bachelor programs in construction related fields. A key question being asked was whether a course in the program is completely devoted to construction safety. Additionally, if such a class is offered, what portion of the class is devoted to the discussion and understanding of OSHA regulations. A copy of this survey is provided in Appendix A of this paper.

2.1 *Intent of Survey*

The survey of four-year college/university level construction programs in the United States was conducted in order to find out how many programs were teaching construction safety to their students, and to what extent the program covered this subject. Furthermore, the survey was intended to discover what portion of the construction programs focused on OSHA regulations. Another focus of this study was to ascertain if construction safety was incorporated into other courses within their program.

2.2 *Sample Selection*

All colleges and universities offering at least a four (4) year bachelor's degree in a construction related field were contacted and asked to participate in this survey being conducted by the Center for Construction Safety and Loss Control at the University of Florida. The contact persons, along with their telephone and fax numbers, were obtained from the Associated Schools of Construction (ASC) 1996 membership directory. Multiple attempts, through mail, facsimile, email, and phone calls were made when necessary to contact all representatives listed in the ASC directory.

2.3 *Procedure*

All representatives were contacted first by facsimile and/or mail. All those who did not respond were contacted a second time by telephone or email, if available. If there was no response, then repeated attempts were made by telephone and email wherever possible. Fifty-eight percent (58%), or 55 out of 95, of the institutions surveyed provided a response to this survey.

2.4 *Analysis*

The reasons for some programs not participating in the study varied, including lack of interest in the survey, lack of an established safety course at the corresponding institution, or absence of an instructor or informed individual at the time this questionnaire was introduced. The recipients who did respond were generally positive about the issue of construction safety in the college classroom, and were very eager to discover the current and future safety education trends of other U.S. colleges and universities.

All results were analyzed in a scientific manner, producing graphic results from a calculated distribution of yes and no responses. Although some responses contained data that was not easily quantified in descriptive statistics, binary responses to yes and no questions are included in charts and graphs as separate variables.

3 RESULTS

In response to Question #1, forty-five percent (45%), or 25 out of 55, of the responding institutions have a course within their curriculum which is wholly devoted to construction safety issues. As summarized below in Figure 1, some of the respondents who answered 'no' to this question cited either that safety is addressed in a more generalized manner through other courses in the program, or that a certain group of required classes provide a satisfactory discourse on the topic of safety as it relates to each individual class. Sixteen institutions reported that one-half (1/2) of the entire class is devoted to construction safety issues, while the other half covers estimating, materials and methods, or soil mechanics.

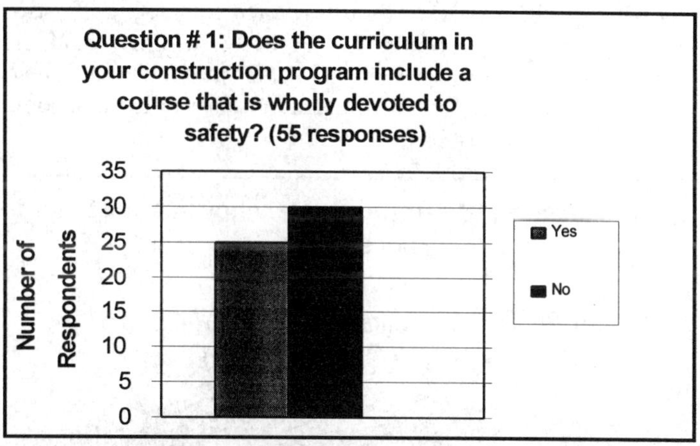

Figure 1: Question #1

The second question asked at what level the safety course (if available) was taught. Most institutions indicated that a safety course is taught predominately at the junior and senior levels. Some schools responded that this course was available to all academic levels within the program, though most tend to complete the course during the last two years of the program. Another question asked if the class is required or if it is an

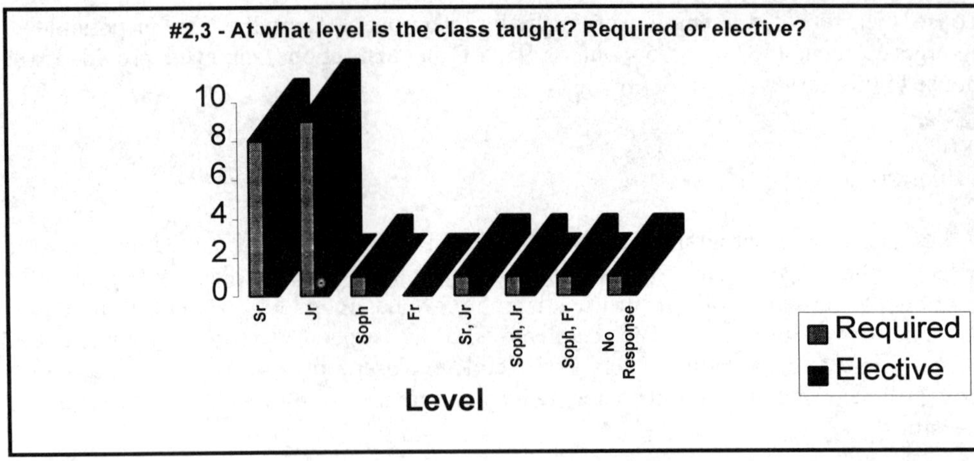

Figure 2: Questions #2 and #3

elective. Senior and junior levels have the highest levels of respondents who indicate that the class is required. Because some of the institutions do not offer a class which is completely devoted to construction safety, but do offer a fairly substantial inclusion of safety in a portion of one or more classes, only the yes respondents are presented in the graph. Many of the school representatives made it clear that a definite effort was being made by the department to include a mandatory construction safety class or seminar for graduating seniors. Questions #2 and #3 are grouped together in Figure 2 in order to illustrate the relationship between the level at which the class is offered, and whether or not the class is mandatory.

Figure 3 provides a range of how many students take the safety class per year, in schools which offer a course wholly devoted to safety. Fifty-two percent (52%) of all schools who responded reported that between thirty and fifty students take the construction safety class per year. On this question, many respondents from schools where construction safety is not a mandatory class noted that some students take elective

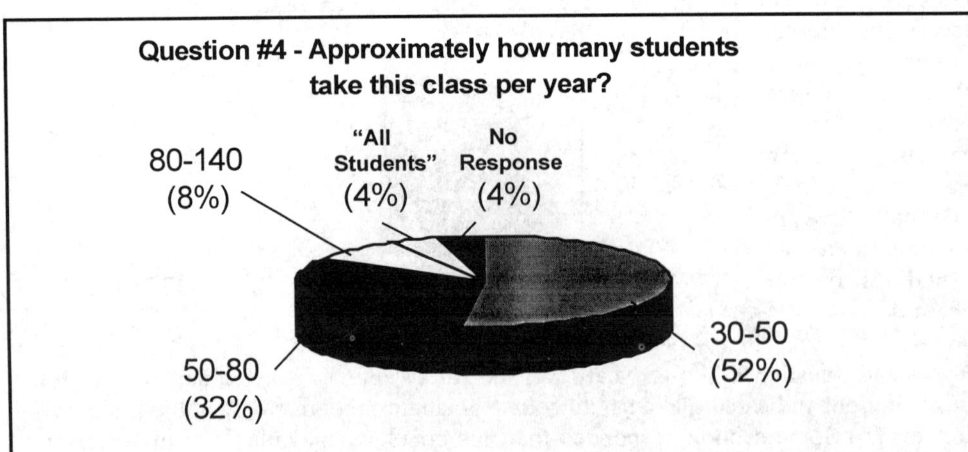

Figure 3: Question #4

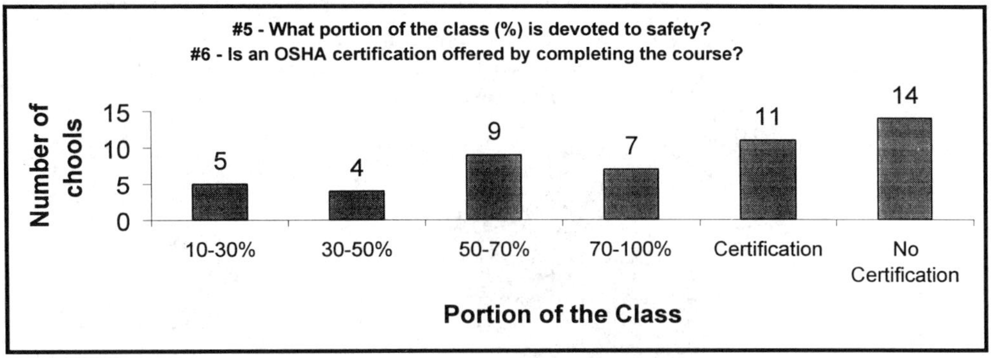

Figure 4: Questions #5 and #6

courses outside their department which may relate to safety. Four percent (4%) who did not respond to this question either did not have access to class registration data or did not wish to disclose this information.

Questions were also asked about the instructional content that related directly to the OSHA regulations. All of the schools that offer a separate safety class responded that they do cover OSHA regulations to varying degrees. Forty-four percent (44%), or 11 out of 25, of these institutions with a separate safety course include OSHA Outreach Training (i.e., OSHA ten-hour, OSHA thirty-hour, or OSHA 500 program), with certificates and wallet-sized cards awarded upon successful completion of the class. The results are summarized in Figure 4. Some of the respondents who stated that no OSHA Outreach Training was offered within their program cited that steps were being taken to include OSHA Outreach Training by the next academic year.

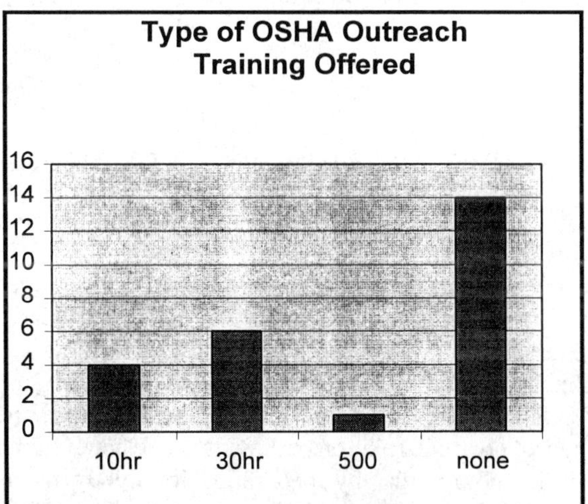

Figure 5: Type of OSHA Outreach Training Offered by Participating Schools

Six of the eleven respondents whose institution offers a separate safety class stated that the thirty-hour OSHA Outreach Training is included in the course, while four offer the ten-hour training (see Figure 5). The OSHA Administration offers an Outreach Program to colleges, universities, and technical or vocational institutions who provide a thorough discussion and demonstration of OSHA regulations for the construction industry. Passing students are issued a certificate and wallet-sized card, which recognizes their successful completion of the OSHA training program.

Questions #7, #8, and #9 will only be addressed briefly here. Answers to Question #7 indicated the use of OSHA regulations and/or safety management texts. Results of

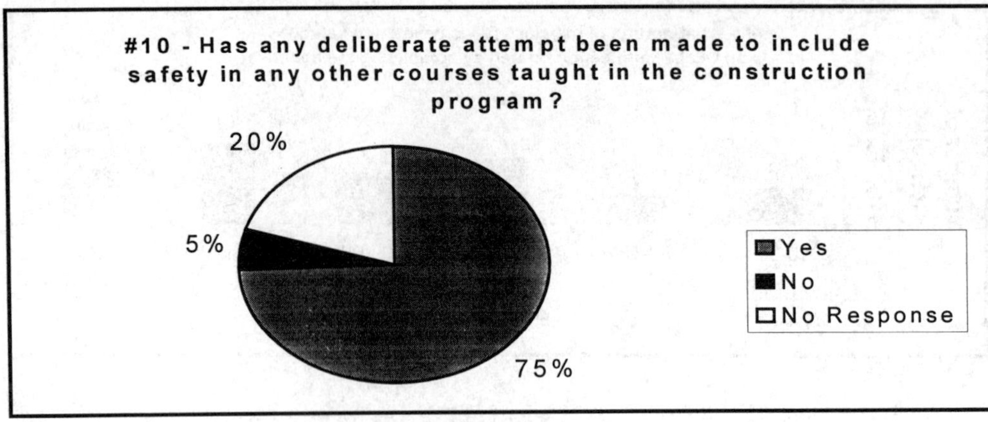

Figure 6: Question #10

Question #8, regarding when the safety courses were first taught, show a wide distribution of years, ranging from the early 1970's up to 1997. In Question #9, most programs with a safety course were willing to provide a copy of their syllabus.

Question #10 asked if a deliberate attempt has been made to include safety in any other courses taught in the construction programs in these participating schools. The results are presented in Figure 6.

Figure 6 provides responses from all participants in the survey, without regard to whether or not they have an individual safety course in their curriculum. The five percent who responded 'no' to Question #10 indicated that they either offer a class devoted to safety and/or guest lecturers who discuss construction safety issues at the schools. Some indicated that time constraints on the delivery of other course materials often do not allow for a thorough discussion of construction safety issues.

4 CONCLUSION

The future for an improved, safer working environment for the construction industry is integral to the training that is done in the classrooms of colleges and universities. Students who graduate from a construction science, engineering, or architecture program will eventually find themselves involved in construction safety issues on the job. Therefore, it is important that the construction programs in the United States colleges and universities incorporate safety into their curriculum, not only as a momentary topic of discussion in each class, but as a separate class which focuses on all areas which will help the graduates in their careers. Since the OSHA regulations dictate the minimum construction safety standards on all U.S. job sites, it is encouraging to see from the survey that some colleges have formed a positive relationship with the Occupational Safety and Health Administration, most commonly by becoming involved in the Construction Safety Outreach Program. The closer the relationship between college level construction safety programs and the OSHA Administration, the more prepared students will be when entering the modern construction management workforce.

This study shows that some colleges and universities are very proactive in their safety education, but the authors recommend that all construction science programs seriously consider specifically addressing safety in their curriculum. With the importance of this

UNIVERSITY OF FLORIDA

M.E. RINKER, Sr. SCHOOL OF BUILDING CONSTRUCTION
Survey of Safety Education in Construction Programs

1. Does the curriculum in your construction program include a course that is wholly devoted to construction safety?

If yes, answer the following:

2. At What level is this class taught?
____ freshman ____ soph. ____ junior ____ senior ____ graduate

3. Is this class an elective or is it required? ____ elective ____ required

4. Approximately how many students take the class each year? _____

5. What portion of the class is devoted to coverage of the OSHA regulations (as opposed to coverage of safety management issues)? ____%

6. Is OSHA Outreach Training earned by completing the course?
____ yes, 10-hour ____ yes, 30-hour ____ no

7. What text is utilized in this class? ____

8. When was this class taught for the first time (which year)? _____________

9. Could you provide a copy of the syllabus? _____ yes, attached _____ no

10. Is any deliberate attempt made to include safety in any of the other courses taught in your construction program? ___ yes ___ no
If yes, please describe the course and describe what safety material is covered:
____
____
____

Please provide your name and address below, and a summary report will be sent to you. Thank you for your assistance. Incidentally, your comments will remain anonymous.

Name: ____ DATE: ___________
Address: ____
University: ____
City, State: ____
Fax: ___________________________ e-mail________________________

Please return to: Center for Construction Safety and Loss Control
M.E. Rinker, Sr. School of Building Construction
Gainesville, FL 32611-2032

Or Fax to:
(352) 392 –9606

subject being emphasized by federal mandate and also by construction management, all graduates from construction science programs can benefit in their career by being safety educated. The job site should not be the place where graduates of construction science programs have their first exposure to OSHA and construction safety.

Implementation of Safety and Health on Construction Sites, Singh, Hinze & Coble (eds)
© 1999 Balkema, Rotterdam, ISBN 90 5809 036 1

Safety education expectations for construction engineering and management students

G.R.Smith
Iowa State University, Ames, Iowa, USA

T.M.Arnold
Pennsylvania State University, Indiana, Pa., USA

ABSTRACT: Implementation of safety and health programs on construction projects requires an educated workforce that is knowledgeable in safety requirements and procedures. Education of construction engineering and management students in safety has not often been a high priority. Except for a widely distributed instruction module designed by the American Society of Civil Engineers in 1994, safety curriculum content has not been given much consideration. This paper discusses the results of a survey conducted in one region of the United States to determine what safety knowledge practitioners would expect in newly hired engineering graduates. The second phase was to develop a simple analysis tool that could be used to review existing instructional materials on construction safety to determine which most closely matched to industry expectations. The survey identified twelve key areas where the majority of the responding contractors felt were important knowledge areas for new employees. A comparative analysis is shown, although not detailed, to illustrate the author's application to instructional material evaluation.

1 INTRODUCTION

In 1996 safety education at the Pennsylvania State University Department of Civil and Environmental Engineering was being conducted in a fairly disjointed manner with topics being addressed haphazardly within several courses. There was little continuity in the material and level at which it was being taught. An independent study course on safety management had been developed for graduate students, but the materials in the course were generally outdated. Within the university an undergraduate safety and health curriculum was being developed that offered a bachelor's degree in industrial safety and health. The new degree program faculty identified the need for an elective course on construction safety. A new course for this degree program could be developed, but it was felt that the course should also benefit students in civil engineering.

Searching for relevant instructional material revealed that only a few references were available as potential textbooks, and only a few articles were found that addressed safety education specific to construction. One solution would have been to obtain curriculum information and outlines from other schools. However, it was decided that the construction industry could provide excellent guidance on the appropriate content for such a course, particularly if the information was obtained from the safety directors in the companies.

Suckarieh and Diamantes (1995) noted in their work that safety has not influenced construction education to any significant extent. They indicated that the time dedicated to construction safety is lacking, but formal training could have a significant impact on their performance as soon as they graduate. They reported that only about fifty percent of the construction management programs have courses that are dedicated to construction safety. The common elements of the lower division safety courses were as follows:

- Introduction to Occupational Safety and Health Act
- OSHA standards
- Administration in the field
- Craft education requirements
- Filing forms and accident reports
- Keeping hazardous materials information
- Preparing for OSHA inspections

Upper division courses reported to include the following elements:

- Motivational factors
- Incentive programs
- Safety program effectiveness
- New initiatives

Their paper concludes that a plan for formal education in construction safety can be either a stand-alone course or, where appropriate, integrated into all elements of the curriculum by covering the material in several courses in a coordinated fashion. Their fundamental goal was to assure that their students accomplish the basic outline of the U.S. Department of Labor, Occupational Safety and Health Administration 10-hour safety training. The program content described above exceeds the minumum10-hour course requirements.

While the 10-hour course is an admirable target, a fundamental question should be asked about the remaining course content. What do the contractors, the primary employer of graduates, think is important for graduating engineers or expect the to be capable of performing in regard to safety?

1.1 *Objective*

Finding an answer to the question of contractor expectations was considered to be a fundamental step to determining the baseline for course content. A survey technique was chosen to solicit information on construction safety knowledge or responsibilities. The industry safety managers were asked to identify the career development stage at which employees were required to have certain knowledge. Primary interest would be the entry-level employee through the fifth year of employment. The questionnaire was developed to solicit data on what knowledge a newly hired construction engineer is expected to have. Responses to questions related to responsibility and knowledge at the fifth year or experience were also considered important feedback for course content selection.

A secondary objective of the survey was to develop an evaluation tool based on the survey results for evaluating existing instructional materials. The source that most closely matches the needs identified by industry would be a likely choice as the primary reference in the development of a new course.

2 SURVEY METHODOLOGY

Since the purpose of the survey was to investigate course content for construction students at Penn State University, the survey targeted known employers and likely employers of Penn State students. These employers predominately operate in the Northeast United States. 123 contractors were identified as working in the northeast region at the time of the survey initiation. Each of these contractors reported work volume in excess of $100 million in contract revenues per year. 123 questionnaires were mailed, but 18 surveys were returned due to faulty address information. Therefore, the baseline was reduced to 105 possible survey responses. Twenty-seven contractors completed the survey for a 26% completion rate. This return rate considered satisfactory for course development input from industry. The survey document is provided as an appendix to this paper. Several respondents commented on the importance of the content and that they appreciated the opportunity to provide their comments.

The first section contains 14 safety responsibility areas identified from the literature and other sources. The respondent was asked to identify if a person was responsible for that area or not. The respondents were not restricted to a single choice, allowing an a responsibility to be carried forward in the career path. The career stages were limited as follows:

- as a field engineer (1st year employee),
- field or project engineer (up to 5 years experience),
- project manager (at least 10 years experience),
- the safety manager, or
- some other personnel (they were asked to list titles)
- Respondents were also permitted to respond that an item was not applicable.

The second question asked for a brief description of the safety orientation and training given to newly hired engineers.

The third section identified five, non-regulatory aspects of safety, but common in many upper division courses on safety. The respondents were asked to indicate if they expected engineering graduates to have knowledge of Experience Modification Ratings, Incident Ratings, Incentive Schemes for Safety Accident Costs , or Accident Impact on Morale and Productivity.

The fourth and final section listed the Code of Federal Regulations 1926 subparts with titles and asked which of these newly hired graduate engineers would be expected to know. Respondents could select all, none, or choose individual subpart requirements.

3 SURVEY RESULTS

Table 1 summarizes the responses to the first section of the survey. Responses to any question will total more than 100%, since all applicable responses were to be chosen. Given the wide range of responses, the decision was made that only those areas receiving more than 50% of the respondent support for the first year and 5th year would be considered as important content. These significant categories are highlighted on the table in bold print and listed as follows:

- Conduct pre-project hazard analysis
- Prepare accident reports
- Conduct tool box meetings
- Participate in project safety meetings
- Perform hazard analysis
- Recognize common hazards
- Conduct site safety audits
- Maintain MSDS files
- Permits (confined spaces, etc.)

The responses to the question on safety orientation indicated that 12% required a 30 hour OSHA course, 28% required a 10 hour OSHA course, 80% provided a company-specific safety program and 4% required no safety orientation. The total sums to greater than 100% percent because some companies require and OSHA course supplemented by company training. One firm considered their site safety meetings to be their safety orientation. Other company programs were 30 to 60 minute reviews of the company's written safety policy. After the review the new employee signs the document acknowledging having read and understood the document.

Table 2 identifies four of five additional safety topics were considered important. Only incentive schemes had less than 50% support from the surveyed group. This may seem low, but it is consistent with findings by Kranz (1997). EMR and accident ratings were both selected as important elements. The impact of accidents in terms of cost, morale and productivity were also identified as important information. Comparison of these results with those in Sucharieh and Diamantes (1995) shows a great amount of consistency, with the obvious exception of the incentive schemes not desired by industry. Speculation on this difference is beyond the intended scope of this work.

Table 3 is a summary of the major subparts identified as important to construction. Overall more than 60% of the respondents selected all subparts. Table 3 summarizes those subparts that received 80% or more responses. Column 3 identifies the percent of industry citations categorized by these subparts. The selected subparts address 80% of the most frequently

Table 1 Percent of Responses Yes Responses to Question 1

Responsibility (1)	Field Engineer (1st year employee) (2)	Field or Project Engineer (5 years experience) (3)	Project Manager (≥ 10 years experience) (4)	Safety Manager or Equivalent (5)	Other Personnel (6)
Conduct Preproject Hazard Analysis	**(52)**	**(84)**	48	92	4 - supervisor 28 - superintendent
Prepare Project Safety Manual	8	24	36	96	8 - superintendent
Jobsite posters	48	28	28	80	4 - Safety Tech 4 - Office Manager 16- Superintendent
Maintain OSHA 200 Log	40	36	24	92	4Expediter 4Office Manager 20Superintendent
Prepare Accident Reports	**64**	**60**	48	76	4 Office Manager 20Foreman 48Superintendent
Conduct Tool Box Meetings	44	**56**	40	56	4Supervisor 8Foreman 44Superintendent
Participate in Project Safety Meetings	**96**	**96**	96	96	4Supervisor 8Foreman 44Superintendent
Perform Hazard Analysis	48	**68**	44	92	8 Foreman 40Superintendent
Recognize Common Hazards	**88**	**96**	92	92	8Foreman 4Everyone 36Superintendent
Conduct Site Safety Audits	**52**	**72**	56	100	8Foreman 44Superintendent
Maintain MSDS Files	**52**	**52**	24	88	4Office Manager 4Foreman 36Superintendent
Maintain Field Personnel Safety Training Records	44	44	28	76	4Personnel Office 4Office Manager 4Foreman 36Superintendent
First Aid Inventory and Supplies	52	36	24	60	4Safety Tech 4Office manager 12Foreman 40Superintendent
Permits (confined spaces, etc)	44	**60**	36	76	12Foreman 36Superintendent

cited areas. Although all subparts should be addressed, this table provides guidance to the that could be given more attention than others, if instructional time is available.

Table 2 Percent Yes Responses to Question #3

Do you expect newly hired graduates to have knowledge of:	Yes (%)
EMR	64
Incident Ratings	68
Incentive Schemes	36
Accident Costs	84
Accident Impact on Morale and Productivity	92

Table 3 OSHA Regulation Subparts with Greater Than 80% Support

SUBPART TITLE (1)	Yes % (2)	Percentage Of Citations (a) (3)
Subpart A – General	80	
Subpart B – General Interpretation	80	
Subpart C – General Safety and Health	92	5
Subpart D – Occupational Health and Environmental Controls	88	15
Subpart E – Personal Protective and Life Support Equipment	88	3
Subpart F – Fire Protection and Prevention	84	3
Subpart K – Electrical	88	20
Subpart L – Scaffolding	84	15
Subpart M – Fall Protection	88	7
Subpart P – Excavations	88	4
Subpart X – Stairways and Ladders	88	8
Subpart Z – Toxic and Hazardous Substances	80	Na

(a) Most Frequently Cited Standards Distributed by Subpart

Table 4 Analysis for Instructional Material Source

TOPIC	Source 1	Source 2	Source 3	Source 4
Pre-project Hazard Analysis		x		
Prepare Accident Reports		x	x	
Conduct Tool Box Meeting	x	x		
Prepare Task Hazard Analysis	x	x		
Recognize Common Hazards/ 29 CRF 1926 all Subparts		x		x
Conduct Safety Audits		x		
MSDS			x	
Permits				
EMR	x	x		
Incident Ratings	x	x		
Accident Costs	x	x	x	x
Impact on Morale, Productivity	x	x		
SCORE	6	10	3	2

4 INSTRUCTIONAL MATERIAL ANALYSIS

By concentrating on the safety knowledge elements required by 50% or more of the responding contractors, the general objectives and content of a 'safety curriculum' could be determined. These elements appear in column 1 of Table 4.

The approach, although highly simplistic, matches the 12 significant areas from the survey to the content of reference sources available for teaching. Participation in safety meetings is not indicated in the topics. It was felt that this was a policy issue and not an instructional issue. However, the data did indicate that most entry-level personnel would be participating in project safety meetings. The reference source list was not a comprehensive list of safety resources. Titles of the materials are not provided, since the selection choices may not include all the known references. Source 2 matched 10 of the 12 criteria and would likely be chosen as the primary course reference.

5 CONCLUSION

The development of a course on construction safety can be accomplished by simply teaching the fundamental requirements of the OSHA 10-hour course. However, this approach does not address all the expectations of contractors. The authors identified the12 topic areas that had the greatest amount of agreement among the contractors surveyed. The author's feel that the information recovered from the limited data is generally representative of the industry expectations, not only the northeast region of the United States. However, increasing the number of contractors would very likely improve the consistency of the information. The survey concentrated on contractors reporting $100 million or more in annual revenues. An important area to consider would be the contractor responses in an equivalent sample size for smaller volume contractors where it is more likely that management employees may be required to handle more responsibilities in safety.

6 REFERENCES

Suckarieh, G. and J. Diamantes 1995. "Educating construction management students In safety," *Safety and Health in Construction*, CIB Publication 209, Published by CIB Working Commission 99, pp. 154-164.

Kranz, Joseph E. 1997. *A Safety Management System for Electrical Contractors*, Master of Science Thesis in Architectural Engineering, The Pennsylvania State University.

7 APPENDIX

1. In the following table, please check yes if this employee is responsible for the item in the left hand column. The first row and column would read *"Does a field engineer (1st year employee) have responsibility for conducting preproject hazard analysis?"* Check the appropriate response: Yes or No. If none of the personnel categories are appropriate, indicate in the right hand column who has responsibility for the item.

Responsibility	Field Engineer (1st year employee)	Field or Project Engineer with 5 years experience	Project Manager with at least 10 years experience	Safety Manager or equivalent	Not Applicable (NA)	Superintendent Other Personnel (please list)
Conduct Preproject Hazard Analysis	Yes No	Yes No	Yes No	Yes No	NA	
Prepare Project Safety Manual	Yes No	Yes No	Yes No	Yes No	NA	
Jobsite Posters (OSHA and Emergency Contacts)	Yes No	Yes No	Yes No	Yes No	NA	
Maintain OSHA 200 Log	Yes No	Yes No	Yes No	Yes No	NA	
Prepare Accident Reports	Yes No	Yes No	Yes No	Yes No	NA	
Conduct tool box meetings	Yes No	Yes No	Yes No	Yes No	NA	
Participate in Project Safety Meetings	Yes No	Yes No	Yes No	Yes No	NA	
PerformTask Hazard Analysis	Yes No	Yes No	Yes No	Yes No	NA	
Recognize Common Hazards	Yes No	Yes No	Yes No	Yes No	NA	
Conduct Site Safety Audits	Yes No	Yes No	Yes No	Yes No	NA	
Maintain MSDS Files	Yes No	Yes No	Yes No	Yes No	NA	
Maintain Field Personnel Safety Training Records	Yes No	Yes No	Yes No	Yes No	NA	
First Aid Inventory and Supplies	Yes No	Yes No	Yes No	Yes No	NA	
Permits (Confined Space, etc.)	Yes No	Yes No	Yes No	Yes No	NA	

2. Please describe the safety orientation and training that is given to all newly hired engineers or provide an outline of the training materials.

3. Would you expect newly hired graduates to have knowledge of (Check all that apply):

☐EMR
☐Icident Ratings
☐Incentive Schemes for Safety
☐Accident Costs
☐Accident Impact on Morale, Productivity

4. With regard to OSHA Regulations, would you expect newly hired graduate engineers to have knowledge of:

☐all sections of the 29 CFR 1926
☐None of 29 CFR 1926
☐Only these selected subparts:

☐ Subpart A	General	☐ Subpart N	Cranes, Derricks, Hoists, Elevators, and Conveyors
☐ Subpart B	General Interpretation	☐ Subpart O	Motor Vehicles, Mechanized Equipment, and Marine Operations
☐ Subpart C	General Safety & Health	☐ Subpart P	Excavations
☐ Subpart D	Occupational Health and Environmental Controls	☐ Subpart Q	Concrete and Masonry Operations
☐ Subpart E	Personal Protective and Life Saving Equipment	☐ Subpart R	Steel Erection
☐ Subpart F	Fire Protection and Prevention	☐ Subpart S	Underground Construction, Caissons, Cofferdams, and Compressed Air
☐ Subpart G	Signs, signals, and Barricades	☐ Subpart T	Demolition
☐ Subpart H	Materials Handling, Storage, Use and Disposal	☐ Subpart U	Blasting and Use of Explosives
☐ Subpart I	Tools-Hand and Power	☐ Subpart V	Power Transmission and Distribution
☐ Subpart J	Welding and Cutting	☐ Subpart W	Rollover Protective Structures
☐ Subpart K	Electrical	☐ Subpart X	Stairways and Ladders
☐ Subpart L	Scaffolding	☐ Subpart Y	Diving
☐ Subpart M	Fall Protection	☐ Subpart Z	Toxic and Hazardous Substances

Implementation of Safety and Health on Construction Sites, Singh, Hinze & Coble (eds)

Reducing construction-related injuries and illnesses through education

M.H.Sweeney, C.Jeff Bryant & John Palassis
Centers for Disease Control and Prevention, National Institute for Occupational Safety and Health, Cincinnati, Ohio, USA

P.Becker
West Virginia University, Safety and Health Extension, Morgantown, W.Va., USA

ABSTRACT: In the United States, construction workers experience one of the highest occupational fatality rates as well as rate of injuries and illnesses resulting in lost work days. Young workers, particularly those under 18 years and those with less than two years of experience in the trade are at high risk for injury or death. Training programs in occupational health and safety are seen as an effective means of increasing the awareness of workers and managers alike of work hazards and of methods of preventing injuries and illnesses. A variety of approaches are being used to develop training programs to address the many health and safety risks encountered on the construction site. Programs to prevent falls, musculoskeletal disorders and respiratory diseases, three principal causes of morbidity and mortality among construction workers are described.

1. BACKGROUND

More than 7 million workers (about 6 percent of the total U.S. labor force) are currently employed in the construction industry. Compared with other industries, construction workers experience one of the highest occupational fatality and injury and illness rates resulting in lost work days (BLS, 1996). Of all work-related deaths in 1996, 16.9% (or 1,039) occurred among construction workers; falls were the leading cause (31%). By trade, ironworkers and roofers accounted for more than 75% of deaths due to falls in the industry (CPWR, 1998).

Nonfatal injuries also occur frequently among construction workers. In 1995, construction workers experienced more than 182,000 illnesses and injuries causing lost work days. Having contact with or being struck by an object and musculoskeletal disorders account for more than 50% of all traumatic injuries; backs, hands/fingers, and eyes are the body parts most affected (BLS, 1998; Lipscomb,1996).

In partnership with researchers throughout the United States, NIOSH is developing and evaluating methods to reduce work-related injuries and illnesses among construction workers. One approach is to develop and disseminate educational programs, training materials, and methods that address the needs of construction workers and the industry as a whole.

1.1 *Future construction workers*

In the United States, the injury rate among adolescent workers is 5.8 per 100 full-time equivalents. To reduce this rate and increase the safety awareness and education of vocational, technical, and post-secondary industrial school teachers and administrators, NIOSH is creating a comprehensive set of trade-specific teaching materials with an occupational safety and health emphasis. Currently, six trade-specific curricula are planned: electrical work, drywall and plastering (both installation and finishing), and carpentry.

Thus far, the developed material contains thirteen modules on general safety and health topics relating to vocational shops: noise and hearing conservation, personal protective equipment overview, employee right-to-know/MSDS, PPE—hearing conservation, PPE—eye, hand, foot, disconnecting, locking and tagging-out, respiratory protection, dangers and prevention of silicosis, lead hazards, ergonomics, compressed gases safety, safety and health committees, and first aid. These modules will be evaluated for their effectiveness. This program has the potential for reaching more than 11 million students enrolled in more than 20,000 programs throughout the country. Through such programs, NIOSH will help safeguard the health and life of young workers and begin to create a culture that values safety and health on the job.

1.2 *Respiratory protection program handbook for small businesses and contractors*

Pulmonary diseases and cancers, particularly asbestosis, lung cancer, mesothelioma, bronchitis and occupational asthma consistently cause disability, morbidity and death among U.S. construction workers. These conditions may be attributed to on-the-job exposure of construction workers to asbestos, crystalline silica, diesel fumes, wood dust, paint fumes, resin adhesives and other respiratory hazards.

In 1998, OSHA implemented a new respiratory standard to reduce such exposures (29 CFR 1910 and 1926 [1988]). This standard requires that both large and small employers have respiratory protection programs if their work sites pose specific respiratory hazards to workers. Whereas large companies will most likely have a well-established respirator protection program, small contractors may not.

NIOSH has designed a series of easy-to-read handbooks that will provide small contractors with the tools for establishing and maintaining good respiratory protection programs. The first handbook in the series will target drywall finishing contractors. Drywall finishers may be exposed to high concentrations of dust that may contain silica and may exceed current OSHA standards for total and respirable dusts. A draft outline of the handbook is included in Table 1. The handbook may be accompanied by other materials such as videos and CD ROMs.

1.3 *Fall-Safe: an intervention program*

In the United States, more than 30% of all fatalities and 20% of injuries with days off work are caused by falls. The Fall-Safe Construction Program was designed by researchers, educators, and

Table 1: Respiratory Protection Programs Handbook for Drywall Contractors (DRAFT)

I. Introduction.
   - A. Health Hazards of the Trade.
   - B. Legal—Introduction to the Respirator Program Required by 1910.134.
   - C. Liability.

II Requirements for a Respirator Program.
   - A. Procedures : Selecting Respirators.
   - B. Employee Medical Evaluations
   - C. Fit Testing Procedures.
   - D. Procedures :Proper Respirator Use
   - E. Procedures : Cleaning, Storage, Inspecting, Repairing, Discarding and Maintaining Respirators.
   - F. Procedures to Ensure Adequate Air Quality, Quantity, and Flow for Atmosphere-Supplying Respirators.
   - G Training of Employees in the Proper Use of Respirators.
   - H. Procedures: Program Evaluation.

II. Checklists, Forms, Useful Tools.

practitioners at the West Virginia University Health and Safety Extension (Fall-Safe Construction Program, 130 Tower Lane, P.O. Box 6615, Morgantown, WV 26506-6615) in collaboration with NIOSH and the Center to Protect Workers' Rights. The program was created as a comprehensive fall hazard management system to (1) increase fall safety knowledge, awareness, accountability, and communication among different levels of workers in an organization or at a job site, and (2) provide a systematic approach to training and managing fall-prevention activities. The Fall-Safe system includes specific modules for training all levels of the workforce: contractors and safety coordinators, competent persons or field supervisors, and workers. At the end of the program, contractors participating in Fall-Safe are awarded the status of Fall Safe Contractor. West Virginia University audits their performance by quarterly site audits. Poor performance on two successive site audits results in withdrawal of Fall-Safe status. Currently, the program is being evaluated for effectiveness in increasing awareness of fall safety and the reduction of falls among workers employed by trained contractors..

1.4 *Reducing musculoskeletal disorders through training*

Musculoskeletal disorders among construction workers account for 22.5% of injuries resulting in lost work days (BLS, 1998). In a recent survey of 3,000 construction workers in Eastern Iowa, more than 70% reported having symptoms related to musculoskeletal pain in the 12 months before the survey (Cook, 1992).

Through its extramural program with universities, NIOSH currently sponsors five research projects to develop training programs and materials that teach workers and managers how to identify risk factors for musculoskeletal disorders and how to change the work environment to reduce or eliminate them. Due to the large variety of hazards on the construction site, most projects concentrate on trade-specific risk factors. Steps for developing training materials include symptom surveys of workers and work-site assessments to identify and quantify risk factors associated with musculoskeletal disorders. These risk factors include repetitive tasks, awkward postures, forceful exertions, lifting, vibration and cold work environments. Detailed discussions with retired and current workers, supervisors, trainers and other knowledgeable individuals detail job activities, stresses and possible solutions to identified job and task-related problems and concerns.

Using this information, researchers at the University of Massachusetts-Lowell (UML), University of Iowa (UI), University of Oregon (UO), University of Cincinnati (UC), and the United Brotherhood of Carpenters and Allied Trades (UBC) have taken a variety of approaches in the development of course work and training materials. Most of these approaches encourage participatory training and include sections on the identification of musculoskeletal disorders and the associated risk factors and methods on prevention.

To date, the programs target carpenters, drywall installers, bricklayers, and operating engineers. Two programs target carpenter groups. The UBC created a module to be added to their apprenticeship program; UC produced tool box talks for on-the-job training and instructor's guides for carpenter apprentices for drywall installation and scaffold erection, two jobs identified jobs which place considerable stress on the body. Two additional jobs performed by carpenters and identified as physically demanding include pile driving and concrete form work were addressed by the UBC program.

The UML developed a train-the-trainer course with emphasis on operating engineers, but which may be applied to other trades. This course has been so well-received that they anticipate that more than 2500 workers and trainers will have taken the course over the next year. Using results of a musculoskeletal symptom survey of approximately 3000 construction workers in Eastern Iowa, researchers conducted in-depth analyses of job sites of bricklayers, sheetmetal workers and operating engineers and presented class-based lectures to union members. In addition, they are working with a large contractor to modify excavation equipment to make the equipment more ergonomically sound. Finally, researchers at UO are working with a variety of contractors in the Portland, Oregon area in an attempt to develop site-specific interventions to reduce musculoskeletal disorders. Their approach varies depending upon the type of work being performed, but includes regular on-site audits of ergonomic conditions on the worksite and the use and refinement of standard training modules that integrate ergonomic principles, body mechanics and exercise training for delivery to foremen and craft workers on construction sites.

Each of these programs have been immensely successful in that they, by default, increase the awareness of workers and management regarding musculoskeletal disorders and associated risk factors. However, more work is needed to evaluate the effectiveness of the various approaches in the long-term prevention of work-related musculoskeletal disorders among construction workers.

2. WHERE DO WE GO FROM HERE?

Although many safety and health training programs and courses are designed for construction workers, more research is needed to determine their effectiveness in reducing work-related injuries and illnesses. Furthermore, more work is needed to identify the safety and health training needs of each employee in the construction industry (including young adults who are not yet in the industry) and to determine how the information is most effectively translated and transmitted to them.

REFERENCES

Bureau of Labor Statistics 1996. Fatal Workplace Injuries in 1994: A Collection of Data and Analysis. Washington, D.C.: U.S. Government Printing Office.

Bureau of Labor Statistics 1998. Occupational Injuries and Illnesses: Counts, Rates and Characteristics, 1995. Washington, D.C.: U.S. Government Printing Office.

Cook, T.M. & C.L. Zimmermann 1992. A symptom and job factor survey of unionized construction workers. In: Kumar S (ed.), *Advances in Industrial Ergonomics and Safety IV*. New York: Taylor & Francis.

Code of Federal Regulations 1997. 29 CFR 1910.1000. Washington, D.C.: U.S. Government Printing Office, Office of the Federal Register.

The Center to Protect Workers' Rights 1998. The Construction Chart Book: The U.S. Construction Industry and Its Workers. Washington, D.C.

Lipscomb, H.J. Kalat, J. & J.M. Dement 1996. Workers' Compensation claims of union carpenters 1989–1992: Washington State. *Appl Occup Environ Hyg* 11(1):56–63.

Implementation of Safety and Health on Construction Sites, Singh, Hinze & Coble (eds)
© 1999 Balkema, Rotterdam, ISBN 90 5809 036 1

Continuous enhancement of health and safety awareness in undergraduate civil engineering courses

M.A.Al-Mufti
Department of Civil Engineering, The Queen's University of Belfast, UK

ABSTRACT: Civil engineering degree courses should provide students with adequate knowledge and an enhanced awareness of H&S in construction. Students in the department of civil engineering at Queen's University of Belfast are therefore made aware of H&S issues throughout their years of undergraduate study through lectures, projects, and practical sessions. In recent years innovative teaching / learning exercises related to construction disasters and hazards have been conducted. All students at both undergraduate Levels 1 and 2 take part in these active learning exercises which lay emphasis on feedback and learning from failures through case studies and role-play. Final year projects have included reviews on current construction H&S practices in the province and devising systems for costing accidents. The overall approach is therefore based on examining and referring to safety at relevant points throughout the undergraduate course rather than to cover the topic in a separate module.

1 INTRODUCTION

Civil engineers have a pivotal role in ensuring the health and safety (H&S) of construction site personnel. They are expected to be both well educated and qualified to play that important role. This is a challenging role at the best of times but much more so to young graduate engineers who may be faced with major responsibilities for considerable human and material resources on site without necessarily having the full experience and competency at the early stages of their careers. It is therefore necessary that they are provided with adequate knowledge on H&S during their degree studies. Time constraints in such courses will always restrict the quantity, and perhaps the quality, of H&S knowledge and training that students may have access to during their studies.

2 CIVIL ENGINEERING COURSE STRUCTURE AT QUB

Similar to many other universities in the United Kingdom, there are currently two streams of degree studies within the Department of Civil Engineering at Queen's University of Belfast (QUB). A three-year BEng course and a four-year MEng course. The former degree requires the successful completion of eighteen modules and the latter twenty-four modules over the entire period in each case. Following the Standards and Routes to Registration (SARTOR) prepared by the Engineering Council in the UK; requirements for Chartered Engineers will include four years of academic study instead of three as the educational base. The extended degree should allow more adequate coverage of topics which broaden students' appreciation of work in the construction industry and the interaction of the engineer with business and social environment in which construction projects are carried out.

Consequently the teaching of H&S, which is mandatory in civil engineering degree courses, may be enhanced and continue to be an integral part of all subject areas with particular emphasis in design and management studies. Examining and referring to H&S at relevant points throughout undergraduate courses appears to be preferred in most UK universities rather than to cover the topic in separate courses.

As illustrated in Fig.1, in the civil engineering degree course at QUB, H&S on site is first referred to at Level 1 under *Construction Practice 1*. This module cites among its primary objectives the *teaching of the implications of safety on site*. It also deals with items such as site organisation, specifications, programming of construction works, and construction techniques. The Construction Industry Training Board has been involved in the past in organising practical sessions and demonstrations related to safety on site at their training facilities.

Further teaching on safety is included in Level 2 under *Construction Practice 2* and this includes problems associated with movement of materials on site. Staff from Northern Ireland H&S Executive are involved in teaching this module and students are thus kept aware of the latest developments particularly from the legal and regulatory points of view.

The *Civil Engineering Management and Economics* module at Level 3 includes H&S in its syllabus. In recent years there has been much emphasis on increasing students' awareness of the Construction (Design & Management) Regulations, which came into force in 1995 and are considered as the latest and most important development in H&S law covering the construction industry in Europe. Further, H&S is presented as an important component area of construction management and the management *Body of Knowledge* (BoK). In this respect BoK documents produced by the American Association of Project Managers and the United Kingdom's Association for Project Management are referred to in this module.

H&S topics dealt with at Level 3 are further enhanced at Level 4 in the MEng degree under the *Construction Management* module. Students on this module are also required to make presentations on recent research and publications in management areas including H&S.

Complementing the above coverage of H&S, the topic is being given increasing prominence in other subjects being taught such as structural analysis and design projects.

In addition to H&S covered in taught modules, the list of honours projects at Level 3 which both BEng and MEng students have to carry out often includes projects on H&S. Dissertations submitted in recent years have included such topics as review of health and safety practices in the local construction sector, cost of implementing H&S legislation, and cost of construction-related accidents.

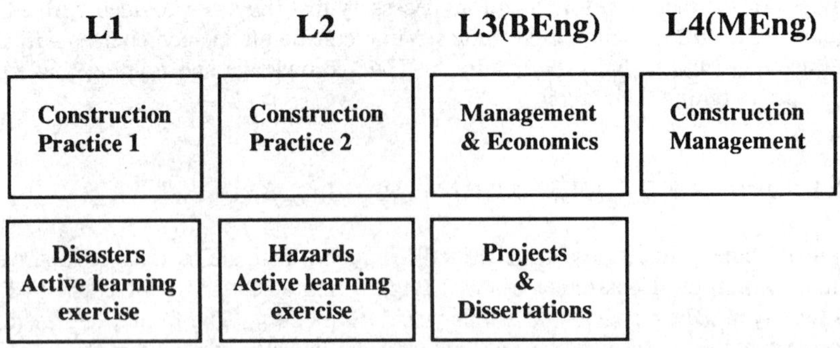

Fig.1 : Health and safety coverage in civil engineering degree courses

3 STUDENT CENTRED INVESTIGATIONS (INNOVATIVE TEACHING / LEARNING)

During the period 1994-97, three student centred investigations entitled *Learning from Disasters* were conducted at Level 1. These exercises were conducted with students working in groups during one week dedicated to this purpose. Each group was assigned a particular engineering disaster to investigate and report on. The various groups then made presentations on their assigned disasters elaborating on the technical and human factors involved as well as the lessons to be learnt. Sources of information utilised for this purpose included engineering news magazines, inquiry reports, journal papers, and books (Jennings, 1995 & 1997).

A further Level 2 exercise entitled *Hazards Investigation* was also developed and run for the first time in January 1997 in parallel with the Level 1 exercise. Here, Level 2 groups were asked to investigate a specific hazard by examining at least four disasters which have occurred and in which the hazard was a contributing factor (Jennings & Mackinon, 1997).

Some factors highlighted in the above mentioned presentations included historic progression, safety versus finance, scheduling pressures, politics, innovation, care in design, careful checking, communication problems, and adoption of quality assurance schemes.

Such innovative teaching / learning exercises can be very demanding in terms of staff time and information requirements particularly when involving large classes as it has been in this department. Moreover, time and space has to be found for students who are already under pressure from overcrowded syllabus. Several members of academic staff have had to be involved every time such exercises are carried out. Students from senior levels were also involved both to assist and benefit from these exercises. For example, Level 3 students undertaking research projects provided valuable assistance with the investigation of available literature and possible assignments. Level 4 MEng students and some postgraduate research students have also been involved as *proctors* thus providing them with an interesting experience in observing the workings of small teams. In addition to academic staff from within the department itself, leading civil and structural engineers from industry have also been involved in these exercises. They assisted in the initial planning stages when students were deciding on their disaster case studies, preparing draft documents and making arrangements for consultation sessions. In these sessions, each engineer from industry would be paired up with a member of academic staff to consult with specific groups. Later they would be involved in questioning the groups during presentations and assist in their assessment of the groups. Finally they would attend *post-mortem* meetings of the groups to review the outcome and make suggestions for improvements. The input from engineers from industry was a valuable contribution both from the standpoint of the students and the staff.

The objectives set out for the disasters and the hazards exercises included students improving their knowledge of disasters and the lessons to be learnt, the need to pay great attention to detail in design, the implications and possible consequences of design changes, accessing and making full use of existing sources of information, and the importance of good communication and group skills. Further, one of the main outcomes has been students' increased awareness of the importance of safety management.

Replies to questionnaires about the exercises indicate that students have improved their understanding of engineering as well as their appreciation of the responsibilities of engineers. Just as significant, the feedback also confirms that these exercises have provided them with an enjoyable learning experience.

4 H&S COVERAGE IN CIVIL ENGINEERING COURSES

Reduced contact hours in modularised engineering courses, larger class sizes, and reduced staff resources are all constraints restricting increased coverage of any topic in the various modules being taught. This invariably applies to the teaching of H&S in civil engineering. It is therefore more necessary than ever that a focused approach is followed in this respect. Such an approach may comprise of the following:

(a) First and foremost H&S must be an integral part of all other teaching rather than being presented as an individual module separate from other modules. A separate module devoted solely to H&S will have problems of timing, fragmentation, and possible duplication with other modules. More important however, such an approach may lead engineering students to develop a passive view of the subject as a non-technical, unimportant, and a subject to *pass, forget, and never revisit* in their professional career. It is this latter perception that is dangerous if unintentionally formed in engineering students' way of assessing the subject as it may continue to dominate their thinking when they are practising the profession.

(b) In an integrated approach the critical importance of H&S needs to be emphasised within the context of the various modules being taught. Some more prominently than others. For example particular attention must be paid to H&S within the context of subjects such as structural design and construction management.

(c) It must be emphasised that approaching H&S in construction projects must be based on consideration of complementary elements or components. These can be based on technical aspects such as those dealt with in design, including for example safety factors, permissible stresses, load factors, and failure modes. Other technical aspects include detailing, specifications, and risk analysis. Non-technical H&S aspects may come under the general umbrella of management. They include human, organisational, and corporate considerations. They are concerned with the psychological aspects related to the engineer's views and convictions, as an individual, regarding H&S. They are also concerned with the social aspects related to H&S including the welfare of personnel, peer pressure and apathy on site, and the overall socio-economic and cultural environment in which any particular construction project is being carried out.

(d) The importance of feedback and learning from failures must be emphasised to students. Major failures generally result in inquiries to investigate and often publicises the details of such cases and the lessons to be learnt. The outcomes of such investigations in the form of inquiry reports or papers published in conferences and journals will be of great benefit to students and young graduate engineers on two accounts. Firstly they will highlight the technical aspects of such failures. These include design deficiency, inaccurate detailing, incorrect specifications, etc. Secondly they may provide valuable insights into the management related aspects. These will include decision-making processes, communication problems and the roles and responsibilities of engineers in the context of the projects where the major failures have occurred.

(e) Safety needs to be perceived as part of quality and the general motto that *safety and quality go hand in hand* should be promoted in engineering courses. In construction, it is unreasonable to expect high quality workmanship and craftsmanship under conditions that are perceived to be unsafe by the employees concerned. The Total Quality Management concept (TQM), now that it is well established, incorporates safety as a major component, and management for quality is seen as part of management for safety. In a recently published paper on the Glaxo project, one of the largest and successful building projects in the UK in recent years, a primary project objective identified in the context of Total Project Quality was for it *to be the safest construction site in the UK* (Winch, 1998).

(f) It is generally agreed that the construction sector has a poor safety record and that construction is still perceived to be a more hazardous profession than many other industrial professions. It is important therefore that new graduates are fully aware of this fact and that they are about to pursue a career in a profession which still has this negative attribute associated with it. The situation is bound to improve for a variety of good reasons and young graduate engineers must see this as a challenge to which they can respond to positively aided by their awareness, knowledge and conviction, all of which would have been instilled in them during their degree studies.

(g) As already pointed out in a previous section, due to modularisation of courses,

Fig.2 : Health and safety profile for site engineers

reduced contact hours and staffing resources, there will always be a limit to how much can be taught by way of regulations, codes of practice and techniques pertaining to H&S. The overall objective should therefore be to inculcate an awareness of the continued and increasing importance of H&S in construction. The fact that it is do with safeguarding the life and health of fellow human beings in the first instance. Damage to materials, plant and properties, important and significant as they are nevertheless secondary. This awareness needs to be supplemented by the graduate engineers' belief in the importance of H&S in construction as well as the confidence to convey this belief to the workforce and finally put it to practice. This amounts to planting the seeds of a positive H&S culture in the mind of graduate engineers as they take responsibilities on site. This culture can then be enhanced through training and practical experience throughout their professional career (Fig.2).

5 CONCLUSIONS

Fortunately H&S appears to be acquiring more prominence in the construction industry. It is now viewed as an integral component of the overall function of project management. It is also becoming accepted that attention to H&S matters should span the life cycle of projects and should be the joint responsibility of all parties involved throughout the various project phases. Graduate engineers need to have an appreciation of this as they start their professional career and often having to assume early responsibilities for considerable human and material resources. They must therefore be equipped with adequate knowledge of the subject as well as being fully aware that H&S constitutes an important consideration in the background of most if not all project activities. The integrated approach to H&S adopted in civil engineering degree courses at QUB is aimed at fulfilling these objectives through highlighting the H&S issue in the various modules spanning the duration of the degree courses and through adopting different teaching and learning modes.

REFERENCES

Jennings, A. 1995, Disasters and their role in education, *The Structural Engineer*, Vol.73, No.18, pp297-300.

Jennings, A., & Mackinon, P., 1997, Disaster Awareness – a necessary ingredient to engineering education, *Second International Conference on Teaching Science and Technology at Tertiary level*, KTH, Stockholm, Sweden.

Jennings, A. & Mackinon, P., 1997 : Student Centred investigation of hazards, *Mouchel Centenary Conference : Innovation in Civil and Structural Engineering*, Cambridge, England.

Winch, G., Usmani, A., & Edkins, A. , Towards total project quality : a gap analysis approach, *Construction Management and Economics*, Vol.16, No.2, pp 193-207.

9 Design engineering safety

Implementation of Safety and Health on Construction Sites, Singh, Hinze & Coble (eds)
© 1999 Balkema, Rotterdam, ISBN 90 5809 036 1

On the factor of safety for the shear strength of reinforced concrete beams

T.Tamura
Department of Civil and Environmental Engineering, Nagaoka University of Technology, Japan

T.Shigematsu
Department of Civil Engineering and Architecture, Tokuyama College of Technology, Japan

K.Nakashiki
Kyuken-Sekkei Corporation, Fukuoka Branch, Japan

ABSTRACT: Recent design codes for the shear strength of reinforced concrete members are based on many experimental research works. However, in the case of the member subjected to axial tension, the design equation on that concept has inadequate safety margin because of the insufficient study of this problem. The purpose of this study is to investigate how much is the factor of safety of the ACI (American Concrete Institute) code and the JSCE (Japan Society of Civil Engineers) code for such cases. To verify the factor of safety for the nominal shear strength of the codes approximately 380 specimens were provided and were tested.

1 INTRODUCTION

In the designing of the structure the engineers are required many design concepts. The factor of safety for the shear strength of the member is also one of the important concept. The failures of the reinforced concrete members are classified into bending failure or shearing failure. The bending failure of the member progresses slowly and toughly. However, the shear failure occurs suddenly and explosively. Therefore, engineers have to maintain the design concept that the reinforced concrete member fails not in shear failure but in bending failure when the collapses are inevitable. (In spite of such a concept, on January 17, 1995, the news of Hansin Earthquake shocks the civil and architectural engineers in Japan. They observed many wrecks of the reinforced concrete members in shear.)

The study on the shear strength of the reinforced concrete member was started about 1900 (Ritter 1899), (Mörsh 1903). Since then, it has been studied by many researchers. A.N.Talbot suggested that the shear strength of reinforced concrete beam is determined by concrete strength, reinforcement ratio, shear reinforcement, and length of shear span (Talbot 1909). However, the problem of reinforced concrete in shear was not so simple. It is reported that a factor of the partial collapse of the Willkins Air Forced Depot warehouse in Shelby, Ohio (on August 3, 1955), is tensile stress due to the restraint caused by shrinkage and thermal movements (Anderson 1957), (Elstner 1957). Since then, from the 1950's to the 1960's, Mattock et al. proposed the fomula of the shear strength including the influence of an axial force in the experimental study (Haddadin, Hong, Mattock 1969). Subsequently Leonhardt analyzed the T-beams and plates subjected to axial tension and bending experimentally (Leonhardt 1965). Those reports show the effects of the compressive strength of concrete, shear span-depth ratio, reinforcement ratio and axial force for the shear strength of reinforced concrete member.

The present-day engineers well known the axial tensile stress due to the restraint of shrinkage, thermal movements, creep and applied load is avoidable problem in the statically indeterminate as rigid frame. Also it can be predicted that these initial axial tensile stresses reduce the shear capacity of the members. Recently, nominal shear strength of reinforced concrete beam without shear reinforcement is generally determined by concrete strength, reinforcement ratio, effective depth of cross section, shear span to depth ratio and applied axial force. However, in spite of the fact men-

tioned above, the ultimate strength and failure mechanism are not clearly decided under various conditions. In a recent paper, Collins applied the modified compression field theory to predict the strength of biaxial stressed structures (Vecchio, Collins 1986), (Bhide, Collins 1988). He suggested that the modified compression field theory is applicable to accurately predict the shear capacity of reinforced concrete member subjected to axial tension. Also using this theory, it is conclude that the shear capacity of such members is influence strongly by the amount and distribution of the longitudinal reinforcement. The other hand, recently researchers have investigated the development of computational models for nonlinear analysis of reinforced concrete members by finite element method (ASCE 1982), (ASCE 1993).
Definitely, these analytical manners are applicable to accurately predict the shear capacity of reinforced concrete member in the practical use. However, at present time, any code has not adopted these methods. (In the near future it will adopt the methods.)

From above mentioned point of view, this paper discussed the review of previous investigations and the factor of safety of the shear problem of reinforced concrete beams.

2 NOMINAL SHEAR STRENGTH

For the calculation of the nominal shear strength of the concrete nonprestressed member both ACI and JSCE codes take the same concept.

In ACI code the nominal shear strength V_n is computed by:

$$V_n = V_c + V_s \tag{1}$$

where V_c is nominal shear strength provided by concrete, and V_s is nominal shear strength provided by shear reinforcement.

In JSCE code the designed shear strength V_{yd} is computed by:

$$V_{yd} = V_{cd} + V_{sd} \tag{2}$$

where V_{cd} is designed shear strength provided by concrete, and V_{sd} is designed shear strength provided by shear reinforcement.

2.1 *Nominal shear strength provided by concrete*

2.1.1 ACI Code

The latest ACI code provides that the nominal shear strength V_c provided by concrete for nonprestressed members shall be computed as follows.

For members subject to shear and flexure only,

$$V_c = \left(\frac{\sqrt{f_c'}}{6}\right) b_w d \tag{3}$$

For members subject to axial compression,

$$V_c = \left(1 + \frac{N_u}{14A_g}\right)\left(\frac{\sqrt{f_c'}}{6}\right) b_w d \tag{4}$$

Where A_g is gross area of section and N_u are factored axial load normal to cross section occurring simultaneously with factored shear force V_u. Quantity N_u/A_g shall be expressed in MPa.

However, the code provides that the basic expression for shear strength of members without shear reinforcement is given by,

$$V_c = \left(\sqrt{f_c'} + 120 p_w \frac{V_u d}{M_u}\right)\frac{b_w d}{7} \tag{5}$$

but not greater than $0.3\sqrt{f_c'}b_w d$, which shall be used for the more detailed calculation of the V_c.

For members subject to axial compression, it shall permitted to compute V_c using Eq.(5) with M_m substituted for M_u and $V_u d/M_u$ not then limited to 1.0, where

$$M_m = M_u - N_u \frac{(4h-d)}{8} \tag{6}$$

For members subject to significant axial tension,

$$V_c = \left(1 + \frac{0.3N_u}{A_g}\right)\frac{\sqrt{f_c'}}{6} b_w d \tag{7}$$

but not less than zero, where N_u is negative for tension.

2.1.2 JSCE Code

The JSCE code provides that designed shear strength of members without shear reinforcement is calculated by concrete strength, reinforcement ratio, effective depth of cross section, shear span to depth ratio and applied axial force as follows.

$$V_{cd} = \beta_d \cdot \beta_p \cdot \beta_n \cdot f_{vcd} \cdot b_w \cdot d/\gamma_b \tag{8}$$

Where, $\gamma_b = 1.3$ and $f_{vcd} = 0.19\sqrt[3]{f_{cd}'}$ (MPa)

$\beta_d = \sqrt[4]{100/d}$ (d:cm) β_d is 1.5 when $\beta_d > 1.5$

$\beta_p = \sqrt[3]{100 p_w}$ ($p_w = As/b_w d$) β_p is 1.5 when $\beta_p > 1.5$

$\beta_n = 1 + M_o/M_u$ ($N_d' \geq 0$) β_n is 2 when $\beta_n > 2$

$\beta_n = 1 + 2M_o/M_u$ ($N_d' < 0$) β_n is 0 when $\beta_n < 0$

where N_d' is design axial load, which is taken as positive for compression and negative for tension. M_u is ultimate moment and M_o, named decompression moment, is the bending moment when the axial stress is calculated in bending one. If M_o has the same signs M_u, it is taken as positive.

2.2 *Nominal shear strength provided by shear reinforcement*

In design of shear reinforcement, basically both codes provide that it is based on a modified truss analogy. When shear reinforcement perpendicular to axis of member is used,

$$V_s = \frac{A_v f_y d}{s} \text{ (ACI), and } Vsd = \frac{A_v f_y z}{1.15s} \text{ (JSCE)} \tag{9}$$

where A_v is the area of shear reinforcement within a distance s and $z=d/1.15$.
When inclined stirrups are used as shear reinforcement,

$$V_s = \frac{A_v f_y (\sin\alpha + \cos\alpha) d}{s} \text{ (ACI), and } Vsd = \frac{A_v f_y (\sin\alpha + \cos\alpha) z}{1.15s} \text{ (JSCE)} \tag{10}$$

3 THE FACTOR OF SAFETY FOR THE SHEAR STRENGTH OF THE CODES

The factor of safety of the both codes were verified by the experimental data with various cross sectional parameter and various loading conditions. To verify the factor of safety for the nominal shear strength by concrete 278 specimens were provided and were tested. Also to verify the factor of safety for the nominal shear strength by shear reinforcement 106 specimens were provided and were tested.

3.1 *Experimental program*

3.1.1 Test beams

The dimensions and the each cross sectional parameters of rectangular beam, T-shape beam and I-shape beam are shown in Fig.1. The length of all specimens is 180cm and effective span is 150cm. The ratios of the flange width to the web width are from 2.3 to 3.0 in T-beam, and are from 2.0 to 6.7 in I-beam. Also the reinforcing ratios in rectangular, T-shape and I-shape are 0.0119, 0.0162

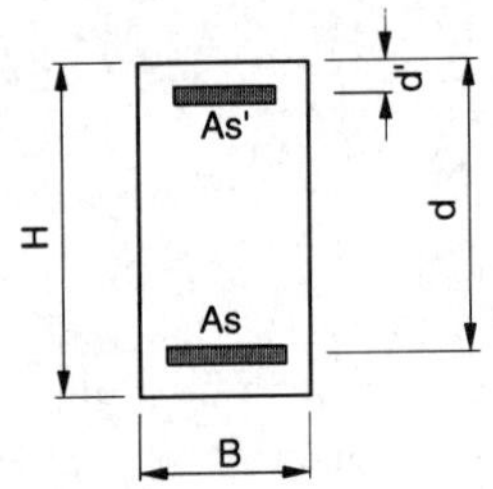

Rectangular beam			
B	10.0cm	d	18.0cm
H	20.0cm	d'	2.0cm
p	0.70%~3.0%	p'	0.34%~2.6%
Total	172 specimens		

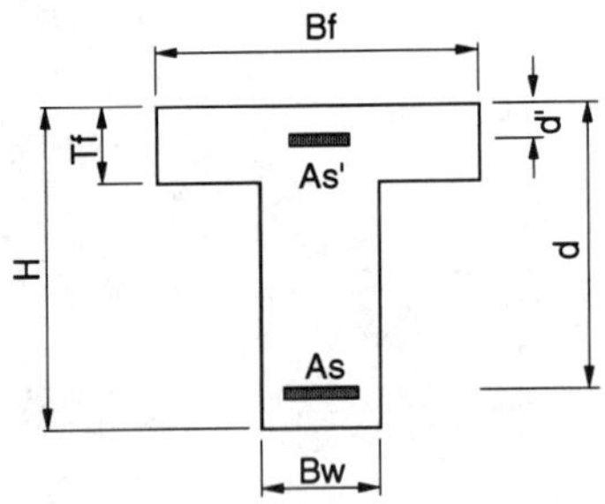

T-shaped beam			
Bf	20.0~24.0cm	d	18.0~23.0cm
Bw	8.0~10.0cm	d'	2.0cm
H	20.0~25.0cm	Tf	4.0~6.0cm
p	0.40~0.86%	p'	0.22~0.29%
Total	39 specimens		

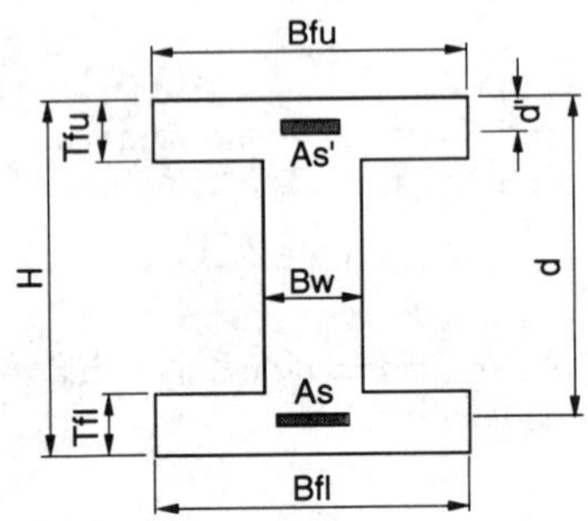

I-shaped beam			
Bfu	5.0~30.0cm	d	28.0cm
Bfl	15.0~30.0cm	d'	2.0cm
Bw	5.0cm	H	30.0cm
Tfu	4.0~8.0%	Tfl	4.0~8.0%
p	0.45~2.87%	p'	0.20~1.28%
Total	67 specimens		

Fig.1 Cross sections and test parameters

and 0.0147, respectively. For the specimens to verify the shear strength by concrete, no stirrups are placed for shearing reinforcement except the supporting points in every beam. Also for the specimens to verify the shear strength by shear reinforcement, shear reinforcement perpendicular to axis of member is used in various spacings as shown in Fig.2. At the ends of all of the specimens have holes to introduce the axial tensile force. The maximum axial tensile stress is from 400KPa in rectangular beam.

3.1.2 Test apparatus and procedure

Testing apparatus with axial tension are shown in Fig.3. It is composed of two oil pressure actuators controlled by the electrohydraulic servo mechanism. Also both supporting points are hinged by bearing joints. The axial tension is introduced into the both ends of the beam via longitudinal actuator and is held constantly after reaching of the expected tension. Then the transverse loads are provided by transverse actuator which can introduce the load into two points by the loading beam. The transverse load increases continuously until the beam failures under displacement controlled system. At each loading stage new cracks are marked on the faces of the beam. Dial gauges are placed at the loading point and center of the span to measure the deflection of the beam. Also the bending strain are measured by wire strain gages at the center of the tensile reinforcement and the top of the flange.

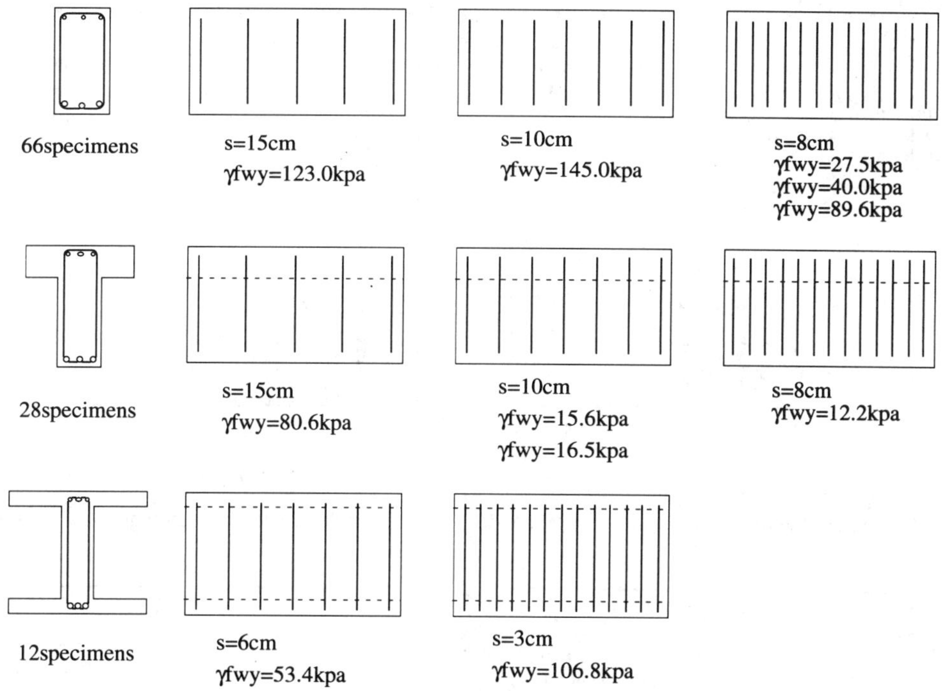

Fig.2 Shear reinforcement ($\gamma f_{wy} = A_v \cdot f_y / (b_w \cdot s)$)

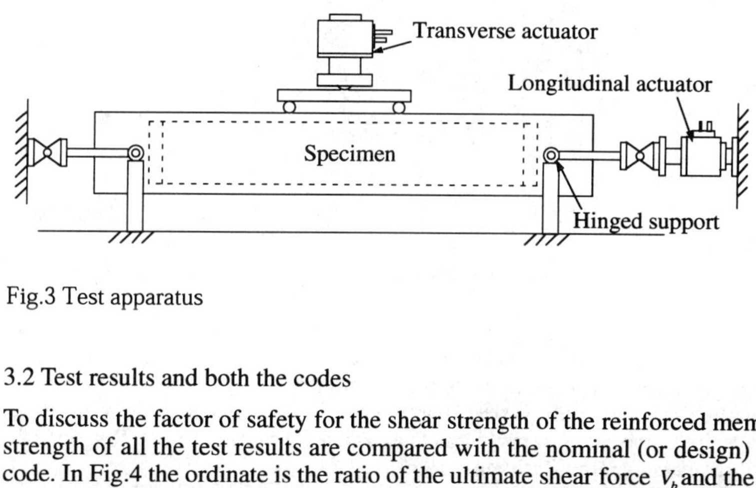

Fig.3 Test apparatus

3.2 Test results and both the codes

To discuss the factor of safety for the shear strength of the reinforced member the ultimate shear strength of all the test results are compared with the nominal (or design) shear strength of each code. In Fig.4 the ordinate is the ratio of the ultimate shear force V_b and the nominal (or designed) shear strength by concrete V_c (or V_{cd}) of each code, and abscissa is the shear span to the depth ratio. Also in Fig.5 the abscissa is a factored axial stress (MPa). Consequently, in this figure, the value of each mark means the factor of safety of each member. From these figures both codes keep the factor of safety for the almost all of data. However, the factor of safety of the ACI code spread over a wide range because it doesn't consider the influence of the size effect, and it is confirmed that the shear capacity of the I- or T-shape beam tends to grow doe to the extended flange width. Also ACI code considers too hard the influence of the axial tension. For the axial force the JSCE code has adopted the conception of decompression moment that is originally developed in CEB/FIP model code. The concept gives good agreement with test results when the reinforced concrete members

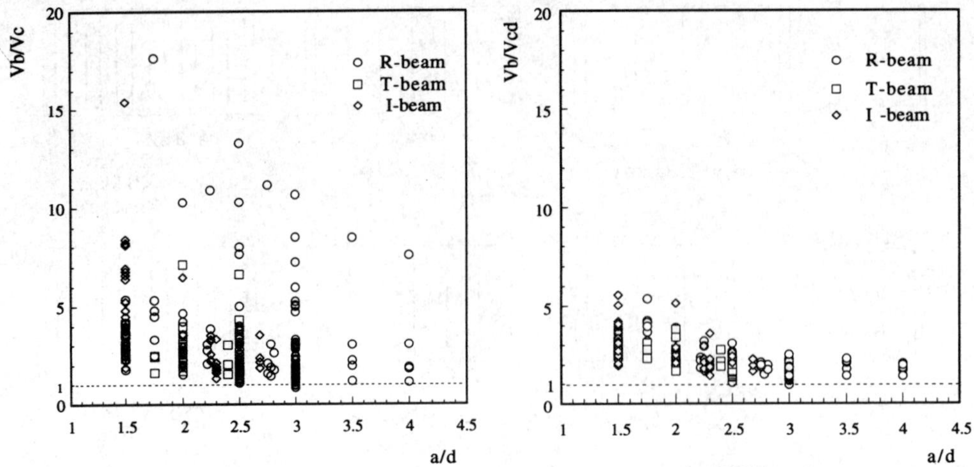

ACI code JSCE code

Fig.4 Relation of the test results and each code on the nominal (or designed) shear strength by concrete

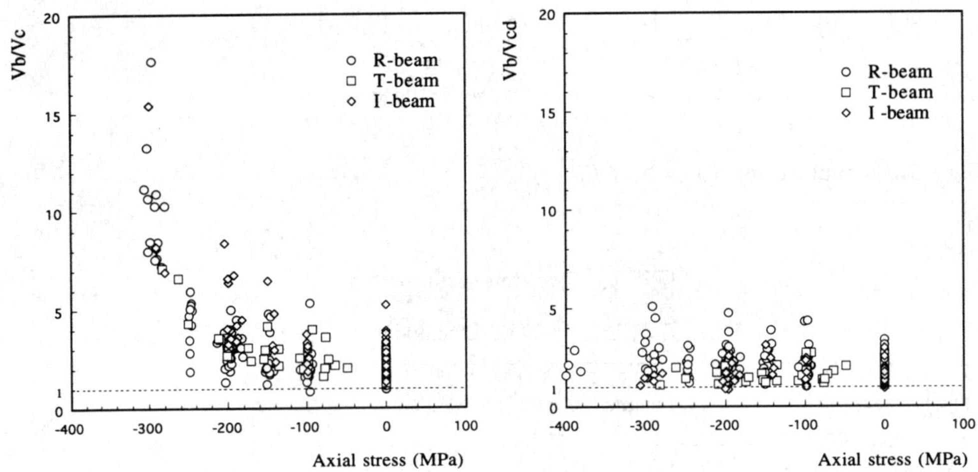

ACI code JSCE code

Fig.5 Relation of the test results and each code on the nominal (or designed) shear strength by concrete

are subjected to axial compression. However, in the case of axial tension, the design equation on that concept has inadequate safety margin because of the insufficient study of this problem. Furthermore, from the numerical analysis with various parameters, it was clear that the reinforcing ratio affects the shear capacity of the member subjected to axial tensile force (Tamura, Shigematsu, Hara 1994), (Tamura, Shigematsu, Hara, Maruyama 1996).

In Fig.6 the ordinate is the ratio of the ultimate shear force V_b and the nominal (or designed) shear strength by shear reinforcement V_n (or V_{yd}) of each code, and abscissa is the shear span to the depth ratio. Some results are less than the safety margin 1.0 as shown in Fig.6. It meams that the concept of the equation (1) and (2) includes some problem. In ACI code, the nominal shear shear strength V_n is reduced by strength reduction factor ϕ.

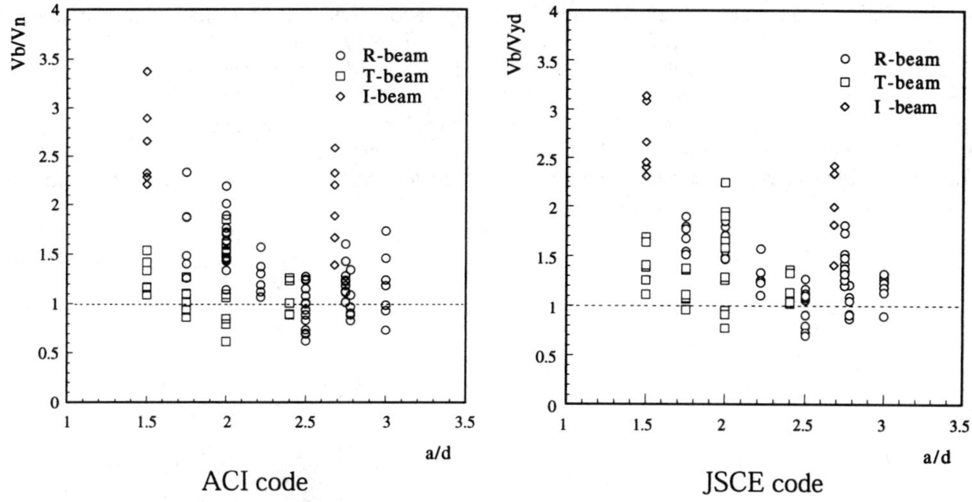

Fig.6 Relation of the test results and each code on the nominal (or designed) shear strength by shear reinforcement

4 CONCLUSION

Experimental results of approximately 380 test specimens indicate the magnitude of the factor of safety of both the ACI and JSCE codes on the nominal shear strength.

The shear strength of I-or T-shape beam is influenced by the ratio of the web width to flange width. Also, definitely, it can be predicted that these initial axial tensile stresses reduce the shear capacity of the members. However, in case of the reinforced concrete member subjected to axial tension, that design equation has an unreasonable factor of the safety due to the insufficiency of the study of this problem.

A simply design equation is generally preferred, and it have to keep the factor of safety sufficiently. However, if it take the factor of safety too much, that is bad economy. Also designer should clear the factor of the safety of the structure for person's request. So we have to clear the failure mechanism of the reinforced concrete member in shear in more detail.

REFERENCE

American Concrete Institute Building Code Requirements for Structural Concrete (ACI 318M-95) and Commentary-ACI318RM-95.

Japan Society of Civil Engineers 1996. Specification for Concrete.

Ritter, W. 1899. Die Bauweise Hennebique, *Schweizerishe Bauzeitung*, Feb., 41-43.

Mörsh, E. 1903. Versuche über Schubspannungen in Betoneisenträgern, *Beton und Eisen*, IV. Heft., 269-274.

Talbot, A.N. 1909. Test of Reinforced Concrete Beams, Resistance to Web Stress, *Bulletin No.29, Engrg., Experiment Station, Univ. of Illinois*.

Anderson, Boyd G. 1957. Rigid Frame Failures, *ACI Journal*, Proceedings V.53, No. 7, 625-636.

Elstner, Richard C., and Hognestad, Evind 1957. Laboratory Investigation of Rigid Frame Failure, *ACI Journal*, Proceedings V.53, No. 7, 637-668.

Haddadin, M.J., Hong, S. and Mattock, A.H. 1969. A Study of the Effectiveness of Web Reinforcement in Reinforced Concrete Beams Subject to Axial Forces, *Structures & Mechanics Report SM 69-2, University of Washington*, Sept..

F. Leonhardt 1965. Die verminderte Schubdeckung bei Stahlbeton-Tragwerken, Begründung durch Versuchsergebnisse mit Hilfe einer erweiterten Fachwerkanalogie, *Der Bauingenieur*, Heft 1, 1-15.

Vecchio, F.J. and Collins, M.P. 1986. The Modified Compression-Field Theory for Reinforced Concrete Elements Subjected to Shear, *ACI JOURNAL, TECHNICAL PAPER*, Title no. 83-22, 219-231.
Bhide, S. B. and Collins, M.P. 1988. Influence of Axial Tension on the Shear Capacity of Reinforced Concrete Members, *ACI STRUCTURAL JOURNAL*, Vol. 86, No. 5, Sep., 570-581.
ASCE 1982. State of the art Report on Finite Element Analysis of Reinforced Concrete.
ASCE 1993. Finite Element Analysis of Reinforced Concrete Structures II.
T. Tamura, T. Shigematsu, T. Hara 1994. Schubtragverhalten durch auf Längszug und Biegung beanspruchte Stahlbetonbalken, *Beton- und Stahlbetonbau*, Heft 11, 289-294.
T. Tamura, T. Shigematsu, T. Hara, K. Maruyama 1996. A Study of Design Equation for the Shear Strength of RC beams subjected to Axial Tension, *Concrete Library International Proc. of JSCE*, No.27, 209-220.

Implementation of Safety and Health on Construction Sites, Singh, Hinze & Coble (eds)
© 1999 Balkema, Rotterdam, ISBN 90 5809 036 1

On the safety of pre-cracked reinforced concrete member

H. Matsuo & T. Shigematsu
Tokuyama College of Technology, Japan

T. Tamura
Nagaoka University of Technology, Japan

K. Nakashiki
Kyuken-Sekkei Corporation, Fukuoka, Japan

ABSTRACT : This paper reports the safety of reinforced concrete member under axial tensile stress and pre-cracked reinforced concrete member experimentally and numerically. It is predicted that the cracks on the reinforced concrete members affect the ultimate strength. Two type of reinforced concrete beams are carried out experimentally. One is the pre-cracked reinforced concrete member and the other is the member which is subjected to the axial tension and bending simultaneously. For the comparison of the cracked reinforced concrete behaviors, non-cracked reinforced concrete member is tested. In numerical analysis, two type of reinforced concrete member are carried out in the same conditions as the experimental conditions. Cracked elements are modeled as the thin elements which possess the weaker tensile and compressive strength than the normal concrete element.

Both from the experimental results and numerical results we obtained the influence of crack on the reinforced concrete.

1 INTRODUCTION

R/C members subjected to axial tensile stress show the bending failure and/or the shear failure depend on the various axial tensile stress level. Similarly, pre-cracked R/C member shows the bending failure and/or the shear failure depending on the degree of cracks and the loading and supporting condition. These conditions are classified into some items, i.e., initial compressive strength, reinforcement ratio, shear span ratio, effective depth and axial tensile force. The failure mechanisms of the R/C beams subjected to axial tensile stress are different on the conditions whether pre-cracked elements are exist or not. In the experimental results, it is evident that the axial tensile force and the pre-cracked elements influence on the ultimate strength of the member. In the numerical predictions, the same tendencies are obtained. In the numerical procedure the finite element approach is adopted using of three dimensional solid hexahedron elements which is adopted for the R/C member model. It is indicated that the approach is a good way to investigate the influence of axial force on the strength of the R/C member. The numerical results are compared with the experimental results.

2 FINITE ELEMENT METHOD

In the finite element formulation, the three dimensional solid element shown in Fig. 1 is adopted and the geometric and the material nonlinearities are taken into account. The propagation of cracked zones, as well as the compressive behavior of the concrete is analyzed.

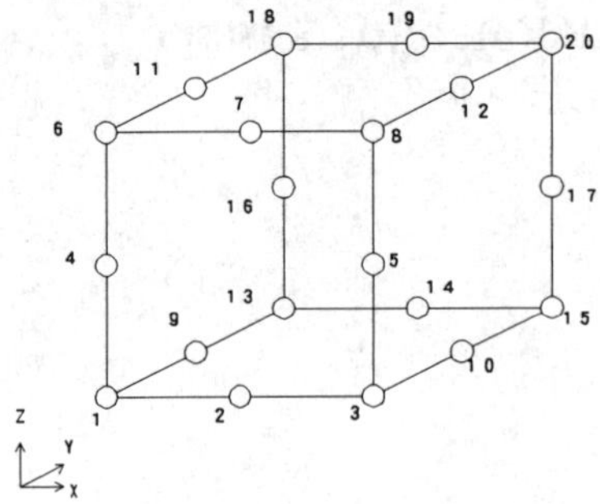

Fig.1 Three dimensional solid element

2.1 *Concrete material model*

2.1.1 *Behavior in compression*

To evaluate the stress state of the concrete for given strain state, following four conditions must be defined.

2.1.2 *The yield criterion*

Under a triaxial stress state, the yield function depends not only on the mean normal stress I_1 (σ_{oct}) but the shear stress invariant J_2 (τ_{oct}). The yield condition of triaxial compressive concrete is defined by Druker-Prager yield criterion.

$$f\,(I_1, I_2) = \{\beta\,(3\,J_2) + \alpha\,I\}^{1/2} = \sigma_0 \qquad (1)$$

2.1.3 *The flow rule*

The normality of the plasticity deformation rate vector to the yield surface is commonly assumed. In plastic area, the flow rule of Prandtle-Reuss is adopted.

$$d\,\varepsilon_{ij}^{p} = d\,\lambda\,((\partial f\,(\sigma\,))/\partial\,\sigma_{ij}) \qquad (2)$$

2.1.4 *The hardening rule*

To define the hardening rule, the relationship between the accumulated plastic strain and the current 'load surface' is assumed to be the conventional 'Madrid Parabola'.(Hirton et al. 1984)

2.1.5 *The crushing condition*

The crushing condition of concrete is decided whether the magnitude of the equivalent strain, which substitute the strain components of defining the yield surface, exceeds the ultimate strain or not.

$$\{\beta(3J_2') + \alpha I_1'\}^{1/2} = \varepsilon_u \qquad (3)$$

2.1.6 *Concrete behavior in tension*

In this paper the response of concrete under tensile stress is assumed to be a linear elastic

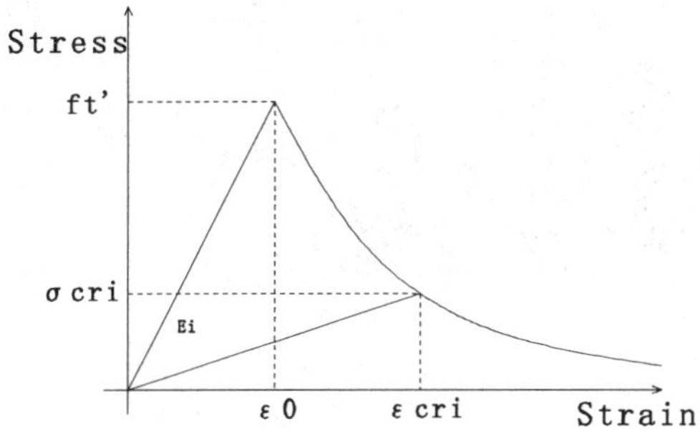

Fig. 2 One dimensional stress strain relation of concrete in tension

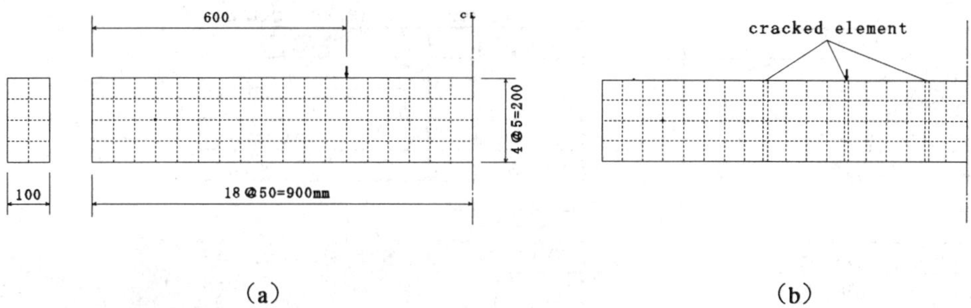

Fig.3 FEM mesh for analyzed R/C beam models

until the fracture surface is encountered and its behavior is just the same as the previous response of concrete under compressive stress. The smeared crack approach is adopted to represent the cracked element behavior.

2.1.7 *Tension Stiffening*

Fig. 2 shows the stress-strain diagram of R/C in tension and represents the cracked concrete modulus of loading and reloading: The tension stiffening is assumed as following equations:

$$\sigma = f_t' \exp\{-(\varepsilon - \varepsilon_0)/\alpha\} \tag{4}$$

$$\alpha = \left(G_f - 0.5 f_t' \varepsilon_{cr} l_c\right) / \left(f_t' l_c\right) \tag{5}$$

where f_t' is maximum tensile stress of concrete, ε_0 is cracking strain, G_f is the fracture energy of concrete and l_c is the characteristic length of the sampling point.

2.2 *Material model for the reinforcement*

The reinforcing bars are considered as steel sheets which possess equivalent thickness and uniaxial behavior resisting only against the axial force in the bar direction. The bilinear idealization is adopted in order to model the elasto-plastic stress strain relationship and both the tensile and the compressive state are governed by this relationship.

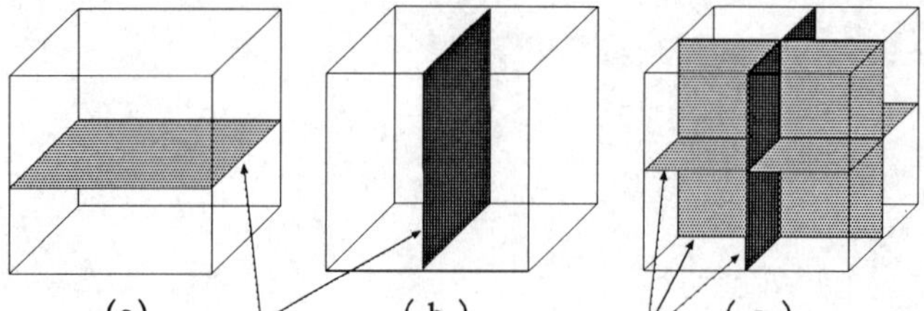

Fig.4 Solid elements pattern

Table.1 Material Properties

	Concrete	
	non-cracked element	cracked element
Elastic modulus(E_c)	28.42 (GPa)	28.42(GPa)
Poisson's ratio(ν)	0.2	0.2
Density(ρ)	2.45	2.45
Compressive strength(σ_c)	32.23 (MPa)	32.34(MPa)
Tensile strength(σ_t)	2.36 (MPa)	2.36(MPa)

Steel	
Elastic modulus(E_1)	205.80 (GPa)
Hardening modulus(E_2)	2.06 (GPa)
Yield stress(σ_y)	356.72 (MPa)

3 NUMERICAL MODEL

Fig.3 shows the dimensions of specimen of R/C beam. Numerical model is a half of beam considering the symmetry of the structural and loading condition. The span of R/C beam is 1,500mm. The test beams have no shear reinforcements. The ratio of the shear span to the effective height is 2.5. Two type of specimens is considered. One is 144 element and 973 the material constants of analyzed models. In the analysis for the member cracked by axial nodes.(Fig 3(a)) The other is 152 element and 1129 nodes.(Fig 3(b)) Table 1 represents tensile loads (pre-cracked member), the finite element model includes the cracked element under the assumption that the tensile strength σ_t and the compressive strength of cracked concrete is 0MPa and σ_c =22.56MPa respectively, as shown in Fig.3(b). That is the compressive strength of concrete is reduced 30% of the non-cracked element. The steel layers are included to the analyzed structures.(Fig.4)

4. EXPERIMENTAL PROCEDURE

4.1 *Test Beams*

In the experimental analyses the same dimensions of specimen and the same loading conditions as the numerical model are used. The length of all specimens is 1800mm and effective span is 1500mm. The dimensions of each specimen are shown in Fig.5. The reinforcing ratio is 1.19%. No stirrups are placed except the supporting points to clarity the structural behaviors. Holes are placed at ends to introduce the axial tensile force.

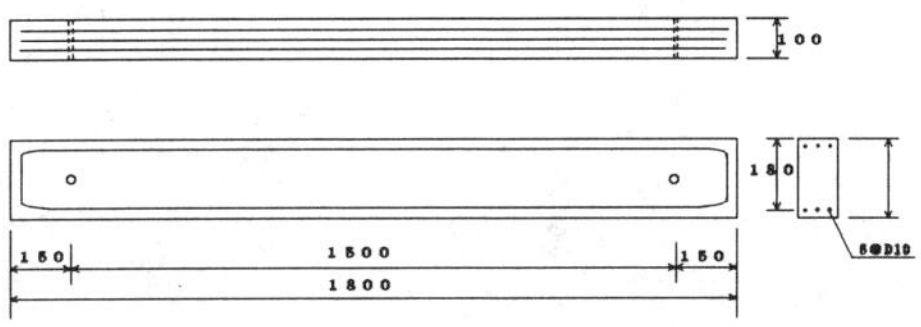

Fig. 5 Specimen

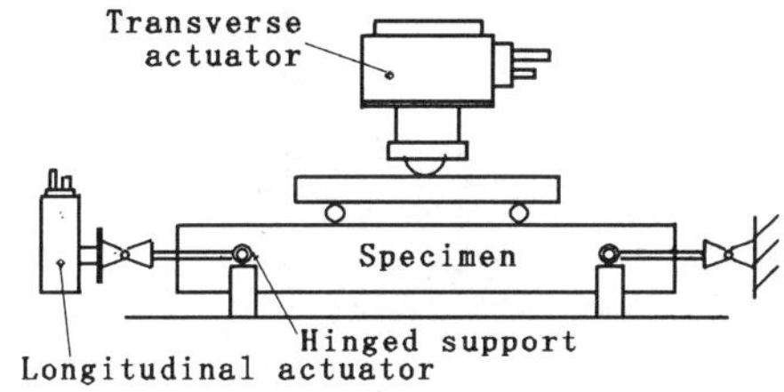

Fig. 6 Test apparatus hinged by bearing joints.

Table.2 Ultimate strength

type	①	②	③
condition	N=0kN	pre-cracked	N=39.2kN
Experimental results(kN)	51.83	49.51	39.54
Numerical results (kN)	58.48	49.51	50.46

4.2 *Test Apparatus and Procedure*

A testing apparatus with axial tension is shown in Fig.6. It is composed of two oil pressure actuators controlled by the electro-oil servomechanism. And both supporting points are hinged by bearing joints. To discuss the shear strength depending on the pre-cracks, we carried out three types of loading tests. ① type is subjected to no axial tension. ② type is subjected axial tension N=39.2kN and then the axial tensile force is released. ③ type is subjected axial tension and bending simultaneously. In the test of the ② type and ③ type, the axial tension is firstly introduced into both ends of the beam via longitudinal actuator till reaching the expected tension and is held constant. Then the transverse loads are provided by transverse actuator that can introduce the loads into two points by the loading beam. The transverse load increases continuously until the beam fails under displacement controlled system. In the test of ② type, the axial tension is firstly introduced until the tensile cracks occurs in the beam. Secondly, the axial load is released from the beam. Then the transverse load applied continuously until the beam fails under displacement controlled system.

① type test is provided to compare with these two tests. In the test, new cracks are marked on the surface of the beam at each loading stage. Dial gauges are placed at the loading point and the center of the span to measure the deflection of the beam. Then the bending strains are measured by wire strain gauges at the center of the tensile reinforcement and the top of the beam.

5 RESULT AND DISCUSSION

5.1 *Ultimate strength*

Table 2 shows the ultimate strength of the beam obtained from the experimental study and the numerical results. Numerical results show that the ultimate strength of the pre-cracked beams is quite smaller than that of non-stressed ones. Also those values approximately show the same ultimate strength of the beam subjected to axial tension N= 39.2kN

5.2 *Load Deflection Relationships*

Fig.7 shows the load deflection curves obtained by the experimental results. Fig.8 shows the the numerical results. In Fig7 and 8 the abscissa indicates the deflection at the center the beam. The ordinate indicates the applied load to the beam. By using displacement incremental method, the load reduction caused by the extension of cracks is detected under the additional

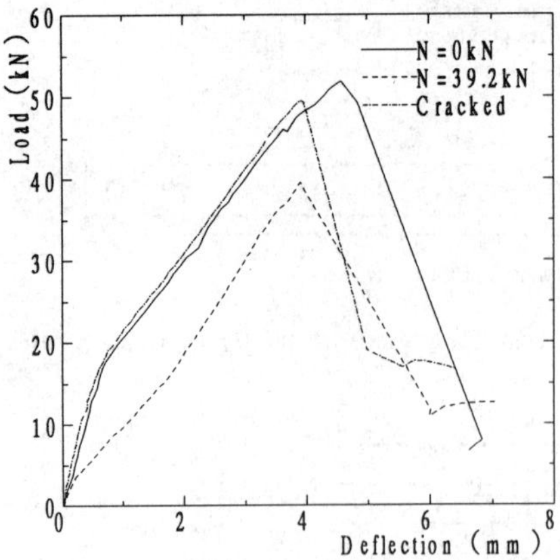

Fig.7 Load deflection curve(Experiment)

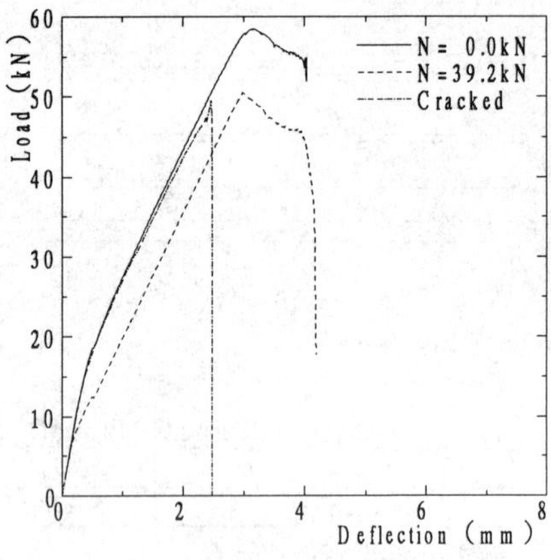

Fig.8 Load deflection curve(Numerical)

cracked elements by the constant displacement increment 1/100mm. In the experimental analysis, the displacement incremental control is also adopted. The loads are applied by each approximately 1/10mm displacement increment. Figure shows that the numerical predictions and experimental results are in good agreement up to the ultimate strength in ① type. The initial gradients of each load deflection curve are almost the same. However, after the occurrence of the initial cracks, the deflection of the beam subjected axial tension N=39.2kN is larger than the other results. In the pre-cracked beam, the load deflection relation is almost equal to that of non-stressed beam until near the maximum loading stage.

6 CONCLUSION

In this paper, we investigated the relationship between the various type of R/C members concerning the strength and the cracks. Pre-cracked R/C member and R/C member subjected axial tension are reduced the ultimate strength for the non-cracked R/C member from the experimental and numerical results. From these results cracked R/C member shows brittle behavior. From numerical results by finite element method using of three dimensional solid hexahedron elements and experimental results, the following conclusions are obtained.

(1) Cracked R/C member is less safety than non-cracked R/C member.

(2) ② type and ① type are plotted approximately same line in load deflection curve. ② type failed faster than ① type.

(3) Crack is influenced for the ultimate strength of the R/C member both from experimetal results and numerical results.

(4) The finite element method reveal that cracks have a bad influence on R/C member.

7 REFERENCE

Hinton,E. and Owen, D.R.J. 1984. Finite element Software for Plates and Shells. Pineridge Press Swansea U.K.

Kupfer,H. and Hilsdorf, K.H. 1969. Behavior of concrete under biaxial stresses, *Proceedings of ACI.*

M. Beppu, S Katsuki, N.Ishikawa and A. Miyamoto, 1971 "An Experimental Study on Circumferential Crack of RC Pier Model by Push-up Impact Test", *Proceedings of The 2nd Symposium on Hannsinn-Awaji Earthquake, pp.213-218.*

Implementation of Safety and Health on Construction Sites, Singh, Hinze & Coble (eds)
© 1999 Balkema, Rotterdam, ISBN 90 5809 036 1

Earthquake technology for American and Japanese educational facilities

V.E.Garland
College of Liberal Arts, The University of New Hampshire, Durham, N.H., USA

ABSTRACT: The damages caused by major earthquakes in Japan and the United States within the past five years have hastened the development of advanced technology systems for disaster mitigation in the construction of public buildings. Computers are now widely used in both countries to mitigate structural and non-structural hazards of earthquakes. With the new technology, old buildings can be retrofitted and new facilities can be designed to resist major temblors. The safety of personnel and material on construction sites during emergencies can be greatly enhanced by the use of telecommunications and computer based crisis planning. Because of their special functions in serving large numbers of children and in being designated as post-earthquake relief and recovery centers, educational institutions should be constructed beyond the minimal seismic building code.

1 INTRODUCTION

On January 17, 1995, at the one year anniversary of the Northridge, California earthquake, which killed 61 people, destroyed 5900 buildings, and resulted in $15 billion in damages, the "Japan/US Workshop on Urban Earthquake Hazards Reduction" was being held in Osaka, Japan. That morning, a temblor measuring 7.2 on the Richter scale hit neighboring Kobe. This "Great Hanshin Earthquake" killed nearly 6500 people, made an additional 300,000 homeless, demolished over 100,000 buildings, and caused almost $100 billion in damages. In the aftermath, Japanese and American structural engineers surveyed the destruction and recommended more technologically advanced systems in the design of earthquake resistant buildings as well as more computer use in relief and recovery efforts.

Educators in Kobe, Japan's foremost international city, quickly mobilized their school and university resources to serve the needs of survivors. The author, who was a visiting professor at Kobe University during the 1995 earthquake and who had also experienced the 1987 Whittier, California earthquake, conducted a three year study of earthquake mitigation in the design and construction of educational facilities in Japan and America.

Results of this study include the need for computer integrated seismic safety and emergency plans for construction site personnel. An assessment of structural and non-structural hazards in both the 1994 Northridge and the 1995 Kobe earthquakes informs architects, engineers and crisis managers on future standards in the construction of public buildings in fault zones with high levels of seismic activity. Recommendations for safety in the building and reconstruction of schools and universities include special considerations of the laws in the United States affecting educational institutions in earthquake zones and in their role as refugee and relief centers under the Federal Emergency Management Agency. Additional analysis is made of current disaster prevention plans in Japan.

2 SEISMIC MITIGATION IN JAPAN AND AMERICA

The United States and Japan share the intense earthquake activity of the "ring of fire" Pacific Rim area which extends from California north to Alaska, and then south to Japan and to New Zealand. In the "subduction zones" of this region, the thinner Pacific plate is often forced beneath the thicker continental land masses, accounting for three fourths of the world's seismic energy (Economist 1995). The biggest recent earthquake in this area occurred in 1976 in Tangshan, China. That 7.8 magnitude temblor killed 240,000 people.

Americans and Japanese have also experienced major earthquakes in this century, including the 1906 San Francisco earthquake which killed 524 and the 1923 Great Kanto earthquake in the Tokyo area, which left 140,000 people dead. As populations increased and urban areas expanded, seismic building codes were codified in the 1920s in Japan and in the 1930s in the United States. In the 1990s, American and Japanese engineers learned that earthquakes could still topple and create major structural damage to buildings which had been "built to code" as "earthquake resistant." As a result, advanced technology systems are being developed to assist buildings, especially high rises in earthquake prone cities such as Los Angeles and Tokyo, to absorb the shock waves of magnitudes greater than 7 or 8 on the Richter scale.

There are some philosophical differences between the Japanese and the Americans in how seismic technologies should be developed and implemented. The government of Japan has a $500 million earthquake research program, whereas the Federal Emergency Management Agency (FEMA) has only allocated $50 million for similar studies in the United States. Japan, which is geographically about the same size as California, is a mountainous landscape dotted with seismic monitors which were can be used to shut down the 250 mph bullet train. In both countries, these monitors cannot give the populations adequate response time in case of dangerous tremors, but they are used in Japan to signal landslides which could be triggered by earthquakes (Ashiya City Hall 1995). Although earthquakes still cannot be predicted with any degree of accuracy anywhere in the world, the Japanese have had some success with submarine seismometers, "It is possible to forecast earthquakes in the Tokai region, since submarine seismometers have been placed there and the mechanism of the quakes in the area is clearly known, but forecasting in other areas remains impossible at present" (Ninomiya 1995).

In the aftermath of recent earthquakes, United States agencies appear to have more useful technology than the Japanese government officials. After Kobe, emergency teams had far less fiber optics cameras for searching in rubble and fewer fire fighting helicopters than American rescue operators after Northridge. Although the Japanese also lag behind the United States in the use of computers and the internet, Kobe University officials used the World Wide Web and dedicated government phone lines to help survivors after the 1994 earthquake (Garland & Morimoto 1996).

FEMA authorities in the United States have recently developed an array of communication technologies to deal with natural disasters such as earthquakes. Federal support occurs when normal communications systems, such as telephones and some computers, are crippled by power outages and downed lines. Mobile communications systems, which can establish open emergency telephone lines, are brought in to stricken areas. Through an agreement with the United States Joint Chiefs of Staff, FEMA can field "an airborne disaster operations center - a communications-equipped 747 military jet - to allow immediate response following a catastrophe" (FEMA 1996). In addition, FEMA technocrats have created advanced computer models to predict the nature and extent of earthquake damage, a World Wide Web site that provides information on disaster assistance, and global links to other emergency management officials.

3 EARTHQUAKE RESISTANT CONSTRUCTION DESIGN

There are also differences between Japanese and American structural engineers in the construction of buildings which should withstand earthquakes. Engineers in the United States tend to emphasize strength in building construction, whereas their counterparts in

Japan focus on flexibility (Zorpette 1995). Japanese buildings do generally well in moderate earthquakes, whereas American buildings are more resistant to larger temblors. However, it is difficult to determine how well United States cities would have fared in the 7.2 magnitude "Great Hanshin Earthquake" since Kobe has the dubious distinction of being the only urban area of the two countries to have had the direct impact of a temblor greater than 7 on the Richter scale in the past fifty years. Computer simulations of earthquakes are helpful, but seismologists concur that "the only way to get a full picture of how buildings react in an earthquake is to have one" (Heaton & Hall 1995). Hence, there are lessons from the rubble in Kobe.

Dr. Ueno, assistant professor of forensic medicine at Kobe University, was one of the medical examiners who determined the exact cause of earthquake deaths. He indicated that most died of suffocation resulting from compression of the chest and abdomen (Ueno 1996). According to the Hyogo Prefecture's medical records, over seventy percent of the victims had been crushed to death, "The primary cause of death had been the collapse of wooden structures built according to conventional methods. The worst hit were those built before 1981, when new earthquake-resistant standards were adopted" (Nakaho 1996).

In addition to the collapse of older wooden and non-ductile reinforced concrete buildings during the initial Kobe earthquake, many other severely damaged structures collapsed in the numerous aftershocks. The local and national governments' failure to act quickly in removing people from cracked buildings in Japan caused further injuries (Garland & Tsuda 1998). Similarly, after the 1994 Northridge earthquake American engineers found major cracks in the steel welds of 120 buildings which had been built to California's stringent 1991 seismic code (Zarghamee & Ojdrovic 1995).

An American engineer who inspected the damage to Kobe's buildings the day after the earthquake observed, "Japan's steel-framed buildings are somewhat stiffer than their American counterparts. Although numerous multi-story modern engineered buildings partially collapsed and several high-rise buildings were seriously damaged, the lack of widespread damage to buildings built in approximately the last 15 years is encouraging, in that our building code standards are similar to Japanese codes...Damaged small-scale buildings with heavy tile roofs and no lateral resisting systems posed obvious continuing collapse hazards" (Kornfield 1995).

Destruction to buildings in the 1989 Loma Prieta earthquake in the San Francisco region indicates other types of weaker structures, such as the pre-code unreinforced masonry buildings, concrete frame structures with unreinforced masonry infill, "tilt-up" buildings, and "apartments with 'soft stories' from large openings on the ground floor for garage doors or windows" (Nemetz 1994). Numerous of the latter type of building collapsed in the Kobe earthquake, causing considerable loss of life of apartment dwellers.

However, both nations are discovering and testing new "earthquake resistant" construction techniques, which are intended to support buildings in these larger earthquakes. More recently, earthquake mitigators have been implementing "seismic isolation" technology to maximize a building's protection from seismic waves. Rubber and steel pads are used to isolate the base of a building from ground motion. A seismically isolated hospital withstood the Northridge earthquake unscathed, whereas a nearby hospital built by traditional techniques suffered almost $400 million in damages.

Engineers at the National Center for Earthquake Engineering Research at the State University of New York at Buffalo designed a network of sensors and computers that activate 'tendons' that brace buildings as they sway, allowing structures to bend instead of toppling in an earthquake. The computerized machinery 'adjusts the building's structure hundreds of times a second to offset the effects of ground vibrations - so-called active designs for earthquake resistance" (Spotts 1995). This type of "active technology" that allows buildings to react to earthquake forces has been installed in several buildings in Japan (Naj 1994). The world's first "smart building," which can reduce seismic vibrations by eighty percent, is the Kyobashi Seiwa building in Tokyo. A computerized system is key to its success, "the sensor network would carry information from seismic vibrations to a central computer, which in turn would analyze the data, programming the weights, braces, or rockets to respond accordingly" (Senders 1992).

Many buildings can be effectively retrofitted to resist considerable earthquake damage.

Kornfield believes that wooden buildings in Japan and the United States can be successfully upgraded, "The Kobe earthquake showed, as did the Loma Prieta earthquake, that large benefits can be gained by modest structural improvements to buildings. In Kobe, some wooden buildings with only a few solid wall sections were seen to have avoided collapse. As most of San Francisco's buildings are wood, a strong push for retrofit with foundation bolts and plywood shear walls could go far in reducing major building failure. No other building construction type seems to provide such cheap, easy opportunities for increased seismic safety" (Kornfield 1995).

One California law mandated retrofitting for certain earthquake prone buildings. After the damages suffered by concrete buildings in the Northridge earthquake, the Los Angeles City Council passed an ordinance requiring that owners of pre-1976 "tilt-up" concrete wall buildings retrofit the structures to make them earthquake resistant (Rackham 1994).

In addition, measurements from both earthquakes taken from "strong motion" recorders in the areas affected indicate that vertical shaking appeared to equal or exceed the force of horizontal vibrations. This suggests that structural engineers will have to revise their construction codes to account for the possibility of strong vertical accelerations (Monastersky 1995).

4 COMPUTER USE ON CONSTRUCTION SITES

Crisis management plans in natural disasters include the use of networked computers for telecommunications when traditional telephone lines are not in use (Garland & Morimoto 1996). During the 1995 Japanese earthquake, networked computers were key links to the outside world for Kobe University students, faculty members and administrators (Garland & Tsuda 1998).

Computers are also vital in personnel safety and in hazard mitigation on construction sites during earthquakes. Architects, engineers and construction site supervisors should be equipped with portable, battery operated computers for communication with outside agencies in case of emergencies. A prompt connection over the Internet could potentially reduce injury and loss of human life after a destructive earthquake.

Technology is also important in the mitigation of structural and non-structural hazards on building construction sites. Some "smart buildings" are equipped with computerized systems, which activate the structural elements needed to resist strong shock waves. Non-structural hazards, such as lighting and machinery, should be inventoried on the computer and secured to withstand major earth movements. Computerized systems can also be used to shut off gas, water and electrical lines on construction sites during earthquakes.

After the 1994 Northridge earthquake, contractors were effective in stabilizing and in saving damaged structures. They gave first priority to "life safety situations;" then the contractors tested construction performance, restored service where appropriate to do so, and made permanent repairs to salvageable buildings (Dooley 1994).

5 EDUCATIONAL FACILITIES

Seismic building codes in the United States and Japan establish the minimum standards which determine how buildings should be designed and constructed to resist earthquake ground-shaking. Despite the existence of building codes in both countries covering construction standards, many structures "built to code" toppled or were irreparably damaged in the Northridge and Kobe earthquakes.

These codes may be more stringent for certain buildings, such as hospitals, fire stations and schools. In light of the recent non-structural and structural earthquake damages to schools, it is imperative that Japanese and American lawmakers, school board members and educational administrators build their schools "beyond the code" of seismic resistance. An American expert on the planning of educational facilities states, "The concern for safety is of primary concern, being a higher priority than even the curriculum and instruction. Some old buildings are unsafe...period. Because of their basic design, they are unsafe for children and adults" (Holcomb 1995). A 1996 report by the United States General Accounting Office confirms this need for improved school construction methods, indicating that "one-third of the schools serving about 14 million students nationwide needed extensive repair or at least partial replacement" (Hardy 1997).

While seismic codes require specified construction methods, "severe earthquake damage can occur even if building structures remain essentially intact...the 1983 6.5 magnitude Coalinga, California, earthquake resulted in non structural damage at Coalinga's schools totaling $2 million and that the 1987 5.9 magnitude Whittier Narrows, California, earthquake caused almost $16 million of damage to elementary and high schools, most of which was non structural" (FEMA 1990). The 1994 6.7 magnitude Northridge earthquake seriously affected 300 schools and forced the closure of the California State University campus at Northridge. School district officials estimated damages to school facilities at $700 million (Dillon 1994).

According to earthquake hazards reduction authorities, "The non structural components with both life safety and major property damage consequences include exterior nonbearing walls, exterior veneers, infill walls, interior partition systems, windows, ceiling systems, elevators, mechanical equipment, and electrical and lighting equipment" (FEMA 1990). In Kobe and in Northridge, building occupants were fatally victimized by these falling non structural objects. Fortunately, those devastating earthquakes occurred before schools were in session. Otherwise, many children and school officials would have been killed from unsecured non structural elements. According to the United States Department of Housing and Urban Development, if the Northridge earthquake had happened later in the day, "thousands of children would have been injured or killed by falling debris, furniture, and lighting fixtures" (Spotts 1995).

Schools in Kobe which survived the quake suffered massive non structural damages, yet they also became refugee and relief centers. After Northridge, the city of Los Angeles, in collaboration with FEMA, implemented a seismic retrofitting to remove the danger of falling lights in the schools (FEMA 1998).

Because they are buildings which serve dense populations of children and which could potentially support earthquake evacuees, it is essential that educational facilities be constructed with maximum resistance to seismic activity.

6 CONCLUSION AND RECOMMENDATIONS

There are an estimated 100,000 earthquakes each year around the globe. The probability of earthquakes of over 7 magnitude on the Richter in the Tokyo and San Francisco areas in the next 30 years is over 65% (Economist, 1995; Sereno 1996). Not only the United States and Japan, but all nations should develop earthquake resistant buildings in order to mitigate the terrible effects of major temblors.

A proactive approach is needed before the next big earthquake topples more buildings and kills more people. Building codes for seismic safety need to be updated to account for structural and non-structural hazards as well as for vertical and horizontal earth shaking movements. The primary goal should be the saving of lives. The use of modern technology, in constructing new high-rises or in retrofitting old buildings, and the implementation of computer aided disaster relief plans are steps in the right direction.

The conclusion is inescapable: engineers, educators and policy makers must learn from the recent disasters in Northridge and Kobe how to protect people and buildings from seismic activity of large magnitude. Preventive actions now will reduce the devastating impact of earthquakes in the future.

REFERENCES

Dillon, S. 1994. New concerns loom as Los Angeles schools reopen. *New York Times*. 143 (49588): A1.

Dooley, T. 1994. Earthquake creates building-saving role for contractor. *Los Angeles Business Journal.* 16 (9): 30.

Earthquake Restoration Plan. 1995. Ashiya, Japan: Ashiya City Hall.

Evacuation in accordance with earth and rock avalanche sensors (wire sensors) and heavy rain warning display panels. 1995. *Ashiya Newsletter.* 3 (3): 3.

Garland, V. & Morimoto, M. 1996. The Kobe earthquake: Telecommunications survive at Kobe University. *Technological Horizons in Education.* 23 (8): 79-81.

Garland, V. & Tsuda, Y. 1998. Japanese educational leadership in natural disasters. *Society of Educators and Scholars* (in press).

Hardy, L. 1997. Covering construction costs. *The American School Board Journal.* 184 (10): 20-23.

Heaton, T. H. & Hall, J. F. 1995. Response of high-rise and base-isolated buildings to a hypothetical Mw 7.0 blind thrust earthquake. *Science.* 267 (5195): 206-212.

Holcomb, J. H. 1995. *A guide to the planning of educational facilities.* New York: University Press of America.

Kornfield, L. 1995. Our experiences in Kobe Japan. In *Comprehensive study of the great Hanshin earthquake.* Nagoya, Japan: United Nations Centre for Regional Development.

Monastersky, R. 1995. Kobe disaster offers lessons for U. S. *Science News.* 147 (4): 54.

Naj, A. K. 1994. Engineers want new buildings to behave like human beings. *Wall Street Journal.* 223 (14): B1.

Nakaho (ed.). 1996. *Chronicle: The great Hanshin earthquake.* Osaka, Japan: The Yomiuri Shimbun.

Nemetz, P. N. 1994. Estimating potential capital losses from large earthquakes. *Urban Studies.* 31 (1): 99-122.

Ninomiya, K. 1995. In T. Kajimoto, Predicting quakes. *The great Hanshin quake.* Tokyo, Japan: The Japan Times, Ltd.

Project impact: Building a disaster resistant community. 1998. Washington, DC: Federal Emergency Management Agency.

Rackham, A. 1994. City orders retrofit of 'tilt-up' concrete wall buildings. *Los Angeles Business Journal.* 16 (5): 8.

Seismic considerations for communities at risk. 1990. Washington, DC: Federal Emergency Management Agency.

Senders, C. 1992. Shock steady: Smart buildings guard against bad vibrations. *Omni.* 14 (8): 26.

Sereno, L. B. 1996. Building owners on shaky ground. *Business Journal Serving San Jose & Silicon Valley.* 14 (14): 5B.

Shaky ground. 1995. *The Economist.* 335 (7911): 5-8.

Spotts, P. N. 1995. How engineers are making buildings more quake-proof. *Christian Science Monitor.* 87 (40): 1.

This is FEMA. 1996. Washington, DC: Federal Emergency Management Agency.

Ueno, Y. 1996. In A. Nakaho (ed.), *Chronicle: The Great Hanshin Earthquake.* Osaka, Japan: The Yomiuri Shimbun.

Zarghamee, M. S. & R. P. Ojdrovic. 1995. Northridge postscript: Lessons on steel connections. *Civil Engineering.* 65 (4): 68-72.

Zorpette, G. 1995. Bracing for the next big one. *Scientific American.* 272 (4): 14-15.

Implementation of Safety and Health on Construction Sites, Singh, Hinze & Coble (eds)
© 1999 Balkema, Rotterdam, ISBN 90 5809 036 1

Analytical study on shear strength of R/C column under axial force

K. Yamazaki, T. Shigematsu & T. Hara
Tokuyama College of Technology, Japan
T. Tamura
Nagaoka University of Technology, Japan

ABSTRACT: Engineers have to design the reinforced concrete structure under the conception that the reinforced concrete member should not fail in shear failure but in bending failure when the collapses are inevitable. However, the shear failurc has not confirmed when the structures are subjected to vertical earthquake motion. In this paper, the strength of reinforced concrete column under axial force was examined using the finite element method based on three-dimensional solid element. From the numerical investigation, it is concluded that the axial force strongly influences on the response of reinforced concrete structures.

1 INTRODUCTION

The failures of the reinforced concrete members are classified into the bending failure and the shear failùre. The member under bending shows the ductile behavior. On the other hand, the shear failure shows the brittle behavior. Therefore, engineers have to maintain the design concept that the reinforced concrete member should not fail in shear failure but in bending failure when the collapses are inevitable. The shear strength of reinforced concrete member is evaluated by concrete strength, reinforced ratio, effective depth of cross section, shear span-depth ratio and applied axial force. However, in addition to the above factors, it has been confirmed that the shear failure occurs when the structure is subjected to vertical earthquake motion. Especially, on Hanshin Earthquake, many reinforced concrete columns which compose the flat slab with piloti in the base story had failed in the shear.

In designing the column, the member is merely assumed to be subjected to axial tensile force. However, when the structure is subjected to unusual loading condition such as a vertical earthquake motion, it is possible that the members are subjected to axial tensile force. When the members are subjected to axial compressive force, the shear strength may increases more than double for the no axial stress condition. In contrast, it is confirmed that the shear strength decreases extremely when the members are subjected to axial tensile force.

In this paper, shear strength of reinforced concrete column subjected to axial force was examined analytically using the finite element method based on three-dimensional solid element. The axial stress varies from 3.6MPa (compressive stress) to 3.6MPa (tensile stress) and we examined the effects of axial force.

2 FINITE ELEMENT METHOD

In this paper, we use three dimensional solid element which could be applied to any shapes. An element adopted is twenty nodes isoparametric three dimensional solid element. (see Fig.1)

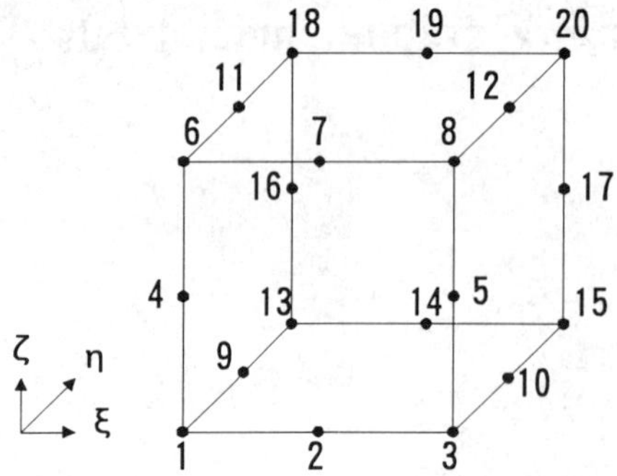

Fig.1 Three dimensional solid element

2.1 *Material model for concrete*

2.1.1 *Behavior in compression*

The inelastic behavior of concrete possesses the recoverable strain components and irrecoverable strain component. In this paper, elasticity theory is used for recoverable strain components, and the strain hardening plasticity approach is used for irrecoverable strain components. To evaluate the stress state of the concrete for given strain state, following four conditions must be defined.

(i)Under triaxial stress state, the yield function depends not only on the mean normal stress $I_1(\sigma_{Oct})$ but also on the shear stress invariant $J_2(\tau_{Oct})$. The yield condition of triaxial compressive concrete is expressed as follows;

$$f(I_1, J_2) = \sqrt{\beta(3J_2) + \alpha I_1} = \sigma_0 \tag{1}$$

where α and β are material parameters and σ_0 is the equivalent effective stress. When the results by Kupfer's experiment (Kupfer 1969) are employed, α =0.355 σ_0 and β =1.355 are defined. Fig.2 shows the three dimensional stress space representation of the concrete stress combination.

(ii)The normality of the plasticity deformation rate vector to the yield surface is commonly assumed.

(iii)To define the hardening rule, the relationship between the accumulated plastic strain and the current 'loading surface' is assumed to be conventional 'Madrid Parabola'. Fig.4 shows equivalent stress strain relation of the concrete.

(iv)The crushing condition of concrete is described as a strain control phenomenon and the crushing condition is defined as like as yield condition Eq.(1).

2.1.2 *Behavior in tension*

Cracking is one of the major cause of nonlinearity of the reinforced concrete structure because cracking stress in tension shows lower than the failure stress in compression. In this paper, the response of concrete in tension is modeled as a linear-elastic brittle material and maximum tensile stress criteria is employed. The smeared crack approach is adopted to represent the cracked element behavior. The crack criterion is defined as follows;

$$\sigma_{i0} = f_t \qquad \text{for tension-tension zone}$$

$$\sigma_{i0} = f_t\left(1+\frac{\sigma_j}{f_c}\right)\left(1+\frac{\sigma_k}{f_c}\right) \qquad \text{for tension-compression zone} \tag{2}$$

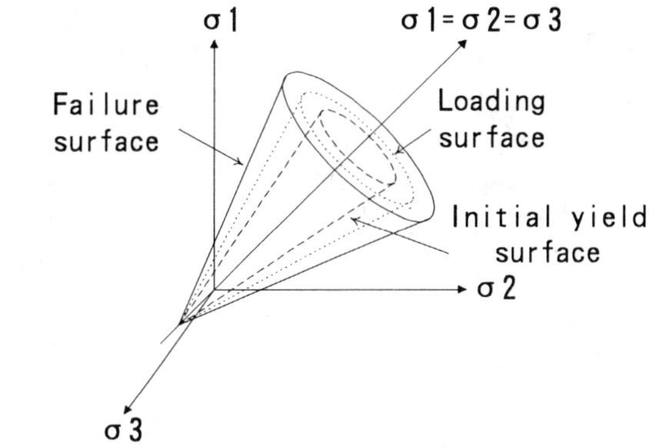

Fig.2 Three-dimensional stress space representation of concrete

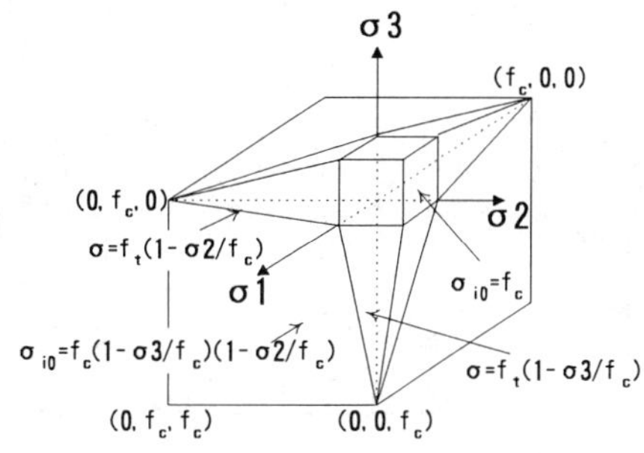

Fig.3 Three-dimensional stress space representation of concrete (tension)

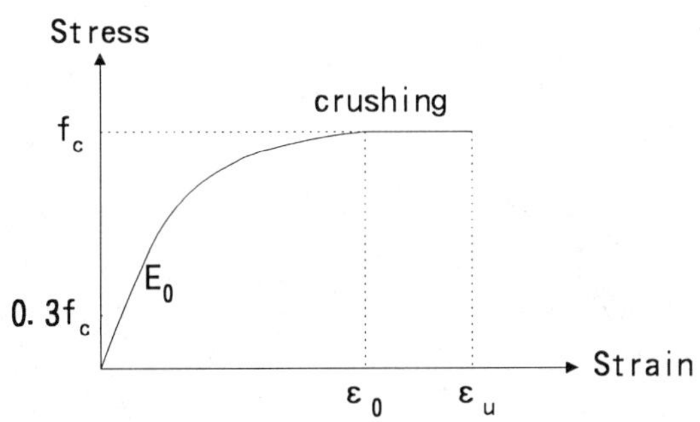

Fig.4 Stress strain relation of concrete in compression

where σ_{i0} is the cracking stress (i=1,2,3) and σ_j and σ_k is the principle compressive stresses perpendicular to the principle tensile stresses, respectively. Three-dimensional stress space representation in tension or tension-compression is shown in Fig.3.

After cracking, the concrete is still partially able to carry the tensile stresses normal to the cracked plane due to the concrete adhesion to the reinforcing bars and the aggregate interlocking. To evaluate the tension stiffening of reinforced concrete, the stress reduction of the concrete normal to the cracked plane is assumed as shown in Fig.5. In Fig.5, the stress in the degradation zone is defined.

$$\sigma = f_t' \exp\left(-\frac{\varepsilon - \varepsilon_t'}{\alpha}\right)$$

$$\alpha = \frac{G_f - 0.5 f_t' \varepsilon_{cr} l_c}{f_t' l_c} \qquad (3)$$

where α =0.2 in this paper.

2.2 *Material model for the reinforcement*

The reinforcing bars are considered as steel sheets which possess equivalent thickness and uniaxial behavior resisting only against the axial force in the bar direction. The bilinear idealization is adopted in order to model the elasto-plastic stress strain relationship and both the tensile and the compressive state are governed by this relationship.

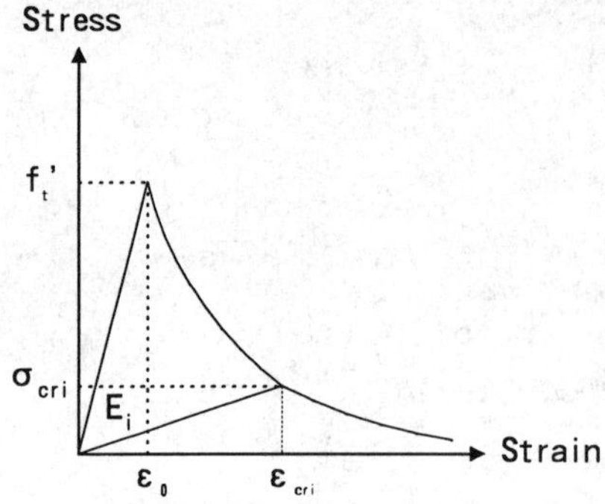

Fig.5 Stress strain relation of concrete in tension

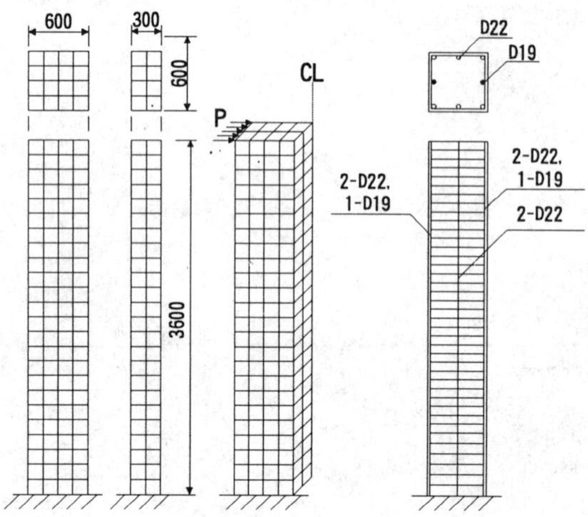

Fig.6 Numerical Model (unit: mm)

3 NUMERICAL MODEL

Fig.6 shows the dimensions of reinforced concrete column model. We assumed that the height of reinforced concrete column is 3,600mm, the column width is 600mm × 600mm based on Building Standard in Architectural Institute of Japan. The column is divided into three dimensional solid elements. Since both the shape and the loading condition are symmetric about the center of the reinforced concrete column, we take the half of reinforced concrete column as a numerical model. Column model is divided into 24 × 2 × 4 meshes (192 elements) and 1285 nodes. The applied distributed load is idealized as the concentrated load, and also the reinforcement is modeled as the steel sheets defined in the local coordinates. The material properties are shown in Table 1.

We calculate the shear strength of no axial stress condition. Then we calculate the effect of the axial stress. We calculated shear strength of column under the axial stress σ_{0c}=1.2MPa. Also, we calculate the shear strength of the column subjected to axial compressive stress σ_{0c}=2.4MPa, 3.6MPa and axial tensile stress σ_{0t}=1.2MPa, 2.4MPa, 3.6MPa. In the numerical analysis, we examine the relationship between the axial force and the shear strength.

Table1 Material Properties

Concrete	
Elastic modulus (GPa)	30
Poisson's ratio	0.3
Density (kg/m^3)	2.4
Compressive strength (MPa)	21
Tensile strength (MPa)	2.0
Steel	
Elastic modulus (GPa)	210
Hardening modulus (GPa)	2.1
Yield stress (MPa)	340

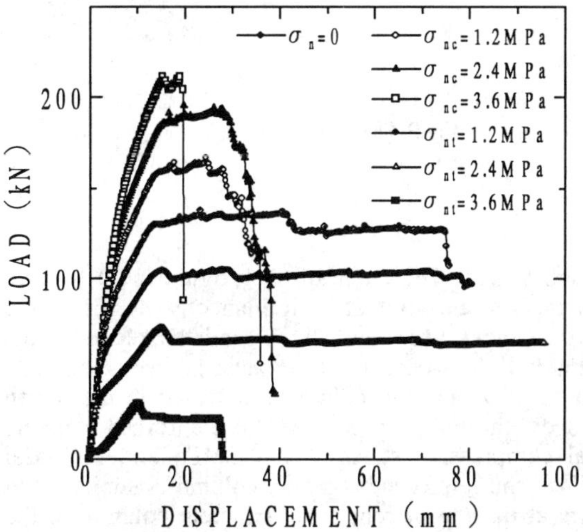

Fig.7 Load-displacement curve

4 NUMERICAL RESULTS

4.1 *Load-displacement response*

In the numerical analysis, the displacement is controlled by 0.2mm(downward) at the top of column shown Fig.6 using the displacement increment method . The load-displacement curve is shown in Fig.7. Each curve shows the numerical results of the model that is subjected to various axial compressive stress σ_{nc}=3.6MPa, σ_{nc}=2.4MPa, σ =0 and σ_{nc}=1.2MPa and to the axial tensile stress σ_{nt}=1.2MPa σ_{nt}=2.4MPa σ_{nt}=3.6MPa. The initial gradients of each load-displacement curve are almost the same. However, after the occurrence of the initial cracks, shown in Fig.7, as the reinforced concrete column is subjected to larger axial tensile force, the maximum loading capacity become smaller. In contract, as it is subjected to larger axial compressive force, the ultimate strength become larger. Therefore, it is expected that the axial force strongly influences on the response of reinforced concrete structures. However, the reinforced concrete column subjected to axial compressive force shows brittle behavior. On the other hand, the reinforced concrete column subjected to axial tensile force shows the ductile behavior. Therefore, the reinforced concrete column that is usually failed by bending shows the shear failure under axial compressive force.

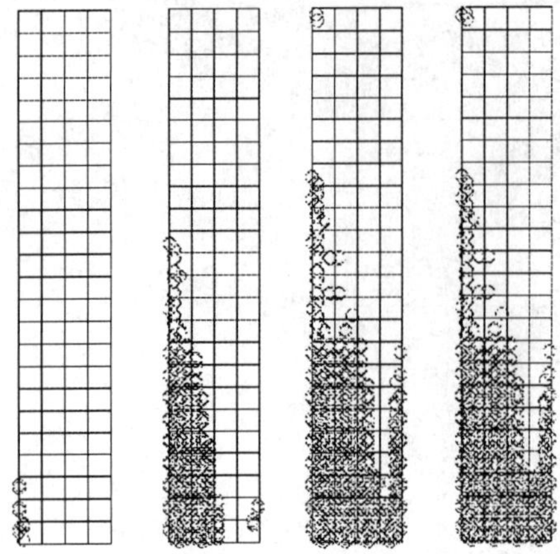

I II III IV

Fig.8 The failure process ($\sigma_{n}=0$)

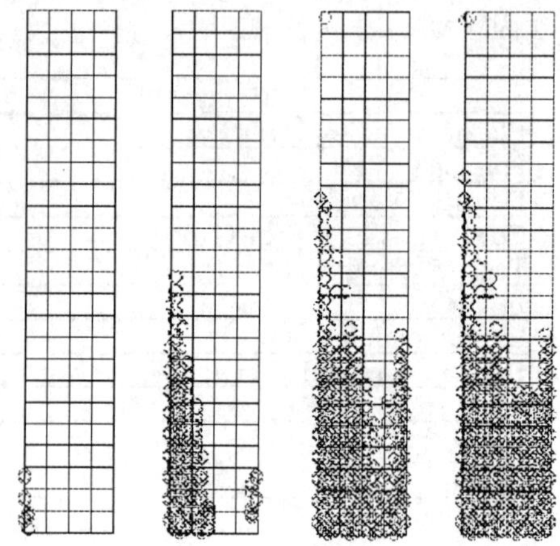

I II III IV

Fig.9 The failure process ($\sigma_{nc}=1.2MPa$)

4.2 *Failure process*

The cracking patterns shown in Fig.8 through Fig.11. The circle of each figure indicates that concrete stress exceeds the tensile strength in Gassian point of each element. The first stage of each figure shows the generation of initial cracks (see mark I), and the second stage shows the progress of bending crack(see mark II). Also, the third stage indicates the state of crack after the maximum loading(see mark III), and the failure is indicated in the fourth stage(see mark IV). Fig.8 through Fig.10 show the failure process of the reinforced concrete column with the no axial force, the axial compressive stress $\sigma_{nc}=1.2MPa$ and the axial tensile stress $\sigma_{nt}=1.2MPa$, respectively. When the reinforced concrete column is subjected to axial tensile force, the bending crack occurs at the top of reinforced concrete column on the first stage as show in Fig.10. From figures, the bending crack is starting from the bottom of reinforced concrete column.

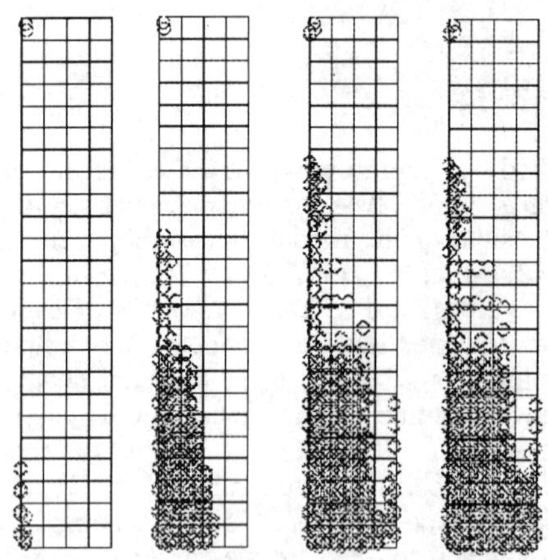

I II III IV

Fig.10 The failure process ($\sigma_{nt}=1.2MPa$)

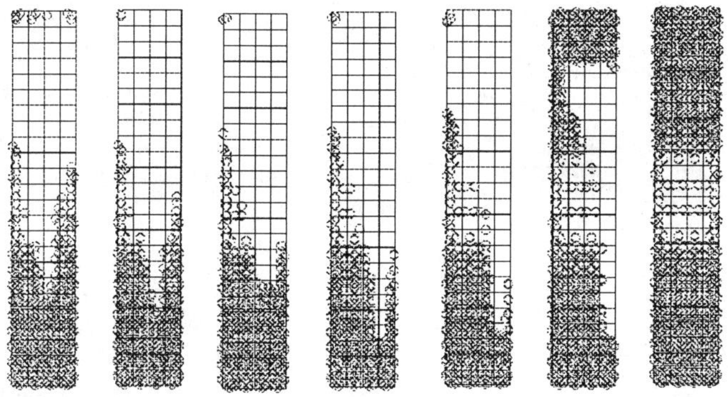

compressive force tensile force
3.6 2.4 1.2 0 1.2 2.4 3.6 (MPa)

Fig.11 The failure condition

4.3 *Effect of the axial force*

Fig.11 shows the failure conditions under different axial stress. The reinforced concrete column subjected to axial compressive stress tends to fail at the bottom of reinforced concrete column. Also, the reinforced concrete column subjected to axial tensile stress tends to fail at the bottom through the top.

5 CONCLUSIONS

The finite element analysis of reinforced concrete column subjected to axial force is examined considering geometric and material nonlinearities. The conclusions are as follows;
(1)The ultimate strength of the reinforced concrete column subjected to axial tensile force drop compare with one subjected to no axial force. Also, the failure becomes ductile.
(2)Engineers have to maintain the design concept that the reinforced concrete member should not fail in shear failure but in bending failure when the collapses are inevitable. However, in this paper, the reinforced concrete column subjected to axial compressive force tends to fail in shear. Therefore, in order to overcome this difficulties we should place the closer spaced hoop reinforcement under the axial compressive stresses.
(3)The axial force influences on the failure process after generation of cracking. It is expected that the axial force influences on the response of reinforced concrete structures.

6 REFERENCE

Architectural Institute of Japan, 1988. Standard for Structural Calculation of Reinforced Concrete Structures.

Hinton, E. and Owen, D.R.J..1986. Numerical Methods and Software for Dynamic Analysis of Plates and Shells. *Pineridge Press Swansea U.K.*.

Kupfer, H. and Hilsdort, K.H..1969. Behavior of concrete under biaxial stress. *ACI JOURNAL*, pp.656-666.

T.Tamura and T. Shiba. 1996. Shear Strength of Reinforced Concrete Pillar Subjected to Axial Force. *Our World in Concrete Structure, Vol.15.* pp.233-238.

Implementation of Safety and Health on Construction Sites, Singh, Hinze & Coble (eds)
© 1999 Balkema, Rotterdam, ISBN 90 5809 036 1

Repair and improvement of safety for the cracked cooling tower

T. Hara & T. Shigematsu
Tokuyama College of Technology, Japan
S. Kato
Toyohashi University of Technology, Japan

ABSTRACT: The cooling tower repair systems are evaluated considering the safety and the improvement of structural strength. Two repairing systems have been ordinary adopted for the old cooling tower. One is adding the concrete layers on the shell surfaces. The other is stiffened by the R/C ring stiffeners. In this analysis, former procedure is adopted and is analyzed by focusing on the effect of the additional thickness. The finite element procedure based on the isoparametric degenerated shell is adopted to evaluate the effects of repairing the old cooling tower. Numerical parameters are the shell thickness and the amount of additional reinforcement. From the numerical analysis, the initial crack load, the reinforcement yield load and the failure load are detected and the safety of the old cooling tower is discussed.

1 INTRODUCTION

Many cooling towers are used for cooling down the hot water produced by the power plant and the industrial plants. The cooling towers are made both from concrete and reinforcements. Both concrete surfaces suffered from the penetration on inner surfaces and from the carbonation in outer surfaces. Therefore, old cooling towers are subjected to such aging effects. To repair these damaged structures is important not only to maintain the plants but also to prevent both thermal drainage and air pollution against the environment. However. cooling towers are the huge structures. Therefore, the numerical evaluation is one of the best ways to assure the safety of these structures.

In this paper, the cooling tower repair systems are evaluated considering the safety and the improvement of structural strength. Two repairing systems have been ordinary adopted for the old cooling tower. One is adding the concrete layers on the shell surfaces. The other is stiffened by the R/C ring stiffeners. In this analysis, former procedure is adopted and is analyzed by focusing on the effect of the additional thickness. The cooling tower modeled in this analysis is Lagisza cooling tower in Poland. The height is about 100m. The finite element procedure based on the isoparametric degenerated shell is adopted to evaluate the effects of repairing the old cooling tower. In the numerical analysis, both material and geometric nonlinearities are taken into account. Also the layered approach is used to evaluate the stress redistribution after cracks through the thickness. Applied load is the self weight and the wind pressure recommended by IASS (IASS 1977,1979).

Numerical parameters are the shell thickness and the amount of additional reinforcements. From the numerical analysis, the initial crack load, the reinforcement yield load and the failure load are detected and the safety of the old cooling tower is discussed.

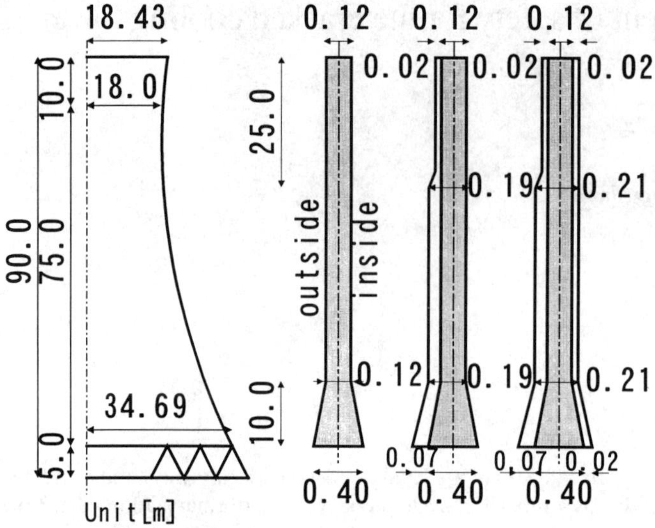

Figure 1 Dimensions of Lagisza Cooling Tower

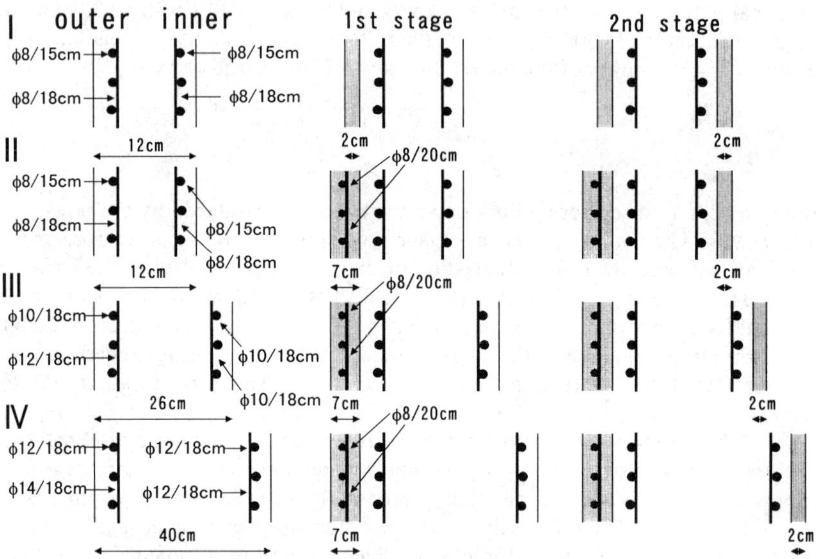

Figure 2 Cross Section of the Cooling Tower

2 NUMERICAL MODEL

In this paper, the cooling tower at Lagisza Power Plant (Golczyk 1996) is adopted as the numerical model. The height is about 100m and suffered the damages from the temperature effects, the penetration of the water and the carbonation of the concrete. Lagisza cooling tower was repaired by adding both the concrete thickness and reinforcement mesh. Additional reinforcement mesh was placed around the middle height of the tower. Figure 1 shows the dimension and the shell thickness of Lagisza cooling tower shell. Figure 2 shows the cross

Table 1 Shell Thickness and Cross Section

Lagisza Cooling Tower		
cross section	height z[m]	thickness t[cm]
I	-15.00~+10.00	12.0
II	-15.00~-60.00	12.0
III	-67.50~-60.00	12.0~26.0
IV	-75.00~-67.50	26.0~40.0

Table 2 Material Properties

Concrete		Steel	
Elastic Modulus E_C	28.3GPa	Elastic Modulus E_s	210GPa
Poisson s ratio ν	0.175	Hardening Parameter H'	0.01
Compressive Strength f_c	25.0MPa	Yield Strength f_y	232MPa
Tensile strength f_t	2.50Mpa	Tensile Strength f_b	394MPa

section of the cooling tower and the repair schemes through the shell height.

The shell thickness throughout the height is shown in Table 1.

Lagisza cooling tower is stiffened under two stages repair. At the first stage, the concrete covers of 2cm and 7cm are added on the outer surface of the shell. In the region of the 7cm additional concrete layer, the hoop and the meridional reinforcements are placed. At the second stage 2cm concrete layer is added on the inner shell surface.

The material parameter of the cooling tower is shown in Table 2.

3 NUMERICAL PROCEDURE

3.1 *Finite Element Representation*

In the numerical analysis, the finite element method is used (Mang 1983, Hara 1994). The shell elements are derived based on the isoparametric degenerated element discretization. In the numerical analysis, both material and geometric nonlinearities are considered. To represent the strains in the element, Green Lagrange strain definitions are used. R/C shell element contains both concrete and steel material. Therefore, to show the behavior of the shell element, each element is divided into some layers.

To represent the concrete behavior, the Drucker Prager criterion is defined for the biaxial compressive stress states and the flow rule is adopted for the plastic behavior. In the behavior of the concrete in tension, tension stiffening is introduced (Hara 1994). To represent the behavior of reinforcement, it is assumed that the reinforcing steel has a uniaxial bilinear constitutive relation in both compression and tension.

3.2 *Numerical Procedure*

The finite element mesh division of the model is shown in Figure 3. The model is divided into 8 elements in the hoop direction and into 14 elements in the meridional direction. Numerical model is a half of the cooling tower because of the symmetry of the configuration, the supporting and the loading conditions. The cooling tower is simply supported at the base and free at the top. In this analysis the deformation effects of the supporting V columns are not considered. Each shell element is divided into 8 concrete layers and some steel layers under non-repaired conditions.

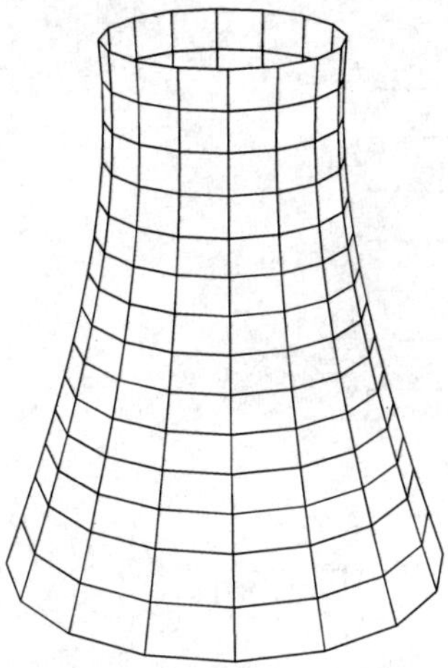

Figure 3 Finite Element Mesh

To investigate the R/C cooling tower responses, both the load increment and the displacement increment are used to represent the nonlinear behavior and to avoid the numerical instability caused by the sudden changes of the stress states due to the concrete cracks.

In the numerical calculation, the self-weight is loaded gradually under the load increment control. After the applied load reaches the self-weight, the wind pressure is applied quasi statically under the displacement increment control. In this analysis wind pressure recommended by IASS (1977, 1979) is applied. In this analysis, we assumed that the wind velocity at 10m above the ground is 44.7m/sec (100mph).

In the numerical calculation, the concrete has the material properties that are considered as the deteriorated concrete material on both surfaces. Therefore, the concrete possesses 50% tensile strength and the 80% compressive strength.

4 NUMERICAL RESULTS

4.1 *Load deflection behavior of repaired cooling tower*

Figure 4 shows the load deflection behavior of Lagisza cooling tower. The displacement is measured at the throat on the windward meridian. The abscissa and the ordinate denote the displacement at the throat and the load factor, respectively. The load factor denotes the ratio of applied wind pressure to the designed wind pressure.

Solid line denotes the load deflection behavior of Lagisza cooling tower without stiffening. The curve concerning the shell stiffened by only concrete cover (dashed line) is coincide with that by both concrete cover and the additional reinforcements (dotted line). The load factor grows up until the tensile cracks occur on the windward meridian. After the load factor drops due to the stress redistribution, the load deflection curve shows the zigzag behavior and the load factor converges to almost constant value due to the yielding of the meridional reinforcements. Figure 5 shows the deformation pattern of the cooling tower near the failure. The windward meridian of the shell surface shows the concave deformation. Therefore, in the lower height of the windward, meridional reinforcements yield. The cooling tower fails followed by the yielding of the hoop reinforcements.

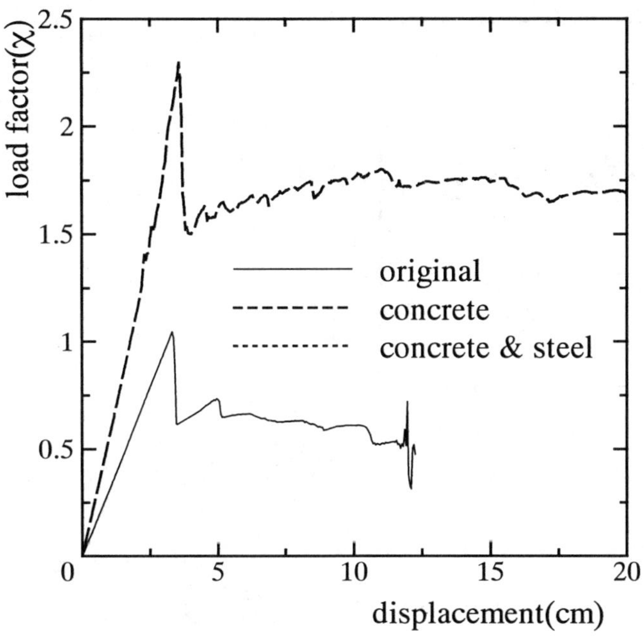

Figure 4 Load Displacement Behavior

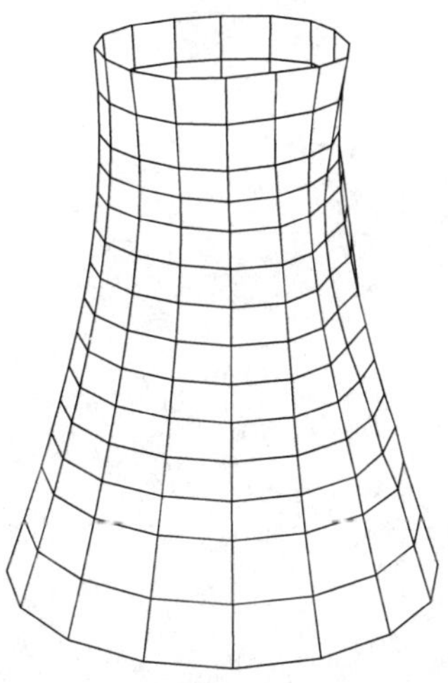

Figure 5 Deformation Pattern near the Failure

4.2 *The effects of reinforcing position*

To investigate the effects of the reinforcing position, three types of reinforcements are examined. In the first model, the stiffening zone on which the reinforcements are placed (see Table 2 I,II,III) is the designed position (whole stiffening). In the second and third model, reinforcements are placed on lower and upper half of the designed stiffening zone, respectively. Figure 6 shows the load deflection curve of these three models. The dashed line, the dashed dot line and the dashed doted dot line denotes the load deflection curve concerning the whole

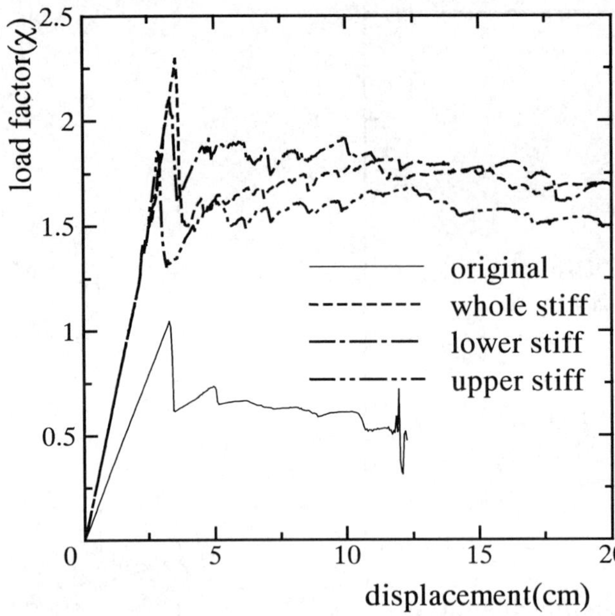

Figure 6 Comparison of the Stiffening Position

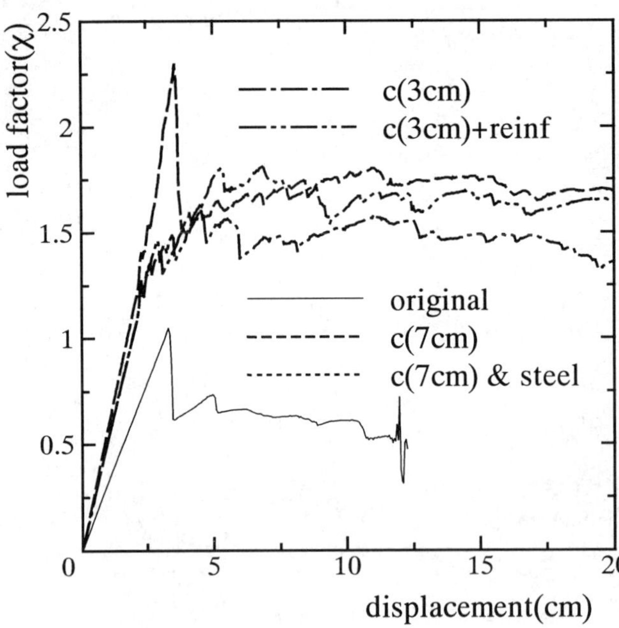

Figure 7 Effect of the Concrete Thickness

stiffening, the lower stiffening and the upper stiffening condition, respectively.

From the figure, the results of the upper half stiffened condition shows the lowest strength $\chi=1.55$. However, the results of both the whole stiffened condition and the lower half stiffened condition show the same ultimate strength χ=1.75. Therefore, the shell, which is stiffened around the lower height of the cooling tower by the additional reinforcements, is effective.

4.3 *Effects of concrete cover thickness*

Figure 7 shows the load deflection behavior of Lagisza cooling tower which has different additional concrete cover thickness. In the case of 7cm concrete cover (designed thickness), the load deflection curve of the shell stiffened by the concrete cover (dashed line) and that stiffened both by concrete cover and reinforcements (dotted line) show the same load deflection curve. However, in the case of 3cm concrete cover, the load deflection curve of the shell stiffened by concrete cover (dashed dot line) is different behavior from that of the shell stiffened both by concrete cover and the reinforcements (dashed dotted dot line). The effects of reinforcements are influenced by the shell thickness.

5 CONCLUSIONS

In this paper the load deflection behavior and the ultimate strength of the repaired cooling tower are examined. Lagisza cooling towers is adopted as the numerical model. The general material parameters and the IASS (1977, 1979) recommendation are adopted to evaluate the cooling tower behavior.

The conclusions obtained in this analysis are as follows:

1. In the numerical model, the final failure behavior is caused by the yield of the meridional reinforcement on the windward meridian. Therefore, ultimate strength is determined by the amount of meridional reinforcement. However, as the thick concrete cover layer is placed on the shell outer surface, the influence of the stiffening reinforcement is small.
2. The stiffening position by the additional reinforcements influences the increasing strength of the cooling tower. It is efficient to stiff the lower portion of the cooling tower by additional reinforcement.
3. It is useful that the deteriorated cooling tower is repaired both by the concrete cover and the additional reinforcement considering the deformation patterns and the stress redistribution of the cooling tower.

REFERENCES

Golczyk M., Abramek W. and Centkowski J. Analysis and safety of repaired reinforced concrete cooling towers , *Proceedings of the 4th International Symposium on Natural Draught Cooling Towers* May 1996 pp277-281.

Hara T., Kato S.. and Nakamura H. (1994). Ultimate strength of RC cooling tower shells subjected to wind load , *Engineering Structures* 16(3), pp.171-180

Hara T., Kato S. and Ohya M. Numerical simulation of repairing the cooling tower shell, *XIII Polish Conference on Computer Methods in Mechanics* May 1997 Vol.2 pp517-524.

Hara T., Kato S. and Ohya M. Behavior of Repaired Cooling Tower Shell, *The Sixth East Asia-Pasific Conference on Structural Engineering and Construction* January 1998 pp2075-2080

Hinton, E. and Owen, D.J.R., (1984). *Finite element software for plates and shells*, Prineridge Press, Swansea, UK.

IASS Working Group 3 (1977). *Recommendation for the design of hyperbolic or other similarly shaped cooling towers*

IASS The Working group on recommendations of IASS (1979). *Recommendations for reinforced concrete shell and folded plates*

Mang,H.A. and Floegl,H. (1983). Wind-load reinforced concrete cooling towers: buckling or Ultimate load ? , *Engineering Structures*, 5, pp.163-180.

10 Fatigue and human error

Implementation of Safety and Health on Construction Sites, Singh, Hinze & Coble (eds)
© 1999 Balkema, Rotterdam, ISBN 90 5809 036 1

Implications for construction safety of studies of the management of human error

Andrew R. Atkinson
South Bank University, London, UK

ABSTRACT: Recent developments in research and practice related to the avoidance of accidents and failures in the construction industries world-wide has emphasised managerial forces rather than the actions of an individual. This paper reports research related to human error and quality assurance which has implications for safety. A model of the pathology of error, related to construction projects and which emphasises managerial and systemic aspects, is conceptualised and evaluated.

1 INTRODUCTION

Researchers from the field of psychology, driven by several major accidents in the 1970s and 1980s, have given a major impetus to the study of human error (Rasmussen, 1983, Rasmussen et al 1987, Lourens, 1990, Reason 1990), but their perspective is largely limited to the actions of the individual. Further, many errors are seen as inevitable, or even beneficial, leading to the conclusion that they cannot be removed at source and must be managed. Reason (1990) gives some attention to wider causes related to the management of organisations and makes a distinction between *latent* and *active* errors. He develops the distinction between latent and active errors in proposing a "resident pathogen metaphor" - latent defects in a system waiting for a combination of circumstances to precipitate a failure.

The distinction between latent and active errors is echoed in major reports on human error in several industries (HSE 1976, ACSNI 1993, Bea 1994) as well as in monographs directed to human error (Petroski, 1985, 1994, Kletz 1985, Blockley 1992). From this distinction some researchers attempt models of the error process within complex systems. Embrey (1992) classifies causes of errors as *"actual errors"*, *"error inducing factors"* and *"higher level policy factors"* and Whittington et al (1992), in specifically considering construction accidents, classify causes as operating on four levels - *"company policy"* level, *"project management"* level, *"site management"* level and *"individual"* level.

Few models give detail of the causes of errors within these general levels, but authors writing from several industries cite similar causes sufficiently regularly to begin to form categories. Analysis of literature identified thirteen categories of causes of errors in project work covering *primary* actions of the individual, *managerial factors* and wider *global factors*. At the primary level, lack of *knowledge*, poor *selection* of personnel and failures in *self-checking* work are widely cited (Ransom 1994, Jones and Nathan 1990). At the managerial level, lack of *checking* (Stewart 1993, Kaminetzki 1991), poor *division of responsibilities* (Korman 1991), lack of *change control* (Brown and Xiaochen Yin 1988, Daoud and Hamdani 1988), poor control of *concurrent working* (Ellingwood 1987), poor *communications* (Rollings and Rollings 1991) and poor *organisational culture* (Turner 1978) are widely cited. At the wider global level, environmental factors of *political pressures* (Brown and Xiaochen Yin 1988), *societal pressures*

(Peperell 1987) *cost pressure* (Hadipriono 1985) and *time pressure* (Allen 1986) are emphasised.

In addition, several authors emphasise the socio-technical or systemic nature of error incubation where one cause could impinge upon or drive another (Perrow 1984, Fortune and Peters 1995).

2 CURRENT RESEARCH

2 1 *A model of errors in construction projects*

The literature review led to the construction of a conceptual model of the nature of errors in construction projects. The components of the model are listed in table 1. Subsequent field research tested this model with respect to three defining characteristics, its *content*, *form* and *operation.* Research was directed primarily to defects rather than accidents or failures in use, but the literature upon which the model was based also related to accidents and failures. The model is therefore broadly applicable to outcomes other than defects.

2.2 *Tests of the model*

Four empirical studies were carried out, a general survey and three studies of speculative house-building involving a statistical, interview and observational components. The studies combined quantitative and qualitative research methods and relied on an overall triangulation methodology for verification of findings. This paper concentrates on the statistical and interview studies.

3 GENERAL SURVEY

The general survey is reported in detail in Atkinson (1998). 107 UK based construction industry practitioners were asked to rate the importance of factors in the model in a detailed questionnaire and were also given the opportunity to supply further factors. *Communications* was the highest rated factor, followed by *time* pressures, avoiding *concurrency*, controlling *changes* and *education and training*. The lowest rated factors were *economic*, *societal* and *political* pressures. The form of the model was tested in two respects, Respondents were invited to supply additional factors they considered important in the generation of errors. In response there was some indication that factors related to the individual qualities of the manager, (such as leadership

Table1-Model Of Errors In Construction Projects

LEVEL OF THE MODEL	FACTOR
PRIMARY	Knowledge (education/training)
	Selection of knowledgeable personnel
	Self inspection of tasks
MANAGERIAL	Checking work
	Dividing responsibilities
	Controlling change
	Controlling concurrent working
	Communications
GLOBAL	Organisational culture
	Economic pressure
	Time pressure
	Political pressure
	Societal pressure

and motivation) and to "systems" were important. Regarding the layers of the model, factor analysis was used to investigate whether responses loaded on the levels of the model. Factor analysis showed a fairly clear split between global factors *outside* the control of the project participants and factors *within* their control, but not between managerial and primary factors.

4 STATISTICAL STUDY

4.1 *Background*

The statistical study investigated whether there was any statistically significant relationship between factors in the model and the performance of construction projects. 23 housing projects were analysed and data were primarily collected verbally in an interview with the site manager. The study compared a dependent variable of the level of errors leading to defects with independent variables constructed from the factors in the model. "Primary" factors were represented by a rating for the skill level of the trade operative for each site. In addition the influence of *primary factors related to the site manager* on performance was rated using three variables representing the *experience* level, *background* (trade or management) and *qualifications* of the manager.

"Managerial" factors were represented by five variables. *Communications* was divided into two categories, *informal* and *formal communications*. For informal communications, the number of contacts between site managers and *non-line* staff per month was assessed and sites were divided into two categories, those with a high as opposed to those with a low level of communications. For formal communications, the site manager rated the quality of project documents (drawings and specification) and sites were divided into those with high as opposed to those with low quality documents. *Organisational culture* was treated as a managerial factor in the analysis and, following literature equating organisational culture with the extent of managerial participation (ACSNI 1993, Turner 1978), the number of contacts between site managers and *line staff* per month was assessed and sites were divided into those with a high level as opposed to a low level of participation. Divisions of *responsibilities*, control of *concurrency* and control of *changes* were combined into one overall *planning* factor for the purpose of statistical analysis on the basis (Morris 1994, Hinze and Raboud 1988) that control of these factors is represented by good planning. Sites were divided into two categories, those with a high level of planning as opposed to those with a low level of planning. Additionally, the overall influence of the *project management* of sites was assessed by taking the product of scores for planning, informal and formal communications. Sites were subsequently divided into intensively as opposed to less intensively project managed.

"Global" factors were represented by two variables. For *economic/cost* pressures, the site manager was asked whether the site was pressurised in terms of cost and sites were divided into two cost categories, tight/not tight. For *time* pressures, the site manager was similarly asked whether the site was pressurised in terms of time and sites were divided into two time categories, tight/not tight. Two factors, *checking* and *societal* pressure could not be included in the statistical analysis because data were not suitable for constructing variables.

4.2 *Analysis*

Independent variables were compared with dependent variables for strength of association and tested using chi-square computed with SPSS for Windows. The null hypotheses were that there would be no significant association between independent and dependent variables. Results are as shown in table 2.

Table 2 - Chi-Square Association - Reported Defects (P<0.1 Shown Shaded)

p=0.510		**Skill level of trades-people**		
		High	Low	Total
Level of reported defects	Low	6	6	12
	High	7	4	11
	Total	13	10	23
p=0.059		**Experience of manager**		
		Experienced	Inexperienced	Total
Level of reported defects	Low	4	8	12
	High	8	3	11
	Total	11	11	23
p=0.019		**Background of manager**		
		Managerial	Trades	Total
Level of reported defects	Low	8	4	12
	High	2	9	11
	Total	10	13	23
p=0.033		**Qualifications of manager**		
		Qualified	Unqualified	Total
Level of reported defects	Low	6	6	12
	High	1	10	11
	Total	7	16	23
p=0.855		**Level of informal communications**		
		High	Low	Total
Level of reported defects	Low	5	7	12
	High	5	6	11
	Total	10	13	23
p=0.057		**Quality of formal communications**		
		High	Low	Total
Level of reported defects	Low	10	2	12
	High	5	6	11
	Total	15	8	23
p=0.827		**Organisational culture (level of participation)**		
		High	Low	Total
Level of reported defects	Low	6	6	12
	High	6	5	11
	Total	12	11	23
p=0.049		**Planning/programming**		
		Local and central	Central	Total
Level of reported defects	Low	7	5	12
	High	2	9	11
	Total	9	14	23
p=0.007		**Intensity of project management**		
		High	Low	Total
Level of reported defects	Low	10	2	12
	High	3	8	11
	Total	13	10	23
p=0.304		**Cost pressure**		
		High	Low	Total
Level of reported defects	Low	9	3	12
	High	6	5	11
	Total	15	8	23
p=0.552		**Time pressure**		
		High	Low	Total
Level of reported defects	Low	4	8	12
	High	5	6	11
	Total	9	14	23

From table 2 it can be seen that the null hypothesis of no association between dependent and independent variable can be rejected for the following associations:-

At P<0.05

- The background of the site manager and the reported level of defects (A managerial background is associated with a lower level of defects)(p=0.019)
- The qualifications of the site manager and the reported level of defects (Qualified managers are associated with a lower level of defects)(p=0.033).
- The level of planning and programming and the reported level of defects (More detailed planning is associated with a lower level of defects)(p=0.049)
- The intensity of project management and the reported level of defects (More intense management is associated with a lower level of defects)(p=0.007)

At P<0.10

- The quality of formal communications and the reported level of defects (Better quality formal communications are associated with a lower level of defects)(p=0.057)
- The experience of the site manager and the reported level of defects (<u>Inexperienced</u> managers are associated with a lower level of defects)(p=0.059)

4.3 *Discussion*

The statistical study shows that significant results were both concentrated on managerial factors and, in particular, those of a more tangible nature - formal communications, planning, the experience, background and qualifications of the site manager and the overall intensity of project management. The findings support two indications from the general survey. In that survey, *communications* was rated as the most important factor in avoiding errors and the significant association between formal communications and performance in relation to defects supports that finding. The general survey also indicated that *primary* factors, but at a managerial level were important and significant associations between the qualifications, background and experience of the site manager supports this.

An interesting finding is that *inexperienced* managers are associated with better performance. A further test between the two independent variables, *experience* and *qualifications* of the site manager also shows a significant association indicating that inexperienced managers are also better qualified (p=0.016). This suggests that qualifications more than make up for lack of experience in avoiding errors.

5 INTERVIEW STUDY

The interview study examined the content and operation of the model in a series of interviews of 40 managers in the house-building industry. Interviewees included the 23 site managers for the sites in the statistical study, eleven more senior managers within the two firms involved in the statistical study and a further independent group of six managers from building guarantee insurance companies. The primary method of interviewing was to ask open questions concerning recollections of specific errors or general problems either leading to defects or requiring corrective action. Content analysis (Crano and Brewer 1973) was then used to extract and enumerate factors related to the model.

5.1 *Content of the model*

In total 220 causes of errors were categorised as shown in table 3.

Table 3 - Interview Study Record Of Errors

LEVEL OF THE MODEL	FACTOR	NUMBER
PRIMARY		47
MANAGERIAL	Checking work	17
	Dividing responsibilities	5
	Controlling change	12
	Controlling concurrent working	2
	Formal communications	61
	Informal communications	9
	Planning*	5
	Manager*	11
	Systemic*	7
GLOBAL	Organisational culture	7
	Economic pressure	16
	Time pressure	19
	Political pressure	0
	Societal pressure	2
		220

* Factor added as a result of general survey or statistical study.

5.2 *Operation of the model*

Analysis of the accounts of interviewees revealed three types of systemic interaction, simple *self contained errors*, *two way interactions* and *complex interactions*. Simple self contained errors when related to the operative, would directly result in a defect or failure. In contrast, simple errors at a managerial level often surfaced as operative errors or did not cause a problem as they were corrected by the operative. In two way interactions, interviewees acknowledged that two errors (such as bad work and poor supervision) led to a defect. Complex interactions of two types were illustrated in the interviews. Firstly, some incidents involved complex multiple errors leading to defects, which spread beyond the unit or site initially involved. An example of such an interaction was an incident where a bricklayer built fixings for first floor timbers at too high a level because clear instructions were not provided. The carpenter "following on" subsequently over-notched the timbers to correct the levels. Electricians in an independent incident also over-notched the timbers for cabling and this resulted in the floor being of inadequate strength. The series of defects was repeated throughout the site as a result of the site managers failing to give clear instructions to the carpenters and electricians not to cut into timbers in future. Thus, in one incident, multiple errors related to *communications*, *checking* and operative *primary errors* were involved.

The second type of complex interaction recorded was systemic multiple errors *driven from the top*. An incident of this type involved a series of "show houses" on one site containing several defects. The defects were effectively planned by senior managers who admitted that the houses were built too quickly in order to meet sales targets. This was needed to improve year end accounts and reduce the threat of take over of the company on the Stock Exchange! The "pathology" of errors appears as operative *primary errors* forced by *concurrent working* and over-pressurised managers not *checking* adequately in turn caused by *speed of working* driven by *cost/profit* targets and the need for survival of the firm. The problem with this type of incident is that the defects "appear" to be caused by operatives and not by senior managers.

6 CONCLUSIONS

6.1 *Generally*

The studies reported here give support for the proposed model of errors in construction projects. In particular, they emphasised the importance of managerial factors. Within these factors, communications stood out as of primary importance. The statistical study supported some indications in the general survey, that error incidence was associated with the primary individual characteristics of the key manager for this type of project (the site manager). The operation of the model was demonstrated in the interview study, where a number of incidents showed how errors in one part of a project or organisation can modulate or disguise errors elsewhere. The implication of this is that error occurrence cannot be treated in isolation and that analysis of cause *must treat the whole project as a system.*

6.2 *The implications for construction site safety*

Although the "outcome" studied in this research was defects rather than accidents, the "cause", human error, is common to both outcomes and the links between outcomes are strong. In particular, many actual or potential defects identified in the research could have led to a dangerous failure *in the permanent finished work.* Whether an error caused a minor defect or led to disastrous failure was simply a matter of chance and the "riskiness" of the technology (It should be noted that many major failures have been precipitated by relatively inconsequential errors). By extension, the research also has direct application to failures of *temporary works*, which are subject to the same requirements in terms of organisation, planning and control as finished work (HSE 1976).

The application of the research to operational site safety is less obvious, but many authors (ACSNI 1993, Whittington et al 1992, Turner 1978) have claimed a link between quality and safety. Feedback from the interviews suggests that there is a difference between safety and quality objectives *in the short term.* Some personnel still appear to see benefits in compromising safety (for a variety of reasons) in order to maintain production and good site managers considered operational safety "as a given" rather than directly aiding quality or production.

6.3 *Further work*

Attention should be concentrated in future research in three areas. Firstly, the potential differences between errors having quality implications as opposed to safety implications should be investigated by replicating some of the work (particularly the statistical study), but using measures of safety. Secondly, the strong indictment of communications in the research suggests that a comprehensive examination of patterns of communications in the construction industries is needed. Thirdly, the associations between performance and *individual* managerial qualities revealed in the statistical study suggests that further detailed work on what makes a good manager would pay dividends. Concentrating on managerial qualities may pay greater dividends, in terms of safety, quality *and* business performance, than concentrating on (often ignored) procedures.

REFERENCES

ACSNI 1993. *Advisory Committee on the Safety of Nuclear Installations - Study Group on Human Factors Third Report: Organising for Safety.* HMSO.

Allen D E 1986. *Human Error and Structural Practice*. National Research Council Canada, Institute for Research in Construction, Paper No. 1429.

Atkinson A R 1998. Human Error in the Management of Building Projects. *Construction Management and Economics* 16, 339-349.

Bea R G 1994. *The Role of Human Error in Design, Construction and Reliability of Marine Structures*, US Ship Structure Committee (National Technical Information Service, USA).

Blockley D I (Ed.) 1992. *Engineering Safety*, McGraw Hill.

Brown C B and Xiaochen Yin 1988. Errors in Structural Engineering. *ASCE Journal of Structural Engineering* 114 Nr11: pages 2575-2593.

Crano W D and Brewer M B 1973. *Principles of Research in Social Psychology*, New York: McGraw-Hill.

Daoud O E K and Hamdani S K 1988. Human Mistakes in Construction - Case Study in the Gulf Region. *Forensic Engineering* 1 No.3: pages 175-184.

Ellingwood B 1987. Design and Construction Error Effects on Structural Reliability. *Journal of Structural Engineering* 113 No.2 February: pages 409-422.

Embrey D E 1992. Incorporating management and organisational factors into probabilistic safety assessment. *Reliability Engineering and System Safety* 38: pages 199-208.

Fortune J and Peters G 1995. *Learning from Failure: The Systems Approach*: Wiley.

Hadipriono F C 1985. Analysis of Events in Recent Structural Failures. *ASCE Journal of Structural Engineering* 111 Nr7, July: pages 1468-1481.

HSE - Health and Safety Executive 1976. *Final Report of the Advisory Committee on Falsework* (the Bragg Report): HMSO.

Hinze, J and Raboud P 1988. Safety on Large Building Construction Projects. *Journal of Construction Engineering and Management* 114 No 2 June: pages 286-293.

Jones P and Nathan N D 1990. Supermarket Roof Collapse in Burnaby, British Columbia, Canada. *Journal of Performance of Constructed Facilities* 4 No.3 August: pages 142-160.

Kaminetzky D 1991. *Design and Construction Failures - Lessons from Forensic Investigations*: McGraw-Hill.

Kletz T A 1985. *An Engineer's View of Human Error*: Institute of Chemical Engineers.

Korman R 1991B. Defense Attacks Splintered Process. *Engineering News Record* 226 29/4/1991: pages 8-9.

Lourens P F 1990. Theoretical perspectives on error analysis and traffic bahaviour. *Ergonomics* 33 Nr10: pages 1251-1263.

Morris P W G 1994. *The Management of Projects*. London: Thomas Telford.

Pepperell N 1987. *Professional Indemnity*. Building Technical File No.19 October: pages 53-56.

Perrow C 1984. *Normal Accidents*. New York: Basic Books.

Petroski H 1985. *To Engineer is Human. The Role of Failure in Successful Design*. London: Macmillan.

Petroski H 1994. *Design Paradigms - Case Histories of Error and Judgement in Engineering*: Cambridge University Press.

Ransom W H 1994. *Building Failures - Diagnosis and Avoidance*: E and F N Spon.

Rasmussen J. 1983. Skills, rules, knowledge: signals, signs and symbols and other distinctions in human performance models, *IEEE Transactions*: Systems, Man and Cybernetics.

Rasmussen J. Duncan K. Leplat J (Editors) 1987. *New Technology and Human Error*. Chichester: Wiley.

Reason J 1990. *Human Error*: Cambridge University Press.

Rollings R S and Rollings M P 1991. Pavement Failures: Oversights, Omissions and Wishful Thinking. *Journal of Performance of Constructed Facilities* 5 No.4 November: pages 271-286.

Stewart M G 1993. Structural Reliability and Error Control in Reinforced Concrete Design and Construction. *Structural Safety* 12: pages 277-292.

Turner B A 1978. *Man Made Disasters*. London: Wykeham.

Whittington C, Livingston A and Lucas D A 1992. *Research into Management, Organisational and Human Factors in the Construction Industry*. HSE Contract Research Report Nr. 45/1992: HMSO.

Implementation of Safety and Health on Construction Sites, Singh, Hinze & Coble (eds)
© 1999 Balkema, Rotterdam, ISBN 90 5809 036 1

An investigation of job related discomfort and layout preferences of masons

Rasha M. Stino & John G. Everett
Department of Civil and Environmental Engineering, University of Michigan, Ann Arbor, Mich., USA

ABSTRACT: Construction activities, such as bricklaying, are characterized by a constantly changing workplace layout. The construction of a brick wall consists of repetitive, short cycles which involve a wide range of working postures for the mason. This study indicates that the postures assumed by masons during their work has resulted in musculoskeletal problems which are most pronounced in the low back (81.3%), shoulders (56.3%), elbows (52.1%), and knees (41.7%). As part of a questionnaire, masons were asked to rate the "degree of physical discomfort" and "speed of work" for construction of each of five zones of a brick wall. Both the masons' perceived speed and comfort levels were rated the highest in the "Wrists/Hands to Elbows" zone of a brick wall, where minimal bending and extension of the back and minimal motion of the arms is required. 87.2% of the masons answered that it made a difference to them how high the brick supply is stacked on the scaffold. When given the option to specify how high they wanted the materials supply to be located on the scaffold, 66.7% chose the "Wrist/Hands" level, 18.5% chose the "Knees" level, and 7.4% chose the "Elbows" level.

1 INTRODUCTION

Construction activities, such as bricklaying, are characterized by a constantly changing workplace layout. The construction of a brick wall consists of repetitive, short cycles which involve a wide range of working postures for the mason. The various positions a mason assumes are defined primarily by the workplace layout: location of the materials (brick/block) at the supply end or "origin", and, position of the materials (brick/block) at the wall or "destination". The destination of materials in the wall and the supply size changes from cycle to cycle as the brick wall is built up and as the materials are depleted. Working height of the hands varies from foot level to shoulder height and overhead.

It is expected that each brick's unique location at the origin and destination will, to varying degrees, affect the mason's productivity and level of discomfort. This is due in part to the fact that a mason must assume a wide range of postures while constructing the wall. Repeated flexion and extension of the back and upper extremities is experienced by the mason to varying degrees, depending on which section of the wall is being built and how high the material supply is. In addition, visibility and control vary as the wall is built up to the mason's head level. A brick placed at the mason's waist level, for example, would probably not require the same amount of time as a brick placed at the mason's ankle level. Also, the extreme postures assumed by the mason at very low and very high levels of the wall contribute to the job related discomfort.

This paper is part of a larger study which investigates the productivity and ergonomic implications of bricklaying as a function of the changing workplace layout. The objectives of this paper are to present and evaluate the: (1) prevalence of musculoskeletal complaints among masons, (2) masons' perceived degree of discomfort at different heights of a brick wall (destination), (3) masons' perceived productivity at different heights of a brick wall (destination), (4) preferences of masons for materials supply height (origin), both in general and as it relates to the changing height of the brick wall, and, (5) recommendations of masons with regards to materials, equipment and methods, to increase comfort and/or speed of work.

2 METHODOLOGY

One hundred and fifty questionnaires were mailed to three masonry locals in Michigan and were distributed on various construction sites among masons. The four page questionnaire consists of 10 questions.

Questions 1 and 2 comprise the first page of the survey and are based on the Nordic questionnaire (Kuorinka 1987) which addresses musculoskeletal symptoms. First, mason-specific data such as age, weight, height, gender and years of experience is gathered. Next, the mason is asked if there have been any job related aches or pains during the last 12 months in different parts of the body. In addition, the mason is asked if the job related aches or pains (if present) have prevented him/her from doing the job "at any time" and if he/she has seen a doctor/chiropractor as a result.

Questions 3 and 4 were graphically designed to capture the mason's perceived level of discomfort and perceived speed of work as a function of layout. The brick wall that a mason constructs is illustrated in the questionnaire and is divided into five vertical zones: (1) ankles/feet to knees, (2) knees to wrists/hands, (3) wrists/hands to elbows, (4) elbows to shoulders, and, (5) above shoulders. The mason is asked to indicate on a scale the "degree of physical discomfort" for construction of each of the five zones of the wall. Physical discomfort ranges from "No discomfort" (numeric value of "0") to "Extreme discomfort" (numeric value of "10"). The mason is then asked to indicate on a scale the "speed of work" for construction of each of the five zones of a wall. Speed of work (productivity) ranges from "Slow to build" (numeric value of "0") to "Quick to build" (numeric value of "10"). Masons are asked to explain both their selections of the most uncomfortable and the slowest zones of the brick wall.

Questions 5, 6 and 7 address the mason's preferences for the materials supply height, particularly as it relates to each of the five brick wall zones. First, the mason is asked to select a single, optimal height for the material supply. Next, the mason is given the option of adjusting the material supply height relative to the changing wall height and is asked to select a preferred combination of material supply vs. wall heights.

Finally, questions 8 through 10 are designed to capture the masons' observations regarding materials, equipment, tools and methods and their suggestions to increase the comfort level and/or speed of masonry work.

3 RESULTS AND DISCUSSION

A total of forty-eight questionnaire responses were collected from across Michigan. The average age, weight, height and work experience were 42 years (SD 9), 189 lb. (SD 32), 5.9 feet (SD 0.3), and, 18.7 years (SD 9.8), respectively. All masons who responded were male. 85% of the masons were right handed, meaning that the right hand is the trowel hand and the left hand is the bricklaying hand.

Based on these responses, Table 1 shows that the low back (81.3%), shoulders (56.3%) and elbows (52.1%) appear to be the regions with notable musculoskeletal problems among masons. This is in line with a study by Cook et al. (1996), where the percentage of musculoskeletal complaints of thirty-nine bricklayers was 72.5%, 54.1% and 41% for the low back, shoulders and elbows respectively.

The results of the masons' perception of the degree of discomfort and speed (productivity) as a function of the five defined brick wall zones are presented in Tables 2 and 3. The respondents in this study indicated that the brick wall zones of least discomfort and most discomfort are "Wrists/Hands to Elbows" and "Ankles/Feet to Knees" respectively. Working above the shoulder level also ranks highly on the discomfort scale for masons (Table 2).

Table 1. Percentage of musculoskeletal complaints in 12 month period among masons

Body region	Musculoskeletal complaints (%)
Neck	33.3
Shoulders	56.3
Elbows	52.1
Right Wrist/Hand	39.6
Left Wrist/Hand	37.5
Upper Back	27.1
Low Back	81.3
Hips/Thighs	25.0
Knees	41.7
Ankles/Feet	27.1

Table 2. Masons' average rating of discomfort (10 is extreme discomfort) as a function of the five brick wall zones

Wall Zone	Average rating of discomfort
Ankles/Feet to Knees	6.5
Knees to Wrists/Hands	3.1
Wrists/Hands to Elbows	1.8
Elbows to Shoulders	3.6
Above Shoulders	5.9

Table 3. Masons' average rating of speed of work (10 is quick to build) as a function of the five brick wall zones

Wall Zone	Average rating of speed of construction
Ankles/Feet to Knees	2.8
Knees to Wrists/Hands	7.1
Wrists/Hands to Elbows	8.2
Elbows to Shoulders	5.4
Above Shoulders	3.0

Masons indicated that the brick wall zones slowest to build and quickest to build are "Ankles/Feet to Knees" and "Wrists/Hands to Elbows" respectively. It is worth noting that, according to the masons, the speed of work is generally higher around the middle section of the body and slows down at the upper and lower extremes of the brick wall (Table 3).

Masons commented that bending over to construct the ankles/feet to knees zone of the wall was extremely stressful on their backs and that working at the other extreme, above the shoulders, hurt their elbows and shoulders due to the reaching and lifting involved. Several explained that the discomfort they were experiencing combined with the awkward line of sight at extremely low elevations and high elevations slowed down their work.

With regards to materials layout preferences, 87.2% of the masons answered that it did in fact make a difference to them how high the brick supply is stacked on the scaffold. When given the option to specify how high they wanted the materials supply to be located on the scaffold, 66.7% chose the "Wrist/Hands" level, 7.4% chose the "Knees" level and 11.1% chose the "Elbows" level (Table 4).

When the masons were asked if their preferences for material supply height changed as the wall height changed, 50% responded "yes". When given the option of adjusting the material supply height relative to the changing wall height, the majority of masons still chose to maintain the material supply height at the wrist/hand level for all the wall heights. The only variation appeared at the ankles/feet to knees portion of the wall, where some masons preferred to have a matching material supply height at the ankles/feet level.

Table 4. Masons' preference for material supply height

Alternative material supply heights	Preference of masons (%)
Ankles/Feet	0.0
Knees	18.5
Wrists/Hands	66.7
Elbows	7.4
Shoulders	1.8
Other	5.6

A wide range of recommendations were made by the masons with regards to the methods, materials, equipment and tools they used. According to the masons, a working environment that is clean, safe, well organized and a scaffold that is stable and level boosts their confidence and allows them to focus more on their work. Maintaining a layout that eliminates extremes in elevation (high and low) both at the supply and at the wall, and consequently reduces bending and reaching is also high on their list of suggestions. Several masons pushed for the use of adjustable or tower scaffolds and it was suggested that another lift bracket be added to the 6 foot, tubular scaffolding. Lighter, more uniform masonry units are being strongly requested by masons. Correct consistency and properly tempered mortar takes the stress off the wrist. Suggestions for hands tools included using special coatings to prevent mortar from sticking, making tools out of more durable material and incorporating rubber handles to reduce shock. An additional notable comment was made regarding the fact that mortar dries out and chaps the skin of the hands severely to the point of bleeding but that it is difficult to wear protective gloves and still be productive.

This study indicates that the postures assumed by the masons during their work has resulted in musculoskeletal problems which are most pronounced in the back, shoulders, elbows and knees. Not surprisingly, speed and comfort levels are the highest in the "Wrists/Hands to Elbows" zone of the brick wall, where minimal bending and extension of the back and minimal motion of the arms are required. Furthermore, most of the masons preferred to have the materials supply at the "Wrist/Hand" level. There appears to be a sincere desire on the part of masons to do their work well, as one mason put it, "[I] take pride in the material, the product and the way it is installed". The masons' suggestions and recommendations should be taken as a positive step toward an open dialogue between the masons, designers and contractors that can improve the quality and efficiency of masonry work.

The data presented in this paper can be used to assess existing work conditions in the light of masons' needs and preferences. Ultimately, it is hoped that this effort will lead toward future research, development and selection of better ergonomically suited and productivity enhanced workplace layouts, equipment, tools and materials that still preserve the unique character of masonry work.

REFERENCES

Kuorinka, I., Jonson, B., Kilbom, A., Vinterberg, H., Biering-Sorensen, F., Andersson, G., and Jorgensen, K., 1987. Standardised Nordic questionnaires for the analysis of musculoskeletal symptoms. *Applied Ergonomics* 18(3): 233-237.

Cook, Thomas M., Rosecrance, John C., and Zimmermann, Chris L., 1996. *Work-Related Musculoskeletal Problems in Bricklaying*. Injury Prevention Research Center, The University of Iowa.

Implementation of Safety and Health on Construction Sites, Singh, Hinze & Coble (eds)
© 1999 Balkema, Rotterdam, ISBN 90 5809 036 1

A behavior-based approach to construction: Applied behavior analysis for construction safety

W.F. Maloney
University of Kentucky, Lexington, Ky., USA
G.R. Smith
Iowa State University, Ames, Iowa, USA

ABSTRACT: This paper examines the characteristics of the construction industry, especially in terms of safety and safety programs. The focus of construction safety must shift from the top-down imposition of rules and programs to a bottom-up focus in which workers and their work crews are responsible for performance including productivity, quality, and safety. The work crew, using a continuous process improvement paradigm, will track at-risk behaviors. The reduction in the occurrence and frequency of at-risk behaviors will result in the reduction of accidents and injuries.

1 INTRODUCTION

In the United States construction industry, safety has been a system developed by management and staff personnel and imposed upon the workforce, if it is formally addressed at all. The larger companies, because of their greater resources, are able to develop a safety program that is implemented on each of the companies' projects. Company safety and/or management personnel monitor each project. Thus, safety has been totally driven by management with a top-down perspective.

Construction accidents, like accidents in any other industry, stem primarily from either unsafe working conditions or unsafe employee behavior. Most safety management schemes focus on the physical conditions at the worksite, because that is the major focus of the Occupational Safety and Health Administration's (OSHA) compliance requirements (another example of the top-down perspective on safety). However, unsafe behavior has been shown to account for 50% to 90% of all accidents, depending upon which study is referenced. Other studies report that unsafe behavior occurs with 7 times the frequency of unsafe conditions. This is highly subject to the definition of unsafe behavior. While there is a wide variance around that number, there is clearly a need to direct attention to an area that has great potential to impact worker safety.

Despite the success of the use of intervention techniques to improve safety performance by changing worker behavior in other industries, a preliminary literature search reveals few controlled studies documenting or demonstrating the effectiveness of various intervention techniques designed to improve safety performance in the U.S. construction industry. The 1980s saw the development of behavior-based safety programs in many industries, but not in construction. Behavior-based safety refers to the use of applied behavior analysis methods to achieve continuous improvement in safety performance. The methods of behavior-based safety include "identifying and operationally defining critical safety-related behaviors, observing to gather data on the frequency of those behaviors, providing feedback, and using the gathered data for continuous improvement." As such, behavior-based safety is very similar to the continuous process improvement concept and techniques that are a major component of Total Quality Management. A critical dimension of behavior-based safety is employee involvement. The sustained engagement of employees in the improvement of the work process, especially in terms of safety, is the primary driving force in improving safety. Thus, the adoption and adaptation of behavior-based safety in construction will necessitate a refocusing from a top-down perspective to a bottom-up one.

Safety systems, to be widely used in the construction industry, need to be effective for large and small firms alike. The large contracted dollar volume of the major contractors often overshadows the preponderance of small firms. Hinze indicates that small firms consistently have lower accident rates than larger firms. Although no suggestions are offered regarding the cause of this relationship, the smaller firms are more likely to operate as teams and the employees are more likely to be permanent employees. Larger firms hire much of their labor on a temporary basis.

Interventions such as safety audits, training programs, and many incentive schemes can result in safety improvements. However, the time required to implement and monitor these types of intervention systems requires full-time supervision. This is impossible for most medium and small firms where full-time safety personnel do not exist. Therefore, an approach to construction safety intervention that focuses on changing worker behavior and worker involvement is a rational approach for small to medium firms, because they retain many of their workers for long periods of time.

2 CONSTRUCTION INDUSTRY CHARACTERISTICS

The American National Standards Institute has established Standard A10.33, Construction and Demolition Operations, Safety and Health Program Requirements for Multi-Employer Projects. In this standard, the following selected terms are defined:

- *Contractor Safety and Health Program*: the written program developed by the contractor specifically for the work to be performed on the jobsite by the contractor. The program meets the requirements established by the Project Safety and Health Program, and covers activities and conditions under the responsibility of the contractor.
- *Project Constructor*: a person, firm, or corporation, i.e., the Construction Manager, General Contractor, Prime Contractor, or other entity, as designated in the project documents, responsible for supervising and controlling all construction work performed on the project.
- *Project Safety and Health Program*: the written program developed by the Project Constructor specifically for the project that describes the requirements and procedures for implementing an employee safety and health program. This program applies to all contractors and/or individuals.
- *Senior Contractor Supervisor*: the individual designated by the contractor who has overall authority and responsibility for work performed by that contractor, including the Contractor Safety and Health Program.
- *Senior Project Supervisor*: the individual designated by the Project Constructor who has final authority and overall responsibility for all construction and related activities, including the Project Safety and Health Program.

This standard has not been widely adopted by the industry, but it does provide a basis for developing an understanding of the process of safety in the construction industry.

The unique characteristics of the construction industry make developing a safety program extremely challenging. These characteristics include:

- A construction project has a finite life, which varies significantly from project to project.

- On a typical project, the Project Constructor is a general contractor who performs approximately 15-20% of the project work and subcontracts the remainder of the work to specialty contractors.
- Multiple employers are employed and working on the site at any point in time.
- Contractors may perform work onsite and offsite such as fabricating HVAC duct at the shop and installing it at the project site.
- Under Occupational Safety and Health Administration rules, each construction employer is responsible for the health and safety of its employees, regardless of who created the unsafe condition or who might be contractually liable for correcting the unsafe condition to which their employees are exposed.
- The employer has full responsibility safety with the worker having a passive role. Compliance with employer mandated procedures is the worker's sole responsibility.
- Using the ANSI A10.33 standard definitions, the responsibility for safety on a construction project can be depicted as follows:

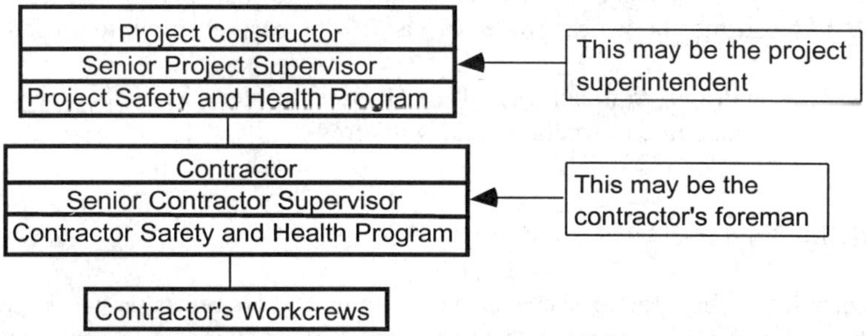

- Construction workers are employed on a project for a finite period, which is equal to or less than the period the workers' contractor is on the project.
- The tasks that a worker is required to perform vary throughout the worker's time on the project.
  Construction work is unstable because a worker is employed to work on projects. When a worker completes the work for which he/she was hired on a project, the need for the worker on that project is eliminated. The contractor will move the worker to another project or lay the worker off. The length of time that a worker is a member of a workcrew and the length of time that the worker is employed on a project have an impact on that worker's performance and the potential for effecting behavioral change through training.
- Similarly, the composition of workcrews is unstable because of voluntary and involuntary turnover.
- The movement between projects reduces the worker's familiarity with specific jobsites.
- Workers hired on a construction project may have an extremely varied background in terms of safety. There is no requirement that workers be trained in safety, Some may have received formal safety training through a union or employer. Others may have received informal training on a construction project.
- The worksite is in a constant state of flux as work is begun and completed and contractors and their employees move on and off the site. The work conditions that are present today may be very different from those that will be in place in a week.
- Contractors range in size from those who do home remodeling to ones that build dams and power plants. However, the average contractor employs approximately 10 people, who with the exception of the owner and possibly a clerk are field craft personnel.
- On a typical project, a superintendent who has overall responsibility for the execution of the project, including administration of all subcontracts and monitoring of performance of subcontractors represents the general contractor in the field. For larger projects,

there may be a clerk and an assistant superintendent. A subcontractor is normally represented by a foreman who is responsible for the performance of the work contained in the scope of work in the subcontract. The subcontract foremen report administratively to the general contractor's superintendent. The superintendent is responsible for safety for the general contractor's employees and the subcontractor foremen are responsible for the safety of their workers.

- Typically, there is not a full-time safety person assigned to a project. Larger contractors may have a full-time person who will oversee all of the contractor's projects. There may be a full-time safety person on a project if the project is large enough, dangerous, required by contract, or provided by the project owner.
- There has been recognition that the roles of labor and management must change if performance is to be improved. The traditional battle line between labor and management has been drawn with workers focusing on job security and management focused on performance. There has been a realization that labor can do very little to improve job security and management can do very little to improve performance. Labor, because it performs the work processes, is in the best position to improve performance.

The conclusion that must be drawn from analysis of the above characteristics is that the key to improving safety performance in construction is the workcrew.

3 CREW PARADIGM

The model for improving safety performance is based upon the concept that work is a process and that the individuals performing the process are its owners. As such, they determine how the work is to be performed, including a determination of what constitutes safe behavior and performance. They also have the responsibility for investigating near-miss situations and accidents to determine the real cause(s). The model will be an adaptation of the employee-driven, behavior-based system developed and advocated by Krause. The group of workers executing a process will follow a structured teamwork approach to the: (1) identification of problems and causes, (2) examination and measurement of processes, and (3) solution of problems.

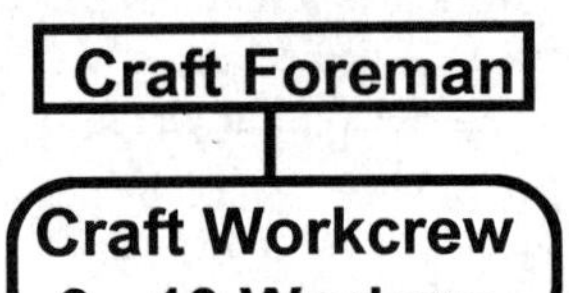

The model may be depicted as follows: The foreman is assigned responsibility for the accomplishment of a specified scope of work on the project that is to be completed using the workcrew. The workcrew performs the required work using a specific work process for which the workers will have to be involved in determining the means and methods to be employed in performing the work. This includes the determination of the safety procedures and behaviors to be followed.

Training will be a major element of the model. ANSI Standard A10.33 specifies that each contractor shall be responsible for the safety and health training of their employees. Several different types of training will be required:

- Supervisory training: employees assigned to supervisory positions shall be trained to carry out the safety and health responsibilities of the positions to which they are assigned. This training will be different than that which is normally done. For the employee-driven, behavior-based paradigm, the supervisor must function as a facilitator and coach. Thus, craft foremen will be trained to facilitate the functioning of the workcrews and to coach the crews to improve their functioning.
- Craft personnel: training in safety and health requirements shall be provided to each non-supervisory craft worker. The objective of this training will be to make each craft worker a competent person. The ANSI standard is based upon the idea that each

project should have a competent person, who is a member of management, assigned to it. The employee-driven paradigm is predicated on the belief that just as every worker is responsibility for quality, a basic tenet of TQM, every worker is responsible for safety. Therefore, each worker should be qualified as a competent person. Training programs of the following types will be utilized to do this:

- ♦ New Hire Orientation: training for newly hired employees oriented toward the contractor's safety and health policies, program, and rules. This will also include an orientation about the workcrew, its functioning, and its responsibilities.
- ♦ Job Specific Training: training in safety and health requirements and rules specific to the necessary knowledge, skills, and abilities to safely perform assigned work.
- ♦ Site Specific Training: training in safety and health requirements and rules unique to the particular work site involved.
- ♦ Safety Meetings: training conducted on a regularly scheduled periodic basis during the progress of the work to reinforce the requirements of the safety and health program, review compliance and incidents or near misses caused by non-compliance and establish any special procedures for present or upcoming activities.

In addition, training programs will be required to equip craft workers with the skills to conduct pre work planning in the following areas.

- Project Survey: prior to the start of work, each crew shall conduct a physical survey of the jobsite and make a survey of the work to be performed by reviewing the drawings and conducting discussions as applicable with one of more of the following: owner, engineer, general contractor, construction manager, and other contractors.
- Hazard Analysis: at the initiation of a construction project and for critical stages of work, a hazard analysis shall be conducted and implemented by the crew describing potential hazards and actions required to provide a safe and heathy work place. This will involve an analysis of each activity; the identification of each unsafe condition, action or hazard caused by the activity; and preventative or corrective action needed to eliminate the unsafe condition, action, or hazard.
- Pre-phase Planning Meeting: a meeting of affected crews shall be held to coordinate and assign responsibility for all items identified in the hazard analysis.
- Daily Production Review Meeting: a meeting shall be held prior to each workday to review the work activities to be performed that day, the safety considerations for each activity, and the specific worker responsibilities for safety.
- In addition to safety training, programs should be provided to equip workers with the skills necessary to undertake process analysis and improvements and to improve the group dynamics and functioning of the crew. These will include: process diagramming and analysis, process improvement. critical incident investigation, behavioral analysis, statistical analysis, group functioning and dynamics, problem solving and decision making, and metrics and performance measurement.

4 BEHAVIOR-BASED PARADIGM

Antecedents → Behavior → Consequence

Krause has pioneered the field of behavior-based safety. His two books are the primary literature in the field and are based on the following model. In this model, antecedents are events that precede or trigger behaviors which have direct and indirect influences on behaviors. An antecedent could be a supervisor who allows workers to engage in unsafe behaviors. Behaviors are observable actions such as not using shoring when working in a six foot deep trench. Consequences are states or events that follow behaviors and have a direct influence on behavior. Those consequences that happen soon after the behavior, are certain to happen, and are viewed positively have the most powerful effect on behavior. Supervisory praise for safe performance would be such a consequence. Krause views behaviors in terms of critical behaviors, i.e., behaviors that are critical to safety. When performed in a safe manner, critical behaviors prevent injury. Performed in an unsafe manner, critical behaviors result in exposure to injury.

Although Krause's framework is easy to understand, the expectancy model of motivation provides a richer model for understanding behavior. This model has been examined by Maloney in the context of the construction industry.

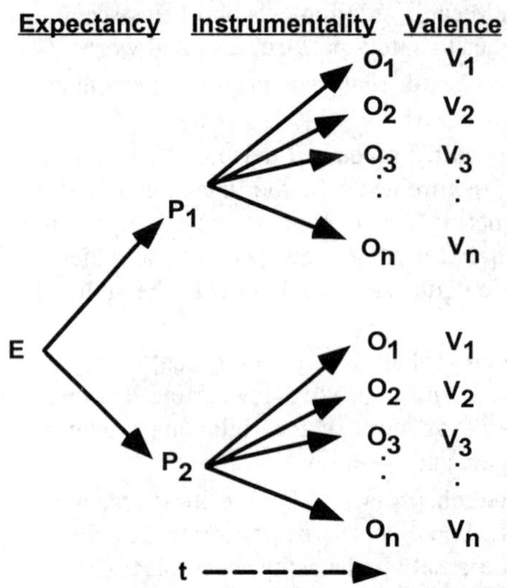

In this model, E represents the effort expended by a worker to engage in a behavior B. B1 and B2 are mutually exclusive behaviors, such as behaving in a safe manner in performing a job and behaving in an unsafe manner. In the model, expectancy is the subjective probability of the worker that exerting effort E will result in a particular behavior and can range from 0 to 1.

O represents a specific outcome. Instrumentality is also a subjective probability in this model represents the likelihood that by engaging in a specific behavior the worker will receive a specific outcome. For example, engaging in behavior B_1 results in the receipt of outcome O_1. Each outcome will have a different instrumentality. The variable t represents the elapsed time from the performance of the behavior to the receipt of the outcome. Each outcome has an importance to the worker, which is its valence. Valence is the anticipated satisfaction the worker associates with the specific outcome and its value ranges from -1 to +1. For example, praise from coworkers will have a positive valence while a reprimand from a supervisor will have a negative valence.

A behavior-based analysis of a work process will result in the identification of critical behaviors, i.e., those that have a strong influence on safety performance. If it is assumed that B_1 is performing the behavior in a safe manner and that B_2 is performing it in an unsafe manner, the following analysis can be conducted:

- Expectancy $E \Leftarrow B_n$: the probability that the worker can exert the effort and engage in the specified behavior. This is a function of the worker's knowledge, skills, and abilities. Does he have the capability to behave that way? Has he had the requisite training to equip him with the knowledge of what is the behavior? Has he had experience in performing the task in that manner? If the worker has the requisite knowledge, skills, and abilities and has experience in performing the task in that manner, the expectancy should be high.
- Instrumentality $B_n \Leftarrow O_n$: the probability that by engaging in behavior B_n, the worker will receive outcome O_n. There is a set of feasible outcomes associated with each behavior. The set of outcomes is relatively stable between behaviors. What differs is the instrumentalities, i.e., the probabilities associated with the receipt of a specific outcome. For example, the probability associated with receiving praise from a superior for performing a task in a safe manner should be higher than the probability for receiving praise from that supervisor for performing the task in a unsafe manner.
- Valence V_n: may be viewed in terms of the importance the worker associates with the particular outcome. An outcome with a positive valence is desired by the worker while one with a negative valence is to be avoided. Recognition, a bonus, or performing a

difficult task well are typically viewed as having positive valences. Getting injured or being fired are perceived as having negative valences.

- Elapsed time t: the time between the performance or behavior and the receipt of an outcome is critical. The sooner the outcome is received the more strongly it is associated with the behavior, which increases the motivational value of the outcome.

A motivational score for a particular behavior can be calculated using the following equation:

$$\text{Score for } B_1 = \text{Expectancy}(E \Leftarrow B_1) \times \sum^{n} [\text{Instrumentality } (B_1 \Leftarrow Oi) \times \text{Valence } (V_i)]$$

The higher this score, the greater the worker's motivation to engage in this behavior. When faced with the choice of behavior B_1 or B_2, the worker will engage in the behavior with the higher score.

Changing a worker's behavior so that he engages in the safe behavior, B_1, rather than the unsafe behavior, B_2, requires management to change the culture of the organization with regard to outcomes by acting to reduce the instrumentalities associated with unsafe behaviors. For example, a worker may engage in an unsafe behavior if it allows his crew to finish arduous task faster and his fellow crew members give him an "atta boy" for letting them get done sooner. Management must take actions to reduce the receipt of positive outcomes associated with unsafe behaviors if it hopes to eliminate unsafe behavior.

5 BEHAVIOR MANAGEMENT

The management of behavior to improve safety performance begins with the establishment of a workcrew, which is responsible for the design of the work process. The design of the work process involves the determination of the means and methods to be employed in accomplishing the work. An integral part of the crew's design process is the identification of behaviors that are critical to safety within the work process and, especially, those that put the worker at risk of an injury.

As with the continuous process improvement paradigm, data are collected. Instead of collecting data on defects or product characteristics, incidence rates for at-risk behaviors are collected. In Krause's model, an individual external to the workcrew is used to collect the data. This is very similar to the process of work sampling, which can be defined as

> ...a statistical sampling procedure that involves collecting a large number of random samples or observations of the activities of people or equipment. When an observation is made, the state of being or condition of the subject being studied is noted, and this state is classified into predetermined categories of activities that have been selected because of their pertinence to the work activity. From the proportion of observations in each category, inferences are made regarding the total work activity.(Thomas, 1981)

The process of work sampling involves an observer walking through a construction project at random times on random routes observing workers and categorizing what he sees each worker doing into a predetermined set of categories. For behavior analysis for safety purposes, the observer would be looking for critical behaviors, as identified by the workcrew, and determining whether the behaviors were performed safely or unsafely.

The problem with this approach is that the outside observer is overlaid on the existing organization and is not really a part of the operation. The authors propose the use of the crew foreman to collect the data necessary for the behavioral analysis. The foreman will be actively involved with the crews, familiar with their work design, and able to work with them to improve performance. In addition, the added expense of using an outside observer is eliminated. The behavior data collected by the foreman is provided to the appropriate workcrews and the crews use it in analyzing their performance.

6 Conclusion

The behavior-based crew approach to safety is depicted in the figure below. As shown in the figure, the crew is responsible for the design and execution of the work process including safety. Acting upon behavioral data provided by its foreman, the crew

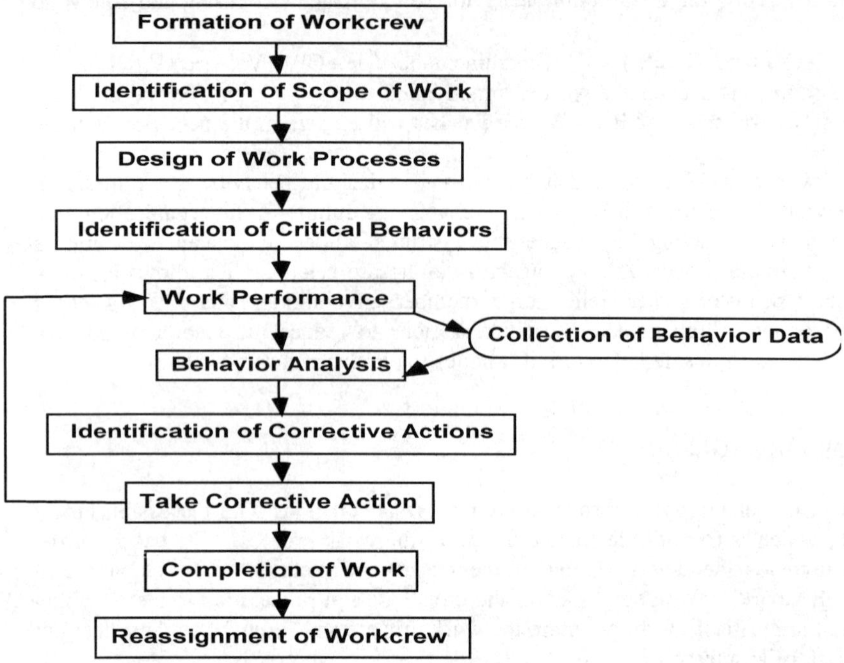

determines the causes of at-risk behaviors, identifies potential corrective actions that will eliminate the at-risk behaviors, and takes those actions that have the greatest likelihood of eliminating the unsafe behaviors. For example, the crew may determine that it needs more training in the safe performance of a task or that more safety equipment must be made available so that workers do not give up out of frustration and perform the task in an unsafe manner. Changing from a top-down to a bottom-up focus on safety will improve safety performance in construction. Having the people performing the work responsible for safety will result in better safety.

REFERENCES

Krause, Thomas R. (1997) *The Behavior-Based Safety Process - Managing Involvement for an Injury-Free Culture*. Second Edition. New York: Van Nostrand Reinhold.

Krause, Thomas R. (1995) *Employee-Driven Systems for Safe Behavior – Integrating Behavioral and Statistical Methodologies*. New York: Van Nostrand Reinhold.

Maloney, William F. (1986) "Understanding Motivation," *Journal of Management in Engineering*, American Society of Civil Engineers, Vol. 2, No. 4, October, 1986, pp. 231- 245.

Thomas, H. Randolph, Jr. (1981) "Construction Work Sampling," Department of Civil Engineering, The Pennsylvania State University, University Park, PA.

Implementation of Safety and Health on Construction Sites, Singh, Hinze & Coble (eds)
© 1999 Balkema, Rotterdam, ISBN 90 5809 036 1

Concrete slab placing and finishing: Physiological demands

Tariq S. Abdelhamid & John G. Everett
Department of Civil and Environmental Engineering, University of Michigan, Ann Arbor, Mich., USA

ABSTRACT: With the exception of earthmoving equipment, cranes, and other machinery, construction craft work remains a physically strenuous and demanding occupation. This paper investigates the physical demands of concrete slab placing and finishing operations in construction. Physiological measures of energy expenditure, including oxygen consumption and heart rate data, were collected for a crew of construction workers while performing actual construction work. The results indicate that these construction workers routinely exceed one or more published guidelines for acceptable levels of energy expenditure, oxygen consumption, and heart rate. The methods described in this paper have wide applications in identifying excessively demanding construction tasks so the work can be modified to accommodate the abilities of all workers, to reduce physical demands, and resulting physical fatigue, and improve productivity.

1. INTRODUCTION

Construction work is physiologically demanding. Except for earthmoving equipment and cranes, highly capital intensive automation and robotic equipment that has become widespread in many manufacturing industries has not gained acceptance in construction. The culture of the construction industry has evolved such that contractors rely heavily on hand labor with small, relatively inexpensive, multipurpose tools.

Physically demanding work leads to physical fatigue. Bonjer (1971) suggested that "the longer a body has to work, the lower its power to convert chemical energy into mechanical energy." In general, physical fatigue leads to decreased productivity and motivation, inattentiveness, poor judgment, poor quality work, job dissatisfaction, accidents, and injuries (Brouha 1967, Janaro 1982).

Many investigators, dating back to Frederick W. Taylor (the father of Scientific Management) in the early 1900s, have argued that since fatigue affects performance, then performance can be improved by either eliminating the causes of fatigue or by at least finding ways in which to combat the effects of fatigue.

Clearly, physically demanding and physically fatiguing tasks have to be identified first before any performance improvement efforts may start. Most of the data for construction work activities was collected during the 1950s and 1960s. The work and workforce have changed since then. Moreover, the equipment available in the 1950s and 1960s did not allow for accurate assessment of construction activities, which is particularly difficult because of the variety of individual operations involved (Astrand and Rodahl 1986). The equipment available today is vastly superior to that used in the 1950s and 1960s, and enables researchers to efficiently measure metabolic and physiologic response to work and exercise in a manner that is non-intrusive to the work itself.

The objective of this paper is to investigate the feasibility of measuring the physical demands of construction work and whether the demands are physically fatiguing to the workers. To keep the scope reasonable, a common construction operation was chosen: concrete slab placing and finishing operations. An 8-member concrete placing and finishing crew engaged in real-life work situations were selected for a field study. For each worker, oxygen consumption and heart rate values were measured and used to characterize the physical demands of work based on work physiology standards. This paper is not aimed at generalizing any findings to all construction tasks nor to the general construction craft worker population.

2. BACKGROUND: WORK PHYSIOLOGY

The most common methods of measuring physical demands are those of measuring the oxygen consumption or oxygen uptake, VO_2 (usually measured in liters of oxygen per minute), during work or exercise, and/or recording the heart rate, HR (measured in beats per minute), associated with the performance of an activity. VO_2 estimates from HR or from direct measurements have also been used to indirectly estimate energy cost of performing various human activities, which in turn is used to assess potential for physical fatigue.

By measuring the amount of oxygen consumed before, during, and after performance of work or exercise activities, the total energy expended by a human can be estimated, because the anaerobic energy yield is quite small under such conditions (McArdle, Katch and Katch 1996). This technique is referred to as indirect calorimetry. For every liter of oxygen consumed, on average 4.83 kilocalories (kcal) of energy are produced. The exact amount varies slightly depending on a physiological attribute termed the Respiratory Quotient. Thus, measured VO_2 may be converted to energy expenditure.

The energy cost of a task as a measure of physical demand is only one side of the story, the so-called assessment phase. The other side is the evaluation phase. The objective of the evaluation phase is to determine whether the physical demands of a certain task are excessive, and whether the worker performing the task may suffer from physical fatigue. Absolute energy cost values, oxygen uptake values, and/or heart rate values are all used to perform these evaluations.

A widely used rule of thumb is that activities requiring less than 5 $kcal \cdot min^{-1}$ can be performed continuously for a work shift without overly taxing the worker. An activity requiring more than 5 $kcal \cdot min^{-1}$ can be performed for a limited time before the worker needs a rest to recoup energy from stores within the body (Oglesby et al. 1989). Table 1 illustrates the use of oxygen uptake and heart rate to classify work intensity (Astrand and Rodahl 1986).

The most reliable evaluation technique is to express the energy cost of a task relative to a worker's capacity, i.e., as a percentage of the worker's maximum aerobic capacity. Unfortunately, techniques for determining the maximum aerobic capacity are rather impractical and may even be dangerous for unfit construction workers. Moreover, prediction techniques are unreliable for data collected in-situ. Therefore, it was not possible to obtain VO_{2max} data for the eight workers in this study.

3. METHODS

The AeroSport KB1-C ambulatory metabolic analysis system (AeroSport, Inc., Ann Arbor, Michigan) was used for indirect calorimetry assessment of energy expenditure in this study.

Table 1. Severity of prolonged physical work and cardiovascular response

Work severity [1]	Oxygen uptake (VO_2) $liters \cdot min^{-1}$ [2]	Heart rate response ($beats \cdot min^{-1}$) [3]
Light work	up to 0.5 liter/min	up to 90 beats/min
Moderate work	0.5 - 1.0 liter/min	90 - 110 beats/min
Heavy work	1.0 - 1.5 liter/min	110 - 130 beats/min
Very heavy work	1.5 - 2.0 liter/min	130 - 150 beats/min
Extremely heavy work	over 2.0 liter/min	150 - 170 beats/min

This device measures VO_2 (in addition to many other physiologic attributes) during work or exercise. This device is approximately 3 x 6 x 2 inches (7.5 x 15 x 5 cm) in size and is worn in a chest-vest containing electronic instrumentation, battery, oxygen and carbon dioxide sensors, and telemetry connections to a microprocessor that permits radio transmissions of up to 3 miles (5 km) to a receiver and computer. The metabolic system weighs approximately 1.13 Kg (2.5 lb) and is easily transported during physical activity.

Another appealing feature of the AeroSport KB1-C system is that it allows collection of HR data via its telemetry technology from a separate heart rate monitor system known as the Polar Vantage XL. The HR is measured using a chest band fitted with a sensor/transmitter which measures the HR and transmits it to a microprocessor in the AeroSport KB1-C unit.

Oxygen uptake and heart rate measurements for an 8-member concrete slab placing and finishing crew were collected while *actual* construction work was being performed. Measurements were taken under normal work conditions. The average age, height, and weight for the 8 construction workers was 31 years, 179 cm, 84 kg, respectively. For each subject, VO_2 and HR data were collected at 20 second intervals for a period ranging from 30-60 minutes. For each worker, data was collected long enough to ensure that steady state VO_2 and HR had been reached and that several typical work or work-rest cycles had been completed.

4. RESULTS AND DISCUSSION

The activities observed for the 8-member concrete slab placing and finishing crew are summarized in Table 2.

The average and standard deviation for collected VO_2 and HR data, as well as estimated energy expenditure values are shown in Table 3. By closely examining Table 2 and the results in Table 3, it is clear that the workers who were using machines (walk-behind trowel, concrete buggy, front-end loader, and Bobcat tractor) had, on average, lower energy expenditures than the workers who worked manually.

Classification of the physical demand based on Table 1, and evaluation of physical fatigue potential for each worker based on the 5 $kcal\cdot min^{-1}$ limit, are shown in Table 4. These results indicate the inconsistent nature of these guidelines. For example, subject 3 was performing "heavy work" according to his HR, but was performing "moderate work" according to his VO_2, and was working at a rate that is not fatiguing when considering the 5

Table 2. Observed construction activities

Subject [1]	Trade [2]	Description of work [3]
1	Cement Finisher (Foreman)	Used two 270 lb, 45" gas engine walk-behind trowels with float and trowel blades to finish a 400 sq.ft. area
2	Cement Finisher	Hand screeding using a 2x4 for a 200 sq.ft. sidewalk slab
3	Laborer	Transported concrete a distance of 40 ft from concrete truck to sidewalk placing crew on a gas powered "concrete buggy"
4	Laborer	Clean-up: brooming, lifting and transporting wood, welded wire fabric, paper, etc. to dump truck. Distances traveled ranged between 40 and 110 ft.
5	Operating Engineer	Earthmoving; part of preparation work for concrete floor slab placement. Used a front-end loader and a "Bobcat" tractor
6	Laborer	"Come-Along" during placement of a 100 sq.ft. concrete floor slab
7	Carpenter (Foreman)	Preparation work for concrete placement, mostly involved measurements, and hammering expansion joints along the walls (120 ft).
8	Cement Finisher	Hand troweling (mostly on knees) a 350 sq.ft. concrete floor slab.

Table 3. Physiological demand data

Subject	Test duration (min)	VO2 (liter.min^{-1}) mean±SD	Peak VO2 observed (liter.min^{-1})	HR (beats.min^{-1}) mean±SD	Peak HR observed (beats.min^{-1})	Estimated energy expenditure (kcal.min^{-1})
[1]	[2]	[3]	[4]	[5]	[6]	[7]
1	60	0.59±0.16	1.24	86±7	111	2.78
2	60	1.08±0.31	1.91	139±10	162	5.34
3	30	0.89±0.12	1.23	113±5	129	4.28
4	58	1.48±0.35	2.20	131±11	156	7.15
5	28	0.46±0.23	1.41	106±6	126	2.18
6	23	1.24±0.34	2.28	108±16	150	6.12
7	30	1.25±0.3	2.07	118±8	138	5.79
8	33	1.27±0.29	1.89	155+10	171	6.19

Table 4. Evaluation of work intensity and physical fatigue potential

Subject	Work intensity based on mean VO_2	Work intensity based on mean HR	Work intensity based on energy expenditure	Worker perception of work
[1]	[2]	[3]	[4]	[5]
1	Moderate work	Light work	Not fatiguing	Tough, I sometimes fall asleep while driving home
2	Heavy work	Very heavy work	Fatiguing	Tiring especially when you go back home
3	Moderate work	Heavy work	Not fatiguing	Kills you
4	Heavy work	Very Heavy work	Fatiguing	Hardest job, and you go home wanting only to sleep
5	Light work	Moderate work	Not fatiguing	It's very tough and depends on day, some are worse than others
6	Heavy work	Moderate work	Fatiguing	Very tiring
7	Heavy work	Heavy work	Fatiguing	Tough
8	Heavy work	Extremely heavy work	Fatiguing	Very hard, but it is what we do for a living

kcal□min^{-1} limit. Another example is that subject 8 was performing "heavy work" according to his VO_2, but was considered to be performing "extremely heavy work" according to his HR. These limitations of the guidelines are due to the wide variations among individuals.

In any case, it is clear that these construction workers routinely exceed generally accepted thresholds for sustained energy expenditure, VO_2 and HR. It should come as no surprise that these workers are exhausted at the end of the day and may not be fully recovered at the beginning of the next work shift. Few workers can sustain this level of performance. Many burn out and seek alternative, less demanding, work. If alternative work cannot be found, the worker faces the dilemma of continuing at a job that causes excess fatigue, or perhaps dropping out of the workforce.

5. CONCLUSION

This paper has shown that it is now technically feasible to measure physical work load for individual construction craft workers performing their normal work in the field. As an example, the physical demands of concrete slab placing and finishing work were investigated in order to characterize work intensity and determine whether the demands are physically fatiguing to the workers. The results reveal that concrete slab placing and finishing craft workers routinely exceed generally accepted thresholds for three physiological indicators of work load: oxygen consumption, heart rate, and energy expenditure and are at risk for developing physical fatigue. This may lead to decreased productivity and motivation, inattentiveness, poor judgment, poor quality work, job dissatisfaction, accidents, and injuries.

Examples have been presented that indicate that use of any one physiological measurement does not account for individual differences and response to strenuous work. Future research should investigate how oxygen uptake and heart rate relate to the individuals' maximum values for each measure, which would lead to a better prediction of physical fatigue.

Other factors such as neuro-muscular functions disturbance, water balance, drop in blood glucose concentrations, depletion of glycogen depots in working muscles, are also good predictors of physical fatigue or its perception by workers. Studying these factors requires specialized laboratory procedures, and highly trained technicians. On the one hand, the results and findings from such controlled laboratory studies would have high internal validity and would be quite beneficial from a physiological standpoint. On the other hand, since real construction environment and tasks would be quite difficult to simulate in the lab, the results and findings of such studies would lose ecological (external) validity, which renders their use strictly academic. Moreover, such studies would involve invasive techniques such as drawing blood samples, which, as our experience suggests would not be welcomed by a vast majority of construction workers.

Through the replication of this study on other types of work, management and labor would have better chances to identify opportunities to reduce physical demands and physical fatigue, and cooperate in introducing changes in work procedures and methods to accommodate the abilities of all workers while improving productivity.

REFERENCES

Astrand, P., and Rodahl, K. (1986). *Textbook of work physiology, physiological bases of exercise*. McGraw-Hill, New York.

Bonjer, F. H. (1971). Temporal factors and physiological load. In *Measurement of man at work, an appraisal of physiological and psychological criteria in man-machine* system. (Singleton et al) Taylor and Francis, London.

Brouha, L. (1967). *Physiology in industry*. Pergamon Press, New York.

Janaro, R. E. (1982). *The development and implementation of optimal multi-rest break scheduling models.* Ph.D. Dissertation, Florida State University, Florida.

McArdle, W. D., Katch, F. I., and Katch, V. L. (1996). *Exercise physiology*. Williams & Wilkins, Baltimore.

Oglesby, C. H., Parker, H. W., and Howell, G. A. (1989). *Productivity improvement in construction.* McGraw-Hill, New York.

Implementation of Safety and Health on Construction Sites, Singh, Hinze & Coble (eds)
© 1999 Balkema, Rotterdam, ISBN 90 5809 036 1

Stress and construction

J.J.Smallwood
Department of Construction Management, University of Port Elizabeth, South Africa
R.Ehrlich
Department of Community Health, Medical School, University of Cape Town, South Africa

ABSTRACT: The construction industry is a stressful industry in which to work. Construction managers, supervisors and workers often have to deal with shortened contract periods, reduced margins and poorly defined problems. These factors and long-term shortages of adequately qualified staff result in enormous pressure and stress. The unique nature of the construction industry exacerbates the problem.

Stress can have serious implications for both the individual's health and performance. Besides heart disease, depression, anxiety, and low self-esteem, burn-out and reduced job performance constitute some of the negative outcomes of stress. Reduced job performance includes among others, faulty decisions.

The causes of stress in construction can be classified according to nine categories: physical work environment; organisation itself; way in which organisation is managed; individual's role within the organisation; relations within the organisation; career development; personal and social relationships; equipment, and the individual's anxieties.

However, stress can be managed, there being three levels of interventions. First, focus on the individual, e.g. employee assistance programmes (EAP); second, individual – organizational interface e.g. person-environment fit, and third, organizational e.g. training.

To date limited stress related research has been conducted in South Africa among members of the Building Industries Federation of South Africa (BIFSA) to determine the status quo with respect to stress. Findings include, among others: the incidence of stress increased commensurately with job responsibility, and the five most important causes of stress are: project schedule; inadequate staff; incompetent staff; job complexity and overburdening.

The need for interventions to reduce and/or eliminate stress is amplified by the direct cost of stress to both the individual and the organisation. However, there is also an indirect cost of stress, unattractiveness as an industry in which to pursue a career, being just one of the indirect costs.

1 INTRODUCTION

The construction industry is a stressful industry in which to work. Construction managers and supervisors often have to deal with shortened contract periods, reduced margins, poorly defined problems and long-term shortages of adequately qualified staff, all of which result in enormous pressure and stress (Morton, 1990). The construction industry is also unique: 'once-off' product; nomadic; relies on a transient workforce; occurs in the elements; the factory floor' changes as frequently as weekly on a multi-storey project, and the extent of subcontracting (Smallwood, 1997).

2 NATURE OF STRESS

Selye describes stress as "the rate of wear and tear" caused by life, and Cox considers stress to be a by-product of the many demands made on the individual which they feel they are unable to cope with adequately (Sommerville and Langford, 1995). However, there are two types of stress: eustress which is good and distress which is bad. Eustress can be a motivation for growth and development, a positive stimulus for change. Fryer (1997) maintains moderate levels of stress produce the best performance.

3 REACTION TO STRESS

According to Selye (Ross & Altmaier, 1994) a person's response to stress follows a universal pattern, namely the General Adaptation Syndrome (GAS). The first stage, alarm, is the body's initial response. There is a brief period of lowered resistance followed by a period of heightened resistance e.g. increased heart rate and blood pressure and a release of glucose to provide energy. If the stressor is prolonged, the second stage is reached and the stressors are replaced with responses that promote long-term adaptation. The third stage of collapse is reached when the energy required for continued adjustment becomes depleted and is characterised by a loss of resistance to the stressor. Exhaustion, collapse and even death can occur.

4 CONSEQUENCES OF STRESS

Stress can have serious implications for both the individual's health and performance. Beside heart disease, depression, anxiety, and low self-esteem, burn-out and reduced job performance constitute some of the negative outcomes of stress. Reduced job performance includes among others, faulty decisions (Djebarni, 1996). Stranks (1994) cites absenteeism, accidents, erratic job performance, loss of concentration, loss of short-term memory, mistakes, poor personal appearance and poor staff relations as the consequences of stress.

According to Fellows, Langford, Newcombe and Urry (1983), the Swedish equivalent of the Building Research Establishment (BRE) identified stress as the major contributor to accidents. The stress being caused by the acceleration of activities as a result of too short project periods, and trades following closely on others, creating stressful, and consequently more hazardous working conditions. Fryer (1997) maintains alertness drops when an employee has been doing a job for too long under low stress conditions. According to Hinze (1997) stress imposed on an individual either by the internal environment (fatigue, loss of sleep, or psychological stresses e.g. worry or anxiety) or by the external environment (noise, illumination, temperature or excessive physical strain) will increase accident occurrence. Stress diverts a worker's attention, which diversion increases the susceptibility of injury. Hinze (1997) cites pressure to keep costs within budget and pressure to meet unrealistic deadlines as two common sources of mental diversions.

5 CAUSES OF STRESS IN CONSTRUCTION

According to Sommerville and Langford (1995) the causes of stress in construction can be classified according to nine categories.

5.1 *Physical work environment*: temporary office accommodation characterised by inadequate temperature control, inadequate ventilation, poor levels of illumination and lack of privacy; excessive noise, and mud/dust and grime on site.

5.2 *The organisation itself*: inadequate staffing; poor coordination between subcontractors; inappropriate communication channels, and insufficient training for the individuals.

5.3 *The way in which the organisation is managed*: crisis management; overtime; undue emphasis on competitiveness, and shift-work.

5.4 *The individual's role within the organisation*: role ambiguity and conflict; overburdening, and inadequate managerial support.

5.5 *Relations within the organisation*: interpersonal conflicts; poor or failing relations with superiors, peer groups and subordinates and lack of feedback.

5.6 *Career development*: perception of job insecurity and inadequate remuneration.

5.7 *Personal and social relationships*: inadequate opportunities for social contact.

5.8 *Equipment*: unsuitable and unreliable equipment which is unable to meet job demands or which frequently breaks down respectively contributes towards unsafe working practices.

5.9 *The individual's anxieties*: inherent difficulty in managing change; poor personal time management, and fear or concern over safety on the site.

6 COPING STRATEGIES

Research conducted in the United Kingdom determined that site managers adopt coping strategies which include those which are not conducive to good health and those which aid a healthy lifestyle. Time management was used on a relatively small percentage of occasions. The healthy coping strategies, namely, maintaining prime time with family and friends and a substantial range of sports/hobbies were adopted by the greater percentage of site managers. A significant percentage of site managers consumed excessive quantities of coffee or alcohol as a primary coping strategy (Sommerville and Langford, 1994). Fryer (1997) recommends that managers must learn to say 'no' i.e. tactfully decline requests which do not contribute to their achieving their primary objectives.

7 PREVENTIVE MEASURES

Stress can be managed (Morten, 1990). Djebarni (1996) advocates three levels of interventions based on Defrank and Coopers prior research: focus on the individual; individual – organizational interface, and organizational. Focus on the individual includes the following interventions: relaxation techniques; cognitive coping strategies; biofeedback; meditation exercise; employee assistance programmes (EAP), and time management. Individual – organizational interface interventions include: relationships at work; person – environment fit; role issues, and participation and autonomy. Interventions in the organization include: organizational structure; selection and placement; training; physical and environmental; nature and characteristics of job; health concerns and resources, and job rotation.

Sommerville and Langford (1994) advocate a greater emphasis on the individuals within the project, i.e. on the composition of the team, as a possible way to realise a significant reduction in one of the primary stressors on any project, people.

8 PERSONALITY TYPES AND STRESS

According to Stranks (1994) seven personality types, A to G, have been classified by psychological researchers over the last 30 years. The type A personality characterised by: ambitious; active and energetic; impatient; conscientious; maintenance of high standards, and intolerant, is most at risk to stress.

9 RESEARCH

Recently completed occupational health research by the authors includes two descriptive surveys and consultations of construction workers. The stress related findings of this study are presented below.

1111 metropolitan based general contractors (GCs) who were either members of the Building Industries Federation of South Africa (BIFSA) or the South African Federation of Civil Engineering Contractors (SAFCEC) were surveyed using a written questionnaire. 48 Questionnaires were returned which constituted a response rate of 4.3%. 790 Non-student members of the South African Institute of Building (SAIB) were surveyed using an identical written questionnaire. 27 Questionnaires were returned which constituted a response rate of 3.4%. The low level of response is in itself a finding in that normally approximately a 10 to 15% level of response to a comprehensive postal survey is received. The response is likely attributable to the populations not being familiar with the subject. 145 Construction workers were consulted using a structured questionnaire.

The majority of GCs undertook industrial (66.7%) and commercial (66.7%) work. Only 25% undertook domestic and 39.6% infrastructure work. On average the GCs subcontracted 11.6% of the value of work to labour only subcontractors (LOSC's) and 26.1% to 'full' subcontractors.

On average 32.5% of SAIB members became involved with industrial work, 38.2% with commercial, 15.5% with domestic, and 3% with infrastructure work. The greater percentage of SAIB members became involved with work between '0 & 2 floors' (37%), '0 & 10 floors' (29.6%) and at 'ground level' (14.8%).

The greater percentage of workers (49.7%) worked in commercial construction followed by industrial (23.4%), domestic (10.3%), infrastructure (7.6%) and other (8.9%). The majority of workers (71.7%) worked for a main contractor, 13.1% for a 'full' subcontractor (labour, material, etc.), 8.3% for a labour only subcontractor, and 6.9% for other types of contractors. Concreting (45.5%) predominated in terms of the work workers were mainly involved with, the next common trades being excavating (31.7%), formwork (31%) and bricklaying (29.7%). The mean age was 38,8 and the median 38 years, the youngest worker being 20 years and the oldest being 65 years of age. On average the workers had worked 14,8 years in construction and 7.4 years for their current employer, the median being 12 and 5 years respectively. 30.3% were general workers, 13.8% semi-skilled and 28.3% artisans (craft workers).

An importance/incidence index was computed to enable the occupations and causes to be ranked in terms of severity based on a value of 4 derived from weighting the range of responses. The following formula was used:

$$\text{Importance/Incidence Index} = \frac{4n_1 + 3n_2 + 2n_3 + 1n_4}{4(n_1 + n_2 + n_3 + n_4)}$$

where n_1 = frequency of 'Often'
n_2 = frequency of 'Sometimes'
n_3 = frequency of 'Rarely'
n_4 = frequency of 'Never'

Contracts Management was identified as the most stressful occupation, the incidence of stress increasing commensurately with job responsibility (Table 1).

The top six causes of stress, although not in the same order, are common for both the GCs and SAIB members: project schedule; job complexity; crisis management; inadequate staff; overburdening, and incompetent staff.

Project schedule, which attained a ranking of first, is a time related variable. Overtime, another time related variable achieved a joint ranking of thirteenth for both populations.

Three of the five organisational related variables achieved a ranking in the top six: crisis management; inadequate staff, and incompetent staff.

Inappropriate communication channels achieved a ranking of ninth and seventh for the GCs and SAIB members respectively, and change ninth and fifteenth respectively.

Table 1. Extent of stress among construction management/supervision and workers (construction management).

Level	Importance Index		Ranking	
	GC's	SAIB	GC	SAIB
Contracts Management	2.21	2.33	1	1
Operational Management:				
Site Manager/Agent	2.17	2.26	2	2
Foreman	2.06	2.07	3	3
Junior Foreman/ Section Foreman	1.65	1.56	4	4
Leading Hand	1.40	1.48	5	5
Production Workers:				
Skilled	1.35	1.33	6	6
Semi-skilled	1.10	1.00	7	7
Unskilled	0.98	0.81	8	8

Table 2 indicates that according to workers 35.9% experienced stress daily and 20.7% weekly, which collectively constitutes 56.6% on a weekly or less basis.

Table 2. Extent to which workers experience stress.

Frequency	Response %
Daily	35.9
Weekly	20.7
Fortnightly	11.0
Monthly	11.7
Never	10.3
Don't know	10.4
Total	100.0

Table 3. The causes of stress constituting a problem to construction management / supervision, production workers and others (construction management).

Cause	Incidence Index		Ranking	
	GCs	SAIB	GCs	SAIB
Physical environment:				
Dirty' conditions	1.08	1.63	19	12
Excessive noise	1.15	1.52	18	18
Inadequate ventilation	0.92	1.41	23	21
Lack of privacy	0.79	1.26	26	25
Poor illumination	0.88	1.44	24	20
Temperature	1.35	1.67	10	9
Temporary accommodation	1.00	1.22	21	26
Unsafe & unhealthy site	0.85	1.33	25	23
Organisational:				
Change	1.38	1.56	9	15
Crisis Management	1.46	1.85	6	3
Inadequate staff	1.54	1.77	2	6
Inappropriate communication channels	1.35	1.74	9	7
Incompetent staff	1.54	1.81	4	5
Individual:				
Delegating	1.31	1.63	12	11
Inadequate managerial support	1.19	1.67	16	8
Job complexity	1.54	1.92	3	2
Job insecurity	1.25	1.50	15	19
Lack of training	1.38	1.63	7	14
Overburdening	1.49	1.85	5	4
Relationships with superior	1.17	1.56	17	16
Relationships with subordinates	1.33	1.33	11	22
Role ambiguity	1.02	1.28	20	24
Time variables:				
Overtime	1.29	1.63	13	13
Project schedule	1.88	2.04	1	1
Shiftwork	0.79	0.99	27	27
Off-the-job:				
Family	1.27	1.67	14	10
Life events	1.00	1.56	22	17

Two of the nine individual related variables achieved a ranking in the top six: job complexity, and overburdening. Lack of training also achieved a ranking of seventh for GCs and inadequate managerial support a ranking of eighth for SAIB members. Delegating achieved a relatively high ranking of twelfth and eleventh for GCs and SAIB members respectively.

Although temperature achieved a ranking of tenth and ninth for the GCs and SAIB members respectively, generally physical environment related causes did not achieve high rankings.

Similarly, off-the-job related causes did not achieve a high ranking.

An importance index based on a value of 3 was also computed for 27 causes of stress (Table 3).

In a study conducted by Sutherland and Davidson (Djebarni, 1996) work-load-constant time pressure, lack of competent staff and inadequate number of staff were ranked 3rd, 4th and 5th respectively as sources of stress among construction site managers.

Although table 4 schedules the worker responses to the extent to which occupational health interventions are required, it is important to note that with the exception of PPE and physical demands, they are all 'physical environment' related factors. The latter mentioned being one of nine categories of causes or stress identified by Sommerville and Langford (1995).

10 CONCLUSIONS

Construction by its very nature tends to be stressful. The literature and surveys of both management and workers substantiate this. However, a moderate level of stress enhances performance, but excessive stress can have serious implications for both the individual and the organisation, inter alia, physical injury as a result of accidents, and decreased productivity. After the time related variable, project schedule, the organisational related variables: crisis management; inadequate staff; incompetent staff, and inappropriate communication channels, are the predominating category in terms of causes of stress. Thereafter the individual related variables: job complexity; overburdening, and inadequate managerial support, feature prominently. A number of interventions can reduce or eliminate the stress. These interventions are far reaching and include acceptance by clients of, inter alia, the reality that there are optimum durations for construction projects, regardless of interest rates. Consequently interventions need to be undertaken at national, industry, organization and individual level. Initially these interventions may seem ridiculous to some, but the cost of stress is more far reaching than realised by most. Unattractiveness as an industry in which to pursue a career being just one of the indirect costs of stress.

11 RECOMMENDATIONS

Clients and designers should determine optimum project periods.

Contractors should ensure that both the management and worker environments on-site are complementary to the physical and mental wellbeing of employees. This will also enhance the image perceived by the public. Optimum organisation structures should be evolved for projects. This implies adequate, trained and experienced personnel with suitable personality types.

Table 4. Actions which should be taken to improve health on site (workers).

Action	Importance index	Ranking
Improve facilities (toilets, etc.)	2.52	2
Improve lighting	1.94	6
Improve site tidiness	2.54	1
Provide more PPE	2.30	4
Reduce dust	2.14	5
Reduce physical demands (make job easier)	2.39	3
Reduce noise	1.93	7

Contractors should have systems, procedures and checklists in place to empower management and workers to undertake work in the correct manner, which will avert crisis management and consequently stress. Sound management, manifested in, inter alia, planning will reduce stress during the construction process. Communication is also essential to complement systems, procedures and checklists.

The level of awareness of stress and its consequences should be raised among contractors and in industry. Contractors should introduce stress management programmes.

Given that the descriptive findings contained in this paper emanate from an occupational health study, a more in-depth study of stress in South African construction should be conducted.

REFERENCES

Djebarni, R. 1996. The impact of stress in site management effectiveness. *Construction Management and Economics,* (14) 4: 281-293.

Fellows, R., Langford, D., Newcombe, R. & Urry, S. 1983. *Construction Management in Practice.* New York: Construction Press.

Fryer, B. 1997. *The practice of construction management.* Third edition. Oxford: Blackwell Science Ltd.

Hinze, J. 1997. *Construction Safety.* New Jersey: Prentice-Hall, Inc.

Morton, C. 1990. The cost of stress. *The Chartered Builder,* December: 13.

Ross, R.R. & Altmaier, E.M. 1994. *Intervention in occupational stress.* London: Sage Publications Ltd.

Smallwood, J.J. 1997. Construction is a stressful industry. *SA Builder/Bouer*, October: 30.

Smallwood, J.J. & Ehrlich, R. 1997. Occupational health in construction. *Proceedings of the 1st South African Construction Health & Safety Conference. Health & Safety in Construction : Current and future challenges*, Cape Town, 7-9 October: 171-187.

Sommerville, J & Langford, V. 1995. Stress in the construction industry. *Construction Papers, 51,* Ascot: The Chartered Institute of Building.

Sommerville, J. & Langford, V. 1994. Multivariate influences on the people side of projects: stress and conflict. *International Journal of Project Management.* 12(4): 234-243.

Stranks, J. 1994. *Human factors and safety.* London: Pitman Publishing.

11 Ergonomics

Implementation of Safety and Health on Construction Sites, Singh, Hinze & Coble (eds)
© 1999 Balkema, Rotterdam, ISBN 90 5809 036 1

Ergonomic design considerations in construction

C.L.Zimmermann & T.M.Cook
Injury Prevention Center, The University of Iowa, Iowa City, Iowa, USA

ABSTRACT: The following project was undertaken to identify the presence of work-related musculoskeletal disorder (WMD) risk factors among sheet metal workers, electricians, plumbers, and pipefitters (mechanical trades). WMDs account for a significant percentage of injuries among the mechanical trades. Many of these WMDs can be associated with exposures to repetitive motions, high forces, awkward postures and vibration while in the workplace.

Mechanical trades workers were evaluated while on a construction site. Visual observation of the workers and work-site were performed to identify WMD risk factors. Several of the risk factors identified were attributed to the presence of other installed materials (i.e. sheet metal ducts in the way of electrical conduits or plumbing, etc). In these and other situations, consideration of construction techniques during the facilities' design stage appears to be lacking. "Design for production" in the construction trades could eliminate many risk factor exposures and needs to occur.

1 INTRODUCTION

In 1996 the construction industry ranked second behind manufacturing in the rate of lost workday injury and illness cases and cases without lost workdays (BLS 1997a). Sprains and strains are the leading cause of these missed work injuries among construction workers (BLS 1997b). According to a recent NIOSH review (NIOSH 1997) of the epidemiological data regarding the association between workplace factors and musculoskeletal disorders, there is strong evidence to support a relationship between working posture and increased risk of neck and neck/shoulder disorders. A strong relationship also exists between combinations of repetition, force and posture and an increased risk for elbow disorders, carpal tunnel syndrome and tendinitis. Increased risk for low back pain has been strongly associated with exposure to lifting/forceful movement and whole body vibration exposure.

A recent survey of workers in the mechanical trades (sheet metal workers, plumbers and pipefitters and electricians) has identified the low back, knees, wrists and hands, shoulders and neck as the five most symptomatic areas for these workers (Cook et al. 1997). This same survey identified the low back, neck, upper back and shoulders as the regions most often associated with physician visits regarding work related pain. The low back, upper back and shoulders were identified as the regions most often associated with missed work. In the same survey workers were asked to provide self-perceptions of the association between workplace factors and work-related aches and pains. The workplace factors ranked as most related to their aches and pains were:

1) Working in the same position for long periods
2) Bending or twisting your back in an awkward way
3) Working in awkward or cramped positions
4) Carrying, lifting or moving heavy materials or equipment
5) Reaching or working over your head or away from your body.

These self-reports combined with the epidemiologic data reviewed by NIOSH (NIOSH 1997) appear to be in strong agreement.

The following paper describes problem areas identified in the evaluation of a construction site. Specifically, problems areas related to building design characteristics that negatively influenced those job factors rated as most problematic by both the NIOSH review and the workers themselves.

2 METHODS

In this informal worksite evaluation, one of the authors spent approximately 100 hours on a local construction site. The site involved the production of retail sales space in a shopping mall (both the mall area and anchor stores). Initial permission to work on the site was provided by the general contractor for the entire site and permission was later granted by three of the five anchor stores. Evaluation was preformed through visual observation and videotaping. Due to the scope of the project, only two of the anchor stores and the mall proper were evaluated. During the course of site visits, work practices and techniques were compared between the subcontractors employed by the mall and anchor store general contractors. This group included two sheet metal, two plumbing and pipefitting and two electrical contractors.

The investigator watched workers and conversed with them regarding the methods, tools and materials used on the various sites. This provided an employees view on the rationale for construction methods.

3 FINDINGS

In many instances where ergonomic risk factors were identified, the primary rational for work practices provided by the workers was "That's the way it is on the blue print." Listed below are examples of situations where design criteria had a significant effect on workers' postures, force requirements and general exposure to WMD risk factors.

1. Installation of a ceiling mounted heater.
2. Placement of electrical drop boxes relative to an electrical bus system.
3. Height of the electrical bus system on the mall roof.
4. The proximity of HVAC, electrical and plumbing systems.
5. Height of gas lines on the mall and anchor store roofs.

Regarding the installation of a ceiling mounted heater, the following is a description of the heater's placement. The heater was located above a hanging fluorescent light fixture and directly under a run of electrical conduit. The proximity to these other fixed structures created an increased risk for awkward postures. The need to work in awkward postures resulted in greater force requirements for the musculature and an increased potential for muscular fatigue. The heater was also hung approximately 15 feet from the vent in the roof. This required the sheet metal worker to install between 15 and 20 feet of pipe from the roof to the heater, resulting in increased materials and work time. There were no other structures directly below the roof vent. Changing the placement of the heater to directly below the vent would have eliminated the increased risks of awkward postures and muscular fatigue and reduced the materials and labor costs as well.

The placement of drop boxes relative to an electrical bus system also presented increases in exposures to risk factors. The design of the electrical bus system was such that there were no convenient mechanisms for moving the electrical drop boxes from the outer edge of the roof to the region enclosed by the bus system. The drop boxes had to be physically lifted from one side of the bus to the other. This required the use of three to four electricians and installation required awkward postures, including kneeling and bending over and twisting the back. The first problem was the logistics of moving drop boxes. This could have been simplified with the inclusion of ramps over the bus as part of the design or the elevation of sections of the bus to allow materials to pass under the bus system. The second option may be preferred, as this would have reduced the awkward postures associated with installation of the bus system.

Installation of the electrical bus system was another task that was strongly influenced by the system's design. The bus system was placed approximately 10 inches off the roof. This resulted in considerable bending and kneeling during installation of the support structure and electrical wires. An increase in the elevation of the bus system as part of its design could considerably improve the work practices involved in its installation. A reduction in the need for forward bending of the trunk and kneeling postures during construction could be easily achieved with a slight change in design. This, if following the example set by bricklayers, should result in a significant reduction in energy expenditure, exposure to awkward postures, and increased productivity (Vink and Koningsveld, 1990).

The final design criterion, evident in all of the sites, was the proximity of installation for all of the mechanical trades. In several occurrences, it was noted that the electricians, plumbers and pipefitters and sheetmetal workers were required to install their systems in the same vertical plane. This practice had a significant impact on the adoption of awkward postures on at least two of the three trades. The first trade to occupy a space worked with the least restrictions (Figure 3). However, each of the trades that followed were typically required to install their conduits, duct work or pipes while working around the other trades materials. This resulted in a significantly greater need to work in limited space using awkward postures. Although nearly all of the trades were working just below the roof and overhead work was required for all of the trades, the restrictions caused by the presence of other materials in the work space provided greater need for the adoption of awkward postures and resulted in the potential for significantly greater force requirements on the part of the workers. Reconfiguration of the location of duct work, plumbing and electrical runs in the facility could greatly reduce the need to work around other materials and its associated awkward postures. These awkward postures not only produce greater physical demands on the workers, they also require greater time for the installation of materials.

Thus, the problems associated with the above examples include:

1. Heavy lifting.
2. Awkward postures including; bending and twisting of the spine, and kneeling
3. Prolonged and awkward postures of the head, neck, upper extremities and low back.

4. SUMMARY AND CONCLUSIONS

Of the situations mentioned, all could be considerably improved or eliminated by integrating ergonomics into the design of the building site. For example, raising the level of the bus system a few inches could significantly reduce the musculoskeletal demands placed on workers backs and knees during installation. Placement of mechanical systems within designated elevations and separate regions would reduce the need for awkward postures and high forces.

"Designing for production" could have a significant impact on decreasing the presence of risk factors in the mechanical trades. The prevalence of awkward postures and cramped working conditions can be reduced through foresight and planning that considers the requirements placed on tradesmen and workers. By considering construction methods, and work processes or work organization during the design process, WMD risk factors may be significantly reduced. Designing with consideration of ergonomic principles should lead to a healthier workforce, more efficient work methods and improved productivity.

REFERENCES

Bureau of Labor Statistics (BLS) 1997b. Workplace Injuries and Illnesses in 1996. Table 7. Incidence rates of nonfatal occupational injuries and illnesses by industry division, 1993-1996. Washington, DC: US Department of Labor, Bureau of Labor Statistics, USDL 97-

Bureau of Labor Statistics (BLS) 1997b. Occupational Injuries and Illnesses By Selected Characteristics. Table 5. Percent distribution of nonfatal occupational injuries and illnesses involving days away from work by selected injury or illness characteristics and industry division, 1996. Washington, DC: US Department of Labor, Bureau of Labor Statistics, USDL 97-

Cook, T.M., Rosecrance, J.C. and Zimmermann, C.L. (1996). *The University of Iowa Construction Survey*. Iowa City, IA: The University of Iowa, Ergonomics and Biomechanics Facility. Report E1-96.

National Institute for Occupational Safety and Health (NIOSH) 1997. Musculoskeletal disorders and workplace factors: A critical review of epidemiologic evidence for work-related musculoskeletal disorders of the neck, upper extremity, and low back. Cincinnati, OH: U.S. Department of Health and Human Services, Public Health Service, Centers for Disease Control and Prevention, National Institute for Occupational Safety and Health, *DHHS (NIOSH) Publication No. 97-141.*

Vink, P., & Koningsveld, E.A.P 1990. Bricklaying: A step by step approach to better work. *Ergonomics* 33(3):349-352.

Implementation of Safety and Health on Construction Sites, Singh, Hinze & Coble (eds)
© 1999 Balkema, Rotterdam, ISBN 90 5809 036 1

Ergonomic and medical analysis of plasterwork a basis for ergonomic redesign

J.Wakula
Institute of Ergonomics, University of Technology Darmstadt, Germany
F.Wimmel
Arbeitsgemeinschaft der Bau-Berufsgenossenschaften, Frankfurt am Main, Germany

ABSTRACT:Quantitative and qualitative knowledge of stress-strain-situations by different tasks of plasterwork with respect to postural load and manual materials handling was obtained. It could be demonstrated, that stress generated by working postures and working conditions are a major problem of plasterwork. The analysis of more than 2300 working postures revealed, that many plastering tasks contain more ergonomically awkward working postures than harmless ones. The statistical analysis of medical questionnaire results of 204 plasterworkers showed significant correlation between musculoskeletal disorders and handling of heavy materials. Personal factors, such as age and overweight played also an important role.
Some proposals for ergonomic redesign of certain plastering tasks are made.

1 INTRODUCTION:

It has been frequently hypothesized, that heavy physical work with awkward working-postures and manual materials handling in the building and construction industry can cause a variety of musculoskeletal pains and disorders.

Musculoskeletal disorders and especially back pain and lower back disorders are a major health problem in the working force. They put a considerable financial burden on the systems of social security. These ailments are responsible for early retirement and they cost in terms of sick leave compensation and lost working hours. Over 20% of medical sick- leave certificates and about 25% of premature retirements in Germany are due to degenerative diseases of the spine. Reducing the suffering of people and the financial burden on insurances and taxpayers are of outstanding importance and a challenge for occupational physicians, engineers and the social security administration alike.

Complaints of lower back pain and aching joints are common in plasterworkers (e.g. SCHNIEDERS, 1984). Nevertheless, scientific knowledge of workloads, handling of tools and working-postures in plasterworkers is still limited.

The main purpose of this study was to advance knowledge of loads in plasterwork with respect to postural load and manual materials handling, its influence on musculoskeletal disorders and on the lower back pain and to propose ergonomic redesign of certain aspects of plasterwork.

2 METHODS AND MATERIALS

The study included ergonomic and medical parts (s. WAKULA&ROHMERT, 1996). At the beginning of the ergonomic part all purpose job analysis and posture study methods (MAS and

Table 1: Methods used in the study

Methods for rating of:	Ergonomic part	Medical part
Stress:	All purpose job analysis and posture study methods (AET/MAS; BVA; OWAS/ KhS); Compressive forces on L5/S1 (biomechanical model implemented in man-model HEINER)	Occupational history
Strain: - Cardiovascular system	Pulse rate recording	
- Muscles	Recording of muscular activity with EMG	
- subjective rating	BORG rating- scale	
Complaints		Clinical history
Medical findings		Orthopedic examination

OWAS/KhS, see Table 1) were used to analyse specific tasks and work situations of plaster workers.

14 different plastering tasks (e.g. putting up plaster on walls or ceilings, polishing walls or ceilings, roughcasting) were distinguished. The results of the job analysis with MAS showed, that the loads generated by working postures and external forces rewarded further investigation.

For this investigation the "Posture Study Method" (Körperhaltungs-Studie -KhS- ROHMERT et al., 1993) was used. Today, KhS is realised as an enhancement of the Finnish OWAS-method and it still retains an OWAS compatibility.

Working-postures of plasterers were observed at different construction sites (e.g. apartment houses, office buildings) and documented with a videocamera. The personal data were documented separately. The rating of working-postures was made in a videolaboratory. In addition to video-recording, the observers registered the weight of specific loads (e.g. tools, materials). 35 plasterworkers were observed during a normal 8-hour working day doing different tasks of plasterwork. Thus, more than 2300 working-postures were recorded and analysed.

For the assessment of stress effects on the plasterer, a strain rating procedure to analyse sequences of operations (BVA-method,) and electrophysiological methods were used. The strain rating procedure - BVA (WAKULA&LANDAU, 1997) - analysed detailed sequences of different plastering operations and identified their effects by means of a rating scale of strain (BORG scale), which the worker perceived. Sequences and duration of elementary tasks were analysed to reveal relations between stress and strain.

Effects of different plastering tasks on the cardiovascular system and the workload for certain muscles were assessed with heart rate measurements and measuremenzs of the electrical activity of the m. erector spinae and the m. trapezius. The data were digitally recorded with a Par-Port-System and analysed.

The medical part included a medical and occupational history and a short orthopaedic examination of the lower back. A questionnaire, based on a standardised form introduced by KUORINKA (1987), for taking an occupational and medical history was used to acknowledge stress factors and complaints, which were taking effect over longer periods (years). With this questionnaire the long- term effects of stress factors of work on plasterworkers were assessed.

Stress factors were the frequency of lifting and handling of heavy materials (more than 20 kg), working in awkward postures (kneeling, crouching, stooping), the real weekly working time, forms of payment (piecework or hourly), grade of specialisation and environmental factors,

Since sports, gardening and other activities might contribute to lower back pain and other musculoskeletal disorders, we asked for possible private stress- factors as well.

All 204 plasterers were examined and medical parameters such as flexibility of the spine, anatomic abnormalities and muscular status and so on were recorded.
All stress factors were checked for possible relations with harmful effects on the examined persons and statistically evaluated with the χ^2-Test.

For analysis and evaluation of stress on the lower back (L5/S1) a biomechanical model implemented in computer aided man model HEINER (SCHAUB, 1989) was used. Based on the obtained data of working postures and weight of materials or forces, the compressive force on the intervertebral discs of the lower back were calculated. The resulting pressure forces on the disc L5/S1 due to different plastering tasks were evaluated according to results of NIOSH (1991) and JÄGER (1994).

3 RESULTS

3.1 Job analysis

With the job analysis method MAS 14 different plastering tasks on construction sites were distinguished and evaluated. This job analysis revealed, that working with uplifted arms was frequent, especially during "smoothing plaster ceiling" and "finishing plaster ceiling". Moreover, the adverse influences of coldness dampness and draughts were found. Static work were stress-factors for the hand/arm system and the back.

3.2 Working postures

Figure 1 compares normal and harmful working postures from the ergonomic point of view, according to OWAS action categories. It can be discerned that from 6 plastering tasks only "polishing wall (outside)" and "polishing (ceiling)" contain more ergonomic awkward working postures than harmless ones.

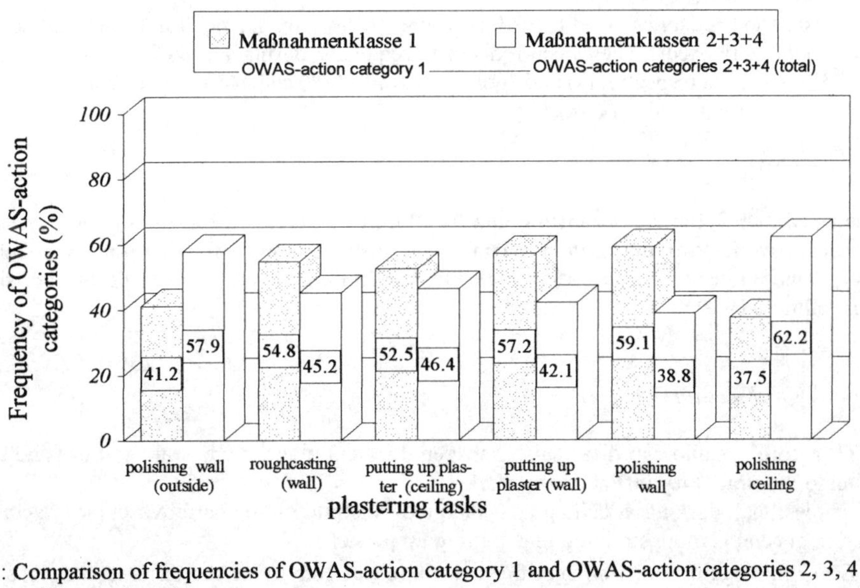

Fig. 1: Comparison of frequencies of OWAS-action category 1 and OWAS-action categories 2, 3, 4 for different plastering tasks.

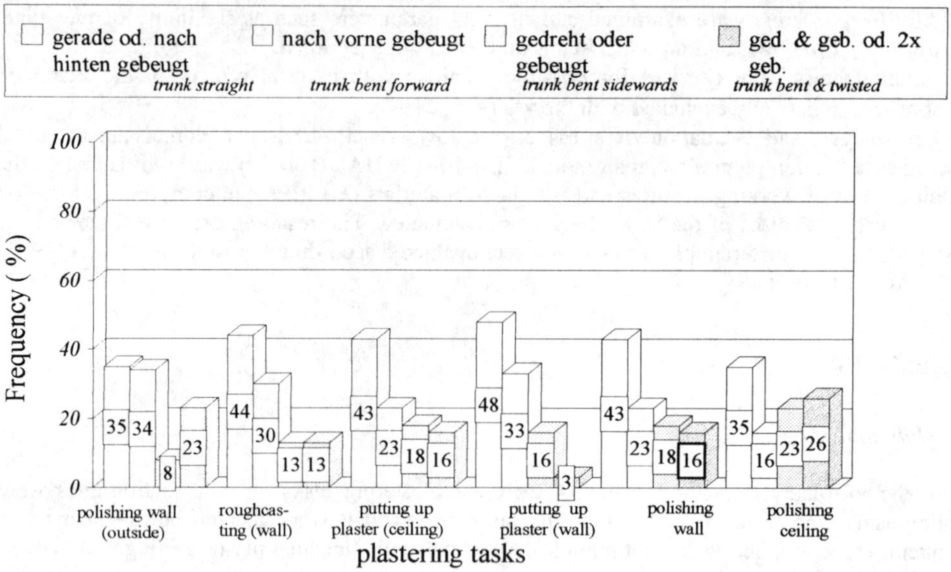

Fig. 2: Frequency of trunk-postures according to OWAS-analysis

According to OWAS-action categories normal or slightly harmful working postures prevail (OWAS-action categories 1 and 2).

Detailed analysis revealed that most working postures requiring ergonomic improvements belonged to OWAS-action category 2 ("slightly harmful"). The postures of this action category is dominant and ranged from 33,8% to 58,3% of total analysed postures in all different tasks. The postures of OWAS-action categories 3 and 4 rated below 6%.

OWAS- analysis of postures of the trunk showed that an upright, backward bent and forward bent position prevailed (Fig. 2).

Detrimental trunk postures from a medical point of view would be "trunk bent sidewards" and "trunk bent & twisted". They occurred most frequently during plastering tasks "polishing ceiling" (23% and 26% respectively) and "polishing wall"(18% and 16% respectively).

3.3 *BVA-analysis*

It can be shown (s. Table 2), that lifting and handling of weights >20 kg is no decisive stress factor in plasterwork. Working in double bent trunk postures is most frequent during polishing plaster at walls. Subjective strain rating with the "BORG" scale was highest during polishing plaster at ceilings (14,9).

3.4 *Physiological measurements*

The ECG- recordings allow to distinguish between different objects (e.g. wall, ceiling) and tasks in relation to the strain they inflict on the worker.

Table 3 shows, that working at ceilings put the greatest strain on the cardiovascular system and polishing in general is more straining than putting up plaster.

The EMG findings revealed no significant variation of activity of the m. erector spinae and the m. trapezius during any specific plastering task, when compared with the mean electric activity over a complete working day (see table 4).

The EMG findings showed no fatigue in the respective muscles regardless of the plastering task.

Table 2: Results of strain rating procedures of different plastering tasks

Task	Prevailing elementary plastering task (in %)	Dur. of working with double bent back	Handling of weights >20 kg (in %) Freq.	Dur.	Strain rating according to BORG
Putting up plaster (ceiling)	Working with power driven tools on scaffolds: 67% Carrying weights < 10kg : 64% Working with uplifted arms: 58%	3%	5%	3%	13,4
Putting up plaster (wall inside)	Working with power tools: 57% Working with manual tools: 43% Finish: 31%	27%	11%	9%	13,2
Polishing plaster (ceiling)	Polishing/smoothing (scaffold: 90% Working with uplifted arms: 85% Finish: 29%	7%	5%	3%	14,9
Polishing plaster (wall inside)	Polishing/smoothing : 90% Working with uplifted arms: 34% Finish: 25%	32%	10%	6%	13,1
Roughcas-ting plaster (outside)	Working on scaffolds +ladders:86% Manual roughcasting: 86% Exposure to dust: 86%	24%	4%	2%	12,0
Work prepa-ration	Working with power tools: 51% Scaffolding: 42% Packing up tools; 33%	25%	10%	8%	11,7

Dur. = Duration

Table 3: Order of tasks in relation to strain (measured as heart- rate) of 8 plasterers

Tasks	Object	Plasterer 1	2	3	4	5	6	7	8
Putting up plaster inside	Wall	2		4	2	2			
Putting up plaster inside	Ceiling			3					
Putting up plaster outside	Wall						2		2
Polishing plaster inside	Wall	1	2	2	1	1			
Polishing plaster inside	Ceiling		1	1					
Polishing plaster outside	Wall						1		1

3.5 *Medical Questionnaires*

The 204 questionnaires were statistically evaluated. 68 % of the plasterers complained of continuous or episodic back pain and / or painful joints during the last 12 months. 45% located the pain in the lower back, 24% in the shoulders, 17% in the elbows and 15% in the knees.
36% of the plasterers said, that weekends, days off or the annual leave did not lessen their

Table 4: Exemplary results of the muscular electric activity of three plasterers

	Tasks	MEA m erector spinae (mV)				MEA m. trapezius (mV)			
		X	S	X_0	X/X_0	X	S	X_0	X/X 0
Plasterer 1	complete recording	24	18,3	13	1,85	10	18,6	7	1,43
	Putting up (wall)	25	14,1		1,93	22	20,1		3,14
	Putting up (ceiling)	28	12,8		2,15	38	27,6		5,43
Plasterer 6	complete recording	11	8,5	6	1,83	10	6,7	3	3,33
	Putting up (wall)	17	7,7		2,83	9	5,8		3,0
	Polishing (wall)	13	8,8		2,17	14	6,9		4,67
	Polishing (ceiling)	15	8,3		2,5	13	7,8		4,33
	Finishing (wall)	9	6,7		1,5	9	6,3		3,0
Plasterer 7	complete recording	12	11,2	8	1,5	29	11,9	11	2,64
	Polishing (wall)	14	10,2		1,75	31	12,6		2,82
	Finishing (wall)	10	9,4		1,25	27	9,4		2,45

X - mean MEA (mV); S - Standard deviation; X_0 - Refernce- MEA (mV);

Table 5: Correlation between complaints and stress and other factors in plasterworkers

Stress and other factors	Complaints in the lower back	Complaints in the knees	Complaints in the shoulders	Complaints in the neck
Age	0,04500	n.s	0,0001	0,00099
Years of professional experience	0,01795	-	0,00284	0,02451
Broca-Index	n.s.	n.s.	n.s.	n.s.
Lifting and handling of heavy materials > 20 kg	0,01604	-	-	-
Working in awkward postures	n.s.	n.s.	n.s.	n.s.

Legend: Significance on a 5% level n.s.: not significant

problems. Moreover, they had to consult a doctor at least once during the last 12 months. 20% were absent from work for at least one day due to pain in the back or joints.

Summarising high prevalence rates for lower back pain and shoulder problems were found. Significant correlations were found between complaints in the lower back, neck and in the shoulders on one side and age and years spent in the profession on the other (s. table 5) .

Lifting and handling of heavy materials > 20 kg showed a significant correlation only with lower back complaints. Surprisingly, working in awkward postures was not significant at all.

3.6 *Compressive force on the disc L5/S1*

Table 6 shows the results of calculated pressures on the intervertebral disc L5/S1 through prevailing posture with or without external forces for different plastering tasks. It could be shown, that critical threshold limits (e.g. NIOSH, 1981) were not reached. Thus, as far as loads on the lower back are concerned, plastering tasks do not exert critical pressure forces on the lower back.

Table 6: Calculated pressures on the intervertebral disc L5/S1 through prevailing posture for different plastering tasks

Task; prevailing posture	Calculated pressures on the intervertebral disc L5/S1			
	through posture alone	through posture + external force per hand (vertical-F_{ver} und horizontal - $F_{hor.}$)		
Putting up plaster (wall inside) (Standing in upright position, both arms below shoulder level and middle distance from the trunk; external force 30 N to 100 N)	F_{ver}=0, F_{hor}=0	F_{ver}=15N, F_{hor}=0	$F_{ver.}$=50N,$F_{hor.}$ =10;	$F_{ver.}$=100N, $F_{hor.}$=0
	1133	1281	1772	2167
Polishing plaster (wall inside) (Standing in upright position, both arms on shoulder level and middle to far distance from the trunk; external force up to 80 N)	Fver=0, Fhor=0	Fver=20N,Fhor=0N	Fver=20N, Fhor=50	Fver=20N, Fhor=80
	1433	1530	1740	1866
Putting up plaster (ceiling) (Standing in upright position, one arm on shoulder level and middle distance from the trunk; external force 30 N to 100 N per hand)	Fver=0, Fhor=0	Fver=15N, Fhor=0	Fver=50N, Fhor=10	Fver=100N, Fhor=10
	1133	1281	1772	2167
Polishing plaster (ceiling) (Standing in upright position, both arms on or above shoulder level and far distance from the trunk; external force 50 N to 100 N per hand)	Fver=0, Fhor=0	Fver=20N, Fhor=0	Fver=20N, Fhor=50	Fver=20N, Fhor=80
	1583	1680	1890	2016
Roughcasting plaster surface (outside) (Standing in upright position, one arm on shoulder level and middle distance from the trunk; external force 50 N to 80 N)	Fver=0, Fhor=0	Fver=15N, Fhor=15	Fver=20, Fhor=50	Fver=40N, Fhor=40
	1133	1347	1540	1700

4 DISCUSSION AND CONCLUSIONS

Our analysis of working postures in general and with special emphasis on trunk position demonstrated the influence of lifting and handling of heavy materials and postural load on the frequency of lower back pain. The occurrence of shoulder and neck complaints can be easily explained with the work loads through the frequent working with uplifted arms and the static position of the hand-arm-system. Some critical tasks of plastering work with respect to loads on the lower back could be recognised.

However, the calculated compressive force on the intervertebral disc L5/S1, resulting from working postures and external force, was nowhere near critical threshold values.

The most important stress factor in plastering work are working postures with uplifted arms above shoulder level. We observed, that bags with cement and mortar are unnecessarily heavy.

Some strategies to facilitate plastering work can be recommended and easily implemented. Others reward further research.

- Ready mixed plaster in silos, which are pumped to the respective workplace should preferably be used instead of bagged material.
- In case bagged materials are necessary (small sites, restoration projects in old city centres) trolleys should be used for moving them around. Equally important is the reduction of bags in size and weight. Smaller sizes can be handled more easily and much closer to the trunk. Bag weights of 20 to 25 kg should not be exceeded.
- Pump hoses for putting up plaster should be kept close to the trunk.
- Rods or planks used for polishing plaster should be no longer than 1 to 1.5 meters to reduce forces resulting from long levers.
- Roughcasting is a relatively rare working procedure. However, it requires considerable strength. It can be mechanised, although further research for that is necessary.

- Organisation of work should be redesigned in a way, that strenuous tasks are distributed more equally among the work force.

REFERENCES

JÄGER,M., LUTTMANN, A. 1994. Biomechanische Beurteilung der Belastung der Wirbelsäule beim Handhaben von Lasten. Sonderdruck aus: *Der medizinischer Sachverständige*, Gentner Verlag Stuttgart, , Heft 5, S. 160-164

KUORINKA, I., JONSSON, B., KIBBOM, A., VINTERBERG, H., BIERING-SÄRENDEN, F., ANDERSSON, G., JÖRGENSEN, K. 1987. Standardised Nordic questionaires for the analysis muskuloskeletal symptomps. *Applied Ergonomics* 18 3, S.233-237

NIOSH, Work Practices Guide for Manual Lifting, NIOSH Technical Report No.81-122 (1981), US Department of Health and Human Services, National Institute for Occupational Safety and Health, Cincinnati, OH

ROHMERT, W.; SCHAUB, KH.; WAKULA, J. 1993. Entwicklung und Anwendung der Körperhaltungs-Studie (KhS) als Methode zur differenzierten Körperhaltungsanalyse, dargestellt am Beispiel von Fliesenlegertätigkeiten. In: Rohmert, W. (Hrsg.): Stand und Zukunft arbeitswissenschaftlicher Forschung und Anwendung. REFA, München: Hanser,

SCHNIEDERS, H., AMAROTICO, E., BAUER, E., BÖTTGER, M., STARK, K., URBAN, K.-D., VALENTIN, H. 1984. Ergebnis einer arbeitsmedizinischen Untersuchung von 24892 Beschäftigten in der Bauwirtschaft. *Arbeitsgemeinschaft der Bau-Berufsgenossenschaften, Frankfurt am Main*

SCHAUB, Kh. 1989 Biomechanical modelling and work evaluation with the man model HEINER. In: *Physical strains and manual handling of loads: Epidemiologic, Prevention and Normative aspects in the European countries.* Mailand: Proceedings .., pp.15-30

WAKULA J. and ROHMERT, W. 1996. Analyse und Bewertung von Belastungen und Beanspruchun-gen bei der Bauarbeit als Determinanten berufstypischer Rückenbeschwerden bei Betonbauern und Verputztern. Schriftenreihe Arbeitssicherheit und Arbeitsmedizin in der Bauwirtschaft,

WAKULA J. & LANDAU, K. 1997. Method Battery for investigating stress and strain in building work. Abstracts from the 1.International Symposium on Ergonomics in Building and Construction, Finnland, Tampere,1997, Center to Protect Workers' Rights, Washington, 110-117

Implementation of Safety and Health on Construction Sites, Singh, Hinze & Coble (eds)
© 1999 Balkema, Rotterdam, ISBN 90 5809 036 1

Participatory approach to ergonomic risk reduction: Case study of body harnesses for concrete work

B.Gibbons & S.Hecker
University of Oregon Labor Education and Research Center, Salem, Oreg., USA

ABSTRACT: The purpose of this project was to use a participatory process to develop a system for carpenters and laborers to obtain body harness equipment satisfying both safety and ergonomic concerns. Focus groups with workers identified poorly fitting and inappropriate body harnesses as one contributor to discomfort and pain among concrete form workers. There was no consistent ordering system foremen followed, and workers were unaware of available body harness options. There were also concerns about whether employees were adequately trained in techniques for inspecting and wearing harnesses. An ordering system and selection guide was developed and implemented working with suppliers, the safety department, field crews and foremen. The system will be piloted at new sites. Foremen will implement it, but project researchers will perform the evaluation. Initial and follow-up interviews with workers will be conducted to learn whether the process led to actual comfort, usability, and ergonomic improvements for workers.

INTRODUCTION

Numerous construction tasks pose significant ergonomic risks to workers. Many of the tasks involved in the preparation, pouring, and finishing of concrete are very demanding physically and reducing these demands through traditional control methods is particularly challenging. The weight and limited "shelf life" of the material, work location, whether ground level or at heights, and environmental conditions all contribute to the physical demands. The project described in this paper began as an effort to address the incidence of musculoskeletal injuries incurred by the work force of a large general contractor performing concrete work. Researchers working with the contractor applied a participatory ergonomics approach to identifying high-risk tasks and developing potential solutions. When promising solutions were identified, a team consisting of an ergonomist, contractor safety personnel, and an equipment manufacturer developed a process to implement the solution through the contractor's existing structures.

IDENTIFICATION OF THE PROBLEM

The contractor is a large general contractor in the Pacific Northwest. A subsidiary of the general contractor performs concrete and structural work, mostly on the general contractor's own projects. The contractor determined that the largest portion of its injuries and claim costs occurred during the concrete phase of construction projects. These are primarily injuries to the back and shoulder. The authors previously worked with this contractor on a large microelectronics construction site where the focus had been on the mechanical trades. The contractor was now interested in looking at the ergonomics of the concrete phase. In January 1997, an ergonomist conducted walkthroughs at the site of a new medical office building, consisting of a five-story parking structure with seven floors of offices above it. A variety of types of concrete work were being performed including the pouring of slabs at grade, columns, walls, and the decks of each floor. Workers were interviewed and some tasks were videotaped.

Following the walkthroughs, separate focus group discussions were held with carpenters and laborers, two of the three trades involved in concrete work (the third trade is cement finishers). A variety of issues were identified as contributing to physical stress in this work for each group. For laborers the most stressful tasks were those involving handling the concrete itself, including:

1. Holding and manipulating the hose through which concrete is pumped;
2. Holding the 45-lb. vibrator which eliminates voids in the wet concrete;
3. "Mucking," or raking the concrete after it's poured; and,
4. Screeding to level the concrete.

This group also performs other physically stressful tasks on existing concrete, using tools like rock drills, chipping hammers, and jackhammers.

Carpenters main tasks involve the building and stripping (dismantling) of forms into which the concrete is poured. Tasks and conditions they identified as physically stressful included:

1. Carrying heavy tool bags all day;
2. Lifting and carrying heavy materials, including lumber, plywood, and forms;
3. Climbing and hanging on wall and column forms wearing full-body harnesses;
4. Frequent bending when lifting tools and materials;
5. Awkward postures while accessing points of operation, e.g. inserting snap-ties in wall forms; and,
6. Repetitive hammering.

It was clear that there were numerous serious ergonomic concerns, but only a limited number of these could be addressed at once. The decision to pursue the body harness issue was made for the following reasons:

1. Body harnesses concerns were raised repeatedly in the focus group meetings;
2. Little or no attention had been paid to body harness issues previously;
3. It seemed to offer the potential for solutions without great expense or major modification of the way work was done;
4. Other problems, while perhaps more pressing like the sheer force required to muck concrete or handle the hose, did not seem to have ready, practical solutions; and,
5. The contractor made a commitment to address this issue.

ERGONOMIC CONCERNS RELATED TO BODY HARNESSES

Body harnesses have become a common piece of personal protective equipment on construction sites. They are used for two main purposes: as part of personal fall-arrest systems when working at heights and, in conjunction with body belts, for positioning when working on walls. Most of the concerns centered on carpenters who are required to wear harnesses continuously throughout the day because they are periodically working at heights which require fall protection. Laborers who wear harnesses while working at heights also experience similar problems. This case report focuses on carpenters assembling and dismantling concrete wall forms. The targeted groups included apprentices, journey-level, foremen, and supervisors.

The usual process by which workers obtain a body harness is through their foreman at the beginning of their assignment to the project. Foreman or superintendents generally order one-size-fits-all harnesses at the beginning of a project with consideration for the number of workers they anticipate and the type of work being performed. For example, carpenters would get a positioning belt included with their harness for building and stripping forms and columns, whereas laborers would only need a harness periodically and usually just for fall protection.

Carpenters who wear harnesses throughout the day reported poor fit as causing major discomfort. Performing tasks with an ill-fitting harness causes strain from pushing and pulling, especially in conjunction with material handling tasks. A poor fitting harness limits functional range of motion leading to awkward postures and excessive forcefulness in working tasks. Wearing a body harness with tool bags and other accessories can compound the problem.

Carpenters also reported the length of positioning chains as a common problem. Positioning chains allow the worker to hang on the wall. If the positioning chains are too short, the worker is pulled too close to the work, reducing leverage and causing strain against the chain. If the chains are too long, the worker must lean forward and reach, creating static loads on the back muscles that hold the worker in position. Individual attempts at correcting the problem involve removal of chain links, but this can compromise the integrity of the chain. When workers found a comfortable harness, they often commandeered it for use at other sites. Additionally, there were no inspection procedures to ensure that a used harness continued to provide reliable fall protection. Most workers were not aware of the options available in the way of size, type, application, and accessories available for harnesses.

PARTICIPATORY PROCESS

With this background the participatory process was expanded to include workers on other sites managed by the same contractor. The following steps were outlined:

1. Conduct an equipment needs assessment involving affected workers and supervisors;
2. Draft a procedure for meeting the equipment needs of the workers that fit with the operations of the general contractor;
3. Pilot the procedure for obtaining appropriate equipment;
4. Modify the procedure based on feedback;
5. Establish a method and develop measures for evaluating the effectiveness of the new process and its outcomes;
6. Implement the procedure on selected sites;
7. Evaluate outcomes and make recommendations for modification; and,
8. Implement the procedure company-wide.

THE NEEDS ASSESSMENT

The first step of the needs assessment was to assure that the body harness issue was of importance to workers beyond the original focus group site. Focus groups were conducted on three additional sites where concrete work was being performed, two downtown high rises and a hospital addition. Each group was composed of 8 to 10 carpenters, their foremen, and a few laborers. Along with group discussion, participants also completed a needs assessment form. Based on the discussions and needs assessment responses, it was established that body harnesses were a legitimate concern for the majority of workers. The needs assessment also ascertained what body harness equipment these target groups felt they needed, and what training and communication systems needed to be in place to obtain and properly use the equipment.

Equipment: Workers responded to questions on the needs assessment form such as:

1. What type of work they usually performed;
2. What types of harness and accessories best fit their needs for this work;
3. What other factors placed stress on their bodies; and
4. How much did they know about specific types of equipment that are better than others.

The form included a figure of a worker with instructions for the participants to draw a body harness that best suits their needs. Respondents also were asked which lengths, sizes, materials, features and accessories they preferred. Specific issues included placement of D-rings, how well the belt and harness accommodated tool bags, ability to obtain specific sizes instead of "universal" fit, and ability to obtain flexible webbing material.

Many workers felt that the standard webbing impairs their movement and ability to use good body mechanics. Some workers preferred "D" rings placed at the hips for positioning, but they weren't aware of harnesses that allowed space for tool bags in the same location. Cutting through their tool bags to be able to put the D-rings through often solves this problem. Most workers preferred "tongue" type buckles around their legs instead of the standard "friction" type, for both security and comfort. Only a small number of workers indicated their preference for back and shoulder pads. However, many couldn't state whether they preferred pads because they had never tried them. A small number who had tried shoulder pads in the past didn't like them because the particular brand they used slipped down the back, forcing them to reach to continuously adjust them. Equal numbers of workers preferred saddle straps, which form a seat, as did not, but again many had never seen this feature.

Training in proper use and inspection: Workers knowledge about types of equipment was based on their experience on different sites. Carpenters were very aware of how to don harnesses, but few workers or foremen understood the importance of inspecting harnesses on a regular basis.

Communication: Preliminary investigation revealed that communication barriers at multiple levels in the company contributed to body harness problems at the work sites. There were different perceptions among workers, foremen, safety personnel, and equipment yard personnel about harness availability. The focus groups revealed poor communication about harness options from management and central office to the field, or about equipment needs from the workers in the field back to management. Some workers believed that even if they were aware of their options they wouldn't be able to get what they wanted. Some foremen confirmed this while others indicated that workers could get whatever they thought they needed. Some of this was self-fulfilling: if workers believed their needs would not be met they didn't bother asking and took whatever was given to them.

Another level of communication is between the project site and the equipment yard. The yard is where all equipment and materials are ordered and purchased for company sites. The perception at the site level was that the yard sends whatever is sitting around or left over from previous jobs before ordering new equipment, regardless of the specific request made or the condition of the equipment. The yard has a reputation of sending out equipment and materials that are inoperable or inappropriate.

Finally, another level of miscommunication was between the safety office and the yard and sites. The safety office has traditionally dictated to the yard what brand and types of safety equipment will be purchased. These decisions are made based on compliance with OSHA standards, best pricing for bulk orders, and fringe benefits provided by vendors such as free equipment training on request. Few attempts have been made in the past to identify individual site and trade needs.

DRAFT PROCEDURE FOR MEETING EQUIPMENT NEEDS OF THE WORKERS

Two parameters had to be defined before a procedure could be developed. First, researchers investigated harnesses, features, and accessories to compare available products with needs expressed by the workers. Second, these findings were communicated to company managers and the safety director for feedback as to how much they were willing to commit to the acquisition of new and different equipment, training connected with the equipment, and improved communications in general.

The safety director expressed strong preference for two specific vendors. Each vendor offered specific things the safety director wanted: good prices and ready availability in one case and complimentary training in the other. The safety director maintained an interest in standardizing the equipment. However, the necessary addition to the process was to make the vendors understand the ergonomic concerns related to the harnesses. The existing practice of a one size-fits-all "universal" harness would not address the ergonomic issues. We concluded that workers need different sizes, different types coinciding with specific uses, e.g. positioning versus fall arrest, and a flexible ordering system to initiate and maintain availability of equipment. The two vendors were well known and reputable with a broad range of styles, sizes, and fabrics to meet these needs. The question was whether a system could be developed to get workers what they needed and whether the contractor would agree to the flexibility needed to meet ergonomic needs. The ensuing discussions among the safety office, vendors, and sites were all part of the evolution of their better understanding of ergonomics in the construction work environment.

An ordering process was developed to ensure communications both up and down the company organization. The needs assessment indicated that both workers and contractor would benefit by incorporating other personal protective equipment needs into this procedure. The ordering procedure was envisioned as a process in which the worker would obtain the necessary equipment but also receive training if needed on proper use and inspection. A form was developed that the worker would fill out in the presence of the foreman when the worker first reported to the site. Questions included:

- What kind of safety equipment do you have?
- What kind of safety equipment will you need (including glasses, hard hat, harness, body belt, gloves and a variety of items necessary to do the job)?
- Do you know how to inspect a harness (your own, or one issued to you)?

The workers would need to demonstrate their ability to inspect the fall protection equipment at that time, and foremen would demonstrate the correct method, if necessary. If the workers brought their own harnesses onto the site, they would need to demonstrate how to inspect it. Once completed, the workers and foremen would both sign the form.

If workers did not bring their own equipment, they would complete the order form for specific equipment in the presence of the foremen. This was to insure that foremen and workers both understood what is needed and is being ordered. The form listed and illustrated a selection of harnesses and accessories available from the two vendors. The research team selected the items to be listed based on discussions with the workers and vendors. Once selections were made, foremen would send the orders to the yard and the yard would provide the equipment as soon as possible.

A mechanism was needed to address the delay between the time the worker filled out the form and the equipment arrived and was issued. A basic "starter kit" containing universal safety equipment would be issued the first day, and then retrieved when the specialized equipment arrived.

PRETESTING THE PROCEDURE

A draft of this procedure and forms was pretested at two sites. One site was the medical office building where the crews had participated in the original focus groups and the second was the hospital addition from the second round of focus groups.

We introduced the procedure and forms to groups of foremen and some superintendents at lunch-hour meetings. Foremen were most critical to the process because they would ultimately implement the procedure on all sites. During each 45-minute meeting we distributed the ordering forms, outlined the procedure and their role in it, and responded to questions.

Their assignment during the following two weeks was to go through the procedure with their current workers and others who might come to the site. They were to fill out the forms with the workers and talk about the equipment issues, finding out if workers had what they needed or could get it. The foremen were also to tell us if they had sufficient training and knowledge to facilitate the procedure. Some foremen had already expressed the need for more training, particularly on proper inspection procedure for harnesses.

We returned after two weeks to receive feedback from the two sites. At the first site issues arose concerning the variety and availability of the specific items on the forms. Several foremen felt that not enough choices were provided to address what workers were requesting. Many workers wanted the "Duraflex" strap, a stretchy web material, but it didn't appear on the form.

Some foremen had used the order forms as if they were already adopted and sent them to the yard to obtain the equipment they wanted. The yard had sent them the standard universal fit body harnesses. The problem was that they had ordered based on an actual harness they had on site (size "small"), not realizing that this had been a special order and was not a standard item the vendor provided to this contractor. Since the yard knew only about the universal size that's the one they sent. At the second site workers and foremen ordered only from the items available on the form and were happy that they received what they ordered.

The pretest taught us several lessons. Simply listing and illustrating available equipment was not enough. A series of prompts raising issues of ergonomics, type of work being performed, size, and type of material was needed to lead both workers and foremen through the process. It was clear that we had narrowed the list of available equipment prematurely. Since this had been done in large part to satisfy the safety director's desire to standardize, we needed to reach an understanding with him. A compromise was that we would start with a fairly broad range and, after using the form for a while, the workers would have a better idea of what they wanted most of the time. We could then narrow the list to the most frequently chosen items.

A further concern that arose during the pretest was how to track the body harness equipment. Identification tags could assure that the equipment got to the right worker, but the issue of ownership went beyond this. Since the same general contractor managed all the various sites, taking equipment from one site to the other was permissible. And those who had worked for the contractor for a long time often assumed the harness was now theirs, even if they ended up working for another contractor. Tracking was a problem yet to be solved.

FINALIZING THE PROCEDURE

To solve the problem of too few choices, the primary vendor developed a "selection guide," a mini-catalog, especially for this particular contractor and its employees. The guide incorporated our language concerning selection of various features, which attempts to direct workers and foremen through the process, anticipating common questions. Much of this language prompts workers on ergonomic issues we wanted them to consider in their selections. It is extensively illustrated so workers and foremen will have a better chance of knowing what they are getting before it arrives.

FINDINGS FROM THE PARTICIPATORY PROCESS

The pilot process identified training needs. Both foremen and workers wanted more training. We identified the equipment and material needs and learned that more choices of harness equipment were needed.

We identified numerous relationships that affected the project: between workers and foremen, foremen and the superintendents, the superintendents and the yard, safety director and foremen, safety director and workers, safety director and the yard, and safety director and the vendor. Any of these could determine whether the right equipment would ultimately be available to the workers. Sometimes the issue would be economic, other times ergonomic and/or safety. These are relationships that need further exploration.

The safety director wanted a consistent process providing standard equipment to all sites at an inexpensive price. The idea of bulk buying with price breaks and complimentary training seemed to dominate the thinking of the superintendents, the yard, and the safety director. The safety director became educated to realize that addressing ergonomic issues required worker involvement to learn exactly what their needs are rather than dictating those needs through purchasing decisions. The safety director was eventually able to say "We expect to make a customized body harness purchase for each person."

Another lesson is that this project is more viable because it addresses not only ergonomic issues, but also safety issues. It addresses not only material issues, but on-the-job relationship issues as well. In essence, it is multi-faceted in a way that draws all of the active players into the participatory process.

ERGONOMIC OUTCOMES

The process itself has identified numerous obstacles to workers obtaining appropriate and necessary personal protective equipment. The new procedures are intended to overcome these barriers. However, the ultimate goal is improved comfort, safety, and ergonomics. There are specific outcomes for these objectives that will be measured to assess the effectiveness of the revised procedures. Worker outcomes include:

- Did the equipment you received actually match what you thought you had ordered?
- What effect does a customized harness have on
  - Comfort
  - Fatigue
  - Ability to perform the task
- Did the procedure increase workers' ability to properly inspect their fall protection equipment?

Three new construction sites were selected to collect data on these outcomes. Each new worker coming to the site will be interviewed within a week of coming on site and going through the ordering and training process. Data collected will include satisfaction with current and past fall protection equipment in terms of comfort, musculoskeletal symptoms, specific features, appropriateness to the task, and condition. The efficacy of the initial foreman-worker meeting will be assessed in terms of the worker's obtaining the needed PPE, training in PPE use, and usefulness of the session as an orientation. Each worker will be surveyed at four-week intervals for several months to determine longer-term impact of the harness on comfort, symptoms, and ease of performing the worker's tasks.

Foremen will also be interviewed to collect their perceptions of the procedure as a training tool and its impact on safety and ergonomics.

CONCLUSION

The focus of this project was to use a participatory process to develop an improved system for carpenters and laborers to obtain body harness equipment that satisfied both safety and ergonomic concerns. Focus groups with craft workers identified poorly fitting and inappropriate body harnesses as one contributor to discomfort and pain among concrete form workers. There was not a consistent ordering system that foremen at all sites could follow, and workers were unaware of some of the options that were available for body harnesses. There were also concerns about whether employees were adequately trained in proper techniques for inspecting and wearing harnesses. Working with suppliers and the safety department, we developed an ordering system and selection guide that would go into the foremen's manual.

We will pilot the new process at three new construction sites as they start up. Site foremen will have actual implementation responsibility but project researchers will perform the evaluation. All workers who come on site will be interviewed shortly after their arrival concerning their previous body harness equipment, what they received on the new site, and how the ordering and training process worked for them. Each site will then be visited at 4-week intervals for four months to conduct follow-up interviews with each worker to learn whether the process led to actual comfort, usability, and ergonomic improvements for those workers. This process introduced a participatory ergonomics model into the company's organizational structure.

REFERENCES

Schneider, S, Susi, P, 1994: Ergonomics and construction: A review of potential hazards in new Construction, American Industrial Hygiene Association Journal, 55(7): 635-649

Cook, T., Rosecrance, J.C., Zimmerman, C., 1996: The University of Iowa Construction Survey, (University of Iowa Biomechanics and Ergonomics facility, Iowa City).

Hugentobler, M.K., Israel, B.A., Schurman, S.J., 1992. An Action Research Approach to Workplace Health: Integrating Methods. Health Education Quarterly Vol. 19(1):55-76

Basch, CE., Focus Group Interview: An underutilized research technique for improving theory and practice in health education. Health Education Quarterly 14:411-448, 1987

Miles, M.B. Huberman A.M., Qualitative Data Analysis: A Sourcebook of New Methods. Beverly Hills, CA. Sage, 1984

Implementation of Safety and Health on Construction Sites, Singh, Hinze & Coble (eds)
 ISBN 90 5809 036 1

Technological interventions to reduce the incidents of back injury in building construction

M.L. Davis, S.J. Lorenc & L.E. Bernold
North Carolina State University, Raleigh, N.C., USA

ABSTRACT: Occupational back injury represents an amazingly large segment of worker injuries and results in more lost work days than any other illness, second only to the common cold. Between 1985 and 1987, twelve states were surveyed and the results were astounding. Occupational back injury was found to be the most frequent work related disorder, accounting for 22% of all the reported cases and 32% of all workers' compensation dollars. The construction industry has consistently been considered a high risk industry for low back injuries. This paper presents the first results of a research project that has as its goals to identify hazardous tasks and to redesign existing workplaces, jobs, and equipment to fit the worker. The main portion of the paper is dedicated to discuss the steps that led to an ergonomic nailgun system that has been tested in the field.

1 INTRODUCTION

The most frequent injury to the common worker is lower back pain, generally the result of cumulative damage done to the spinal column area. Back pain is nothing new. It has afflicted man from the beginning of time; there is even evidence that primitive man suffered from some form of low back disease. Backaches, as common as they are, are very rarely lethal, but statistics show that they cause an extreme amount of misery. They tend to persist for years, sometimes even decades, flaring up from time to time, often for no apparent reason, and then subsiding again, just as mysteriously. These recurrent attacks of pain generally occur during a person's adult years, often interfering with the prime years of life.

Most of the advances and understanding of back injuries have occurred in the last 20 years. Enough books and articles have been written on the topic to fill an independent library. Along with the increase in interest and concern, the social and economical costs have risen as well. Back pain has, indeed, developed into man's most important non life-threatening disease. Despite all this, we still lack in knowledge about both cause and cure. In addition, we as humans lack in the necessary appreciation of its danger to a normal life.

1.1 *Occupational Back Injuries*

Occupational back injury is the leading cause of activity limitation among young adults, is the second most common reason for work absenteeism and is the third leading cause of work disability (Deyo and Diehl, 1988). Over the past 20 years, the number of persons permanently disabled as a result of low back pain has proportionately exceeded the population growth and the disability from other chronic health conditions . Between 1985 and 1987, twelve states were surveyed and the results were astounding. Occupational back injury was found to be the most frequent work related disorder, accounting for 22% of all the reported cases and 32% of all workers' compensation dollars (Federal Register, 1992). Over the past decade, workers' compensation benefits have increased 157% and employer costs have risen 143% (Nelson, 1993).

According to Malcolm Pope, the greatest potential for low back pain (LBP) prevention also exists in the workplace (Pope, M. et al. 1984). Incidence, severity, and potential disability are all related to the demands on the individual on the job.

1.2 *Back Injuries in Low-Rise Residential Construction*

Workers in the construction field are exposed to a number of recognized occupational risk factors for back pain including heavy work, materials handling, pushing, twisting, frequent lifting over 25 pounds, requirements for sudden unexpected maximum effort and a number of awkward postures (Bernold, 1993; Pope, 1984). Despite this, however, there has been very little research done relating specifically to occupational back problems among various construction trades. Interventions dealing with the prevention of construction related back problems is also a topic not often addressed.

During a four-year time frame in Ontario, Canada, data was collected by the Construction Safety Association of Ontario on lost time due to back injuries. It was found that the number of back injuries in the low-rise residential construction business greatly exceeded the injuries incurred by workers in other construction sectors. Low-rise residential building construction averaged about 22% of all back injuries in the construction industry. This distribution of back injuries by construction sectors can be seen in Figure 1. One explanation for this is the fact that the low-rise residential field employs more workers compared with the other sectors. Another explanation is that smaller contractors use less equipment and their material handling capabilities are limited.

Despite this fact, however, there is little literature specifically related to occupational back problems in construction. As a result, a growing concern for back injury in building construction has arisen, stemming from both the loss of productivity and the high cost of medical bills associated with the treatment of back injuries. The problem, thus, is how to decrease the incidents of back injury for the average worker in the construction industry while increasing productivity.

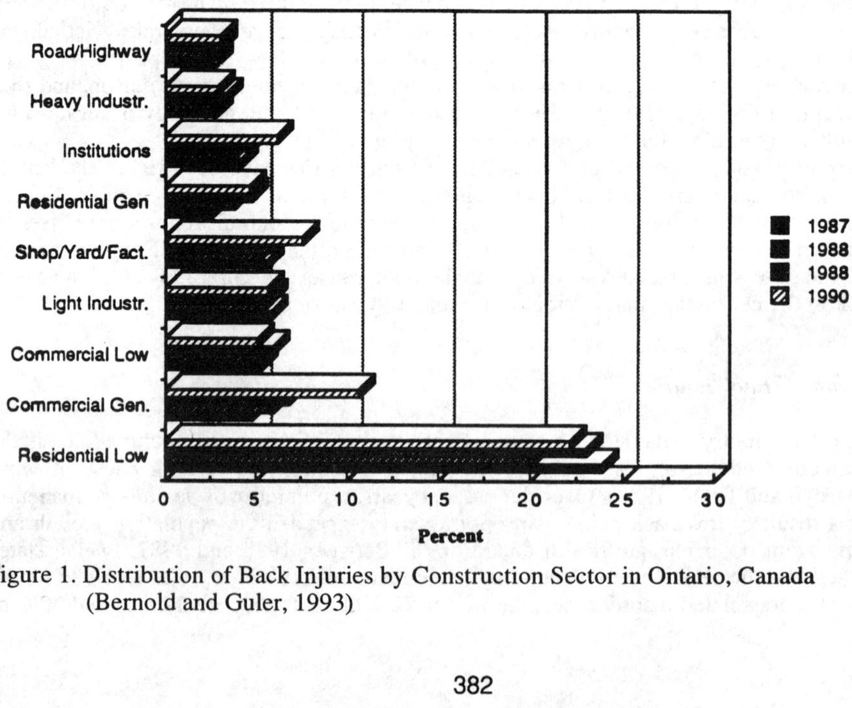

Figure 1. Distribution of Back Injuries by Construction Sector in Ontario, Canada (Bernold and Guler, 1993)

2 PROBLEM IDENTIFICATION AND ASSESSMENT

There exist several approaches to midigate the risks of back injuries on the job. One of the most promising are technological interventions that involve introducing ergonomics by redesigning existing workplaces, jobs, and equipment to fit the worker.

2.1 *Defining the Dangerous Tasks*

Three cooperating partners have joined forces to conquer this issue. Duke University Medical Center is responsible for conducting a symptom survey of a large sample of workers consisting exclusively of masons, carpenters, and roofers. The second partner, North Carolina State University's Industrial Engineering Department (Ergonomics Lab), is currently conducting extensive ergonomic analysis of various worker tasks (Mirka et. al., 1998). Biomechanical back stress associated with these specific tasks will also be assessed using the Lumbar Motion Monitor (LMM). The Construction Automation and Robotics Laboratory (CARL) at NC State University is the third participant in this research endeavor.

Construction workers are constantly exposed to a variety of differing job situations and must modify their behavior in accordance with the ever-changing circumstances. No two jobsites are ever exactly alike and even the same jobsite can change on a daily basis. Thus, it is essential to monitor jobsite conditions in order to identify "hazardous" tasks. First, a motion analysis of carpentry work on two residential construction sites was conducted; fifteen experienced workers from framing subcontractors were the chosen subjects involved in the study. Data was collected via continuous video analysis for the period of several workdays. After the videotaping was completed, each tape was analyzed with a video coding system. Using this approach, it was possible to continuously describe the biomechanical stress on each worker throughout the workday (Mirka, 1998).

Using this method of analyzing biomechanical stress, it becomes clearly apparent that certain activities are indeed back laboring and stress inducing. From the entire list of observed activities, six tasks were identified by ergonomic analysis as the greatest contributors to low back injury for carpentry workers. These tasks are:

1. Lifting raw materials from ground level
2. Working with power/air tools on ground level
3. Manually lifting trusses to the roof level
4. Carrying plywood flooring
5. Raising walls
6. Moving materials to the second floor

2.2 *Task Analysis*

The task of working with power/air tools on ground level was chosen for further analysis. In particular, the task of fastening plywood flooring to floor joists using a pneumatic nail gun, otherwise referred to as the subflooring task, became the focus of this research endeavor.

The subflooring task cycle basically involves three workers and tracks subflooring plywood from its initial location in the stockpile to its final placed position on the floor of the house being constructed. The first worker fetches the plywood sheet from the stockpile and transports it to his sawhorse cutting location. Here, Worker #1 measures the plywood and cuts it to its necessary size. After this task is complete, Worker #1 picks up the cut plywood and carries it to the construction area. More often than not, Worker #1 completes his task at ground elevation and the plywood is transferred to Worker #2, who is on the building foundation. Worker #2 bends over and picks up the sheet of subflooring plywood. He then carries the plywood to the location that it is needed and places it horizontally on top of the floor joists. Using a rubber mallet, Worker #2 guides the plywood into position by hitting its edges. His job with this section of subflooring is now complete and Worker #3's job begins. The plywood

Figure 2. Laboratory Test with Lumbar Motion Monitor

must now be fastened to the floor joists. This task is generally done with the application of nails by a pneumatic nail gun. Worker #3 must bend over at the waist and apply numerous nails into the sheet of plywood, taking care to position his nail gun in such a manner as to attach the plywood to the underlying floor joists. The subflooring cycle is now complete.

2.3 *Measurements of Stresses on Back*

For the purpose assessing the stress that the nailing task exerted on the back of a human a Lumbar Motion Monitor (LMM) system was used. It is data collection device that allows objective and quantitative measurements of the lumbar range of motion, velocity, and acceleration while a human is performing specific tasks. The LMM is an exoskeleton of the spine that has been instrumented with sensors and fitted to the subject as shown in Figure 2. The outputs of the sensors are transmitted through an umbilical cable to the A/D board in a personal computer. The computer then calculates the instantaneous position, velocity, and acceleration of the lumbar spine during the engaged activity.

Three subjects were chosen to perform the task of nailing plywood sheets to floor joists in a controlled laboratory setting. All subjects were familiar with the use of the pneumatic nail gun and all were of approximately mean stature for the male population. In this fashion, each worker nailed 2 sheets of plywood to the floor joists; each sheet required 5 rows of 4 nails, totaling 20 nails per sheet of plywood. Table 1 present the results of the averages of all the conducted tests.

Table 1: Results of Lumbar Motion Monitor for Traditional Nailing

	LP	LV	LA	SP	SV	SA	TP	TV	TA
Trad. Nailing	-4.9	5.5	26.7	66.9	6.4	28.8	-9.7	2.4	14.0

(L = lateral; S = sagittal; T = transverse; P = position; V = velocity; A = acceleration) (Position measured in degrees; Velocity measured in degrees per second; Acceleration measured in degrees per second^2)

As expected, the largest motions are measured in the sagittal position (SP), which is the forward bending of the trunk. With 66.9 degrees, it represents the quantitative measure of the large bending motion, while the -9.4 degrees of TP represents the much smaller rotation of the trunk during bending. This data demonstrates that nailing with one hand is not just a simple bending operation, Due to the fact that the worker must extend one hand further away from his body then the other, the upper body has to rotate above the hip. As a result, the total stress on the spine is the product adding bending- and torque-stresses together.

Since the small weight of the nailgun itself does not result in large stresses from nailing one nail, the cumulative or repetitive motion of nailing is significant. In fact, the video analysis of the site operations showed that the nailing of the subfloor will take a significant amount of time, followed by the assembly and nailing of wall frames, done on the finished floor. The extended duration of operating the nailgun in a bend-over posture was the main reason why this operation was identified as one of the most "hazardous" tasks.

After the nailgun operation on the ground level was identified and its impact on the spine studied, an technological intervention was thought.

3 DEVELOPMENT OF A TECHNOLOGICAL INTERVENTION

In order to reduce the risk of getting a back injury duc to operating a nailgun, a variety of approaches can and have been taken. They include, exercise, setting material on work horses, etc. Due to a variety of reasons, the research team decided to develop a technological solution to the problem by redesigning the nailgun in such a way that it could be operated from a standing position. The resulting device was called the Ergonomic Nailing System (ENS).

The ENS is an attachment that serves as an extension device that is able to handle any commercial pneumatic nail gun. The ENS is attached to the nail gun with two Velcro straps on the handle and two adjustable braces. Key features which make the system unique are its:

- Ergonomically correct design
- Lightweight aluminum structure
- Height and angle adjustment capability
- Trigger operation at waist level
- Adaptability to any pneumatic nail gun
- Mobility with the use of multidirectional, height adjustable casters
- Safe operation by utilizing the safety mechanisms of the nail gun itself
- Quick and efficient method of attaching and detaching nail gun.

The Ergonomic Nailing System, as pictured in Figures 3 and 4, has a lightweight aluminum tubing handle that is both height and angle adjustable. The trigger is located on the right handle grip and is operated with two or more fingers. A cable connects the finger trigger and the actual trigger housing mechanism. This housing mechanism engages the trigger of the pneumatic nail gun. This special housing mechanism allows only one nail to be shot at a time and is only operational when the nail gun is in the correct position for nailing. The entire device is completely mobile in all directions, with the addition of two multidirectional, height adjustable casters.

In order to disengage the first safety, the worker must push the handle away from his body to depress the front safety at the point of impact. With the front end down, the worker can now squeeze the trigger and discharge a nail. He must, at this point, raise the front of the nail gun by pulling the handle back towards his waist. This can be done while gliding the nail gun to its next position. Because the casters are multidirectional, the worker is not limited to a specific sequence of nailing and can go back and forth with ease.

Figure 3. First Prototype During Laboratory Testing

4 COMPARATIVE EVALUATIONS

In order to assess quantitatively the effect of the ENS, the controlled laboratory tests using the LMM were repeated using the new prototype.

4.1 *Controlled Laboratory Tests*

Again, each participant wore the Lumbar Motion Monitor (LMM) on his back during the laboratory testing period, in order to quantify the amount of bending involved.

The results of the controlled laboratory testing, (Table 2) revealed that the Ergonomic Nailing System drastically reduced the average trunk sagittal position (SP), which is the forward bending of the trunk. Trunk bending was decreased from 66.9 degrees to 12.4 degrees. This reduction resulted from the elimination of forward bending usually involved in the task of nailing. In addition, the trunk transverse position (TP) or the rotation of the trunk, was reduced from –9.4 degrees to –4.4 degrees. Since the worker no longer has to rotate his body while nailing, the rotation of the trunk is greatly reduced. Both of these differences were statistically significant at the $p<0.0001$ level.

After analyzing the videotapes of the controlled laboratory testing and the accompanying LMM data, it was noted that the use of the Ergonomic Nailing System required the subjects to rotate their bodies to the side to view the exact point of impact of the nail gun.

Table 2: Lumbar Motion Monitor Results

	LP	LV	LA	SP	SV	SA	TP	TV	TA
Trad. Nailing	-4.9	5.5	26.7	66.9	6.4	28.8	-9.7	2.4	14.0
ENS	-6.9	6.3	27.1	12.4	6.3	29.7	-4.4	3.0	15.8

(L = lateral; S = sagittal; T = transverse; P = position; V = velocity; A = acceleration)
(Position measured in degrees; Velocity measured in degrees per second; Acceleration measured in degrees per second per second)

Figure 4. Pneumatic Nailing of Subfloor for Manufactured Houses

In order to solve this problem, a mirror mounted on the frame of the ENS was added to the design. This mirror allows workers to view the exact point of impact without rotating their bodies unnecessarily.

4.2 *Comparative Field Tests*

In order to prove that the ENS would actually work in the field, a comparative field experiment was designed. It was decided to focus first on the nailing process during the prefabrication of manufactured houses. Since this process is highly repetitive inside a covered space a true comparison was possible. Figure 4 shows the present nailing of the subfloor.

As depicted in Figure 4, the worker operates the large nailgun by bending over. Each time one nail is "shot", he experiences a recoil effect in his arm and upper body. While the same worker operates the nailgun during the day (e.g., dedicated nailgun operator), he has to interrupt frequently and sometimes for a long time. Some of the reasons include: 1) Refilling the nailgun with a strap of new nails, 2) lack of plywood sheets, 3) adjusting plywood, 4) pushing completed floor to next work station, and 5) breaks during the day. While the nailgun is in use, however, the operator is capable of nailing 2 nails a second. As can also be seen on the picture, the handle of the nailgun is drastically different then the nailguns commonly found in building construction. Before any field test could be conducted it was necessary to modify the ENS. The resulting second prototype, as shown in Figure 5, was eventually used for the test.

The experimental tests, with and without the use of the ENS, were videotaped for later analysis. A quantitative comparison of the two system did not show any reduction in the amount of nails the the operator was able to "shoot". The comparative tests allowed the researchers to make following important observations: 1) The new tool was immediately embraced by the nailgun operator because he wanted to get rid of his back pain, 2) no learning effects could be assessed, 3) the recoil effect on the body was eliminated and immediately recognized by the worker, 4) the selected trigger mechanism required too much movement of the right hand, and 5) wheel assembly created difficulties when the operator had to nail around protruding (pre-installed) conduits.

Figure 5. Second Prototype of the ENS During Field Testing

5 SUMMARY AND CONCLUSION

Occupational back injuries are on the rise and seem to be universal,. Due to its large impact on the life of people and the related cost, it *should* be prevented at all costs. This paper presented one approach, technological interventions, namely a simple and advanced engineering solution to the operation of nailguns in building construction. After an extensive period of site observation and analysis, the process of nailing sub-floors was identified as a high-risk operation. Based on the nature of the work, the Ergonomic Nailing System (ENS) was designed, prototyped, and tested in the field. The first comparative tests showed that the proposed simple system works well in the field of manufactured housing, is immediately embraced by workers that suffer from constant bending during nailing, and does not result in loss of productivity. Laboratory tests also showed the drastic change of this innovative system on the sagittal position of the human, who is not required to bend anymore.

6 ACKNOWLEDGEMENTS

This publication was supported by Grant Number 5R01/CCR413061-02 from the National Institute of Occupational Safety and Health. Its contents are solely the responsibility of the authors and do not necessarily represent the official view of the National Institute for Occupational Safety and Health.

7 REFERENCES

Bernold, L.E. and Guler, N.1993. Analysis of Back Injuries in Construction. ASCE, *J. Construction Engineering and Management*, Vol 119, No. 3, Sep., pp. 607-621.

Deyo, R. and Diehl, A. 1988. Psychosocial Predictors of Disability in Patients with Low Back Pain. *J. of Rheumatology*, 15, pp. 1557-1564.

Federal Register 1992. Department of Labor, Part V. Occupational Safety and Health Administration, *Ergonomic Safety and Health Management*, Proposed Rule, Aug. 3.

Mirka GA, Kelaher DP, Nay, DT, Lawrence, BM and Reese, JS.1998. Continuous Assessment of Back Stress in Framers in the Home Building, Proceedings of the Annual International Occupational Ergonomics and Safety Conference, Ypsianti MI, June 11-14, pp. 293-296.

Nelson, W.J. 1993. Worker Compensation: coverage, Benefits, and Costs, 1992. *Social Security Bulletin*, Volume 56, 3, pp. 68-74.

Pope, M., Frymoyer, J. and Andersson, G. 1984. *Occupational Low Back Pain*, Praeger Publisher, New York, NY.

Implementation of Safety and Health on Construction Sites, Singh, Hinze & Coble (eds)
© 1999 Balkema, Rotterdam, ISBN 90 5809 036 1

Ergonomics, economics and operating engineers

C.L.Zimmermann & T.M.Cook
Injury Prevention Research Center, The University of Iowa, Iowa City, Iowa, USA

ABSTRACT: The following paper provides a case study evaluation of an earthmoving and underground utilities contractor. This contractor has implemented numerous changes to both equipment and facilities. Facilities changes include the placement of overhead lifts in each mechanic's bay (decreased lifting) and the installation of two-hand control switches for the overhead doors (decreased potential for pinched or severed fingers). Modifications to existing equipment include additional steps and grab bars (reduced potential for slips and falls during ingress and egress). Cable controls on scrapers have been converted to hydraulic controls, resulting in decreased demands on operators, and increased load capacity and work availability. Equipment maintenance points have also been relocated; preventing slips and falls by decreasing climbing on equipment. Most of these interventions cost less than $100. Yet, it is estimated that these interventions have resulted in reduced workers' compensation claims and a four-year premium savings of over $380,00.

1 INTRODUCTION

Recently there has been an increased awareness of ergonomics and its beneficial role in reducing the effects of workplace factors on musculoskeletal disorders. In the summer of 1997, NIOSH published a critical review of the epidemiological evidence associating workplace factors with musculoskeletal disorders. (NIOSH 1997). This document, along with the success of ergonomic interventions in the manufacturing sector, has provided a basis for the current renewed interest an Ergonomics Standard. NIOSH's review has identified strong evidence to support a relationship between working posture and an increased risk of neck and neck/shoulder disorders and strong relationships between combinations of repetition, force and posture and increased risks for elbow disorders, carpal tunnel syndrome and tendinitis. More pertinent to the present study, NIOSH has identified strong evidence to support an increased risk for low back pain associated with exposure to lifting/forceful movement and whole-body vibration.

Numerous examples exist in the manufacturing sector where ergonomic intervention has lead to greater productivity (Hendrick, 1996; GAO, 1997), decreased injury rates (CTD News, 1995, Hendrick, 1996; GAO 1997), decreased workers' compensation costs (Hendrick, 1996; GAO 1997), decreased maintenance costs (Hendrick, 1996) and increased profits through a reduction in exposure to the risk factors of force, posture, repetition and vibration. However, the acceptance of ergonomics' beneficial role in the construction industry is moving at a very slow pace.

When discussing ergonomic interventions with members of the construction community, many contractors would agree that enhanced worker safety and ergonomic improvements would be beneficial. However, the major objection to the implementation of ergonomics as part of standard work practice is usually related to perceived initial costs (the cost of new equipment, costs associated with equipment modification or the cost of additional equipment).

Unfortunately, the costs associated with lost work time injuries (workers' compensation premiums, medical expenses, lost productivity etc.) and decreased productivity (compared to ergonomically improved work environments or methods) are seldom considered. The following is a case study of ergonomic and safety modifications and their associated benefits experienced by a contractor in the earth moving and underground utilities trades.

2 MODIFICATIONS

In an interest to improve productivity and reduce the physical demands and risks of injuries placed on equipment operators and mechanics, this contractor has made significant modifications to his dozer / scraper combinations. Initially the scrapers were designed with cables to manipulate the cutter and load. Operation of the cable control levers required awkward postures of the shoulders, upper back and neck and high force demands on the operator when emptying the scraper. It also involved high force and time demands on the mechanics when providing maintenance (replacing the springs, cables and rollers in the scraper).

Modifications to these dozer / scraper combinations have included additional weights to the dozer (increasing the contact area between the tracks and earth) and replacement of the cable and spring mechanisms with a hydraulic system. Since the dozers were originally designed with a blade on the front, the owner has attached additional grab bars and steps to the equipment (allowing operators to maintain three points of contact during ingress and egress). Lubrication points on the equipment have also been moved, allowing maintenance to occur from the ground rather than from the top of the scraper where it previously was performed.

In the shop area, overhead door switches (originally single button) have been replaced with double button switches that require the use of both hands. There are also steps, fabricated in-house, which allow the mechanics to climb onto and off the equipment while carrying small items. A rather unique in-house modification is the development of a device to remove the connecting pin from excavator tracks. This system involves the use of a ram to drive the connecting or link pin out. However, the most significant modification in the shop area is the placement of overhead lifts in each of the mechanic's work bays.

3 HEALTH AND SAFETY BENEFITS

These modifications have resulted in numerous benefits. In terms of the dozer / scraper modifications, the following benefits have been observed.

1) The addition of weights to the tractor provides a smoother ride for the operators. This is reported to be due to a lack of porpoising and an associated reduction in exposure to shocks.

2) The addition of grab bars and steps on the cabs allows the operators to climb on and off the equipment while maintaining three points of contact at all times. These three points being either one hand and both feet or one foot and both hands. This results in a decreased risk of slips and falls.

3) The movement of lubrication points from the top of the scrapers to the sides reduces the risk of slips and falls. In addition, this change provides a decrease in maintenance time and increased ease of maintenance.

4) The conversion to hydraulics has probably been the most significant modification. In addition to replacing the cables, the controls for the scraper (which were originally two handles located to the side and rear of the operator) have been replaced with a single pistol grip control. This modification has significantly reduced the postural requirements for operation of the scraper. Instead of working with the arms elevated to or slightly below shoulder level, the operator now controls the scraper with one hand and the shoulder only slightly abducted. This significantly reduces the muscular effort required to maintain the operator's arm posture. It also has a significant impact on the force required to operate the controls. The original controls required considerable force when the scraper was being emptied. Now this force is provided by the dozer's hydraulic system.

In the shop area, the most notable change occurred following an accident. One of the employees was lowering an overhead door with one hand on the switch and the other on the door's tracks. The door was lowered and pinched the employee's finger causing a laceration. Within approximately two weeks following this incident, the owner had a double button system installed to control all overhead doors. This two button system is designed such that neither of the operator's hands can be in proximity of the door's tracks while the door is being raised or lowered. This has eliminated the possibility of a similar occurrence.

Additionally in the shop area, steps have been fabricated which allow the mechanics to climb from floor level to the level of the equipment's tracks. The use of these steps eliminates the need to climb on and off the equipment using grab bars and frees up the mechanics hands to carry tools or small parts.

A rather unique tool in the shop was designed in-house to assist in the removal of excavator tracks (both in the shop and field). This tool allows the use of a lifting device and a hydraulic ram to remove the link pin from excavator tracks after they have slipped off the rollers. This system replaces the use of a sledgehammer, punch, and two mechanics. It provides increased safety through a reduction in the potential for struck by injuries and decrease maintenance time and effort.

As part of the design of the shop, overhead hoists were installed in each of the mechanics bays. This has greatly reduced the requirements for heavy lifting and forceful exertions. Part of the evidence for the success of this intervention is the lack of any OSHA recordable incidents in the shop over the past year.

4. PRODUCTIVITY BENEFITS

In terms of productivity, the greatest changes are evident in the conversion from cable to hydraulic controls. The owner estimates that one additional load can be moved per hour with the converted scrapers. In addition, each of the loads carried by the converted scrapers is approximately 100 yards greater than that carried by a non-converted scraper. The conversion also allows the dozer / scraper combinations to be used in wet conditions where conventional cable scrapers would require such great force to empty that it is not economically feasible to work.

The placement of overhead lifts in each of the shop's bays has also had a significant impact on productivity. Having these lifts allows the mechanics to continue to work at all times. There is no need to wait because a lift in unavailable. With the conversion to hydraulics and in-house shop (with lifts), the contractor reports a 91-93% equipment availability status. This allows the company to obtain return on equipment investments nearly all of the time resulting in quicker pay-off and increased profit over the equipment's life cycle.

5. WORKERS COMPENSATION BENEFITS

A quantifiable outcome of the health and safety benefits mentioned in section three is the effect that these modifications have on injury rates and worker's compensation premiums. Based on estimated rates of claims (calculated using NCCI data and payroll information) this company has experienced a nearly 60% reduction in the number of workers' compensation claims when compared to state averages. (Table 1)

Table 1. Workers' compensation history

Workers' Compensation Job Class Code	Claims/ $100,000	Payroll $	Estimated Claims	Actual Claims
5606	0.25	2,229,039	6	*
6217	0.523	27,110,885	142	*
6306	0.5698	5,808,185	33	*
8227	0.326	4,281,588	14	*
Total			194	83

*Data unavailable

Table 2. Workers' compensation premium savings

Year	Experience Modifier	Standard Compensation Premiums ($)	Actual Compensation Premiums ($)	Compensation Premiums Savings ($)
1993	1.00	327,157.95	327,157.95	0.00
1994	0.95	463,153.47	439,995.80	23,157.67
1995	0.70	442,240.43	309,568.30	132,672.13
1996	0.68	503,300.68	342,244.46	161,056.22
19971	0.84	418,060.62	351,170.92	66,889.70
Total Savings				383,775.72

This reduction in claims has had a significant impact on the company's workers compensation premiums. Compared to the standard rating for the state, this company has saved over $380,000 in workers' compensation premiums over the past four years. (Table 2)

6. SUMMARY AND CONCLUSIONS

The previous sections have discussed modifications to equipment, work areas and work practices implemented by an earthmoving and underground utilities contractor. As evidenced by the reduction in injuries and savings in workers' compensation insurance premiums, the changes have had a significant impact on the corporate bottom line. In addition to the changes discussed there are some intangible factors associated with this corporation. Many would consider the management style participatory. The owner, when on job site and shop visits, frequently asks his employees what they might do to make their jobs better in terms of productivity, efficiency or decreased physical exertion. This has fostered a commitment on the employee's part to look for ways to improve their work environment. It may also play a part in the decreased number of workers' compensation claims filed over the last five years. The second intangible is the contractor's commitment to safety and health. "No matter how you look at it, safety pays." These intangibles along with the modifications discussed result in significant cost savings and increases in productivity. Taking into account the NIOSH document on workplace factors and musculoskeletal disorders and the results provided above, it is hard to argue that the implementation of ergonomics practices can't have a positive effect in the construction industry.

REFERENCES

Center for Workplace Health Information, 1995. An ergo process that runs like a Deere. CTD News 8, August, 6-10.

General Accounting Office (GAO) 1997. Worker Protection: Private Sector Ergonomic Programs Yield Positive Results. GAO/HEHS-97-163.

Hendrick, H.W. 1996. The ergonomics of economics is the economics of ergonomics. *Proceedings of the Human Factors and Ergonomics Society 40th Annual Meeting.* Philadelphia, PA: Human Factors and Ergonomics Society, 1-10

National Institute for Occupational Safety and Health (NIOSH) 1997. *Musculoskeletal disorders and workplace factors: A critical review of epidemiologic evidence for work-related musculoskeletal disorders of the neck, upper extremity, and low back.* Cincinnati, OH: U.S. Department of Health and Human Services, Public Health Service, Centers for Disease Control and Prevention, National Institute for Occupational Safety and Health, *DHHS (NIOSH) Publication No. 97-141.*

12 Safety assurance

Implementation of Safety and Health on Construction Sites, Singh, Hinze & Coble (eds)
© 1999 Balkema, Rotterdam, ISBN 90 5809 036 1

Integrating construction worker protection into project design

Jimmie Hinze, Richard J.Coble & Brent R.Elliott
M.E.Rinker, Sr. School of Building Construction, University of Florida, Gainesville, Fla., USA

ABSTRACT: This paper discusses how construction worker protection can be improved by addressing safety concerns during the design process. Historically, the constructor has been the party responsible for worker safety and health. Owner and designer responsibilities have been recognized in recent years, but often not until a catastrophe has occurred, in which case the courts are usually left to determine liabilities of the various parties involved in a project. Designers play a key role in improving worker protection, and this paper provides direction regarding what the designer's role is and how it can best be developed. Specific suggestion are given for incorporating safety into the design process. Since design choices often determine which construction methods will be used, it is crucial that such choices are made with worker safety and health in mind. In order to make design decisions that will improve worker safety, designers need to be educated to recognize how their design will impact construction worker safety. If the designer does not have such education or knowledge, the design team should include individuals with knowledge of construction safety. In addition, a design component needs to be developed which specifically addresses construction worker protection on each project. Owners are in an ideal contractual position to establish the priority of worker safety and health on their projects, and such priority needs to be upheld during the design phase. Owners also benefit from safe projects, for although contractors bear initial cost when workers are injured, owners ultimately bear the cost associated with injuries since it is part of the cost of construction.

1 INTRODUCTION

The delivery of a completed facility occurs at the culmination of several phases of work effort. These efforts can be broadly categorized as taking these phases from design through construction. It is generally agreed that successful projects are those which are delivered on time, within budget, and in conformance with the requisite specifications. Traditionally, the issue of safety is focused on only after the facility is completed. That is, the completed facility should be safe for use by the facility user. This interpretation is too narrow as it does not consider the issues of safety as it relates to the construction workers. The construction workers are also facility users, albeit temporal ones. This makes construction workers a unique group that warrants particular attention in the design phase. This responsibility has historically been relegated to the construction parties, but the designer must also address this issue. The issue to be addressed then is the means by which this responsibility can become an integrated component of the designer's work effort. To accomplish this end, the U.K.'s construction, design, and management (CDM) regulation is viewed as a model for U.S. application.

2 OWNER'S COMMITMENT TO CONSTRUCTION WORKER PROTECTION

Based on a purely humanitarian principle, no one would want any construction workers to be injured on the job. Unfortunately, this is not a sufficiently strong principle to make safety a top priority in all firms. However, it is ultimately the owners who bear the costs associated with injuries. The construction firms do incur the initial cost, but this cost is eventually incorporated in the cost of construction. With this recognition, it would stand to reason that owners should play a leadership role in construction safety, which starts in the design and continues during construction.

Historically, owners have generally not been active in the area of safety, especially during the design phase. Owners are, of course, in a prime position to dictate contractually established priorities for safety. Owners can begin by insisting that architects and engineers address or consider safety in their design decisions. While this mandate might be made in the contract, it must be recognized that added compensation would appear to be appropriate to address the added work effort. Compensation commensurate with the effort requested is imperative if safety is to be addressed fully.

Design professionals on the owner's team typically try to distance themselves from any responsibility for the safety of construction workers. Today's design codes and standards reflect this attitude, and consequently construction worker safety typically rests only on the constructor's shoulders. Therefore, it is very important that designers include construction worker protection as a design component. It is important that the designers realize that the safety performance on a project can be dictated to a large extent by decisions made by the designer during engineering or design.

3 CHANGES IN THE DESIGN TEAM

Although requiring and paying for the designer to address construction safety in the design establishes the foundation for change, the issue of knowledgeable input in the design process is not assured. Since designers have historically not addressed construction safety in their designs, it is not a topic on which they have a great deal of expertise. This knowledge must be acquired. There are several steps or phases by which this can occur.

Perhaps the simplest way to have knowledgeable input on safety in the design is to include in the design team an individual or individuals with knowledge about construction safety. These team members might be contractors, consultants, or other parties possessing the requisite skills in safety. These persons will function similar to consultants who regularly conduct constructability reviews during the design phase. These team members could be employed as consultants or they could even be hired by the designer.

The consultant's selection will be important. This suggestion must be addressed seriously by the entire design team. The practical approach would be to conduct a thorough risk assessment of each component of the facility being designed. This would include the risks associated with different design approaches. The relation risks to construction when safety associated with each alteration approach will then provide valuable information with which the appropriate design can be made.

4 THE CONSTRUCTION (DESIGN AND MANAGEMENT) REGULATIONS 1994

For worker protection to become an active part of project design, the designer must assume a new role. The Construction (Design & Management) Regulations used in the United Kingdom are an excellent model to accomplish this end. Known as CDM, these Regulations emphasize the fact that "the designer's duties are essentially to ensure that health and safety is given proper consideration from the outset" (Williams, 1995).

The CDM Regulations as they relate to designers, as detailed in Regulation 13, requires the following four areas of emphasis:

1) Designers should take reasonable steps to alert clients to the existence of the regulations.

2) Designers should ensure, so far as is reasonably practical, that any design either avoids foreseeable risks or includes control measures for reducing that risk at its source in a way which will protect everyone subsequently carrying out the work.

3) Pass on information about aspects of the structural design and materials which subsequently might be important to the health and safety of any person carrying out the work.

4) Cooperate with the planning supervisor and other designers in carrying out their function, which is a simple requirement to make the regulation work" (Williams, 1995).

All of the above requirements for the project designers are conditions which make for safer working conditions for construction workers.

5 DESIGNING FOR CONSTRUCTION WORKER SAFETY

The main objective in designing for safety is to improve worker safety and health. Accomplishing this, however, requires that designers' awareness of construction safety be increased in order to compel them to address construction safety in their design. Designers can become more aware of construction safety through continuing education, considering their previous formal education was void of such education. The safety education program for designers should include activities that are basic to the design and construction process, proper handling of materials, education on hazardous substances and safety specification writing.

Designers need to incorporate a component for protection of construction workers into their designs. They need to identify possible design decisions that will ensure construction worker protection and eliminate, or at least minimize, possible hazards to workers on site.

5.1 *Suggestions for Incorporating Safety into Design*

The following suggestions should help designers to incorporate construction worker safety into their designs:

1. Use of materials:
Design engineers need to understand not only the safe design limits of a single material, but the physical interactions involved in the combined use of multiple materials, and make allowance for them. They need to understand the process and sequence of their construction, and design for safe construction as one follows the other. Designers need to identify hazardous materials and dangerous conditions and either eliminate or reduce the hazards to workers.

2. Working Areas:
One of the most important design suggestions to improve construction worker safety is to have design solutions that will provide an uncongested work site, and distance the workers from moving traffic. Design permanent stairs and walkways to be constructed early in the construction phase so that temporary structure use is minimized. When used, temporary structures should be designed for construction worker safety.

3. Slip and Fall Hazards:
The design should reduce worker exposure to falls from elevation and being struck by falling objects. Reduce slip hazards due to snow and ice and prevent fall hazards by simplifying the work area for the construction workers.

4. Cost Effective Safety Design:
Integration of prefabricated and precut components into the design result in reduced worker exposure on site. Specific safety measures can be implemented into the design as well, such as predrilling for guardrail holes.

5. Determination of risk to construction workers:
Design engineers need to recognize that safety does not fall into the hands of the general contractor, and that risk assessment needs to begin with the design of the structure and its members. These risks to construction workers and the criteria on which they are based must be made known to and understood by the owner and contractors.

5.2 *Requirements of the Design Component*

When developing a design component for construction worker protection, the design team should be made responsible for coordination of recognized construction methods so that unnecessary hazards are not created. Design choices often determine construction methods, which directly impact worker protection, so the design engineer should understand construction. One of the most important requirements of the design component to protect construction workers should be a system review. System review is a vital process that ensures that the design creates no hazards to the construction worker during the construction phase.

The following guidelines should help the design engineer in developing a design component for construction worker protection.

1. The design component should define pertinent reliable safety requirements, including applicable consensus standards of the American National Standards Institute (ANSI), the Occupational Safety and Health Administration (OSHA), and other applicable standards such as the US Army Corps of Engineers' Safety and Health Requirements Manual.

2. The design component needs to list potential high-risk hazards, such as falls from elevations, excavations, powerlines, underground utilities, soil conditions, toxic materials, wastes and/or contaminants, fumes, fire, common work areas and any other hazards inherent to the construction worker on the project.

3. It should list design concepts governing the construction methods to be used, such as tilt-up, lift-slab, prefab concrete beams, decks, etcetera, so that special conditions to ensure safe construction methods can be included in contract specifications.

4. The design component should require that the entire design team receive appropriate hazard prevention training.

The design component for protecting construction workers should be used as a design tool to enhance the safety of the worker during the construction phase by providing design solutions for a safe and hazard free working area for the workers. The following suggestion should be included in the design component for construction worker safety:

1. Establish procedures for identifying possible hazards that could be created by the designs and establish hazard prevention issues that would concern worker safety.

2. It should provide information on the techniques, sequences or procedures that are to be employed by the constructors. Any structure that will have significant construction loads imposed on it should be designed for stated construction loads, and in some cases the construction process should be outlined.

3. It should consider designing structures in such a fashion that permanent stairways can be completed early in the construction stage, permitting these stairs to be used by the construction workers.

4. The design tool should ensure that controls, valves, switches, cutoffs, etcetera, are placed where accessibility is assured. These locations should be clearly noted on the drawings.

5. Schedule the underground firewater system to be constructed early in the construction phase. Similarly, in high-rise construction, schedule the firewater protection system to be operational early in the construction phase. Permanent emergency exit signs should be required to be erected as early as possible.

6. It should be specified that it is necessary to design steel structures so that construction guard rails can be fabricated integrally with the structure. When the steel structure is erected, the guardrails will not have to be erected as a separate component, further reducing fall hazards.

7. The design component should require concrete test results to be obtained prior to removing forms and shoring on selected structural members. Where concrete placement loads impact the design, specify the procedure for placing the concrete. Similarly, in piping systems and structural steel systems, perform weld tests at the critical locations.

8. The design component should consider maintaining a constant depth in the excavated portions of the projects. To the extent possible, facilitate minimizing the amount of time excavations are open. Keep all below grade work as simple as possible. Allow adequate clearances to accommodate shoring, forms and workers in the excavation.

9. The entire design should reflect a sensitivity to the need to reduce falls resulting from irregular surfaces, steep slopes on ramps and roofs, and slippery walking surfaces.

10. It should emphasize on designing trenches at minimum feasible depths and detail methods to be used in trenches. Avoid high fill and embankment heights. Design trench requirements to match shoring requirements.

11. The design component should necessitate substituting hazardous materials with less hazardous materials for cleaning, paints and castings. Specify natural gas to be purged from manholes.

12. It should use CAD programs to plan safe work sequencing and to identify and mediate hazards before construction begins. Conduct a hazard assessment for construction during design.

13. To improve construction worker protection the design component should indicate designing higher parapet walls or an integrated guardrail system along all roof edges.

14. It should emphasize laying out the piping and mechanical systems so workers will have more than one exit, thus avoiding the need of passing through hazardous areas to escape.

15. The design component should ensure that designs avoid overhead powerlines at the work site by relocating them (i.e., underground or rerouting), by altering the point of access to the work area, or by installing barrier cables.

5.3 *The Need for Education and Training*

Incorporation of construction safety concerns in the design effort requires the application of knowledge that has not historically existed in the design teams. While the use of consultants does provide a means of acquiring such knowledge rapidly, this is not the ideal mechanism. Ideally it should be an inherent consideration of every designer to address construction safety issues in all design decisions. To date, most designers do not have the academic background or requisite understanding of construction safety to adequately address worker protection.

Higher education, at both the college and university levels, has a tremendous challenge to take deliberate steps to include construction safety training in their design curriculum in architectural and engineering degree programs. The challenge lies in the fact that the professors in these design disciplines do not generally have the necessary knowledge to effectively satisfy this need. Construction professors at most academic institutions are housed in either the college of architecture or the college of engineering. Unfortunately, these programs have historically had considerable lack of safety education in the design phase. This must change if designers are to have increased sensitivity to construction safety. Of course, one problem is that in some academic programs of engineering, architecture, and construction management, construction safety is not even taught as a part of the curriculum.

Thus, in some programs the need to incorporate construction safety in the construction curriculum will be the first objective. In others, the need will be to devise a mechanism of establishing a healthy exchange of ideas between the designers, engineers, and constructors, since some designers are very reluctant to have their creativity compromised by the added constraints imposed by the need to address construction safety. These concerns should be alleviated once they realize or accept the fact that construction safety is an obligation of designers. This obligation is both moral and legal.

There is at least one aid that has been developed which can enlighten designers on how to address construction safety in the design phase. This is a software program entitled "Design for Safety Toolbox," which was the result of an extensive research effort sponsored by the Construction Industry Institute (Haas & Hinze, 1996). This program takes general information about the project that is input and then offers a variety of choices of how a potentially unsafe condition can be addressed.

Once designers have begun to address construction safety in their designs, whether though consultants or other design assistance, they will hopefully become more proficient in construction safety. Ideally, addressing safety in the design phase will be an ingrained ability. It should be a natural matter for the designer to address construction safety, but this can only occur by consciously doing so on a series of project designs.

6 CONCLUSION

It is a foregone conclusion that we need to seek a means for sensitizing designers to the need to provide for construction worker safety as a design component. Safety is not an issue which can simply be left to the constructor. It would be naive to suggest that designers have no role in construction safety; decisions of the designers have a direct impact on construction worker safety. One of the means by which designers can become more responsive to the safety of construction workers is through education. Designers must be aware of the various means by which their design decisions impact the job-site safety conditions for construction workers. Such an educational process can begin with creating a design tool which will successfully address construction worker safety. Once the designers are educated in design decisions that could impact construction worker safety, they can successfully include construction worker protection as design component.

The issue of reduced cost related to designing for safety needs to be emphasized. Cost saving, which is often a design motivator, is also an issue that can involve safety. Using better means and methods can start in the design process, which may include the use of prefabrication in a shop or job site situation.

By reviewing the suggestions in "Design for Safety Toolbox," designers can gain some insights about how they can address construction worker safety in their designs. It is not sufficient to simply accept these design suggestions with the intent of utilizing them when the need arises. The designer must begin to think about construction worker safety whenever design decisions are made. Designers must make an effort to address construction worker safety in all aspects of the project design. Many design suggestions offered in this paper may appear simple or even matters of common sense. These suggestions are designed only as a guide in making further design decisions. Once the designer considers the design examples that were provided, the designer will hopefully attempt to generate additional ideas that are unique to a given project.

REFERENCES

Churcher, D.W. and Alwani-Star, G.M. (1996). Incorporating Construction Health and Safety into the Design Process, *International Council for Building Research Studies and Documentation (CIB)*, Proceedings of Implementation of Safety and Health on Construction Sites. Alves Dias & Coble (Eds.), Balkema, Rotterdam, pp. 29-39.

Haas, Carl T. and Hinze, Jimmie (October, 1996). Design for Safety Toolbox (IR101-2S), *Construction Industry Institute (CII)*, Austin, Texas.

Hinze, Jimmie and Gambatese, John A. (February, 1996). Design Decisions that Impact Construction Worker Safety, *International Council for Building Research Studies and Documentation (CIB)*, W99 Report, Publication 187, pp. 219-231.

Hinze, Jimmie and Weigand, F. (1992). The Role of Designers in Construction Workers Safety, *Journal of Construction Engineering & Management, American Society of Safety Engineers (ASCE).*

Williams, Stephen (March, 1995). Construction Safety (The Team), *The Safety and Health Practitioner.*

Implementation of Safety and Health on Construction Sites, Singh, Hinze & Coble (eds)
© 1999 Balkema, Rotterdam, ISBN 90 5809 036 1

Owner involvement in construction at a national laboratory

G. Lipka
Sandia National Laboratories Facilities ES&H, Albuquerque, N.Mex., USA

ABSTRACT: In a construction project, the contractor and the owner each have a responsibility for ensuring the health and safety of personnel on a project site. The contractor has the responsibility for ensuring that the provisions of OSHA's safety and health regulations are followed and that the work is conducted in a safe and well thought out manner (Kohn 1996). The owner has a responsibility for disclosing to the contractor those owner-controlled hazards that are present in the work area due to ongoing and past operations (OSHA 1997). With the owner taking an active role in disclosing the potential hazards, the contractor is able to account for, plan, and mitigate potential health and safety issues during the performance phase of the project. At Sandia National Laboratories, this disclosure is made early in the project through the use of processes developed specifically for this purpose.

1 INTRODUCTION

Sandia National Laboratories, New Mexico (SNL/NM) is a multi-program laboratory that undertakes fundamental and applied research activities, with its primary mission being national security with emphasis on nuclear weapon research, development, and engineering. Sandia Corporation, a wholly owned subsidiary of Lockheed-Martin, under contract to the Department of Energy (DOE), operates the laboratory for the DOE. Sandia Corporation oversees approximately $55M in new and renovation construction every year to meet the mission and infrastructure needs of the laboratory. Several buildings serving this mission have been in use for over 40 years, with a long history of varied projects and programs potentially involving the use of hazardous and/or radioactive materials.

In a multi-program environment, test sites, buildings, and individual rooms within buildings may have been used for a variety of projects. Because individual projects supporting the programs may have used hazardous and/or radioactive materials, there is a DOE and corporate driven requirement to assess the potential risks associated with personnel safety and health prior to working in any area (DOE 1995). As a result, it is incumbent upon persons managing construction projects within the laboratory to provide and disclose as much information as possible to the contractors coming onto the site before any construction work is performed.

To assist in meeting this requirement, an Environmental Safety and Health (ES&H) support team, comprised of industrial hygienists, waste disposal specialists, construction and electrical safety, asbestos consultants, and radiation protection personnel, was formed in the early 1990's to address building hazard and construction safety issues. Using the framework provided by the DOE's Integrated Safety Management System (ISMS), processes were developed to assess the potential risks and regulatory requirements before work is conducted in or around any project site. Among these processes, the Building Modification and Hazard Assessment (BMHA) process was created to assess issues related to personnel safety and health in an effort to identify and mitigate potentially hazardous exposures to the contractors coming onto the site. This process, along with a program to ensure contractor compliance in the areas of safety and health during construction, have helped maintain Sandia's construction contractor OSHA recordable rate in construction safety to approximately one half the national average.

2 INTEGRATED SAFETY MANAGEMENT SYSTEM (ISMS)

Fundamental to Sandia's hazard assessment process is the DOE's Integrated Safety Management System. This system provides a framework for management of all laboratory work, including construction on the Sandia site. The basic philosophy of the ISMS process is defined in a set of guiding principles, which ensure that:

- All personnel know and understand their roles, authorities, and accountabilities
- All work is planned with appropriate rigor
- All hazards are identified and controlled through engineered or administrative processes
- The correct standards are identified and applied to safe work practices
- Work is performed by trained, qualified personnel
- Safe work practices are evaluated and improved through timely feedback

The management of work performed under the ISMS process, whether employed early during the conceptual stages of a project or later at the task-specific activity levels, can be categorized into five functional areas of: planning; identification of hazards; controlling hazards; performing work; and feedback. As depicted in Figure 1, it can be seen that this is a continuous process that can be applied at any phase of a project or activity.

2.1 *ISMS at the Project Phase*

Applied to a construction project, the ISMS model provides a simple, methodical roadmap for hazard evaluation and mitigation. Each point on the star triggers processes and decisions initiated by those having responsibility for a specific phase of a project. For instance, at the Planning phase of the star, the project engineer consults with the team of ES&H subject matter experts to decide whether a particular project can proceed or if it requires a full hazard

Figure 1. Integrated Safety Management System

evaluation prior to proceeding with the work. The decision to obtain or not obtain a full hazard evaluation is based primarily on the complexity and location of the project. For instance:

- Is the project located in a new building or an old building?
- Is this project a simple office modification or a renovation in a wet chemistry laboratory with fume hoods and acid drains?
- Will the project disturb any areas with known soil contamination?

Once the decision is made to proceed with a full hazard evaluation, the next point of the star, the Hazard Identification phase, is initiated. The ES&H Support team reviews the project information and conducts surveys to determine if any hazards will be encountered. These hazards include, but are not limited to, chemical hazards from existing or past operations, radiation hazards either in the form of residual contamination or existing sources, asbestos and lead hazards, or underground / hidden utility hazards. Using past documentation, site surveys, instrumentation measurements, and analytical results, the information gathered by the ES&H support team is forwarded back to the project engineer in the form of a report.

Using the report during the Control Hazards phase, the project engineer again has a decision to make. Knowing the hazards which may be encountered, the engineer can design around the hazards, proceed with the design as planned, or, if the identified hazards create a significant problem, stop the project. If the project proceeds as planned, the information gathered during the hazard evaluation phase is accounted for in the design and disclosed in the construction documents. Armed with the hazard information, the contractor assigned the work can then plan for the safety and health of its crew.

It is the functional area of Identifying Hazards and Controlling Hazards where the hazard assessment process takes place within the confines of a construction project. The remaining portions of this paper will concentrate specifically on that process and the integration of the hazard identification results into the contractor's work plan.

3 BUILDING MODIFICATION AND HAZARD ASSESMENTS

The purpose of the Building Modification and Hazard Assessment (BMHA) process is to identify and evaluate environment, safety and health hazards associated with current job-site conditions prior to performing Facilities-managed construction or maintenance activities. When appropriate, recommendations for safe work practices, environmental protection measures, and personal protective equipment are included in the hazard evaluation to provide protection from the pre-existing hazards.

This process evaluates job-site conditions and hazards associated with: chemicals, confined spaces, lead-based paints, shielding, and dust, asbestos, environmental restoration sites, radiological activities, hazardous, radioactive, and mixed wastes, and general construction and maintenance activities. This process is not intended to address activity-specific hazards, such as machine guards, electrical safety or crane safety concerns. Activity-specific hazards are addressed through operating procedures, job safety analysis, or in the contractor's site-specific health and safety plan.

3.1 *ES&H Support Team and Responsibilities*

The persons responsible for conducting the hazard evaluations are subject matter experts in the fields of industrial hygiene, radiation protection, construction safety, asbestos and lead management, and waste management. They are co-located within Sandia's Facilities Engineering Department, providing for easy access by the engineers and designers to consult on the hazards that may be encountered in a project. Information on any health and safety concerns is gathered using existing maps, site surveys, analytical laboratory services, and the process knowledge of the owners of the space where the work is to be conducted.

A BMHA coordinator is the initial point of contact for the engineer. This coordinator provides consultation as to whether a hazard assessment is necessary. In addition, he is responsible for logging and tracking each hazard assessment request from the initiation of the request through completion. BMHA requests, which include a detailed project description and location, are logged into a database and are forward electronically to each of the subject matter experts by the coordinator. Individual reports on specific hazards are completed by the support team members and then compiled into a single summary report by the coordinator. This summary report is then forwarded to the engineer. The individual reports are then kept in a file, which can then be used for reference on future projects.

Presently, there are twenty-six members in the support team, including twelve people who are on-site contractors assigned to implement Sandia's asbestos and lead management program and six persons responsible for Sandia's fire protection program. The team also consists of a Certified Industrial Hygienist, a radiation protection technician, a Certified Safety Professional, an electrical safety expert, and two people with responsibilities for waste management and disposal. The remaining members on the team are comprised of administrative support staff.

3.2 *Determining the need for a BMHA*

Stated previously, the project engineer must decide if the hazard evaluation process is warranted. In some circumstances however, the BMHA process is a mandatory activity. These circumstances include all Facilities projects or work activities performed in environmental restoration sites, when hazardous chemicals are known to be present, intrusion of laboratory exhaust or drainage systems, or construction projects performed in posted radiological areas.

Most renovation projects do not require a full suite of hazard assessments, due to the nature and scope of the work. The project engineer can select specific hazards for review based on his understanding of the work to be performed. In these cases, the coordinator will only route the BMHA request to those subject matter experts indicated. If, however, it is determined that a more extensive review is needed, the project engineer will be contacted and a more extensive review will be done.

3.3 *The BMHA Process*

The process for obtaining a hazard evaluation involves several steps with specific responsibilities by individuals, depending upon where they fit into the overall project scheme.

Specifically, the requestor of the hazard assessment:

- Completes the top portion of the request form with a requestor name, project number and description.
- Identifies the necessary services (i.e. radiation, industrial hygiene, asbestos)
- Provides design details with the request, such as drawings, maps, equipment lists, description of work activity, and any process knowledge that may indicate the type and location of any hazards.
- Forwards the BMHA form to the coordinator.

The BMHA Coordinator:

- Receives the BMHA request form and attachments from the requester.
- Negotiates the schedule for the BMHA.
- Forwards copies of the request to the appropriate ES&H Support team subject matter experts for evaluations as required by the requester and project description.

- Receives the completed evaluations from the ES&H Support team subject matter experts.
- Returns completed BMHA form and other applicable documentation to the evaluation requester.
- Tracks BMHA information in database.

The ES&H Support team subject matter experts conduct evaluations as required, following their evaluation process and forms. The basic hazard identification methodology is as follows:

- Contacts and interview the space/equipment owner, Facilities contact, and the employees familiar with the site and/or equipment.
- Reviews past investigation reports and BMHA's that provide information relevant to the evaluation request.
- Performs an on-site inspection, if required, of the site/equipment to identify concerns associated with the evaluation request.
- Performs sampling and/or monitoring as required to fully characterize or determine the extent of chemical, radiological, asbestos, or other contamination.
- Communicates with other subject matter experts on the team when appropriate to identify and verify absence or presence of other hazard impacts (i.e. radiological concerns on ER sites).
- Documents existing hazards, work practice/engineering controls, and sampling data associated with the project/work site on the BMHA Form.
- Documents any recommended worker PPE and/or additional recommendation guidance, or work practices necessary to safely perform the work on the project/work site.
- Identify regulatory requirements/guidance on the BMHA Form.
- Returns results of the assessments or evaluations with recommendations to the coordinator.

The last step in the BMHA process is the implementation phase, where it is necessary to ensure that the information gathered is then transmitted to the work force. In this phase, the engineer, maintenance planner, work leader, or supervisor completes the project design or maintenance work package, and incorporates the BMHA guidance into the work documents either directly on the drawings or as attachments to the project documents. The construction inspector or construction manager conducts pre-job briefings to review BMHA guidance with the workers as appropriate. In this last step any required training is verified and any required permits are coordinated and obtained.

3.4 *Controlling hazards*

Upon receiving the summary report, the project engineer must make a decision on how best to proceed with the project. The summary report will identify the hazards that will be encountered and contain information on how to control those hazards. For instance, for intrusive work on an exhaust system, the industrial hygiene report may require the use of chemical resistant clothing and a respirator. Work in an environmental restoration site may impose restrictions on excavations and soil removal. Given this information, the engineer can design around the hazards or provide engineering controls within the project documents. Another option is to completely remove the hazard by using specialty contractors prior to performing the general construction work. This is most often seen in asbestos abatement projects, where abatement contractors come in ahead of the scheduled project and remove any asbestos that may be impacted by the general work.

In those cases where the hazards cannot be avoided, the contract documents provided to the contractor will disclose the hazard information. This disclosure also contains the restrictions that are placed on the project.

These restrictions may include:

- The need to obtain Sandia-issued permits, based on a specific task to be performed i.e. hot work, confined space entry.
- Additional oversight by one or more subject matter experts
- Additional worker training specific to a unique hazard expected to be encountered in the work area

The hazard assessment information is included as a contractual requirement necessitating the need for the construction contractor to account for the identified hazards in its site-specific safety plan. This safety plan is a requirement for each project and must be submitted prior to performing any work. This plan must acknowledge that the hazards are understood and include a method for working around the identified hazards. This plan is then transmitted to Sandia in the form of a submittal that is reviewed by the support team to ensure that the contractor's plan is adequate and fully addresses the hazard information provided.

4 SYSTEMS APPROACH TO WORK

Sandia's involvement in construction safety does not end once an authorization to proceed is issued to the contractor. The BMHA is one of several processes within an integrated system of processes that is used to manage construction safety.

Sandia's ES&H Specification for construction provides the contractor with guidelines and procedures for construction activities at the Sandia site. Specific items in the specification include provisions which are more stringent than the OSHA 29CFR1926 Regulatory Guidelines, as well as site-specific guidelines for waste disposal, confined space entry, lockout / tagout, mobile crane use, hot work and torch cutting on lead-based painted surfaces, and electrical safety. For example, a hidden hazard / digging permit process must be completed prior to excavating any soil or penetrating any concrete wall or slab. This permit process seeks to ensure that utilities, which may be buried in these locations, are identified to the best of Sandia's ability, using as-built record drawings and instrumentation. Strict provisions for personal protective equipment use are provided and enforced under this permit process.

Sandia employs a construction inspection team comprised of twelve people who are experienced in the structural, mechanical, and electrical construction trades. In addition to their work experience, Sandia provides additional training to subsets of the inspection team by outside agencies. Competent person training has been provided in the areas of crane inspection and rigging, fall protection, scaffold erection, and confined space entry. Members of the support team having certifications in OSHA's Construction Outreach Program offer monthly in-house training on individual OSHA subparts in a manner consistent with OSHA's 10 and 30 hour voluntary compliance program. This in-house training is also extended to the contractor's employees on a voluntary basis. Training topics are scheduled to coincide with known future activities, with emphasis given to those areas where an increase in accidents or noted deficiencies has been identified.

On a weekly basis the inspection team, along with Sandia's construction safety experts, performs site-wide audits of all active construction projects. Deficiencies found are corrected immediately, if possible, and tracked in a database. The database is used to identify any trends, which may be indicative of a need for future training sessions. The database is also used to monitor contractor performance in construction safety. This information can then be used, in addition to other criteria, for future evaluation and selection of construction contractors.

5 SUMMARY

Sandia's engineering department averages approximately 3600 projects annually. In the first seven months of Fiscal Year 1998 312 BMHA's have been processed with an average turn around time of one week. In the previous year, a total of 373 BMHA's were processed within an average time of eight days. To ensure that cost is not a prohibitive factor in obtaining a hazard evaluation, the cost of the BMHA is borne by the department responsible for managing the assessments and not by the customer.

Both the BMHA and ISMS processes, coupled with the other safety and health related processes, have demonstrated success in managing construction projects at Sandia National Laboratories.

Sandia management believes the cost associated with the program to manage health and safety in construction in this manner is a good business decision. The program seeks to ensure the health and safety of the contractors coming on to the Sandia site, limits liability associated with potential legal actions resulting from accidental exposures and release of hazardous materials to the public, and demonstrates good community relationships. Specifically:

- The BMHA process, as an integral part of ISMS and the construction program, effectively identifies and discloses ES&H hazards present at a construction site in a proactive manner.
- The programs and subsequent processes provide the engineers with tools to engineer out the hazards and provides them with timely information to make sound economical decisions.
- The contractor is provided with information that can be used to reduce exposures to its employees and lost work time for ES&H related issues resulting in greater efficiencies and lower workers compensation insurance premiums.

The resultant decrease in the construction accident / injury rate to one half the national average (BLS 1997) and a zero fatality record for Sandia-managed construction projects demonstrates that Sandia's role in construction safety is a positive one.

REFERENCES

Bureau of Labor Statistics (BLS) 1997. *Table 7. Incidence Rates of Non-fatal Occupational Injuries and Illnesses, by Industry Division, 1993-96*

Department of Energy (DOE) 1995. DOE Order 440.1. *Worker Protection Management for DOE Federal and Contractor Employees.* Attachment 1: 2

Kohn, J.P., M.A. Friend, & C.A. Winterberger 1996. *Fundamentals of Occupation Safety and Health.* Rockville: Government Institutes, Inc.

Occupational Safety and Health Administration (OSHA) 1997. *Code of Federal Regulations Title 29.* Part 1910.120.3(b)

Implementation of Safety and Health on Construction Sites, Singh, Hinze & Coble (eds)
© 1999 Balkema, Rotterdam, ISBN 90 5809 036 1

Construction accidental injury insurance in China

Fang Dongping, Cao Xiaoling & Huang Xinyu
Department of Civil Engineering, Tsinghua University, Beijing, People's Republic of China

ABSTRACT: Based on the analysis of current practices and development tendencies of construction management in China, and referring to foreign methods and experiences, a calculation method dealing with construction accidental injury insurance premiums suited for China is presented in this article. The core of the method is to adjust the insurance premiums of contractors according to their safety performance. The establishment of this system will be an important factor to encourage contractors to improve their safety management. Following analysis of examples, further prospects concerned with this method are also shown in this article.

1 INTRODUCTION

The Construction Law of P.R.China came into force on Mar. 1st 1998. In Article 48 of *the Law*, it is clearly stipulated that contractors must insure for workers involved in dangerous jobs and must pay the insurance premium. It is the first and only stipulation on compulsory insurance in construction regulations and laws in China. Although the industries of both construction and insurance are not yet fully developed in China, the rights and interests of construction workers are well protected with the enactment of *the Law*, the risks taken by the contractors are distributed. Further, *the Law* will lead to significant commercial opportunities for insurance companies in China.

In China, the insurance industry still lacks experience and technical expertise because of its short period of development (Fang 1997, Fang 1998). For example, even in Shanghai, an economic developed region in China, only nearly 30% construction companies insure for the projects they contract, in comparison with nearly 100% in some developed countries. Despite progress in the construction industry, many problems still exist in management, especially in safety management. Without reasonable and appropriate safety management policies and regulations, the objectives of construction safety management cannot be fulfilled. Instead, haphazard safety management policies may simply create more problems. Furthermore, the wrong idea that the insurance companies will take all the risks of accidents may lead to the ignorance of construction safety management by the contractors. For these reasons, in China it is necessary to improve construction safety management by referring experiences in some developed countries, such as advanced technologies and regulated management methods. However, the most essential countermeasure at present is to establish a reasonable insurance system to provide incentive for the improvement of construction safety management of contractors.

2 CURRENT SITUATION AND PROBLEMS CONFRONTED IN CONSTRUCTION ACCIDENTAL INJURY INSURANCE

For years, the injured in construction accidents could only get limited compensation. Because of the lack of related laws and regulations, the chances of getting compensation through prosecution have been very small. In 1996, a case in which a construction worker disabled by serious injury in construction accident got 270,000 RMB (32,530 USD) in compensation through prosecution was the first case in China's construction industry.

In 1997, construction accidental injury insurance was first carried out in Shanghai City, Zhejiang and Shandong Provinces. The objects of construction accident injury insurance are limited to certain projects of building works, piping works, installation, dismantling and decoration of buildings. The contractor should pay the insurance premiums, while the insured are all the people working on site. The beneficiary are the injured themselves or the legal or designated bodies of the fatalities. The insurance policy should be effected since the commencement of the works until the complement of the works.

Practically, the insurance premium can be calculated either by determination of the annual premium per capita according to 40～50% the number of workers involved at the peak time of construction, or by a certain ratio of contract price or by the unit construction area. The rating depends on the accidental risks of the project, the average local living level and the maximum insurance premium thereof. For example, some insurance company charges 3 RMB per capita for every 1,000 RMB of insurance annually. Insurance compensation for construction accidental injury includes the cost caused by diagnosis, medicine, emergency care, hospitality and other sorts of compensations. The insured will obtain the insurance compensation once the insurance policy is valid, regardless of whether or not they have been compensated for any other accident before. The insurance premium should be accounted into the construction cost. The cost of construction accidental injury insurance should cover cost and profit of the insurer besides the insurance to be paid. Costs caused by the safety supervision and safety training of construction industry and the reserve funds required are also included. To encourage the contractors with good safety performance, the insurance company can decrease the insurance premium rating in the next year or return some of the insurance premium to those without fatality and whose accidental injury rate this year is under the provided level (For example, those without accidents in a duration of 3 years are free of insurance premium in the fourth coming year).

Although great progress has been made in China, there are still many problems such as: (1) the insurance premiums depend on rough judgement of experiences instead of a reasonable calculation method; (2) no modification rating is provided to encourage the contractors with good safety records and punish those with bad; (3) the determination of the insurance premium based on the number of workers involved in is not feasible for construction industry in which the site and position of the personnel change frequently.

3 THE METHOD TO CALCULATE CONSTRUCTION ACCIDENTAL INJURY INSURANCE PREMIUM

For data source and statistical reasons, it is difficult to research and calculate construction accidental injury insurance premium precisely and properly. Based on the current situation and limited documentation in China, and referring to foreign experiences, especially those of the USA (Levitt 1987, Hancher 1997, Williams 1985 and Everett 1995), a method to calculate construction accidental injury insurance premium, with the payroll of the contract price as a factor, is shown below,

$$IP = LC \times PR \times MR$$

in which,

$PR = PPR / (1\text{-}CR)$

$PPR = TL / LRU$

$MR = EMR \times SAR$

where

IP—Insurance Premium

LC—Payroll

PR—Premium Rating

MR—Modification Rating

PPR—Pure Premium Rating

CR—Cost Rating

TL—Total Losses

LRU—Loss Risk Unit (Total Payroll)

EMR—Experienced Modification Rating

SAR—Safety Assessment Rating

3.1 *Calculation of Insurance Premium Rating*

3.1.1 *Estimation of classified accident losses*

The construction accidental injury losses mainly consist of compensation for the victims, which includes various costs of losses. At present in China, the loss caused per fatality is estimated as 80,000 RMB(9,640USD), per serious injury 48,800 RMB and per slight injury 400 RMB (CPIC 1997).

3.1.2 *Determination of the number of casualties*

According to Heinrich 1:29:300 principle (Cao 1998), in every 300 cases which may cause accidents there are 29 slight injuries and 1 serious injury. Following Chinese statistics, the ratio of death to serious injuries is 1:4, so totally the ratio of death to serious injuries and slight injuries shows 1:4:116. As the statistics of national construction accidental casualties only show the precise number of deaths, the calculation of insurance premium ratio followed is carried out by this ratio.

Table 1. Calculation of probability of losses

Accident Category	Annual fatalities			Annual Average fatality	Serious Injury	Slight Injury	Losses (10,000 RMB)	Payroll (LC) (100,000,000 RMB)	Loss Proba-bility (‰)
	1995	1996	1997						
Earth work	139	106	90	111.7	446.7	12953	3591.20	320.321	1.1211
Formwork	102	79	60	80.3	321.3	9319	2583.52	219.235	1.1784
Scaffolding	270	242	164	225.3	901.3	26139	7246.72	329.104	2.2020
Edges and openings	317	280	254	283.7	1135	32905	9122.72	461.148	1.9783
Portal frame	215	221	140	192.0	768.0	22272	6174.72	289.524	2.1327
Tower crane	85	89	77	83.7	334.0	9705	2690.72	146.827	1.8326
Construction machine	177	114	130	140.3	561.3	16279	4513.12	234.198	1.9271
Temporary electric	152	196	89	145.7	582.7	16897	4684.64	198.301	2.30624
Existent electric	56	64	41	53.7	214.7	6225	1725392	182.239	0.9471
Dismantling	83	54	36	57.7	230.7	6689	1854.56	264.519	0.7011
Others	273	216	199	229.3	917.3	26603	7375.36	894.671	0.8244

3.1.3 *Calculation of the pure premium rating*

The pure insurance premium is the cost caused by the compensation when accidents happen in a certain period of time. i.e. pure insurance premium equals total sum of compensation, which theoretically requires that pure premium rating equals probability of loss. We classify the accidents into 11 categories and calculate their probability of loss respectively (see Table1).

The expectation value of the probabilities of losses, whose values are calculated in the table above, turns out to be

$E(X) = 1.564276‰$.

Required by the financial stability of insurance companies, the pure premium rating equals the expectation value plus standard deviation which is,

$\sigma = 0.613226‰$

and the value of pure premium rating is

$PPR = E(X)+\sigma = 2.177502‰$

3.1.4 *Calculation of premium rating*

Besides the PPR, the premium rating takes additional cost, profits and provision for accidents into account. The cost rating is estimated to be 40%(Tao Jun 1991), then,

$$PR = PPR / (1-CR)$$
$$= 2.177502‰ / (1-40\%)$$
$$= 3.629170‰$$

3.2 *Modification rating*

3.2.1 *Calculation of EMR (Experience Modification Rating)*

EMR is actually the comparison of the safety level of a certain contractor with the national average level in the past 3 years. The equation below is suggested to calculate EMR,

$$EMR = (APL + AEL \times W) / (EPL + EEL \times W)$$

in which,

$EPL = EL \times D$

$EL = LC \times PPR$

where

APL—Actual Primary Losses

AEL—Actual Excess Losses

W—Weighting Value

EL—Expected Losses

EPL—Expected Primary Losses

EEL—Expected Excess Losses

D—Discount Ratio

LC—Payroll (Labor Cost)

PPR—Pure Premium Rating

The above equation is familiar to that used in the USA to calculate EMR(Everett 1995, William 1997), except that neither of the terms B(ballast) and EEL×(1-W), which are to limit the variation of EMR, are involved here. The purpose is to enlarge the difference of premium ratings for different contractors to improve their safety management, which is necessary for the current situation in China.

The national safety supervision organizations are responsible for the determination of EMR, by means that losses caused in each accident are recorded and summarized annually for each project. And EMR of a contractor this year are determined by the records of the last 3 years and is valid for each project the contractor contracts to this year.

D-ratio is a ratio of national average expected primary losses and expected losses. The factors are determined as,

APL is supposed to be 15,000 RMB, actual losses less than 15,000 RMB are identified with its own value, while excess is regarded as AEL. For project with the contract price to be 50,000,000 RMB, the weighting value is 0.35 and D-ratio is 0.4 (Cao 1998). The example below shows how EMR works, with A, B, C representing 3 different construction companies that have only 1 safety accident respectively in the last 3 years with the value of SAR to be 1. Table 2 shows the injury records and losses classification of each company. Table 3 illustrates the calculation of EMRs, while Table 4, that of premiums.

Table 2. Injury records and losses classification

Co.	LC (10,000 RMB)	fatalities	Serious Injury	Slight Injury	AL	APL	AEL	EL	EPL	EEL
A	1250	2	3	6	308800	15000	293800	45375	18150	27225
B	1250	0	1	6	51200	15000	36200	45375	18150	27225
C	1250	0	0	8	3200	3200	0	45375	18150	27225

Table 3. Calculation of EMRs

Co.	APL	AEL	EPL	EEL	AEL×W	EEL×W	APL+EPL×W	EPL+ EEL×W	EMR
A	15000	293800	18150	27225	102830	9528.75	117830	27678.8	4.2570
B	15000	36200	18150	27225	12670	9528.75	27670	27678.8	0.9997
C	3200	0	18150	27225	0	9528.75	3200	27678.8	0.1156

Table 4. Calculation of Premiums

Company	Payroll (¥10,000)	Fatalities	Serious Injury	Slight Injury	AL	Premium	EMR	Premium modified	Premium difference	Percentage in invest
A	1250	2	3	6	308800	45375	4.2570	193161	147786	3.863
B	1250	0	1	6	51200	45375	0.9997	45361	-14	0.001
C	1250	0	0	8	3200	45375	0.1156	5245	-40130	0.803

Obviously, the safety performance of company A is worse than the others, so its premium is 4 times larger than that calculated without EMR and covers 4‰ of the total investment. For company B whose safety performance is average, the premium is slightly modified, while the premium of company C decreases to only 1/10 of its previous value because of its excellent safety performance. One has no difficulties in finding out that a lower EMR value means a larger potential income to the contractor while a higher EMR value means greater losses. With no upper limitation for EMR, a high premium of a contractor will cause refusal of insurance by insurance companies or further increases of premium, which will result in loss or even failure in bidding of the contractor. If no insurance companies would accept the insurance, the contractor cannot be allowed to bid on new projects.

3.2.2 *Discussion on SAR (Safety Assessment Rating)*

EMR efficiently modifies the construction injury insurance premium of the contractors by their safety records over the last 3 years. However, no consideration has been taken to such factors as the safety condition on site, safety planning and project risk, so the following equation is suggested,

$$MR = EMR \times SAR$$

in which EMR mainly indicates the assessment of the safety performance records of the contractor in the past. With the rapid development of China, contractors are improving quickly, so it's far from satisfactory to evaluate their present safety management and ability only by their past performance. Therefore, it's more reasonable to determine MR of the project insured by the appraisal of the safety performance with both EMR and SAR of the project.

SAR includes safety on site assessment, safety planning assessment, project risk assessment and assessment of other factors.

Safety on site assessment is accomplished according to JGJ 59-88(the National Construction Safety Scoring Standards, enacted in 1988 and came into force in 1989), in which safety on site assessment standards of safety management, outer scaffolding, construction electricity, portal frames and cranes are provided.

Safety planning assessment consists of assessment of safety on site planning and safety education and training.

Project risk assessment includes assessment of such factors as project characteristics, building height, construction method, geological conditions of surroundings, duration and season of construction.

Assessment of other factors includes qualification of the contractor, type of the contract, financial state of the contractor, loss caused by delay of works and construction season.

Specific items and standards of SAR and their shares in the total safety assessment of the project are being investigated. However, SAR can be 1 or determined by the insurance company at present.

4 CONCLUSIONS

Based on the analysis of current practice and development tendency of construction management in China, and referring to the foreign methods and experiences, a calculation method dealing with construction accidental injury insurance premiums which is suitable for China is presented in this article. The core of the method is to adjust the insurance premium of the contractors according to their safety records and safety management.

Constrained by factors such as statistical methods and caliber, precise and further research proved to be difficult. Yet we are confident that the conclusion reached, although far from perfection, and the method we apply in this article are suitable for the current situation in China. With further experience and practice, more reasonable and precise construction accidental injury insurance systems are expected to be presented in the future.

REFERENCES

Williams, C. Arther & Heins, Jr. Richard M. 1985. *Risk Management and Insurance*. New York: McGraw-Hill.

Cao, Xiaoling 1998. *Construction Safety and Its Related Insurance* (Graduation Dissertation). Beijing: Department of Civil Engineering, Tsinghua University. (in Chinese)

China Pacific Insurance Co.(CPIC) 1997. *Insurance Regulations on Construction Accidental Injury*. Beijing(in Chinese)

Hancher, Donn E. Jesus M. de la Garza, Gregory K. Eckert. 1997. Improving Workers' Compensation Management in Construction. *Journal of Construction Engineering and Management*. 9: 285-291.

Fang, D., Qin, C. and Jin, L.1997. Providing the incentive to improve construction safety management by the modification of construction safety insurance premium rating. *Biulding*, 12: 19-21. (in Chinese)

Fang, D., Cao, X., Huang, F. and Fong, P. 1998. Construction accidental injury insurance. *Shanghai Insurance*. 5: 13-15. (in Chinese)

Everett, John G. & Thompson, Willard S. 1995. Experience Modification Rating for Worker's Compensation Insurance. *J. of Construction Engineering And Management*. Vol.121, No. 1.

Levitt, Raymond Elliot & Samelson, Nancy Morse 1987. *Construction Safety Management*. New

York: McGraw-Hill.
Tao, Jun & Yin, Chunhua 1991. *Modern Insurance*. Nanjing: Nanjing University Express. (in Chinese)
William, L. 1997. Factors of the Experience Modification Rating and What They Mean. *Professional Safety*. 9:27-29.

Implementation of Safety and Health on Construction Sites, Singh, Hinze & Coble (eds)
© 1999 Balkema, Rotterdam, ISBN 90 5809 036 1

Communication of health and safety in the design phase

J. MacKenzie, A.G.F.Gibb & N.M.Bouchlaghem
Loughborough University, UK

ABSTRACT: Poor investment of management practices, particularly within the design phase, have been identified as a prime cause of the unacceptable accident rate within the construction industry. The Construction (Design & Management) Regulations, which became law in March 1995 as the UK response to EC Directive 92/57/EEC, require designers to design in a way that avoids, reduces or controls risk to the workforce. However, for many, the adoption of such practices is still unfamiliar and there is little information on procedures or best practices. Thus, to a large extent the application of such regulations is intuitive and relies on the attitude, behaviour and experience of the designer and their ability to communicate the identified risks effectively. This paper presents the preliminary findings of a major research project to assess safety communication in the design phase of construction projects.

1. INTRODUCTION

The design process and its management offer complex problems and challenges that are not well defined or understood. Project and contracting uncertainties have increased at the same time as industrial fragmentation and professional institutionalisation. There have been many government and stakeholder sponsored studies on the culture and operation of the UK construction industry. These reports have acknowledged that communication and information flow during design have a major impact on the performance of construction projects. The Emmerson report (1962) stated that "in building there is all to often a lack of confidence between architect and builder amounting at worst to distrust and mutual recrimination. Even at best, relations are affected by an aloofness which cannot make for efficiency, and the building owners suffer. In no other important industry is the responsibility for design so far removed from the responsibility for production". Back in 1964, Banwell found that "insufficient regard is paid to the importance or value of time and its proper use in all aspects of a project, from the client's original decision to build, through the design stages and up to final completion. Time well spent can mean time and money saved". Faster Buildings for Industry (1983) acknowledged that the design must account for buildability and that specialist consultants and contractors with design input must be properly co-ordinated as soon as is reasonably practicable. Work on site must be based on full information, communicated clearly and in adequate time. This is facilitated by high levels of site management with strong links with the design team and client. The report also found that the form of contract does not primarily determine performance but the attitude of the contributing parties may. The Latham Report (1993) suggests clients must allow adequate time for the preparation of the brief and should seek advice. " Even the best clients are likely to benefit from some advice on alternative methods of achieving their aim, which may produce better value for money". In relation to the design process, Latham noted that "effective management of the design process is crucial for the success of the project". Lack of sufficient management during the design phase results in insufficient information for completing designs or resolving conflicting construction details. The importance of improved design management has been widely recognised (Austin et al 1993, 1994, 1995, Grey *et al* 1994). A report by NEDC showed that more than fifty per cent of problems on building sites were related to poor design (NEDC 1987). More recently, a study by Boudjabeur and Skitmore (1996) found that in a survey of design and build managers, client and information related factors were the most frequent and had the biggest impact on the performance of construction projects. These problems are the probable root cause of many of the accidents and near misses experienced by the industry. In summary, these reports have identified good practice which has long been advocated but rarely followed. It is depressing to read the Emmerson Report and the Banwell Report to find that the 'same' problems persist a generation later. While there is "a widespread acceptance that change is necessary, and that it is long overdue" (Latham 1994), the culture change that the industry is waiting for is not evident.

2. WORRYING TRENDS

Over the last twenty years the UK construction industry has witnessed a steady decline in the number of fatal and non-fatal accidents. Unfortunately statistics for 1996/97 have seen a increase across the range, with fatal accidents up 12.2 % and major/non-fatal accidents up nearly 17.5 % on previous annual figures (HSE 1998). Whether this increase is a 'sinister blip' as a result of the increased workload or better reporting as a result of more stringent legislation, remains to be seen. However, what is known is that the European construction sector accounts for thirty percent of all fatal workplace accidents while employing less than ten percent of the working population (Bishop 1993). Similarly the US construction industry accounts for twenty percent of workplace fatalities while employing five percent of the labour force (Ehmer and Jaselskis 1998). Alternative management practices must be sought in order to redress this unacceptable trend. Until recently most emphasis had been placed on analysing the construction phase, where it was commonly perceived accidents had their root causes. Previous research into design, some of which has been summarised in this paper, suggests there is considerable scope for improvement in analysing procedures in the pre-construction phase in order to reduce the risk of accidents.

3. DESIGNING FOR SAFETY

'Prevention is better than cure'. This well known proverb suggests that the best way of preventing an accident is by eliminating the possibility of it taking place. The reduction of risk should begin with the conception of a new building, designing intrinsically safe and easy to build projects. A report entitled 'Blackspot Construction' (HSE 1988) concluded that positive action by managers within the UK construction industry could have prevented 70% of the fatalities. One of the main causes identified was the lack of co-ordination between the professional team members at the design stage. A survey in South Africa by Smallwood (1996) found that fifty percent of respondents identified design as a factor to negatively affect safety. These views are supported by the pre-amble to the European Temporary and Mobile Sites Directive (1992) which suggests that unsatisfactory organisational options and poor planning at the preparation stage of projects influenced more than half of the accidents occurring on construction sites in the community. The root cause of accidents can be broadly split into three classes: those due to design decisions, those due to lack of planning and those due to construction methods and processes. A study by Churcher and Alwani-Star (1996) found that sixty three per cent of fatalities are traceable to pre-construction activities. Similar results were also found by Jeffry and Douglas 1994. Studies by Barnard (1996) and Hinze (1992, 1998) have emphasised the role of safety in design and more recently, Gambatese (1998) has developed a computer based design tool which enables identification of safer design solutions. However, before safety information is implemented, designers must be willing to use it. Gambatese argues that by incorporating safety knowledge, a designers exposure to liability increases, which may deter the designer from implementing knowledge. "Traditionally, unless specifically written into a contract, a designer is not responsible for overseeing construction worker safety." (Gambatese 1998)

The early stages of design are particularly important. In complex decision making situations the ability to affect project outcomes diminishes exponentially over time. Brandon (1978) suggests that, by the time sketch design is formulated, the major decisions affecting management costs, structure, envelope and quality standards have been made and the ability to influence remaining decisions is probably limited to twenty percent of the final cost. It is highly probable that this is also true of safety. If we accept that safety, like quality, can not be 'inspected' into a project and that 'bolt on' safety measures are seen as a reaction to potential hazards then surely safety must be implemented in early design.

4. ANTECEDENTS TO THE RESEARCH

The Construction (Design and Management) Regulations 1994 (CDM) were introduced in March 1995 as the UK's response to EC Directive 92/57/EEC, in an attempt to address the absence of safety management during the pre construction phases. As a result, regulations that ensure health and safety is co-ordinated and managed effectively throughout a project's life cycle have been enforced for the last three years. However, for many designers, the adoption of such practices is still unfamiliar and there is little information on procedures or best practices.

A recent report titled "Experiences of CDM" (CIRIA 1997), found that many designers are unsure of their duties and the extent of risk assessment required and that as a result, the practices of designers are affected by their lack of familiarity with the requirements. " There are undoubtedly problems in understanding and applying the regulations and as yet, only a minority of participants are fully competent at their roles", (CIRIA 1997). More concerns about the implementation of the CDM Regulations quickly emerged and a study was commissioned by the HSE (1997) to examine the impact of the regulations. The principle conclusion was that the regulations had led to a increase in awareness of health and safety issues and that the Health & Safety Plan was an important and useful document. Although it failed to quantify the cost and benefits of implementation, the report suggested that the long term application would yield both efficiency benefits and improvements in H&S. Specifically the report established that the professionalism (knowledge) of the client has an affect on the implementation. Clients, not so much the larger more frequent players but infrequent small and medium enterprises, are more likely to be dismissive of the perceived benefits and discharge their duties to others. Most see the CDM Regulations as an additional burden in bureaucracy and cost, a view shared by the Institute of Directors (1997). The ability of clients was also questioned in relation to the appointment of consultants and contractors, on the basis that they assumed competence rather than determining it. The review observed that larger contracting organisations have incorporated CDM into their management, quality and environmental procedures which has enabled an effective and efficient approach. Unsurprisingly, contractors have adapted to the regulations better than other parties, with over fifty six percent of those surveyed found to be in full compliance. Contractors have expressed concern over the competence of Planning Supervisors and the added bureaucracy in respect of competence assessment of them and their requirement to assess appointees. In relation to designers the report found that engineers had a more structured approach to design and more familiarity with H&S issues, which resulted in a higher level of competence in the identification and elimination of risk, so much so that they too are more likely to question the competence of Planning Supervisors. Architects however, like their American counterparts, are concerned with professional indemnity and criminal liability. However, the lack of knowledge of construction materials, processes and techniques used in the industry is of major concern. This is compounded by the inability to identify and eliminate risks. Thus, to a large extent the application of such regulations is intuitive and relies on the attitude and behaviour of the designers and their ability to communicate the identifiable risks effectively.

In summary the report indicates that there is increased awareness of H&S issues as a result of increased planning, control and behavioural changes. There was however no substantive evidence of improved health and safety. In positive reflection the report's claim that it is too early to tell, may prove to be right. Although time will not address the reluctance and apathy of some parties in the acknowledgement of safety in design. The writer agrees with the views of Clark (1998) in that the key to the regulations lies in continuous design assessment and the planning supervisor's ability to ensure designers co-operate and communicate with each other.

5. ASSESSMENT OF RISK

The construction industry is no stranger to the apportionment of risk. Risk permeates through construction projects from conception through to demolition. However, safety risks have not received the same amount of detailed attention as those associated with financial considerations. Reasons for this include:

- The identification and assessment of safety risk is more difficult to address, often subjective and it's root causes are unclear in origin.
- Safety risks tend to be analysed later in the project development, resulting in 'bolt-on' safety systems, rather than projects being designed safely from the outset.
- Comprehensive data on accidents and accident reporting techniques simply do not exist in many cases resulting for example, in occupational related statistics being underestimated.
- Safety performance is a minor consideration compared to the tender price, fitness for purpose, completion time etc.

Central to the UK's Management of Health and Safety at Work Regulations (1992) is that sufficient and suitable risk assessments are made and maintained in order to identify measures to be taken to satisfy health and safety law. The CDM Regulations expand on these requirements and place specific duties on those who can contribute to health and safety which include the client and design consultants. As part of these duties designers are required to assess risks so that they can be avoided where necessary or combated at source i.e. alter the design. Where risks can not be avoided or combated they must be assessed in terms of their potential to cause harm and included in the health and safety plan. The health and safety plan is essentially a medium to communicate health and safety issues. The plan must be made available to any contractor before arrangements are made to manage or carry out construction work, thus providing a link between design and construction. The risk assessment technique generally used in design is probabilistic risk assessment (PRA) where the likelihood of a hazard occurring is multiplied by the severity of harm caused if the hazard manifests itself. Therefore designers must ask themselves what is the probability of a hazard occurring and if so how severe will it be? There are no figures on the probability of outcome of potential hazards in the construction industry. Therefore the probability is an estimate of occurrence and severity based on the assessors own perception and experience. Thus the person who assesses the risk is insulated from the working conditions and hazards. Lichtenstein et al (1978) has shown that the risk of rare adverse occurrences tend to be over estimated, while the risk of more common events are likely to be underestimated. People differ in their reaction to perceived risk. Also people will be motivated differently in response to the same hazard. As a result PRA tends to be highly subjective in nature. Designers require more information in order to assess risk in the design phase, this could be facilitated by the use of contractor feedback on design issues.

6. AIMS AND METHODOLOGY OF EXPLORATORY STUDY

The overall aim of this research programme is to determine whether communication between the client, planning supervisor, designers and principal contractor has an effect on the health and safety performance of the project. Within that research boundary, the aims of this exploratory study are to:

- Examine the importance attributed to safety in the design phase.
- Assess the time during the design phase when safety is analysed.
- Determine factors that affect communication in the design phase.
- Identify priorities for further research.

A comprehensive literature search reviewed related books, professional journals, conference papers, and publications concerning the nature and management of design, generation and dissemination of information and current health and safety design techniques. The literature search was supported by informal interviews and briefings with construction industry and allied professionals to identify current best practices for communicating safety during the design process and the problems associated with design management. On the basis of information developed through the literature search, a exploratory questionnaire was utilised to carry out semi structured interviews with professional design team members. It was felt that the five organisations selected would be sufficient to enable initial exploration of the issues within the time available and the sensitive nature of the enquiry. The organisations represent:

- a regional repeat-order end user client with a newbuild and maintenance budget
- a national consulting engineering company with a planning supervising facility
- an international consulting engineering company
- a international construction company
- an international construction company specialising in the plant and process industry

7. ANALYSIS OF EXPLORATORY STUDY

7.1 *The importance of safety during the design phase*

The study found that the importance attributed to safety in the design phase varied. The offshore and process industries identified worker safety, occupational health of the workforce and environmental issues as important to their organisation. Their documentation and approach reflected a culture which regards safety as a priority during all aspects of design, construction, operation and decommissioning.

The construction industry's perception of safety was somewhat different, in that:

- The most important project criteria for the design team were: 1) Level of Specification
2) Suitability of Purpose
3) Date for Completion
- Safety was not a critical issue for the design team
- Contractual arrangements do not affect safety during construction
- No benchmarking or safety performance measurement was evident
- Safety documentation is linked to the organisational quality management system
- Value judgements and cost / benefit analysis are based on the experience of the designer
- The Principle Contractor has the most influence on safety during construction

In summary the study revealed that while safety is deemed to be of value to the construction industry, it is not as important as other project criteria. This view is supported by the lack of performance measurement and established procedures. It was also evident that the Principle and Trade / Sub Contractors were deemed to have more control of construction safety, which suggests a reluctance to accept proportional responsibility for safety performance. It was agreed that safety during the design phase could be improved by; simplified documentation, improved communication and improved auditing to ensure implementation. It was also noted that; improved culture would lead to designing safely as opposed to bolt-on reviews, and more information and feedback on construction issues would aid in designing safely. How this would be achieved was not forthcoming or evident.

7.2 *Time during the design when safety is analysed*

Timely consideration and application of procedures has been identified as of prime importance to the success of a project's performance. It was therefore premised that early application of safety procedures would have the most beneficial impact on reducing potential accidents. The study identified that:

- No additional time has been allowed in the design phase for the application of the CDM Regulations
- Approximately 1-2 percent of the design phase is committed to the implementation of the CDM Regulations
- Feasibility stage is the best time for the appointment of the Planning Supervisors. However, this was rarely achieved in practice
- Some clients are reluctant to engage parties too early in the design phase for fear of wasting money.
- Scheme stage has biggest impact on safety while detailed design was highlighted as important.
- The technique utilised by all design team parties (including the process industry) for identifying risk was Probabilistic Risk Assessment.
- Review of in house documentation including Risk Assessments is done as the need arises
- The design tasks that were most difficult to assess were those associated with unfamiliarity or minor changes (e.g. change from expansion bolts to resin anchors in a confined space).
- Designers were sceptical on the use of risk assessments.

During the study it was cited that there was no proof that risk assessments carried out by designers actually improved health and safety. This view may be justified in light of the amount of time dedicated to identifying potential hazards. It was also acknowledged that the CDM Regulations have increased awareness, but not to extent that designers are committing adequate resources. The responses during the study have clearly indicated that insufficient time is dedicated to the implementation of safety procedures during the design phase. What was found to be of use was a list of potential hazards which was passed on to the principle contractor via the health and safety file.

7.3 *Factors affecting safety communication*

The communication and dissemination of information between project stakeholders is important throughout the projects life-cycle. From the literature it has been established that during design, communication is critical to the projects success. The study found that:

- The communication of safety information was unaffected by the procurement route chosen or the time in the design phase when safety is analysed.

- Obtaining a realistic design brief in the allotted time was found to be the most difficult task for a designer.
- Method statements were indicated as the key document for all elements of work especially those identified as containing potential risk.
- Information Technology would aid in the communication of risk in design. However it was not rated highly in the identification and assessment of risk.
- IT communication between organisations was hampered by system and software compatibility
- The use of 3D visualisation and Virtual Reality has little practical use in identifying potential hazards in the construction industry.

The study has identified a number of factors which affect the communication of safety during the design phase. Most are associated with lack of clear lines of communication and leadership. It was felt that increased use of IT would aid in the dissemination of information and that it's future potential was tremendous. Currently, inter and infra organisational compatibility is seen as the main problem. It was established that measures taken to ensure compatibility at the outset of a project would enable increased information transfer. Also recommended were common computer servers and the increased use of Local Area Networks (LAN's).

8. CONCLUSIONS

There has been much literature published in relation to health and safety in the construction industry. More recently, attention has been focused on the design phase, where early implementation has been premised to result in positive, proactive results. Surveys carried out in relation to the CDM Regulations suggest that while legislation has increased safety awareness and will in time have a beneficial effect, many especially those in small and medium enterprises are sceptical on the cost/benefit results and regard their application as bureaucratic.

This exploratory study has confirmed that the CDM Regulations have increased awareness. However it's application is not considered to be of paramount importance compared to other performance criteria e.g. specification, suitability of purpose and completion time. Most feel that they are implementing properly, although none are actively comparing or benchmarking against set targets. The study has clearly indicated that insufficient time is dedicated to the implementation of safety procedures during the design phase. The study concludes that there is much need for improvement. Designers need more information about the projects' potential hazards at a time when information is traditionally scarce. The contractors input into the design as been identified as beneficial to buildability, which in turn has positive links to safety. Design risk assessment is subjective and based on the experience of the designer, who is insulated from the actual risks. It is therefore apparent that the use of contractor feedback in identifying potential hazards would aid in the proactive reduction of their occurrence. It would also close the communication loop that exists at the end of construction projects .

REFERENCES

Banwell, Sir G. H. 1964. *The Placing and Management of Contracts for Building and Civil Engineering Works*. Ministry of Public Buildings and Works. London: HMSO.

Barnard, M. J. 1996. Learning to live with major changes in the UK construction safety legislation. In L.M. Alves Dias & R. J. Coble (eds), *Proc. Int. Conf. on Environment, Quality and Safety in Construction*, Lisbon, 4-7 September 1996: 61-66. Rotterdam: Balkema.

Bishop, D. 1993. *The Professionals' view*. A paper given to the joint Health and Safety/ Construction Industry Council Conference. 11th February. London: HSE/CIC.

Brandon, P. 1978. A Framework for Cost Exploration and Strategic Cost Planning in Design. *Chartered Surveyor, Building and QS Quarterly*, 5(4) 60-65.

Churcher, D. W. and Alwani-Starr, G. M. 1996. Incorporating construction health and safety into the design process. In L. M. Alves Dias & R. J. Coble (eds), *Proc. Int. Conf. on Environment, Quality and Safety in Construction*, Lisbon, 4-7 September 1996: 29-39. Rotterdam: Balkema.

CIRIA 1997. *Experiences of CDM*. CIRIA Report 171. London: Construction Industry Research and Information Association.

Clarke, T. 1998. Planning Supervisors within the construction process. *Construction Safety Briefing*. Croner Publications Ltd. Issue no 30. 13 March.

Council of the European Communities. 1992. *Council Directive 92/57/EEC of 24 June 1992 on the Implementation of Minimum Safety and Health Requirements at Temporary or Mobile Construction Sites*. Brussels: European Commission.

Ehmer, J. and Jaselskis, E. J. 1998. Technological Advances in the Construction Industry. *Proceedings of the International Conference on Environment, Quality and Safety in Construction*. Lisbon: CIB Working Commission W99.

Emmerson, Sir H. 1962. *Survey of Problems Before the Construction Industry*. Ministry of Works. London: HMSO.

Gambatese, J. A. 1998. Liability in Designing for Construction Worker Safety. *Journal of Construction Engineering and Management. ASCE*. Awaiting Publication

Gray, C., Hughes, W. and Bennett, J. 1994. *The successful management of design*. Reading: University of Reading,

Hinze, J. W. and Wiegand, F. 1992. Role of Designers in Construction Worker Safety. *Journal of Construction Engineering and Management. ASCE*. Vol. 118 Part 4: 677-684.

Hinze, J. W. 1998. Addressing Construction Worker Safety In the Design Phase. Technological Advances in the Construction Industry. *Proceedings of the International Conference on Environment, Quality and Safety in Construction*. Lisbon: CIB Working Commission W99.

HSC 1988. *Blackspot Construction*. London: H.M.S.O.

HSC 1992. *Management of Health and Safety at Work Regulations 1992* and *Approved code of Practice*. London: H.M.S.O.

HSE 1998. *Key Facts: Injuries in the Construction Industry 1961 to 1995/96*. Operations Unit Merseyside: Health and Safety Executive.

HSE 1997. *Evaluation of the Construction (Design and Management) Regulations (CDM) 1994*. The Consultancy Company Ltd. London: H.S.E.

Institute of Directors 1996. *Health and Safety*, A discussion paper. London: Institute of Directors.

Jeffrey, J. and Douglas, I. 1994. Safety Performance of the UK Construction Industry. *5th Annual Rinker International Conference focusing on Construction Safety and Loss Control*. University of Florida. Gainesville: 233-254.

Latham, Sir M. (1994) *Constructing the Team*. Joint Review of Procurement and Contractual Arrangements in the United Kingdom Construction Industry. London: HMSO.

Lichtenstein, S., Slovic, P., Fischhoff, B., Layman, M. and Combs, B. 1978. Judged frequency of lethal events. *Journal of Experimental Psychology. Human Learning and Memory*, 4: 551-578.

NEDO. (1993) *Faster Buildings for Industry*. National Economic Development Office. London: HMSO.

Smallwood, J. J. 1996. The influence of designers on occupational health and safety. In L.M. Alves Dias & R. J. Coble (eds), *Proc. Int. Conf. on Environment, Quality and Safety in Construction*, Lisbon, 4-7 September 1996: 203-214. Rotterdam: Balkema.

Implementation of Safety and Health on Construction Sites, Singh, Hinze & Coble (eds)
© 1999 Balkema, Rotterdam, ISBN 90 5809 036 1

Application of owner controlled insurance programs to construction

J.J. Lew
Department of Building Construction Management, Purdue University, West Lafayette, Ind., USA
M. Overholt
American Contractors Insurance Group Incorporated, Dallas, Tex., USA

ABSTRACT: An owner controlled insurance program (OCIP), or controlled insurance program (CIP), has recently become commonplace in the construction industry, and is expected to be used more frequently in the future. This paper presents an application process for the use of CIP in the bidding and construction processes currently being utilized in the construction industry.

The guiding principle of a CIP is to provide broad and consistent insurance coverages to all contractors at a project site that are in the program. CIP produces worksites characterized by consistent safety and loss control programs. The result is an improved safety culture, reduced claims and expenses, and additional profits for participating contractors. This paper presents CIP in an applications format so that the working manager of construction operations can gain a working knowledge of CIP.

1 INTRODUCTION TO CONTROLLED INSURANCE PROGRAMS

An owner controlled insurance program (OCIP), or controlled insurance program (CIP), is a method of providing various insurance coverages including worker's compensation, general liability, auto liability, builder's risk, and other coverages as applicable, for all the contractors and approved subcontractors on a major construction project or on approved, selected projects. (Herke, 1994) The projects are normally confined to a single location, but can be applied to multiple projects within a geographic region. A CIP is generally in force for a finite duration. The common application is a two-to-five-year project, where at the end the project, accounts are terminated and the files closed. The benefits are several and are presented here.

Lew (1998) reported information relating to CIP that significant insurance cost savings and broader insurance coverage can be produced through a well-designed and administered CIP. Contractors generally must make significant adjustments in their business practices to successfully participate in a CIP construction project. CIP creates an economy of scale by centralizing the purchase of insurance coverages. CIP streamlines insurance project management by placing a number of insurance functions such as loss control, safety management, security, and recordkeeping under a single authority. Cost savings of 1 to 2 percent of the total project costs have been and can be realized with the use of CIP.

1.1 *Terms*

Following are the terms used in CIPs:

- CIP (Controlled Insurance Program): A centralized and controlled insurance and loss control program authorized by the project manager and applicable to a defined worksite.

The program is purchased and administered by the CIP manager and covers all participating contractors and subcontractors on any tier.

- Owner: The body or entity that has the contractual agreement with the General Contractor and who provides the direction to the General Contractor and the Project Manager.
- Project Manager: The person or entity charged with the overall responsibility for the project and worksite where the CIP is in effect. This responsible person or entity may be the project owner, a representative of the owner such as a management or engineering firm, or a general contractor.
- Contractors/Subcontractors: All of the contractors and subcontractors in any tier performing work at the defined worksite under the direction of the project manager, and who are participating in the CIP.
- CIP Administrator: The person or persons who have the responsibility for the administration of the program. This person(s) resides on the project. The CIP administrator's responsibilities include payroll administration, premium collection, coordination of claims handling with the carrier, and interaction with the subcontractors, general contractor, and the owner regarding the insurance administration of the program. The administrator handles all of the day-to-day functions that the manager does not coordinate.
- CIP Manager: The person or entity responsible for the design and administration of the CIP, and resides in the project manager's home office. This maybe a broker, carrier, account executive, consultant, or a member of the project manager's staff. Under normal circumstances, the CIP Manager reports to the project manager.
- Project Safety Representative, (PSR): The person or persons responsible for the coordination of the day-to-day safety and health functions on the project. Responsibilities include safety training, accident investigation, jobsite inspection, conducting safety meetings, and development and implementation of the safety program.

2 BENEFITS

As a major benefit, subcontractors are able to take the experience of the project as a whole with them to their next project. Generally, for a subcontractor, the benefits of the CIP will be much greater regarding the aspects of risk management and safety, resulting in improved EMR and other safety-based statistical data. The principle of the CIP is to provide broad and consistent coverages of insurance for all contractors in the program. (Alexander, 1995)

For the owner, the coordinated approach of contractor management is appealing for several reasons:

- The sponsoring contractor or owner earns underwriting profit on the subcontractor premium.
- The sponsoring contractor or owner earns investment income on the subcontractor premium.
- The subcontractor insurance markup is eliminated.
- A coordinated approach to project safety and risk management is enabled.
- All onsite contractors can benefit from an efficient and cost-effective safety program.
- The CIP assures that claims-handling techniques are uniform.
- The CIP provides a system in place for uniform tracking of insurance credits, EMRs, payroll, and premium figures on a routine basis.

For the owner and the general contractor, there are two primary benefits: (1) There is only one policy for the project, whereas in the standard construction documents, every contractor must provide the owner with copies of certificates of insurance; and (2) broader coverages can be purchased by the project with additional limits of liability based on the purchasing power of all approved contractors on the site.

The cost of the CIP is loss sensitive, thus allowing for a substantial premium savings if losses are controlled on the project. On the other hand, premiums have the potential to increase greatly if losses on the site are not controlled, therefore, the objective of a successful project in terms of schedule, productivity, and finances has to be the safety and loss control program. Normally, the general contractor and the owner share the premium savings and adjustments to the program. The breakdown of percentages are agreed upon by both entities through their contractual agreement.

For owners, a great benefit is the administrative ease that is gained through uniformity.

For minority business enterprises (MBE) contractors, and women's business enterprises (WBE) contractors, opportunities are more available. At times, these contractors might have an advantage given that contract letting will only accept bids from a MBE/WBE contractor (the thought here is that minority contractors would not otherwise be able to provide the necessary coverages because of their size, and the inability to purchase higher limits on their policies).

Direct control of exclusions and fewer disputes are additional benefits for everyone concerned. The emphasis in CIP programs should be on the owner—in other words, the owner must accept accountability and stipulate cooperation from all involved parties.

CIPs provide the lowest cost of risk which results in cost reduction, and CIPs provide protection from catastrophic loss by providing higher liability.

There are fewer intermediaries taking a piece of the action.

It is easier to secure discounts for large accounts. The premise here is the ability to have *group* purchasing power.

3 ADMINISTRATION OF CIPs

The administration of the CIPs is as follows.

3.1 *EMR and safety performance*

A contractor's experience modifier (EMR) encourages the contractor to improve safety performance while allowing the insurance industry to collect the required premium to pay for the incurred losses. The insurance premium savings offered through the experience rating plan almost always outweigh the expense needed to improve safety performance. In other words, safety does indeed pay.

Owners use EMR to gauge the safety performance and experience of the general contractor or the prime contractor. The general contractor then uses the experience of the subcontractors to gauge the safety performance through the pre-bid documentation. An experience modifier of .80 means the contractor will receive a 20 percent discount on its worker's compensation premium. A contractor with an experience modifier of 1.20 will pay a 20 percent surcharge on its worker's compensation premium. A contractor's EMR is only one measurement of past safety performance. Integrating the experience modifier with other key measurements will provide a better decision-making model.

3.2 *Minimum coverage*

For a typical project with a parameter of $100 million in work scope, at a minimum, the following coverages should apply:

- Specific coverages: The CIP provides worker's compensation, general liability, and builders risk insurance for all approved subcontractors (of any tier) who have operations on the project. Any subcontractors that might pose any undue risk to the CIP may be declined enrollment in the program at the discretion of the owner.
- Worker's compensation: Each approved subcontractor is issued an individual worker's compensation policy covering that individual subcontractor's operations in support of the work at the site.
- General Liability: A *master* general liability policy is issued in the name of the general

contractor. The policy is site specific. Each approved subcontractor is added to the master policy as an insured, effective with the date that the approved subcontractor begins operations at the site.

- Builder's Risk: A builder's risk policy is effective for the project as a whole. Each approved subcontractor is named as excess of the primary coverages. The subcontractors are added, effective the date they begin work at the site.

3.3 *Additional considerations*

Additional considerations are high risk, administration, on the job injuries, documentation, and claims management.

- High Risk: In addition to the low bid, consideration is given for previous safety performance, and EMR is included in the selection process for subcontractors to be included in the CIP. Subcontractors with above-average injury rates and a modifier in excess of 1.0 are considered *high risk.* Subcontractors in the high risk category will be required to submit a safety plan that outlines how they will execute the work in an injury-free manner. Special emphasis will be placed on their performance both from a project and from a home office perspective. Should frequent accidents, non-compliance, or indifference become a factor, extreme measures will be considered, up to removal from the project. These actions include personal contact with the owner of the company, stoppage of the work in progress, and possible revocation of the subcontract. Early warning signs of non-compliance will not be ignored, and all subcontractors will be held to the same standard of performance. Since subcontractors comprise a large portion of the premium base for the CIP, the overall success of the project effort hinges heavily on the ability of subcontractors to control their actions and limit their losses.
- Representatives and Administrators: The CIP utilizes a Project Safety Representative (PSR) and an onsite CIP administrator to assist management in the overall coordination of the safety, loss control, and risk management aspects of the project. Among the duties of these individuals will be pre-job/pre-task planning, review of the subcontractor's safety plans, auditing of the subcontractor's work, conducting new-employee orientation, conducting accident/incident investigations, reviewing and inputting the monthly payroll and premium figures, and oversight of the drug and alcohol program and functions.
- On the Job Injuries: All subcontractors must be informed that they will be expected to make reasonable accommodation to return injured employees to work as soon as possible following an on-the-job injury. The work assignments for employees on modified duty will be monitored by the PSR to ensure that the activity is not aggravating the condition or prolonging the healing process of the employee. Medical providers willing to work with the General Contractor in the modified duty program must be identified in the early stages of the project and should be fully informed of the program.
- Documentation: Among the $100 million parameter, other parameters include an annual premium of $1 million or more, and a typical labor component of 25-30 percent of the total project cost. Specific language will be used in the AIA documents and in the bid documents. Some of the key assumptions will be an average modifier of .80 based on the trade classifications of the project and an estimated worker's compensation premium of $2.6 million with the average project term being 30 months. The cumulative benefit to the owner and the general contractor will be directly related to the loss performance. Loss performance is determined by the loss ratio, which is the ratio of loss costs to paid insurance premiums. Lew (1998) reported typical loss ratios in the range of 60 to 65 percent at non-CIP construction sites, while loss ratios on CIP sites were in the 30 percent range. The calculations of loss ratios for the project in its entirety will be demonstrated on a monthly basis to review any adverse trending, or pinpoint any subcontractors that are not performing safely. An on-line reporting system is established for information, and for financial and

safety reporting functions. Among other items, the data to be reported will be premium/credit reports, loss/safety data, OSHA safety statistics, project schedule and cost reports, payroll information, loss development data, and claims closing ratios.

4 CIP MANAGEMENT

The CIP team is made up of the owner and owner representatives, the general contractor, all of the subcontractors, the insurance company, and the broker for the owner.

The safety program management portion of the project includes:

- Weekly safety meetings to be held by all contractors
- A monthly CIP team meeting
- Regularly conducted audits of the project by supervisors to identify safety hazards
- A quick response effort to all safety-related problems
- Close coordination with the project manager and the safety staff
- Compliance with all owner and federal safety regulations and standards
- A progressive discipline program
- An incentive program that rewards safe work habits by employees

All of the project safety elements, if defined, coordinated, and implemented properly, will ensure a successful and safe project.

The intent of the owner and the general contractor is to provide the best medical care available and to establish an effective return-to-work program for all employees if an injury does occur. In providing these handling techniques and programs, as well as having professionally developed claims management procedures, claims are expected to close within an expected time frame and not reopen. To ensure this portion of the program, all available resources are utilized. A partnership is created to establish the necessary parameters of communication and decision-making techniques, and is comprised of the onsite administrator, the project manager, the PSR, medical professionals, and the insurance carrier.

4.1 *Claims management*

The three criteria to be used in the claims management performance measurement are:

- Accident-to-report-date time frame
- Claims closing ratio
- Loss development

5 CONCLUSION

Streamlining the project management functions of loss control, claims management, safety, *etc.*, under a single program will provide greater control and administration in cost accounting, record retention, and overall project control, and reduce litigation between the parties. Administration of the program is much easier under the direction of one entity. Trying to coordinate the many aspects of an insurance program that has multiple contractors contracting their own insurance carriers and brokers results in a wide variety of safety and loss control programs that vary in sophistication. The general contractor's insurance carrier or broker does not need to follow up on certificates of insurance, safety programs, or acceptability of carriers; and does not need to deal with a myriad of agents, brokers, onsite safety representatives, and claims adjusters. It is vitally important that the project team has the support of the owner and the commitment from all subcontractors to provide a healthy and safe working environment for all of the employees on the worksite.

In a properly run and administered CIP, the safety and loss prevention applications should be emphasized, and the efforts of a safety-driven contractor amplified. As mentioned earlier, the subcontractor's modifier rate may decrease, which in turn will produce a positive effect on the insurance program. As the modifier rate decreases, the loss ratio also drops, thus additional profits are available in the form of premium rebates or lower premiums. In the long run, the subcontractor's ability to competitively bid and win projects will be enhanced. Thus, the most likely result of a CIP is an improved safety culture, and reduced claims and expenses, which will ultimately benefit the subcontractors and others working in association on the project.

6 REFERENCES

Alexander & Alexander, Inc. (1995, Fall). *Manual of insurance procedures for a wrap-up insurance program for IPALCO [Indianapolis Power and Light Company) Enterprises.* (Available from Alexander & Alexander of Indiana, Inc., 251 North Illinois, Suite, 1500, Indianapolis IN 46204)

Herke Insurance Center. (1994, Fall). *Owner controlled insurance manual for contractors and subcontractors; for work performed on Indiana sites for Eli Lilly and Company.* (Available from Herke Insurance Center, P. O. Box 40309, 727 E. 86th Street, Indianapolis IN 46240)

Lew, J. J., & Overholt, M. (1998). Managed contractor's insurance programs. In C. Berryman (Ed.), *Proceedings of the Thirty Fourth Annual Associated Schools of Construction Conference* (pp.205-215). Lincoln NE: University of Nebraska-Lincoln.

13 Law and regulation

Implementation of Safety and Health on Construction Sites, Singh, Hinze & Coble (eds)
© 1999 Balkema, Rotterdam, ISBN 90 5809 036 1

The role of the planning supervisor under new health and safety legislation in the United Kingdom

C.N.Preece, K.Moodley & C.Cavina
Construction Management Group, School of Civil Engineering, University of Leeds, UK

ABSTRACT: The construction industry has always had one of the highest incident rates of fatal accidents and serious injuries of all industries. Safety professionals and legislators recognised that the accident records could be improved if obligations were placed on those parties indirectly involved in construction as well as those actually carrying out the work. In compliance with European Directives, the Construction (Design and Management) Regulations (CDM) were introduced on 31 March 1995 in the United Kingdom. These regulations placed new duties on all parties involved to plan, co-ordinate and manage health and safety throughout all stages of the construction process. To date little research has been conducted on the impact of the legislation, and literature on the subject is minimal. This paper will outline the results of research including surveys and interviews conducted with clients, designers, planning supervisors, contractors and independent specialists, and question the roles and competence of the new duty holders in particular, the planning supervisor.

1 INTRODUCTION

The Construction (Design and Management) Regulations (CDM Regulations) (Appendix 1) were introduced in March 1995 as a result of an European Union Directive and were designed to provide a legislative framework aimed at achieving co-operation and co-ordination in an effort to improve construction health and safety. The importance of health and safety is often overlooked and when tendering is often the first item to be cut in the budget. Improvements in health and safety management could not only save lives but have additional further reaching benefits in areas such as quality and productivity and in turn reduce the need for civil litigation and the hidden costs of accidents and ill health.

The construction industry is notoriously one of the most dangerous industries in terms of its safety record. Since the introduction of the first piece of health and safety law in 1802 there has been steady flow of new and backup legislation. In the last twenty five years there has been an emphasis on individuals and their duties compared with previous legislation. Rather than a prescriptive approach it was based on principles designed to bring about greater awareness of the problems associated with health and safety issues.

This philosophy has carried through to the CDM Regulations where every party has been given individual and specific duties. It is a new system to bring about a more efficient and structured approach to health and safety management. The single most significant change has been the introduction of a new role, namely the Planning Supervisor, whose main task it is to co-ordinate the parties and documents to bring about the successful management of health and safety on a project and beyond.

There are many issues and questions that arise from the introduction of such legislation and this paper intends to provoke ideas and investigate the important issues.

2 BACKGROUND

From the introduction of the first piece of health and safety law in 1802 there has been a steady flow of new or backup legislation ever since. The development of Factory Legislation and the following Construction Legislation led to the most radical piece of health and safety legislation the Health and Safety at Work Act 1974. There was now an emphasis on individuals and their duties compared with previous legislation. Rather than a prescriptive approach it was based on principles designed to bring about greater awareness of the problems associated with health and safety issues.

This philosophy has carried through to the CDM Regulations where every party has been given their individual duties. New features have been created such as the health and safety plan and the health and safety file to allow better management of health and safety in construction. The single most significant change has been the introduction of a new role, namely the Planning Supervisor, whose main task is to co-ordinate the parties and documents involved to bring about the successful management of health and safety on a project.

The CDM Regulations is the new system to be adopted by all parties involved in a construction project to bring about a more efficient and more structured approach to health and safety management. Along with the introduction of the CDM Regulations the HSE gave an indication that enforcement would be carried out sympathetically in the first nine months of operation. This implies that the response after this 'honeymoon' period would be to monitor carefully the progress of projects using the CDM Regulations. It must always be remembered when considering the CDM Regulations that they cover the whole of the construction industry in all its complexity and diversity. The HSE has said that the construction industry has adopted the CDM Regulations faster than was initially expected. This must give those involved in health and safety management some encouragement. However there are still a number of 'grey' areas that need to be considered and some thought needs to be given to them in order to aid the progress of the CDM Regulations.

It is only natural to assume that there will be problems when introducing some new legislation and that some members of the construction industry will greet it with open arms while others will try to brush it aside. In order to assess this research needs to be done to ascertain the problems and possible solutions can be derived from this.

2.1 *Review of the Important Areas in the CDM Regulations*

The CDM Regulations have been introduced in a attempt to improve the management of health and safety in the UK. They provide a system for the individuals and companies involved in a project to follow in order to structure the way in which they approach health and safety management. During the design and construction life of the project it helps to promote the area of teamwork as highlighted in the Latham Report (1994).

This section will look at the important areas in the CDM Regulations and those relating to some of the duty holders.

2.1.1 *The Client*

It is true to say that without a Client there is no project and this demonstrates the importance of them in the process. Once a Client has taken the decision to commission a project the initiative lies with them to apply the CDM Regulations.

The appointment of the Planning Supervisor is central to the duties of the Client. Regulation 6(1)(a) provides:

> *'Subject to paragraph (6)(b), every client shall appoint -*
> *(a) a planning supervisor... in respect of each project.'*

This immediately raises problems of how the Client will select the Planning Supervisor with respect to their competence. This however will be discussed in greater detail in the section dealing with the Planning Supervisor.

Another important decision that the Client must make is when they should employ the services of the Planning Supervisor. The timing of this can be very important so that the Planning Supervisor is involved from the initial stages of the project. Regulation 6(3) provides:

> *'The planning supervisor shall be appointed as soon as is practicable after the client has such information about the project and the construction work involved...'*

From this it can be seen that the Client has to appoint a Planning Supervisor 'as soon as is practicable', which could be interpreted as with out unreasonable delay. However without a more specific time in the project many Clients may appoint them too late, either due to the fact that they are ignorant or want to cut costs.

The Planning Supervisor must ensure that the Client receives the health and safety file on the completion of the project. The Client must take reasonable steps to ensure the health and safety file is kept available for inspection:

'... by any person who may need information in the file for the purpose of complying with the requirements of and prohibitions imposed on him by the relevant statutory provision'

The requirement in the CDM Regulations to hand over the health and safety file, which should include 'as-built' drawings, is therefore an extension of existing practice. The Client cannot simply file the health and safety file away never to be seen again and should be treated in a similar manner to other important documents such as the lease and the mortgage. This requires the Client to be aware of the importance of the health and safety file.

2.1.2 *The Designer*

The CDM Regulations acknowledges that risks should not always be dealt with at the construction stage and realises that the Designers can design out many of the high risk activities with some careful thought. The challenge facing Designers is the ability to seek out and discover or develop other techniques or construction methods to produce the same or similar results than a more inherently high risk option. At each stage the Designer is required to ask of himself, 'Can this design be constructed more safely without undue detriment to the other important factors including buildability, environmental impact and aesthetics?'

The ACOP requires the Designer to balance the risk of a certain feature of the design against the cost of excluding that feature by:

1. designing to avoid risks to health and safety;

2. tackling the causes of risks at source; or if this is not possible,

3. reducing and controlling the effects of risks by means aimed at protecting anyone at work who might be affected by the risks and so yielding the greatest benefit.

This is a very big change in the duties of the Designer, who used to leave the problems of risk assessment to the Contractor, but who now has a major role in the improvement of health and safety.

2.1.3 *Principal Contractor*

With the creation of the health and safety plan in the tender documents all Contractors can allow for managing health and safety in their prices. This, in theory, will allow the management of health and safety to be taken seriously and to have its own budget.

Regulation 16(1)(e) provides:

The Principal Contractor appointed for any project shall -
(e) promptly provide the planning supervisor with any information which -
(ii) it is reasonable to assume the planning supervisor would include in the health and safety file...

This provides that the Contractor passes on information to the Planning Supervisor in order for them to compile the health and safety file. This is a change in their role for which they are used to producing their own Operation and Maintenance Manuals.

2.1.4 *The Planning Supervisor*

This new role of Planning Supervisor introduced by the CDM Regulations can be a company, partnership or individual. There is no requirement for the Planning Supervisor to be from one professional background or another - engineers, architects, surveyors, safety professionals or others will all be eligible to fulfil the role. There is no requirement as to any particular level of qualification, so that technician grades and tradesmen could equally be suitable subject to the assessment of competence which is discussed below. However, it should be recognised that the Planning Supervisor needs familiarity and knowledge of the design process which will eliminate a number of these from taking on the role.The competence of the Planning Supervisor is covered in Regulation 8. It is important to recognise here that it is only the Client who can appoint the Planning Supervisor and consequently must assess the Planning Supervisor's competence. The remaining duty holders can be appointed by others and consequently qualified persons can assess their competence. Although the guidance to Regulation 8 contained within the ACOP suggest that 'reasonable enquires only are required' there is little in the way of further guidance as to what is reasonable. For a Client unfamiliar with the construction industry their assessment of competence may vary greatly with that of a Client with experience in construction. This shows how difficult to pin-point competence and its elusive qualities and demonstrates how difficult it will be to prove or disprove competence. The requirements upon the Planning Supervisor are outlined in Regulation 14. In the ACOP relating to Regulation 14 the first paragraph explaining the Planning Supervisor's role states that:

'The Planning Supervisor appointed by the Client, has to CO-ORDINATE the health and safety aspects of the project design and planning.'

Throughout the CDM Regulations and the supporting guidance the role of the Planning Supervisor as a 'CO-ORDINATOR' is stressed. In their role there is little in the way of 'supervising' to be done. Therefore there seems to be an error in the naming of the Planning Supervisor and a name that better reflects their role should have been attributed to them. Titles such as 'Health and Safety Co-ordinator' or even 'Planning Co-ordinator' would have been more descriptive. It is therefore difficult to understand why the title of Planning Supervisor, which can confuse the issue, has been used. It is important to realise that whether they are supervisors or co-ordinators they hold no powers over anyone on the project and as such are named correctly from this point of view.

2.1.5 *A need for adequate Resources*

The CDM Regulations require that the persons fulfilling the different roles demonstrate that they have allocated or intend to allocate adequate resources to health and safety. This area of the CDM Regulations is less than clear because it does not specify who or what should be allocating the resources. Does this refer to the individual or the company or region that the individual is employed by? It needs to be more project specific and clearer as to where the resources should be coming from.

3 RESEARCH FOCUSING ON THE CDM REGULATIONS 1994

The areas that have been researched regarding the CDM Regulations are explained as follows Attitudes to introduction. This will look at the views of those involved with health and safety management as to whether they regard the CDM Regulations as a useful tool in combating injuries and accidents on the construction site. Also at the introduction were the duties and expectations clear to the parties involved?

Training - An important issue with the introduction of new regulations is the area of training. Can compliance with the CDM Regulations been done satisfactorily with out some degree of formal training? What are the views of the duty holders and if they wish to undertake some training are their satisfactory courses available provided by relevant and qualified professionals?

Time/Communication - Inevitably the introduction of new legislation will bring about an increase in paperwork in order to comply. If this paperwork is a necessary and worthwhile part of the process then it is a significant aspect. However there may be some parts that are not required and are wasting the time of the parties involved. This can be linked in to the communication between the parties that are working together and if the introduction of the CDM Regulations has helped in this area at all.

The Health and Safety Executive - The HSE are the only enforcement authority and therefore have a key duty in the application and monitoring of the CDM Regulations. Is their targeting and guidance relating to specific duty holders sufficient and as an organisation do their have the resources to adequately enforce the CDM Regulations?

Designers - Under the CDM Regulations the Designer has major new duties to incorporate safety in to their design wherever possible. Is this being heeded by the Designers and are they meeting their obligations?

Planning Supervisor - The creation of this new health and safety co-ordinator is the most significant change brought about by the CDM Regulations and brings with it a new set of duties and responsibilities. There are a large number of aspects connected with this new role and the second model has been devoted to this.

3.1 *The Role of the Planning Supervisor*

The areas to be researched regarding the Planning Supervisor and their new role are explained below:

Client - With the introduction of the CDM Regulations the Client is one party that has new duties. They are different from the fact that on many projects they will not be construction professionals and therefore in many cases may not be aware of the duties required of them. One of the Client's most important duties is the appointment of the Planning Supervisor and the timing of this can be very important to the rest of the project.

Contractor - At present the Planning Supervisor is not required to go on to site unless at the request of the Client. As mention before many of the Clients are not construction professionals and having the Planning Supervisor going on to site to check progress and then reporting back to the Client as an advisor may be a useful function.

Position and Size - The appointment of a Planning Supervisor is generally corporate and the day to day operations are down to one person. On a multi-million pound project can a single person perform the functions of Planning Supervisor satisfactorily or would the creation of a team be a better alternative? The role of the Planning Supervisor can be performed by a number of people involved in the project. The two most popular are as part of the design team or as an independent party. Is either one of these better?

Health and Safety Plan - One of the Planning Supervisor's duties is to ensure that the health and safety plan is prepared. Should the health and safety plans prepared, especially the pre-tender health and safety plans, be project specific rather than containing general health and safety information?

Health and Safety File - The preparation of the health and safety file is currently the duty of the Planning Supervisor. As part of this duty they have to collect all the information relevant to the health and safety file from the Principal Contractor who it could be said is in a better position to assemble the file. The Principal Contractor is closer to the construction and could possibly have a better idea of what is required by the file. The Planning Supervisor, when the file is complete, has to pass it on to the Client for them to safe keep as you would other important documents such as the lease or mortgage. This is were the Client must realise the value of the health and safety file.

Enforcement Powers - Currently the Planning Supervisor has no powers and acts in a role co-ordinating the health and safety management of the project. As the duty holders go the Planning Supervisor is the only one with no powers or authority. For example the Client can hold back payment and the Principal Contractor can hold back payment to the subs or hold up the works. It may be that the Planning Supervisor should have more enforcement powers or at least a link with the HSE to inform them of anything untoward.

The Association of Planning Supervisors (APS) - The APS is the advisory and representative body to assist in the development of Planning Supervisors. Being a young organisation is it able to provide those who require it with adequate and useful information?

Competence - This is an area that is often very difficult to measure in any context. What skills and qualities does the ideal Planning Supervisor need to possess to perform the duties required of them? What mixture of construction and health and safety experience and knowledge is needed? With the possible introduction of NVQ's in the role of Planning Supervisor who is aware of them and are they a good idea.

4 METHODOLOGY

The groups and individuals used to obtain information are listed below:

1. Contractors - On a project Contractors are used to dealing with health and safety issues especially before the introduction of the CDM Regulations when they had to produce an Operations and Maintenance Manual. Therefore they are in a prime position to assess the differences and changes that have taken place.
2. Clients - Before the introduction of the CDM Regulations the Client of a project had little or nothing to do with the management of health and safety. Their role has therefore changed dramatically and their views are very important.
3. Consultants - On a project Consultants can be involved in a number of areas including design and supervision of the whole project. Given this role in a project they may be able to provide a more overall view of the CDM Regulations.
4. Planning Supervisors - Being a completely new role created by the CDM Regulations they are a vital source of information.
5. Independent Specialist - It is also important to consider the view of those not directly involved in the day to day operation of the CDM Regulations. They may be able to express opinions and judgments given the fact that they are not so closely involved.
6. The Health and Safety Executive - Being the enforcing body their views on how they perceive the CDM Regulations at their introduction and since is a vital source of information.

With these groups and individuals it was felt that they would cover the majority of the areas that needed to be studied. The interviews also helped to highlight a number of other areas that could be incorporated in the survey.

4.1 *Survey*

Forty eight questionnaires were sent out, the majority to Planning Supervisors, and a few to Clients and Contractors. The Planning Supervisors' names were obtained from the Association of Planning Supervisors. The companies that these Planning Supervisors work for came from a number of different backgrounds such as Civil Engineering, Architecture, Water Engineering and Demolition

4.2 *Background and experience of respondents*

The sample consisted of Directors/Chairman, Safety Advisors/Executives and Project Managers. These are all senior enough positions to assume that they are very involved in the running of the CDM Regulations either in their own firm or for another. Approximately one third had a professional background in Civil Engineering and a quarter trained as an Architect or Designer. Only about thirteen percent had a background in health and safety. Just over seventy percent had had only between 0-10 years experience in health and safety. Of these, 74% of them were members of a health and safety professional body.

5 RESULTS & ANALYSIS

5.1 *CDM - a natural progression to help reduce the accidents and injuries sustained on site.*

In the survey ninety one percent of the respondents believed that the CDM Regulations were a natural progression in health and safety management.Eighty two percent of those asked felt that the CDM Regulations would reduce the number of accidents in construction. Only nine percent believed that the number of accidents would not be reduced. They felt that Contractors were still not aware of their role in the regulations and the responsibilities that they now had. They also felt that structures were not being designed with safety at the forefront and that designers were not adjusting to their new role. One respondent claimed that most accidents that occurred were slips, trips and falls and this area has not been addressed by the regulations.

The CDM Regulations have taken the idea that safety should be thought of not just in the construction life of the project but in to the operation and maintenance period. Of those questioned seventy eight percent thought that the regulations had highlighted the area of safety in the operations period an concept that in the past was so often overlooked.

5.2 *Formal training*

In the survey ninety five percent of those who responded had had some type of training related to the CDM Regulations. The two most common ways that they had received this training was either in-house and company training or at lectures and seminars. However, just over a quarter felt that the courses provided were inadequate. This possibly meant that some of the people that had undergone training thought that it was insufficient.

It appears from most of the evidence in the research that the majority of the parties involved in the CDM Regulations have undertaken some type of formal training although less are satisfied that there are enough courses available provided by relevant and qualified professionals. The importance of the training can only be recognised when some parties lack it sufficiently and cost others time and money.

5.3 *The duty holders have a closer working relationship and communication has improved.*

Virtually all the evidence collected from the interviews and the survey strongly support this proposition. Contractors have found that there has been a significant increase in the number of meetings that they were attending. This was thought to be useful as it developed in to more communication between the parties even at the extra cost and time. They were being kept up to date and were able to keep the other parties up to date with any information or changes.

Clients feel that they have been kept in touch with events going on at the regular meetings that take place and their communication with all the parties has been good. The more experienced Client, currently coming to the end of a project, feels that he has become far more involved in the project especially regarding the future use and plans for the structure.

Planning Supervisors see it as an important part of their role to improve communication between the duty holders. They have found it a useful aspect of the regulations to involve everyone, especially the Client and Designer, in the project team and share the responsibility rather than load it on one party's shoulders. Even in light of the improvements they still find problems in acquiring information off some parties which in turn makes his job more time consuming.

Eighty seven percent of the respondents believed that communications had been improved under the CDM Regulations.

A major reason why they felt this was the case is the placement of responsibility on the parties that had previously avoided it. People are being forced to communicate with the other parties on the project and then realising that it can be a useful tool to understanding and smoother completion of the project.

A framework has now been created for construction health and safety within which communication is likely to be more effective. It has set formal communication channels for the flow of safety related information and planning which encourages people to involve the other parties. If the Planning Supervisor does his job properly then this should improve awareness and communication.

It can be seen that there have been major improvements in the working relationships of the duty holders and most seem to value this highly even at the increase in cost or time spent on it. It allows for the smoother running of projects if any problems that arise can be talked through and resolved early and in an efficient manner. Not only can it help the health and safety aspects of a project but many other areas if the communication links are strong.

Virtually all the evidence from the results collected strongly support the proposition that since the introduction of the CDM Regulations the duty holders have a closer working relationship and communication between them has improved.

5.4 *The Health and Safety Executive, is failing to provide sufficient guidance to the specific duty holders and has inadequate resources to enforce the CDM Regulations.*

All the views and results show that the HSE is failing to provide sufficient targeting and guidance and as an organisation is lacking in resources to enforce the regulation.

In the survey, when asked if they thought that the HSE's guidance for Planning Supervisors was sufficient, approximately one third said no. They feel that there should be more detail in the ACOP regarding training , experience and qualifications and there should be minimum standards of these that the Planning Supervisors have to follow.

In the survey seventy eight percent of the respondents believed that the HSE needed to target Clients in a better manner to help increase their awareness of the existence of the CDM Regulations and the role that they play in them. A number of suggestions were given as to ways in which the HSE could achieve improved targeting of the Client.

The key is that Clients need to be targeted before they have even thought of a project or very early on in the project life. Advertising through direct mail to health and safety officers in business and commerce would be a possible way to make them aware even before the start of a project. As soon as it is known that, for example, a company intends to expand through construction the HSE could send them information to notify them of there duties and the possible consequences that they will suffer if they fail to comply with them. This could also relate to the Client when they have just applied planning permission the HSE could write or visit them to inform them of their duties.

Possibly the most realistic and most commonly suggested approach that the HSE could take would be to concentrate on the duty holders that are in the construction industry. As they are the first point of contact for the Client the HSE can stress to them the need to inform them of their duties and if they are unsure to contact the HSE directly. This requires good coverage of the duty holders in the industry so that they inform the Client to a sufficient level.

In the survey, eighty seven percent felt that the HSE lacked the resources to adequately enforce the CDM Regulations. Only four percent were confident in the ability of the HSE. This shows a serious lack of confidence in the HSE and this may lead to other problems. People may feel that if the HSE don't care about the CDM Regulations then why should they and they may try to cut corners and get aware with breaching the regulations confident that no one is monitoring them.

5.5 *The health and safety plan should be project specific*

Of the respondents to this question sixty eight percent felt that the health and safety plan should be project specific. Many people say that health and safety plans are filled out with information regarding the companies safety policy and has no relation to the project. They try to pass this off so that they can give the same plan out for each job that they do. There is a call for all the paperwork in it to be job specific.

This evidence seems to support this proposition that the health and safety plan should be project specific and in fact all paperwork should be job specific.

5.6 *The best position for a Planning Supervisor is in the design team*

Results from the survey show that sixty five percent of the respondents believed that the best position for the Planning Supervisor is as an integral part of the design team. Ninety one percent of them thought that the independent Planning Supervisors were in the role in order to capitalise on the business opportunities that are available.

From the evidence that has been collected it seems that the best position for the Planning Supervisor is as part of the design team. Even though many of the independent organisations have their advantages many people feel that when things settle down they will not receive as much work as at present many are only in the role because of the monetary benefits.

5.7 *The role of Planning Supervisor is vital to successful implementation of the CDM Regulations.*

In the questionnaire eighty three percent of the respondents felt that the role of Planning Supervisor is vital to the successful implementation of the CDM Regulations. It must however be remembered when looking at these figures that the majority of the respondents are in fact Planning Supervisors themselves.

Those who claimed that the role was not vital came up with some interesting arguments. They felt that as a co-ordinating role someone else could take responsibility for it such as the Principal Contractor who they feel completes the majority of it anyway. On small to medium projects the role of the Planning Supervisor is not vital because many parties co-ordinate well together already. Also the successful implementation of the CDM Regulations lies with the ability of the Client to understand their role and not on the Planning Supervisor.

It seems that different duty holders view the role of the Planning Supervisor differently. The Planning Supervisors and the Clients view the role as vital while the HSE and Contractor see it less importantly.

5.8 *The 'ideal' Planning Supervisor will have experience in construction and health and safety.*

In the survey ninety one percent of the respondents felt that the 'ideal' Planning Supervisor should have experience in both construction and health and safety. A number of definitions for competence for a Planning Supervisor were suggested by those who completed the survey.

The most popular was for someone with formal training, knowledge and experience in construction and health and safety. Some individuals thought that it was enough to have experience in just construction to be a competent Planning Supervisor. Other suggestions included someone with sufficient resources and understanding to implement the CDM Regulations on a large project or someone with awareness of all aspects of any one project and knowing the limitations of his/her knowledge seeking competent assistance as necessary.

Most of the evidence so far supports the need for a Planning Supervisor to have experience in construction and health and safety. However this is contrasted by the sixty eight percent of respondents who felt that a construction professional could complete a safety course and become a competent Planning Supervisor. Most of the people who felt that this could not be done gave the reason that a Planning Supervisor can only become competent with extensive training and experience coupled with existing knowledge.

Virtually all the evidence from the interviews and results support the proposition that competence for the 'ideal' Planning Supervisor should include experience in construction and health and safety. It should be included with training and knowledge of construction and health and safety.

5.9 *Amendments are needed to the CDM Regulations for health and safety management to function better in the future.*

Seventy percent of respondents to the survey felt that amendments needed to be made to the CDM Regulations in order for them to function better in the future and eighty seven percent were unaware of any amendments that may be made in the future.

The amendments suggested were divided in to three categories, areas that should be changed, added or removed. Firstly the important areas that were suggested to be changed will be looked at. The duties of the Planning Supervisor should given to the Principal Designer and the role of Planning Supervisor as a separate function should disappear. The amount of paperwork, such as the number of health and safety files produced, should be reduced as much of it is instigated by the Planning Supervisor in an attempt to justify their role. The health and safety plan should be smaller for smaller contracts and the general applicability of the regulations to smaller/shorter needs clarification.

Areas to be added include a clearer definition of the responsibilities of the duty holders. Pre-construction health and safety plans need to very specific with regards to the hazards on the job and when people fail to abide by the regulations there should be imprisonment for people at the head of the businesses concerned.

One area that an individual feels should be removed from the CDM Regulations the role of the Planning Supervisor until a clear definition can be found and their role can be set out more effectively.

It is clear from these suggestions that the evidence supports the proposition and there is a need for some amendments to be made to the CDM Regulations in order for health and safety management to function better in the future.

5.10 *The future for the CDM Regulations*

In order to ensure success in health and safety management in the future there may need to be changes to the CDM Regulations, the Health and Safety Executive and the attitudes of people in the industry.

From the interviews both the Contractors believe that the HSE must play a much bigger role in the enforcement of the CDM Regulations. This is backed up by many who completed the survey and feel that there needs to be high profile activity and stricter enforcement by the HSE. As with many areas of life the majority conform and the minority cause the problems. Most Contractors and Consultants take the CDM Regulations seriously while others particularly small Contractors don't. These are the companies that the HSE need to focus on. The targeting of certain duty holders needs to be improved. Clients need to be made more aware of their responsibilities from an early stage in a project while Designers need a greater education in the role that they are to provide.

The Specialist Inspector from the HSE believes that the CDM Regulations fit in with the ideas of the Latham Report (1994) regarding teamwork rather than the traditional adversarial approach. Some people from the survey feel that the spirit of the regulations, especially communication and teamwork, needs to be accepted by the industry.

One of the Contractors feels that the CDM Regulations may be the beginning of a new culture in health and safety by seeing a new facility as being more than a building project and looking in to the future beyond the construction and in to the operations period.

6 CONCLUSIONS

The CDM Regulations have been created under the philosophy of giving every party individual and specific duties. It is a new system designed to bring about a more efficient and structured approach to health and safety management. The single most significant change has been the introduction of a new role, namely the Planning Supervisor, whose main task it is to co-ordinate the parties and documents to bring about the successful management of health and safety on a project and beyond.

These conclusions on health and safety management and more specifically the CDM Regulations and the Planning Supervisor help to demonstrate the value of the report as very little active research has been completed on this subject. It is hoped that these conclusions will help to highlight the areas and issues that need to be addressed if the CDM Regulations are to change the face of health and safety management in the construction industry. Each paragraph contains a separate conclusion.

Before the introduction of the CDM Regulations health and safety in construction was managed unsuccessfully due to the role played by Contractors, the lack of responsibility accepted by other parties and the general lack of awareness of health and safety highlighted by the high accident rate in the industry. Contractors used to be solely responsible for health and safety and the other parties on a project were content to leave it in their hands. However when tendering a number of Contractors would cut their budget on health and safety to create a more attractive tender price.

The CDM Regulations are a natural progression in health and safety management and have highlighted safety after the construction stage and in to the operations period. Even though the regulations were introduced in line with an EU Directive they are an evolution not a revolution and have emphasised the importance of the new concept of considering the safety elements of a structure into its operations and maintenance period.

The introduction of the CDM Regulations has been welcomed by all parties involved and has helped to improve awareness of health and safety and in the process create a more structured and formalised approach to health and safety management concentrating on involvement from all parties to a project.

The majority of individuals involved in projects complying with the CDM Regulations have undergone some type of formal training although many are not fully satisfied that there are enough courses available provided by relevant and qualified professionals. However there are still a number of individuals who have failed to obtain any training and are therefore causing the main problems when it comes to operating the regulations.

Since the introduction of the CDM Regulations working relationships and communication between the duty holders has improved considerably primarily due to the placement of responsibility on to parties that had previously avoided it. A framework has now been created for construction health and safety within which communication is likely to be more effective. It has set formal communication channels for the flow of safety related information and planning which encourages people to involve the other parties.

The Health and Safety Executive lacks the resources to adequately enforce the CDM Regulations and the majority of people in the industry are less than confidence in their ability to perform their role. Their guidance for and targeting of the Client and Designer is particularly insufficient regarding their awareness of the regulations and education in them. High profile and well publicised prosecutions of a Client, Designer and Planning Supervisor are required in the future for people to take the CDM Regulations more seriously and for them to act as a deterrent. To date there have been a number of prosecutions but the parties involved in the regulations are totally unaware of them highlighting their lack of publicity.

Substantial amounts of paperwork completed in compliance with the CDM Regulations are not project specific. For example, the pre-tender health and safety plan in many cases is just completed with the safety policy of the company with no reference to the particular risks on the job.

Clients are unaware of the value and importance of the health and safety file. They fail to realise that the health and safety file should be treated with the importance of a lease or mortgage document mainly due to their lack of knowledge and general ignorance of the CDM Regulations.

The Planning Supervisor and the Client view the role of Planning Supervisor as vital to the successful implementation of the CDM Regulations. Other parties in the CDM Regulations did not view it as importantly as they felt that as a co-ordinating role someone else could take responsibility for it such as the Principal Contractor.

The best position for the Planning Supervisor is as an integral part of the design team to utilise their knowledge of the design process and to allow the easy passage of information. It is felt that the majority of the independent planning supervision organisations are in the role for the monetary gains and may phase out over the years to come.

The 'ideal' Planning Supervisor should have formal training, knowledge and experience in construction and health and safety. This will provide the most rounded Planning Supervisor capable of completing their duties to a satisfactory level.

Amendments, such as the clearer definition of duties and a reduction in the amount of paperwork, may need to be made to the CDM Regulations in order for them to function more effectively and improve the management of construction health and safety in the future.

6.1 *Recommendations to the Construction Industry*

The introduction of the CDM Regulations has helped to improve awareness of health and safety and it is important that this momentum is retained with health and safety continually highlighted by all parties in the industry. There must be monitoring and an improvement in the courses that are on offer provided by relevant and qualified professionals so that all individuals who undertake such a course are confident that they are receiving the best guidance and support.

The Health and Safety Executive must reassess its guidance and targeting of Clients and Designers to improve their awareness and education of the CDM Regulations. All duty holders need to target the Client to make them more aware of the value and importance of the health and safety file at an early stage in a project.

Individuals who set themselves up as Planning Supervisors in the future should do so in the environment of a design team and feel confident that they have adequate training, knowledge and experience in construction and health and safety to comply with their duties. The Association of Planning Supervisors should take responsibility not to award the title of Planning Supervisor to anyone without the relevant training, knowledge and experience.

REFERENCES

Latham, Sir Michael (1994) Constructing the Team, HMSO

Implementation of Safety and Health on Construction Sites, Singh, Hinze & Coble (eds)
© 1999 Balkema, Rotterdam, ISBN 90 5809 036 1

The integration of construction (design and management) regulations into small and medium companies

A.H.Tyler & A.Pope
Department of Civil and Building Engineering, Loughborough University, UK

ABSTRACT: Employers are duty bound under various legislation and codes of practices to provide for the health and safety of employees. The construction industry has one of the worst records for heath and safety and as a result has a bad reputation. In 1994, the Construction (Design and Management) Regulations were introduced to combat the problem. Industry knew about the well-publicised CDM regulations, which were introduced on 31st March 1996 but while some took them very seriously, others seemed very hazy in their knowledge. The bespoke nature of the construction industry causes many problems with integration of new ideas and trends and this paper will analyse the problems caused by the integration of the regulations for small and medium sized companies alike.

1. INTRODUCTION

The CDM Regulations were introduced as a result of an EU directive 'Implementation of minimal safety and health requirements at temporary or mobile construction sites'. It became the responsibility of member states to adopt the directive into national law and the instrument that achieved this in the UK was the Construction (Design and Management) Regulations 1994. They carry criminal sanctions – up to two years imprisonment and unlimited fines.

The regulations apply to widely defined construction sites and are applicable when the project is 'notifiable' to the HSE. This occurs when the construction phase is longer than 30 days or will involve more than 500 person days and where there are more than 5 persons carrying out construction work at any one time.

Their objectives are to ensure proper consideration of health and safety issues throughout each stage of the construction process – from inception to demolition including maintenance. Essentially it is a management solution - involving co-ordination in a fragmented industry but also by integrating the major parties in the construction process.

- The client will have direct responsibility to appoint competent planning supervisors, designers and contractors who must be properly resourced. Fundamentally if safety costs money – which it does – then it is right that the client should pay.
- The designer must take account of health and safety issues throughout his design including maintenance.

- The planning supervisor who must co-ordinate the health and safety issues and prepare a pre tender health and safety plan for the use of the contractor before the work commences, and upon completion prepare a file for the owner of the building.
- The contractor must prepare a health and safety plan and abide by the terms of the plan and ensure that his sub-contractors do likewise.

Against this background, a pilot study of small to medium sized building construction companies in the West Midlands was carried out to assess the impact and effects of the regulations. Only a limited sample of such firms were studied and it follows that only preliminary and tentative conclusions can be drawn from this initial research.

Selective interviews were also undertaken to further clarify the industries opinions of the regulations and discussions held with various professionals in the Construction Industry and an Official of the Heath and Safety Executive.

The objectives of this research are therefore to:-

1) Assess the effect of the regulations on a sample of small to medium sized West Midlands Building Construction Companies.
2) From the information obtained, to critically examine the regulations to establish what improvements could be made.
3) From published data to examine the views of a range of construction professionals and establish whether the regulations have improved site safety.

Research has been carried out from published sources across the spectrum of the construction process – to obtain a balanced view. The views of clients, consultants, planning supervisors, contractors, maintenance supervisors, demolition contractors and our tentative conclusions drawn as follows:

2. THE SCOPE OF THE REGULATIONS

The definition of 'construction works' is comprehensive and the reason for the inclusion of demolition and maintenance works are well founded. There are some conflicts regarding the extent of demolition works that could be clarified – on the other hand minor demolition works sensibly should not need to be notified – but it is difficult to envisage a safe satisfactory definition.

3. NOTIFIABLE PROJECTS

The definition of 'notifiable projects' is somewhat arbitrary and there appears to be no valid justification in HSE literature for the periods quoted in the regulations. Continental practice is different. In any event the requirements as between the limits of 30 days, 500 person days and maximum number of persons working on site at any one time seems incompatible. There is scope for further research on this subject.

Dramatic changes can be applied to scheme notifications merely by defining a series of projects as a Contract, but this seems a contrived arrangement.

4. THE ROLE OF THE CLIENT

We are persuaded by the argument that the principle of ultimate responsibility for health and safety should rest with the client as this will force the client to ensure adequate provision of fees. There is undoubtedly a cost to implementing safety – as 75% of the respondents thought – but it is right that the costs should apply to the project as a whole.

The requirement that the client should divulge whatever information he may have at the pre-tender stage to the Principal Contractor – must be beneficial. It ensures that contractors tender on the same basis and are saved the individual costs of collection of this information.

The philosophy of distributing responsibility for health and safety throughout the project team is sound and specifically that the client now has a share of responsibility. A clients main concern is the contract sum but he will now be forced to ensure that there are enough fees to cover the duties, and as a result health and safety on site is certain to improve.

5. THE ROLE OF THE DESIGNER

Designers have not previously been extensively involved in giving advice about systematic consideration of health and safety issues. This has been seen as exclusively the responsibility and domain of the Contractor. It was the expertise of the contractor to win contracts by not 'risk assessment', but 'risk taking'. Attitudes will surely change fundamentally now that designers have the new specific duty.

The role of the designer to 'avoid foreseeable risks' as a duty for all construction projects – whether notifiable or not – is a very onerous and responsible task. Yet nothing in the regulations formalises this requirement. This is surprising and unsatisfactory. The solution would be to include a requirement that the Designer should prepare a form listing parameters taken into consideration for safety purposes and concluding with a signed certificate/declaration. The notification of project form F10 should include details of the designer as equivalent status to the client, planning supervisor and principal contractor.

Will the regulations inhibit innovative and bold designs? There is the responsibility to design in a way that 'avoids reduces or controls risk to health and safety in reasonably practicable' – so that buildings can be both constructed and maintained safely. If risks remain – due to some designed architectural or engineering feature perhaps – they have to be stated – so that a reliable and competent contractor can execute the works. As a consequence it is likely that the average designer, mindful of his duties, will seek to minimise risks and it follows that bold imaginative design will be the province only of large consultancies and heavily resourced clients because this element of design will be expensive to execute. Thus, in our view, designers will suffer, but this is the price of safety.

Fundamentally designers will have to reconcile conflicting requirements – the clients brief, aesthetic considerations, building regulations, budgetary constraints and now CDM.

6. THE ROLE OF THE PLANNING SUPERVISOR

This has been misinterpreted by many. There is much evidence that planning supervisors are preparing safety plans that are frequently verbose, over complicated and therefore the plans are not fully respected. All threads of the research confirmed this to be so.

The issue of 'competence' is on balance a difficult one. The lack of definition in the regulations is unsatisfactory.

But the issues of safety are not complex. They need to be readily understood by all engaged in the construction process from Director to labourer.

There is a concern that the Planning Supervisor, if lacking professional status, could be resented by other professionals.

The regulations make no attempt to specify the qualifications of the planning supervisor, merely requiring of the client to be 'reasonably satisfied' that the person he intends to select is 'competent to perform the functions of planning supervisor'.

Intrinsically safety issues are not complex ones requiring persons of professional standing to administer. They are essentially simple structured processes capable of comprehension by sensible, not only intellectually gifted people, and the HSE's principle of 'proportionality to the risk and complexity of each project' applies. Therefore it is only on large complex projects that Chartered status of the planning supervisor may be legitimate, with sound practical people able to deal with the smaller projects without Chartered status.

7. THE ROLE OF THE CONTRACTOR

The introduction of the CDM regulations will be beneficial for the principal contractor. No longer will the issue of safety be his sole responsibility. Although still responsible for safety on the site he will have a specific document which will guide him through the health and safety requirements. The mere existence of the document will act as an 'aide memoire'.

Of considerable assistance to him will be the fact that he has won the contract in the knowledge of the requirements of the regulations and will have built into his prices the cost to cover them.

The regulations will also assist in the control of other contractors on the site.

8. THE HEALTH AND SAFETY PLAN

There is much evidence to demonstrate that most planning supervisors have made the safety plan unnecessarily complicated and too verbose. This is a serious fault and by adopting such wrong practices the danger is that the real safety issues will become obscured and forgotten. The firms who advertise computer packages have a tendency to further complicate the simplicity of the process.

The reason for this appears to be that a number of planning supervisors are attempting to make a profession out of the relatively straight forward job of planning supervisor and surrounding the task with a mystique.

There is ample expertise in the large consultancies to prepare the correct sort of plan for the more complex projects, but it is in the mundane building construction work in which difficulties have arisen. The HSE have distributed much information and advised upon simplicity and brevity but this is consistently being ignored – the time has come therefore for the HSE to reinforce the message.

9. THE HEALTH AND SAFETY FILE

There can be no doubt that a formalised record of what is in effect a normal maintenance manual of a building or structure, enlarged to alert those who will be responsible for future maintenance upon any risks that may be associated with the building, can only be beneficial. There is a good case for a legal requirement to ensure that clients update the file throughout the life of a building.

10. THE ROLE OF THE HSE

The introduction of the regulations will in itself have had a profound and lasting effect on construction site safety. It has focussed all construction professions minds upon the issues and this can only be beneficial. There has been extensive publicity and support provided by the HSE.

There was unanimous agreement that HSE were understaffed and this is an issue for central government to address It seems clear that the Executive are not intending to police the Regulations by any form of positive control. The implementation of the Regulations has caused a considerable increase in the Executive's workload without an appropriate increase in staffing levels. However, no doubt prosecutions will, as at present, follow when there have been serious accidents consequent to poor on site safety standards.

11. THE VIEWS OF SMALL TO MEDIUM SIZED COMPANIES

It has got to be emphasised that the sample of companies from whom responses were received was very small and therefore only preliminary and tentative conclusions can be drawn. Bearing that in mind the following conclusions are drawn from the survey of West Midlands building contractors in the small to medium category but excluding large public companies.

There is almost universal awareness of the existence of the regulations except in respect of firms with less than 5 employees.

Seventy five percent of the firms agreed that projects they dealt with were notifiable, which suggests that the efforts of CDM are being felt across a wide sector of the construction industry.

Many firms undertake their own health and safety plans to save costs. Invariable the companies worked as principal contractors.

Seventy five percent of respondents reported that the contract sum increased when tendering for projects under the CDM regulations, implying that a sum has to be included for safety considerations.

Tender sum was believed to be the main determinant when the client considers which firm is to be awarded the contract.

Only a few respondents thought that the client was concerned about the contractors safety record.

Generally safety procedures on site were perceived as being better as a result of CDM, and the existence of a health and safety plan was welcomed.

The regulations themselves were not considered to be easily interpreted. Overall, although CDM is helpful in the construction business it is not considered beneficial to the client.

12. THE VIEW OF SMALL COMPANIES

As with the analysis of small to medium sized companies only a small sample of results was obtained and conclusions drawn must be considered tentative.

There is considerable ignorance about even the existence of the Regulations for this sector of the construction industry. This is so despite the existence of the domestic exemption. It was evident that notifiable work is being done without the knowledge of the HSE and this must be a cause for concern.

The question of the legitimacy of the domestic exemption arises. It is likely that many small contractors are likely to continue their working practices without ever experiencing the rigours of the regulations.

For those contractors who did apply CDM, it was unanimously considered that the imposition of the regulations has increased the contract sum, in other words health and safety plans have a real cost. All the other evidence confirms this to be so – but this is fully to be expected.

It was disappointing to see that Contractors believe that the clients' first concern is the tender price and not the safety reputation of the firm, but this will doubtless change over time.

The ultimate test of how successful the introduction of the regulations has been will be the extent to which construction site safety has improved nationally. Unfortunately after such a short time since their introduction there is insufficient evidence to enable a meaningful conclusion. The results are obscured by other issues such as the upturn in construction activity. We have no doubt that when further research has been done there will be every likelihood of substantial improvements.

The value of CDM is that it imposes requirements on all concerned in the construction process in a structured way – and despite inevitable teething troubles – the goal of reducing dangers to health and safety will, we are sure, be achieved.

BIBLIOGRAPHY

Barnard, M. 1996. *Making CDM an Asset not a Liability.* ICE Introductory notes.

Bielby, S. 1992. *Site Safety.* CIRIA Special Publication 90.

Birkby, G. 1996. *CDM in practice and legal responsibilities.* ICE

Bone, S. 1995. *HSE Report No 72/95.* HMSO

Burkett, J. 1993. *Spreading the Net.* Building Magazine 8/1/93.

Cawthorne, R. 1996. *CDM Regulations on the Ministry of Defence Estate.* Thomas Telford.

Fear, K. 1994. *The Views and Experience of the Designer Consultant Eur Ing Kevin Fear.* Posford Duvivier (Introductory Notes).

HSE, 1994. *Construction Design and Management Regulations 1994.* HMSO

HSE, 1995. *Health and Safety for small construction sites.* HMSO

HSE, 1995. *Managing Construction for Health and Safety, CDM 1994* ACOP. HMSO.

HSE. 1995. *Information Sheets No 38-44.* HSE Publications.

King, A. 1996. *Applying CDM in a District Council.* Introductory notes, pp 2.

Minogue, A. 1993. *HSE Law to Penalise the Client Hardest.* Building Magazine 8/1/93.

Nattrass, S. 1996. *CDM: Better Design and Management – or Just More Paper?* ICE Conference

Scholey, J. 1996. *Competence – are you reasonably satisfied?* ICE Conference.

Implementation of Safety and Health on Construction Sites, Singh, Hinze & Coble (eds)
© 1999 Balkema, Rotterdam, ISBN 90 5809 036 1

Construction shortcuts: An invitation to disaster

August W. Domel Jr
Engineering Systems Incorporated, Aurora, Ill., USA

ABSTRACT: The competitive nature of the construction industry necessitates the contractor complete a project in the shortest time as a matter of financial survival. Safety may be sacrificed when speed of construction is the top priority. Shortcuts taken during construction leading to disaster can be as simple as a laborer not wearing personal protection equipment or as complex as a failure to build a structure in the proper sequence. There are several technically induced failures that plague the construction industry on a regular basis. These failures could be easily avoided, if field personnel understood the potential for disaster of some of their shortcuts. The shortcuts discussed here are: absence of permanent bracing, failure to install construction bracing, and improper planning.

1. INTRODUCTION

The construction industry is a slave to time. Schedules must be followed and deadlines must be met to make a project financially successful. An elongated project not only increases direct field costs (increased labor hours, additional wear and tear on equipment etc.), it also increases those costs not as obvious (insurance and bonding premiums, cost of administrative staff, etc.). Usually the faster a contractor completes the project the greater the profit. Also, many construction contracts provide for penalties for late completion of a project.

Speed of construction can be the nemesis of safety. Rushed laborers may make errors in judgement causing injury to themselves, other workers, and the general public. This discussion does not focus on this type of error by the individual worker. Rather, this discussion presents common reoccurring errors in judgement when choosing shortcuts that have ramifications on the integrity of the structure.

Project superintendents and other site management personnel make decisions on the means and methods for construction. Decisions to take shortcuts to save time and money are always welcome on a project provided that safety and structural stability are never compromised. However, there are mistakes related to means and methods that the author sees repetitively in the construction industry. These reoccurring mistakes cause structural collapses and, subsequently the dangers associated with a collapse.

This discussion will present three types of technical decisions that produce unnecessary structural failures. These three types are: absence of permanent bracing, failure to install construction bracing, and improper planning. These three types of failures were chosen since they represent a significant number of failures that can be prevented with little effort. Each discussion will include suggestions on administrative procedures to avoid these problems.

2. ABSENCE OF PERMANENT BRACING

First, it is important for this discussion to distinguish between permanent and construction bracing. Permanent bracing is needed for a structure to perform as designed by the engineer, both during its service life, and possibly during construction. Temporary bracing is needed to prevent a member or members from collapsing during erection. This section will discuss permanent bracing. Construction bracing will be discussed in the next section.

Generally, permanent bracing does not participate as a main load carrying member in the structure as do the beams, walls and columns. The main load carrying members are designed to be in the direct load carrying path. Permanent bracing function is to buttress these main load carrying members. This bracing is often referred to as a secondary member but one should never underestimate its significance.

Examples of permanent bracing include bracing between bridge girders (Figure 1), cross bracing for buildings (Figure 2), and x-bracing for wood floor joists. Contrary to what some contractors believe, this bracing is often needed during construction, as well as during the life of the structure. Contractors often will omit permanent bracing during construction to allow for adjustments to the main members or to provide unencumbered access in and around the structure. This time saving maneuver can result in temporary instability of the structure. The example that will be presented will be the impact of the absence of permanent bracing during construction of a bridge.

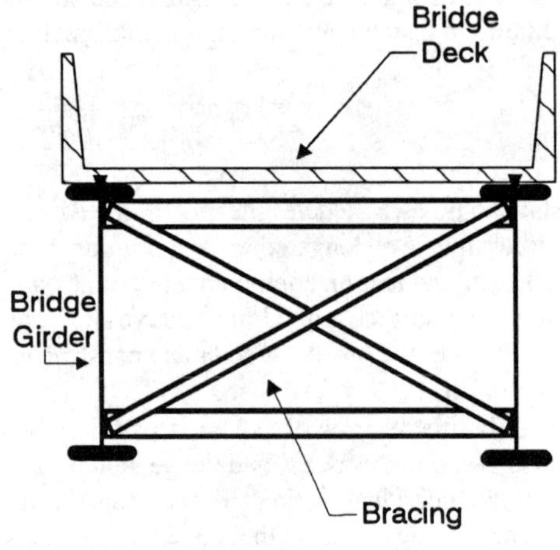

Figure 1. X-bracing for bridge girders.

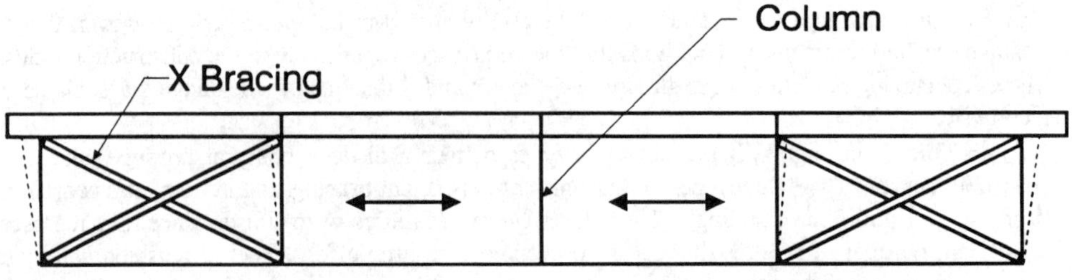

Figure 2. Wind Bracing in a building frame system.

The cross bracing shown in Figure 1 provides lateral bracing for the compression flange of a steel beam, which in most cases will be the top flange. As a beam bends, the compression flange has a tendency to bow laterally, as shown in Figure 3. This condition exists during the life of the structure when the bridge is carrying traffic and during erection under the beam's own weight. If the lateral bowing is not restrained, the girder will become unstable and deflect possibly until it reaches failure.

Figure 4 shows the results when bridge beams had their permanent bracing removed after erection. These beams were in excess of 14 feet deep. During fabrication, holes had been mis-drilled on the connection plates for the permanent bracing. This resulted in the permanent bracing not fitting properly between adjacent beams, and distortion in the beams themselves. The field personnel decided to remove practically all the bracing to facilitate the quick redrilling of holes. In the absence of the permanent bracing, the bridge beams became unstable, and began to bow laterally at the compression flange until they collapsed. Figure 5 shows a portion of the failed beam, and the massive distortion that resulted. This could have been avoided if the bracing were removed one at a time.

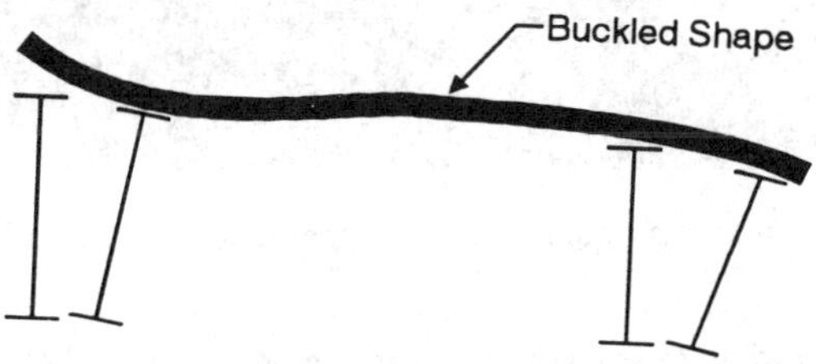

Figure 3. Girder displacement when no x-bracing is utilized.

Figure 4. Collapse of bridge structure during construction due to inadequate bracing.

Figure 5. Severe deformations of a bridge girder that failed due to inadequate bracing.

The solution for avoiding this type of collapse can only be achieved by proper education of those working on site. This education has to come from personnel with a greater understanding of structures than the typical field personnel. To avoid these type of failures it is suggested that the technical staff do the following:

1. Review the plans and specifications and make a list of all members that function as permanent bracing.

2. Determine if the permanent bracing is needed for stability during construction.

3. Review the construction schedule to determine if the installation of permanent bracing, along with the members that are being braced, occurs at the same time.
4. If permanent bracing is not installed at the proper time, the stability of the affected members needs to be reviewed to prevent a collapse.

3. FAILURE TO INSTALL CONSTRUCTION BRACING

While a building is under construction, it lacks the structural integrity of a completed building. Beams, columns and walls that will be fully braced, secured, and supported when the building has been completed are only temporarily connected, lack load carrying redundancy and are absent of supports during the time it is under construction. To avoid collapse during this vulnerable stage in their life, structures must be provided with temporary bracing.

The amount and location of temporary bracing are unique to the type of structure that is being erected. The temporary bracing needed for a steel frame building is much different from that needed for wood roof trusses. Each type of project requires that the construction sequence be analyzed to determine the stages when the structure will be vulnerable to collapse due to the potential for temporary instability. The example presented here will be the installation of prefabricated wood trusses.

Prefabricated wood trusses are utilized in residential construction for spans from 10 feet to 40 feet and in commercial buildings for spans up to 80 feet. Their wide spread use, economical costs and ease of installation make them a very viable structural system. Despite their wide use there is a lack of knowledge among contractors regarding the proper bracing methods. This lack of knowledge results in a significant amount of collapses each year with the accompanying potential for personal injuries.

Wood trusses are highly susceptible to collapse during construction. These long span members, which can be in excess of ten feet high on pitched roofs are constructed of 2 x 4 and 2 x 6 lumber. This results in a member that is prone to tipping. When the structure is complete, the trusses will be part of a rigid roof system that will be comprised of a series of trusses connected rigidly by the plywood roof sheathing. Until the roof sheathing is installed, the trusses have a tendency to tip to the side even with some bracing.

Requirements for the temporary bracing of wood trusses and other pertinent information is provided by the Truss Plate Institute located in Madison, Wisconsin. Proper bracing of wood trusses requires braces in many different locations. Figure 6 provides a bracing layout that is required by Truss Plate Institute document "Handling and Erecting Wood Trusses."

Workers often omit the web bracing and only install the lateral bracing. Figure 7 shows the consequences of this omission. In the absence of the web bracing the trusses, although connected by lateral bracing, tend to flop to one side.

To understand what typically transpires at the site one needs to understand how the temporary bracing information is provided to the field personnel. The standard procedure for passing on the bracing information to the field workers is to transmit it with the truck load of trusses. Upon arrival on site, the workers do not have the luxury to sit back and read the truss bracing instructions. Rather, scheduling pressures require installation begin immediately. For this reason it is incumbent on the contractor to have a procedure in place to insure that bracing will be properly installed.

Transferring knowledge of temporary bracing to the laborers should be initiated by the technical staff. Once this policy is set in motion it can be used on all similar projects. The extent of preparation may be limited by the size of the project. A possible procedure can be as follows:

1. Determine sequence of erection of the structure.

2. Identify elements which will be temporarily unstable during construction.

3. Determine the proper bracing methods for each instability scenario

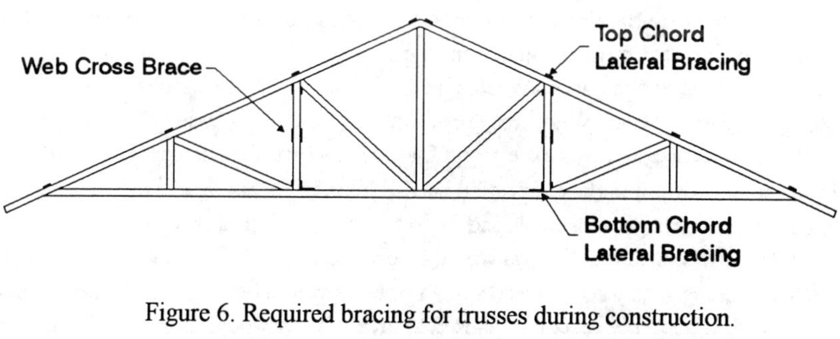

Figure 6. Required bracing for trusses during construction.

Figure 7. Wood trusses that toppled during installation due to lack of proper bracing.

4. Provide checklists, diagrams or other information to field personnel.

5. Specifically assign task of installing, or directing installation of bracing to an individual who will be involved in the work.

4. IMPROPER PLANNING

Improper planning for supplying manpower and material has a detrimental impact on the financial health of a project. It can also result in field personnel making dangerous decisions to accommodate the conditions caused by the lack of planning. Injuries and property damage resulting from these dangerous errors in judgement can be the death knells for a project.

During the estimating phases of a project, the means and methods that will be employed must be determined as well as the associated costs. The estimator must determine a practical and economical way to perform the construction tasks. In the event that the method assumed cannot be implemented the contractor must deviate from the original plan in a cost effective manner. One inappropriate way for the contractor to adjust is by using equipment ill-suited for the task in an effort not to exceed the budget. The adjustment may result in the sacrificing of safety. An example that will be discussed in this section involves excavation for the construction of a tank.

In the subject example the excavation needed to be in excess of 25 feet deep per the project drawings. To achieve this depth a steel pile and wood planking system was to be installed to retain the soil. The estimator envisioned that the soil would be excavated partially by laborers, and partially by the use of lightweight machinery such as a bobcat. The latter would be used after laborers' excavation proceeded to a depth where a bobcat could be used to break the soil loose and fit under the bracing, as shown in Figure 8.

The field personnel could not perform this task within the financial parameters of the original cost estimate if the bobcat was used. Therefore, they utilized a larger piece of machinery, a front end

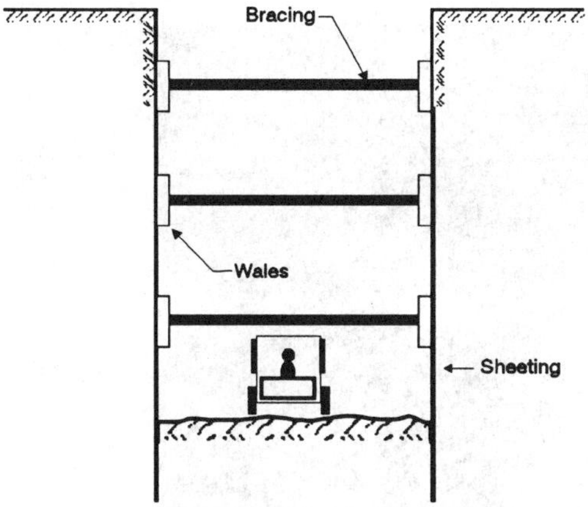

Figure 8. Excavation with small equipment

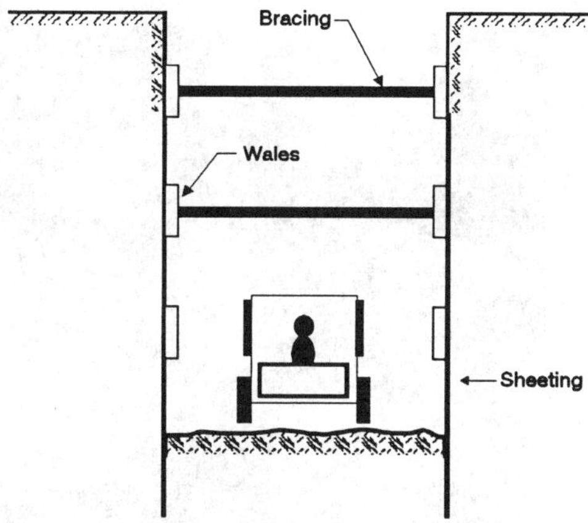

Figure 9. Excavation with large equipment

loader, which could only be used if the bracing was omitted, as shown in Figure 9. The absence of this bracing increased the loads on the piles beyond their capacity, and caused a failure of the soil restraint system. Figure 10 shows workers removing soil that bulged inward when the support system failed.

These types of failures are caused by a combination of poor planning of the office personnel coupled with pressure for cost reductions on the field personnel. To avoid these type of failures it is suggested that the following be done:

1. An estimator should determine the equipment necessary to complete the task.

2. Verify the equipment is available and can be used within the necessary parameters.

3. Train field personnel to recognize that it is inappropriate to modify structural systems to accommodate equipment. In fact, the opposite approach should be employed.

Some other areas of construction where poor planning can lead to a disaster are:

1. Demolition: The demolition of a structure can only be safely performed if there is a fundamental understanding of the structural system. The removal of structural members out of sequence can result in a premature collapse of the structure.

2. Leveling: It is not unusual that a floor of an older building may require jacking during renovation to level the floors. Jacking a single member results in a transfer of stresses and allows for overloading. Also, the slight movement of jacking a structure can bring out the worst in previously undetected construction or design flaws.

3. Sequence Performing tasks out of sequence or leaving work temporarily uncompleted to save time in an effort to keep up with the construction schedule can lead to problems. A classic example of this is an iron worker using a minimum amount of bolts to make a connection so that erection can proceed rapidly. The iron worker plans on finishing all of the connections at a later time to consolidate the task. In the meantime, the connections can become overloaded causing a sudden and unexpected failure.

Figure 10. Failed soil support system due to unapproved changes to allow larger equipment.

5. SUMMARY

The preceding discussion presented three types of structural failures that continue to occur even though their avoidance could be easily achieved. The failures are not the result of any global lack of knowledge in the construction or engineering industries. Rather, these failures occur because the global knowledge is not transferred to those controlling the work at the site.

Workers take shortcuts in construction. Shortcuts help in meeting deadlines and reducing physical labor. These shortcuts can often be an invitation to disaster as illustrated in the examples.

The importance of communication between those with technical knowledge and the workers can not be over emphasized. Management must make field personnel aware of what work is critical during construction to avoid creating an unstable structure. A shortcut should save time and money and avoid being an invitation to disaster.

Implementation of Safety and Health on Construction Sites, Singh, Hinze & Coble (eds)
© 1999 Balkema, Rotterdam, ISBN 90 5809 036 1

Historical and philosophical evolution of Italian laws about safety and health in construction sites

F.Terranova
University of Rome 'La Sapienza', School of Architecture, Department of Technological Innovation of Architecture and Culture of Environment, Italy

S. Bartolomeo
Sanitary National Service, Maternity and Child Department in Latina's Township Sanitary Firm, Italy

ABSTRACT: The management of safety in construction sites, due to the peculiarity of this kind of working place, represents the paradigm of the relationship between an "hard" science (engineering, architecture) and the labour factor, between technological innovation and muscular labour, between past and future. The concept of prevention, as a right and proper undertaking of measures meant to protect the safety and health of workers in all production fields, grows in Italy in the second half of the '50 years and since then a long way has been covered. Today, the reception of some European Community's policies has let Italy follow the same rules about safety in buildings sites available for all other E.U. countries, even if in the past each country has lived his own reality.

1 CONSIDERATIONS ON THE WORKERS "RIGHTS" ABOUT SAFETY

The Legislative Decrees about safety in the work sites and about the protection of the workers' health appears in the Italian normative framework twenty years after epic struggles for safety, in working places, in a political situation dominated by the occupational stalemate, and by a kind of fatalistic disengagement of those same forces that in the past had been more engaged in the battles for health that marked the '70 years.

Other countries had envied us the creativity and originality which those battles had been led.

The Legislative Decrees at which we referred, the 626/ 94 (general Act about safety) and the 494/ 96 (safety in construction sites), followed by the unavoidable Ministry explanatory Circulars, had to be applied in a situation politically stagnant, in which the lack of ideal strains and of battle spirit, and in which the resignation to the globalisation of the economy, had been imposed as the new religion of the humanity, thus suffering real risks.

Risks can he found not only and not as much in the emptying of innovative contents of the laws itself, but mostly, and above all, in their bureaucratic application and sanctionatory orientation, exactly the opposite of the objectives and willing of those crowds of workers who previously imposed and vindicated rules for the safety and the healthiness of the places of work.

However it remains a normative system that will have, according to the future political situations, more or less advanced applications.

The wish is a resumption of the collective conscience in the organizations (mostly in labour unions and political parties) that represent the lifeblood of any democracy.

suitable tools in order to keep the safety of the work environment and workers health protection as the thread of social negotiations, and labour collective agreements.

The management of safety in construction sites, due the peculiarity of this kind of working place, represents the paradigm of the relationship between an "hard" science (engineering, architecture) and the work factor, between technological innovation and muscular labour, between past and future.

A future pervaded by utopian fantasies and a reality made of accidents, invalidity, professional illness and death.

This is the tragic fate, more than in other categories, of the building industry workers. But is it then fate, fatality, and not unconcern and exploitation?

An endless controversy continuing since the land revenues joined the financial capital. The assignment is an honest contribution at the knowledge of the problem and to the correct use of technical instruments.

Knowledge and information represent milestones of the emancipation process of man (in this circumstance of the worker) and of the humanity as collective representation of the individuals.

The juridical system of the modem States, have, till now, registered events and facts, opinions and common sense.

It could be said that they have operated by consequence of the phenomenon, only after they have consolidated, and giving them a legitimization.

Till now, even in Countries where the transformations happens almost in real time, laws and rules are approved, jurisprudence is activated, only after the facts to which they are referred to have become a part of t he common aspirations of individual e and communities.

This, as more happens in Italy, as the history of our Country is deeply marked by a juridical legislative approach.

The legislative procedures and the juridical interpretations have "historic" times that contrast with a productive reality, and financial and trading trends extremely fast in transformations.

But it is necessary to assume that the things go this way, and if such procedures have not been modified, it means, as we live in a system of representative democracy, that the majority of the citizens approves such a system and, however, consider it more guarantistic than a “deregulated" environment.

In the specific the relationship "Production - Safe Work," that produces numerous specific and general juridical provisions, represent the level of update, deflected during the time but recurrent, of the state-of-the-art (innovation and conquests, or defeats, on the salary and on the normative about workers rights and duties) of two fundamental factors: the technical progress and the development of the trade-union relationships.

If we examine these two factors, and their relapses over the workers health, we have the following picture: the technical progress has marked an historical passage, from the mechanized productive cycle to the automated productive cycle.

The process is not yet fully completed and consolidated. But in the more anciently industrialized areas, and in the production strategic sectors, the process is accomplished, even if, as we will see, new problems are set, and their solution appears harder also because of a fall of values and ideals if compared with the past.

2 EVOLUTION OF THE "PRODUCTION - SALE WORK " RELATIONSHIP AND TECHNOLOGICAL INNOVATION

The mechanized productive process that leads its origin in the scientific organization of the work (Taylor) and in the Human Relations School (Mayo) is characterized by a decomposition, fragmentation and shattering of the human labour.

The consequence of this way of performing the work is a continuous repetition of the worker's actions, that you can shape as real micro-actions.

Whoever doesn't remember C. Chaplin 's "Modern times " labour?

The consequences, at individual worker's level of the personality are alienation, estrangement, anonymity, feeling of out-of-control circumstances, poor motivation, anxiety for reaching production targets, penalties for not maintenance of predetermined rates, and so on.

All the variables above exposed are factors of risk for the health of t he worker.

The intensity of the job and the muscular fatigue are two main reasons of accident and invalidity.

The use of certain substances in the productive cycle and their association, combined with the lack of suitable aspirant cloaks are common causes of professional illness.

Work in automated production processes is basically centered on the function of control.

That involves the presence of a quite different kind of worker if compared with the mechanized factory's one.

The worker has an a high school degree, a strong motivation, a contractual strength that makes him feel irreplaceable, a strong company spirit (particularly exasperated in Japanese industry).

Obviously mental fatigue is as dangerous as physical fatigue.

But, someone assumes, they exist all the conditions in order to carry out a line, still admittedly a trend, of a “zero” accident risk.

Even if theoretically things were this way, mental fatigue and VDT stress, stress by permanent adjournment, by recurrent selection tied to the evaluation of the performed job quality, by the implementation of new procedures of production control, by possible loss of varying salary (being this tied to productivity and quality), are all factors that certainly can predispose to our days' typically illnesses, the so-called "a-specific" ones.

The more remarkable problem is the management of the transition from the mechanized to automated product ion, where the coexistence of old and new production methods creates manifest contradictions, that is even more emphasized by the current globalisation process, where competition in the markets and competition between manufacturers in a financial virtually context is more and more determinant.

During this transition, the must, is increasing the firm productivity (the so-called rate of profit) by a reduction in expenditures, to be obtained operating upon the technological innovation and by increasing the production rates, according to the incoming orders.

The probability of human error, merely caused by fatigue, when increasing production rates and of overtime working and with the discontinuity of a no more scheduled but implemented on the orders basis production, becomes evidently common matter.

At this point opens itself the great chapter, rather the favourite of Confindustria. that at the basis of an accident there is always an human error.

The production is neutral.

The once faced solution was to set out a great campaign against the careless workers, ignoring some incontrovertible data such as: a) the prevention must be implemented in the moments of the planning and of the realization of productive process, b) labour illnesses are to a large extent imputable to the forcing of biological rhythms and to the exceeding of individual adaptation capacities; c) the charges attributed to the "human factor " are imputable to high work rates, to inadequate professional training, to heavy working times, etc., and also by external individual (as diet and others) and social (as transportation problems and others) conditions.

The labour union action follows step by step the production systems' evolution, is conditioned by them, and suffers of a limit: of no more bearing an innovative culture of work.

In fact, until '50 years, the labour union accepts the fatality ideology: "the accident is a price to be paid to economic and social growth of the Country".

The central point of the negotiation is salary and the so-called "risk indemnity", subsequently denominated, in contempt of this vision "magnetization for risk".

Health and money are not interchangeable goods; indemnities though high, don't pay back the faced risk and the suffered damage.

The goal is "safety in work". Still in not posed the safety "of" work goal.

As a matter of fact, without high employment levels, measures of prevention can unlikely be imposed.

The strategy, this time, is not centered on wages, but on legal rights of changing the working conditions inside the factory, such as: abolition of overtime hours; abolition of second job; compulsory turnover for those who carry out repetitive tasks , and so on.

This strategy needs strong alliances in order to succeed, above all within the scientific and technological research context, and in machines design.

In the while, a public action is necessary, that - through tools of control and/or vigilance - directs the Companies' choices in the direction of a correct localization, suggesting them some unbreakable technical features, that point out solutions for a production process looking at the human factor besides the firm arguments (working hours, turnover, pauses, medical preventive controls, etc.). The required level of close examination of the productive process and of the human factor is remarkable.

A labour "culture" of the prevention is drafted, that represents a redemption from the subjection suffered by some technician inside the factories.

Paradigmatic is the role of the work physician, that from selector becomes an ally of this potential victims, drawing new ways for the manufacturing process.

Only for memory we remember as meaningful steps for the formulation of the four factors of risk: 1.physical-technical risk (noise, lighting, micro-climatic conditions, etc.) 2. chemical risk (dusts, smog, smokes, toxic substances released as aerosol, etc.); 3. muscular work related risk and, above all, this is the newness, mental fatigue related risk.

An important scientific team, coordinated by the Professor Cesare Musatti elaborated a paper on "Stress and work " that was the basis for a re/negotiation of the "dangerous and harmful" jobs.

Another fundamental step is the risk mapping of each production department, done onto experience basis, by the same workers of each department (homogeneous group), an important phase for a engineering re-drawing of the production process itself, in terms of modalities, of environments, of paths, of materials used for containers and so on.

The importance of these facts, is not perhaps sufficiently valued and realized today, and not last, the remaking of the "risk theory" itself, projected and experimented for the nuclear industry, has allowed to the process industry (chemistry and petrochemistry) a remarkable progress "in safety".

So much commitment and fervour received an important acknowledgment, by the legislative power, with the approval, by the national Parliament of the 300/ 70 Act, also called "Brodolini Act", from the name of the syndicalist at the time Minister of Labour and Social Security.

At article 9 of the Statute a principle is expressed, sustained with peculiar tenacity as an alibi by the confindustrial background: the industrial secret, that will have subsequently devastating effects, with the 833/78 Act, for the institution of the National Sanitary Service (SSN).

This principle could be synthesized in t is way: "no limit could be set to the knowledge and to the control of the health state of the labours".

A strategic role is undertaken by the sanitary territorial services of the Local Administrations, that are authorized to investigate inside the firms, at firms' charge, but, chiefly, by the Magistracy, with the task of attending directly to the work organization, including production rates, the legitimization of individual and collective transfers, the means of control on working activity, individual and collective transfers, the means of control on

Table 1 Operative schedule for safety work

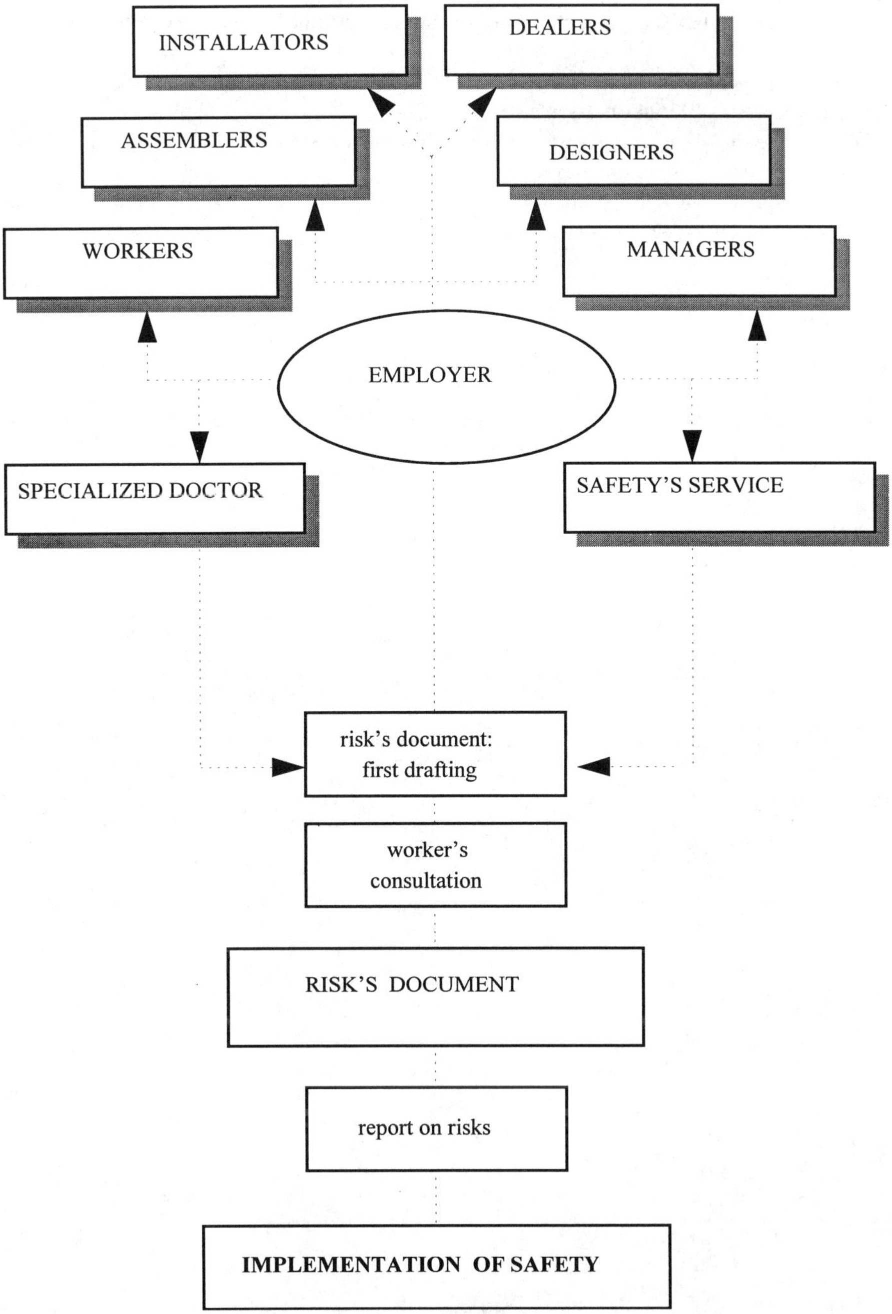

working activity, individual and collective firings, the suspensions and reductions of working time.

In the 1978 the Govern approved the law 833: it is a key law to prevention suggest a new model of society.

The period between ‘80 and ‘90 was negative for our Country (everyone remember *tangentopoli)*.

The national body laws is changing for the reception of some European Community’s directives:

- Legislative decrees n° 626/94 and 242/96 about "safety in every working sites";
- Legislative decree n° 493/94 on “sign's security ";
- Legislative decree n° 494/96 on "safety on construction sites";
- Legislative decree n° 459/96 on "device’s security ".

The goals of new laws are the labour’s free circulation and the implementation of safety and health in every working sites.

REFERENCES

Romeo C. 1994. *Manuale degli ambienti di lavoro*. Roma : Isedi.
Lepore M. 1996. *Prevenzione infortuni ed igiene del lavoro*.Roma : Epc.
Giofrè F., Piendibene C. 1998. *Costruire in sicurezza*. Roma: Officina della Carta.

Implementation of Safety and Health on Construction Sites, Singh, Hinze & Coble (eds)
© 1999 Balkema, Rotterdam, ISBN 90 5809 036 1

The new construction safety regulations in Spain

M.Casals, P.Alavedra & D.Giménez
Department of Construction Engineering, Technological University of Catalonia, Terrassa, Spain

ABSTRACT: On the 29th December of 1997 the Decree-Law RD.1627/1997 about Minimum Safety and Health Requirements on Construction Sites took effect. This was the transposition of the EU Directive 92/57/EEC.

Originally this Directive was inspired by the Spanish Decree-Law RD.555/1986. For this reason, in principle, it should be easy to adapt RD.1627/1997 into Spanish law.

In this paper we compare the EU Directive 92/57/EEC with the Spanish transposition and its application on construction projects. We remark on the differences in prevention processes, in the new figures with legal powers, etc,... and the new particular Spanish aspects in order to disseminate these aspects and differences.

1 INTRODUCTION

In Spain, the process of adapting the EU regulations about Safety & Health on Construction Sites was long and difficult. So, it was necessary in this process to accomplish two main objectives.

The first one was the necessity of adapting the Spanish regulations to the EU Directives.

The second one was the desire to maintain the successfully operating aspects of the Spanish regulations already in existence.

Finally, in October 1997 the Decree 1627/1997 about Safety & Health on Construction Sites was published. This took effect on 29th of December of that year.

The particular characteristics of this Decree, combined with some new aspects in relation with the EU Directive, necessitate discussion about it, comparison with the Directive and the giving of opinions on this new Construction Safety law.

2 HISTORICAL EVOLUTION OF SPANISH REGULATIONS

Safety on construction has historically been a preoccupation in Spain due to the importance of these activities in the national economy. Curiously we have even found a Royal Proclamation dating from the 18th Century on this topic.

In modern times, the great body of regulations starts in the thirties due to the recommendations of International Labour Office (OIT in Spanish) of 1937. In January of 1940 the Spanish Regulation of Safety and Health on Work took effect. Several chapters of these regulations were strongly related with construction (eg. Chapter VII dealing with scaffolding).

In 1952 Regulations of Safety on Work in Construction Industry focussing on these activities was redacted (BOE 1952). These Regulations were adapted in 1970 into Safety Regulations on Work in Construction, Glass and Ceramic Industries (BOE 1970).

In 1971 the General Regulations of Safety and Health in Work took effect (BOE 1971). These Regulations plus the Construction Regulations of 1970 were the Spanish framework for safety in construction until the well known Decree 555.

Finally in 1986 the Decree 555/1986 about Safety and Health on Construction was created (BOE 1986). This Decree forced the inclusion in projects with budgets over 100 M pesetas (around 650.000 US$) of the Safety & Health Project and the Safety & Health Plan in Construction Projects. In 1986 the Incidents Book was also included in construction in order to register any event that happens on site.

Recently in accordance with EU Regulations, the new Decree 1627/1997 was written which is explained in the next pages.

3 CURRENT SITUATION BEFORE NEW REGULATIONS

The Spanish construction sector had usually presented high accident rates. In order to reduce this, the Spanish government published the Decree 555/1986 in 1986. This Decree, well known by those who are working in Construction Safety issues, was partially the origin of the EU Directive 92/57/EEC.

With this laws the labour accident rates in Construction were decreasing continuously with some significant exceptions until the 1997 situation. For example the years immediately prior to 1992 (Olympic Games in Barcelona, EXPO in Seville...) the rates worsened. It is already known that the labour accident rates are related to economic activity (Alves Dias 1996).

The three years after 1992, show stabilised labour accident rates in construction. For example in 1995, 117.490 accidents were reported in construction, that comprises nearly 20% of total labour accidents in Spain. It is necessary to remark that a large economic crisis in general and in construction in particular affected the activity in Spain in these years.

Another example is that in 1997, in a period of increasing construction activity in Spain (in part due to the proximity of the EU monetary union), the labour accident rate comprised 137.068 reported accidents, around 20.3% of the total as well .

From this data, we can observe the continuous repetition of this 20% of labour accidents in construction with respect to the total number of accidents.

The causes of these high rates could be anyone of the following: the difference in treatment between little projects and bigger ones, the difficulty in identifying responsibilities, the high number of construction firms in Spain that leads to a high fragmentation of the sector with the correspondent difficulties of site control and checking, and the over-use of subcontractors with their associated difficulties (Hinze 1996), etc. The aim of the new Decree 1627/1997 is to help to solve these problems.

4 THE EU DIRECTIVE 92/57/EEC AND THE SPANISH DECREE-LAW 1627/1997

Following article 118 A of the constitution treaty of the European Economic Community, in June of 1989 the Directive 89/391/EEC was published. As defined in its preface, the main objective is to improve the safety and health of workers in all the sectors of activity, public and private, through adequate measures which cannot be subject to purely economic considerations. This framework Directive promoted the development of other focused regulations in this area. Particularly the Eight Special Directive, named 92/57/EEC, establishes the minimum provisions about Safety and Health on construction sites.

The Spanish transposition framework directive was the Law 31/1995 about Prevention of Labour Risks, and during 1997 various related Decrees were developed. The Decree-Law 1627/1997, about Minimum requirements of Safety and Health on Construction Sites, and in line with 92/57/EEC was published.

In Figure 1 we show the Three Prevention Tools in line with the Directive 92/57/EEC (CEE 1993).

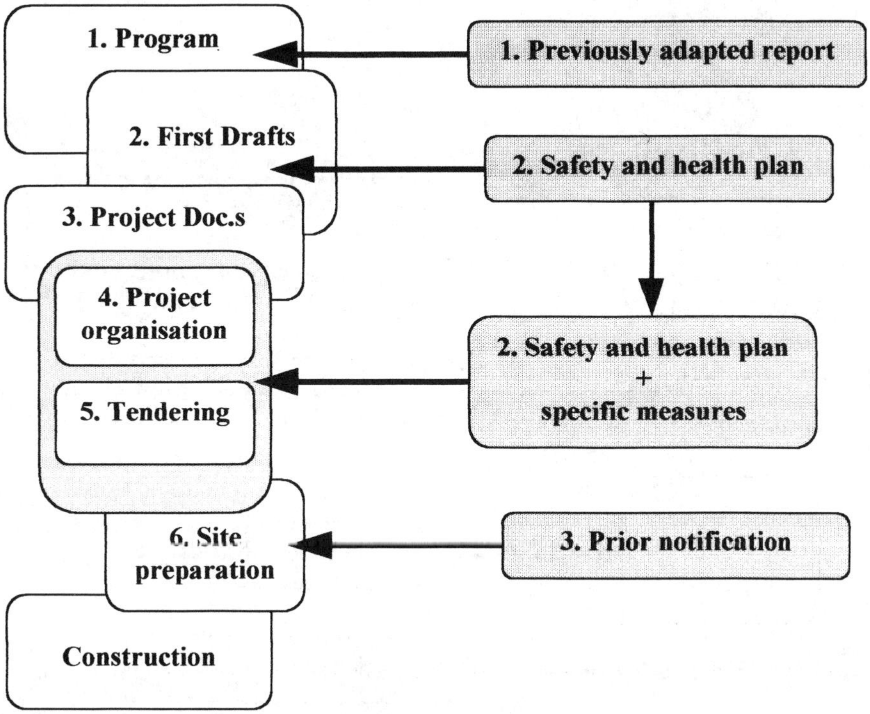

Figure 1. The Three Prevention Tools (CEE 1993).

In this Figure we can observe the accordance of the prevention tools established by the Directive with the construction process flow.

The Previously Adapted Report and the Safety and Health Plan is under the supervision and responsibility of the *Safety and Health Supervisor during the design phase* and the adaptation of this Plan and the realisation of the Prior notification is under the supervision and responsibility of the *Safety and Health Supervisor during the construction phase.*

The most relevant aspects introduced are the extension of the prevention to all the construction process including the design phase and the creation of the supervisor figures.

The transposition of these Directive into Spanish law, in accordance with the aim of preserving those aspects already common in Spanish construction projects, necessitate the extension of the Prevention Tools. The prevention tools in accordance with the Decree 1627/1997 are shown in Figure 2.

In this Decree the existing Safety & Health Studies are maintained. This removes the necessity of developing the Previously Adapted Report. But, a big contribution of this new law is that the necessity of producing this study has been extended to all construction projects. Obviously, the possibility of making a difference between little construction projects and bigger ones exists. So, the barrier is (i) a budget of over 75 M pesetas (500.000 US$), (ii) work of over 500 man-days, (iii) a construction phase longer than 30 days involving more than 20 workers at any one time or (iv) special construction sites like tunnels, dams, etc.

If any one of these four aspects is present in a construction project it is necessary to develop the *Safety & Health Study* (similar to the old Decree 555). If it is not necessary to complete this study, Decree 1627 forces one to develop the *Safety & Health Basis Study.* The Design Process Supervisor (the designer if he is working alone but otherwise the supervisor) is responsible for producing the Basis Study and must identify labour risks and develop corrective measures in order to eliminate these or, if this is not possible, to reduce these risks.

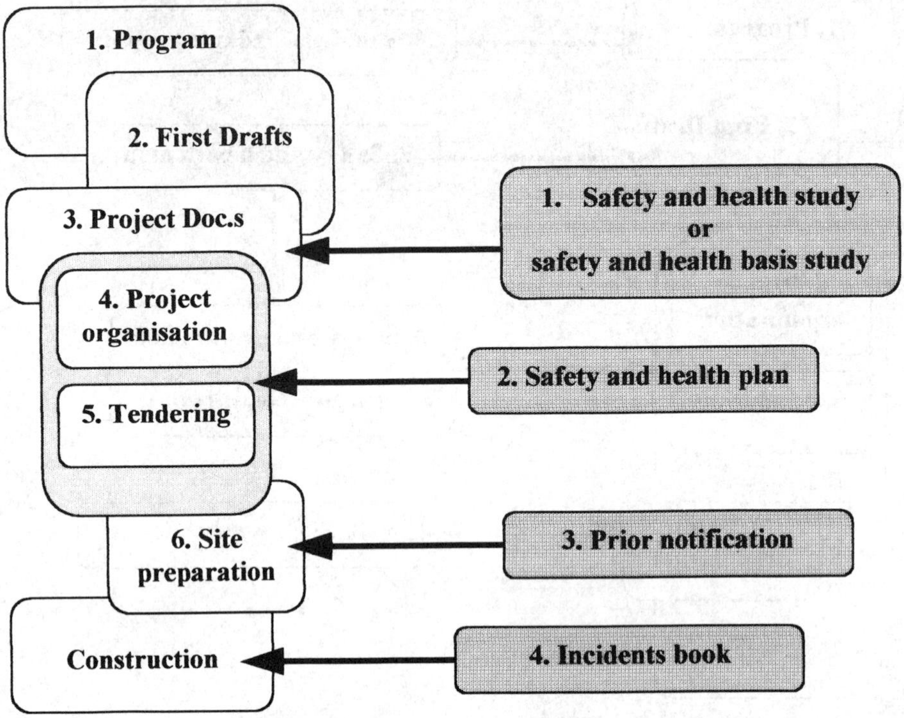

Figure 2. The Prevention Tools in accordance with the Spanish Decree.

If it is obligatory to realise the bigger Study, as a minimum it must include:

- description of working procedures
- description of technical equipment to be used
- identification of labour risks, with removing measures or reducing measures
- preventative actions to be developed
- description of sanitary services and common services on site
- justification of fulfilment of general and particular regulations on safety and health
- prescriptions of maintenance of equipment, machinery, etc
- drawings concerning the implementation of preventative measures
- safety & health budget. This amount is restricted and can't be reduced on site.

These Safety and Health Studies must be completed by the Supervisor of Safety and Health during the design phase under the responsibility of the owner. The existence of the correspondent study is necessary to obtain the project validity stamp from professional bodies (called Colegios in Spain), and obviously, it is absolutely necessary to obtain the authorities permission to start the construction.

As well as this, before the construction phase starts, the contractor must give the *Safety & Health Plan* to the Supervisor of Safety and Health during construction phase. The process is always the responsibility of the owner. After the acceptance of this plan by the supervisor, this plan together with the Prior notification should be delivered to the correspondent Labour Authority. In theory, the Plan is the real application of the Study on-site, depending on the final working procedures and the final site management, and it is strongly based on the Study. Although the Plan can change the safety rules to improve them, the final planned safety budget can't be reduced with respect to the initial study budget.

The Spanish Decree introduces the possibility of not having a Supervisor in the construction phase, if only one contractor firm works in the site. In this case, the project manager automatically acts like a safety supervisor.

Also, the Decree drives the accomplishment of the general rules of Safety and Health developed in the framework regulations (89/391/EEC and Ley 31/1995). As well as this it establishes particular obligations and responsibilities for the supervisor figures, for the contractors and sub-contractors, and for the individual autonomous workers highly common in Spanish construction sector.

Finally, the Decree obliges one to use and maintain the *Incidents Book* similar to the Decree 555/1986. This book is a register where whatever incident that happens in the site related with the safety and health is written. Note that an incident includes the correction of imminent dangers (including the stops of the site if it is necessary) even though the incident proper does not occur. This book should be always on site, in the manager's hut ready to be checked by the supervisor, the workers safety representative, the labour authorities, etc. If one of these writes on the book, the annotation must be delivered immediately to the labour authorities in order to report the irregularity.

5 SUMMARY AND CONCLUSIONS

In this paper, attention has been drawn to the presentation of the Spanish regulations about Safety and Health on Construction Sites and the additions that have been introduced with respect to the EU Directive 92/57/EEC.

To make objective comments we have encountered some relevant difficulties. So, in respect to the data evolution, due to the slow data speed recollection in this field, until late 1999 it isn't going to be possible to have accurate statistics about labour accident rates under the application of the new regulations. Another added difficulty is the big increase in construction activity during 1998, that probably will produce an increase in the total number of accidents. So, the recollected data for this year will be difficult to analyse from an objective point of view. The last big difficulty is the increase of works in building rehabilitation, an aspect strongly promoted by the Spanish governments (State and Autonomous) in order to re-use for housing the large number of buildings existing in town centres. Obviously this work is more difficult than the construction of new buildings and a high percentage of fatal injuries happen in these works as can be seen in current news although we can't find statistics on this issue. Furthermore at the moment clear data to measure the accident rates doesn't exist, so we can just measure the reported accidents, and it is extremely difficult to compare data with the situation in other countries.

In relation with the general aspects like the acceptation of the Decree for workers, professional designers, professional associations, contractors, etc,... we should mention that at first the Decree was poorly received due to some lacks, for instance who is the professional responsible if Spain doesn't have these studies and training or the necessity to add the Safety Basis Study to extremely small projects (e.g. many local councils request this Study for a domestic bath or kitchen renovation!). Although these initial problems, at the moment, it has been possible in practice to achieve a good level of compliance with the initial objectives.

Another inconvenience is that the Decree fixes a bigger role for the Labour Inspection Authority but at the moment it continues with the same structure as before, so inspections follow depending in general at random or due to complaints.

Probably the best aspect is the clear delineation of responsibilities including those of the owner and the existence of the supervisor figures that establish a clearly defined rules for construction professionals. Positive implications of this delineation is great. Another aspect will be the jurisprudence of the courts in case of liabilities determination. Depending on this, the point of view of the different participants can probably be changed.

Furthermore the existence of the supervisor as a reference figure is basically to build a good safety structure on construction sites.

The inclusion of the sub-contractors and autonomous workers treatment in the Decree is also a new positive aspect to be take into account because it is the most common situation presently in Spain. Obviously this treatment must be associated with a good contractual definition between participants and probably as well a good quality system in order to improve the communication, performance and relations between the different levels of firms and workers. These new figures in the site management structure mustn't be separated from the body of the work-force. So, it is very necessary to develop and introduce management systems to obtain a good level of integration of the normal management, the quality management and the safety management. As a conclusion of this, the existence of more than one management line must be avoided in a construction site and the integration of the different managerial structures is fundamental.

The final conclusion is that the general perception of this new decree is positive although some aspects must probably be improved in practical application.

REFERENCES

Alves Dias, L. 1996. *Construction Safety in Portugal.* Safety & Health on construction sites. Publication 187. CIB Report. Rotterdam

BOE, 1952. Reglamento de Seguridad e Higiene en el Trabajo. *Boletín Oficial del Estado*: June 15th. Madrid [in Spanish].

BOE, 1970. Ordenanza de trabajo de la Construcción, Vidrio y Cerámica. *Boletín Oficial del Estado*: August 28th. Madrid [in Spanish].

BOE, 1971. Ordenanza General de Seguridad e Higiene en el Trabajo. *Boletín Oficial del Estado*: March 16th and 17th. Madrid [in Spanish].

BOE, 1986. Real Decreto 555/1986 about the inclusion of Safety and Health Study in Construction Projects. *Boletín Oficial del Estado*: February 21st. Madrid [in Spanish].

CEE, 1993. *Seguridad y Salud en construcción.* Oficina de publicaciones Oficiales de las Comunidades Europeas. CECA-CEE-CEEA, Brussels.

Hinzie, J. 1996. *The role of project co-ordinator in construction safety.* Safety & Health on construction sites. Publication 187. CIB Report. Rotterdam

14 Innovative safety technology

Implementation of Safety and Health on Construction Sites, Singh, Hinze & Coble (eds)
© 1999 Balkema, Rotterdam, ISBN 90 5809 036 1

The use of photogrammetry in accident investigations: A case study

N.Holland
Department of Construction Science, Texas A&M University, College Station, Tex., USA
R.Warden
Department of Architecture, Texas A&M University, College Station, USA

ABSTRACT: One major problem that is prevalent in forensic accident investigation is the disturbance of the accident scene, such as changes in the structure, between the time of the accident and the time litigation is brought forth. Digital photogrammetry technology allows one to graphically recreate the accident scene at a later date for investigating accidents. This paper presents a case in which photogrammetry was successfully used in the investigation and preparation of evidence for use in court. Resolution of the case required the three dimensional location of several points on a high, steeply sloped roof (12:12 pitch) and accurate measurements of the distances between the points. One of the first principles of safety management is to engineer the hazard out of the work. Verification of the dimensions using traditional measurement techniques would have put the investigators in a hazardous position. This technique allowed the investigators to perform the work from the ground with a high level of accuracy.

1 INTRODUCTION

Courts admitted photographs as evidence as early as 1859 (Nickell, 1994). Color photography, though common for many years, was not admissible until 1943. As technology advanced, the courts have more readily accepted it with proper back up documentation. In the construction industry one of the incidents that may lead one to the courts is the construction accident. A thorough investigation of the accident is imperative to determine the cause of the accident so that measures may be taken to prevent its reoccurrence and to provide a permanent record of the incident for potential use in litigation at a later date.

2 THE INVESTIGATION

When investigating accidents, time is of the essence. The investigation of an accident should begin immediately upon notification of an accident by securing the site. Securing the site helps to assure that the injured party can be rescued and be given medical attention without others being injured in the process. Securing the site also helps to preserve valuable evidence that will lead to a determination of the root cause of the accident.

The following is a summary of guidelines to be followed when investigating an accident. The investigation should be conducted without bias and the report prepared with as much detail as possible. It is important to assure all that blame is not trying fixed but rather cause is trying to be determined. The accident investigation and report should include (National Safely Council, 1993) (Hinze, 1997): 1) general information about the accident;

2) information about the nature of injury, the background of injured party, and an interview with the injured party as soon as possible; 3) the identification of witnesses and their interviews; and 4) the identification of equipment and materials associated with the accident and a photographic record of the these items and the accident scene.

2.1 *General information about the accident*

In general, the investigation and report should answer the questions who, what, where, when, and why. In all aspects of the investigation these questions should be kept in mind. As much of the following information as possible should be recorded immediately following the incident, before witnesses leave the scene or it is altered. The initial investigation should begin with obtaining the following information:

1) the name(s) of the injured, the task being performed, and if the activity was being performed alone or with another employee;
2) a description of how the incident occurred, the exact location of the accident;
3) the task the injured person was performing when the accident occurred;
4) the sequence of events leading up to the accident;
5) the date and time of day the accident occurred;
6) weather conditions at the time of the accident;
7) whether or not established procedures were followed;
8) a list of witnesses, what they were doing at the time of the accident and their phone numbers and addresses;
9) a dimensional sketch of the scene locating the point of the accident, reference points, equipment and material locations, and the location of witnesses; and
10) the identification of equipment and materials involved in the accident.

2.2 *Interview of the injured party(s)*

As soon as the injured person(s) is able, he should be interviewed while the facts are fresh in his memory and before it is influenced by the recollections of others. His account of the accident and the events leading up to the accident can provide invaluable information in determining the cause of the accident.

One should also document the injured party's technical training, how long he has performed the work, safety training, and other information relating to the person's past performance and physical condition prior to the accident.

2.3 *Witness interviews*

Interviews with witnesses should take place as soon as possible after the accident before memories fade. It is important that each witness be interviewed individually in a private location and before the witnesses can discuss the incident with each other and their memories of the incident are subconsciously merged with those of others. The witness should be assured that the information provided is necessary to determine the precise cause of the accident so that corrective action can be taken and the workplace made more safe. The interview's notes should distinguish between that which the person actually saw and heard and that which is speculation. It is also important to note what the witness was doing prior to the accident and immediately after the accident. When applicable the witness should be asked to make diagrams that indicate what happened. Insofar as possible, direct quotes should be used. The interviewer should repeat the witness's statement to assure that it was understood correctly and that quotes are accurate. Again, the questions to be asked and answered are: who, what, where, when, how, and why.

2.4 *The identification of equipment and materials and the photographic record*

In the initial investigation all equipment and materials associated with the accident should be identified, the condition verbally described and photographed in-situ. In addition, some items may need to be photographed under more conducive conditions. If destructive testing is required, it is imperative that all necessary photographs required for documentation be taken prior to testing. Many times the court may require that the photographs be made with a scale to indicate size and without an aid to indicate scale to show in site relationships.

In the event of a later court case, care should be taken to assure that the photographic evidence meets the court's requirements for admissibility. For each photograph, the following information should be recorded in the event it is requested in court (Nickell, 1994):

1) data to identify the photographs with the incident;
2) the identification of the photographer;
3) the location the camera position with respect to the accident location;
4) the date and time each photograph was taken;
5) lighting and weather conditions, *f/* number and exposure;
6) the type and make of camera, film, and any special photographic equipment used;
7) focal length of the lens;
8) data regarding the developing, printing, specialized laboratory techniques and who developed the photographs; and
9) a record of the complete chain of custody of the photographs and negatives.

3 THE CASE STUDY

In this case, the owner/general contractor hired a mechanical subcontractor to install the heating, ventilation, and air-conditioning system (HVAC). In turn, the HVAC subcontractor hired another subcontractor to install the heating and air-conditioning system and associated roof vents. The owner/general contractor had supplied roof anchors that were installed by the framing subcontractor. The subcontractor to the mechanical subcontractor had installed a vent on one side of the roof, after which he crossed the roof ridge and was attempting to go to a second location, when he slid off the roof of the one story residence and fractured multiple vertebrae. The only possible witness to the accident was the installation subcontractor's employee, his father. The installation subcontractor had no insurance or workers' compensation and subsequently filed suit against the mechanical subcontractor and the owner/general subcontractor. He claimed that the owner was negligent in the placement of the roof anchors and the design of an unusually dangerous steep pitched roof. He also claimed that the owner should have provided him with a safety harness and instruction in its use.

When the attorney of the owner/general contractor brought the authors into the case, several years had passed since the accident had occurred. The information provided were depositions of the parties involved, that were taken one or more years after the accident and a set of plans that did not match the configuration of the roof system as constructed. The only photographs available were those taken by the injured party's (plaintiff's) attorney. This plaintiff's photograph, similar to the one in Figure 1, showed vents and roof anchors on the side of the roof from which the plaintiff fell. In his testimony the plaintiff stated that he had waited at the chimney for his father to throw him a rope and was going to the roof anchor on the side of the roof to tie off when he slipped and fell. The plaintiff's photograph did not show and neither did the plaintiff's nor the defendant's attorney know, until our investigation, that there was a roof anchor located behind the chimney where the plaintiff claimed to be standing.

To demonstrate that the roof anchors were properly located, it was necessary to create an accurate model of the existing roof and locate all roof protrusions such as anchors, vents and chimneys. The existing roof differed from the roof shown in the construction documents for the house so the model could not be constructed solely from the information on those documents. The steep pitches of the roof (12 in 12 and 8 in 12)

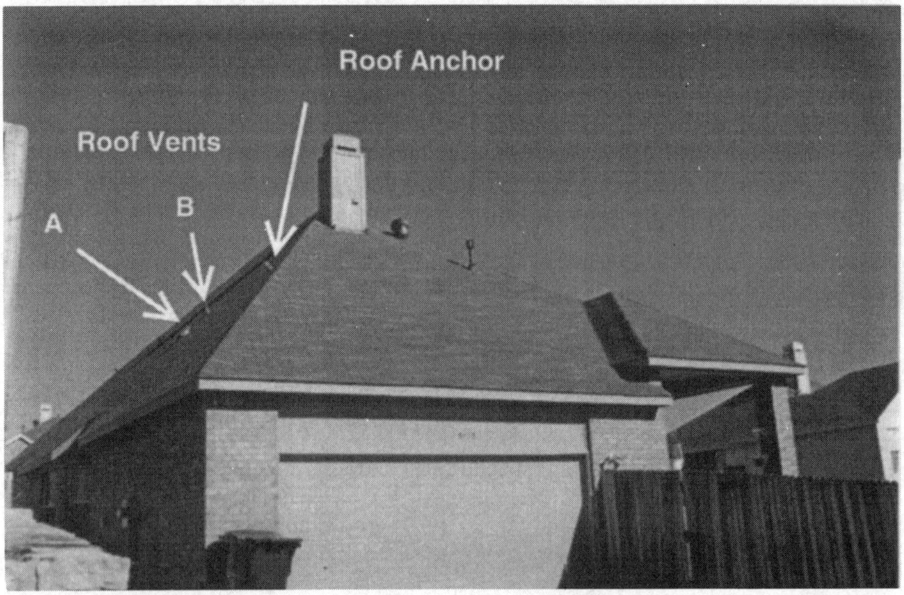

Figure 1. Accident scene: image of roof structure showing roof anchor and roof vents (At the time of the accident roof vent A was existing and B was to be installed.)

prevented safe access for direct measurement with tapes or electronic distance meters, so a remote sensing system of measurement was required. These factors contributed to the decision to use close-range photogrammetry as the primary generator of the 3D model.

3.1 *Photogrammetry and development of the 3-D model*

Close-range photogrammetry is a method of measurement that uses photographs in conjunction with computer software to calculate the coordinate positions of various points in the photographs. It utilizes the age-old theory that the coordinates for any point in space can be calculated by triangulation using angles and distances. Surveying teams have used this method for decades to calculate positions with surveying instruments. If one knows the relevant information about the camera and film used, then the camera can be utilized in place of surveying instruments. This is often helpful because one can minimize the time spent in the field gathering data since there are no particular setup requirements for taking the photographs.

The particular software used for this project is called Photomodeler 3.0 which is produced by EOS Systems, Inc. The program allows the user to extract measurements and develop 3D models by processing photographs. One must also know the focal length of the lens and film type. Several photographs of a scene or an object are taken from different locations and at 90° angles. The photographs are converted to digital images and displayed on a screen and the user marks each image with the mouse, marking points of interest. The same points are marked on different images. The more points that are marked and the greater the number of photographs used, the greater the accuracy, especially on complex models. The Photomodeler was then used to generate the 3D position of points by projecting straight lines from the camera position, through the marked point on the image and out into space. The position of a point is found by the intersection of two or more projecting lines. This processing of images involves the use of an iterative algorithm for determining the position of the marked points and the camera position. An iterative algorithm is one in which a number of steps are repeated until the

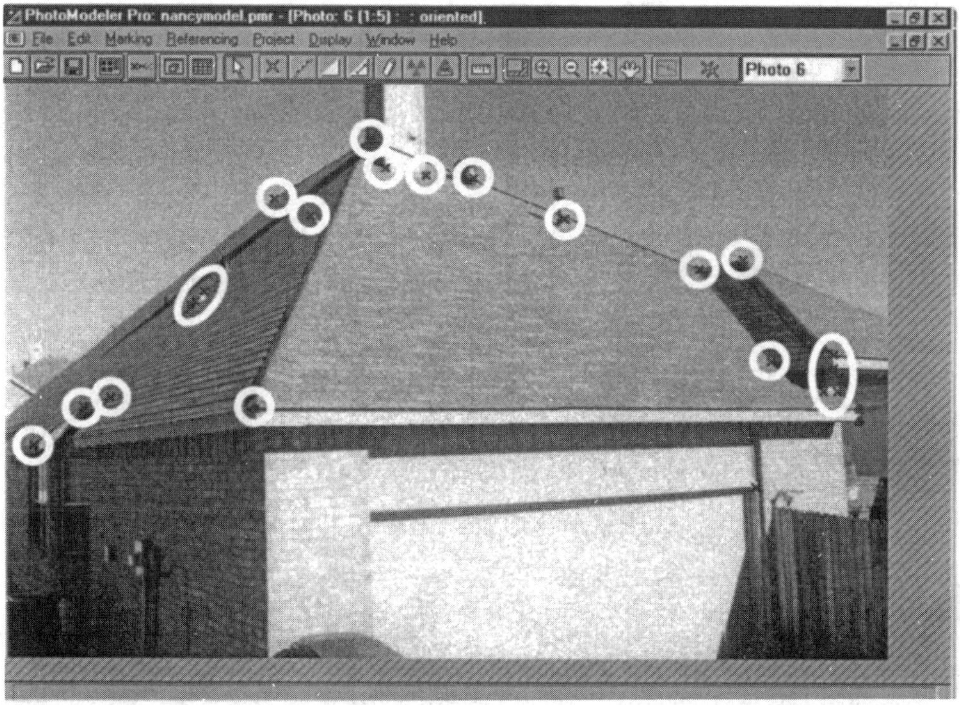

Figure 2. Accident scene with reference points marked (with "x" and highlighted in white circle) for use in Photomodeler

best solution is obtained. In this case, the best solution is the model that gives the "best fit" for positions of marked points.

The scale of the model is controlled by supplying either dimensional information taken directly by hand measurements or coordinate information supplied by surveying instruments. Since most of our photographs contained the walls of the house as part of the image, we were able to measure these accessible parts of the building by hand and thus use this information to control the model. Control points were placed at discernable corners of the masonry, garage door and windows.

The accuracy of these models depends on many factors: the quality of the lens, the resolution and process of digitizing the image, the number of good known control points, the clarity of the photographs, and the coverage of the photographs. The program utilizes camera information, such as, focal length, film type, and fiducial inserts input into the program by the user to calculate the position of the camera for each of the photographs. With a good fixed focal length lens and a high resolution (2700 dpi) film scanner, the quality of the digital images can be quite high. The linear dimensional accuracy can run from 1:500 up to 1:10,000. For typical careful work with 35mm cameras the accuracy runs about 1:2,000. This means that for an object like this house with a length of 18.1 (60 ft) each point is accurate to within 0.18 cm (0.33 in).

While the Photomodeler alone could have been used to create a model of the house scene, it was decided to use a combination of Photomodeler and Auto Cad because of time constraints. Although AutoCad models are accepted in court proceedings, this was the first time Photomodeler was used for this purpose. Thus, it was necessary to verify the model of the house obtained by the combined methods. In the generation of the model of the house by the combination of methods, the first step consisted of verifying the

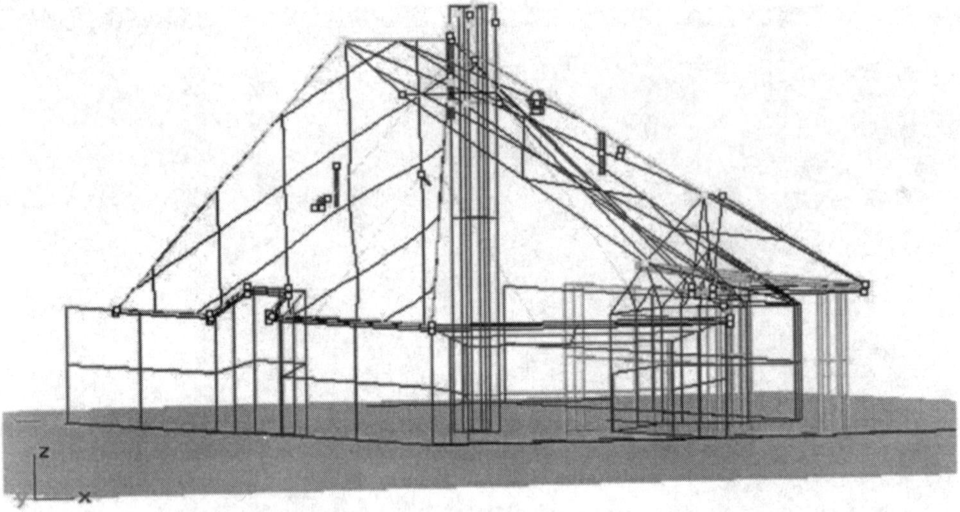

Figure 3. Framework for the 3D model using AutoCad and Photomodeler

dimensions of the original drawings including the house perimeter, wall heights, and eave dimensions. AutoCad was then used to construct the plan and elevations. Since the as-built roof structure was different from the plan, the roof was constructed using geometry, the location of known points, known slopes and the photographs were used to recreate the roof structure. Photomodeler was then used to locate the roof anchors and roof vents; thereby resulting in the model of the house by the combination of methods, Photomodeler and AutoCad.

To verify this model based on the combined methods, a check of the locations of roof anchors and vents was made using material dimensions, such as counting shingles. This is a common procedure used to check building surveys. The agreement between this model and the one generated by the combination of Photomodeler and AutoCad was very good, thereby verifying the use of the program Photomodeler for this purpose.

Figure 3 presents an image of the merged construction of the structure using AutoCad and Photomodeler. The 3D model presented in Figure 3 was then exported into a solid modeling package called Rhino 3D for refinement and rendering. Since the time-frame for completing the project was quite short it was decided not to generate texture mapped images, but to supply images that could be queried dimensionally in order to easily demonstrate the location of roof vents and roof anchors. Figure 4 shows the completed model in Rhino 3D. The model is capable of being queried for any dimension so that the accuracy of the final model can be checked with accessible points on the real structure. The program can then be queried to determine the distances between the physically inaccessible points (the location of and distance between the roof vents and anchors, and the chimney) and thus, determine whether there is compliance with the roof anchor specifications.

Within eighteen hours after the deposition where the dimensioned computer generated model and new photographs were entered into evidence, the case was settled out of court. The guidelines for photographic evidence, presented in Section 2.3, were followed when presenting these materials into evidence.

Figure 4. Created 3D partial model of the structure with located roof anchors and roof vents using AutoCad and Photomodeler and surfacing using Rhino 3D

4 CONCLUSIONS

This paper presents the use of close range photogrammetry for creating 3D computer models of existing structures for the purpose of analyzing situations involving construction accidents. This use of the program is a new application of Photomodeler. Without the use of the program Photomodeler as an aid in the investigation and preparation of images for use as evidence, the authors would have had to either put themselves in harms way by making hand measurements on the steeply sloped roof or investing considerably more time and effort by using survey equipment. In this case the program was used to generate a model of the roof with anchors and vents located and dimensions computed.

Thus is has been demonstrated that photographs taken at the accident scene can be used at a later date, with the program Photomodeler, to generate a model of the accident scene as it was at the time of the accident. The model generated can be helpful in accident investigation and use as evidence in litigation.

Photomodeler has proved to be an excellent means of creating a model of an accident scene using the guidelines for accident investigation and photography set forth in this paper.

REFERENCES

Hinze, Jimmie 1994. *Construction Safety*. New Jersey: Prentice Hall.

Nickell, Joe 1994. Camera Clues. Kentucky: The University Press of Kentucky.

Photomodeler 2.0 Users Manual 1995. British Columbia: EOS Systems Inc.

Supervisor's Safety Manual, 8th Edition 1993. National Safety Council.

Implementation of Safety and Health on Construction Sites, Singh, Hinze & Coble (eds)
© 1999 Balkema, Rotterdam, ISBN 90 5809 036 1

Towards a virtual environment for safety-integrated site layout design and organisation

G.A.John & B.Hobbs
Construction Research Unit, University of Teeside, Middlesbrough, UK
C.J.Anumba
Department of Civil and Building Engineering, Loughborough University, UK

ABSTRACT: This paper describes a research project that addresses a new approach to the organisation and layout of temporary construction facilities on the site, with safety as the main objective. The research effort involved two major activities: (a) the compilation of a knowledge base from in-depth review of published literature on site layout design, review of safety manuals, and elicitation of expert knowledge from construction practitioners; (b) the development of a virtual environment for site layout design and organisation. The virtual environment incorporates a knowledge-based component that encapsulates expert guidelines and 'rules of thumb' within a constraint management system which ensures that safety is adequately considered in meeting site layout design and organisation. The virtual environment offers a unique opportunity to visualise the safety implications of different site layout configurations. The prototype system being developed will be applicable to both new site layout designs, as well as evaluating the safety implications of existing site layouts. The paper concludes that the system demonstrates a proactive approach towards construction risks reduction and also a new way of visualising and solving construction site layout problems.

1 INTRODUCTION

Construction safety includes the safety of the facility to be constructed as well as safety in the process of constructing the facility. Safety of the structure to be constructed is usually dealt with during the planning and design stage of the facility, while safety in the construction process was mainly the responsibility of the general contractor. Safety may include better quality control, avoidance of human error, robustness of the structure, design of the temporary facilities and the construction process against unexpected catastrophe, as well as safe site operations. Safe site operation involves the selection of temporary facilities; equipment and their operations; the selection of well-trained construction personnel; and the planning, organisation and design of an effective and efficient site layout design.

Construction safety is, for the most part, managed re-actively and is the sole responsibility of the contractor (Hinze and Wiegand 1992). The construction industry has for a long time been plagued by poor safety records that show figures that are high when compared to other industries. In addition, there are huge cost implications that the construction industry can ill-afford. Factors contributing to the poor safety records include: the nature of the construction process; the threat posed by hazardous substances; lack of effective management control; inadequate training of site operatives and supervisory staff in health and safety matters; inadequate supervision and monitoring of site activities; and inappropriate site layout and organisation (Anumba and Bishop 1995). To achieve safe site operations, construction practitioners need an effective means by which to monitor and control safety during construction and to assess safety plans as a criterion for selecting contractors (Kartam 1997). Contractors also need a system by which to actively monitor their site operations and implement measures to eliminate unsafe acts and practices. Before presenting the outline of a prototype system that will help in this regard, the framework under which site operations are undertaken needs to be examined in order to establish gaps, inefficiencies and inadequacies.

2 CONSTRUCTION SAFETY FRAMEWORK

2.1 Safety regulations

The legal framework within which the UK contractors operate is replete with rules and regulations relating to health and safety (Anumba and Bishop, 1997). Those that have a significant impact on site layout design and organisation are Common Law, the Health and Safety at Work Act (HMSO, 1974), the Control of Substances Hazardous to Health (COSHH) Regulations (HSE 1988), and the Construction Design and Management (CDM) Regulations (HSE 1994).

Under common law, an employee injured at work may sue the employer for the tort of negligence. The employer is directly liable if there is a breach of a duty of care owed to the employee, or vicariously liable if the injuries are caused by the negligent act of one of his/her employees acting in the course of employment.

Under the Health and Safety at Work Act (1974), it is the duty of the employer to ensure, so far as is reasonably practicable, the heath, safety and welfare at work of all his/her employees. In addition to the duty of care owed to employees, employers must also ensure their own safety and that of third parties.

The COSHH regulations were introduced in 1989 and contain guidelines on the control of virtually all substances hazardous to health (except asbestos, lead, and other substances that are covered by special regulations). These substances occur naturally or may be by-products of the construction process. In construction, they include dust fumes, saw-dust, concrete dust, welding fumes, paints, solvents and wood preservatives.

The CDM regulations were introduced in 1995 and provide for the client to appoint a Planning Supervisor who has overall responsibility for co-ordinating all aspects of health and safety at the pre- and post-construction phases. Duties are placed on all participants in a construction project - clients, designers, and contractors - so as to encourage a proactive approach to construction safety. The imposition of duties on both clients and designers served to plug a loophole in the Health and Safety at Work Act. Thus, the CDM regulations recognise, as do others (Hinze and Wiegand 1992), that construction safety is influenced to a large extent by design decisions.

Clearly, the regulatory framework discussed above is adequate to enable a methodology for proactive safety management to be developed. In many cases, strict compliance with the regulations facilitates the effective consideration of health and safety issues in site layout design and organisation.

2.2 The role of designers

Traditionally, designers have not explicitly addressed construction worker safety and often are not aware of the full extent to which their design decisions impact on construction safety. The lack of designers' involvement in worker safety is attributed to their limited education and experience in addressing safety on the construction site, and their attempt to minimise their liability exposure (Gambetese et al 1997). Construction worker safety is typically overlooked until the start of the construction phase, where the mechanics for the increased involvement of contractors in safety are well established. Where there are recurring site failures involving similar design errors or construction deficiencies, the need exists for more effective feedback to the design professions. As construction project become more and more complex, involving a proliferation of new materials, systems, and construction methods, practice often lags behind current research. There is a great need for strategies that will help designers make the transfer from research into practice of the new information (Carper 1998). Consideration of construction worker safety by design professionals can enhance project safety, instigate common safety hazards methodology, and reduce the number of worker injuries.

2.3 Training of personnel

In order to lower the risks associated with working on the construction site, it is prudent that all staff should have some formal training in the relevant safety topics. It is equally important that they should have a period of working with more experienced colleagues in on-the-job training. However, until recently, initiatives in the construction sector have considered education and training without a clear idea of the safety requirements of designers, suppliers and clients (Carper 1998). Furthermore, while contractors are often prepared to invest in training, sub-contractors are prefer to employ skilled and experienced workers who do not need much training. With these varied interests, the short term benefits and considerations for the different parties involved in the construction process seem to outweigh the long-term goal of achieving a safer site (Hsieh 1998).

2.4 Subcontracting

Essentially, there are four types of services supplied by the subcontract resource market. These are: specialised design input, bulk material supply, component fabrication and/or pre-assembly, and site erection services. On site, most subcontractors are concerned with the detailed layout and internal organisation of the area assigned to them. The general contractor or the construction manager has to consider contractor areas relative to one another and focus on major facilities and shared resources (Tommelein et al, 1992). The benefit of subcontracting is that it not only allows the general contractor to acquire various construction services efficiently from the resource market, but is also an effective means of cost control and risk management. Unfortunately, this practice may pose tremendous barriers to site productivity and site safety as the general contractor may reduce the size of the site management team. This reduction of site management personnel weakens the general contractor's ability to plan for and communicate with subcontractors, and to handle their individual problems or interference. Also, when market forces dictate stiff and intense competition, one way general contractors take risk is by lowering the mark-up in their bids, thus improving their chance of winning contracts. Invariably, when such contracts are awarded, the general contractor spreads out this risk to subcontractors by contractual arrangements and exercising strict cost control procedures upon sub-contractors. If the risk is unbearable, the sub-contractor may compromise safety by violating some safety regulations in order to cut costs.

2.5 Safety management

The managerial problems of safety are mainly concerned with aspects of human frailties and the difficulties faced by modern industrial organisations. It is increasingly recognised that human error in design, construction, or operation of plant, is a major contributor to accident rates. The functionality of the management systems within an organisation (in the broadest sense) is a key determinant of that human performance (Cottam and Maguire 1995). One crucial problem is often unfamiliarity with the project, attributed to the one-off nature of construction projects. Managers also have to function in the face of conflicting pressures, whether it is in a design firm or a contracting firm. One of the most serious is the need to achieve safety at a price, or within defined project time scales. In this regard, deployment of individuals with the right attitudes and skills is a key factor in achieving safety targets for most managers. The skill level of the work force and their familiarity with the technology and methods used, as well as the managerial commitment of the firms to train their workers, is also essential. Where there is a well-structured management system and managers are committed to job site safety through efficient site operations, the construction site is usually less hazardous.

2.6 Site Layout Design and Organisation

The construction site is one of the primary resources available to the contractor. The aim in planning the site layout and facilities is to produce a working environment that will maximise efficiency and minimise risk. It reflects the organisation's commitment to safety and well-being of the work force and its determination to satisfy the needs of its customer (Gibb and Knobbs, 1995). Site layout planning is complicated, time-consuming and is often considered a heuristic skill acquired over a long period of time. In most companies, design procedures are not written down, but are considered 'second nature', remaining locked in the minds of experienced managers and planners. Even where contractors do operate a company policy, they are normally unwilling to disclose the details (Rad and James 1983). The site plan produced usually shows the relationship between the site boundary, the permanent facilities and the temporary facilities. The effective organisation of the project depends on the partial interactions between the static and dynamic resources and the temporary facilities on sites. Thus, it is important that the design of the site layout takes cognisance of the fact that the requirements of the site are time-varying, changing as the project progresses (Tommelein and Zouein 1993). As changes are introduced, their safety implications should always be reviewed, as situations that were not considered dangerous may become so. Aspects of site layout design that need to be addressed are: access and traffic routes; material and storage handling; site offices and amenities; construction plant; fabrication workshops; services and facilities, and the site enclosure (Anumba and Bishop 1997).

3 TOWARDS A SYSTEM FOR SAFETY-INTEGRATED SITE LAYOUT DESIGN

3.1 Review of previous work

Previous studies on site layout design and organisation (Tommelein et al 1992) have been devoted to ways of optimising them for productivity reasons, often without adequate consideration of the health and safety implications of the layout configurations (Anumba & Bishop 1997). Rad and James (1983) suggested three possible methods by which a layout could be represented: drawings, templates and 3-D models. They considered equipment positioning, access, lay down areas, overhead utilities, and topography as being capable of analysis using 3-D models. The knowledge-based system proposed by Rad and James was further developed by Handa and Lang (1988) to establish elements that affect site productivity.

Farrell and Hover (1989) addressed the problem of safer site layout and organisation to a limited extent by developing CRANES - a computer program to assist in the selection and positioning of cranes on the construction site. One of the objectives of the work was the prevention of crane accidents through proper planning. The work by Hamiani (1989) centred on the development of a knowledge-based system for site layout design with a view to satisfying a set of pre-defined objectives. Other site layout design studies include the use of a database management system integrated with a geographic information system (GIS) to lay out the temporary construction site facilities (Cheng & O'Connor 1996). Annealed neural networks have also been used for the layout of temporary construction site facilities (Yeh 1995). None of these have given adequate consideration to the safety implications of the developed site layout configurations.

Although there are a number of systems and methods available to facilitate site layout design and organisation, they are deficient in not providing for adequate consideration of health and safety issues. In such a situation, the tendency is for site managers to adopt solutions that they are familiar with, so as to save time. Invariably, this may not be the best solution from the safety point of view. There is therefore the need to enhance the site manager's capabilities and at the same time provide a system that will meet safety requirements on site. The approach to be adopted in ensuring safer site layout design and organisation should involve an objective analysis of the requirements, elimination of risks through design and forward planning, development and implementation of procedures for dealing with known accidents, setting up a clear safety management structure which all workers can easily identify, implementation of a regime for safety training, and the continuous monitoring and review of adopted measures (Anumba and Bishop, 1997).

3.2 Visualisation and Virtual Reality in Site Layout Design and Organisation

Visualisation is the creative ability to form mental images. In the context of the engineering design process, it refers to the visual thinking and design modelling processes that involve perception, imagination and communication. There are several visualisation techniques, with virtual reality being one of the most powerful. Virtual reality (VR), is a visualisation technique in which a computer is used to form an illusion of reality by simulating a three-dimensional environment. It can be used as an innovative planning system that is able to illustrate planning mistakes at an early stage. It also enables the building of virtual models that are more cost effective than the building of physical models. The efficiency of planning is strongly increased by the flexibility of such virtual systems. A VR system can also recognise various gestures, such as grabbing or picking, which can instigate actions involving the objects within the virtual world (Heger and Betsos 1996, Bouchlaghem et al 1996).

Visualisation using virtual environments extends the ability to explore multi-dimensional spaces and their contents. Teylingen et al (1997) are of the view that the payoff could be greatest for multivariate information spaces whose variables are highly correlated, and spatially complex fields that may also change significantly over time. They also state that consideration of problems that include navigation through complex spaces where the order of tasks in the space is critical, has brought the realisation that engineering design problems can be treated similarly to data visualisation and analysis. It is therefore logical to consider how virtual environments could be utilised to provide new ways of visualising information about a particular application. The application domain in this case is safety-integrated site layout design, which involves the relative location of site layout components (temporary facilities, plant, etc.), taking into consideration site layout constraints as well as safety constraints. The use of a VR system can greatly enhance safety-integrated site layout design and organisation, as unsafe situations can be readily visualised and simulated.

4 THE PROTOTYPE SYSTEM

A prototype system, SAFESITE, is being developed to facilitate adequate consideration of health and safety issues in site layout design and organisation. The system incorporates a virtual environment (to enhance visualisation and communication of safety issues) and a knowledge-based component (for encapsulating, enforcing and managing the rules that govern the safe relative location of temporary facilities on site).

4.1 System Development Environment

The prototype system is being developed on an IBM compatible PC, which provides a cost effective platform for the development. The development toolkit being utilised is the World Up Development System (Version 3) with Sense 8 as the main software. The World Up Development Environment consists of the World Up Modeler and the Universe Browser. The graphical objects (or site components) are constructed using the 3D World Up Modeler. The main software, Sense 8, contains a library of functions grouped into different classes to develop interactive virtual worlds using object-oriented programming features. There are also facilities for linkage with other applications such as spreadsheets, databases and graphics packages. Linkage to graphics packages facilitates the input of the site layout details.

4.2 System Architecture

The architecture for the prototype system is shown in Figure 1 while its three main components are described below.

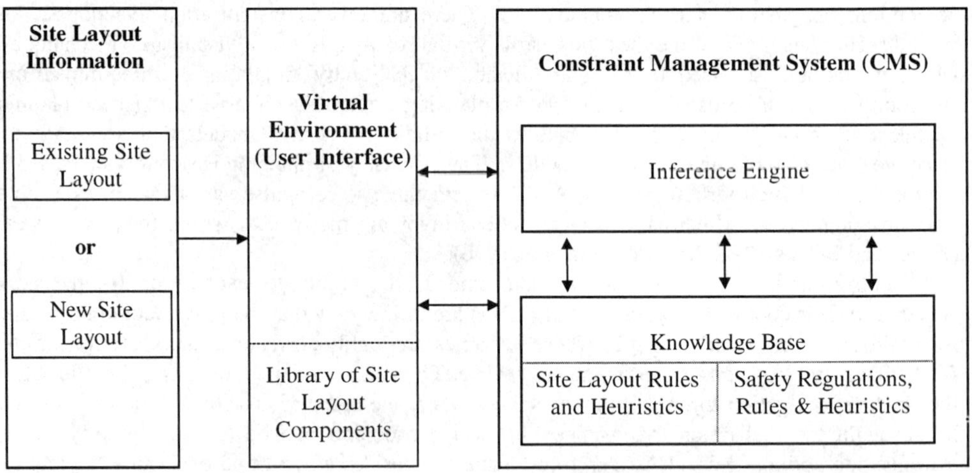

Figure 1: Architecture of SAFESITE

Site Layout Information: This component provides for the design of the site layout within existing 3D graphical modelling packages. The (new or existing) site layout can then be imported into the virtual environment where the evaluation of its safety implications takes place.

Virtual Environment (User Interface): The VR environment allows the user to interact directly with the site layout components (site facilities) and, if desired, to undertake the whole site layout design within this environment. A library of standard site layout components consisting of temporary facilities (such as cranes, site offices, scaffolding, etc.) with variable properties that can be adjusted to meet user requirements is available to facilitate this. Textual safety information and messages that are useful in aiding the site layout design process can be displayed within a window, and there is scope to highlight unsafe situations in a highly visual way within the virtual environment.

The Constraint Management System (CMS): This consists primarily of a knowledge base and an inference engine. The knowledge-base contains both heuristics and rules relating to site layout design and safety while the inference engine reasons about the site layout configuration and determines whether any safety or site layout rules have been violated. If any problem is found, the system requests that the relevant site components are repositioned or the constraints satisfied in some other way. The CMS advises the user on what safety violations have been committed. An iterative process is followed until the site layout configuration is found satisfactory.

4.3 Knowledge acquisition

The knowledge encapsulated in the system knowledge base has been obtained from several sources - literature review, an industry survey, case studies of construction sites, and interviews with domain experts. These sources were considered complementary and provided enough rules and heuristics for developing the prototype system. Knowledge elicitation from domain experts was based on semi-structured interviews, but other knowledge elicitation techniques may be utilised in further development of the system.

5 SUMMARY AND CONCLUSIONS

This paper has sought to address the inadequate consideration of health and safety issues in previous studies of site layout design and organisation. It argues for a proactive approach to safety-integrated site layout design and organisation. The use of a virtual environment that is underpinned by a knowledge-based constraint management system (CMS) is considered useful in this regard and is the basis for a prototype system, SAFESITE, being developed to assist site managers in the design of safe site layouts. The potential benefits of the prototype system include:

- an interactive design tool that can be used to assess a variety of site layout configurations;
- a virtual environment that provides a highly visual demonstration of safety problems that can go undetected when other site layout design techniques are used;
- use of the system as a decision support system for construction managers;
- a training tool for teaching inexperienced engineers aspects of safety on construction sites;
- potential for carrying out dynamic simulations of site operations;
- the facility for site layouts to be developed within existing CAD systems and imported into the virtual environment;
- encouragement of a proactive approach to construction safety management;
- major scope for reducing accidents on site through better site layout designs.

To realise the above benefits, further development of the system will focus on the detailed implementation and testing of the inference mechanisms, and improvements to the user interface. These will be undertaken in close collaboration with the industry partners on the project.

ACKNOWLEDGEMENT – This project is part-sponsored by Birse Construction Ltd., UK.

REFERENCES

Anumba C. J. and G. Bishop 1995. Safety-integrated site layout and organisation. *Proceedings Annual Conference of the Canadian Society for Civil Engineering*, Vol. III, pp. 147-156.

Anumba C. J. & G. Bishop 1997. Importance of safety considerations in site layout and organisation. *Canadian Journal of Civil Engineering*, 24:229-236.

Bouchlaghem N., A. Thorpe, & I. G. Liyange 1996. Virtual reality applications in the UK's construction industry. *Proceedings of CIB Workshop on Construction on the Information Highway*, June 10th -12th, pp. 89-94.

Carper, K. L 1998. Current structural safety topics in North America. *The Structural Engineer*, 76(12):233-239.

Cheng, M. Y, & T. J. O'Connor 1996. ArcSite: enhanced GIS for construction site layout, *ASCE Journal of Construction Engineering and Management*, 122(4):329-336.

Cottam M. P. & H. Maguire 1995. Hazard and risk assessment in other industries. *The Structural Engineer*, 73(21):364-365.

Farrell C. W. & K. C. Hover 1989. Computerized crane selection and placement for the construction site. *Proceedings 4th Int. Conf. on Civil & Structural Engineering Computing*, City University, London, UK, Vol. 1, pp 91-94.

Gambetese, P. T., Hinze, W., & C. T. Haas (1997). Tool to design for construction worker safety. *ASCE Journal of Architectural Engineering*, 3(1):32-41.

Gibb, A. G. F. & T. Knobbs 1995. Computer aided site layout and facilities. *Proceedings 11th Annual Conference of ARCOM*, Sept. 18-20, 1995. Thorpe, A (Ed.), pp 541-550.

Hamiani, A. 1989. A KEE approach to site layout problems. *Artificial Intelligence Techniques and Application for Civil and Structural Engineers*, Topping, B. H. V. (Ed), Civil Comp. Press, Edinburgh, pp. 65-69.

Handa V & B Lang 1988. Construction site planning. *Mag. of Construction Canada*, 5:43-49.

Heger, R. & G. Betsos 1996. Virtual assembly planning system for complex assembly operations - a practical concurrent engineering application. *Advances in Concurrent Engineering*, pp. 280-286.

Hinze, J. & F. Wiegand 1992. Role of designers in construction worker safety. *ASCE Journal of Construction Engineering and Management*, 118(4):677-684.

HMSO. 1974. *Health and safety at work Act*. Her Majesty's Stationary Office, United Kingdom.

HSE. 1988. *Control of substances hazardous to health (COSHH) regulations*. Health and Safety Executive, United Kingdom.

HSE. 1994. *Construction design and management (CDM) regulations*. Health and Safety Executive, United Kingdom.

Hsieh, T. 1998. Impact of subcontracting on site productivity - learned in Taiwan. *ASCE Journal of Construction Engineering and Management*, 124(2):91-100.

Kartam N. A. 1997. Integrating safety and health performance into construction CPM. *ASCE Journal of Construction Engineering and Management*, 123(2):121-126.

Rad P. F., and James B. M. 1983. The layout of temporary construction facilities. *Journal of Cost Engineering*, 25(2):19-27.

Teylingen, R., Ribarsky, W. & C. Mast 1997. Virtual data visualizer. *IEEE Transactions on Visualisation and Computer Graphics*, 3(1):65-73.

Tommelein, I. D. Levitt R. E., & B. Hayes-Roth 1992. SightPlan model for site layout. *ASCE Journal of Construction Engineering & Management*, 118(4):749-765.

Tommelein, I. D. & P. P. Zouein 1993. Interactive dynamic layout planning. *ASCE Journal of Construction Engineering and Management*, 119(2):266-287.

Yeh, I. 1995. Construction site layout using annealed neural network. *ASCE Journal of Computing in Civil Engineering*, 9(3):201-208.

Implementation of Safety and Health on Construction Sites, Singh, Hinze & Coble (eds)
© 1999 Balkema, Rotterdam, ISBN 90 5809 036 1

The requirements of a multimedia system to support health and safety on construction sites

P.Stephenson
School of Construction, Sheffield Hallam University, UK
K.Conheeney
School of Computing and Management Sciences, Sheffield Hallam University, UK

ABSTRACT: This paper describes and illustrates the requirement of a multimedia system development to support health and safety requirements. The aims and objectives of the system are explained in addition to target user groups. Systems development is described, together with contents analysis for requirements capture and system design. Information sources are also discussed, in addition to information updating, availability and cost. System requirements also incorporates codes of practice, material specifications, diagrammatic illustrations, photo-CD and video facilities related to health and safety requirements. This provides a rich and flexible communication environment to promote safe working and the avoidance of accidents during construction.

1 INTRODUCTION

Accident occurrence on construction sites can often be the result of inadequate or insufficient information during the production process. Paper based systems for recording and communicating health and safety information are often inadequate owing to important details being lost, misplaced or ignored through the pressures of work commitments and required project completion times. Multimedia environments provide a rich source of information which can be easily retrieved and communicated to site management and operatives working during the construction process. The use of multimedia during construction provides for the integration of information through text, sound, pictorial representation and animation. Information can be presented in a way that allows health and safety issues in connection with work operations to be identified and communicated effectively. Thus information can be provided in a timely and effective way prior to work being carried out.

2 HEALTH AND SAFETY IN CONSTRUCTION

The construction industry in the UK and throughout Europe has been recognised as having a poor record in health and safety. Accident statistics have prompted the Heath and Safety Executive to warn that those in the industry with responsibility have to take some of the blame (HBS 1997). The major causes of accidents have been identified and can be directly attributed to unsafe design and site practices. The construction process itself is also seen as being poorly planned in terms of both design and construction. The major inadequacies relating to erection, maintenance and demolition of buildings and structures (Cooke and Williams 1998).

Several European directives have been issued in recent years and implemented in the UK. These are represented by several regulations directly related to construction. Reports and consideration of health and safety matters have been highlighted in relation to these regulations and cover a wide range of practical issues. These include health and safety requirements and revised codes of practice for the erection of precast concrete flooring (Harris 1997). Issues related to piling and foundations (Philpot et al. 1996), procedures to reduce noise levels (Wilson 1993) and the safe development of contaminated land (Simmonds 1992). Views have also been expressed about legislation and its impact on safety (Norton 1996), (Birkby 1993). Many within the industry expressing views that certain regulations are cumbersome and time consuming (Mcneil 1996). Additionally, health and safety laws are often seen as complicated and require simplification to be more effective on site (Reading 1996). Furthermore, managers have recognised a need to have a community wide database on health and safety issues including legal implications (Wright 1992). Future education and teaching of health and safety issues are also seen as an important requirement and an essential part of university education (Anderson 1996).

3 MULTIMEDIA DEVELOPMENTS

Many of the concerns related to effective utilisation of regulations, codes of practice and timely information requirements, including education and training can now be addressed by the use of information technology. Several developments have shown to be useful, and in particular, multimedia communications are set to have a major impact on construction (Bunn 1997).

Many information services are being developed, based on multimedia (Swamy 1994), and communication links have been established and have proved to be beneficial on projects worldwide (Thorpe et al. 1995). Interactive images, combined with graphics and animations are being used in several areas (ENR 1995), and multimedia computing is likely to affect the future of engineering software in the civil engineering industry (Vanegas and Baker 1994), in addition to education and training applications (El-Bibany 1996). Other developments have focused on more effective and interactive user interfaces and the development of navigational tools to aid user orientation within an information network.

4 SYSTEM DEVELOPMENT REQUIREMENTS

The aim of multimedia system development for health and safety is to provide a flexible information environment for designers and practitioners within the construction process in order to inform and promote safe working during construction. More specific objectives would include the presentation of information which is visually attractive with an auditory and animated presentation to increase interest and awareness of important issues related to health and safety. The systems should also be structured in way that suits the needs of users involved in different aspects of the construction process in order to present information in such a way that suits a variety of information requirements.

4.1 *Sources of Information*

Information collection for systems development would include sources from:

- technical journals
- material suppliers' specifications
- specifications from suppliers of plant and equipment
- BSI standards

- codes of practice
- health and safety legislation
- health and safety bulletins

Much would depend, however, on the particular type of work or specialisation, and the systems would need to be sufficiently flexible to allow for expansion and inclusion of future work types and related documentation.

4.2 *Information Availability*

The availability of information would be readily available from the above sources subject to permission for its usage. More practical based information related to construction operations on site would be more difficult to obtain where permissions to select and record information would have to be attained.

Other issues concerning accessibility and the timeliness of information collection are also crucial. The recording of work operations and related details would have to be captured at the time the work is carried out, in addition to preparatory work prior to recording. Time restriction may be imposed due to planned work operations, therefore if recordings are not taken at the appropriate time, valuable information may be lost.

4.3 *Cost of Information*

While the collection of published information may be relatively cheap to obtain, the recording of information on specific projects is difficult to determine in terms of cost. Several factors would be influential and include: cost of permission, insurances when recording in dangerous locations, time-related costs, and travel costs, particularly for overseas projects.

4.4 *Updating Information*

The updating of information for the system would have to be carried out on a continual basis and be dependent upon changes in construction methods and legislation. Technical developments take place continuously, including new methods and newly designed materials and components. Provision would therefore have to be made for such developments to be incorporated into the system.

4.5 *Requirements Specification and Design*

The precise requirements of the systems would be dependent upon the extent of work types and related health and safety issues required. Such requirements would vary from company to company depending upon specialisation, but the system should be sufficiently flexible to provide for amendments and the incorporation of additional information in a variety of mediums in order to maintain a high degree of quality, and the communication of important issues.

In order to establish the requirements of the system in terms of content and detail, a contents analysis is required to show the different levels of information and assist in the design of navigational control. Figure 1 illustrates part of a contents analysis for knowledge requirements. Further issues regarding screen design and the representation of information can be provided by the use of storyboards which act as a focal point for discussion, allowing amendments to be made prior to implementation. Figure 2 shows an example of storyboards and the links between proposed screen displays.

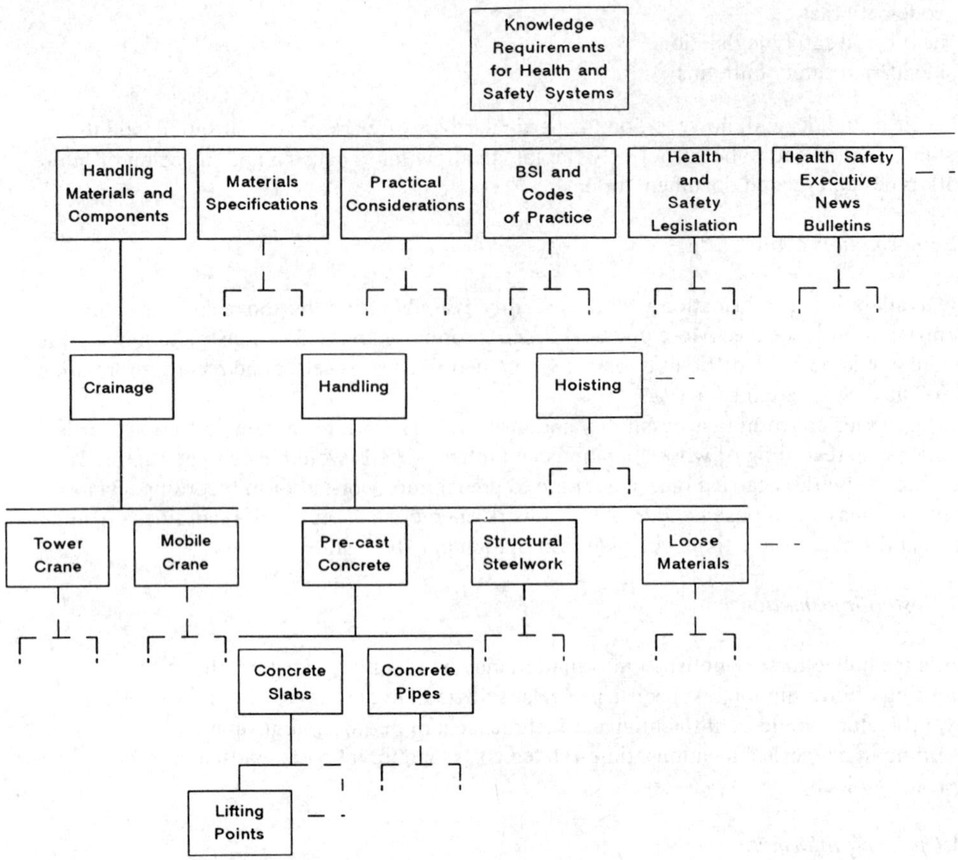

Figure 1. System contents analysis for knowledge requirements

4.6 *Web based development for multiproject use*

Several multimedia development environments can be utilised for system development of dedicated systems for specific project use. Additionally these may be developed for more general company use and accessed via network systems making information available to both designers and site staff within an organisation. Web based developments for systems would also provide the opportunity for utilisation of information on a global scale to support communication links with overseas projects.

The feasibility of the multimedia system has already been proved by its realisation as a working prototype. Essentially the system can deliver a set of content modules supported by different indexes. Content is presented dynamically as HTML web pages via a Cold Fusion server linked to MS Internet Server, itself linked to a back end MS SQL Server database. The database controls both content and indexing, making for a highly flexible, customisable delivery system. Many extensions can be envisaged, including the potential for users to add web page content and personalised indexing. A phased development has been adopted; phase one developed the security and client subsystem; phase two developed the study subsystem of contents and indexing; phase three developed the recurrent task scheduler and phase four provides the ongoing development of the workflow system. The system provides the facility for decision based assessment training which can therefore be focused to individual need.

Additionally, the system could be used to provide a web based operations manual, which could be customised to a specific project. More detail will be provided in a subsequent paper. Currently, the database is being extended to include the domains of contract and tendering, and health and safety. Figure 3 shows an implemented screen design with a display for erecting structural steelwork.

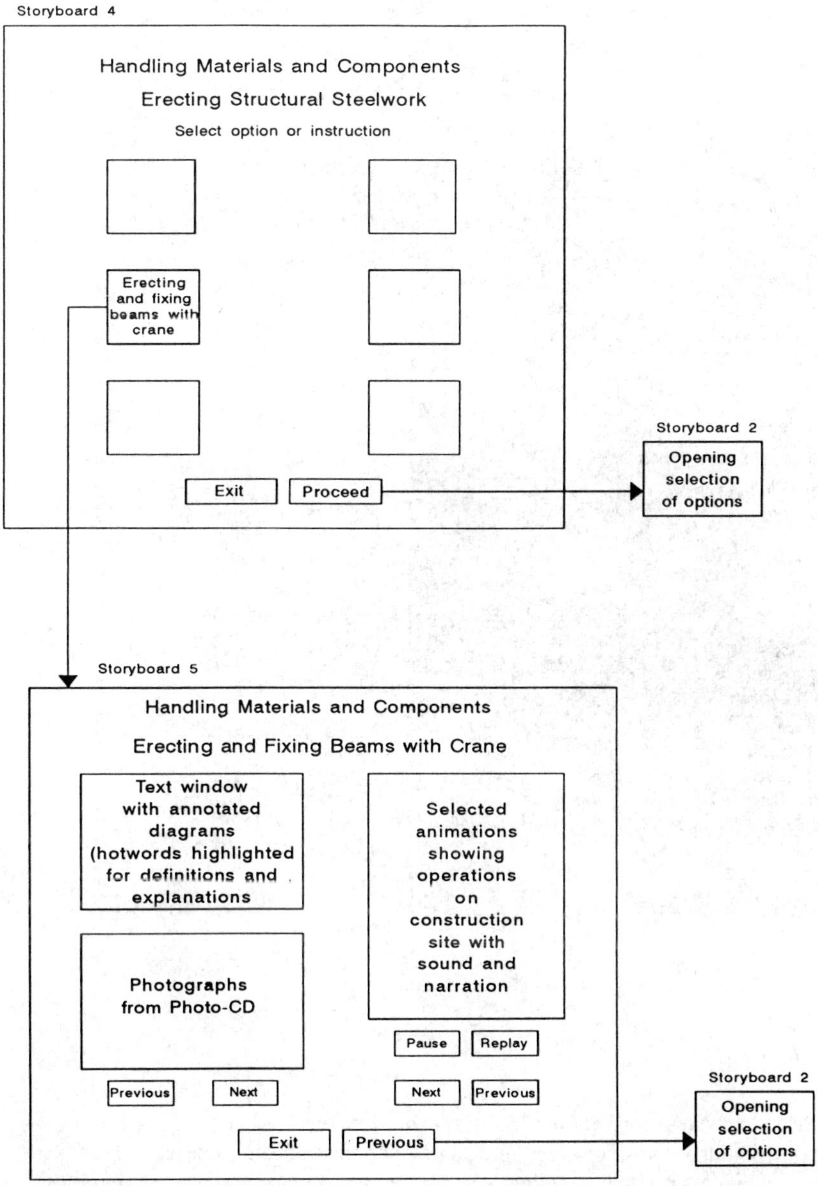

Figure 2. Example of storyboards for screen design

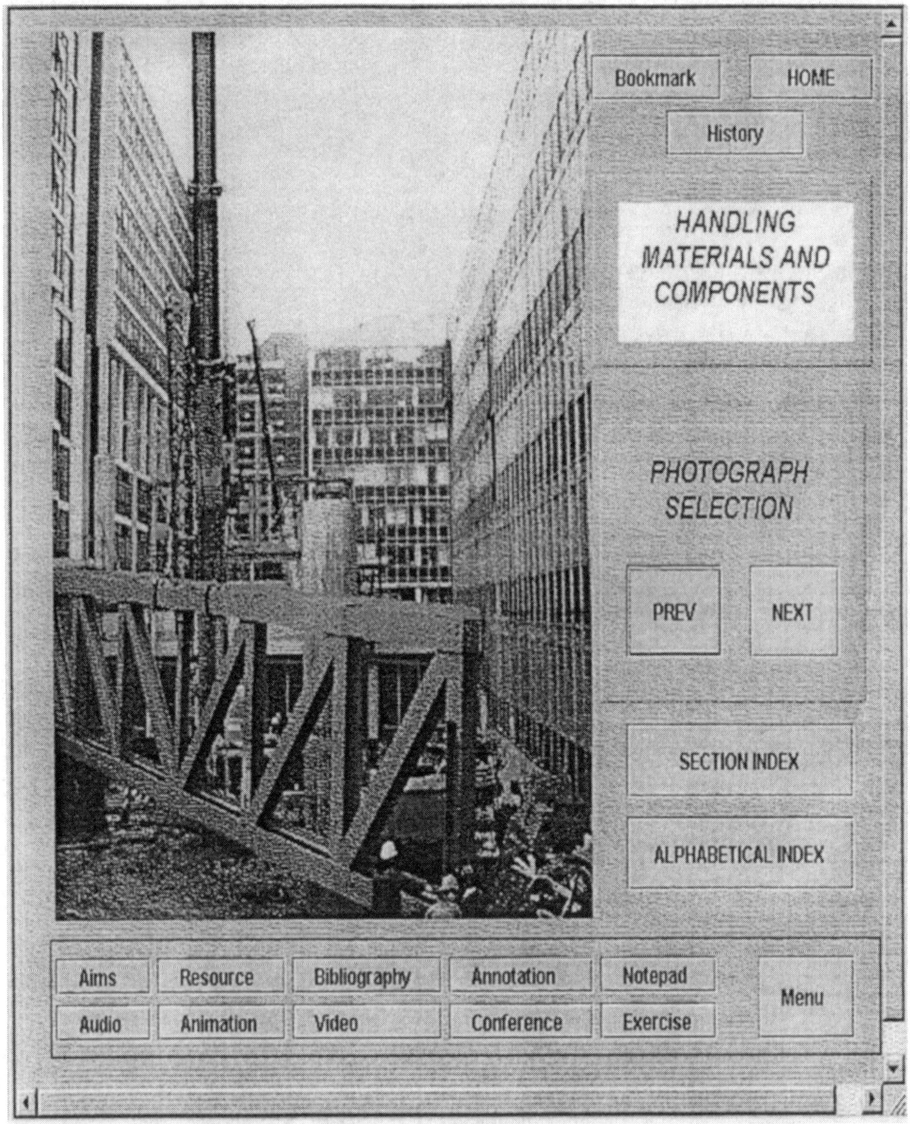

Figure 3. Implemented screen design

5 TARGET USER GROUP

The use of a multimedia system for health and safety would be directed at construction professionals and particularly those involved in design and construction processes. The identified causes of accidents related to design, site practices and planning of work generally would indicate a need for all those involved in the construction process to have a knowledge of the practical issues and legal requirements. A multimedia system would therefore play an important role in promoting awareness, in addition to a learning and training environment to enhance understanding of the issues involved.

6 TRAINING REQUIREMENTS

The training requirements for the system will be largely dependent upon the design and structure of the system. Irrespective of users' prior knowledge of computer systems and their operation, the system should be user friendly and intuitive. Training should be provided so that users of the system will be able to identify the benefits to be gained, which in turn should promote an awareness of the important issues related to health and safety.

7 SUMMARY AND CONCLUSIONS

The requirements for the development for a multimedia system have been illustrated and the important factors considered for successful implementation. Such a system for site use and assist in the design process would enable the integration of text, graphics, sound and narration plus animation of practical situations demonstrating best methods and safe practice. The system should be easily accessible to both designers and site management providing the opportunity for effective communication of health and safety issues. This would also promote a health and safety awareness and help in the reduction of accidents and fatalities during the construction process.

REFERENCES

Anderson, J. 1996. Teaching Health and Safety at University, *Civil Engineering International*, 14, 56-57.

Barnard, M. 1992. Increasing Awareness of Safety in Public Works: is a safe site possible? *ICE Proceedings: Municipal Engineer*, 93(3), 171-173.

Birkby, G. 1993. The Need to Minimise Risk on Site, *Architects Journal*, 198(7), 24-26.

Bunn, R. 1997. Shrinking Distance, *Building Services Journal*, 19(3), 29-31.

Cooke, B. and Williams, P. 1998. *Construction Planning and Control*, Macmillan Press Ltd.

El-Bibany, H. 1996. Multimedia Development Software: Object-oriented Interface-based Simulation, *ASCE Journal of Computing in Civil Engineering*, 10(4), 295-299.

Engineering News Record (ENR) 1995. *Multimedia More than Magic*, 234(22), 65-67.

Harris, P. A. 1997. The Safe Erection of Pre-cast Concrete Flooring, *Structural Engineer*, 75 (23/24), 423-424.

Health and Safety Bulletin (HSB) 1997. *Major injury deterioration leaves HSE chief deeply disappointed and disconcerted*, 261, 13-15.

Macneil, J. 1996. Hazard Warning, *Building*, 261(36), 54-55.

Neale, B. S. 1995. Hazard and Risk Assessment for Construction: A regulator's View, *Structural Engineer*, 73(22), 388-390.

Norton, B. 1996. A Healthy Change of Emphasis, *Contract Journal*, 383(6072), 32-33.

Philpot, T. et. al. 1996. Piling and Foundations, *Contract Journal*, Piling & Foundations 96 Supplement.

Reading, P. 1996. Are Rules a Safe Bet? *Contract Journal*, 383(6076), 24-25.

Simmonds, S. 1992, Safe Redevelopment of Contaminated Land, *Health and Safety Information Bulletin*, 197, 4-5.

Swamy, D. 1994. Market Futures, *Communications*, 21(8), 59-61.

Thorpe, A. et. al. 1995. Multimedia Communications in Construction, *ICE Proceedings: Civil Engineering*, 108(1), 12-16.

Vanegas, J. and Baker, N. 1994, Multimedia in Civil Engineering, *Civil Engineering (ASCE)*, 64(5), 71-73.

Wilson, P. 1993. Sound and Vision, *Health and Safety at Work*, 15(11), 17-18.
Wright, F. 1992, Information on Information, *Health and Safety at Work*, 14(10), 41-43.

Implementation of Safety and Health on Construction Sites, Singh, Hinze & Coble (eds)
© 1999 Balkema, Rotterdam, ISBN 90 5809 036 1

Mini excavators and narrow scissors lifts – Create new safety concerns

Bernard Ross, Brad McGoran & Fiona Walter
Failure Analysis Associates, Menlo Park, Calif., USA

ABSTRACT: The huge increase in popularity of mini excavators and narrow based scissors lifts throughout the construction industry has sparked new job-related safety issues. In particular, the possibility of equipment rollover as well as the serious injuries that can result thereto demand specific attention. The present paper describes the range of safety features currently found on these machines and explores the types of accidents which result inevitably from owner or operator misuse. Case histories are discussed and a novel means of photogrammetric reconstruction is applied to analyze a mini excavator rollover accident.

1 INTRODUCTION

Amazing changes have occurred in the disparate sizes of construction equipment produced during the past decade. The large-scale end of the spectrum has witnessed truck-mounted cranes capable of 800 (mobile) and 1500 (crawler) ton lifts, dump trucks with 200-ton capacities, and the prodigious bucket wheel loaders found in mining applications. The smaller end, more in consonance with construction and building activities, has witnessed a miniaturization of conventional equipment such as hydraulic excavators, aerial manbaskets and scissors lifts. These machines, conceived to take over traditional tasks handled for years by pick and shovel manual labor, incorporate, by necessity, narrow base widths to pass through standard size doorways and work inside building structures. As expected, this class of equipment has brought forth new safety concerns particularly with regard to overturning or rollover accidents.

The mini or compact excavator, traditionally fitted with a dozer blade, can function as either a digger, dozer, or lifting machine. But, it is also vulnerable to lateral upset during these three different modes of operation. For example, the unit can be pushed over by exaggerated downward displacement of the bucket, experience roll instability due to side overloads while craning, and overturn abruptly on moderate side slopes As a dozer, this unit is required to mount a ROPS (Rollover Protective Structure). On the other hand, as a 360-degree rotation hydraulic excavator, the ROPS requirement is excluded. These conflicting demands pose a dilemma for both the manufacturer and user.

An actual case history outlining the reconstruction techniques employed to resolve an insidious mini excavator accident involving roll-pitch overturn is described here and representative images of a 3-D dynamic graphic animation of the event are shown. A comparison of incipient tipping angles for virtually all mini excavators manufactured worldwide as well as other performance parameters are presented.

The overturning problem and other safety concerns for narrow base scissors lifts are also treated. Related advantages of both fixed and automatic pothole protectors are mentioned. Conditions under which a machine will topple are portrayed via graphic animation. Distinguishing features of machine construction which influence the stability problem are discussed.

2 MINI EXCAVATORS

The mini excavator is a marvel of engineering innovation. The machine combines the earth-moving functions of a tractor backhoe, track excavator, and mini dozer. It can also be used for light crane work particularly in sewer and pipeline trenching operations. This flexibility combined with a surprising mobility embrace a range of work sites from underground tunnels to high rise rooftops, and the agility to work in tight corners is remarkable. A typical mini excavator machine is shown in Figure 1 whilst a listing of mini excavator makes, models, and performance data is given in Table 1.

Figure 1. Typical mini excavator.

Table 1. List of typical mini excavator models and stability performance data.

Make	Model	Country of Origin	Engine HP	Calculated Sideroll Tip Angle (Degrees)
Takeuchi	TB007	Japan	8	26
Gehl	GX10	US	12	34
Case	CK13	Germany	13	27
Bobcat	200	US	13.5	25
Bobcat	320	US	13.5	29
Takeuchi	TB15	Japan	14	30
Kubota	KH-35	Japan	15.6	28
Atlas	404	Germany	16	28
Schaeff	HR12	Germany	18	32
O&K	RH1.1	Germany	19	27
O&K	RH1.2	Germany	19	25
Case	CK15	Germany	20	26
Kubota	KX-014	Japan	NA	29
Kubota	KX-41	Japan	22.2	28
Bobcat	225	US	23	30
Bobcat	325	US	23	33
John Deere	30	US	24	39
Case	CK25	Germany	25	32
Schaeff	HR14	Germany	25	37
Kubota	KX-71	Japan	32.2	36
Bobcat	231	US	34	30
Nissan-Kiz.	N-11ss	Japan	~36	35
Kubota	KX-101	Japan	36.6	37
John Deere	50	US	39	38
Komatsu	PC40-6	Japan	39	NA
Kubota	KX-151	Japan	42.7	39
Gehl	GX45	US	46	37
O&K	RH2.4	Germany	46	28
Nissan-Kiz.	N-41ss	Japan	~50	35

Tip angle (degrees) = arctan[1/2*(basewidth-trackwidth)/CG height]. CG height assumed at top of boom platform per Takeuchi 015 and Kubota KX-014 data. CG located on vehicle centerline.

Figure 2. Mini excavator controls showing locking levers.

As for safety equipment, machines sold in the United States are universally equipped with ROPS/FOPS structures and mandatory seatbelts despite the equivocal requirements for ROPS which are required for tractor backhoes, but not for 360° rotation excavators. An important safeguard is a joystick control block activated when the locking lever (Figure 2) is pulled upwards.

These devices provide two related security features. First, they lock out both of the joystick controls which govern boom hoist--unit swing and dipper stick--bucket crowd motions and thereby prevent any inadvertent input by an operator when grabbing a joystick handle to assist in mounting the cab, a clear danger whilst the engine is running at full rpm. Moreover, the down-horizontal attitude of the levers, when the joystick controls are released for working with the excavator, serve to keep an operator's legs enclosed within the cab if a rollover event does occur. As a further safeguard toward keeping the operator belted in his seat and captured within the ROPS structure protective zone, the entire joystick control lever pedestals simulate body block-in arm rests, yet they can be tilted rearward to assist cab access.

It is noted vis-à-vis the joystick controls that the left and right hand functions are different for machine versions produced in Europe and the United States. Specifically, the left-hand joystick in European models controls dipper stick (bucket arm) motion (in-out) and unit swing (slew) rotation (left-right) whilst in the U.S., the boom hoist (up-down) and unit swing (left-right) are activated by the left control. Obviously, the complementary functions for each version are governed by the right hand joystick.

An omnipresent danger with rotating upper works, whereby workers can be crushed against obstructions behind the equipment during swing motion, is minimized by the newer stylized versions of mini excavators with rounded backsides such as the Komatsu PC28UU and the Hitachi EX75UR. These machines manifest turning radii of less than one meter and develop only a 10 cm. counterweight overhang.

The overwhelming mode for mini excavator accidents is overturn. The machine can roll over on a side slope particularly when the boom and bucket are carried transverse to the crawler tracks over the downhill side. Most mini excavators have a side angle working slope limit of 15°. Tilt table tests on at least four different machine makes have established that the actual critical angle for overturn is between 30 and 40°. Overturn can be provoked by just the boom and bucket arm. For example, the unit can be pushed over backwards in pitch or be turned over sidewards in roll by excessive down force of the bucket against the ground.

A more insidious mechanism of overturn can result when a slewed mini excavator is used for craning over-capacity loads, or in a corollary case history, attempting to boost a recalcitrant stump or buried object out of the ground.

If such tasks are attacked properly--that is, the upper works (unit) and boom aligned longitudinally, the crawler base and the dozer blade in front, and compact rigging--little danger exists. The same situation prevails for pure lateral loading where the unit (upper works) is 90° to the longitudinal axis because the dipper stick and bucket will act as a crutch before the point of incipient instability is reached. However, if these same maneuvers are attempted whilst the upper works and boom are positioned at a swing (slew) angle between approximately 20° and 70° off the centerline axis, the machine can keel over rapidly in a combined roll-pitch motion.

Operators should work with the dozer blade in front which is a configuration that provides resistance to forward pitchover during the backhoe digging action. Moreover, the blade serves as a back stop or mold board for earth moved up during excavation particularly while trenching. A dozer blade to the rear can readily create operator confusion since skid steer motion of the tracks is reversed. For example, with the blade in front, the excavator will travel forward when the steering levers are moved away from the operator. Conversely, with the blade in back, the direction of travel is rearward for forward push of the levers--a situation contrary to intuition and usual experience.

3 SCISSORS LIFTS

A representative listing of self-propelled narrow base mobile scissors lifts that meet the size criteria addressed here is presented in Table 2. Typically these scissors lifts have platforms and/or chassis widths between 20 and 36 in. with vehicle weights in the 1,600 to 3,000 lbs. range. Thus, the machines can enter buildings having normal width, standard headroom clearance doorways while also taking advantage of passenger elevator capacities to reach the upper floors of a multi-story building where tenant improvements often necessitate ceiling work.

This genre of scissors lifts incorporates a wide variety of safety related features both as required by governing ANSI Standards and/or furnished altruistically by the manufacturer. Such details include, but are not limited to: automatic side pothole protectors, audible multi axis sensors with alarms and power cutoffs for drive-lift function interrupt at side slope angles greater than 5°, deadman (momentary) controls with compound action joystick handles, and limit switched dual range slow (1 mph) – fast (3 mph) travel speeds depending on whether the platform is elevated above 5-8 ft. (typical) per ANSI and 4.5 m. per CSA Standards. Back-up and platform descent alarms are often standard. Other characteristics include: manual roll out

Table 2. List of typical narrow base scissors lift models and size data.

Slab Scissors Lift Manufacturers	Model	Maximum Platform Height (ft-in)	Stowed Height (in)	Platform Size (in)	Operating Weight (lb)
Access	2033	20'0"	79	30x94	3,400
Air Technical Industries	MLI-1000	12'0"	24	32x60	1,600
Calavar Condor	2032	14'6"	49.5	27.5x60	2,180
Economy	SM30/16	15'10"	39	30x78	2,050
Genie Industries	GS-1530	15'	77	25.5x62	2,563
Get Smart Scaffold	15S	15'0"	39.25	30x72	2,360
Grove	SM2232e	16'0"	76	32X63	2,280
GSI	14ED	14'0"	Not listed	72x24	1,500
JLG Industries	1532E	15'0"	75.7	30x64	2,400
LDC	1433	13'6"	40	33x96	3,590
Lift-A-Loft	MPN 15-10	15'0"	71	30x75	1,960
Mark Industries	J14EP	14'2"	75.5	27x60	1,865
Mayville Engineering	1532ES	14'7"	75	30x66	2,270
Mayville Engineering	1332	13'0"	76.5	30x66	1,990
Simon Aerials	X-14	14'0"	30	30x64	2,230
Skyjack	3015	15'0"	78	26x59	2,155
Snorkel	SL-15	15'0"	78	29x64	2,300
Strato-Lift	TRX-13	13'0"	82	28x56	1,500
Terex Aerials	TS-15	15'	77.5	30	2,000
Time Condor	V1433	19'10"	39.5	30x63	2,350
Upright	MX15S	15'0"	76	28.5x97	1,962
Weber Products	3015	15'	45	26x67	2,780

Figure 3. Lateral (fixed) and side mounted (automatic) pothole protectors.

deck extensions, proportional hydrostatic drive and power steer functions on a single axle, electro- or hydro- mechanical brake systems, and prime movers that are either electrical (battery), propane, or gas engine fueled.

An important requirement found in both the ANSI and Canadian Standard Association (CSA) regulations is a 5° side slope proof test with the unit placed on a tilt table in its least stable configuration under full scissors extension and a prescribed uniformly distributed platform load. Apropos, tests of numerous machines at Failure Analysis Associates have indicated that the typical critical angle for overturn is around 7-8°; thus there is approximately a 60% margin of safety over the alarmed cutout. The CSA standards describe a pothole stability test for two different size cylindrical depressions. At present there is no American test requirement for scissors lift stability vis-à-vis pothole encounter. However, it is understood that a cognizant test protocol will be forthcoming in the next ANSI Standard series.

Among the real life accident case histories encountered with these lifts, the following examples can be cited. First, raising the platform into ceiling trusses, beams, girders, etc. is a well proven threat with obvious exposure to head and shoulder injuries. A more likely consequence of ceiling bashing is the bending, buckling, dislodging of the thin gauge platform structural members which often end up being re-secured by bailing wire whence anyone leaning against the railings can break through and fall.

The omnipresent danger of driving laterally off sidewalks, curbs, and ramps exists even despite the automatic side mounted pothole protectors now found on many machines. In fact, to guard against such accidents, pothole protectors should be attached both transversely and longitudinally as seen in the new model UpRight Series MX15 Scissors Lift (Figure 3). [The fixed (bolted) channel section is actually a forklift blade pocket but it serves as a pothole protector too.]

Fatal accidents have also occurred when scissors lifts were driven off elevated loading docks or upper stories of buildings--a case of pure operator error. In situations where close in, tight maneuvering near free unguarded edges is required, a spotter should be employed to assist the operator.

Freak accidents also happen. For example, a trailing 220 volt electrical line to a power tool was caught under a scissors lift wheel; when the platform was raised, the resultant pull imposed on the line by the wedged tool separated the wires whereupon contact of the live conductor with the metal frame electrocuted a worker.

All of the scissors lifts envisioned here have solid rubber tire wheels and must be used exclusively on grade slabs. None of the machines cited in Table 2 is a rough terrain piece of equipment. Overturning accidents can result if the equipment is operated on unfinished earth basements or outdoors where depressions and ground undulations can wreck havoc. For example, a fully raised lift parked on a 6° compound side slope with one of its wheels over a 2-1/2 in. depression experienced fatal rollover when the occupants moved across the platform, out of the stability triangle, over to the platform corner above the free floating wheel.

Figure 4. Scissors lift tipover due to driving off curb.

The prediction of scissors lift stability performance under an unlimited range of platform heights, deck extensions and occupant loadings is readily accomplished with present-day computer software such as Knowledge Revolution's Working Model 3-D. An image from the 3-D graphics animation of a scissors lift driving off a sidewalk curb is shown in Figure 4.
The composite center-of-gravity for the machine under analysis must be known as a function of platform height. The c.g. location can be determined by rote calculation of individual component-assembly weights with their respective moment arms, or more easily, by tests according to the protocols of SAE J1040.

Of course, most scissors lifts are owned by rental companies and on-the-job servicing, upkeep and worker abuse of leased equipment is always suspect.

4 ACCIDENT RECONSTRUCTION

The engineering investigation of a mini excavator roll-pitch overturn is introduced not only to explain the phenomenon but to illustrate how advanced photogrammetric techniques can be applied in a novel way to reconstruct the accident. The incident at issue occurred whilst an operator was attempting to remove a recalcitrant sapling stump from the ground with the excavator bucket and a chain attachment. Resistance of the stump to its removal, in terms of load, was completely unknown to the operator nor was it even calculable. In effect, the operator was lifting Planet Earth.

The excavator was poorly positioned for the work at hand with both the unit and boom at compound offset angles of 16° and 35° to the longitudinal axis of the crawler base. Moreover, the tracks were reversed with the dozer blade at rear thus providing no support in pitch and very little reaction in roll to counter the offset craning action. Further, it would be expected that the operator abruptly yanked, rather than pulled, on the steadfast stump to cause sudden overturn of the entire unit. The operator was either thrown or bailed from the cab whereupon the exterior handhold on the ROPS post produced a severe crushing ankle injury that resulted ultimately in leg amputation. As usual, in virtually every case where injury or death results from an earthmoving machine rollover, the operator was not wearing his seatbelt.

Immediately following the accident, some amateur photographs were taken of the grass lawn site. A representative picture is shown in Figure 5. To be noted here is the sprinkler head with surrounding doughnut and the palm tree with a prominent knot on its lower right side. The reconstruction investigation was undertaken many years after the accident whence the grass track and dozer blade patterns had completely grown over and the sprinkler head had been removed entirely. Thus, it was necessary to reposition the mini excavator in both linear and rotational planar coordinates vis-à-vis the palm tree and foliage where the stump was located.

To this end, a process known as reverse image projection, or F3 photogrammetry, was employed. First, an existing glossy photo (Figure 5) which contained signatures of the two

Figure 5. Post-accident ground markings.

crawler tracks, the dozer blade, and the handrail contact patch with the ground was blown up to 8-1/2 by 11 size. Next, a full photo size clear plastic sheet was overlaid and marked with black outlines of all prominent features. This transparency was subsequently reduced to the size of a 35-mm photo slide. The 35-mm image was then inserted in a Nikon F3 camera which has a detachable upper body hood and removable viewing screen to assist these purposes. Two such photographs shot from different angles were rendered in this way.

Next, two convergent view F3 cameras were set up on tripods in the backyard of the accident site. The cameras were sighted on the palm tree trunk diameter, as to linear dimension, and the prominent knot, as to azimuth dimension, so as to establish actual camera locations when the original photographs were taken. Then an assistant laid out a series of orange pegs (golf tees) under direction of the cameraman according to the outlines of prominent features as seen through the viewfinder. Once all the pegs were set and strung together (Figure 6), measurements could be made to establish engineering plan and elevation views of the accident scene with all features captured in the original photograph properly identified in situ.

Through computer graphics simulation, the digital rendering of a full side view photograph of the subject mini excavator was superposed on the photogrammetric drawings. Once coordinated in the computer memory file, all manor of zoom and pan views of the excavator and the working site could be obtained and presented either as line drawings (Figure 7) or rendered graphics (Figure 8).

Based on the actual measurements obtained on site, it was possible to establish a quasi exact mini excavator upper works (unit) and boom-dipper stick configuration in 3-D for the intended stump pull (Figure 8). Then, using both analytical calculation, based on center of gravity for the machine in the now determined configuration, and also actual field tests with bucket uplift on a concrete embedded eye-bolt anchor (the machine was tethered to prevent complete overturn), it

Figure 6. Outlines constructed from photogrammetry.

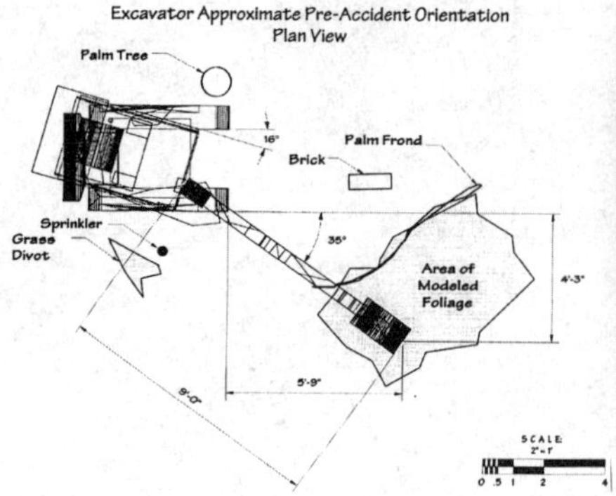

Figure 7 and Figure 8: Line and rendered excavator drawings.

was found that the characteristic loads to provoke roll-pitch instability ranged between 600 and 750 lbs. Conversely, the maximum allowable load found in the Operators Manual under the condition of pure side lift was 420 lbs. Thus, poor machine position, an overload of at least 50%, and failure to observe the mandated use of a seatbelt caused this unfortunate accident.

REFERENCES

ANSI/SIA A92.6-1990—Self-Propelled Elevating Work Platforms.
CAN3-B354.2-M82—Self-Propelled Elevating Work Platforms for Use on Paved/Slab Surfaces.
CAN3-B354.3-M82—Self-Propelled Elevating Work Platforms for Use as "Off-Slab" Units.
OSHA ROPS Standard (29CFR §§ 1926.1000, 1926.1001, 1926.1002).
OSHA Seat Belt Standard (29 CFR § 1926.602).
SAE J1040—Performance Criteria for Rollover Protective Structures (ROPS) for Construction, Earthmoving, Forestry, and Mining Machines, April 1988.
SAE J874—Center of Gravity Test Code, October 1998.

Implementation of Safety and Health on Construction Sites, Singh, Hinze & Coble (eds)
© 1999 Balkema, Rotterdam, ISBN 90 5809 036 1

Developing a proactive Safety Performance Measurement Technique (SPMT) for construction

R. Kunju Ahmad, A.G.F. Gibb & R. McCaffer
Department of Civil and Building Engineering, Loughborough University, UK

ABSTRACT: The construction industry is well known for its poor safety performance and it needs to look into new ways at improving its image. Over the years many changes have evolved regarding safety. With better enforcement of the legislation and commitment from employers and employees, construction safety has received greater attention. To measure the performance of safety on site, a mechanism is prerequisite to identify the elements of safety. In this industry, current measurement of safety performance is based on post-accident data with little focus on pre-accident factors. Safety performance is more than just a measure of lack of safety. Rather it recognises its potential as an effective means of communicating positive safety results. This paper introduces the proposed methodology for developing an effective safety performance measurement tool based on pre-accident measures.

INTRODUCTION

The construction industry has never been a leader in safety. Accidents, injuries and property losses not only cause delays, but also directly and indirectly incur costs. The realisation of the cost of accidents and human suffering that follows has brought about changes towards the attitude of management and employees with regards to safety (Staley 1996). Only with proper management commitment, planning and establishment, is it possible to achieve a safer working environment which is also cost effective. In the period 1992/93 - 1996/97, figures for fatal and non-fatal accidents in the UK construction were:-

The reduction in the figures (Table 1) is mainly due to a smaller workforce or less buoyant industry. Traditionally accident statistics play an important role as a prime indicator of safety performance. The problem with accident statistics is that they will only provide factual information which is analysed later to identify opportunities for improvement (Staley 1996).

It is the purpose of this study to examine new approaches and techniques that will enhance safety performance measurement on construction sites. The paper will discuss the aim, objectives, reasons for the study and methodology to carry out the work. The authors are

Table 1- Fatal and Non-fatal Accidents for the Construction Industry From 1992/93-1996/97 (Employment Gazette 1992/93 and 1998)

Type of Acc	1992/93	1993/94	1994/95	1995/96	1996/97
Fatal	69	73	56	62	65
Non-fatal	13474	11287	11504	10111	12221

currently at the literature stage of the study and no primary data have been collected to date. More data will be available at the time of the conference.

AIM AND OBJECTIVES OF THE STUDY

The aim of this study is to develop a single measurement tool (SPMT) to measure safety on construction sites. This method will not only reflect the level of safety but will evaluate the effectiveness of safety performance; provide continuous information concerning changes in the safety state and enable identification of potential causes of future loss.

The objectives are:-

1. To identify realistic and measurable proactive factors and sub-factors that affect safety performance on site.
2. To develop and test the measurement tool (SPMT) that will be a benchmark for future measurement.
3. To develop a measurement technique that will be able to tell not only how the accident occurred but the remedial effort to be applied.
4. To be able to evaluate over time the degree of progress or retrogression of safety performance on site.

REASONS FOR THE RESEARCH

The authors have identified important and critical reasons to undertake this research (Kunju Ahmad et al 1998). The reasons are:-

- The construction industry has a poor safety record.
- There are good reasons to want to improve the safety performance.
- Accident causation is complex but important.
- It is important to measure performance.
- Existing safety measurement systems are limited in scope and effectiveness.
- Concentrating on proactive measures instead of reactive measures should improve safety performance.
- There is a need to develop a single standardised measurement technique.

METHODOLOGY TO DEVELOP SPMT

The following steps have been chosen to develop SPMT:-

1. *Review of existing literature:* A brief summary of the preliminary findings of this review are summarised here. More information has been provided by the authors elsewhere (Kunju Ahmad et al 1998). Most researchers argue that looking at causes of accidents will definitely reduce the number of accidents (Hinze 1996, Bentil 1995, Staley 1996). Safety efforts should be aimed at all accidents including unsafe acts and near misses rather than targeting serious and fatal accidents (Figure 1). In theory such practice will cause reduction from the base upwards. Factors such as work force, job site conditions and unsafe methods also contribute towards this poor performance as can be seen from Figure 2. Many researchers agree that training is essential as part of the measures (Janandi 1996, Hubler 1995, Tam 1996, Lowery 1988, Wagenaar 1997). What is important throughout the review is that many agree that management involvement is important and essential. Management not only involves by providing the safety manuals but more commitment and accountability for the safety performance (Simon 1991, Jaselskis 1996, Wagenaar 1997, Mattila 1993, Kletz 1993, Booth 1995). Simon (1991) suggests that communication is important to encourage employees to suggest safety improvements and reports near misses as well as unsafe conditions and practices.

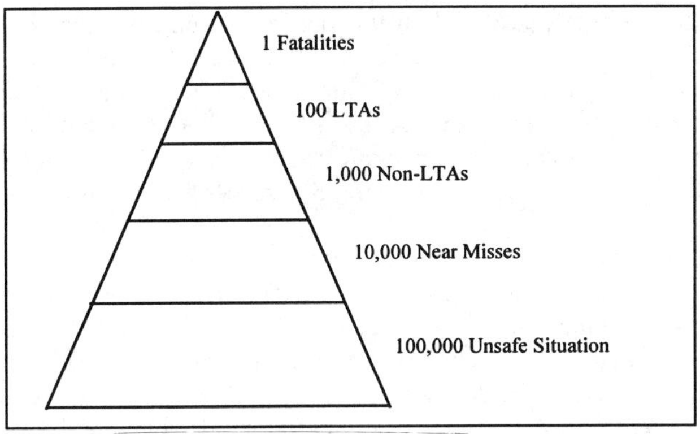

Figure 1 - Unsafe Situation to Fatality Pyramid (Smith 1991)

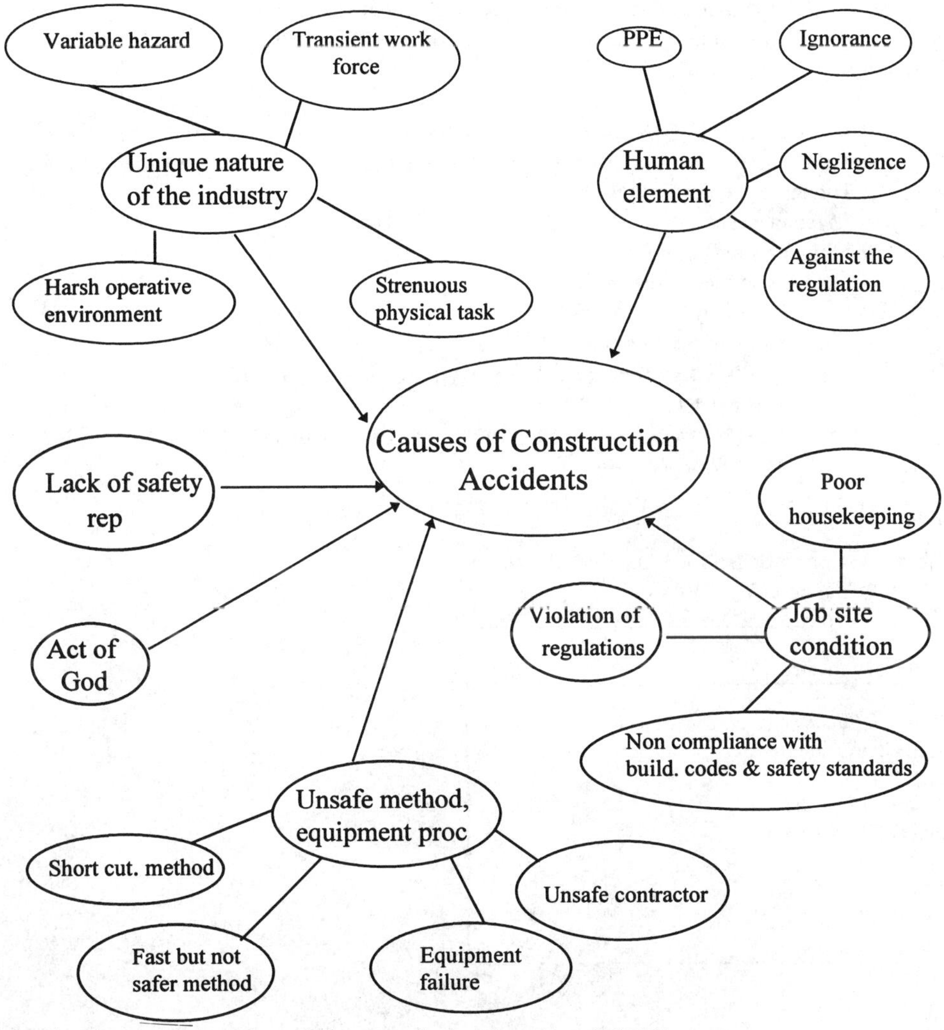

Figure 2 - Causes of Accidents (Diagram adapted from Bentil 1990)

Many studies and models have been developed based on this concept of unsafe behaviour and conditions (Staley 1996, Smith 1991). All these models agree that proactive measures or pre-accident measures are the answer to a better safety performance on site. Unlike reactive measures, proactive measures deal with data from current safety situations. Repeated appraisals provide essential feedback on performance before the injury or the incidents occur. Thus this approach is highly recommended (Staley 1996, Simon 1991).

2. *Isolate proactive measures from the literature reviews*: The authors have identified at least 30 measurement techniques. Many are used in process engineering, mining and construction. All 30 measurement techniques are in-house safety systems designed to their respective needs and objectives. What the authors observed is that the historic and contemporary SPMs have no consistent approach towards safety performance. Figure 3 looks at proactive measures that were derived from the 30 measurement techniques. The vertical axis represent the number of times the same constituent criteria were found. For example, in Figure 3 there are 49 factors. The highest score is 13 and the lowest is 1. Table 2 shows the most frequent criteria based on the 30 measurement techniques. These eight criteria will be considered in designing the conceptual framework which will form the basis for SPMT. The authors intend to develop the SPMT such that it will cause minimal disruption to normal site operations.

3. *Develop a conceptual framework of proactive measures based on Figure 3*: It has been decided that only construction work involving building works will be included for this study. The target respondents are:-
    - Management
    - Supervisors/Foreman
    - Workers/Operatives
        - general operatives
        - plant operators
        - skilled workers (e.g. bricklayers, roofers, painters)
        - steel erectors/fixers
          services operatives (e.g. plumbers, electricians, lift engineers)
        - ground workers

Table 2 - Highest Score for Proactive Measures
(5 Scores and More)

Measuring factors	Units
Organisational/Management control	13
Engineering control	8
Training	8
Behaviour/Attitude	6
Design/Planning	6
Housekeeping	6
Procedural control	5
Supervision	5

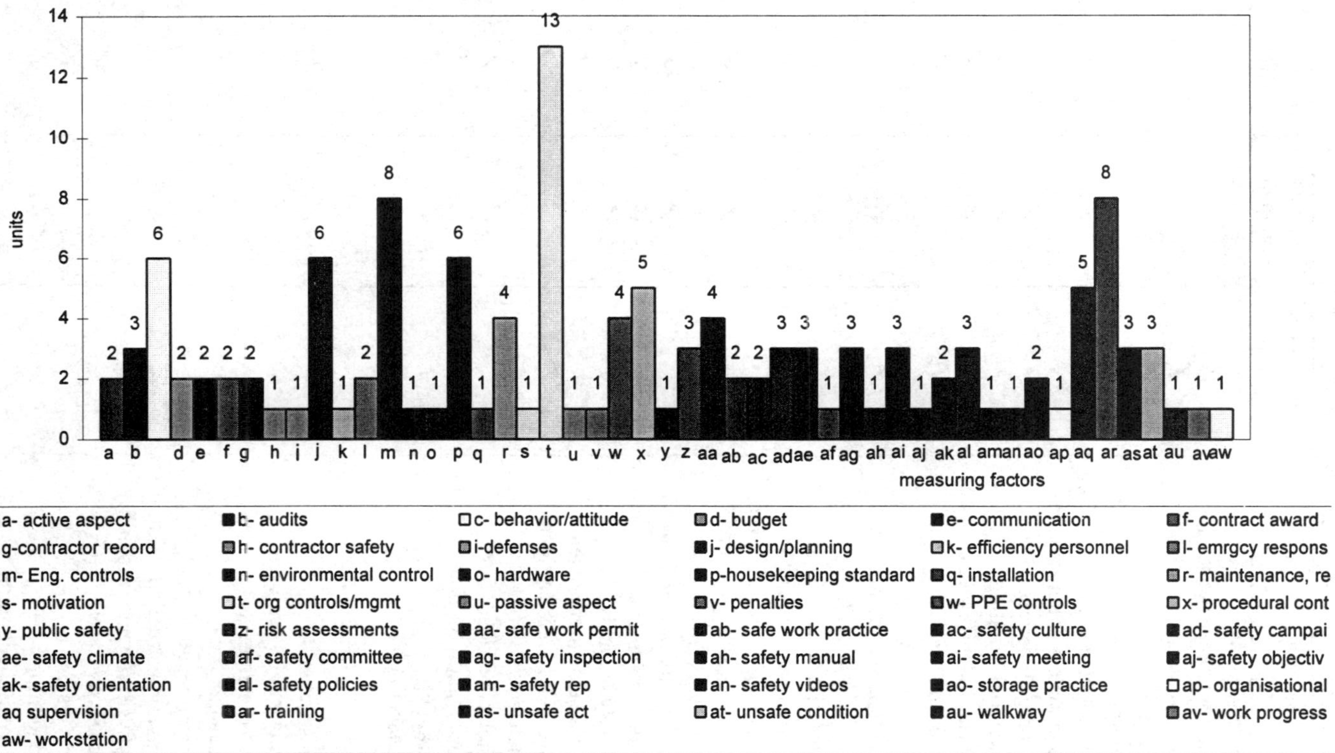

Figure 3 - Proactive Measures

Table 3 - Recommended proactive factors and sub-factors for SPMT

Factors	Sub-factors	Management	Supervisor /Foreman	Worker /Operatives
Management	safety policy	✓		
	safety representative	✓		
	motivation	✓		
	audit	✓		
	budget	✓		
	meetings-contractors, sub-contractors & supervisors	✓	✓	✓
Housekeeping	storage practices		✓	✓
	material handling		✓	✓
	hazard identification		✓	✓
Engineering control	risk analysis	✓	✓	✓
	method statement	✓	✓	✓
	design/planning	✓		
	installation		✓	✓
	maintenance/repair		✓	✓
	adequate knowledge/skills		✓	✓
	procedural/work standards		✓	✓
Safety culture	safe work practice		✓	✓
	safety rules/procedures		✓	✓
	safety standards	✓	✓	✓
	safety manuals	✓	✓	✓
	safe behaviour related to worker, tool & equipment		✓	✓
	safe acts		✓	✓
	safe condition		✓	✓
Training	orientation for new workers	✓		✓
	training	✓	✓	✓
Supervision (formal & informal)	checking	✓	✓	
	inspection	✓	✓	
Communication	channel for reporting			✓
	proper signage	✓		
	channel for recommendation			✓
	giving instruction	✓	✓	
Others	physical/mental capability			✓
	excessive stress			✓
	diversion			✓
	condition of material/equipment		✓	✓

4. *Design appropriate measuring techniques for each level*: For the management levels, the authors will design a semi-structured questionnaire and this will be completed once a month. Whereas for the supervisors and worker/operatives levels, a checklist will be developed as a practical tool for gathering information on a busy site. This checklist will be

used more frequently e.g. once a week.

SPMT will concentrate on measuring safe behaviour instead of unsafe behaviour. This way a worker performing a task will be evaluated in terms of the overall assessment of the potential well-being of the worker. US based researcher Jimmy Hinze has advised the authors to consider using a scale ranging from poor (injury is highly likely) to excellent (safety is highly assured). Table 3 lists all the proactive factors and sub-factors that the authors have identified to date.

5. *Test prototype SPMT*: Once this conceptual framework for all levels is developed, the authors will conduct a 'dry run' to verify the reliability of SPMT.
6. *Review developed SPMT*: Receive feedback and develop further before sending out to a bigger sample.
7. *Validate SPMT on a larger sample*: Receive feedback, review and conduct interviews and workshops to stimulate discussion regarding SPMT.
8. *Trial SPMT*: Use SPMT on a number of case study sites.
9. *Evaluate effectiveness of SPMT*: Analyse and tabulate the data. Compute the percentage of safety improvement.

CONCLUSION

The authors intend to design a common safety performance tool to measure safety on site. The authors have identified the reasons to undertake the study. The industry has a poor safety record and accident causation is important but complex. There is a need to measure safety performance, but currently the many techniques make useful comparison difficult. By having a single measurement tool, then the level of safety should be improved. The SPMT will be a benchmark towards measuring safety performance. The benefit of having a single technique is that it can compare the real progress between construction sites. The result will be easily understood and interpreted by the management, contractors and even the workers. The aim of this single technique is to continuously generate observable improvement in the way people work. Measuring safety performance using accident data or reactive measures does not yield a true measure of safety. It was decided that a measure of safety using proactive measures would be the best strategy. SPMT will aim to move the industry away from a purely reactive response towards a more proactive approach to the improvement of safety performance. This approach, if successful, will contribute to changing the culture of the construction industry with regard to safety.

REFERENCES

Bentil K. K. 1990. Construction Site Safety : A Matter of Life and Cost. *Cost Engineering.* Vol. 32 No 3, pp 7-11.

Booth R.T. 1995. The Role of Human Factors & Safety Culture Safety Management. *Proc of the IME*, Vol 209, No B1, pp 393-399.

Hinze J W. 1996. The distraction Theory of Accident Causation. *Proceedings of the First International Conference of CIB Lisbon W99*, pp. 375-384.

Hubler W.G. 1995. Behaviour Based Approach to Creating A Strong Safety Culture. *SPE/IADV Drilling Conference*. Amsterdam SPE. pp 361-373

Janandi. 1996. Factors Affecting the Safety of the Construction Industry. *Bldg Research & Info*, Vol 24, No 2 pp 108-112.

Jaselskis E.J. 1996. Strategies for Achieving Excellence. Contractor Safety Performance. *Journal of Construction Mgmt & Econ*, Vol 122, No 1 pp 61-70.

Kletz T.A. 1993. Accident Data - The Need for a New Look. *Safety Science* Vol 16 pp 407-415.
Kunju Ahmad R, Gibb
A G F and Ron McCaffer, 1998. Methodology To Develop An Effective Safety Performance Technique (SPMT). *ARCOM Conference*, September 1998.
Lowery R and Moore S L, 1988. Safety Performance: Results Can Be Achieved. *Professional Safety*. August 1988, pp. 72-76.
Matttila. 1994. Effective Supervisory Behaviour and Safety at Building Site. *International Journal Of Industrial Ergonomics,* Vol 13, pp 85-93.
Ngowi A.B. 1996. Culture and Safety at Work Site. *Proceedings of the First International Conference of CIB Lisbon W99.* pp. 417-427.
Simon J.M. 1991. Construction Safety Performance Significantly Improves. *1st Intl Conf of Health,Safety & Environ*, Netherland. SPE Intl. pp 465-472.
Smith G R and Arnold T M 1991. Safety Performance for Mansonry Construction. *Proceedings of the First Intl Conf of CIB Lisbon W99.* pp. 103-114.
Staley B.G. 1996. Investigating Accidents & Incidents Effectively. *Mining Technology*, Vol 78 No895 pp 67-70.
Tam C.M. 1998. Effectivensss Of Safety Mgmt Strategies in Safety Performance in Hong Kong. *Journal of Construction Mgmt & Econ*, Vol 16, No 3 pp 49-55.
Wagenaar W.A. 1997. Accident Analysis The Goal & How To Get There in *Safety Science*, Vol 26, No 1 pp 25-33.

15 Site safety

Implementation of Safety and Health on Construction Sites, Singh, Hinze & Coble (eds)
© 1999 Balkema, Rotterdam, ISBN 90 5809 036 1

Safety program for volunteer construction workers

Boyd C. Paulson, Jr
Stanford University, Calif., USA

ABSTRACT: Volunteer organizations account for significant amounts of construction work. For the most part, volunteers have little or no prior experience in construction and are ignorant of common safety hazards. Even senior safety officials seem uncertain as to how occupational safety and health laws apply to volunteer workers.

This paper describes a safety program designed for a 24-unit condominium project that involved some 6000 volunteers, most just for a day, to perform 80% of the labor. They undertook tasks ranging from concrete foundations through structural framing, roofing, and electrical and mechanical systems to various finish trades. Each was trained in both the safety and technical aspects of their tasks and then spent about six hours to accomplish a good day's work. The safety record, while not perfect, was good. The project confirmed that best practices from contractors with top safety records also apply well in volunteer construction.

1 INTRODUCTION

Volunteer construction efforts – building and rehabilitating homes for poor people, cabling schools for the Internet, painting community centers, restoring historic structures – occasionally make the evening TV news or front-page press. Almost always, however, the glowing reports ignore gross safety violations sometimes evident in the pictures. Indeed, reporters seem to home in on dare-devil acts and even put themselves at risk to get the most dramatic photos or video to illustrate their stories. Such reporters seem to know less than the volunteers.

This paper will briefly discuss unique aspects of volunteer construction and focus on safety considerations for these well-meaning people. It will then describe a safety program designed for a Habitat for Humanity affiliate in California that recently built a 24-unit condominium.

2 VOLUNTEER CONSTRUCTION

Some volunteer organizations account for significant amounts of construction work. Habitat for Humanity has built 20,000 homes in the U.S. and 40,000 in developing countries. Its 1997 domestic production ranked it as the 15th largest home builder (in number of homes) in the U.S. (Professional Builder 1998). Another organization, Christmas in April, rehabilitates thousands of homes and other structures every Spring. Religious and community groups not only build homes but also churches, playgrounds, and other facilities. For the most part their volunteers have little or no prior experience in construction and are ignorant of common safety hazards. Even senior Occupational Safety and Health Administration (OSHA) officials seem uncertain as to how their laws apply, if at all, to volunteer workers.

Common examples of violations observed in volunteer work environments include lack of personal protection equipment (hard hats, ear plugs, safety goggles, etc.), work at heights without fall protection, make-shift platforms (e.g., standing on chair stacked on a table) and unsecured ladders, untrained people using power tools, and working with toxic materials without breathing apparatus (e.g., when removing asbestos or lead-based paint). While these problems are well-recognized and safely handled by responsible construction professionals, volunteers and their supervisors often seem not to even recognize the problems let alone know how to implement the applicable safety measures.

Given that accidents in such environments most likely do not come under standard reporting requirements applicable to occupational health and safety, it is difficult to estimate the number of deaths and injuries that occur each year. However, clues may be found in the most common environment for non-work accidents – the home. In the United States, National Safety Council data show that home accidents are second only to automobiles in accounting for accidental deaths and injuries – 26,400 deaths (7300 from falls alone) and millions of injuries in 1995 (Gibson 1998). This exceeds by a factor of two those for all industries, let alone construction. Many of these accidents result from ill-conceived home maintenance and remodeling situations not unlike those found in volunteer construction. – painting from ladders, working on roofs, mishandling power tools, etc. It is likely that most people who have done such things regularly have sustained at least modest injuries, and most people also know one or more other people who have suffered severe injuries while working at home.

Given the nature of the work undertaken by volunteer construction workers, it would not be surprising to find that volunteer construction fares more poorly than conventional contractors and employees doing similar types of work. How does one set up an effective safety program in such an environment?

3 CONDOMINIUM PROJECT BUILT BY VOLUNTEERS

From late 1995 to early 1998, Peninsula Habitat for Humanity (San Mateo County, California) built a 24-unit condominium project consisting of four 2-story and 3-story residential buildings and six 4-bay garage structures. The design was nearly identical to an adjacent 38-unit non-profit rental project built concurrently, and thus had no concessions toward simplicity for amateur builders. It met all of California's stringent earthquake, energy and multi-family residential codes. Its craftsman style had complex roof lines and numerous architectural details. Prior to this project, Peninsula Habitat had only built nine one-story single-family detached homes and thus had to climb a steep learning curve to build the far more complex multi-family structures.

In this project, some 6000 people – programmers, engineers, biotechnology workers, health care professionals, teachers, homemakers, retirees, students and others – volunteered 80% of the labor. Most volunteers, who came from churches, schools, companies, and community organizations, were on site for only a day in groups ranging from a dozen or less to over 100. Though lacking trade skills, most were intelligent, highly motivated and quick learners. They undertook tasks ranging from concrete slabs and foundations through structural framing, roofing and electrical and mechanical systems to windows, drywall and various finish trades.

While this is not a construction management paper, it is worth noting in passing a few of the differences between the volunteer project and the conventional project next door. First, the bad news. Volunteers took much more supervision. The experienced general contractor, with no direct-hire construction workers, needed only one full-time superintendent and a part-time project manager to marshall its series of subcontractors and suppliers to the site. Habitat, which did most of its work with its own volunteer forces, needed three full-time paid supervisors and also depended on several knowledgeable volunteers who came most Saturdays. The volunteer project also took three times as long and thus its supervisory overhead was far more expensive. Given the drawn-out schedule, the few subcontracted items were chopped up over time rather than tightly sequenced and thus most such trades cost more than did those next door. Habitat's cost was also higher because it elected to require subcontractors to pay prevailing union wages while the adjacent project ran "merit shop."

Now the good news. As expected, volunteer labor made many tasks much less expensive (e.g., framing 60%, electrical about 50%, drywall 50%, etc.). Quality, as judged by city inspectors and other impartial and knowledgeable observers, was superior to that of similar market-rate housing let alone low-income projects. Habitat upgraded some specified products to more durable materials and installed them at or above code standards. (The standing joke in Habitat organizations is that volunteers pound so many nails that if a Habitat house ever did burn down it could be sold for scrap metal!) Most significantly, Peninsula Habitat conscientiously tried to apply OSHA safety regulations conservatively and adhere rigorously to state and federal standards (CalOSHA 1995, Office of the Federal Register 1995). For example, the project was fully scaffolded as soon as the first-floor walls went up and volunteers wore hard hats no matter where they were working. Just to be extra safe, midway through the project they invited a CalOSHA consultation visit in the hope of getting specific advice on further improvements and instead subsequently were given a report that amounted to a letter of commendation. Safety remained top priority throughout the project, but there was always room for improvement.

The bottom line was that Peninsula Habitat's first venture into multi-family housing came in about even in costs per housing unit when compared with the subsidized rental project next door. The latter was built by an experienced non-profit developer (about 4000 units to date) and a respected and professional general contractor who specializes in this type of work. In California's high-cost market, units on both sides of this project sell or rent for about half of equivalent market-rate condominium and rental units not far away. Because home buyers benefit further from Habitat's zero-interest mortgage policy,[1] ownership in this project costs less than the adjacent rentals and builds equity with every dollar paid (in contrast to conventional mortgages where most dollars go to interest in the early years).

4 SAFETY PROGRAM

Knowing that this project would pose far more risks than its previous single-family homes, Peninsula Habitat set out before construction began to develop a formal safety program that would be suitable for its volunteers. Fortunately, the San Francisco Bay Area has several large contractors that have excellent safety programs and outstanding safety records that consistently rank among the best in the nation (Barrie 1992, Levitt 1993). One of these is Rudolph and Sletten (R&S), which does a wide variety of general building, including housing, and is especially well known for its work on the complex technical facilities of Silicon Valley's electronics, computer and biotechnology industries. They made available all of their safety materials and from these their "Code of Safe Practices" and 35-mm safety slide set used for worker orientation were simplified and adapted to the needs of the Habitat project. A copy of Peninsula Habitat's Code of Safe Practices, included at the end of this paper, gives insight into the priorities deemed most essential for typical one-day workers.

Another large local firm, DPR Inc., which works in markets similar to R&S but also does so nationwide, included Habitat's construction supervisors and staff in its own in-house OSHA-certified ten-hour safety course. At no cost they not only provided instruction to each participant, but also the state and federal safety manuals and their own in-house safety course and reference materials. Furthermore, it was most beneficial for Habitat's supervisors, who mainly had a residential construction background, to see just how seriously and professionally truly high-end firms such as DPR and R&S integrate safety into their mainstream operations on a wide variety of projects. They also sometimes build housing, but their main market ranges from luxury office remodels to extremely complex $2-billion electronics plants.

Once the supervisory situation settled down (there were problems initially, to be discussed later), the pattern on Habitat's condominium project became as follows. Even before arriving on site, volunteers and groups who signed up in advance received an orientation letter that, among other things, told them what to wear, how to get to the site, and asked them to read the Code of Safe Practices. The purpose was to distinguish between this type of fun group activity (it is indeed fun to build) and more familiar group experiences, such as weekend softball games, picnics, etc., and to get their attention directed toward safety issues that pertain to construction.

Upon arrival on site, volunteers got together in a 4-bay garage (completed early and set aside for this purpose) for a 30-minute orientation and training session. Here they received some

[1] Habitat's zero-interest mortgage policy is based on several Biblical prohibitions against charging usury to the poor, most notably Exodus 22:25 – "If thou lend money to any of my people that is poor by thee, thou shalt not be to him an usurer, neither shalt thou lay upon him usury." Typical Habitat families are the working poor, falling near what the Federal Government defines as "very low income," which is at or below 50% of the mean for their area. As their "down payment," these families are expected to put 500 hours of their own labor toward their home – about 20% of the overall volunteer effort. All proceeds from their payments go into the "Fund for Humanity," 100% of which is supposed to go toward building more homes.

background about Habitat's program and the role of this project in its community, heard about a 15-minute overview of safety guidelines with emphasis on those applicable to the day's work, and then were divided into teams that would perform specialized tasks.

Team leaders (mostly regular volunteers or AmeriCorps workers[2]) then took their respective teams to the next 4-bay garage where tools and safety supplies were kept. They made sure the volunteers were properly equipped for the tasks they were to perform (all had hard hats; ear plugs, tool belts, glasses and gloves were normally needed as well). Tools in hand, they went from here to their workstations in the buildings under construction. At this point the team leader gave further instruction regarding the technical requirements of the tasks, demonstrated the methods to be used, and reinforced specific safety rules and procedures that applied here.

Only at this stage, about an hour after arrival, were the volunteers ready to start climbing their own personal steep learning curves in the six hours that remained to accomplish a good day's work. Typically, they would spend the morning on a few production iterations of their task (framing a small wall, hanging a few sheetrock panels on the ceiling, pulling electrical circuits, etc.). They usually could do three times as much work in a similar amount of time by the end of the day – if they still had the energy to do so.

At the end of the day, they cleaned up, admired their work, and went away for a month, a year or more. The process started all over again the next morning – more new people, more new things to learn – and it continued this way until 24 families were all in their new homes. This is not a formula for high production, but it got the job done safely and with quality results.

5 PROBLEMS AND RESULTS

As experienced professionals know, even the most complex projects often have their greatest challenges in the organization and people involved. This was no exception.

Surprisingly, the volunteers – even though they numbered in the thousands – were seldom a significant problem once one understood their needs and limitations. This type of activity attracts some of the kindest, most community-spirited and well-meaning people one could hope to know. Working in such groups is its own reward. Most were intelligent, enthusiastic, eager to learn, and willing to follow instructions. Occasionally a high-school group got a little rambunctious and needed to settle down. At times someone started using a power tool without being checked out first. Every now and then someone would have climbed to the top of a three-story roof on a 6 in 12 pitch only to realize that they really should not be there, and needed reassurance to get back down and move to another task. As volunteers go, perhaps a larger problem was groups that insisted on bringing too many people, or to only come for half a day, but these problems normally were handled in the home office. On site, perhaps the main hazard was an occasional home handyman or engineer from another field who convinced site staff that he knew it all already and then proceeded to give conflicting instructions to fellow volunteers. But all in all, these were not the main issues.

The initial crew of three site supervisors (nominally a superintendent and two foremen) had been involved in varying degrees supervising or working on the nine single-family homes Peninsula Habitat had built prior to starting the condominiums. While each had some post-secondary education, they were up-from-the-ranks residential construction craftsmen – two carpenters and a cement mason – and had years of experience with traditional ways of doing things. When working with small groups and with little stress building one single-family home at a time, they seemed to have worked well with volunteers and were well liked. Apart from a volunteer who fell from a ladder and broke his arm in one of those homes, they did not say much about safety on those jobs. Nevertheless, they agreed that the larger project was more of a challenge and two went to DPR's OSHA 10-hour course.

However, in the early months of the project it became evident that they were unlikely to change much from their years of "the way it is always done on these jobs." They mainly were paying lip service to the safety program and seemed to think themselves exempt from the

[2] Like many other non-profit organizations, Habitat benefits from the very low-cost services of the extraordinary people – typically recent high school or college graduates – who volunteer a full year of their lives for public service under the Federal Government's VISTA and AmeriCorps programs. For two years, Peninsula Habitat has had two VISTA volunteers to assist with office administration and four AmeriCorps volunteers to assist in the field, and had smaller numbers before then.

requirements observed by the volunteers. The superintendent delegated the safety orientation to a regular volunteer and it only happened when that volunteer was on site. One foreman, who brought an antiquated and poorly maintained forklift, along with an old dump truck and loader, as part of his hiring "package," was clearly of the hard-driving "cowboy" school of high production at all costs. After his brakes and engine failed and his forklift rolled back to do $2000 damage to a volunteer's car, he received a warning and promised to do better. Shortly thereafter, a large double truss being placed on a second story fell over, pinning his fellow foreman and toppling already-placed trusses like dominos, with the last two falling off the end of the building. He was contrite, but blamed it on a knot in a 2x4 brace and on his goal to finish the wing by the end of the day before more securely bracing it. Contrition notwithstanding, he was terminated at that point.

The other two soldiered on, but their planning in general was as seat-of-the-pants style as their safety and the project fell further and further behind. The superintendent, a good and decent man at heart, pledged to get safety up to par. But he gave priority to cost and production. An example was that he only had enough scaffolding to work from one or two sides of the building at a time, and tried to move it around so that it was under the roof decking and shingling work areas. One day, when the volunteer chairman of the construction committee came up on the roof, he noticed a crew of four volunteers rolling out felt paper near the lower edge of a section of pitched roof. There was nothing between them and a 6-meter (20-ft.) fall to hard concrete below. Three meters (11 feet) is the height at which 50% of such falls become fatal. The chairman immediately pointed this out to the Superintendent, who was just over the ridge, snapped a quick photo to make the point, and stopped the operation. The project remained closed for two weeks while the construction committee and home office staff tried to come up with a better way of managing this site. While this was the most serious situation observed, the pattern was typical of others at that stage. During this shut-down time, the superintendent concluded that it was best for himself and for the project that he resign rather than try to leave the old ways behind to work in the new system – a sad but probably necessary conclusion.

But his resignation left the project in the hands of the remaining foreman. He proved to be the most problematic of all, but for reasons not directly related to safety. Fortunately, a structural engineer and a contractor, who had first come to the project as volunteers with their church group, came to the organization with an offer to take on the supervision. There were three more difficult months of personnel problems with the remaining foreman. He proved unwilling and unable to work with the new team, berated volunteers, harassed prospective owner families, accused city inspectors of collusion, and eventually wound up badgering city hall and picketing the site even while on paid leave. But that is another story. The two new men saw the project to completion and have continued on with subsequent Habitat projects.

It was actually at that stage when the scenario of orientation and training described above really began to become effectively implemented. Indeed, the two new supervisors have wonderful people skills and seem to have what it takes to work with the extraordinary professionals and community-minded citizens who take time from their busy lives in Silicon Valley to volunteer. These volunteers know that their work in turn helps some of their less fortunate neighbors achieve the dream of home ownership in a high-priced market where this dream would otherwise be nearly impossible for low-income and very-low-income workers and their families. Six thousand people leaving their neat middle-class suburbs to cross over the Bayshore Freeway to volunteer their time in a town where many caution their children not to go has helped build broader community understanding, particularly regarding the impact of high housing costs on poor people. The project, like similar efforts around the world, built much more than new housing.

In fairness, by the standards of an R&S or DPR, Peninsula Habitat's organization still has a ways to go to improve its safety program. The condominium project was not accident free. One volunteer, without having been checked out, tried ripping a board on a table saw. It jammed (ironically, in the safety guard), kicked back into her chin, knocked her over and sent her to the clinic for a couple of stitches. Another volunteer, one of our loyal regulars, was walking along the top of a 6-inch curb instead of down on the driveway, tripped, fell, and broke her arm. While not a construction accident as such, it happened on site. The other significant accident was a volunteer who got out on the second-floor joists ahead of the floor decking, slipped, partially fell through, and incurred a doctor visit to treat the resulting bruises and abrasions. Other than these injuries, lesser cuts and slivers were treated with on-site first aid equipment. No doubt at R&S and DPR, any of the three significant injuries mentioned here would have received management attention and probably would have cost the site crew their monthly safety commendation, so again there is little cause for celebration yet. But at the end of the job there was some sense of relief if not satisfaction in seeing 6000 amateur volunteers through a challenging project without anything worse having happened.

6 CONCLUSION

As this is written, Peninsula Habitat has three single-family homes under construction and is about to start a 10-unit town house project. The author just recently returned from a six-month sabbatical at the University of Hawaii and was happy to find that the supervisors had maintained the same safety standards rather than slip back to old ways just because they were back to single-story construction for a few months. These were probably the only fully scaffolded one-story frame homes in the area, but that is the type of measure they take to keep lots of good people safe and healthy – and indeed to live up to the law as it is written. In some respects, this amateur volunteer outfit is ahead of much of the residential building industry.

But all this was not achieved easily and some of the lessons learned extend to the building industry in general. People problems were foremost, especially with the initial supervisors. Without the full understanding and cooperation of those in charge, even the best-designed safety program is of little value. Nevertheless, the project confirmed that best practices from contractors with top safety records also apply well in volunteer construction. With continuing efforts to improve, it should be feasible even for volunteer-based construction organizations to aspire to and achieve the lofty goals set and the commendable results obtained by the leading firms in the commercial and industrial building sectors.

7 REFERENCES

Barrie, D. S. & Paulson, B. C. 1992. *Professional Construction Management*, 3rd Ed. Chapter 17: Safety and Health in Construction. New York: McGraw-Hill Book Co.

CalOSHA October 1 1995. *State of California Construction Safety Orders.* Anaheim, California: BNi Building News.

Gibson, Lydia 1998. How Safe Are Your Houses? *Professional Builder*. (63) 3: 71-74.

Levitt, R. E. & Samelson, N. M. 1993. *Construction Safety Management*. New York: John Wiley & Sons.

Office of the Federal Register July 1, 1995. *Code of Federal Regulations*, Part 1926 (Occupational Safety and Health Standards for Construction). Washington, D.C.: National Archives and Records Administration.

Professional Builder 1998. The Giant 400. (63) 6: 128.

Rudolph & Sletten, Inc. 1995. Code of Safe Practices. Foster City, CA.

PENINSULA HABITAT FOR HUMANITY
-- GLORIA WAY PROJECT

1010 Doyle Way, Suite #6
Menlo Park, CA 94025

Phone: (650) 324-2266
Fax: (650) 324-4375
E-mail: Penhabitat@igc.org

CODE OF SAFE PRACTICES

The following are our site safety rules. Before you start to work, we will request you read or attend a briefing on these safety rules and abide by them and any additional safety requirements of this project. They are intended for your personal safety and for the safety of those around you.

I. IN GENERAL

1. All persons shall follow these site safety rules and take the time necessary to do their work in a safe manner. Failure to cooperate and comply with these rules could result in your being requested to leave the site.
2. All persons immediately shall fix safety hazards within their authority or notify their team leader, Habitat supervisor, or Job Safety Coordinator of the hazard. While the hazard exists, please warn those nearby.
3. All personnel will attend safety meetings when offered, where instructions will be given for the prevention of injuries. Your feedback on safety issues is welcomed.
4. Report all injuries or suspected injuries promptly to you team leader, Habitat supervisor, or Job Safety Coordinator, for arrangements to be made for medical treatment or first aid if necessary. Even splinters can cause dangerous infections.

II. PERSONAL PROTECTION

1. Wear clothing appropriate for construction work:
   a. Do not wear loose-fitting clothing. Shirt (minimum "tank top" or as specified)
   b. Wear a shirt (minimum "tank top" or as specified). Long pants are recommended.
   c. Work or hiking boots are recommended; steel-toed boots are best. No sandals please.
2. Hard hats must be worn at all times, unless otherwise directed.
3. Eye protection (i.e., safety glasses) must be worn when exposed to flying/falling particles, splashing chemicals, and harmful light rays. This includes use while operating power tools, hammering, chiseling, painting, and similar tasks.
4. Additional personal protective equipment such as face, ear, hand, knee and foot protection; dust masks and respirators; and safety belts are to be provided as necessary for your safety. When you need this equipment, ask for it from your team leader, Habitat supervisor, or Job Safety Coordinator.
5. Do not wear musical headphones. They can impair warnings and other communications.

III. COMMON SENSE

1. No alcohol or drugs are permitted on site; do not work under their influence.
2. No one shall work while his/her ability or alertness is so impaired by fatigue, illness, medication, or any other reason that might cause him/her or others to be injured.
3. Horseplay, scuffling, and other similar disruptive acts often lead to injury and are prohibited. Personnel engaged in such acts will be required to leave the site.
4. No task requires running. Walk, do not run.
5. Be aware of work going on around you. Stay clear of suspended loads and traffic areas when possible. Give others room to work and be aware when power tools are in operation.
6. Make sure you have a clear area behind you before swinging sledge hammers or other tools or materials.
7. Know and respect your personal limitations. Don't lift anything or perform any task with which you are uncomfortable. There will be plenty of other tasks available for you to do.
8. If there is other equipment operating on the job site, take personal responsibility to stay clear. Always assume that the operator does not see you.

---

* This document was adapted from a policy used by Rudolph & Sletten, Inc., Foster City, CA

9. Read and follow the health and safety precautions on the containers of chemical products, paint, etc. before you use them. For additional product safety information, please ask the site supervisor.

IV. POWER TOOLS

1. NEVER underestimate the potential for harm resulting from misuse of a power tool. Injuries are swift, unexpected and often severe. Be deliberate and focused in handling power tools, and STOP before you take a shortcut.
2. You must be pre-qualified, or "checked out," by the appropriate personnel before operating any power tool. Wear appropriate eye, ear and hand protection when using power tools.
3. Make sure that all guards and other protective devices are in their place and properly adjusted on all tools and equipment before using. If the guard is missing, the equipment is defective and must not be used.
4. Make sure your power cords are in good condition (no exposed inner wires or spliced cords), that plugs have a ground pin when so designed (3 prongs), and that the prongs are not loose.

V. WORKING AT HEIGHTS

1. Always check a ladder's condition before you use it. Take damaged ladders out of service, tag them, and report them to your team leader, Habitat supervisor, or Job Safety Coordinator.
2. Support all ladders on a stable surface. Straight ladders require a 3-foot overlap at landings and a tie-off. Inclination should be about four-to-one (vertical-to-horizontal).
3. Arrange your work so that you are able to use both hands and face the ladder when climbing or descending; don't rush it. Only one person should be on a ladder at a time.
4. Don't work off the cap or the back of a step ladder. Move the ladder rather than over-reaching for the last bit of work.
5. Before working on a scaffold, check that it has good footing, proper bracing, and full planking, properly overlapped; a railing is required at heights of 7-1/2 feet above the surface below. Report damaged equipment.
6. When railings are not practical, personnel working at unprotected heights above 7-1/2 feet must wear a safety harness with the lanyard tied to a substantial anchorage.
7. Avoid "jump down" shortcuts; use stairs, ladders, ramps, and walkways.
8. Never work in an elevated position above vertical rebar, stakes, etc. unless these protruding objects are properly covered. When working over 6 feet in elevation above these hazards, the use of railings or a safety harness is required.
9. Never take down guardrails or uncover holes without the authorization of the site supervisor.
10. Do not throw materials, debris, or other objects from a building until proper precautions (a spotter, barricades, etc.) are taken to protect others from the falling objects.

VI. CLEAN JOBS ARE SAFE JOBS

1. Do not block stairs or aisles with scaffolds or debris any longer than is necessary.
2. Do not leave loose materials on stairs or in aisles, which create a slipping or tripping hazard. Clean-up oil, grease, or other material spills immediately.
3. Bend over or remove protruding screws or nails in your used materials, crates, etc.; stack materials in an orderly manner and place debris in debris containers or piles.
4. In general, clean up after yourself and put tools and materials back where they belong when finished with them.

Implementation of Safety and Health on Construction Sites, Singh, Hinze & Coble (eds)
© 1999 Balkema, Rotterdam, ISBN 90 5809 036 1

Problems of implementing safety programmes on construction sites and some possible solutions (Hong Kong experience)

S.M.Ahmed, S.L.Tang & T.K.Poon
Civil and Structural Engineering Department, the Hong Kong Polytechnic University, Kowloon, People's Republic of China

ABSTRACT: In Hong Kong, there are many construction sites where contractors are implementing their safety programmes inadequately. There are problems/difficulties causing such inefficiency in implementation. 28 building and civil engineering contractors (mostly large/ international contractors) in Hong Kong participated in a survey in which the contractors' opinions in implementing site safety programmes were solicited. Firstly, the problems/difficulties encountered in their site safety schemes were identified. Secondly, opinions on possible methods for improving the present unsatisfactory situation were obtained. Recommendations are then made based on the results of the survey. Possible further investigation works are also suggested.

1 INTRODUCTION

Hong Kong is a place with world famous infrastructures and construction projects. However, it is notorious for its construction safety records - nearly 300 accidents per 1000 workers per year in the recent few years. Such a poor accident rate is certainly unacceptable. Past studies have focused on safety education or training and the perception of construction professionals on site safety (Kwok et al. 1996). This work revealed that construction professionals in Hong Kong did not pay sufficient attention to safety education/training. Another study (Wong et al. 1996) also revealed that safety education and training is a very important factor for improving the safety records of the Hong Kong construction industry. Different categories of personnel such as project managers, site engineers, safety managers/officers, design engineers and so on must work together closely to ensure effective implementation of construction safety. A similar conclusion was obtained in an investigation (Hinze and Wiegand 1992) that design firms should address construction workers' safety in their design and that co-operation between safety officers (site staff) and designers (office staff) is essential in construction safety.

Some other studies focused on safety management system and safety investment. A research work (Lai et al. 1996) suggested that safety management should be proactive rather than reactive; management should assess the adequacy of its safety management effort through safety audits before works commence for predicting its safety performance based on the audits. As to safety investment, a study (Tang et al. 1997) revealed that the minimum safety investment in a building project in Hong Kong is 0.8% of the contract sum. Another study (Everett and Frank Jr. 1996) recommended that some kind of financial incentives should be provided to contractors for improving construction safety performance.

All the above studies concentrated on one or more special aspects of construction safety such as education or investment. There has been little investigation on a comprehensive or macro basis, that is, the reasons for poor site safety in Hong Kong and the possible solutions to improve the situation. This paper attempts to address these points.

2 OBJECTIVE AND METHODOLOGY

The objective of this work is to find out the problems/difficulties which cause inefficiency in the implementation of construction safety and to recommend some possible solutions for improving the current unsatisfactory situation.

28 building and civil engineering contractors in Hong Kong participated in a survey in which the contractors' opinions in implementing site safety programmes were solicited. The survey was done by asking each contractor to complete a structured questionnaire and then afterwards by interviewing its safety manager if necessary (not all contractors were interviewed). In the analysis of the data obtained, firstly, the problems/difficulties encountered in implementing site safety schemes were identified. They were arranged in an order of decreasing importance. Secondly, opinions on possible methods for improvement were also obtained from the contractors and these methods were also prioritised. Recommendations would then be made based on the results of the analysis.

The sizes of the 28 contractors are shown in Table 1. The nature of work undertaken by these contractors are shown in Table 2. Note that the total count in Table 2 is larger than 28 because each firm may undertake more than one nature of work.

3 PROBLEMS/DIFFICULTIES IN IMPLEMENTING SITE SAFETY SCHEMES

Nine most important factors were identified which were considered to be adversely influencing the implementation of safety schemes on construction sites. Table 3 shows these factors with the number of counts of contractors' opinions.

Table 1. Sizes of the contractors who participated in the survey.

Number of workers employed	Number of contractors
10 to 49	3
50 to 199	4
200 to 999	12
1000 or above	9
	$\Sigma = 28$

Table 2. Nature of work undertaken by the contractors.

Nature of work	Number of contractors
Construction of buildings	25
Construction of roads	8
Construction of transportation systems (e.g. bridge, tunnel, etc.)	7
Construction of utilities	9
Construction of slope and related work (e.g. retaining wall, etc.)	5
Others	5

Table 3. Factors adversely influencing site safety implementation.

Factors	Number of counts of contractors' opinion				
	Very important	Importan t	Fairly important	Not important	Negligibl e
Tight schedule of projects	10	14	1	3	0
Inefficient communication due to multi-level sub-contracting system	8	12	7	1	0
Transient nature of construction workers	4	9	7	3	5
Long hours of work for construction workers	0	7	14	7	0
Use of alcohol by workers	5	6	7	5	5
Inexperienced and/or unskilled workers	5	12	6	5	0
Limited budget on safety investment	8	9	10	0	1
Inability of safety officers to enforce safety regulations	8	13	3	1	3
Insufficient penalty for not implementing safety regulations	6	9	8	5	0

In order to prioritise the above factors so that their degree of importance can be known, a score is given to each factor. The score of a factor is obtained by summing up the multiples of 'opinion mark' and the number of contractors. The 'opinion mark' is defined as follows:

Opinion	Mark
Very important	25
Important	20
Fairly important	15
Not important	10
Negligible	5

For example, for the factor 'tight schedule of projects', there are 10, 14, 1, 3 and 0 contractors who think that the factor is very important, important, fairly important, not important and negligible respectively. The score will therefore be 575 and is calculated as follows:

Score = $10 \times 25 + 14 \times 20 + 1 \times 15 + 3 \times 10 + 0 \times 5 = 575$

According to the scores, the factors are prioritised in an order of decreasing importance as shown in Table 4.

4 POSSIBLE SOLUTIONS

4.1 *Result of questionnaire analysis*

Five possible solutions were identified. They were also prioritised using the same method as discussed in section 3. Table 5 shows the priority and the score of each possible solution.

Table 4. Problems/difficulties of implementing site safety schemes arranged in order of decreasing importance.

Priority	Problems/difficulties	Score
1	Tight schedule of projects	575
2	Inefficient communication due to multi-level subcontracting system	555
3	Limited budget on safety investment	535
4	Inability of safety officers to enforce safety regulations	530
5	Inexperienced and/or unskilled workers	505
6	Insufficient penalty for not implementing safety regulations	500
7	Transient nature of construction workers	440
8	Use of alcohol by workers	425
9	Long hours of works for construction workers	420

Table 5. Possible solutions arranged in order of decreasing importance.

Priority	Possible solutions	Score
1	Compulsory (much stricter enforcement) for contractors to promote safety	530
2	Increased penalty to contractors who do not strictly enforce safety regulations	505
3	Bonus given by promoters (clients) to contractors who make efforts to obey safety regulations for a considerable period of time	500
4	Legal restriction on the number of levels of sub-contracting	440
5	Legal restriction on the number of hours of work for construction workers	430

4.2 *Comments from contractors during interviews*

Besides those as discussed in 4.1, there are other possible solutions suggested by contractors during interviews. They are summarised below.

a) Safety should become a billed item in a contract for ensuring safety investment.
b) Regular audits are needed to review contractors' safety management systems.
c) Site management should prosecute individuals who break safety regulations.
d) The promoter (client) of a project should have the authority to stop work when working conditions become unsafe (such as bad weather).
e) More skillful and experienced workers who know the importance of site safety are needed. Hence, training on the part of workers is essential.
f) Safety culture should be instilled in workers so that they take up the responsibility of protecting themselves.

5 CONCLUSIONS, DISCUSSIONS AND SUGESTIONS FOR FURTHER RESEARCH

In this work, the main problems/difficulties that adversely affect the implementation of construction site safety in Hong Kong have been studied. The most serious problems are the tight construction schedule and the inefficient communication due to multi-level subcontracting system, followed by limited budget on safety investment and inability of safety officers to enforce safety regulations. To reduce the effects caused by the problems, it is recommended that the minimum contract periods for different categories of construction works should be studied so as to avoid site accidents caused by tight schedule of work. For better communication, different channels should be set up and maintained, for instance, compulsory periodical meetings should be held and attended by the main contractor and subcontractors. Also, the percentage of subcontract work should be limited to a certain figure in a contract.

This figure should be found out and it is another interesting piece of further investigation work. As to the limited budget problem, it is recommended that a system be set up to control the money spent on safety so as to avoid contractors to reduce their budget on safety investment. Recently, the Hong Kong Government has proposed a scheme named "Pay For Safety Scheme (PFSS)" (Hong Kong Government 1996) and is now having a trial implementation on it. For the problem of inability of safety officers, stricter requirement of their academic qualification and experience should be needed. For example, all safety officers should be degree holders (or equivalent qualifications) with suitable experience. Their status and authority on site should be upgraded.

Some possible solutions have also been studied. The most effective solutions are mandatory safety promotion of contractors and increased penalty to those who do not strictly enforce safety regulations, followed by giving bonus to those who make efforts to implement the regulations. It is recommended that contractors should provide more safety promotion activities so as to instil a safety culture on the part of the workers. The government may establish some guidelines for contractors to organise such activities. As to increasing penalty and giving bonus, the authors believe that they are effective and feasible. However, before their implementation, further studies are required to determine how and how much the bonus are to be given and how much the penalty is to be increased so that optimum strategies can be established.

REFERENCES

Everett, John G. and Peter B. Frank Jr. 1996. Costs of accidents and injuries to the construction industry. *Journal of Construction Engineering and Management.* 122(2):158-164.

Hinze, Jimmie and Francis Wiegand 1992. Role of designers in construction workers safety. *Journal of Construction Engineering and Management.* 118(4):677-684.

Hong Kong Government 1996. Works Branch technical circular number 4/96: Pay for safety scheme (PFSS). *Works Branch, Hong Kong Government Secretariat, March 1996.*

Kwok, W.K., S.L. Tang and S.W. Poon 1996. Perception of construction professionals on construction safety and imminent need in health and safety training in Hong Kong. *Proceedings of the CITA International Conference on Construction Training, Hong Kong, December 1996:* 248-256.

Lai, K.C., S.L. Tang and S.W. Poon 1996. Management effort and safety performance of construction projects. *Asia Engineer, Journal of the Hong Kong Institution of Engineers, January 1996:* 26-28.

Tang, S.L., H.K. Lee and K. Wong 1997. Safety cost optimisation of building projects in Hong Kong. *Construction Management and Economics.* 15(2):177-186.

Wong, Francis, S.L. Tang and Samuel Lip 1996. Safety education for construction students - Hong Kong experience. *CIB Beijing International Conference on Modernisation and Education, Beijing, China, October 1996.*

Implementation of Safety and Health on Construction Sites, Singh, Hinze & Coble (eds)
© 1999 Balkema, Rotterdam, ISBN 90 5809 036 1

How to manage safety in the construction sites

F.Giofrè & C.Piendibene
University of Rome 'La Sapienza', School of Architecture, Italy

ABSTRACT: In Italy the receipt of some community directives has moved again the attention of the experts of the sector to the problem of safety in the building yards. The building sector is, between the productive compartments, the one with greater risk of accidents. The engineers and the architects are called today to a new professionalism to be able to manage safety: "the design safety manager". The simple conception and design of structure is just a piece of the puzzle and every money spent at the top of the design process contribute to reduce the total cost of construction in economic and human terms. To address these ones is necessary a new approach processing from risk analysis, risk management to risk reduction in the design phase. This paper stresses the necessity to standardise the safety instruments in planning and managing design and construction phases.

1 INTRODUCTION

The construction sector has influenced the economic evolution of Italy strongly playing a key role in the different historical phases.

It is possible to individualise at least three evolutionary periods - denominated building cycles - with precise characteristics:

- 1949/1964: it is the period of the maximum expansion of the construction activity that follows post war condition and the huge migratory flows from the south toward the north of the country;
- 1964/1974: it is the period in which the crisis of the sector starts, due to the generalised increase of the prices, to high level of worker's conflict that has its climax in 1973 with the oil crisis that has been reflected on the building sector with the collapse of the market of constructions in 1974;
- 1975/1980: it is the period in which the end of the crisis is accompanied by a decrease of the construction activity, particularly referred to new building, while the market of recovery is consolidated and it acquires, up to our days, a more and more increasing weight; also, the phenomenon of the illegal house building is developed;
- ''90'ies, starting from 1992 the building sector knows a strong recessive course and the diminution of the constructions sector activity assumes diversified courses according to the productive compartment, in fact the activity of renovation has maintained a continuous positive course even in years of strong crisis.

In this context, a particular role assumes the typology of the building contractor - characterised by a lower number of employees - and the professionalism of the employees of the sector.

Table1. Indexes of frequency of cases of accident in the constructions sector and industry / tertiary - INAIL dates

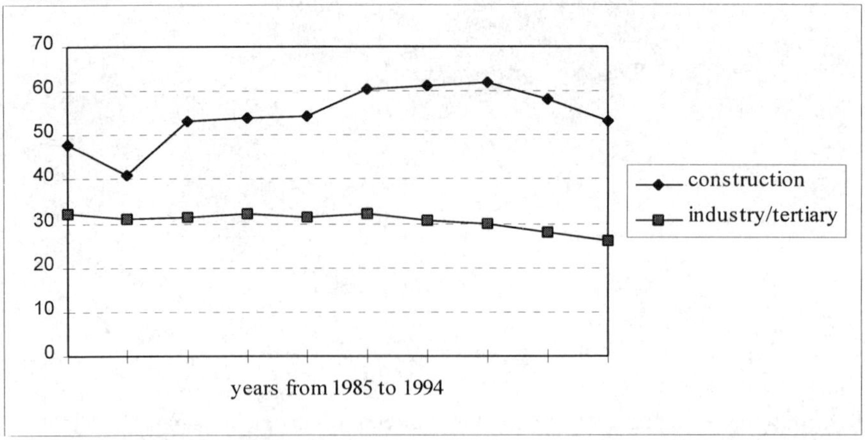

Along with advanced engineering and architectural technologies still today cohabit techniques of antiquated workmanship, levels of insufficient professional qualification, at times, non-existent, unfit or precarious environmental conditions.

Needed to be stressed in fact, that the building sector has always lived with a certain hostility the processes of renewal and modernisation and, with regards to level of tutelage of the workers, it reamins at the edge of legality producing in the same workers a chronic uncertainty on safety in the workplace.

INAIL (National Institute for the Insurance and the Accidents on the Work) furnishes data on the accidents in all the productive compartments.

Readings table 1, in which are shown the indexes of frequency of accident - expressed as the relationship between the number of the accidents and working hours - in three great productive compartments, it is underlined how these are double for the sector of constructions in comparison to those of the industry / tertiary sector.

Yet, as far as the normative is concerned, Italy has shown big attention to the problems of safety at work, from the penal code (1930), to the civil code (1942), to the article 42 of the Constitution to the DPRs n. 547 of 1955 and n. 303 of 1956, to the statute of the workers, to the health care reform (1978).

In the 70'ies and 80'ies the culture of prevention has become fundamental element of the industrial relationships enhancing the role of the person on the workplace. During this period there was a decrease of 50% of accidents at work. There has then been a phase of reflux and, today, the picture is not substantially changed, despite the receipt of the community directives with the legislative decree n. 626 of 1994 and n. 494 of 1996.

The necessity of a different approach to the problem has also been underlined by some definite fiscal initiatives from INAIL: a 10% reduction of the reward expenses for the enterprises that adhere to the territorial joint committees and a further discount of 5% after the check from the controlling organs, of the expenses really sustained for the application of the normative devices on safety.

2 TOOLS OF SAFETY

The legislative decree 494/96 individualises the minimum safety and health prescriptions to realise in the temporary and moving yards. The norm has substantially modified the relationship between the two principal actors of the building process, the employer and the contractor, individualising in the first precise penal and civil responsibility for the safety of the workers. To the goals of a correct management of safety both at the phase of realisation and of maintenance of the building manufactured, it is obligatory to introduce the plan of safety and co-ordination for yards with more than 100 men / day, the general plan of safety for yards with more than 30.000 men / day and however in both of the cases, tools ready to be applied contextual to the executive planning of the work.

Such documents must be compiled from the co-ordinator for safety in the planning phase and managed by the co-ordinator of safety in the realisation phase, he has the power to have them applied and respected. These two new professional figures are named and entrusted from the employer contextual to the charge of the executive planning of the work.

With the coming into force of the legislative decree 494/96 the attention of the safety problem has moved from the phase of realisation to that of planning. In fact from a search developed by our department it has emerged a surprising information, 60% of the accidents are due to causes originating before the opening of the yard, and particularly 35% are caused by errors of planning and inadequate technological choices, 28% from a missing planning of the work, 37% from lack of information and from the missed respect of rules and norms.

3 GUIDELINES FOR THE EDITING OF THE SAFETY PLANS AND CO-ORDINATION

Considered the Italian reality where the sector of the recovery hires the biggest number of the enterprises in interventions of modest dimensions, both in terms of duration and employment of manpower, the safety plan and co-ordination is the most used tool. Insofar in total absence of specific norms, since the law individualises only the general contents, it is useful to trace an operational guideline for the editing.

The choice of furnishing a guideline and not a facsimile, is due to the consideration that every project introduce specific characteristics in relationship to the manifold aspects that characterise it (geographical, temporal, technological, etc.), such to make difficult to standardise a plan that fits to every situation.

The plan must represent a support that guarantees safety for all the categories of workers involved. Therefore the plan can not be a generic document of cataloguing the safety norms but a whole structured information reported to the specific work of which are to be analysed in detail the processes of construction and execution, the formalities of job in relationship to the hygiene and the safety of the workers employed in the yard, the necessary equipment, identifying the predictable risks for every working phase. The plan must point out the anticipated times for the execution of every phase in consideration of the fact that the temporal succession and the simultaneity of different workmanship involves the increase of the risks in the yard. The plan must guarantee the safety of third meaning both persons and things in the neighbouring zones to the yard.

The safety and co-ordination plan is articulated in two parts: the technical relationship and the operational prescriptions - correlated to the complexity of the work to realise and to the possible critical phases of the productive process -.

The technical relationship of the plan is composed by the followings sections:

- register, personal data: it contains the general data relative to the subjects involved in the contract and it allows the immediate individualisation of all the subjects involved in the contract;
- preliminary of the project and of the site: it contains the description of the jobs to be realised, with particular reference to the technological and structural choices, to the used materials and to whatever else can favour a greater comprehension of the problems relevant to the execution of the work; the description of the site before the intervention, with particular reference to the viability, the presence of technological infrastructures, the geologic area, possible areas with specific risk;

- organisation of the yard, contains the layout of the yard or the project of the organisation of the yard;
- planning of the times of workmanship, consists in a diagram that puts in relationship time and workmanship (cronogramma), through which it will be possible to outline contemporaneousness between the workmanship. Such information will be needed to define the prescriptions of the operational cards;
- evaluation of the costs, of which doesn't exist up to date an univocal criterion for the evaluation.

The operational prescriptions is obtained from the general evaluation of the risks that is obtained from the considerations done in the preliminary investigation.

The general evaluation of the risks is a relationship that defines the risks drawn from the execution of the work, in relationship to the adopted technologies and the components that characterise the environment (site, timing conditions, etc.). It also furnishes, the elements for the analysis of the specific risks of every sub phase, described in the tables. For every single phase it is defined, the formalities of execution of the sub phases (description of the jobs), the number of the employees to the workmanship, the used machinery, the analysis of the specific risks, the measures of protection (individual and collective) to adopt and finally the possible interconnections with other workmanship (Table 2).

The number of the cards is varying, since it depends on the division of the workmanship in phases and sub phases, defined in the timetable planning. More phases will be individualised, greater it will be the number of the cards.

Besides the operational cards, integral part of the plan, are the machinery cards where it is pointed out the use of the machinery (Table 3). These are elaborated from the co-ordinator in phase of execution, previous delivery from the enterprise that assigns the contract. Before this the indications on the machinery to adopt will be necessarily generic not being able to know the machinery of the firm that will be been awarded the contract.

Table 2. Example of an operational schedule, to be compiled by the co-ordinator of safety in phase of planning and to be adjourned by the co-ordinator of safety in phase of realisation.

Phaseof work:______________________________				total employed number---------------	schedul number----------------
Description of the jobs	employees number	used machinery and equipment	risk's analysis	safety measures and formality of execution	interconnections with other workmanship
________	________	________	________	________	________
________	________	________	________	________	________

Table 3. Example of machinery schedule (dispositions of use) - Such schedule will be inserted from the co-ordinator in phase of execution previous delivery from the enterprise.

Undertaken performer____________________ Yard______________________________ street__________ zip code_______ city______	
type of machinery____	
Measures of prevention and instructions for the employees	
measures before the use	______________________________
measures during the use	______________________________
measures after the use	______________________________

risks drawn from the analysis of the dangers and the dangerous situations	devices of individual protection

The systematic collection of all the cards, can constitute according to an operational chronological order, a diary of the management of safety in the yard, that will furnish a complete documentation of easy consultation, both in the planning of fulfilment, and in the hypothesis of the necessity of controls afterwards.

The management on the field of the safety plan and co-ordination represents the proper operational phase. Such assignment is up to the co-ordinator in phase of execution whose duties, for the management of the safety plan and co-ordination, in synthesis are: to assure the application of the dispositions, to adjust the plan to the possible intervened changes, to organise the co-operation and the co-ordination, to propose the suspension of the jobs or the firing of the contractor in case of serious non-observances to the norm of plan, to suspend the single workmanship in case of serious danger (Table 4 and Table 5).

Table 4. Schedule for verification of the safety plan and co-ordination to be compiled by the co-ordinator of safety in phase of realisation.

Date of survey	macrofase	phase of workmanship	firm and autonomous workers	found anomaly	dispositions	provisions
________	________	________	________	________	________	________

Table 5. Updating schedule of the safety plan and co-ordination to be compiled by the co-ordinator of safety in phase of realisation.

Macrofase	phase of workmanship	variation	cause	cronogramma of reference	reference schedule number	number new schedule
________	________	________	________	________	________	________

From the analysis done until here of the varied aspects and tools of the safety plan and co-ordination, it is clear that the new approach to the subject of safety introduced from the DLgs 494/96 gives responsibility to all the subjects involved in the realisation of the work and that the necessary competence to the figures preceded to its elaboration and management, can not derive only from a general professional preparation, but from a specific formation and from experience on the field.

REFERENCES

Legislative decrees n° 626/94 and 242/96. *Safety in every working sites.*
Legislative decree n° 493/94. *Sign's security.*
Legislative decree n° 494/96. *Safety on construction sites.*
Legislative decree n° 459/96. *Device's security.*

Implementation of Safety and Health on Construction Sites, Singh, Hinze & Coble (eds)
© 1999 Balkema, Rotterdam, ISBN 90 5809 036 1

Safety management on construction sites in Quebec

J. Brun, M. Arial & C. Loiselle
Université Laval, Québec City, Qué., Canada

ABSTRACT : This presentation examines the results of a study of the implementation of preventive measures in the construction industry. The study began with the observation that there exist numerous preventive measures (technical, organizational or legislative) that are not easily applied on construction sites. The factors studied to understand this situation include the perception of workers and managers, the type of work involved, the characteristics of the contractors, the contractual management of the sites, government inspectors, etc.

The study analysed various construction sites and evaluated whether or not the implementation of preventive measures is conditioned by the relevance or efficacy of the measure itself and/or by factors associated with the workers, companies, job, work-site management or external regulatory agencies. Using an exploratory multi-site study, we attempt to understand, through observations and comparisons of work situations as well as group discussions of these situations, how and why these preventive measures are or are not implemented on various construction sites. We then present a portrait of the conceptions and representations that the different groups involved have of the application of various preventive measures and their relative effectiveness, and likewise discuss the conditions and determinants of this application.

Our results are helping us to draw a portrait of the conditions that positively or negatively influence the application of these measures. They will also help the construction sector better understand the characteristics of the different prevention strategies and the mechanisms that trigger their application.

1. INTRODUCTION

Employment injuries represent a considerable problem in the construction industries of most industrialized countries. In Québec, in 1996, this sector accounted for more than 23% of employment fatalities. Likewise, in 1996, this activity sector gave rise to 16.3% of total payments ($1,179,000,000) of the Commission de la santé et de la sécurité (CSST, Québec's Work Health and Safety Board, 1997). There are a considerable number of preventive measures, such as rules, standards and technical devices, that are intended to lessen this phenomenon and that govern safety on construction sites. However, despite relatively considerable knowledge of construction-related risks and despite the existence of preventive measures to control and limit these risks, we are obliged to admit that these measures are not always applied on the work sites. A good example of this is the fact that between 1995 and 1997, more than 21,000 notices of infractions of the *Construction Work Safety Code* were registered. It would, therefore, seem logical that identifying risks and producing other preventive measures is no longer sufficient; it has now become essential that we understand the conditions and factors that determine whether or not these measures are applied on construction sites. This presentation thus examines the preliminary results of an on-going study of the implementation of preventive measures in Québec's construction industry.

2. THEORETICAL FRAMEWORK

Certain researchers have noted the mixed success obtained in attempts to decrease the number and severity of employment injuries that occur on construction sites (Trinquet, 1996; Hale and Swuste, 1993). In a summary text focussing on the different models and approaches used to examine the phenomenon of employment injuries, Laflamme (1988) noted the many factors that can be involved. For Leather (1987), the factors likely to influence the application of preventive measures are ultimately interdependent. Our study proposes an amalgamation based on a continuum that includes the different dimensions found between a general and local context of preventive-measure application.

DIAGRAM A : FRAME OF ANALYSIS FOR THE IMPLEMENTATION OF PREVENTIVE MEASURES

GENERAL ⟷ **LOCAL**

General Context	Business Context	Work Site	Collective Actors	Individuals
Legislative Framework – laws and regulations Socio-economic context Industrial organisation Technologies	Market Work relations (inter-union, agreements, negotiations, etc) Clients Project owner Architects Civil engineers Tendering companies Project manager General contractor Sub-contractor Contractual system Specific regulations of OHS	Type of work site Physical environment Climatic conditions Work organisation (division of work, supervision, co-activity, etc) Organisation of OHS Suppliers Equipment Temporality Specific risks	Representativeness Involvement in OHS Authority Trade culture	Personal characteristics Qualifications "Current" state Specific working habits Work activity

Required preventive measures ⟷ **Applied preventive measures**

Occupational Health and Safety Act Regulations in effect Work codes	Sector-related work agreements Contractual contents	Specific required measures (eg boats, safety nets) Site-specific system of rules	Social standards General protective equipment	Individual protective equipment « Savoir-Faire de Prudence »

In diagram A, each column and its contextual elements corresponds to a a column below providing examples of preventive measures. Thus, in the general-context column, the Occupational Health and Safety Act constitutes a preventive measure as do the safety regulations that follow from it. In the business-context column, business agreements between contractors may also contain provisions concerning prevention. Each construction site also has its specific risks which give rise to specific preventive measures. For example, repair work on bridges requires that rescue boats be present in the case of a fall. Certain preventive measures also apply to work on high buildings or next to electricity-transmission lines. Several preventive measures also exist for workers. Some of these directly concern individuals. Thus, collective actors such as unions can play a deciding role in the application of preventive measures depending on their representativeness, their authority, their level of commitment to the subject and their promotion of trade cultures. However, when all is said and done, it is up to each worker to ensure his own safety and that of his colleagues. When studying the application of preventive measures, it is also important to consider the personal characteristics and qualifications of the workers, their state at any given moment in time and their prevention abilities.

3. METHODOLOGY

The gathering of data needed to identify and understand the different factors that favour or obstruct the application of preventive measures, as well as the complexity of the relationships between these different factors, has been conducted through continual on-site observations. So as to give more substance to the observed aspects, the research included interviews with groups of workers after work hours. Certain documents, such as the minutes of meetings and excerpts from contracts, were also collected. These observations have made it possible to document work situations in which preventive measures are or are not applied. Observations were made in a total of 16 work sites (commercial, industrial, residential and institutional sectors) situated all over the province of Québec. These observations were conducted by at least two researchers for two consecutive weeks per site.

4. QUÉBEC CONTEXT

In Québec construction sites, the intent of the Health and Safety Act is to make employers and workers responsible for the elimination and control of risks. To this end, the Act counts primarily on the obligation of employers to develop a prevention programme in collaboration with contractors who will be working on the sites. The Act also requires that a health and safety officer be hired when the cost of work on the site exceeds $8 million or when at least 150 workers will be working there simultaneously. The Act also foresees the possibility of workers nominating a representative who can lead or participate in prevention-related activities, such as work-site inspection and health and safety committees. It is worth noting, however, that these provisions have not yet come into effect. Certain legislative provisions which are directly related to the concrete execution of work on construction sites are also foreseen in the safety code for construction work.

5. PRELIMINARY RESULTS

The application of preventive measures is influenced by numerous factors. This section presents some of these factors without claiming to provide an exhaustive list. As these preliminary results are drawn from a study that is still being conducted, we will only touch upon a few of the themes that are emerging from our analyses. This presentation thus does not pretend to examine the overall complexity and subtleties of safety management on construction sites in Québec.

The observations we made on construction sites confirm that a large number of existing preventive measures are still not applied. Factors such as the multiplicity of actors, the sometimes deficient communication between these actors, the difficulty of anticipating certain work as well as its impact on security, the time required to make a situation safe, cooperation and interdependence among contractors, the sharing of responsibility for safety and the essentially reactive role to which health and safety officers are confined seem to be unavoidable determinants of the application of preventive measures on construction sites.

5.1 *Multiplicity of Stakeholders*

One of the characteristics of the construction sector in the province of Québec is the presence of numerous small enterprises (Gervais, 1998). These organizations subcontract part of the work to be accomplished on a work site, which presupposes the simultaneous presence of workers and managers from several companies. As the work progresses, subcontractors finish their part of the work while others begin theirs. In a such a context, the number of stakeholders increases whereas the length of time they spend on the site decreases. This situation represents a considerable challenge to the proper establishment of preventive measures in a business context, such as in the case of a health and safety committee which requires the prolonged participation of its actors. What is more, as these actors are only on the site for a short time, they do not always have enough time to familiarize themselves with the distinctive features of each site, which makes them still more vulnerable to certain dangers.

Another characteristic of the construction sector in the province of Québec is the presence of several trade-based unions. There is, therefore, no homogeneous union allegiance on a construction site, which thereby makes it more difficult to identify representatives for all of the workers. Safety requests by workers which require an action on the part of the site managers thus follow a path laid out by the business structure: the foreman tells the site superintendent or the health and safety officer during a site coordination or safety meeting or during less formal encounters. It is worth noting that the site managers and various subcontractors interact according to a client-supplier logic. Given such a context, workers would not seem to be inclined to make requests. Legal provisions that require the presence of a worker representative at safety meetings for every contractor having 10 or more workers on the site do not sufficiently ensure adequate representation of workers for safety-related matters. On the one hand, contractors who have a small work force on the site are neglected by such a provision. On the other hand, this rule is easily bent, since it is contractors who determine how many workers there are on a site and there is nothing that keeps contractors from hiring less workers on the day when the safety meeting is held. Furthermore, this legislative provision only concerns worker representation at the time of the safety meeting and does not guarantee a continuously open channel for workers to express safety concerns during the execution of a contract.

5.2 *Planning of Work and Anticipation of Potential Danger*

The execution of construction work often entails unforeseen events. Moreover, numerous deviations from the original operational plan occur during execution. The work rate is fast and changes are continually being made to the work environment. The coordination of work must quickly adapt to these unforeseen events, as must health and safety officers. Since work is carried out in a context that is perpetually being transformed, risks and safety issues are likewise transformed as the work progresses. The specific characteristics of each situation demand the flexible application of preventive measures and subsequent adjustments. Numerous contingencies are likely to arise which can greatly modify the configuration of a work site. For example, if an error occurs in the delivery of metal parts required for the assembling of a structure, resources must then be allocated differently. Structural-steel erectors will then be assigned to work on other tasks, creating initially unforeseen zones of co-activity and risks elsewhere on the site. These decisions are made as problems arise and require flexible planning and constant watchfulness on the part of the people in charge of safety. Such a viewpoint means that we must see work-site safety as an on-going process.

5.3 Interdependence Among Contractors

Contractors are sometimes obliged to work in cooperation with each other in order to efficiently execute work-site operations. Operational delays that are run up by one contractor can create delays for contractors carrying out other operations. This phenomenon of interdependence can also affect safety, as is the case with co-activity and the sharing of specific resources that are needed to implement preventive measures. A delay in the assembling of a large skylight, for instance, requires the prolonged maintenance of a safety perimeter on the ground, in places where work normally would have been conducted by other teams. An example of resource sharing would be the use of a steel cable by structural-steel erectors to ensure their safety during welding work on a roof when this cable is also needed by roofers for their work.

Certain contractors also have control of crucial resources on a work site. Access to this type of resource can sometimes be difficult for other contractors due to limited financial resources, a poor relationship with the holder of the resource and so on. This type of situation requires the implementation of alternatives that give rise to additional risks. For example, instead of using a crane to bring up a plumber's tool box, a contractor can ask his workers to raise it up with a hoist winch or even carry up the equipment by hand, thereby contributing to an increase in traffic, fatigue and the risk of dropping equipment.

5.4 Time Needed to Make a Situation Safe

There is always a certain amount of time needed between exposure to a dangerous situation and the moment when a preventive measure is implemented. The corrective work conducted to make a situation safe is also frequently subject to this same phenomenon. The installation of scaffolds or guard rails, the need to complete work on the roof of a building before steel cables can be installed to which workers can attach themselves or the installation itself of these safety devices provide a few examples.

This time period can represent a serious obstacle to the respect of preventive measures. For example, certain operations such as the placing of gypsum plaster board require workers to cover a large area in a short period of time. In such a situation, the setting up of certain preventive measures such as guard rails can require a considerable amount of time with respect to that required to place the plaster board itself.

5.5 Sharing of Responsibility for Safety

A principle that is generally accepted on a construction site is that everyone is responsible for their own safety. For example, structural-steel erectors are in charge of installing the steel safety cables to which they will attach themselves. When a risk is caused by or only concerns a particular group of workers, the responsibility for applying the preventive measures is normally easy to determine. There are, however, cases when preventive measures concern several groups of workers on a work site. This is the case of the layout of common spaces in which workers should be able to move around easily. For example, who is responsible for installing wood panels on the steps of stairways before the cement is poured? The observations we made during our study revealed that often, in such a situation, corrective measures that are desirable and necessary for all concerned are simply not carried out.

5.6 Roles, Constraints and Opportunities of the Health and Safety Officer

In the province of Québec, the logistical organization of a work site is the responsibility of the work superintendent. In turn, the superintendent delegates safety concerns to the health and safety officer. This type of division of work between these two actors tends to separate safety concerns from the operational planning of work. The health and safety officer thus finds himself trying to lessen the gap between the planning and the actual execution of work. The lack of participation of the health and safety officer in the work-planning meetings makes it more difficult to include safety-related concerns when relevant choices are being made, thereby excluding the possibility of eliminating risks before they arise. For example, the choice of scaffolds for the building of an elevator shaft must take into account the weight that these scaffolds must be able to bear but also has a direct influence on the safety of the workers who work on them.

6. CONCLUSION

The elements presented in this text are the results of the preliminary analyses carried out in our study. These analyses made it possible to document certain determinants of the application of preventive measures on a construction site such as the multiplicity of actors, difficulties concerning operational planning and the foreseeing of dangerous situations, interdependence among contractors, time needed to apply preventive measures and problems in sharing responsibilities for safety matters. These analyses likewise allowed us to determine the roles, constraints and opportunities of the health and safety officer. In addition to deepening our understanding of these themes, further pursuit of this study will allow us to document the other aspects mentioned in the analysis framework, thereby merging the general dimensions with the individual ones. Aspects related to the contractual context and to business conditions influencing the execution of operations on a construction site will thus be studied more thoroughly over the next few months. The originality of the elements presented here has made it possible to establish the relevance of studying construction site safety with an approach that takes into account both the specific characteristics of a given work site and the macroscopic characteristics of this activity sector. This approach should be the subject of further studies into the application of preventive measures.

REFERENCES

CSST 1996. *Annexe statistique au rapport annuel d'activité.* Québec : CSST.

Gervais, M., 1998. *La prévention primaire des maux de dos dans le secteur de la construction : une recension des écrits.* Montréal : IRSST.

Hale, A.R., & Swuste, P. 1993. Safety rules : procedural freedom or action constraint ?. *11th network Workshop, Bad Homburg, The Use Rules to achieve Safety,* Netherlands : Delft University of Technology.

Laflamme, L., 1988. *Modèles et méthodes d'analyse de l'accident du travail : de l'organisation du travail aux stratégies de prévention.* Longueil : SyGeSa ltée.

Leather, P.J., 1987. Safety and accidents in the construction industry : a work design perspective, *Work and Stress,* 1:167-174.

Trinquet, P., 1996. *Maîtriser les risques du travail : le cas du bâtiment et de l'industrie française,* Paris : PUF.

Implementation of Safety and Health on Construction Sites, Singh, Hinze & Coble (eds)
© 1999 Balkema, Rotterdam, ISBN 90 5809 036 1

The changing nature of confined spaces

J. Hinze
M.E.Rinker, Sr. School of Building Construction, University of Florida, Gainesville, Fla., USA
S. Baseden
Southwest Division Contracts Office, US Navy, Twentynine Palms, Calif., USA

ABSTRACT: Confined spaces have been the focus of much safety study in recent years. This is because many construction fatalities have resulted from accidents involving confined spaces. Even with the extent of study that has been conducted, there is often a misconception about what constitutes a confined space. The nature of a confined space may vary considerably, typically including trenches, silos, vaults, and other areas in which escape from a hazardous element is limited. To many individuals who would be asked to describe a confined space, the confined space might be envisioned as one in which the physical exit from an enclosed area is limited. Others would describe underground vaults or excavated areas in which toxic gases have accumulated. While such examples might be typical of many confined spaces, the definition is actually much broader. Examples will be provided of several non-typical "confined spaces". These will illustrate the extent of the range the definition must include. It will be shown that even a hallway with unlocked doors can become a confined space.

1 INTRODUCTION

Confined spaces are widely recognized as being the sources of construction worker fatalities. What is probably not as well known is the wide assortment of situations and conditions that are included in the definition of confined spaces. This lack of understanding is perhaps directly responsible for the continuing death toll associated with confined spaces. A full understanding is needed so potential victims begin to recognize, and subsequently avoid, hazardous confined spaces.

It must be recognized that the definition of a confined space is not universal. Some general descriptions can be used to describe many confined spaces. Three characteristics are generally associated with confined spaces. The first is that the confined space is large enough and so configured that an employee can bodily enter and perform assigned work in the space. The second characteristic is that the space has limited or restricted means for entry or exit. Lastly, the space is not designed for continuous employee occupancy. It should be realized that the definition of confined space must be broad. While the three descriptors may engulf many confined spaces in the work place, there are numerous exceptions, i.e., the above descriptors are not valid in all of the cases involving confined spaces.

The need for a broad definition of confined spaces can be clarified by examining a large number of accidents involving confined spaces. This will provide information on the variability that often exists in situations that fall into the category of confined space. This will exemplify the fact that there is no single set of circumstances that can be used to adequately define this common killer.

2 BACKGROUND

The Occupational Safety and Health Administration (OSHA) maintains a record of all jobsite accident investigations that are conducted by OSHA Compliance Officers. These investigation reports document, in general coded form, information about the accident victim (sex, age, craft, type injury), the employer, and the circumstances surrounding the accident (human factors, environmental factors, OSHA regulations violated). Note that while this information is valuable and useful, it does not provide detailed information that can be readily retrieved. For example, the human factor associated with an accident might be coded as "misjudgment" and the environmental factor might be "toxic environment". In this form, the data provide insufficient information by which to accurately depict the specific circumstances surrounding an accident. Even if it were somehow coded as a confined space incident, it would be difficult to ascertain how the accident took place.

The details of accident occurrence are not currently available through the coded information provided in OSHA accident investigation reports. This information is captured in the accident investigations, but it is not recorded in coded form. This information is currently available only in the abstracts that accompany the accident reports. The abstracts typically consist of a single paragraph in which the Compliance Officer describes, in narrative form, the events that presumably led to or preceded the occurrence of the accident. The abstracts are drafted through input from witnesses, from deductions drawn from the existing physical conditions, and, when available, from comments made by the victim(s). The abstracts vary in the value of the information provided, largely due to differences in the availability of information concerning specific accidents and in the differences in the communication skills of the Compliance Officers.

3 RESEARCH METHODOLOGY

The objective of this research was to examine, in greater detail, the circumstances and conditions associated with confined space accidents. The source of this information was the OSHA data acquired through the accident investigations of OSHA Compliance Offices. This information is maintained in OSHA's Integrated Management Information System (IMIS). As public information, this is available to interested parties.

The OSHA IMIS database was utilized to acquire information on confined space accidents occurring between the years 1990-1995, inclusive. The search was conducted by identifying and selecting all cases in which the words "confined space" was included in the abstract. While it was recognized that some cases may inadvertently be omitted because these specific words were not used, it was decided that a sufficient number of cases would be collected with which to conduct a meaningful analysis. This strategy resulted in isolating 120 confined space accidents involving 238 victims. Of the accidents examined, the severity of the injuries were as follows: 106 fatalities, 78 hospitalized injuries, 54 injuries not hospitalized.

4 RESULTS

While the accidents examined did include some incidents that some might consider to be classic confined space accidents, the assumption was made that there would not be general agreement as to what constitutes a "classic" accident.

As described earlier, one of the characteristics is that the space is large enough and so configured that an employee can physically enter the space and perform work in it. The various cases were examined and the confined spaces consisted of the following types of confinement: tanks, manholes, work spaces, truck tanks, vessels, vaults, tanks cars, sumps, silos, trash disposals, swimming pools, pipes, boilers, bins, vent ducts, tubs, pits, elevator cars and draglines. Of these, accidents involving tanks and manholes were most numerous. It is perhaps easy to envision each of these as potential confined spaces.

Of the confined space accidents, it was noted that exceptions did exist to the first criteria (large enough for human entry) given above. One case involved a worker who was not actually performing work while physically inside the space. This consisted of a worker who was spraying water into a "mud" or "frac" tank truck. As the water was being added, methane gas was emitted from the tank hatch. Methane was somehow being released by the drilling or digging operation. The worker was overcome by the methane gas, fell into the open hatch, and drowned in the tank.

The second criteria (limited or restricted means for entry or exit) was also associated with exceptions. One such case consisted of work being performed in a school corridor. In a stepped down portion of the corridor was a drinking fountain that required some repair. The worker entered the hallway and removed part of the cover of the fountain to gain access to the inside of the fountain. The worker worked on the fountain by lying on his back. As work was being done, some of the dense refrigerant escaped and remained near the floor level of the corridor. The worker died in his working position. The stepped down portion of the corridor became a confined space for the dense refrigerant. Unfortunately, the hazardous environment was not recognized by the worker.

The third criterion was that the space is not designed for continuous employee occupancy. This definition is vague in that the timing of the occupancy is not defined. One case was found in which the confined space characteristics were recognized, but they were misjudged. This was a situation in which water was pumped from an excavation by means of a gasoline-powered pump. The operation of the pump was recognized as a source of carbon monoxide generation. Workers turned the pump motor off and did not enter the excavation for several hours. Assuming the toxic gases had dissipated, the workers entered the excavation and, after a few minutes, died of carbon monoxide poisoning. Clearly, the air should have been properly tested rather than making an assumption about the quality of the environment.

There are many substances that can make an environment unsafe for workers. The information in the confined space accident reports was examined to identify the nature of those substances most often involved in confined space accidents. This analysis is summarized in Table 1. The substances are quite varied, but they can all be deadly. With a little thought, it is possible to identify construction operations that will expose workers to each of these hazards.

The cases that have been described seem to indicate that confined spaces are associated with toxic substances in the air. While this is often the case, this is not a prerequisite situation for a confined space. The confined space accidents were analyzed in terms of the nature of the injury actually suffered. The details of these injuries are shown in Table 2.

One drowning accident, associated with methane gas, was already described. In another case, pool maintenance was being performed. When the oxygen tanks for the self contained underwater breathing apparatus (SCUBA) ran low, both workers got out of the pool. One worker left to get fresh oxygen tanks. In the meantime, the other diver, using a mask and flashlight, reentered the pool and went to the pump intake. This worker drowned and was found a few minutes later. This was not an authorized dive by the worker. It is surmised that the diver drowned when he became disoriented and could not find his way to the water surface.

Not all pool accidents involve drowning. In one aluminum pool that was being completed, paint was being applied by workers. Although the pool was quite large, the workers were overcome by toluene fumes in the paint. The workers had not fully evaluated the danger posed by the painting operation.

Table 1. Toxic Agents Associated With Confined Space Incidents

Type of Substance	Number of Accidents
Toluene (in paint)	20
Carbon Monoxide	19
Methane	15
Nitrogen	11
Methyl Chloroform	8
Hydrogen Chloride	8
Fire/Smoke	7
Propane	7
Hexane	7
Carbon Dioxide	6
Petroleum Products	6
Trichloroethylene	6
Gasoline	5
Steam	5
Hydrogen	5
Trifluorethane	5

Table 2. Types of Injuries Sustained in Confined Spaces

Type of Injury	Percent of all Confined Space Accidents
Inhale Toxins	39
Asphyxia	30
Burn, Thermal	12
Poisoning, Systemic	10
Drowning	4
Fracture	2
Other	3
Total	100%

One of the drowning accidents did not involve a traditional body of water. A utility line was being installed in a trench. The trenching operation went through an area that had been previously disturbed, causing a trench collapse. The collapse was a small one and pinned the legs of one of the workers so he could not dislodge himself. Soon thereafter, a water main (in the collapsed region of the trench) broke and caused water to gush into the trench. The water level rose quickly and the worker drowned before he could be rescued.

An example will be described to illustrate an injury categorized as "other" in Table 2. A worker was working above a bin. Somehow, in the process of performing his work, one of his tools fell into the bin. The worker, a man nearly 60 years old, managed to get into the bin to retrieve the tool. The worker attempted repeatedly to get out of the large steel bin but kept falling back in. Eventually, the stress caused the worker to suffer a heart attack. He died in the bin.

From the various descriptions, it should be evident that there are a wide diversity of conditions associated with confined space accidents. Herein lies the problem with confined spaces, i.e., workers must be able to anticipate the many possible hazards to which they may be exposed in the course of performing their work operations. The various accidents were examined to determine the extent that workers were educated about confined spaces. It can be stated that there is generally a low level of training for workers to identify confined space hazards.

5 CONCLUSIONS

Confined space accidents are numerous and they occur in a myriad of work settings. Many confined space accidents occur in tanks and manholes, but many different types of confinement spaces have been associated injuries and fatalities. Most of the atmospheric environments associated with confined space accidents involve either toxic substances or the lack of oxygen.

In many confined space accidents, it appears that the victims failed to recognize that a confined space existed or that the environment in the confined space was life threatening.

When confined space accidents occur, it is common for there to be multiple victims. In addition, the percentage of fatalities, instead of injuries, is generally quite high. So it can be stated that when confined space accidents occur, the situation is often life threatening, but the workers have historically not recognized the potential for injury.

6 RECOMMENDATIONS

It should be an industry objective to drastically reduce the incidence of confined space accidents. This can most readily be accomplished though worker training. This training needs to be diverse. Workers need to be fully appraised of the materials with which work is being performed. The OSHA regulations already require this, but stronger compliance with these regulations is clearly warranted. It is not sufficient to maintain a complete file of material safety data sheets (MSDS) associated with work operations. The information must be communicated to the workers.

The training should extend to the recognition of confined spaces. This is a subject that receives far too little attention. The OSHA regulations do not provide clear guidelines in the identification of confined spaces, largely because this is very situational. Training will be very challenging as a result. Workers must simply be educated about thinking about the work environment being created or in which work is to be performed. Supervisors must similarly be trained to more fully appreciate the broad range of scenarios that are included in confined spaces. Recognition of hazards is associated with the ability to avoid them or to cope safely with them. Because of the serious potential consequences, constant assessment of the workplace is necessary. Without proper attention, they will continue to be potential death traps. Confined spaces can be safer places to work if they are first recognized and adequate precautions are taken.

7 REFERENCES

Pettit, T, & Linn, H. Pre-publication copy. *A Workers' Guide to Confined Spaces*. Morgantown: National Institute for Occupational Safety and Health.

U.S. Department of Labor, Robert B. Reich, Secretary. 1993. *Permit-Required Confined Spaces (Permit Spaces)*. Occupational Safety and Health Administration.

U.S. Department of Health, Education, and Welfare. 1979. *Working in Confined Spaces*. Public Health Service, Center for Disease Control, National Institute for Occupational Safety and Health.

16 Automation and hazard management

Implementation of Safety and Health on Construction Sites, Singh, Hinze & Coble (eds)
© 1999 Balkema, Rotterdam, ISBN 90 5809 036 1

Computer aided construction safety management in 21st Century: A human perspective

A.K.Singh
Department of Commerce, Satyawati Co-ed. College, University of Delhi, India
V.Sachdeva
Integrated Academy of Management and Technology, Ghaziabad, U.P., India

ABSTRACT: The present paper is an attempt to highlight some of the important issues related to the future of CACSM keeping in mind the human factor as the prime area of concern. All the current systems are designed in a manner that suits the organizational goal of maximizing shareholders value using HR as means rather than an end to attain organizational goals whereas human beings should be at the center of all concerns. Hence, there is a need to develop computer based software which should be made keeping the human beings at the center of all concerns both inside and outside the organization instead of the sole emphasis on value maximization of shareholders stock. The authors visualize a new global dynamic and competitive environment where every body is respected and gets his due share right from the conceptual model development stage.

1 INTRODUCTION

Computer aided safety management has gained attention during the last two decades but the focus is primarily on safety management of resources other than Human Resource (HR) knowing fully well that HR is the most important resource of any organization. Almost all the current information systems are designed keeping in mind the prime objective of the organization i.e. value maximization of shareholder's stock, little realizing that all the human beings associated with the organization are equally important (employees, customers, suppliers, etc.). In the closing decades of the 20th century managers started realizing that employees should not be treated as machines and the focus of personnel management got shifted to Human Resource Management (HRM) by imparting the human dimension with the cloak of resource. The authors visualize that in the 21st century the focus is going to be on Human Development (HD) instead of HRM or HRD.

HRM or HRD (Singh 1997) treats human as an input in the production process i.e. a means rather than an end to attain organizational goals whereas HD keeps human being at the center of all concerns. The concern for human beings should not be restricted to employees of the organization only as customers, suppliers and other stakeholders are no less important. HD Index (HDI 1997) is an important indicator in this context. HDI rank of some of the important nations is shown below :

1. Canada	2. France	3. Norway	4. USA	7. Japan
14. Australia	15. UK	19. Germany	108. China	138. India

China and India being the two most populous countries having such a low rank speaks of pathetic situation at this front. Safety of human life is very important component of HD particularly in the construction industry which is the second largest economic activity after agriculture employing 1.5% of Indian population. World wide investment in construction business is more than 1.5 trillion dollar employing more than 100 million workers that almost contributes to 6% to 7% of the total world labor force. The accident rate on an average is 160-250 per 1000 persons in the construction work (Vaid 1986). Realizing the importance of construction safety a complete order 5480.9A was issued by US Department of Energy on April 13, 1994 canceling the earlier order of 1980 which includes program implementation, hazard analyses, training, reporting, project safety and health plan apart from many other aspects (US DOE 1994).

NSA (1995) has defined accident as "An unplanned and something injurious or damaging event which interrupts the normal progress of an activity and is invariably preceded by an unsafe act or unsafe condition." It found that

13-14% of accidents are due to unsafe conditions
85% are due to unsafe act
1-2% are uncontrollable

It is important to note here that 98-99% of accidents are controllable provided there is a will to do that. Hence on the one hand human beings must be educated about safety where HD comes into picture and on the other computer aided construction safety management becomes important assuming that people may commit an unsafe act or may not object for unsafe conditions. The CACSM should be able to not only point out an unsafe condition and/or unsafe act but also be able to rectify unsafe condition and prohibit any unsafe act.

The present paper is an attempt to highligat some of the important issues related to the future of CACSM in the 21st Century putting people at the center of all concerns. The paper has been divided into two parts. Part I briefly deals with issues and developments related to CACSM and Part II provides a detailed vision of CACSM.

2 PART - I - CACSM

There are various issues related to CACSM on which development has taken place in last few decades (NICMAR 1996) keeping in mind the overall objective of the organization - value maximization of shareholders stock. The development has taken place in the areas of computer aided design (CAD), computer aided manufacturing (CAM), calculation and demos of construction sites using simulation modeling techniques. Electronic data interchange, virtual networking through the internet, intranets, extranets, virtualnets, etc. in the field of virtual construction has enabled the experts from different locations of the globe to work on the same project at a time to achieve the best out of the available resources. Thus a full Nth dimensional model is ready before the real execution of the project with flexibility of modifications For the management of these the software like the module of Manufacturing Resource Planning(MRP) in the Enterprise Resource Planning (ERP) have been developed for keeping in view construction as a whole entity keeping in mind the overall objectives of the organization.

Kohli (1998) has described the way Indian construction companies are adopting such software packages. A lot of experts are working in the area of construction safety management and have achieved world class results (Peter 1995) but the box given below shows that there are many more milestones in this journey of total safety so that all the stakeholders are delighted.

It is clear from the above that despite all the developments safety is a casualty in construction even at the end of 20th century. Hence there is a need to develop information systems and software packages keeping in mind the priority of human beings at all levels and putting safety as number one priority (Waterman 1991) for which part II of this paper is devoted.

> Injury Rates Drop Again
>
> The reported rate of serious injuries and illnesses in construction dropped 14% from 1994 to 1995, according to the U.S. Bureau of Labor Statistics (BLS). Injury-and-illness cases with days away from work dropped from 4.9 to 4.2 per 100 full-time-equivalent workers. Since 1990, the rate has dropped by one-third, from 6.2. Also in 1995, for the second year in a row, the rate for all nonfatal injury and illness cases in construction fell below the rate for manufacturing. The construction rate was 10.6 cases per 100 full-time-equivalent workers; the manufacturing rate, 11.6. For 18 years, until 1994, the construction rate was higher than the rate for manufacturing. This is excellent said Robert A. Georgine, president of the Building and Construction Trades Department (BCTD). Our programs in safety and health are paying off in some ways, although construction is still very dangerous. Construction workers suffer more than 190,000 injury or illness cases that require them to miss work well over 750 per workday.
>
> Deaths from injury continue to be 17% of the total, even though construction is 6% of the workforce. BLS announced in August from work dropped from 4.9 to 4.2 per 100 full-time-equivalent deaths from injury continue to be 17% of the total, eve though construction is 6% of the workforce. BLS announced in August that work-related deaths in construction totaled 1,039 in 1996, compared with 1,055 in 1995. Total work-related deaths in the U.S were 6,112 and 6,275 in 1996 and 1995, respectively, BLS reported. The Building and Construction Trades Department consists of 15 international unions and 3 million construction workers. It sponsors a wide range of safety-and-health research and development, mainly through the Center to Protect Workers Rights. Many of the programs are in cooperation with management or government.
>
> Source : IMPACT Newsletter 1997 (impact.html)

3 PART- II - VISION OF CACSM

This part provides a vision for the computer aided construction safety management in the 21st Century. The mission of the organizations in the 21st century should be “stakeholders’ delight” instead of satisfaction. It should include all the stakeholder viz. shareholders, bondholders, employees, customers, suppliers, etc. This requires commitment at the top level of the organization and a different kind of mindset having concern for people to be the most important objective of the organization and it must be shared by using effective communication tools (Davis 1990). This does not mean that the productivity of the organization will be reduced. Mathur (1986) has studied that if stress is reduced by reducing excess work load and safety is increased than it leads to increased productivity. Hence safety increases productivity in the organization and therefore should be dealt carefully. As has been discussed earlier that 98-99% of the accidents are controllable and fall under two broad categories, hence the major focus should be on :

- Providing safe conditions and
- Ensuring safe acts

CACSM should not be considered as the sole method of providing safe conditions and ensuring safe acts. The role of human beings can't be undermined. Nothing is created in this world by human beings unless it is desired by somebody. Computer is the creation of human mind and it will be applied to provide safe conditions and ensure safe acts only if people desire for it. The human beings must be empowered and be conscious of their rights and duties because computer aided systems are designed only if there is a desire or need for the same in the human beings. People should not be merely satisfied with the safety management instead the system should be such that they feel delighted.

The safety should be monitored from the entry point of any organization including homes. Hence there is a need to conduct eligibility tests at the entry point.

3.1 Computer aided entry eligibility test

Only eligible people should be allowed to get entry inside the construction site and computer aided entry eligibility test should be performed on each individual. Further it should be ensured that they are physically and mentally fit to do the assigned work for which the following tests may be performed.

3.11 Computer aided physical and medical fitness test

Both physical as well as medical test must be performed so as to ensure that the employees who are going to work inside the construction site are physically and medically fit. Expert systems should be developed to measure physical strength apart from checking blood pressure, conducting alcohol test, etc. depending upon the organizational requirement. Reddi (1986) has found that money plays an important role in forcing workers to do overtime leading to increased stress which sometimes may prove to be unbearable which this test should be able to detect. This is very important from HD perspective because the probability of accident increases to a very large extent if the person is not physically or medically fit to perform his duty. At present, most of the organizations focus only on computerize time keeping because they have a misconception that punctuality alone can increase productivity little realizing that prevention is better than cure. If a disease is diagnosed at an early stage than it requires less time and money as compared to if it is detected late.

3.12 Computer aided mental stability test

A person commits unsafe act, which is the major source of accidents, mainly because of mental instability. The first author (Singh 1996) conducted a study of 125 organizations spread over India and found that satisfaction level of employees is quite low despite good salaries and concluded that satisfaction is correlated more with the state of mind and how people perceive things. If people are not satisfied or happy than it affect their efficiency and may cause an accident. As evident from the findings of Srivastava (1997), in which he studied multinationals apparently rooted in Japanese, Indian and American culture in India clearly demonstrated that the group of managers(n=150) across the organizations, strongly endorsed the values leading to self realization, that may be due to the fact that managers perceived themselves to be valuing ability utilization, peace of mind, personal development achievement and advancement through their work. Such position provides sufficient foundations to say that organizational functioning is not determined technologically but socially. The importance to conduct computer aided mental stability test can't be undermined but more important than that is to provide some time for mental fitness by allowing them to do some exercises, meditation, yoga, etc., before the commencement of work as some of the organizations are doing in India and Japan (Srinivasan 1990). If a worker is found mentally unstable by the above-mentioned test even after such exercises then he/she should not be allowed to work for that session/shift.

3.2 Restricted security system

Computer aided restricted security system for specific job requirements should be used to ensure that only the eligible person does the requisite work. It has been found (Reddi, 1986) that workers have the tendency to look for more work so that they can earn more irrespective of their skills. This tendency should be stopped by checking at the entry point from the database of the worker that he/she is eligible for the particular job as per the schedule. If the workers are allowed to do a work for which he is not trained then it may prove to be hazardous and they must educated for this safety aspect otherwise they may develop a feeling of being underutilized.

3.3 Computer aided safe working environment

Safe working environment is a very important factor for the success of any organization because if the people don't feel safe in the organization then it reduces the efficiency and effectiveness. There can be various type of problems for which different kind of system would be required depending upon the size of organization. More important aspect is that how safe working environment is created. If people are involved in the decision making process then it becomes very effective because perception about safety is also very crucial. Following is a general framework of computer aided safe working environment.

3.31 Safety through computer aided skin touch sensitive equipment

This removes the possibility of accidents involving machines as sensors are installed to detect entry of human skin in the forbidden area of the machine. If it is detected then the machine is stopped immediately. This concept has gained popularity in manufacturing industry where the labor is organized.

3.32 Closed circuit loop scanners/videos connected to a central monitoring system

Sensitive construction sites require close monitoring to enable immediate and automatic necessary remedial action in case of mishap. To meet this requirement of some specific sites close circuit loop scanners/videos connected to a central monitoring system may be installed. This will avoid the delay in taking the remedial action which otherwise would prove to be very costly.

3.33 Sensors at Construction Site Connected To Simulated Model for Action Oriented systems

This is in contrast with the simulation model being used in the construction for model building of the construction site as the role of simulation model should not end at the point when the construction begins but should continue throughout the construction process as the construction goes on sensors must be installed at various places so as to deal with any variance in the variables to enable necessary action accordingly. This would help preventing various accidents due to collapse of structures, Piping and deep foundation, Concrete framed structure, Structural steel erection etc.

3.34 Fire and mishap controlling robots

NSC (1989) found that about 3.4% of accidents are caused due to fire and 4.3% due to electric mishaps. To save human life from such hazardous situation robots can be used along with fire sensing devices. The computerized check on temporary power supply and wire should be connected with devices which could sense a short circuit, decrease in voltage at a point due to shock or leakage in current, thus blocking the current immediately so as to prevent the accident and thereby saving human life against exposure of such risky situations

3.35 Computer controlled string systems and safe material handling

Computer Controlled String systems are particularly meant for those working positions which are above the ground level. This system would prevent accidents due to fall of persons or objects from a height. NSC (1989) has reported that accidents due to falls account for 12.6% of total accident, hence it becomes more important to implement computer aided string control system. When a person is to work above the ground level than he/she should be provided with a computerized string control systems which when loaded of with pressure of the person jams the string so as to prevent the fall. So far as the object placed at a height must also be inspected properly before keeping for the safe placement and in case of any movement due to mishandling should immediately follow a warning alarm for that area should be activated locating the place of the fault thus preventing such hazardous situations (CSAO 1989).

3.36 Computer aided quality control tests

Due to lack of proper quality control, many buildings have collapsed in the past involving loss of human life apart from investment already committed. The loss is widespread to all the stakeholders including the shareholders as the corporation may have to go into liquidation due to loss of faith in the organization. Quality is the most important aspect for the construction work as the backbone of each construction lies in the raw materials used for the construction thus this should be in the accordance of the simulated model that has been used for the construction. Thus it should not allow use of raw material in variance of its standard which may cause sever condition at a later stage.

3.37 Automatic working and scanning schedule

This corresponds to the database that should be there for the worker at the construction site so as the scheduling of work is done accordingly and the workers are not over loaded with work and should have proper rest between his/her working time. Thus the entry in the site must be in accordance to the work schedule.

3.38 Poisonous and radioactive substances alert systems

The poisonous and harmful substances should be kept at a safe place and only authorized entry to persons is provided who are well equipped and trained in handling such materials. The leakage in case of any poisonous or radioactive substance must be detected through the computer system followed by immediate cap on supply, emergency alarm indicating immediate clearance and necessary remedial action. Bhopal gas tragedy at Union Carbide Plant in India is very important to mention here as the company had to shut down because of one accident.

3.39 Computerized transportation system

A large number of accidents take place during transportation. A study by NSC (1989) has reported that 35.1% of total accidents take place during transportation. There are several causes of accidents and for each one there is a different solution or preventive computerised system. Above all proper and timely maintenance schedules should be adhered to and if the servicing schedule is missed then the computer system should not allow the machine to be used any further without servicing. Database for the service schedules would be required for the same. There should be computer aided inspection after such servicing and before the use of the machine as in some cases a very small lapse may prove to be very costly i.e., aircraft. It remains a challenge for the computer professionals to prevent air-crash caused by a small bird in the air. Similarly there are typical problems on the road leading to such a high rate of accidents. There is some progress in this direction but the systems are still beyond the reach of a common man. However big corporation can afford such transport system which are computer controlled. Further, there is a common problem of overloading for which adequate computers can be used to provide over load warnings and the system should come to a halt if the load increases beyond a limit.

Despite all the above precautions if an accident takes place then there should be an in-built procedure for automatic braking to reduce the impact of the accident on human beings in case the machine or vehicle touches an object in its way and the pressure is beyond a particular limit

4 CONCLUSION

The paper is an humble attempt to provide a vision with a human perspective for CACSM in 21st century. The authors have tried to give a philosophical framework to provide a different keeping in mind the prospective mission and objectives of the future organization instead of concentrating on the technical details of each system and its generic components. Most of the systems suggested are real time systems and their applicability would depend upon the size of the organization and their implementation would depend upon the willingness at top level. However, if the stakeholders become more conscious about the need for safety then the decision makers will have to think and work in this direction. It is suggested that the implementation of any system should not be decided purely on the basis of costs involved, instead a cost benefit analysis (Levitt 1991) should be conducted for taking such a decision and studies have proved that it really pays (COA 1989). It is not necessary to implement all the systems simultaneously, the organization can plan in a phased manner the implementation of CACSM as per the priority fixed in consultation with the stakeholders.

The major conclusion is that prime area of concern should be human beings both inside and outside the organization instead of value maximization of shareholders stock only. The authors visualize a new global dynamic and competitive environment where every body is respected and gets his due share right from the conceptual model development stage. Hence the paper tries to provide a new vision to the policy makers, consultants and computer professionals that any computer aided software should adopt an integrated approach keeping in mind all the stakeholders.

REFERENCES

Construction Owners Association 1989. *The Owner's Role In Construction Safety Management It Really Pays*. Edmmonton, Alberta: Construction Owners Association Of Alberta.

Construction Safety Association Of Ontario 1989. *WHMIS In Construction: Workplace Hazardous Material Information System.* Toranto, Ontario: CSAO.

Davis, Jim 1990. *Hazard Communication In Construction. 10 th Annual Construction Insurance Conference.* Dallas, TX : International Risk Management Institute.

Kohli, Vanita 1998. Corporate Viagra. *Business World,* ,7-21 Aug : 18-26.
Levitt, Raymond E. 1991. *Managing Safety For Profit.* 11th Annual Construction Insurance Conference . Dallas, TX : International Risk Management Institute.
Mathur, G.C. 1986. *Human factor in Construction Safety Management.* Published Seminar proceedings on Construction Safety Management. New Delhi : National Institute of Construction Management and Research.
National Safety Academy 1995. Safety Notes Association Builders and Contractors National Safety Academy, Clemson University.
National Safety Council 1989, *Accident Facts.*32-33 Chicago : National Safety Council.
NICMAR 1996. *A Manual for Project Manager on Safety Management in Construction Industry.* New Delhi : National Institute of Construction Management and Research.
Peter , Newton S, Jerome D. Klug, Larry Guilbault, and Michael C. Hewitt 1995. *Award wining Safety Programs- Achieving World Class Results.* National Construction Conference . Washington, D.C.
Singh, Ajay Kumar 1996. *Accounting for Human Resources : Acquisition, Development, and Retention*, unpublished Ph. D. thesis under the University of Delhi, Delhi.
Singh, Ajay Kumar 1997. *HRD : Vision and Directions in the 21st Century*, in Ajay Kumar Singh (ed), Delhi : Society for Professional Development and Research. pp.1-18.
Srinivasan, A.V. 1990. *Japanese Management.* New Delhi :Tata McGraw-Hill Publishing Company Limited.
Srivastava, Sanjay 1998. A Meeting of Culture in Transitional Organizations. Pranjana: The Journal of Management Awareness, Vol. 1 No. 1, 39-50.
United Nations Development Programme 1997. *Human Development Report-1997,* New York: United Nations Development Programme (UNDP), Oxford University Press.
U.S. Department of Energy 1994. *Complete Order 5480.9*A *Construction Project Safety and Health Management.* Washington, DC : U.S. Department of Energy.
Vaid, K. N. 1988. *Safety Management.* Published Seminar proceedings on Construction Safety Management. New Delhi : National Institute of Construction Management and Research.
Waterman, Eric S. 1991. *Making Construction Safety Priority #1* In The 1990's" Procedding of the 2nd UK-North America Construction Safety Conference Itasca, IL: National Safety Council.

Implementation of Safety and Health on Construction Sites, Singh, Hinze & Coble (eds)
© 1999 Balkema, Rotterdam, ISBN 90 5809 036 1

A knowledge-based approach for automating construction safety management

H. M. Elzarka
Department of Construction Science, University of Cincinnati, Ohio, USA

I.A. Minkarah & R. Pulikal
Department of Civil Engineering, University of Cincinnati, Ohio, USA

ABSTRACT: Concerns about the safety of construction workers have dramatically increased in the past decade. Owners and contractors are continuously working on reducing the unusually high rate of injuries and fatalities in the construction industry. Research has shown that injury reduction is only possible by improving the way safety is managed on site. Currently, managing project safety is a reactive process that is the sole responsibility of the safety manager. The safety manager's main objective is to ensure the contractor's compliance with OSHA regulations. This usually involves spending long hours sifting through volumes of safety regulations to determine what apply to a given project.

To improve safety management, a more proactive approach that integrates safety with other project management functions is needed. Such approach would involve more people in the safety management process, would ensure timely availability of safety information and would utilize safety knowledge in design and in the selection of construction methods. For this proactive approach to be effective, computer based systems that effectively manage safety information are needed.

Some computer-based systems have been developed to support the safety management process. The paper initially presents a discussion of these systems to illustrate their limitations. The paper then proposes an integrated knowledge-based database management system for automating safety management. The proposed system integrates safety and schedule information thereby making safety a more integral part of the overall project planning and control cycle. The proposed system ensures timely access to safety information by associating each schedule activity with required safety actions. Such association is automatically performed by the knowledge base after the user responds to questions related to the specific job conditions and the equipment utilized. Output from the system consists of daily reports that list safety actions to be performed on a given day.

1 CONSTRUCTION SAFETY

In the early 70's, the safety condition in the construction industry was appalling. This has led Congress to enact the Occupational Safety and Health act that established the Occupational Safety and Health Administration (OSHA) in the Department of Labor. OSHA is empowered to develop and enforce detailed safety regulations for all industries including construction. A comprehensive set of safety standards that are applicable to the construction industry was developed by OSHA in order to improve safety.

Recently, many construction companies have realized the importance of safety in their projects. These companies have discovered that safety is as much a part of effective project planning and control as are costs, schedules, procurement and quality (Barrie and Paulson

1992). A good company safety record reduces insurance premiums resulting in a more competitive bid price. Good safety policies that are effectively implemented improve labor productivity thereby increasing profit.

Despite recent efforts to improve safety in the construction industry, statistics show that the accident and injury rate in construction is still significantly higher than most other industries. One factor contributing to such a high rate is that construction includes different activities that require different safety precautions (Oglesby et al. 1989). Although as discussed above, a comprehensive set of safety standards and regulations for most construction activities; was developed by OSHA, many of these standards are vague and ambiguous, and inconsistently categorized. Determining what regulations apply to a given project usually involves spending long hours sifting through volumes of safety regulations.

What construction managers really need is timely and quick access to specific information to protect employees from the hazards associated with materials, equipment and procedures they work with daily. Such timely and quick access to safety information can be achieved through the use of computers. For this reason, several computer-based systems have been developed to support the safety management process. The following section presents a discussion of these systems to illustrate their limitations. The paper then proposes an integrated knowledge-based database management system for automating safety management. The proposed system integrates safety and schedule information thereby making safety a more integral part of the overall project planning and control cycle. The system provides ready reference material to the field engineer for day to day operations and eliminates the need to refer to OSHA safety books for normal construction activities.

2 KBS APPLICATIONS FOR SAFETY MANAGEMENT

A Knowledge based system (KBS), also called expert system is a computer program that emulates the problem solving process used by human experts. A KBS contains the knowledge, experience, and judgment of experts and is well suited to solving problems where judgment and experience play an important role (Bell and Elzarka 1992). At the University of Cincinnati, a wide range of KBS applications have been developed to support the safety management process. Applications have been developed to assist the construction manager to plan and safely execute earthwork, reduce environmental hazards, recommend appropriate personal protective and life saving equipment, and prevent fall accidents (Kak et al. 1995, Minkarah et al. 1996). OSHA safety regulations were incorporated into these systems.

The applications mentioned above recommend safety regulations for a given project based on project specific information. However they are stand-alone applications that do not integrate safety with other project management activities. Because of this, safety recommendations produced by these applications, though complete, might not be disseminated to project participants at the right time. Also in these applications, only OSHA regulations were modeled in the knowledge base. Safety knowledge and experience gained by the company's personnel is not utilized in an organized and efficient manner. From the above discussion, it is clear that the biggest limitations of current computerized safety management systems are the inability to provide timely access to safety information, and to retrieve safety knowledge gained in previous projects.

2.1 *Timely access to safety information*

Timely access to safety information is an important part of the safety management process; Construction managers (CM) need to know what safety precautions are required before the start of construction activities. Early access to data will enable the CM to make informed decisions when it has the most benefit. Thus, the safety management system should not only provide a mechanism for retrieving safety regulation, but should also ensure that the data are

retrieved when actually needed. If safety regulations are not disseminated when required, dangerous and fatal accidents can occur.
Timely access to safety information can be ensured by integrating safety management with the construction schedule. Such integration will provide the construction manager with a scheduled list of safety actions that need to be performed on a given date without overwhelming him/her with massive quantities of safety information required for the entire project.

2.2 *Learning from previous experiences*

Although all construction projects are unique, many operations and processes are common to every project, and require the same safety regulations. An effective safety management system should realize this and should enable its users to share knowledge among projects. Sharing of knowledge and experience from past projects will improve the performance of the management system through continuous learning.

For the continuous learning to occur, a safety management system should be capable of storing the safety data and knowledge gained in previous projects for future access. This will create a knowledge base of all company projects in a central location and will facilitate transfer of knowledge from old projects to current ones. Such a system also makes the implementation of company rules and regulations uniform across all construction projects of the firm.

The discussion above illustrates that the effectiveness of safety management systems can be improved by ensuring timely access to safety information and enhancing the learning mode of the safety knowledge base. To demonstrate the importance of these attributes, an integrated construction safety schedule (CSS) knowledge based system was developed. CSS captures safety knowledge from previous projects into the knowledge base and enables the generation of daily reports of required safety regulations that can be printed out and distributed to managers and formen. The safety management process is greatly simplified and the time that the safety manager spends on sifting through OSHA regulations is greatly reduced allowing him/her to focus more on the important task of ensuring compliance with these regulations.

3 SELECTION OF SOFTWARE DEVELOPMENT TOOLS

CSS consists of two main components: an executable file and a database. The executable file was developed using Microsoft Visual Basic. The database was created using Microsoft Access and is stored separately from the executable file.

Micorosoft Visual basic was selected as the software development tool because it is a mature product that can be used to rapidly develop Windows-based applications. Microsoft Visual Basic is an object-oriented visual development programming environment that makes extensive use of intuitive visual objects, instead of code, to speed development. It has several graphical form capabilities. These forms collect user input through a variety of mechanisms including fields, graphical controls, and scrollable lists. Visual Basic applies several object-oriented techniques to its visual objects. Each object has a set of attributes that control the appearance of the object during the program execution. Examples of these attributes include location, size, font and data type (i.e. string, number, etc.) of the objects.

Each object has its own script that contains a code defining all events associated with the object and the actions to be performed when an event occurs. For example, a script for a field object may contain a code associated with the "Click Field" event. This code is executed when the user clicks on a field. The code can contain either functions that control the graphical user interface (GUI) or control access to the database. GUI functions are used to perform various tasks such as displaying a message, drawing a control, and/or activating a new form.

The Visual Basic event-driven development environment facilitates creating a well-structured application and makes debugging the application much easier. It is more convenient to go directly to the code that is associated with a particular event (Elzarka and Bell 1997).

4 SYSTEM DESCRIPTION

As illustrated in Figure 1, the integrated safety scheduling management process starts with a construction schedule generated by a scheduling software. The schedule should contain the description of all activities, their start dates and end dates. The schedule can be either entered using one of CSS screens (Figure 2) or can be read directly from a file.

Figure 1 also shows that the knowledge base is the central repository of all safety knowledge and is created using Microsoft Access. The knowledge base is divided into two main components: OSHA knowledge base, and company knowledge base.

The OSHA knowledge base contains OSHA safety regulations. The "Occupational Safety and Health Standards for the Construction Industry (29 CFR Part 1926)" promulgated by the Occupational Safety and Health Administration (OSHA) was the basic source for knowledge acquisition. Safety regulations in the OSHA knowledge base are organized by construction

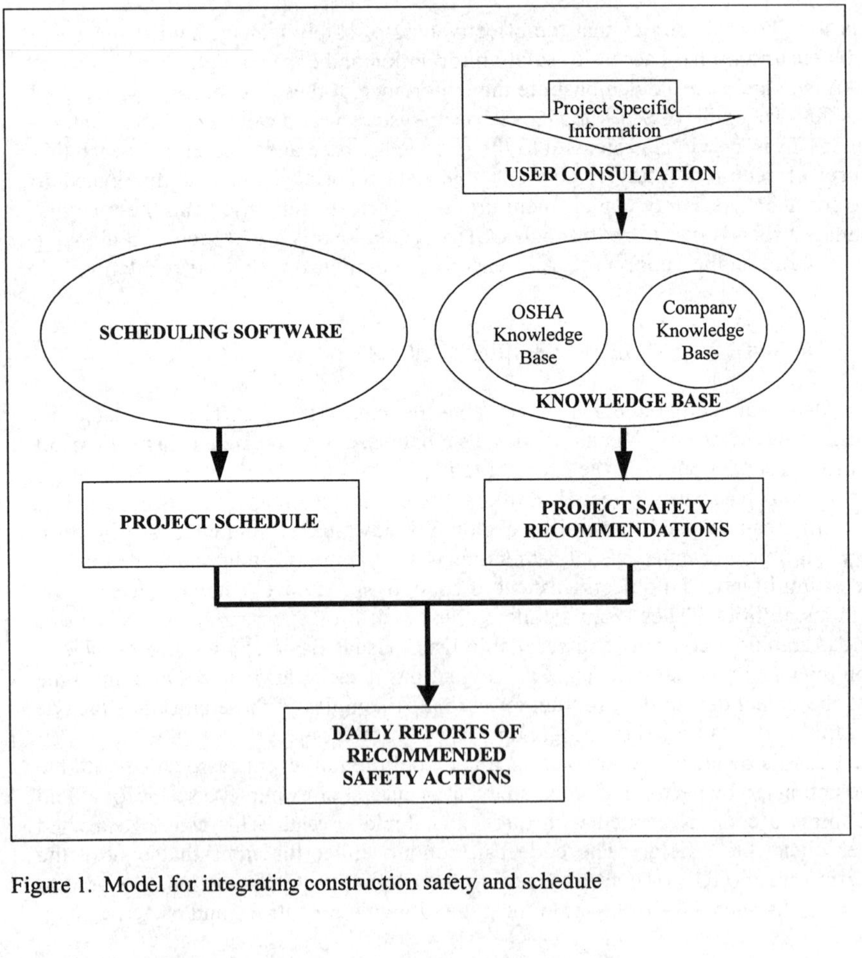

Figure 1. Model for integrating construction safety and schedule

activity (i.e. excavation, steel erection, etc.). Each regulation is presented in a long format and a short format. The long format exactly matches OSHA wordings, while the short format is a one line summary of required safety actions. At the end of the knowledge base consultation, the user is capable of printing the recommended safety actions report in either the short or long format. This capability adds more flexibility for the construction manager in case he/she wants to print safety reports in simpler, easier to understand language. An example of a typical OSHA safety regulation for an excavation with close proximity to existing structures is given in Figure 3.

The company historical knowledge base contains safety actions that have been successfully implemented on previous projects. This portion of the knowledge base allows the construction manager to use the successful experience of other managers from the company that might not necessarily be part of the OSHA regulations. The information in this knowledge base is proprietary in nature and contains the collective experience of all personnel of the company. The use of a central knowledge-base enables the transfer of safety recommendations from old projects to new ones.

The knowledge in both knowledge bases is modeled in the form of production rules that are expressed in an IF-THEN format. An example of a CSS consultation screen is illustrated in Figure 4. During the consultation, the user is asked to enter project specific information. In Figure 4, for example, the user is prompted to answer several questions regarding the excavation activity. If the user thinks that the excavation activity of the current project is similar to that of a previous project, he/she can press the "Suggest" button located at the bottom of the screen. This will list all previous projects stored in the company's historical knowledge base. The user can then select a comparable project and all safety actions performed on this project will be recommended for the new project.

At the end of the consultation, as illustrated in Figure 5, the user can enter a specific date for which a safety action report is to be generated. CSS compares the user provided date with the start date of all project activities to determine which safety actions need to be performed right away. These actions are listed in the "To Do" list (the middle box of Figure 5). The construction manager should review all listed safety actions in the "To Do" list. If some of these actions have already been implemented, he/she can change the status of these actions to "Done" by simply double clicking on the action. This will send the action to the "Done" list

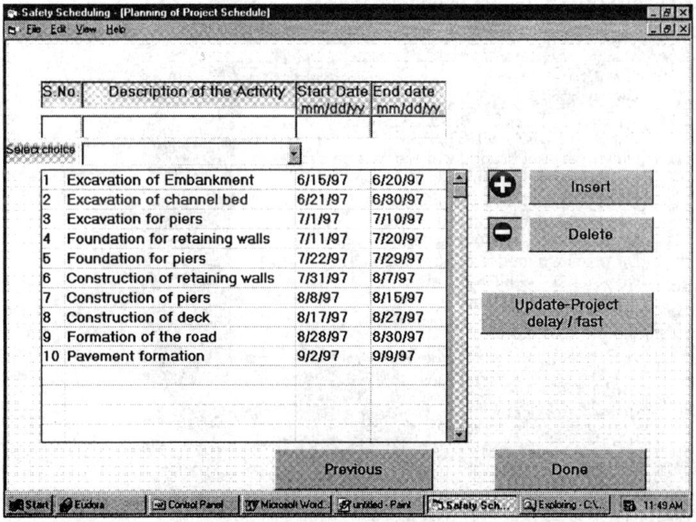

Figure 2- CSS Project scheduling screen

(the upper box of Figure 5). The system also informs the user of any actions that need to occur in the near future. Those are listed in the "Future" List box (the lower box of Figure 5). The text contained in any of these boxes can be copied to any windows based word-processing software where it can be reformatted (if necessary) and printed.

SAFETY ACTIONS NEEDED FOR AN EXCAVATION WITH CLOSE PROXIMITY TO EXISTING STRUCTURES.

1926.651

(1) Where the stability of adjoining buildings, walls, or other structures is endangered by excavation operations, support systems such as shoring, bracing, or underpinning shall be provided to ensure the stability of such structures for the protection of employees.

(2) Excavation below the level of the base or footing of any foundation or retaining wall that could be reasonably expected to pose a hazard to employees shall not be permitted except when:

(i) A support system, such as underpinning, is provided to ensure the safety of employees and the stability of the structure; or
(ii) The excavation is in stable rock; or
(iii) A registered professional engineer has approved the determination that the structure is sufficiently removed from the excavation so as to be unaffected by the excavation activity; or
(iv) A registered professional engineer has approved the determination that such excavation work will not pose a hazard to employees.

(3) Sidewalks, pavements and appurtenant structure shall not be undermined unless a support system or another method of protection is provided to protect employees from the possible collapse of such structures.

Figure 3. Example of safety recommendations included in the knowledge base

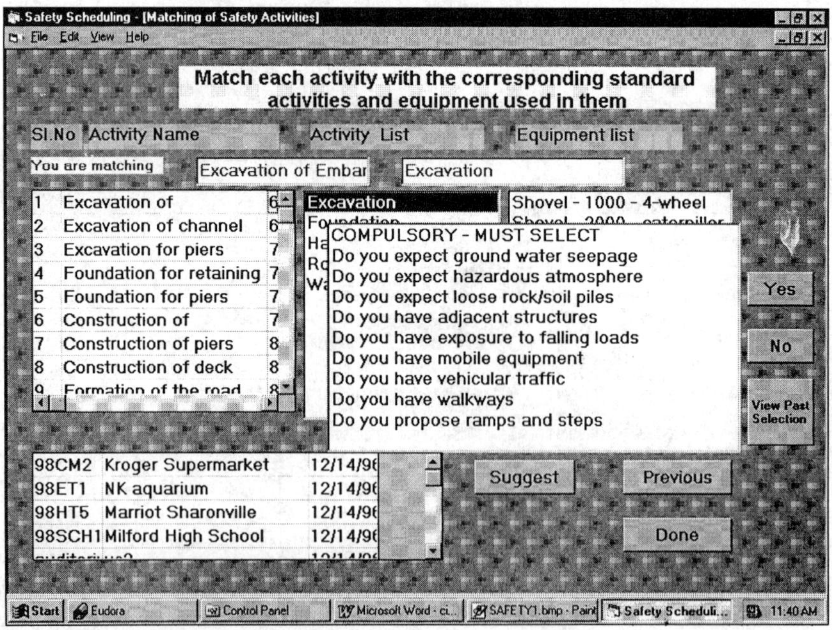

Figure 4.CSS Consultation screen

5 CSS EVALUATION AND FUTURE WORK

CSS was tested and evaluated at two construction sites for the purpose of obtaining subjective information on the benefits associated with the development of an integrated safety and schedule knowledge based system. Feedback from the reviewers indicated that the integrated approach has potential when applied to safety management. Integration of safety and schedule simplifies the safety management process by making safety information accessible when needed and utilizing safety knowledge from historical projects. The reviewers have indicated that the importance of safety and schedule integration becomes more obvious in complex and unique projects where construction managers are not very familiar with required safety precautions.

In its current form, CSS is a stand-alone application that runs on a Pentium personal computer. The personal computer was chosen because of its widespread use and relatively low cost. CSS can be easily expanded in the future. One of the important benefits of using production rules in knowledge representation is that it can be easily expanded and modified. It is easy to add more rules to handle situations not originally contemplated and to modify the knowledge to account for changes in OSHA and/or company safety requirements (Elzarka et al. 1995). This evolutionary improvement is characteristic of knowledge based systems. If many domain experts contribute their knowledge to the system, the knowledge base over time may become more knowledgeable and capable of problem solving than any one of the experts who contributed to its development.

6 CONCLUSIONS

Current computerized safety management systems can be enhanced by creating interface links to scheduling systems and by adding knowledge based modules that capture safety knowledge

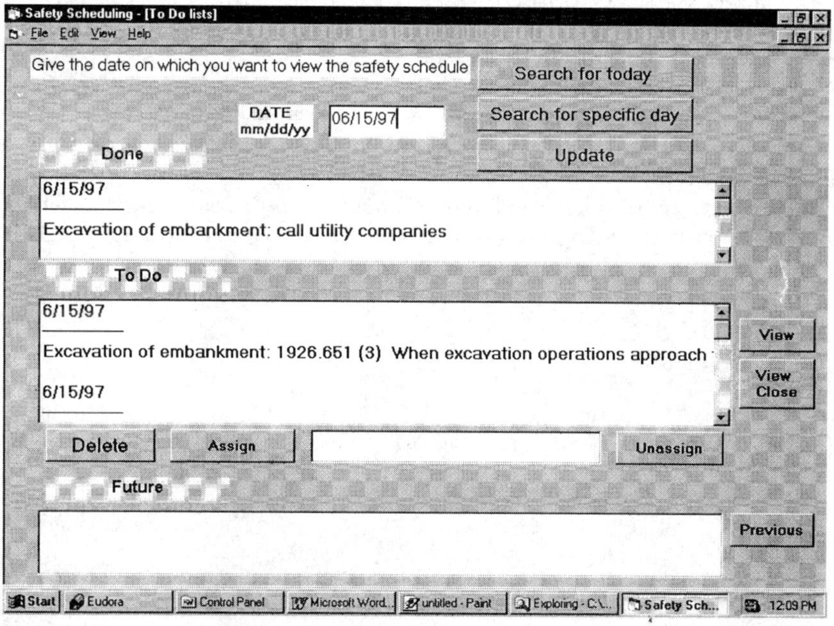

Figure 5. CSS Output screen

and experience gained from previous projects. This paper discussed several capabilities that can enhance computerized safety management systems. The paper also presented a prototype system CSS that was developed to establish the feasibility of incorporating the above mentioned capabilities in current systems. Evaluation of CSS has indicated that the system is an inexpensive tool to improve the existing safety program of any construction organization, provide timely safety standards/regulations, and promote safety awareness among workers. The CSS development process has also illustrated that current Windows software development tools are sufficiently advanced to rapidly create integrated construction management systems.

REFERENCES

Barrie, D.S. & B.C. Paulson 1992. *Professional Construction Management.* New York: McGraw-Hill

Bell, L. C. & H.M. Elzarka 1992. *Expert systems application and tutorial.* Source Document 75, Construction Industry Institute., Austin, Texas

Elzarka, H.M. & L.C. Bell 1997. Development of Pen-Based Computer Field Applications. *Journal of Computing in Civil Engineering,* ASCE, New York, Vol. 11, No. 2, pp.140-144.

Elzarka, H.M. & L.C. Bell 1995. Object-Oriented Methodology for Materials Management systems. *Journal of Construction Engineering and Management*., ASCE, Vol. 121, No. 4, pp. 438-445.

Gargari, M. & H.M. Elzarka 1998. An Integrated CAD/Knowledge Based Approach for Automating Preconstruction Activities. *Journal of Construction Engineering and Management,* ASCE, New York, Vol. 124, No. 4, pp.257-262.

Kak, A., K. Zacharias, I. Minkarah & P. Pant 1995. A Knowledge Based Software for Construction Safety. *Proceedings Second Congress of Computing in Civil Engineering,* ASCE, pp. 997-1002, Atlanta, GA.

Minkarah, I., P. D. Pant, D.S. Rao, & A. Kak 1996. Graphical user interface for site safety management. *Proceedings of the First International Conference of CIB Working Commission W99/Lisbon/Portugal/September 1996.*

Occupational Safety and Health Administration 1991. *Occupational Safety and Health Standards for the Construction Industry (29 CFR Part 1926),* Chicago, Illinois.

Oglesby, C.H., H.W. Parker & G.A. Howell 1989. *Productivity Improvement in Construction,* McGraw-Hill, Inc., New York.

Stanton, W. A. 1990. Conceptual Framework For Computer-Based Construction Safety Control. *Journal of Construction Engineering and Management,* Vol. 116, No. 3, pp.383-398.

Implementation of Safety and Health on Construction Sites, Singh, Hinze & Coble (eds)
© 1999 Balkema, Rotterdam, ISBN 90 5809 036 1

Integrity assessment of steel bridge structures using rough set based knowledge acquisition method

H. Furuta, M. Hirokane & S. Tanaka
Faculty of Informatics, Kansai University, Takatsuki, Japan
Y. Mikumo
Information Engineering Department, NEWJEC Incorporated, Osaka, Japan

ABSTRACT: Many of the existing structures present an ever-Increasing demand for maintenance and repair. It is needed in the first place to diagnose such structures correctly. The diagnoses have been dependent to date upon experts of long experience. To meet the increasing demand for maintenance and repair and perform diagnoses efficiently and correctly, the development of expert systems is hoped for. The present paper discusses the application of the rough set theory to the acquirement of experiential knowledge contained in diagnostic cases. The rough set theory was applied to cases in which experts diagnosed the damage of reinforced concrete bridges, and minimum decision algorithm was derived which could make diagnoses equivalent to those by experts. Also a method of deriving rules from the minimum decision algorithm for the construction of expert systems is discuss.

1 INTRODUCTION

Quite a number of the existing structures have sustained some damage from external loads such a earthquakes, typhoons, shocks, etc. after their construction to date, and now have questionable usability or safety. Nowadays, it has become an important problem how to maintain and administer these existing structures. It does not make much economic sense, however, to pull down and rebuild all such structures for the mere reason that their damage is advancing. Therefore, it is desired at present to have a method which enables us to make correct judgement on which structures should be rebuilt and which ones should be repaired. While such judgement would not be difficult if structures present visually observable, apparent damage, it would be quite difficult if they contain damage inside, showing no trace of it outside.

To make proper judgement on complex damage condition, it is necessary to interrelate visually observable damage condition, various attributes relating to damage condition, and available information (causes of damage, conditions of damage's advancement, importance of structures, surrounding environment, etc.) and assess the integrity of structures from a synthetic point of view. However, because data available for the assessment of integrity are not enough both in quality and in quantity, we have study few, ambiguous data from various aspects estimate the degrees and advancement conditions of damage based on our intuition and experience. Thus, integrity assessment has been dependent to date upon intuition and technical judgement of experts with affluent experience. On the other hand, with the increasing need of maintenance and repair of structures, the shortage in number of experts is ever growing.

To address problems like these, the expert system is drawing attention recently, and many studies have been made to enable even novice engineers to assess the integrity of structures at almost the same level with experts. In building an expert system, most important and difficult is the process of acquiring knowledge about the problem which the system is to address. In the field of artificial intelligence, acquisition of knowledge is one of most important subject. Being studied recently are methods to acquire expertise and empirical knowledge implied in actual cases and to utilize the knowledge; and they expected to offer an approach to the construction of expert systems.

As a fundamental study of artificial intelligence, processing of ambiguous or in accurate, or unclear, or rough knowledge and the like, and then reasoning,
decision, induction, learning, etc. on the basis of the processed knowledge are considered. As the basic technology to handle unclear knowledge, the fuzzy theory has been demonstrating its usefulness for some time, but the basic technology for handling rough knowledge has not been established yet. Rough knowledge means that when one observe a structure and, based on the results of the observation, reviews attributes for the assessment of integrity, the attributes which could not be determined from the observation results or attributes which were determined erroneously may be included in the case. Besides, errors in observation itself may be reflected in judgement on each attribute. Things of this kind are often happening in actual observation, and therefore it is required to remove contradictions from judgement and make as accurate judgement as possible from currently available observation results.

The authors applied the rough-set theory to diagnostic cases in which experts diagnosed the damage causes from the visible damage of panels of reinforced concrete slabs, and removed judgement factors and classes of each judgement factor insignificant in the diagnoses to extract minimum decision algorithm which was still capable of making diagnoses equal to those by factors to which the experts attached particular importance in their diagnoses. Therefore, based on such a minimum decision algorithm, some rules such as " **IF** conditional part (conditional attribute), **THEN** conclusive part (to be determined attribute)" was described.

2 DIAGNOSTIC CASES BY EXPERTS

Used in the present study to extract experiential knowledge were cases in which expert diagnosed the causes from the visible damage of reinforced concrete slabs of bridge structures.

These damage causes are classified into 4 types such as A: loads and the likes, B: design and structure, C: construction work and D: other. Moreover, the details of each damage cause is described, and the external loads such as 1) excessive wheel loads, 2) shocks and 3) relation between wheel load working zones and arrangement of supporting girders are considered as the details of damage cause of A: loads and the likes.

As shown in Table 1, damage was classified into 16 types (judgement factors) such as (1) location of damage and (2) longitudinal crack, etc. Each judgement factors is provided with classes. For example, the judgement factor of (1) location of damage is provided with two classes: 1.haunch and 2.center of effective span, and the judgement factor (2) longitudinal crack is provided with four classes: 1.existing a lot, 2.existing, 3.existing a little, and 4.none.

3 EXTRACTION OF MINIMUM DECESION ALGORITHM BY ROUGH-SET THEORY

The concept of rough-set theory was first introduced by a Polish scholar of calculators, Zdzislaw Pawlak, in 1982, and one of its essential qualities can be said to rest with classification. When one is to make judgment based on several pieces of information obtained by observation, one tries to classify the subjects of the information (to-be-determined attributes) in accordance with various attributes (conditional attributes) contained in the information. If the subjects are classified into one and the same class, one can not distinguish them and treats them as the same things. For instance, when one is supposed to classify cracks on a panel and if a single attribute of "cracking direction" is given, one can not distinguish such cracks as all running in one and the same direction and treats them as the same things. However, when one is given enough additional distinctive attributes such as crack shape, one becomes able to distinguish them. Based on the rough-set theory, minimum decision algorithm capable of making diagnoses equivalent to those by experts can be derived from diagnostic cases.

3.1 Subordination of Damage Causes to Judgement factors

First of all, in Table 2, the decision rules of all the panels were examined to find contradictory rules; i.e., panels which were classified into one and the same class under every judgment factor but the damage causes of which were diagnosed differently. It is necessary in the process of deriving minimum decision algorithm to remove contradictory rules from such a decision table.

No contradictory rules were found in Table 2, and hence all the damage causes proved subordinative to the judgment factors.

3.2 Removal of Judgement Factors

Next, it was necessary to find judgment factors insignificant in the diagnoses of the damage causes. A judgment factor or more were removed each time, and it was checked whether any contradiction occurred or not in the decision table. Every possible combination of judgment factors was removed

Table1. Judgement factors of damage causes

Judgment factor	Classes	Judgment factor	Classes
(1)Location of damage	1. Haunch 2. Center of effective span 3. Cantilever beam	(9)Shortage of covering	1. Existing a lot 2. Existing 3. Existing a little 4. None
(2)Longitudinal crack	1. Existing a lot 2. Existing 3. Existing a little 4. None	(10)Breaking	1. Existing a lot 2. Existing 3. Existing a little 4. None
(3)Lateral crack	1. Existing a lot 2. Existing 3. Existing a little 4. None	(11)Outflow of free lime	1. Existing a lot 2. Existing 3. Existing a little 4. None
(4)Crack in both longitudinal and lateral derections	1. Existing a lot 2. Existing 3. Existing a little 4. None	(12)Water leakage	1. Existing a lot 2. Existing 3. Existing a little 4. None
(5)Grid-like crack	1. Existing a lot 2. Existing 3. Existing a little 4. None	(13)Exposure due to short covering	1. Existing a lot 2. Existing 3. Existing a little 4. None
(6)Hexagonal crack	1. Existing a lot 2. Existing 3. Existing a little 4. None	(14)Exposure due to stripping	1. Existing a lot 2. Existing 3. Existing a little 4. None
(7)Penetrating crack	1. Existing a lot 2. Existing 3. Existing a little 4. None	(15)Rust of reinforcements	1. Existing a lot 2. Existing 3. Existing a little 4. None
(8)Float	1. Existing a lot 2. Existing 3. Existing a little 4. None	(16)Corrosion of reinforcements	1. Existing a lot 2. Existing 3. Existing a little 4. None

Table 2. Diagnostic cases of damage causes by experts

Panel No.	Judgement factor (1)	(2)	(3)	(4)	(5)	(6)	(7)	(8)	(9)	(10)	(11)	(12)	(13)	(14)	(15)	(16)	Damage cause
1	1	2	3	2	2	2	2	2	2	1	4	2	4	1	4	4	A
2	1	3	2	2	2	2	2	2	2	2	4	3	4	2	4	4	A
3	1	2	2	1	1	1	1	2	2	1	2	1	4	1	4	4	B
4	1	1	2	1	1	1	1	2	2	1	2	2	4	1	4	4	B
5	1	1	1	1	1	1	1	2	2	2	1	1	4	2	4	4	B
6	1	1	3	1	1	1	1	2	2	1	4	2	4	1	4	4	B
7	1	4	4	4	4	4	4	4	4	4	4	2	4	4	4	4	B
8	1	4	2	2	2	2	2	4	4	4	4	4	4	4	4	4	B
9	1	4	2	2	2	2	2	4	4	2	2	3	4	4	4	4	B
10	1	4	2	2	2	2	2	4	4	3	3	3	4	4	4	4	B
11	1	3	3	3	3	3	3	3	3	3	4	3	4	3	3	3	C
12	1	3	3	3	3	3	3	3	3	2	4	3	4	3	3	3	C
13	1	4	3	3	3	3	3	2	2	2	2	2	2	2	2	2	C
14	1	3	3	3	3	3	3	3	3	3	4	4	2	2	2	2	C
15	1	4	3	3	3	3	3	2	2	2	4	4	1	1	2	2	C
16	1	4	3	3	3	3	3	4	4	4	4	4	4	4	4	4	D
17	1	4	4	4	4	4	4	4	4	4	4	4	4	3	2	2	D
18	1	4	4	4	4	3	3	4	4	3	1	1	4	3	2	2	D
19	2	1	2	1	1	1	1	1	1	1	4	2	4	2	1	1	D
20	2	4	4	1	1	2	1	2	2	2	4	4	4	4	4	4	A
21	2	4	4	4	4	2	2	4	4	2	4	4	4	4	4	4	A
22	2	2	2	2	2	2	2	4	4	4	3	4	4	4	4	4	B
23	2	1	2	2	2	2	2	4	4	4	3	4	4	4	4	4	B
24	2	1	4	2	2	4	2	4	4	4	4	4	4	4	4	4	B
25	2	4	4	4	4	4	4	4	4	4	4	4	4	4	4	4	B
26	2	4	4	4	4	4	4	4	4	4	2	4	4	4	4	4	B
27	2	2	4	2	2	4	2	4	4	4	4	4	4	4	4	4	B
28	2	2	4	3	3	2	2	4	4	3	2	3	4	4	4	4	B
29	2	4	4	2	2	4	2	4	4	4	4	4	4	4	4	4	B
30	2	3	3	3	3	2	2	3	3	2	4	4	3	3	4	4	C
31	2	4	4	2	2	2	2	2	3	2	4	4	4	4	4	4	C
32	2	3	3	2	2	2	2	3	3	3	3	4	4	4	4	4	C
33	2	3	4	3	3	2	2	3	3	3	3	1	4	3	3	3	C
34	2	4	4	4	4	3	3	3	3	3	4	4	4	2	4	4	C
35	2	3	4	3	3	3	3	2	2	3	4	4	3	3	4	4	C
36	2	4	3	4	4	4	4	4	4	4	3	4	3	4	4	4	D
37	2	4	4	4	4	4	4	4	4	4	4	4	3	3	4	4	D
38	2	4	4	4	4	4	4	4	4	1	3	4	4	4	4	4	D

Table 3. Combination of judgement factors in the decision table without contradictions

Case No.	Judgement factor					Case No.	Judgement factor					Case No.	Judgement factor				
CASE 1	(3)	(4)	(10)	(14)		CASE 2	(3)	(5)	(10)	(14)		CASE 3	(1)	(3)	(4)	(10)	(14)
CASE 4	(1)	(3)	(5)	(10)	(14)	CASE 5	(1)	(4)	(10)	(11)	(14)	CASE 6	(1)	(4)	(10)	(13)	(14)
CASE 7	(1)	(5)	(10)	(11)	(14)	CASE 8	(1)	(5)	(10)	(13)	(14)	CASE 9	(1)	(6)	(9)	(11)	(14)
CASE10	(2)	(3)	(4)	(10)	(14)	CASE11	(2)	(3)	(5)	(10)	(14)	CASE12	(2)	(3)	(9)	(10)	(14)
CASE13	(2)	(4)	(10)	(11)	(14)	CASE14	(2)	(5)	(10)	(11)	(14)	CASE15	(3)	(4)	(5)	(10)	(14)
CASE16	(3)	(4)	(6)	(10)	(14)	CASE17	(3)	(4)	(6)	(11)	(14)	CASE18	(3)	(4)	(7)	(10)	(14)
CASE19	(3)	(4)	(8)	(10)	(14)	CASE20	(3)	(4)	(9)	(10)	(14)	CASE21	(3)	(4)	(10)	(11)	(14)
CASE22	(3)	(4)	(10)	(12)	(14)	CASE23	(3)	(4)	(10)	(13)	(14)	CASE24	(3)	(4)	(10)	(14)	(14)
CASE25	(3)	(4)	(10)	(14)	(16)	CASE26	(3)	(5)	(6)	(10)	(14)	CASE27	(3)	(5)	(6)	(11)	(14)
CASE28	(3)	(5)	(7)	(10)	(14)	CASE29	(3)	(5)	(8)	(10)	(14)	CASE30	(3)	(5)	(9)	(10)	(14)
CASE31	(3)	(5)	(10)	(11)	(14)	CASE32	(3)	(5)	(10)	(12)	(14)	CASE33	(3)	(5)	(10)	(13)	(14)
CASE34	(3)	(5)	(10)	(14)	(15)	CASE35	(3)	(5)	(10)	(14)	(16)	CASE36	(3)	(6)	(9)	(10)	(14)
CASE37	(3)	(6)	(9)	(11)	(14)	CASE38	(3)	(7)	(8)	(10)	(14)	CASE39	(3)	(7)	(9)	(10)	(14)
CASE40	(4)	(6)	(8)	(11)	(14)	CASE41	(4)	(6)	(9)	(11)	(14)	CASE42	(4)	(7)	(8)	(11)	(14)
CASE43	(4)	(7)	(9)	(11)	(14)	CASE44	(4)	(8)	(10)	(11)	(14)	CASE45	(4)	(8)	(10)	(13)	(14)
CASE46	(4)	(9)	(10)	(11)	(14)	CASE47	(4)	(9)	(10)	(13)	(14)	CASE48	(4)	(10)	(11)	(12)	(14)
CASE49	(4)	(10)	(12)	(13)	(14)	CASE50	(5)	(6)	(8)	(11)	(14)	CASE51	(5)	(6)	(9)	(11)	(14)
CASE52	(5)	(7)	(8)	(11)	(14)	CASE53	(5)	(7)	(9)	(11)	(14)	CASE54	(5)	(8)	(10)	(11)	(14)
CASE55	(5)	(8)	(10)	(13)	(14)	CASE56	(5)	(9)	(10)	(11)	(14)	CASE57	(5)	(9)	(10)	(13)	(14)
CASE58	(5)	(10)	(11)	(12)	(14)	CASE59	(5)	(10)	(12)	(13)	(14)	CASE60	(6)	(9)	(10)	(11)	(14)

from Table 2, and it was checked whether any contradiction occurred or not among the decision rules. Table 3 shows 60 combinations of judgment factors, Case1 to 60, each of which consists of the minimum number of judgment factors but still is able to diagnose every bridge without contradiction. Case1 taken as an example, description of how to extract the minimum decision algorithm follows.

The judgment factors other than (3), (4), (10), and (14) of Case1 were removed from Table 2. A pair of the panels 3 and 4, a group of the panels 8, 22 and 23, and so on were governed by one and the same rule each, and the panels of each pair or group were removed except one to obtain Table 4.

3.3 Removal of Classes of Judgement Factors

Lastly, it was necessary to examine the classes of the judgment factors in Table 4. Each class numbers of each judgment factor was removed to see whether any contradiction occurred or not in the table.

Table 4. Decision table based on the judgement factors in Case1

Panel No.	Judgement factor (3)	(4)	(10)	(14)	Damage cause
1	3	2	1	1	, `
2	2	2	2	2	, `
3	2	1	1	1	, a
5	1	1	2	2	, a
6	3	1	1	1	, a
7	4	4	4	4	, a
8	2	2	4	4	, a
9	2	2	2	4	, a
10	2	2	3	4	, a
11	3	3	3	3	, b
12	3	3	2	3	, b
13	3	3	2	2	, b
14	3	3	3	2	, b
15	3	3	2	1	, b
16	3	3	4	4	, c
17	4	4	3	3	, c
18	4	4	1	3	, c
19	2	1	2	2	, `
20	4	1	2	4	, `
21	4	4	2	4	, `
24	4	2	4	4	, a
28	4	3	3	4	, a
31	4	2	2	4	, b
32	3	2	3	4	, b
33	4	3	3	3	, b
34	4	4	3	2	, b
36	3	4	4	4	, c
38	4	4	1	4	, c

If a class number of a judgment factor is removed in such a decision table and contradiction occurs, the class number proves significant in the diagnosis. If not, it proves insignificant. By removing the class numbers one by one as mentioned above, a new decision table, or Table 5 was obtained. Notwithstanding the class numbers insignificant and removed from the sections marked with "-," this table is capable of making diagnoses equivalent to those by Table 4. This decision table contains no single judgment factor or class number removable without causing contradiction and is called "minimum decision algorithm." In the same way, we can find the minimum decision algorithm in each of Case2 to 60 in Table 3. For example, by removing the class numbers of the judgement factors in Case60, the minimum decision algorithm, or Table 6, was obtained. It can be said that the judgement factors of each case shown in Table 3 are all indispensable for its minimum decision algorithm.

4 EVALUATION OF MINIMUM DECISION ALGORITHM

Figure 1 shows the appearance frequency of each judgement factors based on the extracted decision table (Table 3) by a bar chart graph. The judgement factor(14) is needed for many of decision table (Case1~Case60), and the appearance frequency is 61. The judgement factor(15) is necessary for only the decision table of Case34, and the appearance frequency is 1. In Figure 1, describing by order of the appearance frequency, the first is "14: Exposure due to stripping", the second is "10: Breaking", the third is "3: Lateral crack", the fourth is "4: Crack in both longitudinal and lateral directions", the fifth is "5: Grid-like crack".

The evaluation of minimum decision algorithm is available, to verify about the significance of these judgement factors which have high appearance frequency, collating with the diagnosis by experts.

Table 5. Minimum decision algorithm of Case1

Panel No.	Judgement factor				Damage cause
	(3)	(4)	(10)	(14)	
1	-	2	-	1	, `
2	-	2	-	2	, `
3	-	1	-	1	, a
5	-	1	2	2	, a
7	4	-	4	4	, a
8	-	2	4	-	, a
9	2	-	-	4	, a
11	-	3	-	3	, b
12	-	-	2	3	, b
13	-	3	-	2	, b
14	-	-	3	2	, b
15	-	-	2	1	, b
16	-	3	4	-	, c
17	-	-	4	3	, c
18	-	4	-	3	, c
19	-	-	1	2	, `
20	-	1	-	4	, `
21	-	4	2	-	, `
28	-	3	3	4	, a
31	4	2	2	-	, b
32	3	-	3	-	, b
36	3	-	4	-	, c
38	-	-	1	4	, c

Table 6. Minimum decision algorithm of Case60

Panel No.	Judgement factor					Damage cause
	(6)	(9)	(10)	(11)	(14)	
1	2	-	-	-	1	, `
2	-	-	2	4	2	, `
3	-	-	-	2	1	, a
5	-	-	-	1	2	, a
6	1	-	-	-	1	, a
7	4	-	-	4	4	, a
8	2	-	4	-	-	, a
9	-	-	-	2	4	, a
10	-	4	3	-	4	, a
11	-	3	-	-	-	, b
12	-	-	2	-	3	, b
13	-	-	-	2	2	, b
15	-	-	2	-	1	, b
16	3	-	-	-	4	, c
17	-	-	4	-	3	, c
18	-	-	-	1	3	, c
19	-	-	1	-	2	, `
20	-	2	-	-	4	, `
21	-	4	2	4	-	, `
36	4	-	-	3	-	, c
38	-	-	1	-	4	, c

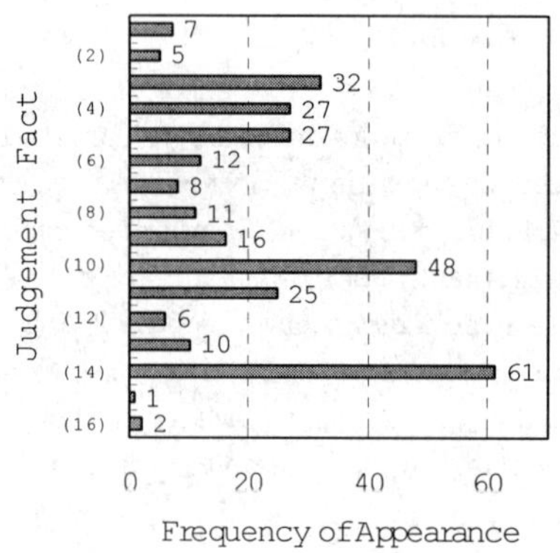

Figure 1. The appearance frequency of judgement factors

5 DERIVATION OF RULES

The minimum decision algorithm mentioned above can be described easily by rules such as "**If** conditional part (conditional attribute) **THEN** conclusive part (attribute to be determined)." The minimum decision algorithm of Case1 shown in Table 5 can be described by the 23 rules shown in Table 7. The rules 1 and 2 express the decision rules of the panel 1 and panel 2, respectively. The rule 23 represents the decision rule of the panel 38.

For example, the decision rule of the panel 1 in Table 5 indicates that its damage cause is "A" when both the conditions are met at the same time, one condition being that a panel is classified into the class 2 under the judgement factor (4) and the other being that the panel is classified into the class 1 under the judgement factor (14). This can be described by a rule below:

IF (4) = 2 *and* (14) = 1 *THEN* Damage cause = A

In the same way, the decision rule of the panel 2 can be described by a rule below:

IF (4) = 2 *and* (14) = 2 *THEN* Damage cause = A

To determine damage causes of conclusive parts by using an expert system constructed based on the above rules, it is necessary to check whether every condition in each rule is satisfied or not. As for the rule 1, it has to be checked whether the panel is classified into the class numbers 2 and 1 under the judgement factors (4) and (14), respectively. As for the rule 2, it has be checked whether the panel is classified into the class numbers 2 and 2 under the judgement factors (4) and (14), respectively. The total number of conditions to be checked in the 23 rules of Case1 shown in Table 7 is

Table 7. The description by rules of the minimum decision algorithm (Case1)

Bridge No.	Rule No.	Rule type description
1	1	*IF* (4)= 2 *and* (14)= 1 *THEN* Damage Factor= ,
2	2	*IF* (4)= 2 *and* (14)= 2 *THEN* Damage Factor= ,
3	3	*IF* (4)= 1 *and* (10)= 1 *THEN* Damage Factor= ,
5	4	*IF* (4)= 1 *and* (10)= 2 *and* (14)= 2 *THEN* Damage Factor= ,
7	5	*IF* (3)= 4 *and* (10)= 4 *and* (14)= 4 *THEN* Damage Factor= ,
8	6	*IF* (4)= 2 *and* (14)= 4 *THEN* Damage Factor= ,
9	7	*IF* (3)= 2 *and* (14)= 4 *THEN* Damage Factor= ,
11	8	*IF* (4)= 3 *and* (14)= 3 *THEN* Damage Factor= ,
12	9	*IF* (10)= 2 *and* (14)= 3 *THEN* Damage Factor= ,
13	10	*IF* (4)= 3 *and* (14)= 2 *THEN* Damage Factor= ,
14	11	*IF* (10)= 3 *and* (14)= 2 *THEN* Damage Factor= ,
15	12	*IF* (10)= 2 *and* (14)= 1 *THEN* Damage Factor= ,
16	13	*IF* (4)= 3 *and* (10)= 4 *THEN* Damage Factor= ,
17	14	*IF* (10)= 4 *and* (14)= 3 *THEN* Damage Factor= ,
18	15	*IF* (4)= 4 *and* (14)= 3 *THEN* Damage Factor= ,
19	16	*IF* (10)= 1 *and* (14)= 2 *THEN* Damage Factor= ,
20	17	*IF* (4)= 1 *and* (14)= 4 *THEN* Damage Factor= ,
21	18	*IF* (4)= 4 *and* (10)= 2 *THEN* Damage Factor= ,
28	19	*IF* (4)= 3 *and* (10)= 3 *and* (14)= 4 *THEN* Damage Factor= ,
31	20	*IF* (3)= 4 *and* (4)= 2 *and* (10)= 2 *THEN* Damage Factor= ,
32	21	*IF* (3)= 3 *and* (10)= 3 *THEN* Damage Factor= ,
36	22	*IF* (3)= 3 *and* (10)= 4 *THEN* Damage Factor= ,
38	23	*IF* (10)= 1 *and* (14)= 4 *THEN* Damage Factor= ,

50. In the same way, the minimum decision algorithm of Case60 shown in Table 6 can be described by the 21 rules, and the total number of conditions to be checked in the 21 is 45.

Thus, it is able to say that the decision table of Case1 is smaller than one of Case60, because the number of the judgement factors in Case1 and Case60 are 4 and 5, respectively. However, the total numbers of rule of Case1 and Case60 are 23 and 21, and the total number of conditions to be checked of their cases are 50 and 45, respectively. In this point of view, it is able to say that the decision table of Case60 is smaller than one of Case1. Rules for the construction of expect system should be described so that their number can be minimized for the sake of knowledge renewal and so on. Beside, conditions to be checked also should be described so that their number can be minimized for the sake of speedy reasoning. Accordingly, as compared with Case1 and Case60, efficient renewal of knowledge and speedy reasoning become possible if an expert system is constructed based on the rules derived from the minimum decision algorithm of Case60.

6 CONCLUSIONS

The basic concept of the rough-set theory was outlined and the application method of the theory to decision-making problems was described. Then, discussed was the method of extracting experiential knowledge of expert from the diagnostic results of damage causes of panels as the decision tables of minimum decision algorithm. Further discussed was a method to derive rule-type knowledge necessary for the construction of expert system from the minimum decision algorithm.

Main findings of the present study are as follows:

1) The decision table stipulates what decision to make when a certain condition is met. Almost all decision-making problems can be formulated by such decision tables.

2) By removing a judgement factor or factors each time in a decision table and seeing whether any contradiction will occur or not, we can know whether the removed factor or factors are significant in the diagnoses or not. In case that subject judgement is required for some judgement factors, the fuzzy theory may be applied, and the basic idea of the rough-set theory of narrowing down judgement factors can be used to solve one of the difficult and important problems in such application of the fuzzy theory.

3) Because rough-set theory start from the problem of how to express the set of conclusions by the set of paired judgement factors and classes, the rough-set theory can be used as a method to acquire knowledge from existing diagnostic cases.

The following are the future subjects:

1) Although at present the classes of each judgement factor are removed one by one to check whether contradiction occurs or not, some other method using genetic algorithms or the like, which can consider the combination of such factors, has to be devised.

2) In the present study, the possibility of applying the rough-set theory to the diagnoses of damage causes of panels was studied with a limited of data, since as the number of data for a problem of combination increases, it becomes more time-consuming. In future, it is necessary to build up case data and study the possibility of the application in a large number of data.

3) In the present study, judgement factors significant in the diagnoses could be identified by removing the judgement factors and checking whether condition occurs or not. However, the degree of effects of the significant judgement factors upon diagnostic results could not be determined. Our study has to be evolved toward a method by which the weights of such factors can be evaluated.

Acknowledgment:
This research was financially supported by the Kansai University Research Grants:
Grant-in-Aid for Encouragement of Scientists, 1997.

REFERENCES

Furuta, H. 1990. Fuzzy Expert System for Integrity Assessment of Bridge Structures, *Structural Engineering Applications of Expert Systems*, Vol.2, pp.133-142. (in Japanese)

Matsumura, H. 1966. Introduction of Set Theory, *Asakura Pub.* (in Japanese)

Mikami, I. Matsui, S. Tanaka, S. and Shinnai, Y. 1988. Knowledge-Based Expert System for Inference of Damage Factors on Reinforced Concrete Slabs of Steal Highway Bridges using Rule and Frame Representation, *Journal of Structural Engineering, JSCE,* Vol.34A, pp.551-562. (in Japanese)

Mikami, I. Tanaka, S. and Kitagisi, H. 1992. An Inference System of Damage Cause on RC-Slabs of Highway Bridges, *Proceedings of the Japan-korea Joint Seminar on Steel Bridges, JSCE,* pp.19-32.

Nishimura, A. Fuji, M. Miyamoto, A. and Ogasa, M. 1987. Basic Study on Application of Fuzzy Sets Theory to Integrity Assessment of Structures, *Journal of JSCE*, No.380/I-7, pp.365-374. (in Japanese)

Pawlak, Z. 1982. Rough Sets, *International Journal of Computer and Information Science*, Vol.11, pp.341-356.

Saitoh, K. and Nakano, R. 1992. RF2 algorithms, *Transactions of Information Processing Society of Japan*, Vol.33, No.5, pp.123-132. (in Japanese)

Implementation of Safety and Health on Construction Sites, Singh, Hinze & Coble (eds)
© 1999 Balkema, Rotterdam, ISBN 90 5809 036 1

Hazard communications using interactive computer based training techniques

P.Jones
Conceptual Arts Incorporated, Gainesville, Fla., USA

ABSTRACT: This paper describes a project with two goals: first, to create an electronic version of a hazardous chemicals database that improves access to information and second, to increase the relevance of the database by linking it to frequently used training modules. *HazCom Interactive* runs on personal computers and delivers information on hazardous chemicals from the International Chemical Safety Card (ICSC) database developed by the International Programme on Chemical Safety (IPCS). *HazCom Interactive* provides specific, high quality information and is linked to relevant training modules on ventilation, respirators and PPE. The training modules are designed to be highly interactive and generally operate like games. In the project's final phase the software will be evaluated for its efficacy in actual training situations.

1 INTRODUCTION

In 1987 the Occupational Safety and Health Administration (OSHA) extended application of the Hazard Communications Standard (HCS) to cover new segments of the American workforce, including the construction industry. As a result, employers in construction were required to develop and implement hazard communication programs appropriate to their industry. In many respects implementation in construction was much more difficult than in manufacturing where the HCS was first applied. Some specific differences are:

- Large to medium sized construction businesses are organizationally decentralized. Administrative headquarters are seldom actually near a company's shifting, multiple worksites,
- The number of company employees (both managers and workers) at any particular worksite is generally small relative to a manufacturing operation,
- In construction a company's employees who work with hazardous chemicals regularly move from one worksite to the next, and
- Construction companies in the U.S. rely heavily on subcontractors whose employees are also very transitory with respect to particular worksites.

Perhaps, the most important difference distinguishing manufacturing from construction operations are their respective safety officers. Construction companies generally have an identified safety officer at their main office, but more often than not the safety officer assignment is secondary to other responsibilities. In most construction companies the person most likely to contend with the practical problems described above are on-site supervisors. Organizationally, a company's project superintendent is directly responsible for implementation of the company's HCS policies on container labeling, maintenance of required Material Safety Data Sheet (MSDS) files, employee training, maintenance of the site's list of hazardous chemicals, and keeping subcontractors informed.

Aside from time management issues, field managers must deal with physical constraints. In practice, construction site field offices are cramped and busy. Although it is possible in a field office to keep files of MSDSs, maintain a list of on-site hazardous chemicals, to have blank, generic hazard warning labels, and to run small training programs, it is a very difficult task. The HCS files required on a commercial construction project can be very extensive and bulky. Establishing a regular filing system and keeping it current is time consuming. Providing training for employees every time a new chemical comes on-site is difficult, especially in a cramped field office.

Workers can also perceive HCS programs as a nuisance, particularly if there are production-based incentive programs and they feel that HCS procedures waste their time. In addition, many workers perceive HCS programs negatively because of problems associated with limited literacy and English language skills (Adult Literacy in America, 1993). Safety materials are often written at levels and presented in ways that are not understandable to average workers (Rabin, 1992). For example, studies indicate that average American workers cannot understand 40% of the information in a typical MSDS (Sattler, 1992). As a result, MSDS-based training can be frustrating or just plain boring to many workers.

In the construction industry, the result is that failures to comply with different aspects of the HCS top the list of OSHA compliance infractions. *The 100 Most Frequently Cited OSHA Construction Standards in 1991: A Guide for the Abatement of the Top 25 Associated Physical Hazards* (OSHA, 1993) identified the lack of a written HCS program, inadequate employee training, lack of MSDSs, lack of OHSA posters, and inaccessibility of MSDSs as the top five citations, respectively. This was still true as recently as 1997, during which OSHA continued "to issue far more citations to employers for violating the Hazard Communications Standard than any other rule" (Karr, 1997). It is noteworthy that the publication provides supplemental information on how to recognize and eliminate physical hazards, but offers no guidance on how to properly institute the HCS Standard. It is a difficult problem.

This paper reports on the development over the last four years of *HazCom Interactive*, an electronic HCS database project, that addresses the issues described above. In its first two years, this project's major goal was to develop procedures for porting information on hazardous chemicals from the IPCS International Chemical Safety Card (ICSC) system into a graphics-enhanced electronic HCS database to serve as a supplement to typical construction industry HCS programs. The database was designed specifically for delivering HCS information on typical construction-site PC computers. The project's specific initial objective was to develop a structured electronic database of multi-lingual ICSC information that included appropriately incorporated icons (or symbols), graphically indicating various types of hazards associated with chemicals and supporting audio (Jones, 1996).

In the last two years the project's emphasis shifted to training. After many demonstrations of *HazCom Interactive* to a wide range of audiences, it was apparent that even though the program was very favorably received, it did not readily fit into existing information delivery programs. In other words it was destined to become a passive resource that might occasionally be used. Apprenticeship educational program coordinators from the International Brotherhood of Painters and Allied Trades (IBPAT) recommended that the *HazCom Interactive* database be directly linked to interactive training programs covering appropriate topics, such as respirators and ventilation. In response to that suggestion, several interactive training modules were produced that cover different presentation styles. Each module links to the underlying database to provide detailed information about specific chemicals. This paper reviews the underlying *HazCom Interactive* database and describes the interactive training modules developed for evaluation.

2 A PARTIAL SOLUTION

By conceptually recreating Hazard Communications Standard (HCS) programs as integrated electronic databases of information for use on field office PCs, many of the inconveniences and inefficiencies of the current system could be greatly reduced. In addition to compact storage, an electronic database makes locating and retrieving specific information easier. Information from an electronic HCS database could be accessed either on the computer screen or printed on demand. Multiple menus based on common names, active ingredients and scientific names would avoid questions about where to find information on a particular material. In addition to the menus, several standard search techniques (such as word wheels, hyperlinks, and full-text indexing) would be applied to the database to facilitate finding specific HCS information.

Likewise, multi-media training programs derived from a HCS database could use the same graphical symbols on screen in conjunction with English (or second language) text and audio. This type of access to multi-media HCS programs would directly address some of the safety training problems discussed in *The Right to Understand: Linking Literacy to Health and Safety Training* (Szudy and Arroyo, 1994). Specifically, the article concluded that safety programming needs to incorporate more "*visually appealing materials*" and opportunities for "*participatory training*" to increase its effectiveness. All of the information delivery capabilities described above are consistent with the presence of PC computers in construction site field offices.

In practice most specific HCS program information currently used in the construction industry comes from Material Safety Database Sheets (MSDSs). Unfortunately, MSDSs are provided on paper (or as facsimile images) and are not generally available in digital form. Even worse, MSDSs are neither consistent nor uniform in their organization or appearance. The sheer volume of MSDSs further complicates the problem. The net result is that MSDSs are not a suitable source for building a HCS database.

At the opposite end of the spectrum, the International Chemical Safety Cards (ICSC) produced by the International Programme on Chemical Safety (IPCS, 1991) are highly structured even to the extent that individual phrases used to characterize chemical properties and hazards are explicitly defined in English as well as a core group of other languages. The ICSC system is fundamentally different from the MSDS system mandated by the HCS standard. ICSCs cover individual hazardous chemicals while MSDSs cover individual products that contain hazardous chemicals. Manufacturers with minimal centralized coordination individually produce MSDSs while the ICSC system is highly coordinated by the IPCS.

3 HAZCOM INTERACTIVE: AN INFORMATIONAL DATABASE

The foundation of the *HazCom Interactive* database is the text information from all currently available International Chemical Safety Cards (ICSC) covering: *Types of Hazards* (*Fire and Explosion*); *Types of Exposure* (*Inhalation, Skin, Eyes* and *Ingestion*); *Spillage Disposal*; *Storage, Packaging & Labelling*; and *Chemical Facts* (*Physical State, Chemical Dangers...*). In the ICSC database, the information is stored as a series of descriptive phrases that are explicitly linked to each information category. For example for the chemical hydrazine under the category "*Physical State: Appearance*" the following descriptive phrases are listed; "*colourless*", "*fuming*", "*hygroscopic liquid*", "*with pungent odour*". Text data (descriptive phrases) are available in many languages. The ICSC text data were imported into an object oriented database management system to support flexible access to the information and forms the foundation for *HazCom Interactive*.

*HazCom Interactive a*llows a user to search for a specific chemical by common name, scientific name or any of five chemical numbers (CAS, RTECS, ICSC, UN, or EC). Once a chemical is selected, a general information screen allows the user to make specific information requests. In addition to various chemical numbers, the opening menu offers access to seven categories of information: *Explosion Hazards, Emergency Medical, Fire Hazards, Exposure Hazards, Spillage Disposal, Storage, Packaging & Labeling, Chemical Facts & Additional Information*.

To obtain more information on inhalation effects, the user selects *Exposure Hazards*; the program moves to a screen that displays the four exposure routes of *Inhalation, Skin, Eyes*, and *Ingestion*; *Inhalation* is chosen and a screen displays *Acute Symptoms*, *Prevention* and *First Aid.* Figure 1 shows the screen that displays for hydrazine if *Prevention* is selected.

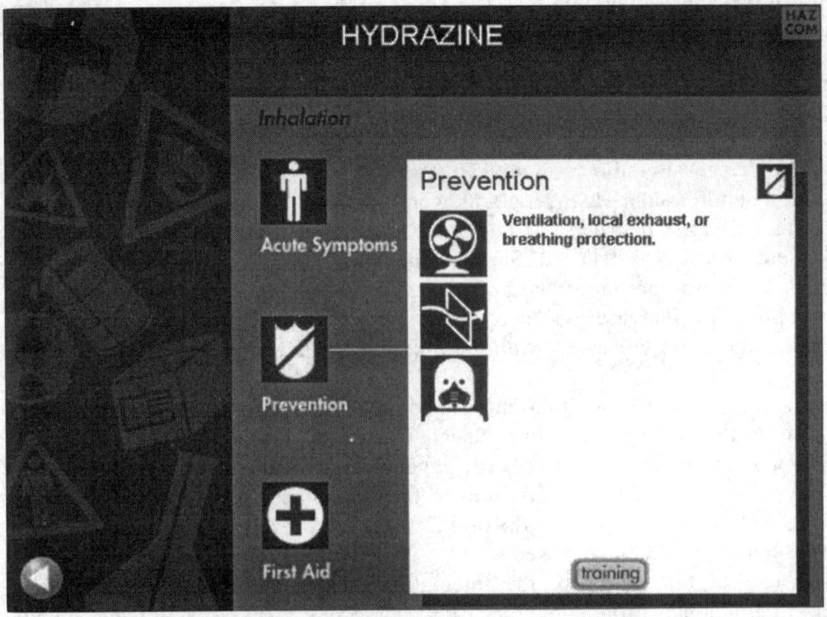

Figure 1. A *HazCom Interactive* inhalation exposure prevention screen with training button that links to specific supplemental training modules.

4 TARGETED TRAINING MODULES

HazCom Interactive supplements text with graphical symbols as shown in Figure 1. Games have been developed to reinforce the meanings of symbols. In one game twelve random pictographs from the total collection are displayed in randomly selected slots. An example opening-screen from the game is shown in Figure 2.

As the game starts, a phrase corresponding to one of the pictographs pops up in the window at the top of the screen, for example "drowsiness". As the words appear the user chooses what they think is the appropriate pictograph and clicks on it. If the selection is correct, the pictograph disappears and a new phrase pops up. A timer challenges the user to move as quickly as possible. If the user hasn't completed the game in 60 seconds, the game stops and the player is asked to play again. If the game is completed under 60 seconds, the player is congratulated and offered the opportunity to either play again to further improve his score or to exit the game. This is a simple, but highly interactive game and in presentations it is enthusiastically received.

The Matching Game shown in Figure 2 directly supports the *HazCom Interactive* database program by increasing users' recognition of the pictographs. It also supports use of pictographs on printed posters, cards and labels. However, it is very tightly linked to the *HazCom Interactive* database and isn't very suitable for use in more general training programs. To further increase use of the database, example supplemental training modules were produced that directly link to and from the database.

An example of how a primarily text based module can be linked to the *HazCom Interactive* database is shown in Figure 3. In this Ventilation training module several widely believed "myths" are presented as true/false questions. After each question is answered an example that demonstrates the correct answer is displayed using text and graphics while a voice-over reads the text. In Figure 2 the text includes a reference to the fact that "carbon monoxide, found in the exhaust from gas-powered compressors has killed many painters." In the program the text phrase carbon monoxide is hyper-linked into the *HazCom Interactive* database where more

Figure 2. An example opening-screen in the Matching Game used to train users in recognizing the pictographs used in the *HazCom Interactive* database.

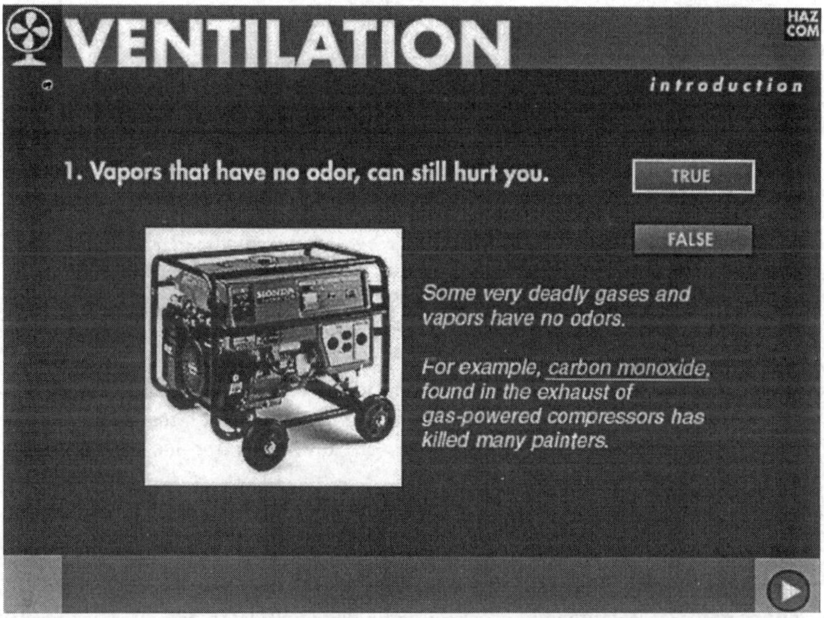

Figure 3. An example screen from a Ventilation training module that includes a reference to carbon monoxide that is hyper-linked to the *HazCom Interactive* database.

detailed information on carbon monoxide (CO) is available. The user has the choice to access the additional information as needed. Ideally, training modules of this type could be developed

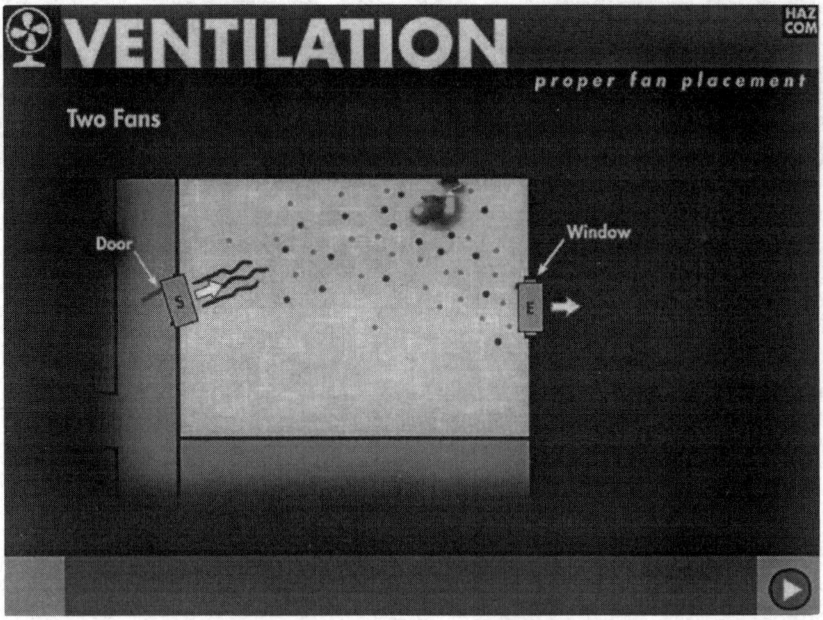

Figure 4. A specific training module on properly placing supply and exhaust fans to minimize solvent concentration in a worker's breathing zone.

to include many references and exercises directly linked to the database.

In addition to being able to link from training modules to *HazCom Interactive*, it is also possible to link from the *HazCom Interactive* database to a targeted training module. For example if hydrazine were going to be used on a site and its inhalation exposure hazards were being reviewed it could offer a direct link to training as shown in Figure 1. The training button could link to a menu of modules or straight to a specific module, for instance the carbon monoxide *HazCom Interactive* entry might link straight back to the true/false quiz described above.

The Prevention measures described for avoiding inhalation exposure to hydrazine (shown in Figure 1), reference ventilation and breathing protection as two requirements. An example of a specific interactive training module that describes inlet and exhaust fan placement to maximize ventilation is shown in Figure 4. The module is presented as a simple game. Initially, both fans and the painter are placed outside the room. The user is asked to drag and drop the exhaust fan and then the supply fan into their proper locations by using the mouse. The dots are two colors, blue for good air and red for solvents. They move around in the room symbolically indicating the fans' effect on local solvent concentrations. Once the fans are positioned, the user is asked to place the "painter" in the correct area to work. If the user tries to position the worker in the wrong location, the "painter" will not drop.

The fan placement module shown in Figure 4 is concerned with ventilation. The inhalation exposure prevention advice for hydrazine also suggested that breathing protection was needed. Another training module covers personal protective equipment (PPE) including breathing protection devices. This module presents the user with a work scenario, such as "You will be spray painting an indoor room with no windows and only one door with latex paint." The user is asked to dress and properly equip him for doing the job. The computer screen shows a supply locker with various PPE categories. The user selects a category and is offered options. In Figure 5 the user has already chosen hearing protection, clothing, shoes and gloves and is preparing to choose from various respiratory protection devices.

Once the user has completed his selections, he can try to go to work by clicking on the arrow to the left. If the user has selected at least minimally acceptable protection then he is congratulated and the game concludes. If the user has made any unacceptable choices then he is offered an

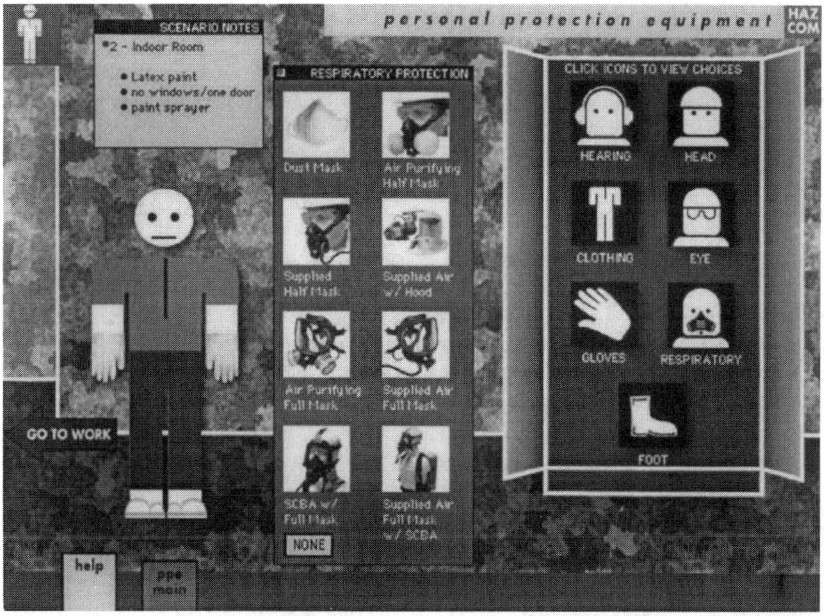

Figure 5. A screen from the PPE game showing the choices of respiratory protection that the user must select from to be suitably equipped for a particular job.

explanation of why particular selections are inadequate and sent back to the locker room to fix the problems. Obviously, the worker shown in Figure 5 will need something sturdier than flip-flops.

5 CONCLUSIONS

HazCom Interactive has all of the advantages of other electronic database systems. Nothing is ever misfiled. The same information can be accessed from many different directions (menus). Version control can be insured. It takes up very little space and is cheap to replicate and distribute. The primary disadvantage is that users need computer hardware to access the information. As computers proliferate that's unlikely to be a serious constraint. The project is expected to continue for one more year during which time its usefulness and practicality are expected to steadily increase until it is delivered to and in use on construction worksites.

The system's navigational schemes are intended to allow users to find specific information quickly (and in a nonlinear fashion). The use of graphics and audio files is targeted to expanding access to semi-literate users. The print-on-demand capabilities are intended to expand the impact into the physical workplace. The system also creates a broad foundation on which training programs can be developed in an environment where a critical mass of information can be found.

Although it is a powerful database, *HazCom Interactive* is a fairly static product. The training modules described above have the potential to integrate the database into more regularly used materials. During the last phase of this project the training modules will be presented to audiences of workers and evaluated for their effectiveness.

6 REFERENCES

Adult Literacy in America: National Adult Literacy Survey. 1993. Office of Educational Research and Improvement. Washington, DC: U.S. Department of Education.

International Programme for Chemical Safety. 1991. *International Chemical Safety Cards.* Commission of the European Communities, Telecommunications, Information Industries and Innovation. Luxembourg.

Jones, P. 1996. Computer-Based Hazard Communication Program Delivery for the Construction Industry. *Proceedings of Implementation of Safety and Health on Construction Sites.* (Alves Dias and Coble). 157-166. Rotterdam: A.A. Balkema.

Karr, A. 1997. Top 10 OSHA Violations: Same Old Tune?. *Safety + Health: Vol. 156(6).*

Occupational Safety and Health Administration. 1993. *The 100 Most Frequently Cited OSHA Construction Standards in 1991: A Guide for the Abatement of the Top 25 Associated Physical Hazards.* U.S. Department of Labor, Washington, DC.

Rabin, A.T. *1992.* Consequences of Adult Illiteracy Reach Far into the Marketplace. *Occupational Health and Safety,* June, pp. 36-42.

Sattler, B. 1992. Rights and Realities: A Critical Review of the Accessibility of Information on Hazardous Chemicals. *Occupational Medicine: State of the Art Reviews,* April-June, pp. 189-96.

Szudy, E. and M.G. Arroyo. 1994. *The Right to Understand: Linking Literacy to Health and Safety Training.* Labor Occupational Health Program, University of California at Berkley.

Implementation of Safety and Health on Construction Sites, Singh, Hinze & Coble (eds)
© 1999 Balkema, Rotterdam, ISBN 90 5809 036 1

Hazard-prediction training (KYT)

Keiichi Kubota
Zero-Accident Promotion Department, Japan Industrial Safety and Health Association (JISHA), Tokyo, Japan

ABSTRACT: The Zero-Accident Total Participation Campaign or ZA campaign is a unique venture that was launched by JISHA in 1973. The ZA campaign takes the form of group participation involving every employee of a company.

1 UNDERLYING THINKING IN KYT

1.1 *What is KYT?*

1. KYT is training to predict hazards and ensure the safety and health of workers.
KYT is an acronym of "Kiken Yochi Training" (hazard prediction training). Its purpose is to ensure industrial safety and health before starting action.

• KYT activities at a work site are usually carried out in small groups.
The group has a leader, who is called various names, such as "shokucho," "kumicho," "hancho" or "boshin" (all mean "foreman").

• Small-group members hold brief meetings (called by the leader) to discuss such activities at various times in the day; day's work starting time, time of arrival at the work site, the time of starting each job, during work and at the close of the day's work.

In order to prevent accidents due to human errors at the work site and ensure workers' health and safety on a daily and hourly, it is necessary to upgrade the team activities for hazard prediction. The upgrade will be aimed at detecting, comprehending and eliminating hazards that lurk at a work site and in a job; sharpening each worker's sensitivity to hazards; increase power of concentration and zeal to resolve problems.

• As part of the zero-accident campaign, KYT is carried out either by all the workers at a work site as a group activity or by each worker separately, and as "short-time training for the elimination of hazards." That is, "short-time hazard prediction training" serves the purpose of "ensuring safety and health to prevent accidents and vocational diseases."

• In order to prevent an accident due to a human error, it is necessary for each worker to detect a hidden hazard and confirm safety through "finger point and call" at each main point in the work process. Incorporating "finger point and chant" and "finger point and call" is integral to a successful KYT program.

2. KYT
KYT is training conducted by:

• using an illustrated sheet that depicts the work site and the job.

• making workers actually work by using real tools or the foreman himself showing an example of actual work.

• discussing, studying and understanding through open and critical discussion in a small work site group the possible hazards and conditions that create those hazards, i.e. actions or unsafe conditions that may cause a labor accident, that lurk at the work site.

• pinpointing the hazard point and confirming the target of safety and health through "finger point and chant" and "finger point and call" either in unison or individually, and

• ensuring safety and health before starting action.

KYT is a method to ensure work site safety before starting any action by bringing together information about hazards, discussing and sharing such information among the members of a group, pinpointing the point of the hazards, and confirming the target of safety and health ensuring work activity. On an individual basis, one should sharpen one's sensitivity to hazards, and input that awareness into the subconsciousness so it becomes second nature to locate and eliminate the hazards which will allow a safe work environment. Retrieving the input from consciousness so one subconsciously carries "point and call" to eliminate hazards is a vital part of KYT.

A human being undergoes the process of learning since birth. As a result, human behavior becomes mostly automatic. He or she judges and acts unconsciously, and his/her body reacts unintentionally. In other words, a human being becomes strongly dominated by one's subconscious. Therefore, it is important to send information about hazards into the subliminal area; transmit such information into the subconscious, in order to be able to prevent hazards in one's second nature. Thus, retrieving information about hazards into the consciousness at each important stage of work is vital.

The information driven into and formed in the subconscious does not fade away easily. Even a wrong piece of information, once accepted as true and inputted into one's mind, makes one's body react unconsciously. In the case of KYT, however, it provides the basis to comprehend the cardinal points of hidden hazards and sets the target of hazard prediction activities. It can never be wrong because it is based on a large amount of data brought together by people, i.e. , professionals. who thoroughly know the job or work site. Therefore, it is important to drive the cardinal points of hidden hazards and the target of hazard prediction activities into the subliminal area of one's mind until one reacts unconsciously and spontaneously, thus making one unconsciously use "finger point and call" at each important stage of work.

1.2 *Goal of KYT*

1. One knows, one can, but...

One has the necessary knowledge and skill. Likewise, one knows the necessary measures to be taken, and the necessary knowledge and skill to carry them out. But if one fails to carry them out, an accident will occur.

So one must ask, "Why didn't one do what one knew one must do and what one could do?" The following three reasons are conceivable:

• One's sensitivity was dull, so one did not notice the hazard as a hazard, thus failing to take the appropriate action. To prevent that, one's sensitivity must be sharpened.

• One was careless or absent-minded, so one failed to take the necessary action. To prevent that, one's power of concentration must be strengthened.

• One didn't have the willingness to prevent a hazard from the very beginning, so one failed to take the necessary action. To prevent that, one's willingness and zeal for practice must be stirred.

2. Strengthening one's sensitivity

KYT sharpens one's sense to feel a hazard as a hazard, that is, the sensitivity to hazards.
If one feels "it's hazardous," "something is wrong" or "there is a smell of a hazard" when one is faced with a hidden hazard, this is the sensitivity to a hazard.

Originally, humans had a very keen sensitivity to hazards. With the rapid progress of material civilization, however, humans became incapable of responding to hazards quickly enough, because their sensitivity to hazards had dulled or was lost. The purpose of KYT is to sharpen that sensitivity and restore it to the original state.

With only one or two KYT sessions a month, it is impossible to maintain one's sensitivity to hazards at a high level for a long time. Therefore, it is necessary to conduct a short KYT activity every day and at every important stage of work in order to maintain the workers' sensitivity to hazards at a high level.

3. Strengthening the power of concentration

In KYT, workers' power of concentration is strengthened by finger "point and call" and "finger point and chant" either as action of each individual or as a group action chanted in unison at each important stage of work. Its purpose is to prevent accidents due to negligence, absent-mindedness or carelessness.

"Each important stage of work" contains a point where a hazard is lurking. By focusing on the point where a hazard is lurking, a worker who implements "finger point and call," increases his/her power of concentration, thus preventing accidents due to human error, e.g., day-dreaming, carelessness, short-cutting. All of these human errors are based on human nature. This is the purpose of KYT.

The "finger point and call" helps one shift one's consciousness to the normal clear level. Uttering the action target in unison and "touch and call," too, also strengthen the team's power of concentration and bolsters its sense of unity and solidarity.

4. Stimulating the zeal to practice

Through frank debate over hazards, KYT stimulates each worker's zeal to practice hazard prediction activities.

Both hazard prediction activities and "finger point and call" produce the desired result only when they are practiced voluntarily, not under coercion.

Willingness and zeal are generated from frank debate over the following questions, "what are hazards?" and "what kind of hazard is lurking and where?" The process of comprehending the condition of a hazard and catching its nature through frank debate produces a strong zeal to practice hazard prediction activities.

Based on a four-round method, KYT generates a strong consensus for practice through a very short (3- to 5-minute) frank debate.

5. Creating a desirable work site environment

The ultimate goal of KYT is creating a bright zero-accident work site where all possible hazards are forecast and prevented, and everybody participates part in the KYT drive.

From a long-range viewpoint, KYT does not mean that safety only is important and other things, for example, the efficiency of production or the quality of products may be sacrificed for safety. Sensitivity which enables one to predict hazards and teamwork which ensures work site safety through prediction naturally contributes to self-initiated resolutions of all other problems.

Based on the above thinking, JISHA has developed a four-round method of hazard prediction training in order to enable one catch hidden dangers before one actually encounters them. Of the four rounds, the first round is particularly important as a method to detect factors of a hazard. Here we will explain it in detail.

2 HOW TO CATCH "THE HAZARD FACTORS" AND DESCRIBE THEM IN THE FIRST ROUND

2.1 *Place yourself in the position of the worker in the illustrated sheet*

In the first round, place yourself in the position of the worker in the illustrated sheet. Look at the sheet as if you were doing the job.

2.2 *Combine "the hazard factors" and "the phenomenon" in time sequence*

The hazard factors are "because the situation is such as such," and the phenomenon is "such and such a thing occurs" or "such and such a thing is done."

2.3 *Describe the "phenomenon" as a "type of accident."*

The "phenomenon"; the hazard which may be caused by a "hazardous condition" and/or "hazardous action" in doing one's job, should be comprehended and described as a "type of accident" as follows:

Fall; tumble; collide with something or somebody; hit one's leg against something; be hit in the head by something; be squeezed; be caught (in a machine); be pinned under something; cut one's hand; suffer abrasion in the hand; suffer a burn or scald; get an electric shock; etc.

• Such expressions as "may be," "there is the hazard of," are "there are fears of" are unnecessary.

• There is no need of mentioning the result of an accident, such as "an injury (a sprain or bone fracture)" or "death."

2.4 *Delve deep into the "hazard factors"*

[Example] Situation: A man is going down the stairs, carrying something.
You would imagine the hazard of "missing one's step and falling."
Why miss one's step?
Because the man cannot see the step so he misses the step and falls.
Why can't he see the step?
Because he is holding something with both hands, he cannot see the step.
The underlined sentences above describe the "hazard factors" that must be detected in KYT.

How to catch the hazard factors and describe them

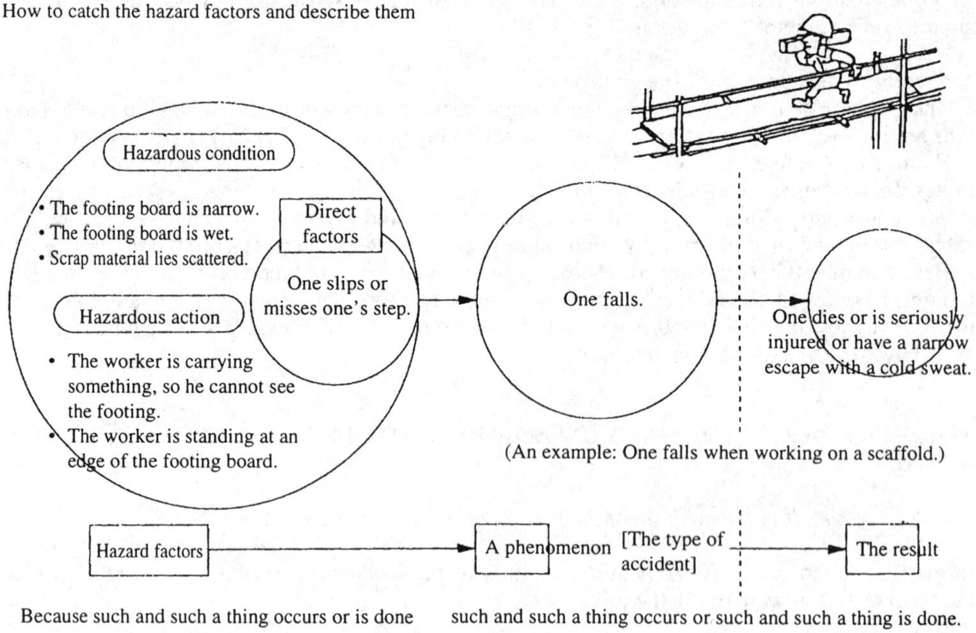

<Figure 1>

2.5 *Describe the "hazard factors" in concrete terms*

You must describe the "hazard factors" in concrete terms to make the point clear and make yourself better understood.

[Example] If you use such abstract expressions as "unnatural posture", "unstable" or "bad," nobody can understand what is unnatural, unstable, or bad, and how.

Therefore, you must say, for example, "because one is in a half-sitting posture" or "because one is standing on tiptoe."

2.6 *Describe the "hazard factors" in a positive, affirmative way*

[Example] Negative sentences like the following are likely to be used:

"Because one is not wearing a safety belt," "because one is not wearing protective goggles" or "because the footing is not fixed."

But such expressions do not reveal the real nature of the hazard. Therefore, you should use a positive, affirmative expression and say, for example:

"Because one is leaning forward," "because one's face is too near to XXX" or "because the footing is not in the right place and is unstable."

A model sheet
< What type of hazard is lurking >

Four basic rounds of KYT
Polishing the door with sandpaper

Situation
You are polishing the door of emergency stairs in order to paint part of the door.

<Figure 2>

No.	Polishing the door with sandpaper

KYT model report

Time	
Place	

Team no. Subteam	Team's name	Leader	Recorder	Reporter	Announcer	Commentator	Other member of the team

1st round (What hazards are lurking?) 2nd round (This is the point of the hazard.)	
○◎	No. Visualize the hazard factors and the phenomenon (the type of accident) and write as "Because..., such and such a thing will happen."
○	1. The door, fanned by a gust of wind, suddenly closes, and the left hand is nipped in by the door.
◎	2. The footstool is near the handrail, and the waist's position is high, so if the man staggers and loses balance, he may fall over the handrail.
	3. If the door is fanned by a wind and moves, the footstool will shake and the man may fall (hitting the handrail).
	4. If the man tries to change the position of his feet while using the sandpaper, he may step out of the footstool and fall.
○	5. When the man is polishing the door, with the door closed, he may fall if the door is suddenly opened by someone from within.
◎	6. Because the man's face is too near to the door, the powder of the sandpaper may enter his eyes.
○	7. If he misses his footing on the footstool and accidentally kicks the paint can, the can may fall and hit a person walking below.
	8.

3rd round (What would you do if you were in the man's position?)
4th round (We would do such and such a thing if we were in the man's position.)

The number marked with ◎	* mark	Concrete steps to be taken
3	*	1. Move the footstool nearer to the wall.
		2. Place the footstool inside the open door.
		3. Put on a safety belt and hook one end of it on the handrail.
		4.
		5.
6		1. Wear protective goggles.
	*	2. Work on the windward side.
		3. Keep the face away from the door and use the sandpaper at a position lower than the eyes.
		4.
		5.

Set the action target for the team. (Let's do such and such things.)	"Move the footstool nearer to the wall and work on the windward side. OK!"
[Confirmation] Item to be confirmed by finger point and call	"Move the footstool nearer to the wall. OK!"

Comment by the boss (coordinator)

How to conduct the four basic rounds of KYT (the process in summary)

Preparations	Team of 5 or 6 members	Division of roles (leader, recorder, reporter, announcer, and commentator); distribution of vellum paper and reporting paper
Introduction	[All the team members stand up.] The leader makes them stand in a row, call the roll, greets and checks their health.	
1 R	[Comprehending the situation] What type of hazard is lurking?	The leader describes the situation aloud. "Hazard factors" and the "phenomenon (type of accident)" that may result are mentioned. "Because ..., such and such a thing happens," or "such and such actions are taken, and such and such a thing results." (At least seven items)
2 R	[Search for the essence of the problem] This is the point of the hazard.	(1) Items considered to be hazards are marked with ○. (2) Select more important ones in the number of items marked with ○. (About two items) → The selected items are marked with ◎ and underlined. = the point of the hazard (3) The point of the hazard is confirmed by "finger point and chant" in unison. The leader says: "The point of the hazard is ..., so we do such and such things, and the result is ... OK!" → All the members of the team follows the leader, saying: "so we do such and such things, and the result is ... OK!"
3 R	[Planning countermeasures] What would you do if you were in the situation?	Concrete and feasible countermeasures against the cardinal points of the hazards are conceived. Concrete and practicable countermeasures against each hazard. → (About three items) for each (Five to seven items altogether)
4 R	[Setting a target] We will do such and such things.	(1) Selecting the more important targets. (About two items) → The targets underlined or marked with * = priority targets (2) Priority items → The target for the team's action is set. (3) The target for the team's action → "finger point and chant" in unison. The leader says: "The target for the team's action... (when we do ...) let's do ... OK!" All the members of the team follows chanting in unison: "The target for the team's action... (when we do...) let's do... OK!"
Confirmation	(1) The item to be confirmed with finger point and call is determined. (One item) The leader says: "Finger point and call XX OK!" → All the members follow the leader by chanting in unison three times: "XX OK!" Three times (2) Touch and call → The leader says: "Let's go for zero accidents. OK!" → All the members follow chanting in unison: "Let's go for zero accidents. OK!"	
Announcement Comments	Two paired teams make comments to each other.	Announcer → read 1R through 4 R Commentator → comment on other team's comments.

REFERENCE

Zero-Accident Promotion Department, Japan Industrial Safety and Health Association 1986. *Zero-Accident Campaign Promotion Handbook*. Tokyo: Japan Industrial Safety and Health Association.

17 Hazard control and indoor pollution

Implementation of Safety and Health on Construction Sites, Singh, Hinze & Coble (eds)
© 1999 Balkema, Rotterdam, ISBN 90 5809 036 1

Robotics applications in hazardous waste sites

Gouranga C. Banik
Construction Department, Southern Polytechnic State University, Marietta, Ga., USA

ABSTRACT: The cleanup of hazardous waste sites is a challenging and complex problem that requires immediate attention for the safety of workers. Each waste site is unique, both in terms of the extent of contamination and the media affected. The great challenge is finding solutions to cleaning up the environment by remediation of hazardous wastes sites and at the same time reducing health risks and improving safety of field worker. Application of robotic technology posses the unique opportunity to this need for its safe, fast and cost-effective operation. This paper addresses the special needs of robotics use, the safety concerns for hazardous remediation operation, the key characteristics of robots and the barrier of its use. This paper also provides suggestions so that the contractors will become encouraged to use more robotics in the near future for the hazardous wastes clean up.

1 INTRODUCTION

It is increasingly apparent that environmental construction markets in the USA and around the globe offer new and emerging business opportunities for the construction industry. Approximately 28,000 hazardous waste sites have been identified in the United States alone. To date, 1200 are on the Environmental Protection Agencies (EPA) Superfund list as because they posses the serious danger to the environment and the human health. (Superfund 1992). Each hazardous site is unique, in terms of the extent of contamination and the media (soil, water, groundwater, surrounding air) affected. The sites are unpredictable because conditions are often unknown prior to the start of remedial action. In addition, the remedial process will also change the site conditions by creating trenches and bulk wastes that in turn cause operation hazards (Andrews 1990).

In 1988, approximately 6000 spills of hazardous materials occur in the U.S., resulting in 15 deaths, over 150 injuries, and damages exceeding $20 million (U.S.DOT). Effective cleanup are difficult to implement, time consuming and expensive. A study by Frost & Sullivan, Inc. suggests that the total cleanup cost for these sites will be about $500 billion and may take 50 years (News 1992). Handling of hazardous waste sites requires a high degree of safety and personnel protection, even brief exposure to these dangerous conditions can have a devastating health effect on unprotected workers. The great challenge to the regulating agencies and engineers, scientists and contractors is finding effective solutions to cleaning up the environment without risking the health and safety of the hazardous waste site field worker.

This paper covers several important topics related to hazardous waste remediation: a) Benefits derived from using robots; b) Safety concerns dealing with current remediation methods; c) Robotic characteristics for hazardous environment; d) Robots which are currently used or under development; and e) Barrier regarding the use of robots on hazardous waste sites.

2 BENEFITS FOR USING ROBOTS IN HAZARDOUS ENVIRONMENT

The threat of hazards exposure during the investigation, mitigation and cleanup of hazardous substances require the acquisition and use of multi-million dollar inventories in protective clothing and equipment by regulating agencies and its Superfund constructors. This high degree of protection is required to implement the levels of personal protection associated with each degree of hazard (EPA 1985). Four such protection levels (Level A, Level B, Level C and Level D) according to the EPA, require different protective equipment and clothing, depending on the site-specific condition. Although the use of protective clothing and equipment provides a certain degree of protection against chemical exposure, it is not foolproof against all exposures. Also, protective clothing and equipment can be a source of potential hazard in itself because it may increase heat, reduce mobility, impair vision and balance, that can drain productivity and safety.

Robotics can also help in improved assessment of hazardous waste sites through automated data acquisition by precisely positioning subsurface imaging. Manual data collection and hazardous investigation are not optimal and generally inefficient, when the human manager is operating under the strain at a hazardous environment combined with cumbersome protection clothing (Esposito et al. 1993). The more tedious and routine a function, the greater the likelihood of the worker becoming bored which resulting in errors and possible accidents. Using robotics to perform tedious functions, such as ground water or air monitoring around the site, not only increases productivity and safety but also help site worker to concentrate on the more technical problems like data analysis and remedial solutions.

Health and safety considerations can increase costs from 10% to 700% over base construction costs for certain type of operations (Lippitt et al.). Furthermore, delays in performing critical tasks affect the entire project. Due to the probable exposure to the worker, short term and long term effect, the environmental cleanup contractors often don't find insurance companies, who can provide adequate coverage for their work (Atkinson 1990). Even if they find the insurance coverage, the insurance premium is unexpectedly high due to the high experience modification rate (EMR), resulting from the probable unsafe operation.

Considering the above safety, cost, and productive disadvantages for working in uncontrolled hazardous sites, robotics can be an effective and optimal option. Arditi (1990) mentioned from his 400 ENR contractor survey that automation and robotics would improve productivity (95%), speed (91%), and quality, safety, and competition (77-79%) in construction work. In this respect, the cleanup of hazardous waste sites provides a unique opportunity for the application of robotics technology. Robotics offers a substantive opportunity for safer, faster and cheaper field operations. Safer operations result from removing the worker from the direct proximity of the hazard and performing work through remote robotic manipulator. Field robotics have already been used during dangerous situation like Three Mile Island and Chernobyl nuclear reactor accidents, less disastrous but similarly hazardous situations encountered in nuclear industries, military operations, deep sea and space exploration. Equally important applications are possible in hazardous waste management and emergency and remedial operations at abandoned hazardous waste sites.

3 SPECIFIC SAFETY CONCERN DEALING WITH HAZARDOUS WASTES

Hazardous waste sites pose a multitude of health and safety risks, any of which could result in serious injury or death. The price for such injury and or death can not be substituted in terms of money. The major hazards encountered include: a) chemical exposure; b) biological (etiologic) hazards; c) fire and explosion; d) ionizing radiation; e) oxygen deficiency; f) electrical hazards; g) heat and cold exposure; h) noise (Martin et al. 1992).

Certain general categories of tasks associated with hazardous wastes remediation has been identified by the author that are impacted by safety concern, resulting lower productivity and higher costs. The most important tasks which are impacted by safety concerns are a) site characterization; b) waste sampling; c) barrel and container handling; d) excavation; e) decontamination; f) transportation of hazardous materials, and g) probable storage.

4 CHARACTERISTICS OF ROBOT DEALING WITH HAZARDOUS ENVIRONMENT

The most important characteristics required for mobile robots dealing with hazardous waste or harsh terrain, comprised of four or five groups of sophisticated capabilities: self-monitoring equipment, vision systems, external environmental sensors, attachments, and other functional devices. The key characteristics or capabilities for hazardous environment robotics, developed by ARD Corp., are presented below:

Rugged Construction
Waterproof
Modular Design
Easy Decontamination
Rugged Electronics
Remote Data Collection System
Remote Operator Console
User-friendly Operator Interface
Tracked Propulsions
Remote Sensing Subsystem
Pay-load Carrying Capacity over 500 lbs
Special Purpose Remote Camera
On-board Power Supply

To achieve these characteristics or capabilities, robotics components include but are not limited to temperature sensors, pressure gauges and recorders, humidity monitors, water intrusion alarms, hydraulic pressure output devices, electrical drain monitors, vibration monitors, neutron detectors, radiation detectors, acoustic sensors, touch or contact sensors, force sensors, proximity sensors, explosive detectors, gas or chemical fume detectors, optical reflective sensors, infrared scanners, and electromagnetic motion detectors. Many units employ television cameras: black and white, color, mini, and stereo, all for visual observation or video recording of data and information (Esposito et al. 1993).

A mobile robot may require to be equipped with one or more of the following, depending on specific environment. This partial list of the tools used for specific functions includes: drills, rakes, bolsters, pipe cutters, sand blasters, water cannons, water jet cutters, pyrotechnic chisels, swipe or smear samplers, radioactivity samplers, air samplers, X-ray analyzers, core boring samplers, spraying for applying water, vacuum suction equipment for sludge and dust, debris and liquid collection etc.

5 DESCRIPTION OF SOME PROMISING ROBOTS FOR HAZARDOUS ENVIRONMENT

Several mobile teleoperated or robotic vehicles have been designed, developed, assembled and operated for hazardous environment in the U.S. and foreign countries in last 15 years. Several authors (Esposito et al. 1993, Lovece et al. 1991) developed a list of robots that were commonly available and the most promising robots of their kind for dealing with hazardous wastes. There is a little coordination among the design activities associated with these units as because the developments are occurring independent of each other. Robotics and the automation industry is developing and changing so fast that it is very difficult to provide a current detailed list, which is in application or in development. The author considers the following robotic systems (partial description) which are already in use or potentially useful under hazardous environment conditions.

Spiral Track Autonomous Robot (STAR): STAR, designed and developed by Lawrence Livermore National Laboratory (LLNL), is an innovative and inexpensive vehicle that uses two screws in contact with the local environment to propel itself. With its low center of gravity, this vehicle can cross-treacherous terrains with sensors for detecting, dislodging, and detonating mines or other ordinance or for detecting hazardous wastes or toxins. The STAR is highly versatile, remotely controlled platform for its modular control architecture and diverse sensor packages. It is equipped with a control onboard electronic control system and wireless data/video links. The control system is responsible for high-level decision making, motion control, autonomous path planning, and execution. Ultrasonic sensors are mounted to provide collision avoidance capability during remote and autonomous operation (Perez & Grasz 1998).

Surveyor: The surveyor, developed by ARD Corporation, is radio controlled, teleoperated vehicle designed to perform remote inspection and surveillance tasks in both indoor and outdoor hazardous environments. The robot is equipped with three cameras, sound detection, radiation detection, lights, motor encoders, equipment diagnostics, temperature, humidity, vibration, pan/tilt for video and for arm, and air sampling. The surveyor operates with real-time three-dimensional vision and is human transportable. The tetherless surveyor's electric manipulator has a vertical and horizontal reach of 14.75 ft, a maximum lift capacity of 9.9 lb. This robot (ARD) is most suitable for this purpose because of its easy modification to an autonomous device that could perform routine monitoring tasks, and would require a minimum of effort to setup and control. Further, this platform shows good versatility of climbing up and descend down.

Scavenger XT 2000, XT 1000, XT 1100 and SS 100: Scavenger XT 2000, developed by ARD Environmental, is a remote controlled heavy duty cleaning system. It is explosion proof, and can be applied for chemical and petroleum industry where large accumulation of contaminant exists. It is fully integrated system, which include multiple tools as cutter head, jack hammer, descabblers, quick look vision system as on-board camera and a crane for deployment and retrieval. It is 4000-lb payload capable (ARD product page 1998).

ARD Environmental also developed scavenger XT 1000 and Scavenger XT 1100. Scavenger XT 1000 is a pneumatically driven and XT 1100 is a hydraulically driven machine. Both robotics systems removes hazardous materials at a rate in excess of 1100 gallons per minute, offer the right balance of efficiency and safety needed to clean today's environment. These system include a wide range of attachments as dredge augurs, bulldozer blades, jackhammer etc. and on-board camera (ARD product page 1998).

Scavenger SS 100 is a telerobotic system removes sludge and solids and contaminants by vacuuming dry or underwater surfaces. It can operate under 100-ft water. It has remotely controlled gripper system, brush cleaning attachments, and auger attachment for hard crust or heel removal.

Haz-Trak: Haz-Trak, developed by Kraft Telerobotics, is a remote-controlled, hydraulically actuated excavator that can be fitted with a bulldozer blade for grading, back-filling and levelling operations. This is most suitable for site remediation, maintenance and repair applications. It is equipped with two fixed-color camera and a pan and tilt mounted color camera. The operator sits at a station with three color monitors, provide a three dimensional view of the work environment. The maximum digging radius is 15.4 ft and its extended lift capacity is 1320 lb. By operating Haz-Trac, the operator is able to 'feel' what is being held by the system manipulator and modify its force of grip as needed for handling a variety of contaminated materials. For example, in handling deteriorating barrels of hazardous waste, the system can gently handle fragile containers. (Kraft 1991).

Stored Waste Autonomous Mobile Inspector (SWAMI): The SWAMI, developed by the University of Southern Carolina and the Westinghouse Savannah River Company (WSRC), is a mobile robot designed to perform autonomous inspection of nuclear and hazardous waste storage facilities. The onboard control system, consisting of three Motorola 68030-based microcomputers, controls a number of subsystem including barcode (6 readers), video (6 camera and a control camera), positioning (rotating laser and control computer), vehicle (itself), and radiation (detector panel, gas containers and control computer) (Fulbright & Stephens 1995).

Remote Work Vehicle (RWV): The RWV, developed by Carnegie Mellon University

(CMU), was designed for heavy recovery work at nuclear accident sites, as washing contaminated surfaces, removing sediments, demolishing radiation sources, applying surface treatment, packaging, and transporting materials. It is waterproofed, stainless-steel, four wheeled multidirectional and teleoperator controlled (via cable) system, and work continuously. It is also equipped with a radiator detector, lights, 10 video cameras, a pitch/roll sensor, a boom tilt sensor, pressure sensors, hydraulic pressure/temperature monitors, a current drain monitor, self diagnostics, a non power shovel, a pry bar, a sledge hammer, a hydraulic grinder, a saw, a water jet cutter, a scarifier, a liquid sprayer etc. It is equipped with a master/salve arm at the end of a tilting/telescoping boom and has a vertical reach of 25 ft and a horizontal reach of 12 ft. The maximum lift capacity of the boom is 1000 lb and of the extended arm is 75 lb.

Intelligent Mobile Sensing System (IMSS): The IMSS, developed by Martin Marietta Aeronautics, for the automated inspection of radioactive and hazardous waste storage containers at DOE sites (Byler et al. 1995). This robot comprised of the vehicle, mission sensors, and the control station. The purpose of the mission executive is to direct the integrated action of various vehicle functions to achieve an efficient and complete site inspection. The functions controlled by this module include mobility and scan platforms, 3-D vision module, color vision module, vision landmark navigation module and the bar code reader. This has the capability of obstacle avoidance by ultrasonic range sensors, geometric inspection, and corrosion inspection.

Advanced Robot Technology Research Association (ARTRA) Robot: The ARTRA robot, developed by Japan's ARTRA, is a highly versatile, sophisticated, remote-controlled demonstration robot aimed at replacing human tasks and is generally used in a radioactive environment (Takehara 1992). This robot is quadruped, capable of smoothly travelling in normal passages, walking up or down stairs, steeping over piping and thresholds, moving around appropriately as the work progresses, and standing up solidly and resisting reaction forces during work. A two-arm four-finger manipulator system provides for overhauling, inspecting and assembling a variety of similar objects using conventional tools and hoists, and also fastening or loosening bolts, attaching or removing parts, and handling soft objects of indefinite shape.

Gyro Runner: Aoki Corp., Japan has developed this system, which is a guiding system for small bore, curved geometry shield tunnels, capable of accurate positional detection and guiding. It is an autotracking total station, a laser gyro and an image processor, and performs accurate shield position/posture detection (Nozawa & Nishino, 1993). The selection of the robot was based on obtaining a readily available, commercial platform that was field compatible and would require a minimum amount of modifications.

6 BARRIERS TO IMPLEMENTATION OF AUTOMATION AND ROBOTICS IN HAZARDOUS REMEDIATION

Although there are many distinct tangible and intangible advantages in using robotics during hazardous wastes remediation, some barriers to the successful automation or use of robotics also exist. These are as follows:

Contracting Approach: Most government agencies in the United States use conventional construction contracting approaches as low bid/lump sum or unit price contracting formats. Although this contracting approach is applicable in most of the Government construction projects, but this approach may not be commonly applicable for hazardous waste remediation projects because of the large degree of uncertainty involved. The uncertainty is relatively high for innovative technologies because of the uncertainty of success in comparison to the conventional methods. As a result, contractors often prefer to use an approved and accepted method that may be less safe, and or more costly because contractual risk forced onto him by the lump sum and unit price contracting formats.

Liability and Risk Allocation Issues: Since remediation starts from feasibility analysis and characterization to the actual construction phase, and it is very uncertain where the exposure or accident may occur. As a result, it is very difficult to allocate the risk associated with specific operation. That means it is difficult to determine who will be responsible for what and in what extent. Although many of the robotics system developed for dealing with specific operation of remediation, but most of them are rarely used under field conditions. Due to the lack of field experience and the problem of risk allocation, contractors, owners and designers are reluctant to accept the risk involved in using these technologies. This leads to 'low tech' use, which ultimately results unsafe and costly operation.

Financial Issues: The construction industry as a whole is an uncertain and volatile market, and comprised of many small companies that find the initial capital required to invest in automated industry difficult. Most contractors always struggle to find enough work for their existence and future growth. The purchase, maintenance and operation costs for robotics are also relatively high. Contractors are reluctant to add another risk by procuring 'high-tech' technology to those already inherent in their industry. For this reason, contractors use low capital intensive technology although this is eventually costly and unsafe.

Robotics itself: Use of robotics is very specific which means certain robotics is suitable for certain operation of hazardous waste remediation. Robotics built for one purpose is difficult to use for another purpose efficiently due to its individual design approach. The device is generally complicated in design, heavy, difficult to operate, and relatively costly which hinders further applications. Many of the devices still need additional modification to specific site conditions considering contaminant characteristics, terrain etc. In the recent years, scientists and engineers are trying to change their design approach by coordinating development among themselves for multiple and economic use, which can eventually be implemented in the field.

Workers Resistance: Most robotic operation requires a high level of human judgement, intelligence and adaptation. Many of the experienced workers who are working for a long time in this industry exert strong resistance in introduction of robotics and automation (Navon & Johnston 1993). Workers, foremen and superintendents offer strong resistance to change due to the fact that they are busy and under pressure to try new methods even though they know that it would cut costs, improve safety, and increase productivity. These workers believe that their job may not be required with the increasing use of automation, and it will change the management-labor equation in terms of job security, control/power, career/promotion opportunities and new requirement.

7 CONCLUSIONS

There is a tremendous potential for the development of user-friendly, robotic devices for use in the clean up and remediation of hazardous waste sites. There is a greater possibility of using more robotics whenever the technology will become mature and cost effective with time. The following steps can be taken to broaden robotics use:

In order to become widely acceptable to the environmental contractor, the robotic system must be simple in design, easy to operate, require a minimum training and be ultimately less costly. For these reasons, it is necessary for the designer, the developer and the user is to work coherently instead of individually.

Insurance companies need to directly participate in the design and the development of robotics so that they better understand the advantages of using robotics in the hazardous environment. Whenever, the insurance company will realize its benefit in respect of worker safety and productivity than the conventional method, they will encourage contractors more to use this technology and will eventually reduce their insurance price.

The government should encourage private-public partnership and provide financial incentives for robotics research and development, and limit liability in the use of this new technology. Government should also be flexible in contract policy in some specific hazard situations. Government should accept bid, even in some cases when the bid price is slightly higher but deemed to be safer due to robotic use and automation for long-term industry interest.

This further possible robotic use in hazardous environment will eventually leads to a win-win situation for the government, contractors, researchers/engineers, workers, and the public in terms of worker safety, cost and productivity.

REFERENCES

Atkinson sells hazardous waste subsidiary 1990. *Engineering News Record (ENR)*, 224(28).

Andrews, L.P. (Ed.) 1990. Worker protection during hazardous waste remediation. Van Nostrand, New York.

ARD Environmental 1998. Brochure materials, Laurel, MD.

Arditi, D., S. Sundareswarm and A.E. Gutierrez 1990. The future of automation and robotics in construction activities. *Proc., 7th Int. Symp. on Automation and Robotics in Construction*, Bristol Polytechnic, Bristol, England 447-454.

Byler, E., W. Chun, W. Hoff and D. Layne 1995. Autonomous hazardous waste drum inspection vehicle. *IEEE Robotics & Automation Megazine*, 6-14.

Environmental Protection Agency. Field Standard Operating Procedure (FSOP) 1985. Number 4-Site Entry. *Office of Emergency and Remedial Response, Hazardous Response Support Division*, Washington, D.C.

Esposito, C., D. Sullivan, U. Frank and R. Cibulskis 1993. Field applications of robotics systems in hazardous waste site operations. *Robotics and Remote Systems for Hazardous Environments, Jamshidi, M. and Eicker, P. (Eds)*, PTR Prentice Hall, Englewood Cliffs, New Jersey, 1-31.

Fulbroght, R. and L. M. Stephens. 1995. SWAMI: An autonomous mobile robot for inspection of nuclear waste storage facilities. *Autonomous robots*, 2, 225-235.

Kraft TeleRobotics, Inc 1991. Brochure materials, Overland Park, KS.

Lippitt, J., J. Walsh, M. Scott, and A. Di Puccio. Costs of Remedial Actions at Uncontrolled Hazardous Waste Sites: Worker Health and Safety Considerations. *EPA No. PB 86-176 344/AS. National Technical Information Service*, Springfield, VA 22161.

Lovece, J.A., and Benson-Walker, G.E. 1991. Military robotics sourcebook: 1991-92 edition, L&B Limited, Washington D.C.

Martin, W.F., J.M. Lippit and T.G. Prothero 1992. Hazardous Waste Handbook for Health and Safety. Butterworth-Heinemann, Boston.

News Update 1992. *Environment Protection*, 3(4), 8.

Nozawa, T., F. Sato, and N. Nishino 1993. The Development of "Gyro Runner" a Totally Unmanned Shield Guiding System. *Automation and Robotics in Construction X.. Walson, R.L. Tucker and J.K. Walters (eds)*, Elsevier, Amsterdam, The Netherlands.

Perez, M. L. and Erna L. Grasz 1998. Using Intelligent Automation and Manipulation to Tame the Hostile Environment. *Proceedings of the 3rd speciality conference and exposition on Robotics for Challenging Environments, Demsetz L.A. R. H. Byrrne and J.P. Wetzel (eds), ASCE,* Albuquerque, NM.

Navon, R., Paul W. Kelly and David W. Johnston 1993. Human factors in introducing on-site construction automation. *J. Constr. Div., ASCE,* 119(4), 801-812.

Superfund site list growing. 1992. *Engineering News Record (ENR)*, 228(7).

Takehara, K. 1992. Japan's ARTRA robot moves forward. *Nuclear Engrg. Int.*, 3f(1), 39-40.

U.S. Department of Transportation (USDOT) 1988. Annual report on hazardous materials transportation, Calendar Year, 1988.

Implementation of Safety and Health on Construction Sites, Singh, Hinze & Coble (eds)
© 1999 Balkema, Rotterdam, ISBN 90 5809 036 1

Hazard surveillance in new residential construction

M.M. Methner
National Institute for Occupational Safety and Health (NIOSH), Cincinnati, Ohio, USA

ABSTRACT: Construction workers suffer elevated disease and death rates. However, with the exception of lead and asbestos, little exposure data among this population have been documented. With this in mind, an observational study aimed at identifying the potential hazards associated with each phase of construction was conducted. Each phase consisted of multiple tasks, each with its own potential for exposure. Observations were made by an industrial hygienist (IH) who visited four different residential construction sites and observed and documented specific work practices and recorded all chemical substances used during each task. The information gathered and examined during this study indicate that a substantial potential for exposure to hazardous agents exists for some workers involved in building a new home. In general, a lack of appropriate/effective personal protective equipment was noted for all trades. Potential ergonomic hazards were noted for 63% of the trades observed. Other hazards associated with different trades are presented in Table 1.

1 INTRODUCTION

Estimates of the number of workers employed in the construction industry range from 6-8 million (BLS 1997). However, these estimates include all workers for many different types of construction. Of particular interest is the new residential construction industry (Standard Industrial Classification (SIC) Code 1521). This interest stems from the fact that virtually no exposure data exists for workers in this sector of the construction industry. According to the Bureau of Labor Statistics (BLS), the construction industry leads all other industries in the number of occupational injuries and illnesses (BLS 1993; Zimmerman, et.al. 1997). The average employment in the residential building construction industry (SIC 1521) in 1995 was 608,600 with an incidence rate of 8.6 total cases of occupational injuries and illnesses (BLS, 1993). These rates were based on logs recorded by employers and are probably influenced by changes in economic activity, working conditions and work practices, worker experience and training, and the number of hours worked. In the past, research to identify occupational hazards associated with construction work has been conducted primarily in the "general" construction sector of this diverse industry. However, workers involved in residential construction have historically been understudied. This may be due, in part, to the fact that most of the residential construction work is often performed by small subcontractors who employ non-union workers. BLS surveillance data suggest that smaller contractors have greater fatal injury rates but lower nonfatal rates when compared to larger contractors (BLS 1993). Since small contractors who often do not have health/safety or monitoring programs comprise a large portion of all contractors engaged in residential construction, the need for hazard identification and task-based exposure assessment studies specific to this industry becomes very clear.

According to the U.S. Department of Labor's Occupational Safety and Health Administration (OSHA), the number one reason for a citation at a construction site was the failure to have a written hazard communication program (OSHA 1997). Other violations frequently cited by OSHA include; lack of adequate fall protection (45%), electrical (20%), and scaffolding (15%). Overall, OSHA citations are more aligned with safety issues than health issues. To further support this trend, an analysis of the OSHA citation data available for construction indicated that health-related citations account for only 4% of the top 100 citations (OSHA 1997). Recently, a pilot program called HomeSafe was initiated in the Denver, CO area in response to 7 deaths that occurred in residential construction. This program focuses on a 10 point list of potential significant hazards on residential construction sites with the ultimate goal of educating workers and reducing injury rates (Gilkey et.al. 1998; Fowler 1997).

Each phase of construction usually consists of multiple tasks, each with its own potential for exposure. Therefore, adopting a task-based exposure assessment strategy is necessary to properly characterize exposure (Susi and Schneider 1995). The purpose of the preliminary study presented here was to identify potentially hazardous physical and chemical exposures among workers involved in the construction of new residential dwellings. The general phases of residential construction observed and evaluated were: excavation, foundation pouring, framing/carpentry, roofing, plumbing, electrical, drywall, painting, interior finishing, and carpet/flooring.

2 METHODS

This study involved the use of a qualitative exposure assessment method (observational) which consisted of an IH visiting four different home sites during various phases of construction. The construction sites consisted of different types of homes (single level and two story). A field data sheet (survey instrument) was developed for the purpose of recording observations of work practices and other qualitative data during the site visits. The survey instrument allowed the collection of information pertaining to a specific phase of construction, self-reported job description, actual tasks performed, and the potential chemical and physical exposures. Once on site, the IH observed workers engaging in their normal job tasks and recorded all products used during each task. Later, information about the materials used was collected from Material Safety Data Sheets.

3 RESULTS

A total of 30 different tasks involving 19 different trades were observed and evaluated for potential exposure to various chemical and physical agents and are briefly outlined here. Table 1 serves as a job-potential exposure matrix according to a given trade. The main potential hazards associated with operators of excavation machinery (e.g. bulldozer, backhoe) for the excavation phase of construction were: vibration, noise and exposure to diesel exhaust. The foundation pouring phase included the pouring of footers, foundation walls, damp proofing the concrete walls, rough plumbing, back fill, and pouring of interior concrete floors. The potential exposures for concrete technicians included ergonomic (awkward bending, pulling concrete), dermal (Portland cement), and respiratory (asphalt emulsion, concrete curing product). The plumbers installing the underground rough plumbing were potentially exposed to the chemicals used to clean and seal the joints of the PVC piping (Methyl ethyl

Table 1: Potential Exposure Hazards Associated with Residential Construction Trades

	RESIDENTIAL CONSTRUCTION TRADES																		
			EARLY PHASE						MID-PHASE						FINAL PHASE				
POTENTIAL	Bulldozer/backhoe	Concrete	Damp-roofing	Plumber	Carpenter (Frame)	Carpenter (Trim)	Roofer	HVAC	Electrician	Drywall Hanger	Drywall Finisher	Insulator	Painter	Siding Applicator	Gutter Applicator	Mason	Floor Prep Tech	Floor Installer	Carpet Installer
Ergonomic	X	X		X	X	X		X	X	X	X		X					X	X
Noise	X				X													X	
Diesel Exhaust	X																		
Portland Cement		X														X		X	
Petroleum Hydrocarbons			X	X															
Mineral Spirits		X	X																
Asphalt			X																
Methyl Ethyl Ketone				X											X				
Acetone				X	X						X								
Tetrahydrofuran				X															
Cyclohexanone				X															
Aromatic Naphtha		X																X	
Calcium Carbonate					X					X	X		X			X	X		
Silica (crystalline)					X		X			X	X				X	X	X	X	
Fibrous Glass					X							X					X		
Hexane					X						X								
Toluene					X													X	
Wood Dust					X	X											X		
Falls					X		X						X	X					
Cuts/Abrasions								X						X					
Gypsum										X	X					X	X		
Aliphatic Petrol. Distillates													X		X				
Ethylene Glycol													X					X	
Volatile Organic Compounds													X						

ketone, acetone, tetrahydrofuran, cyclohexanone).

The carpentry phase involved both framers and trim carpenters. The potential exposures observed for the framers were ergonomic (hand tool vibration), respiratory (silica from marking chalk, acetone, hexane, toluene from sub-floor adhesive), and a safety issue (lack of fall protection). The potential exposures for the trim carpenters primarily involved respiratory exposure to wood dust. The roofing phase was observed to have the greatest potential for falls. Most of the workers wore a body harness, but were not tied off to the roof safety hook. The rough mechanics phase included the trades of Heating, Ventilation, and Air Conditioning (HVAC), Plumbing, and Electrical. The potential exposures for the HVAC technician were mainly safety issues. The potential for injury due to cuts resulting from contact with the sharp edges of the sheet metal were observed during the formation and installation of intake plenums and duct work.

The rough plumbing phase involved the installation of PVC piping and copper water supply lines. The potential exposures observed for the plumbers were mainly due to chemicals present in the materials used to connect the piping. Potential exposure to agents in the PVC pipe cleaner and cement were discussed earlier. The paste flux used in the connection of copper pipes for the water supply lines contained zinc chloride, ammonium chloride whereas the solder contained tin/copper/silver alloy.

The potential exposure for the electricians were ergonomic; the fine hand manipulation required may increase the potential for repetitive-motion disorders or cumulative trauma disorders (CTD). Insulation of the exterior walls was accomplished by first caulking all joints. The caulking compound, a siliconized acrylic caulk, contained aliphatic petroleum distillates and ethylene glycol. The fiber glass insulation contained fibrous glass wools and phenol formaldehyde urea polymers.

The drywall phase involved the installation of gypsum drywall panels by drywall hangers and the subsequent "finish" process performed by drywall finishers. The drywall hangers' potential exposures were respiratory (gypsum dust, silica) and ergonomic (lifting panels). The drywall finishers applied different joint compounds to achieve a smooth finish on the walls. These workers' potential exposures were dermal (vinyl acetate monomer), respiratory (dust), and ergonomic (overhead sanding).

The painting phase involved potential exposure during the application of caulking compound and paint. The caulking compound, a siliconized acrylic caulk, was discussed earlier. The painters were observed to use their bare fingers to smooth the caulk. Painting in an enclosed room with little or no ventilation could result in exposure to Volatile Organic Compounds (VOC's).

The "finish mechanics" phase included the final stages of electrical and plumbing installations. The potential exposures for the electrician were similar to those identified for the "rough mechanics" electrical phase. The plumbers potential exposures were similar to those identified during the rough plumbing phase except for the connection of the PVC pipes underneath cabinetry, which may increase respiratory exposure to the PVC pipe cleaner and cement. The application of siding, gutters, and brick decor were performed during the exterior finish phase. The siding applicators were exposed to fall hazards (no body harness) and other hazards like cuts to the hands and arms from handling vinyl siding. The gutter applicators were observed to sustain injuries (cuts and abrasions) to the hands and arms while handling the aluminum coil stock used to form the gutters. Also, applicators applied a seam sealer to the joints of the gutter using a caulking gun and smoothed the elastomeric aluminum compound into place with their bare fingers.

The potential dermal and respiratory exposures for the mason laying brick occurred primarily during the preparation and application of the mortar mixture. The mason was observed to dump the powdered masonry cement into a portable mixer where a noticeable cloud of cement dust (silica) was released into his breathing zone.

The flooring/carpeting phase included the preparation of the sub-floor and the installation of resilient flooring, ceramic tile, and carpet. The potential exposure for the worker preparing the sub floor was observed to be respiratory (dust) that was re-suspended during sweeping. The potential exposure for the resilient floor applicator involved dermal and respiratory exposure (toluene, ethanol). The worker installing ceramic tile in the bath and shower area used a "wall base" adhesive containing naphtha and ethylene glycol. The grout used by tile installers contained crystalline silica, Portland cement, and titanium dioxide. Potential respiratory exposure to the dry mix was observed when bags of grout powder was dumped into a mixing vessel. The carpet installer was exposed to primarily ergonomic hazards; most of the tasks were performed while working on the hands and knees. The worker wore knees pads while kneeling to installing carpet. During this task, the knee area was subjected to acute impact force while using a "kicker" to stretch and secure the carpet.

4 SUMMARY AND CONCLUSIONS

There were several advantages and limitations of this observational study. The most important advantage of this study was the opportunity to observe residential construction workers performing their jobs. By observing work practices, valuable information was gathered about specific trades and their potential exposure to various chemical and physical agents. This information will be useful in guiding subsequent exposure assessments. Probably the greatest limitation of this study was the lack of participation by more than one homebuilder. Ideally, observations of construction processes would have been more objective if the study included the participation of more than one homebuilder.

Aside from one worker observed to wear safety glasses, leather gloves and a dust mask, virtually no personal protective equipment (PPE) was observed on-site. Often small contractors do not have the financial resources necessary to procure the appropriate PPE and issue these items to the workers. Based on hazard prevalence, professional judgement and the degree of hazardous product use, potential exposures that warrant quantitative sampling efforts are:

Bulldozer/Backhoe Operators - Noise, Vibration, Diesel exhaust.
Concrete Workers - Naphtha, Mineral spirits, Portland cement.
Asphalt Workers - Petroleum hydrocarbons, Asphalt, Mineral spirits.
Plumbers - Methyl ethyl ketone, Acetone, Tetrahydrofuran, Cyclohexanone.
Drywall Finishers - Total and respirable dust, Hexane, Acetone.
Painters - Ethylene glycol, Volatile Organic Compounds (VOC).
Masons - Dust (during the preparation of Mortar).
Floor Preparation Technicians - Total and respirable dust.
Ceramic Tile Installers - Toluene, Naphtha, Silica (from grout powder).

REFERENCES

Bureau of Labor Statistics: Survey of Occupational Injuries and Illnesses, 1995. Table 1. U.S. Department of Labor. Washington, DC (1997).

Bureau of Labor Statistics: Incidence Rates for Nonfatal Occupational Injuries and Illnesses Involving Days Away from Work. Tables R5 and R8. U.S. Department of Labor, Washington, DC (1993).

Fowler, J. 1997. Safety first - we did it our way. *Home Builder*. Denver, CO. HBA of Metropolitan Denver.

Gilkey, D. et. al. 1998. The HomeSafe Pilot Program: a novel approach to injury prevention in residential construction. *Work*: 10: 167-180.

Occupational Safety and Health Administration 1997. OSHA construction industry safety and health outreach program. Des Plaines, IL. US Department of Labor.

Susi, P.; Schneider, S.: Database needs for a task-based Exposure Assessment model for Construction. Applied Occupational and Environmental Hygiene 10(4):394-399 (1995).

Zimmerman, CL; Cook, TM; Rosecrance, JC.: Work-Related Musculoskeletal Symptoms and Injuries Among Operating Engineers: A Review and Guidelines for Improvement. Appl. Occup. Environ. Hyg. 12(7):480-483 (July 1997).

Implementation of Safety and Health on Construction Sites, Singh, Hinze & Coble (eds)
© 1999 Balkema, Rotterdam, ISBN 90 5809 036 1

Issues regarding indoor air quality and safety in residential construction

H. Bashford & K. Walsh
Arizona State University, Tempe, Ariz., USA

ABSTRACT: The desire to make homes more energy efficient has led to tighter construction which decreases air infiltration into the home. It is now understood that this condition can significantly reduce indoor air quality, decrease comfort, and in some cases reduce energy efficiency. Recent studies of 70 homes in Phoenix Arizona have shown that air which is drawn into tightly constructed new homes to replace exhausted air can carry significant amounts of pollutants. This paper discusses these issues as they relate to indoor air quality, comfort, and safety in homes.

1 INTRODUCTION

The efficient use of energy for heating and cooling homes became important in the US housing industry during the energy crises in the 1970's. Prior to that time, readily available sources of inexpensive energy kept energy efficiency as a low priority item. In the early 1970's, energy prices escalated rapidly. Home builders quickly recognized that one of the easiest and quickest ways to make homes more energy efficient was to reduce the amount of air infiltration into the home. The conventional practice which developed was to seal the exterior shell of the home as tightly as possible to minimize infiltration of unconditioned air from outside the house into the conditioned space inside the home. A new test, termed a "blower door" test was developed to measure the tightness of the building shell. A blower door test measures the amount of air infiltration given a known pressure difference between inside and outside, usually 50 pascals.

During this same time period, changes in the materials used to construct homes also began to occur. Prior to 1970 most of the homes constructed in the Phoenix Arizona area were constructed of masonry blocks (CMU). During the 1970's, home builders began to use light wood frame construction. By the early 1980's and continuing to the present time, almost all new homes in the Phoenix area are constructed using light wood frame construction.

2 THE IMPACT OF THE CHANGES

The processes involved in relationship to heating and cooling houses received almost no attention before 1985. Motivation to study these processes began to develop as reports began to be received about "sick house" and "sick building" syndrome, houses which were tightly sealed but were drafty, cold (hot in Phoenix Arizona) and uncomfortable. Records kept by utility companies indicated that the anticipated amount of energy consumption reduction which was expected from increased use of insulation and air sealing was not being realized (Treidler, 1996). Several researchers began to study these issues and quickly focused on the subject of how air moves in and throughout conditioned houses, which gave rise to a new area of inquiry, termed building science.

Researchers quickly found that some very basic principles of physics could be applied to houses, one of these being the law of conservation of mass. They found that tightly constructed wood frame houses seem to naturally operate at a negative air pressure when indoor air pressure is measured in relationship to outdoor air pressure. At first puzzled by this observation, they realized that a number of appliances and a natural phenomenon work to remove air from the house. These include clothes dryers, kitchen exhaust fans, bathroom exhaust fans, leaky air conditioning ducts, and fireplaces, to name a few. Also, stack effect naturally moves hot air to the top of the building. However, with all these methods in operation to remove air from the house, there is nothing which works to move air into the house. Thus, a typical house operates at negative pressure with respect to the outdoors. Air moves into the house because air naturally flows from an area of higher pressure (outside) to an area of lower pressure (inside, in this case). The volumes can be quite significant. A typical electric clothes dryer exhausts 250 cubic feet of air per minute (cfm) when in operation. A kitchen exhaust fan exhausts 300 to 400 cfm, a bathroom fan 30 to 60 cfm. A roaring fireplace can easily move 200 to 300 cfm. The question is, from where does the air come to replace this removed air?

Prior to the 1970's, houses were sufficiently leaky that replacement air was not a problem. Many paths for replacement were available: through or around windows, doors, and numerous other paths. However, as construction techniques focused upon building the exterior shell of the house tighter and tighter, there were fewer and fewer infiltration points for air to enter. Also, because of the growing use of frame construction with the resulting numerous interconnected wall and floor cavities, the actual points of infiltration could be located at a considerable distance from the source of the air. It soon became apparent that several of the most readily available sources of replacement air were sources which were not desirable, such as the garage, the attic, and the under floor area (Cooper, 1997).

In Phoenix, this problem is particularly acute. The typical home constructed in Phoenix has a concrete slab floor constructed on grade, and has wood frame walls with an exterior wall covering of field applied stucco. The floor and walls are virtually air tight. As a part of the Environmental Protection Agency Energy Star certification program, 70 homes have had blower door tests conducted in the past 18 months. The results of the blower door tests are shown on Figure 1. As shown on Figure 1, the homes predominantly perform at about .35 air changes per hour at 50 pascals of air pressure difference between inside and outside. The blower door tests are conducted by depressurizing the house in relationship to the outside. As the homes are depressurized, theatrical smoke is released in the house to

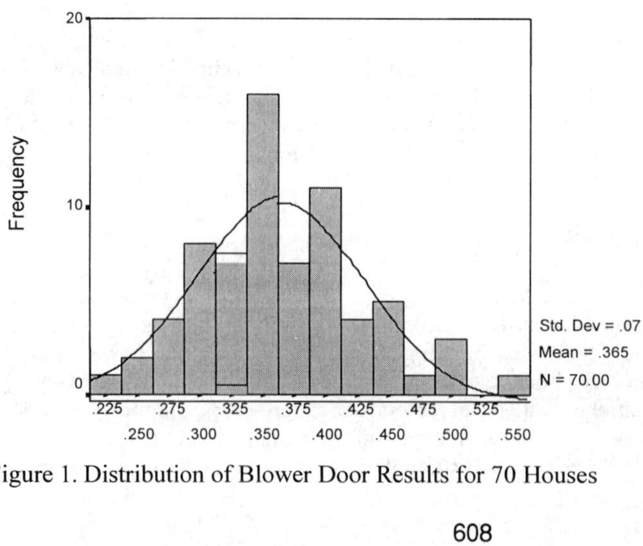

Figure 1. Distribution of Blower Door Results for 70 Houses

determine the source of the air leakage. The predominant source of leakage has been observed to be at recessed ceiling lights, back flow down fireplace chimneys, and leakage around wall penetrations which are connected to framing cavities. The primary issue associated with this is determining what is in the replacement air which infiltrates into the home.

3 AIR INFILTRATING FROM FRAMING CAVITIES AND CHIMNEYS

Framing cavities are generally connected into the attic and into the garage attic. Because little care is taken to seal the garage from the attic, air which has passed through the garage usually ends up in the attic, and hence in framing cavities and in the house. It is not generally recognized that framing cavities are so interconnected. Figure 2 provides a pictorial representation of this difficulty. The figure shows duct leakage of 320 cfm, and an electric clothes dryer exhausting 300 cfm from the house. The replacement air for the 620 cfm is coming through the flue for the domestic hot water heater, or from the garage attic through the interconnected floor joists.

Problems associated with air from these sources, particularly in Phoenix, relate to the temperature of the air and pollutants. If a car has been operating in the garage, high concentrations of carbon monoxide (CO) can be found in the infiltrating air. Mold and dust are frequently found in the attic. Attic temperatures in Phoenix can reach 150° to 160° F (66° - 71° C).

Downdrafts through chimneys, particularly flues for combustion gas appliances, bring combustion byproducts into the house. The byproducts contain substantial amounts of CO and water vapor. In extreme cases, the pressure differences can be substantial enough to cause flame rollout from the combustion chamber of gas fuel burning appliances (hot water heaters and furnaces).

4 RESULTS OF INFILTRATION

The products carried into the home by the infiltrating air can have a substantial effect upon the health and safety and the comfort level in the home. In the Phoenix area, air entering the home

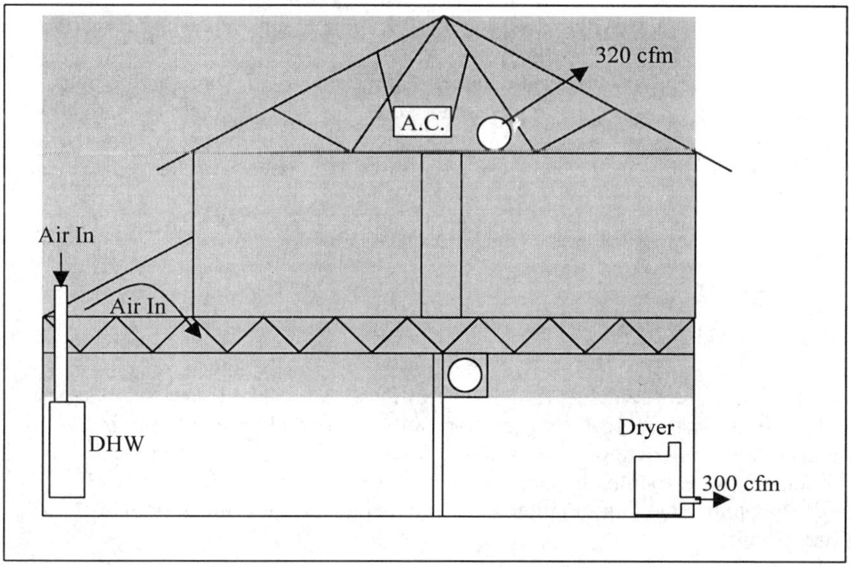

Figure 2. Pictorial Representation of Air Infiltration Locations

from the attic can be very warm, as much as 80° F higher than the air temperature in the home. This infiltrating air is dispersed through the home as it moves toward the point of discharge. Such warm air circulating through the home can result in significant temperature changes in some locations of the home with resulting discomfort. Small rooms with poor circulation are particularly susceptible.

Of course, the other concern is the build-up of hazardous materials in the home. Back drafting flues and air from the garage can contain CO, a poisonous gas even in minor concentrations (Lstiburek, 1998). Dust and mold are allergens. It is estimated that over 1/3 of the residents of the Phoenix area suffer from allergies and/or asthma. Inserting mold and dust directly into the home can aggravate these conditions.

Flame rollout from gas fuel burning appliances is an obvious safety hazard which can lead to a house fire.

The third area of concern is that of energy efficiency. Obviously, exchanging 80° F conditioned indoor air for 160° F attic air is not a desirable exchange. The normal design temperature for refrigerated air conditioning units is an outdoor temperature of 107° F, which is substantially lower than the temperature of attic air in the summer.

These concerns are aggravated by the relatively low natural air infiltration rate of homes in the Phoenix area. The American Society of Heating, Refrigeration, and Air Conditioning Engineers (ASHRAE) standard for residential mechanical systems provides that a minimum of 0.35 air changes per hour is necessary to maintain acceptable indoor air quality (ASHRAE, 1997). The average Phoenix home operates at this level, leaving no room for error.

5 FIXING THE PROBLEM

Several methods are available to control the difficulties associated with infiltrating air when using light wood frame construction. These include:

1. Be certain all framing cavities in light frame construction are sealed in such a manner that they do not interconnect with other framing cavities such as floors to walls, walls to attics, or garage attics to house attics or floor spaces. This involves attention to framing details not usually given.
2. Make sure all air conditioning ducts are tightly sealed, both supply and return, regardless of location.
3. Provide access locations for air to enter the home. All air exhausted from a home must be replaced. Controlling from where that replacement air comes will eliminate concern about what might be entering the home with the replacement air. Controlling where the air enters the home might eliminate much of the discomfort associated with the replacement air.
4. Use gas fuel appliances with sealed combustion chambers, especially in areas subject to depressurization.

6 CONCLUSION

The techniques available for solution to the air infiltration problem are very simple. Unfortunately, the air infiltration problems are not well understood or even recognized by most builders and certainly not by those who actually frame houses. This is compounded by a lack of recognition of the difficulty by the building codes. A large educational effort is necessary to bring this understanding to builders. The educational effort must involve at least the following two groups:

1. Utility companies, particularly those who provide natural gas. Utility companies frequently provide outreach and educational programs for builders. They have access and respect among the builders.
2. Trade school and community college educators involved in the education of builders.

7 REFERENCES

American Society of Heating, Refrigerating and Air Conditioning Engineers, Inc, 1997. *1997 ASHRAE Handbook*, Atlanta, GA: American Society of Heating, Refrigerating and Air Conditioning Engineers, Inc.

Cooper, K 1997. Opportunities for demand controlled ventilation. *Journal of Thermal Insulation and Building Envelopes*, 20 (1), 199-205.

Lstiburek, J. 1998. *Builder's guide:hot-dry & mixed-dry climates*, Minneapolis, MN: Energy Efficient Building Association.

Treidler,B. et. Al. 1996. Impact of residential duct insulation on HVAC energy use and life cycle costs to consumers. *ASHRAE Transactions* 102(1), 881-891.

Implementation of Safety and Health on Construction Sites, Singh, Hinze & Coble (eds)
© 1999 Balkema, Rotterdam, ISBN 90 5809 036 1

Indoor pollution by asbestos and man-made mineral fibers

Ingrid Senitkova & Nadezda Stevulova
Department of Civil Environmental Engineering, Technical University of Kosice, Slovakia

ABSTRACT: Asbestos has been used for many years in building products, mainly because of its excellent insulation properties. Although use of blue and brown asbestos has now been controlled or banned, white asbestos (chrysotile) continues to be used in some products, mostly in the form of asbestos cement. Man-made mineral fibers have similar physical properties and applications, and can superficially resemble asbestos fibers. Exposure to all forms of asbestos and Man-made mineral fibers can occur in buildings, especially if the products containing these fibers are disturbed, such as during home maintenance activities. Indoor asbestos pollution results from various sources, such as asbestos containing insulation products, building materials and heating and ventilation systems. Compared with other indoor air pollutants the concentration of asbestos fibers is very low. Asbestos induced tumours resulting from indoor exposures to friable asbestos are known but the extent of the risk to health from such exposure is still the subject of controversial discussion.

1 INTRODUCTION

The interest in the research of outdoor and indoor environment focused on the air quality and on the effect of air pollution upon health has been growing in recent years. The classic contaminants of the urban atmosphere are the oxides of sulfur and nitrogen, carbon monoxide, Tabasco smoke, house dust, asbestos, radon and volatile organic compounds. At this time, attention is focused on the major indoor air pollution sources because the tendency for an increased proportion of people's time to be spent indoors. This fact affects the control of the indoor air quality.

The industrial exploitation of asbestos began about a hundred years ago, its use peaking around the1960's. Because asbestos is a cheap and effective insulator, fire-retardant and reinforcing material, its use became widespread and it was incorporated into a wide range of products, some of which remain today in the fabric of buildings. The health effects of asbestos were initially unsuspected and workers in the industry were often exposed to massive levels of airborne fibers resulting in serious and often fatal lung disease. Similarly, asbestos-containing products were used in situations where fiber release was more than likely to occur, thereby exposing individuals to potential risk.

Asbestos is one of the components in numerous household products. Indoor pollution from asbestos can result for example, from the use in the home of appliances containing asbestos, such as hair-dryers, toasters and electrical systems, in particular certain types of the night storage heaters. Further hazards are associated with asbestos-containing heating and ventilation systems, thermal and acoustic insulation and vinyl-asbestos tiles. Other known sources of indoor pollution are building materials, such as friable asbestos lagging, sprayed asbestos insulation which becomes brittle, and asbestos fibers release from corroding or weathered asbestos cement sheets, textured paints etc. (Bignon, Doll 1989). Previously came the large-scale exploitation of asbestos. Asbestos is a term for a family of fibrous minerals with biological effects that depend primarily on physical dimension and perhaps also on chemical composition. The best known and the most widely used variety of asbestos mineral is called chrysotile (white asbestos) which is mineralogical a serpentine. The chrysotile crystal has a characteristic appearance. The ultimate fiber has a diameter of about 0,03-0,6 μm. Chrysotile has accounted for about 95 % of all applications in buildings. The others are all amphibole minerals distinguished by their stoichiometry: the commercially important of these is crocidolite (blue asbestos) and amosite (brown asbestos). The amphiboles seem to be far more toxic than white asbestos.

2 HEALTH EFFECTS

There have been many reviews of the health effects of asbestos and other mineral fibers. Occupational asbestos exposure has been associated with mesothelioma, lung cancer and pulmonary fibrosis (asbestosis). Amphibole asbestos is established as the most hazardous form, especially with regard to mesothelioma induction. Although it is widely accepted that there is need for sufficient asbestos exposure to cause frank asbestosis before there is any significantly increased risk of lung cancer (and that this is unlikely to occur in domestic situations), the position with regard to mesothelioma induction is less clear.

Professional interest in the health consequences of asbestos exposure was roused in the industrial countries in connection with reports pointing as a cause of lung fibrosis and its association with lung cancer and mesothelioma of the pleura. The health effects that turned people away from using asbestos were first seen in populations, which were occupationally or environmentally exposed asbestos dust. Mesothelioma as a "signal-tumour", demonstrating previous exposure to asbestos fibers at the workplace, is also regarded as a marker of non-occupational exposure to asbestos fiber dust (Gardner & Saracci 1989). The problem of non-occupational asbestos indoor air exposure as the topical problem was discussed at several indoor air international conferences as well as was reported in the workshops. Asbestos induced tumours caused by indoor pollution and by environmental influences as a result of the occurrence of asbestos or erionite fibers in the soil have been also reported recently (Schneider & Woitowitz 1996). In the UK at least, the incidence of mesothelioma in men has been reported to be increasing rather than falling in line with reduced asbestos usage, and it has been suggested that this is partly due to exposure to amphibole asbestos (especially crocidolite) in the building trades (Peto et al. 1995). This causes concern about domestic exposure to crocidolite and amosite asbestos, especially the higher levels which result from home maintenance activities.

A possible risk from asbestos containing building materials, including asbestos cement products in the immediate living area, e. g. in facades or ceiling panels was recognized. For the initiation of a formation cancer cells in the respiratory tract owing to the inhalation of air polluted with asbestos dust is necessary in order to asbestos fibres have critic dimensions (diameter < 2-3 μm, length > 5 μm; with respect of their ratio greater than 3:1). The threshold limit value of asbestos concentration must be overstepped in the long time exposure. It is now recognized that fibers longer than 8 μm and less than 0.5μm thick have the greatest carcinogenic potential. According to Slovak law of atmosphere asbestos minerals are classed as polluted substances with carcinogenic effect and belong to subgroup 1 of group.

Because some man-made mineral fibers it has been mooted that they may pose similar health risks. However, there is no direct human evidence that non-occupational exposure to man-made mineral fibers are hazardous to humans. The only exception concerns the irritant effects sometimes experienced after skin contact. There are a number of features which distinguish man-made mineral fibers from asbestos, durability or biopersistence is one such factor. Another is the propensity for asbestos fibers to split longitudinally to produce more thinner fibers whereas man-made mineral fibers shear transversely, giving rise to shorter fibers which are likely to be more readily cleared from the lung. Moreover, most man-made mineral fibers have a larger mean diameter than asbestos fibres, which makes them less respirable (Brown et al. 1997).

There are three important properties of fibers which go to determining their pathogenicity, namely dose, dimension and durability. As with any toxic agent, dose is of fundamental importance, although the dose-response relation for fibrous materials remains mostly undefined. The dimension of the fiber is important with respect to both the likelihood of reaching the lung and its biological activity at the target site. Increased attention is being given to fiber durability, or biopersistence, since this determines how long a fiber remains in the lung to exert any potential toxic effect. Fibers of similar size and shape can have quite different toxicities because of differing durabilities, and this may explain some of the observed differences in the pathogenicities of asbestos and man-made mineral fibers. Most man-made mineral fibers are much less durable than chrysotile asbestos, which in turn are less furable and therefore less biopersistent than crocidolite (Harrison & Brown 1997).

The greater toxicity of amphiboles in humans has been explained. Their long straight sharp fibers easily penetrate deep into the lung, while the soft curly chrysotile splits easily into smaller fibrils. This fragmentation, and perhaps also some chemical leaching, causes rapid removal from the lung undo prevents accumulation whereas the amphiboles remain in the lung. The risk of developing mesothelioma is proportional to fiber concentration and is exponentially related to time since first exposure, or alternatively: the increase in mesothelioma incidence caused by each brief period of exposure is proportional to the amount of that incremental exposure and to a power of time since it occurred, independent of age or history of smoking.

The latency for the development of mesothelioma is usually 20-40 years from time of initial exposure. The risk of mesothelioma increases with increasing cumulative exposure. All asbestos fiber types have been implicated as a cause of toumors with the risk being greatest for crocidolite. It is generally accepted that long-thin straight fibers, which are apt to split longitudinally (e.g. crocidolite) are more productive of mesothelioma than shorter curly fibers (e.g. some types of chrysotile).

Generally in Slovakia as in many other countries, there is a tendency towards decreasing the use of asbestos. A cohort of 408 workers exposed to chrysotile was studied through mining and processing the rock for more than 20 years. Only 2 cases of lung cancer and 6 cases of asbestosis has been observed. She concluded that the health risk of exposure to asbestos in this particular plant was relatively small. Across the country, ten years ago, only several cases of asbestosis and cases of lung cancer has been registred in Slovakia.

3 SOURCES

3.1 Main use of asbestos

Asbestos has been used in a wide range of building products, including roofing, cladding, thermal and acoustic insulation and fire resistant internal paneling. Studies of fiber release show that asbestos-containing boards and lagging can release a considerable quantity of fibers when disturbed. White asbestos cement products (particularly flat and corrugated asbestos sheets) account for the greatest volume of domestic usage of asbestos, in normal use these materials will release only very low levels of airborne fibers. However, asbestos cement sheeting has a long life, often over 50 years, and as it weathers fiber release can increase, leading to gradual releases into the environment (West 1997).

3.2. Main use of man-made mineral fibers

The major use of man-made mineral fibers in homes is in loft and cavity wall insulation. Energy conservation drives in the last decade or so have resulted in very widespread use of such products; in Slovakia, for example a lot of loft insulation are installed. The products used are mainly composed of glasswool or rockwool and are able to release respirable fibers if disturbed. Some acoustic insulation products and other boards used in buildings may also contain Man-made mineral fibers, and it has been shown that home maintenance operations such as sawing and drilling can release fibers from these products (West 1997).

4 LIMIT CONCENTRATIONS

The limit value in non-occupational indoor environment is still not established. The original admissible concentration limit established in the previous Czechoslovakia for dust containing more than 10 % asbestos in asbestos plants 2 $mg.m^{-3}$. Consistent with the world-wide trend towards fibre coating, the limit dust standard 1 $fibre.cm^{-3}$ was also adopted in Slovakia. The table 1 shows different values for maximum allowable asbestos concentration in various countries for working environment (Bagel 1997, Dobrovsky 1998). The emission limit value for asbestos fibers concentration in the air was established 0.25 $f.cm^{-3}$ (0.2 mg . m^{-3}) in the Slovak Republic.

5 MEASUREMENT

The intensity of asbestos emission from containing materials as well as its concentration in the indoor air dependents on the degradation of surface structure of asbestos material. In the case of damaged or partly repaired, demolished or deformed units, the release of asbestos fibers is observed. Asbestos fiber is disposed to the fibrillation that leads to the formation of fine particles (asbestos dust) which are dangerous toxic pollutants. A risk from building occupancy arises from inhalation of airborne fibers of suitable dimensions from variety of building products. Some selected asbestos containing building materials are presented in Table 2.

The presently preferred method is transmission electron microscopy with direct transfer of the filter surface to the microscope, thus causing minimal disturbance of the spatial arrangement of particles. The study using this method was carried out in lot of buildings. The data are summarised in Table 3 and with corresponding outdoor samples There can be some discussion whether results should be expressed as average because about 80 % of samples contained no asbestos structures. However it can be say that in these samples the number of fibers were so low as to be below the detection limit.

These results correspond with the most comprehensive study carried out by the Environmental Protection Agency in 50 buildings with extensive statistical analysis published recently showed

Table 1. Maximum allowable asbestos concentration for working environment

Country	Maximal allowable concentration [$fibers.cm^{-3}$]
Canada	0.04
USA	0.1
Finland, Sweden, Denmark	0.5
France, Switzerland, Czech Republic, Slovak Republic	1
Belgium	1 (015 for crocidolite)
Austria, Italy	1 (0.5 for crocidolite)
Japan, Netherlands	2 (0.2 for crocidolite)

Table 2. Percentage of asbestos containing building material

Product	Asbestos percentage
Floor tite and sheet	20 - 30
Cement pipe	20 - 30
Paper pipe wrap	70 - 80
Sprayed, trowelled insulation	50 - 70

Table 3. Concentration of airborne asbestos

Sites	Fibers per Milliliter
administration buildings	0,00082
public buildings	0,00046
schools	0,00073
outdoors	0,00041

that was no significant difference between indoors and outdoors or among the three categories of buildings (with the various damage degree of asbestos containing material) (Cesson et al. 1990). Because of the low fiber concentrations, indoor exposure to asbestos has been, at best, classified as a minimum risk to health. The Environmental Protection Agency at the present time requires bulk sampling and visual inspection of asbestos containing material in building before mandating the need for repairs or reconstruction.

6 DISCUSION AND CONCLUSION

Asbestos-containing materials in poor condition have the greatest propensity to liberate airborne fibers. Man-made mineral fibers in normal use generally produce much lower levels and their inherent toxicity is much less. Indoor situations where asbestos products are exposed and in poor condition, and especially where such materials are inadvertently worked on during home maintenance activities, give rise to highest exposures and most concern. The calculated number of fibers inhaled during a lifetime through "environmental exposures" may appear very high, but this is moderated by the continual processes of clearance and dissolution that take place in the lungs. Moreover, the levels of airborne fibers in buildings are very significantly less that those which have arisen in industrial situations and which have unambiguously been associated with disease.

Existing asbestos in good condition must, where practicable, be managed in situ. Where asbestos materials are damaged and fiber release is occurring, removal is only justified if this would result in a reduction in exposure.

Interference with asbestos-containing materials through home maintenance activities can result in significant intense exposures for a short time. Prudence dictates that such exposures should be avoided. However, unlike exposure to those professionally engaged in working with asbestos, the frequency of such activities is likely to be low, and hence so is the cumulative exposure.

Most currently available man-made mineral fibers products do not readily release airborne fibers, few of any fibers which are released will reach the deep lung and those that do will not persist. Thus the bulk of current man-made mineral fibers products pose little risk. However, vigilance is needed and the properties of new fibers should be monitored. The production and use of fine diameter (<1 μm) fibers should be discouraged, especially where this could give rise to environmental exposure.

While for the individual any health risk is trivial, man-made mineral fibers have become widespread throughout the industrial and built environment. The presence of a vast installed mass of these fiber means that their deterioration into a more dusty condition could cause some exposure during maintenance or demolition of the existing building stock. This exposure would be to relatively thick fibres so that irritancy to skin, eye and upper respiratory tract would be the most likely health outcome. Nevertheless, unnecessary exposure to any fibre should be avoided.

7 ACKNOWLEDGEMENTS

The authors are grateful for supporting as a part of research work on Grant Project of Civil Environmental Engineering Department.

REFERENCES

Bignon, J. 1989. IARC Sci. Publ. 90: 3-29.

Doll, R. 1989. IARC Sci. Publ. 90: 511-518.

Gardner, M.J., Saracci, R. 1989. UARC Sci. Publ. 90: 375-397.

WHO: 1990. Task Group on Indoor Air Quality, Inorganic Fibers and other Particulate Matter. Kingston: 24-28.

Schneider, J., Woitowitz, H.J. 1996. Indoor Built Environment 5: 67-75.

Law No 309/1991 (Law of Atmosphere) in its version No 218/1992.

Dobrovolsky, L. 1998. Indoor Built Environment 7: 122-123.

Bagel, L. 1997. ES 5: 31-36.

Acta hygienica, epidemiologica et microbiologica, enclosure 13: 1986.

Cesson, J., Hatfield, J., Schultz, B. 1990. Environ. Res. 51: 100-107.

Brown, R.C., Harrison, P.T.C., Pigott, G. et al. 1997. Health effects. Proc: IEH (1997) *Fibrous Materials in the Environment*. SR2. Leicester, Institute for Environment and Health.

Peto, J., Hodgson, J.T., Matthews, F.E. et al. 1995. Continuing increase in mesothelioma mortality in Britain. Lancet 345: 535-539.

Harrison, P.T.C. & Brown, R.C. 1997. Discussion and risk assessment. Proc.: IEH (1997) *Fibrous Materials in the Environment*. SR2. Leicester, Institute for Environment and Health.

West, J. 1997. Sources oaf asbestos fibers. Proc.: IEH (1997) *Fibrous Materials in the Environment*. SR2. Leicester, Institute for Environment and Health.

West, J. 1997. Sources of man/made mineral fibers. Proc.: IEH (1997) *Fibrous Materials in the Environment*. SR2. Leicester, Institute for Environment and Health.

Implementation of Safety and Health on Construction Sites, Singh, Hinze & Coble (eds)
© 1999 Balkema, Rotterdam, ISBN 90 5809 036 1

Assessment of hazards for telecommunication tower construction workers

T.J. Lentz
Education and Information Division, National Institute for Occupational Safety and Health, Cincinnati, Ohio, USA

V.J. Casini
Division of Safety Research, National Institute for Occupational Safety and Health, Morgantown, W.Va., USA

ABSTRACT: Growth in the telecommunications industry has been accompanied by concerns about the safety of workers who erect towers to hold transmitting devices and broadcast antennas. These concerns are validated by injury statistics, including 123 tower-related deaths between 1980 and 1994. Researchers at the National Institute for Occupational Safety and Health have been assessing hazards of tower erection while investigating tower-related fatalities. Objectives of the assessment have been to identify hazards and high-risk activities, to promote hazard awareness, and to recommend strategies for their prevention. Activities which contribute to the risk of falls, such as failure to comply with existing standards and the improper or non-use of personal protective equipment, and other factors including exposure to radiofrequency radiation, tower instability, equipment and procedures used to hoist materials and personnel, and inspection and certification of hoist systems were identified for consideration in developing a strategy to minimize hazards during tower construction.

1 INTRODUCTION

The telecommunications industry has experienced rapid growth, fueling the construction of towers to hold transmitting devices for cellular phones, personal communication services, and television and radio broadcast antennas. The Federal Communications Commission (FCC) estimates that no fewer than 75,000 towers have been constructed in the U.S. (Chiles 1997), and industry groups indicate that nearly 25,000 cell sites alone had been constructed through June 1996 (OSHA 1998). As provisions of the Telecommunications Act of 1996 are implemented, the tower construction industry is expected to experience additional growth to accommodate the demand for wireless communication services. This growth has been accompanied by concerns from industry personnel, health and safety professionals and regulatory agencies for protecting the safety and health of tower workers.

In 1997, the Occupational Safety and Health Administration (OSHA) formed a task force to evaluate the hazards of tower construction and develop compliance directives for ensuring the safety and health of tower workers. These directives will provide compliance officers with guidelines for inspections involving tower access, fall protection, and gin pole use. Researchers at the National Institute for Occupational Safety and Health (NIOSH) have been assessing the hazards of tower erection in association with tower-related fatality investigations. Objectives of the NIOSH assessment have been to identify specific hazards and high-risk activities, to promote awareness of these hazards, and to provide recommendations for their prevention. Surveillance activities have included the use of national databases to identify fatalities in tower construction. Additional information has been obtained through collaboration with the OSHA tower task force.

2 INJURIES AND FATALITIES IN TOWER ERECTION

Concern for the safety of tower workers is validated by injury statistics for the tower erection industry. Accident records from the OSHA Integrated Management Information System (IMIS) indicate a total of 108 tower-related accidents (including 89 fatal injuries) occurred between 1984 and 1997 (OSHA 1998); the U.S. Bureau of Labor Statistics (BLS) Census of Fatal Occupational Injuries (CFOI) for the years 1992 through 1996 lists 45 fatalities associated with work on towers (BLS 1997); and a search of the NIOSH National Traumatic Occupational Fatality (NTOF) surveillance system revealed 123 tower-related deaths between 1980 and 1994 (NIOSH 1998a).

Data from the Cellular Telecommunications Industry Association indicate that the number of cell tower sites in the United States has increased from 346 in January 1985 to an estimated 24,802 sites in June 1996 (OSHA 1998). The number of workers involved in tower construction and maintenance work has been estimated to be between 2,350 and 22,560 based on 1993 data collected by the U.S. Bureau of Labor Statistics (OSHA 1998). Because workers who perform this type of work are classified under multiple standard industrial classification (SIC) codes (OMB 1987), it is difficult to determine the exact number of tower workers.

Estimates from the trade group the National Association of Tower Erectors, representing roughly 330 member companies, indicate that the industry currently averages between one and two tower-related deaths each month. This estimate exceeds the annual average determined from data in the NIOSH NTOF surveillance system, which is closer to eight deaths per year for the years 1980 to 1994. Using the National Safety Council figure of $790,000/death in 1995 dollars (NSC 1997), direct costs to the industry from these preventable deaths have been estimated between $6-19 million annually, excluding costs of pain and suffering and other indirect costs.

2.1 *Tower-related fatality investigations*

NIOSH researchers in the Division of Safety Research (Morgantown, WV) perform Fatality Assessment and Control Evaluation (FACE) investigations in response to requests for technical assistance following an occupational fatality which has occurred in one of several participating states in the Southeastern United States. The goal of the FACE investigations is to prevent future fatal work injuries by studying the work environment, the worker, the task the worker was performing, the tools the worker was using, and the role of the employer in controlling other factors which may have been involved. Within the past year, at least three FACE investigations were performed in response to tower-related deaths. Details of these investigations, and one additional tower accident with multiple fatalities, are summarized below.

Case one

On May 15, 1997, a 38-year-old male tower erector/inspector was working with a co-worker connecting antenna-support brackets to a leg of a tower about 67 meters (220 feet) above the ground. The erector/inspector had apparently disconnected or was attempting to reconnect his fall protection equipment and was climbing down the tower to a lower work station when he fell 61 meters (200 feet) to the ground and was killed as a result of injuries sustained from the fall (NIOSH 1997).

Case two (not investigated by NIOSH)

On October 23, 1997, three tower workers (aged 26, 37, and 44 years) were performing minor strengthening tasks on a 31-year-old tower. The workers were positioned at the 457 meter-level (1,500 feet) of the 610 meter-tall (1,999 feet) television transmission tower, and all were using proper fall protection equipment according to a company spokesman, when the tower twisted and the upper 602 meters (1,975 feet) fell to the ground killing all three workers. The exact cause of the tower collapse had not been determined (ENR 1997).

Case three
On December 8, 1997, a 32-year-old male tower erector was working with a crew of two others on a 49 meter-tall (160 feet) cellular phone tower. The crew had completed the tower erection and was in the process of lowering the gin pole (the lifting device used to lift tower sections into place) to the ground. The tower erector had removed two choker cables securing the upper section of the gin pole to the tower and was attempting to ride the hoist cable down to the two lower chokers when the hook on his lanyard slipped off the cable and he fell approximately 40 meters (130 feet) to the ground and was killed by the impact (NIOSH 1998b).

Case four
On March 29, 1998, a 26-year-old male tower erector was working with a crew of five persons erecting a 76 meter-tall (250 feet) cellular phone tower. The worker and a foreman were on the tower working their way down from the top securing flexible electrical supply cables to two of the tower legs. The tower erector and the foreman were free-climbing down diagonal and horizontal tower cross-members on the three-sided tower from the 44 meter-level (145 feet). At the 38 meter-level (125 feet), the tower erector was holding onto a horizontal tower member as he was reaching to secure his positioning lanyard to the tower. The foreman, who was three meters (8-10 feet) from the tower erector, heard him yell and saw him fall 38 meters (125 feet) to the ground, on the way striking a metal electrical switch box and its concrete mounting pad. The tower erector was killed from the impact (NIOSH 1998c).

3 IDENTIFICATION OF HAZARDS

Falls from elevation were determined to be the leading hazard of tower construction; 65-90 percent of tower-related fatalities involved workers falling off the tower structure (OSHA 1998; NIOSH 1998a). The falls ranged in height from six meters to 457 meters (20-1,500 feet). Of those cases (n=47) in which fall height is indicated, nearly 43 percent of falls were from less than 30 meters (~100 feet), and 83 percent were from less than 92 meters (~300 feet).

Multiple factors which contribute to the risk of falls have been identified including: environmental conditions of extreme heat and cold; slippery and uneven work surfaces at elevation; unstable tower structures; electrocution; and potential exposures to radiofrequency radiation. In order to accommodate the demand for tower construction and maintenance services, most tower erection companies and their employees work throughout the year under all weather conditions. Although high winds and precipitation will usually postpone tower work at elevation, construction schedules require many tower workers to continue to work in conditions of extreme heat or cold. Climbing and transitioning on a tower under these conditions places additional stresses on a tower worker, which may result in fatigue or repetitive strain injuries.

The OSHA tower task force, in response to guidance requests from the National Association of Tower Erectors, is drafting a compliance directive which may allow tower erectors to use a procedure for hoisting workers to work stations on the tower under specified conditions. The directive is intended to strengthen the requirements for testing and certification of tower hoisting systems as well as to establish procedures for hoisting employees to reduce the possibility of fatigue, musculoskeletal disorders and the risk of falls. The OSHA task force is also drafting a second directive to establish procedures for the inspection and certification of the gin pole, the temporary hoisting device which is attached to the side of a tower and used during construction to raise successive tower sections into position. The condition of the gin pole and other components of the hoist system, and ensuring that the hoist system is rated for the weight and type of load being lifted (e.g., rated for personnel lifts) are important considerations. Performing activities related to the gin pole, which are an integral part of the tower construction process, have also resulted in fatalities as in *Case three* above.

Exposure to electrical energy and radiofrequency radiation are also potential hazards of tower construction. In many cases, it is possible to eliminate these hazards with an appropriate lock-out program for de-energizing equipment during work on a tower. However, with the co-location of multiple antennas and broadcast devices on a single tower, as well as the grouping of multiple tower structures in areas known as tower farms, it is not always feasible to isolate and de-energize all possible energy sources. Exposure to nonionizing radiation from radiofrequency sources on towers, either by working or climbing in close proximity to an active broadcast antenna or by touching an energized source such as an AM broadcasting tower (which can actually become energized as a receiver of radiation from an adjacent source), is a primary hazard that can cause shocks or burns which might also result in falls.

Fall protection equipment, if used correctly during work on towers, can greatly reduce the risk of tower-related fatalities. In three of the case studies cited above, the tower workers were equipped with fall protective equipment that could have prevented their fatal falls if they had been anchored to the tower. The OSHA tower task force will address the issue of fall protection in drafting a third directive; it is proposed that tower workers should be instructed to maintain 100 percent fall protection during tower construction. One hundred percent fall protection (as defined by the draft directive for tower access) is defined as follows: every employee at risk of falling from work levels greater than about two meters (six feet) above the ground or working surface must be protected by some conventional means of fall protection, which may include an integral fall-arrest system. This applies to ascending, descending, moving point to point, or any tower construction or alteration work activity conducted at an elevated work station. Workers should also be instructed to maintain a minimum of three-point contact (two hands and one foot or two feet and one hand) at all times.

Approximately ten percent (n=12) of the tower-related fatalities identified in the NIOSH NTOF surveillance system were associated with tower collapse while workers were on the structure. In cases such as these, the proper use of personal fall protective equipment does not ensure the safety of the tower worker anchored to the falling tower structure. The possibility of tower collapse suggests a need for additional engineering and inspection guidelines to determine the structural integrity of the tower before work at elevation is performed on the tower.

4 CONCLUSIONS

Activities which contribute to the risk of falls, such as failure to comply with existing standards and the improper or non-use of personal protective equipment, and other factors including exposure to radiofrequency radiation, tower instability, equipment and procedures used to hoist materials and personnel, and inspection and certification of hoist systems comprise the pertinent issues to be considered in developing a strategy to minimize hazards during tower construction. Based on reviews of occupational tower fatalities reported in the NTOF surveillance system and NIOSH investigations of several recent tower deaths, researchers at the NIOSH Division of Safety Research have concluded that employers must continually ensure that workers are provided with appropriate training and equipment for working on telecommunications towers. For each job site, it is important that the employer and employees develop work plans and perform a job hazard analysis of the site prior to starting work to ensure employees' knowledge of new equipment or procedures to be used. Tower workers should be trained on safe climbing procedures, and proper work practices should be enforced, including 100 percent fall protection using equipment that is appropriate and maintained in good condition. Based on the possibility for fatigue with tower climbing, workers should be instructed to take sufficient rest periods during climbs. While these recommendations in general pertain to fall hazards for tower work, other hazards such as exposures to electrical energy and radiofrequency radiation must also be considered. Prior to work at elevation on an existing tower, the structural integrity of the tower should also be determined, preferably by a registered professional engineer.

These NIOSH recommendations and the compliance directives developed by the OSHA tower task force address an obvious need for guidance to protect workers in the tower erection industry. NIOSH will continue to evaluate the effectiveness of this guidance and the need for further intervention with the erection of tall towers up to 610 meters (2,000 feet) for television and high-definition television broadcast, and as many as 100,000 new towers smaller than 122 meters (400 feet) for the next generation of digital cellular communications systems (IBD 1997).

REFERENCES

Bureau of Labor Statistics (BLS) 1997. *Census of Fatal Occupational Injuries, 1992 - 96.*

Chiles, J. 1997. We got us some sky today, boy. *Smithsonian*. July: 44-52, 130.

Engineering News-Record (ENR) 1997. Toppled tower creates mystery. November 3: 14.

Investor's Business Daily (IBD) 1997. Where do you put 100,000 cell phone towers? May 29.

National Institute for Occupational Safety and Health (NIOSH), Division of Safety Research 1997. *Tower erector/inspector dies after falling 200 feet from a telecommunications tower to the ground -- North Carolina (FACE 97-10).*

National Institute for Occupational Safety and Health (NIOSH), Division of Safety Research 1998a. *National traumatic occupational fatality surveillance system, 1980-94.*

National Institute for Occupational Safety and Health (NIOSH), Division of Safety Research 1998b. *Tower worker dies after falling 130 feet from hoist cable to ground -- Pennsylvania (FACE 98-02).*

National Institute for Occupational Safety and Health (NIOSH), Division of Safety Research 1998c. *Tower erector dies after falling 125 feet from cellular phone tower -- South Carolina* (FACE 98-07).

National Safety Council (NSC) 1997. *Accident Facts, 1997 Edition.*

Occupational Safety and Health Administration (OSHA) 1998. *Profile of the Wireless Telecommunications Industry and the Telecommunications Tower Industry (Contract No. J-9-F-4-0013).*

Office of Management and Budget (OMB) 1987. *Standard Industrial Classification Manual.* Springfield (VA):National Technical Information Service.

18 Subsurface safety engineering

Implementation of Safety and Health on Construction Sites, Singh, Hinze & Coble (eds)
© 1999 Balkema, Rotterdam, ISBN 90 5809 036 1

Safety analysis of rock excavations

Michael J.Connolly
Engineers and Surveyors of Hawaii, Honolulu, Hawaii, USA
Amarjit Singh
Department of Civil Engineering, University of Hawaii at Manoa, Honolulu, Hawaii, USA

ABSTRACT: The problem of evaluating the safety of rock faces in rock excavations is compounded by the fact that OSHA regulations are grossly insufficient on this count. While soil is classified in three types, there are no classifications for rock. The Federal Highway Administration adopts analytical criteria to resolve disputes, whereas OSHA regulations leave the determination to the "competent persons" and the inspector to make this determination.

The analytical process of evaluating safety is somewhat tedious and requires special geotechnical expertise. Empirical methods of evaluation were recommended by Terzaghi in 1946 and advanced by Bieniawski in 1973, mostly for tunneling works. This study develops an empirical method of evaluation that pertains closely to volcanic rock and is adapted for construction excavations. The method can be accomplished in the field in about 15 minutes to one hour and can be used by competent persons and site superintendents.

Parameters used in the empirical method are rock strength, rock condition, persistence of discontinuities, spacing between discontinuities, water flow, weathering, roughness, and nature of infilling. Adjustments are made for conditions of dip and dip direction, height of rock face, overhang, and the method of excavation adopted. The result yields a score for Rock Face Safety Rating (RFSR).

The development of RFSR is undertaken through a check sheet using on-site visual confirmation techniques. Since this method can sometimes be subjective, adequate visual evaluation criteria are provided to decide appropriate scores for each parameter, thereby reducing uncertainty in the overall evaluation.

Results of the empirical method were compared for the same rock faces to analytical procedures recommended by Hoek and Bray (1981), in which plane failure, wedge failure, circular failure, and toppling failure analyses were considered. The 3-D stereographic projection method was used to analyze wedge failure. Statistical comparisons indicated that the empirical, visual method of evaluation was significantly identical to the complex analytical method to come to a determination. However, rock excavation safety analysis is not an exact science, still in the development stage, and even analytical procedures yield approximate results.

In light of this, the empirical method is inexpensive, as reliable as the analytical method, and easily accomplished on the construction site.

1 INTRODUCTION

The authors' main objective for this study was to develop a decision system that could be used to determine the stability of exposed rock faces within an excavation. This stability determination would be the decision as to how the excavated faces would or would not be supported to enhance the safety of the workers within the excavation. This system must be user friendly and readily usable on the construction site. After extensive research and field testing, the empirical method proved to be the best system for use on the construction site.

The analytical methods required extensive computing, that required excessive time and expertise to accomplish. Use of this new system will allow the current regulations to be improved to incorporate a better definition of rock stability, particularly within excavations. The contractor will achieve a more complete understanding of rock engineering principles that affect stability, which will allow him to make better determinations, thus improving safety on the job site. Ultimately, as a result of more completely defined parameters within regulations, thereby narrowing the gray area of interpretation of stability, there will be safer job sites and fewer disputes between contractors and OSHA inspectors (Singh, Connolly and Gowda, 1996).

2 OSHA ROCK CLASSIFICATION

Current OSHA regulations classify soils into three types -- Type A, B, and C -- but do not classify rock types. Since the soils have been classified, the types are well known and understood. Rock types on the other hand are not understood nor classified - a result of incomplete definitions within existing regulations. The only classification or definition of a rock condition is as follows: "Stable Rock means natural solid mineral matter that can be excavated with vertical sides and remain intact while exposed."

Under the current regulation, no soil may be classified as Type A if it is cracked or fissured, which has been interpreted by OSHA to mean that if the rock face is cracked or fissured, then that rock was not "intact". With that interpretation, all blasted, fissured rock faces had to classified as a Type B soil and below, thus requiring more excavation and/or support to be adjudged as safe for entry of workers into the excavation. Consequently, the Hawaii Occupational Safety and Health Division (HIOSH) interpreted "stable rock" to mean only rock masses in their natural state (Singh and Connolly, 1998).

3 DEVELOPMENT OF THE ROCK STABILITY DIAGNOSTIC SYSTEM

The Hawaii Rock Face Rating System (HRFRS) was developed from and modeled after several empirical systems already in existence. The starting model began with the Geomechanics Classification System developed by Z. T. Bieniawski (1989), following on Terzaghi's (1946) rock mass classification chart. Bieniawski's empirical system has since become accepted as standard engineering practice used in tunnel, foundation, and slope engineering. As previously mentioned, empirical methods can be (more easily than analytical methods) used quickly at the construction site. Deere and Deere's rock quality designator (RQD) and Wickham et al.'s rock rating structure (RSR) also play a role in the development of the diagnostic, among others. A detailed list of empirical methods developed worldwide are given in Connolly and Singh (1997).

4 ROCK SUBSTANCE, ROCK MASS, AND GEOLOGY

To properly analyze a rock face, the competent person must understand and appreciate the geology of the area as well as the rock composition and properties of the rock mass, The geology of the Island of Hawaii (Big Island) and properties of lava rock are described extensively in Stearns (1985), Hazlett and Hundman (1996), Singh et al. (1996), and Connolly and Singh (1997). The rock substance property of hardness is absolutely crucial in the analysis, but does not tell the whole truth about the stability of the rock mass. Geological properties of dip and dip direction are very important in determining stability. The actual properties of the rock mass, such as discontinuity spacing, conditions of discontinuity, rock condition, and groundwater conditions must be known. Correct analysis of the insitu rock mass requires the correlation of all of these parameters.

5 PROPERTIES OF BASALTIC LAVA ROCK

The hot, low-viscosity basaltic lava produced from volcanic lava flow activity cools to produce two principal types of basaltic lava rock - pahoehoe and aa. Pahoehoe is termed "hard rock", but is not as porous as aa. Different lava flows have intermingled over the centuries, causing weak structural planes at interfaces. There are many lava tubes throughout the island, most of which are unknown until discovered during excavation. The air borne volcanic activity produces less dense and lighter lava such as cinder, pumice, etc., may also be intermingled within the rock strata.

6 ROCK FACE RATING SYSTEM

The Hawaii Rock Face Rating System (HRFRS) diagnostic developed by the authors and described in this paper, reduces the latitude of interpretation of the stability of rock mass. The system is designed primarily to classify lava rock found in the State of Hawaii, but can be easily extended for further application to other rock types. To properly utilize the HRFRS, the material evaluated must be rock, which is defined as a material with the minimum uniaxial compressive strength of 150 lbs per cubic inch or a hardness test result of "very low strength"

6.1 *Parameters and Weights*

Primary parameters and their weights that make up an HRMR (Hawaii Rock Mass Rating) score are:

Primary Parameter	*Weight*
Strength of Intact Rock	15
Discontinuity Density	20
Discontinuity Spacing	20
Discontinuity Condition	35
Groundwater Condition	10
Total HRMR	100

This HRMR score is adjusted using adjustments for the following variables:

Adjustment Parameter	*Value*
F1: Parallelism with slope and dip-direction	0.15 to 1.0
F2: Discontinuity dip in plane of failure	0.15 to 1.0
F3: Relationship between slope angle and dip	-60 to 0
F4: Excavation Method	-8 to 15
F5: Height of rock face	-6 to 0
F6: Overhangs at the rock face	-10 to 0

These parameters provide a value for HRFRS, where HRFRS=HRMR + F1 x F2 x F3 +F4 + F5 + F6. After all adjustments are applied, the HRFRS score is tallied to disclose the classification of the rock mass. The weights assigned were finalized after extensive field testing. Insitu rock conditions in Hawaii plus weights assigned and applied by earlier researchers were considered.

7 USING THE RATING SYSTEM

7.1 *Scoring aids and check-sheet*

The charts and tables developed for use are designed to be user friendly. In addition to the tables and charts, pictures examples are provided within the report to aid the user in determining the insitu condition of that parameter (refer Connolly and Singh, 1997). The user must decide which parameter description or picture best suits the condition or situation, then select the corresponding value. The check-sheet used for determining stability comes in three pages (and are not included in this paper). In spite of the three-page length, it is fairly easy to use at the construction site. Shorter versions were attempted, but without the tables and charts, they were not useful in the field.

7.2 *Apportioning areas for evaluating the rock face*

The user must select areas of similar geological structure and property. The length of the selected areas is not a set parameter for the rock mass, and the length of the selected face ranges from 20 to 50 feet, while the height ranges from 6 to 20 feet. The important consideration in face selection is the insitu rock mass' effect on the rock face as to the actual stability. Evaluating various sample areas is a good practice when the rock face is not homogenoeous.

8 HRMR RATING PARAMETERS

Basaltic rock has the potential to be extremely strong as well as very weak depending upon its weathered condition. This rock has a propensity to form in vertical faces, which may be very high, The insitu condition may be highly fractured or extremely hard and solid. These properties are highly consistent with the OSHA definition of "Stable Rock"but there was not -- until HRMR was developed -- a standard method to prove stability. The parameters for excavation evaluation were selected based upon their influence on the stability of the rock faced. The following parameters are those that most critically influence the behavior of a rock mass, and are described in detail in Connolly and Singh (1997).

8.1 *Intact rock strength*

The strength of the intact rock is significant in the stability the rock mass. When disturbed by excavation, insitu rock may or may not remain stable. The particles in high strength rocks bind tightly, and are more resistance to fragmentation (such as from blasting) or separation (such as from mechanical excavation).

The standard hardness test, which does not require any bore samples, will adequately evaluate this parameter. This manual method to determine the hardness of rock is easy to perform. This test is performed by taking a rock sample the size of one's hand and counting the number of blows necessary from a pick or claw hammer to break it. Once the number of blows has been determined, the investigator uses the check sheet chart to determine strength. Although partially subjective, it can be *easily* done in the field. The user improves in interpreting the results with minimal practice.

8.2 *Condition of insitu rock mass*

The insitu condition of a rock mass is very important to determine stability. The rock condition may vary from solid rock with minimal cracks and fissures (blocky and seamy) to highly fractured, completely crushed rock to being sand and gravel-like. Generally, the more the rock is fractured and the smaller the blocks of rock within the fractures, the weaker the rock becomes. These conditions provide an insight into the inter-connectivity of blocks and the overall resistance that the rock mass has during excavation. The rock mass condition is an overall rating of that evaluated portion. If the Rock Quality Designation (RQD) of the rock mass is known, it will assist in the evaluation. Appropriate tables, charts and pictures, and written descriptions are provided within the worksheet to aid this determination.

8.3 *Discontinuity spacing and condition*

The spacing between the discontinuities (cracks or fissures), which incidentally, can be in any direction, must be measured. Generally, the wider the spacing between discontinuities the more intact and strong the rock mass is.. Each discontinuity must be evaluated as to its condition. In a rock mass, the most critical discontinuities must be evaluated for stability determination.

The Condition of Discontinuity parameter is composed of five variables:

The persistence (length) of the discontinuity.
The separation (aperture) between the surfaces of the discontinuity.
The infilling (gouge) if any is present.
The roughness of the infilling within the discontinuity.
Weathering of the rock face.

The behavior of the rock mass is directly influenced by the location of the discontinuities within the rock mass and the individual condition of each.

8.4 *Condition of ground water*

The presence of moisture or water within the discontinuity is extremely important. Water can be particularly destabilizing with the presence of infilling material that may cause slickensided conditions. A slickensided condition may cause the rock mass to displace along the path of least resistance, possibly causing a cave-in. Always check whether the rock mass is *dry, damp* or has *flowing water.*

9 ADJUSTMENT FACTORS

The HRMR value is based upon the parameters associated with the rock mass with the exception of joint orientation, method of excavation, height and overhang conditions. Corresponding adjustments are required to clearly define and quantify the stability of the insitu rock mass. The HRMR is adjusted to yield the Hawaii Rock Face Rating (HRFR) (see example in Section 11 of this article).

9.1 *Joint orientation*

Adopted from Romana's (1985) work, which simplified the user's ability to determine the impact of joint orientation through simple mathematical computations, factors F1, F2 and F3 for orientation are used. This eliminates the necessity for the user to make a subjective decision for each set of dip and dip direction values. The orientation of joints in the excavation concerns itself with the dip and dip direction of the rock formation and its relative tendency to move into or away from the open excavation.

9.2 *Dip and dip direction*

Four readings required to properly record the dip and dip direction of the slope and the joint, are as (1) Slope dip direction, (As), (2) Joint dip direction, (Aj), (3) Slope dip, (Bs), and (4) Joint dip, (Bj). It is important to note that the dip direction is from magnetic north, and dip is degrees from the horizontal dip direction. The dip direction may be 0 to 359 degrees and the dip 0 to 90 degrees. A common sense check is to note the relative potential movement of the rock mass into or out of the excavation.

9.3 F1: *adjustment for parallelism (check for planar &/or toppling failure)*

The factor F1 is a check for parallelism with the slope and dip directions. This orientation check -- by use of a table -- results in a factor which represents either a favorable or unfavorable condition that influences the stability. For planar failure, F1 is determined from the absolute value, |Aj-As|. For toppling failure , F1 is determined form the absolute value, |Aj-As-180|. If it is not easy to determine which failure is predominant, then calculate both and use the worse case for the adjustment.

9.4 *F2: adjustment for dip angle size in plane of failure*

This factor accounts for the dip angle size. The greater the dip angle, the greater the potential for catastrophic failure. This factor also has a check for both planar and toppling failure. F2 is determined by the value of Bj. The potential of failure approaches maximum effect as the angle increases to and surpasses the angle of friction. Therefore, F2 has a maximum of 1.0 at angles greater than 45 degrees when considering planar failure. The factor F2 is automatically 1.0 if toppling failure is predominant.

9.5 *F3: adjustments for orientation between slope and dip*

The factor F3 is the check for the relationship between the joint and slope dip angles. This is also a check for both planar and toppling failure. When planar failure is the prime consideration, the difference between the joint and slope dip angle is analyzed: a difference greater than 10 degrees is most favorable. The most unfavorable condition exists if the *slope dip is greater than the joint dip* by 10 degrees. The worst case generally indicates that the rock mass has the potential of sliding into the excavation (planar failure).

In the case of toppling failure, the dip and slope angles are added and that value is evaluated: less than 110 degrees is most favorable while greater than 120 degrees is most unfavorable. The most unfavorable condition reflects more vertical dips for the slope and the joint, which could cause a toppling failure.

9.6 *F4: adjustment for method of excavation*

There are various methods used to excavate a rock mass. This adjustment reflects the potential damage the excavation method has to the stability of the insitu rock mass. Random blasting my cause excessive fracture of the mass, thus weakening the structure (worst case). Precision blasting and mechanical excavation methods are as less disturbing to the mass, which usually do less damage to the remaining rock mass, thus allowing it to maintain its insitu structure and strength. With proper excavation planning it is possible to gain stability points for properly executed excavation methods.

9.7 *F5: height adjustment*

Large rock masses with very high vertical faces present a more potentially dangerous state that may contribute to a higher probability of cave in. Many of the stable rock faces, previously seen excavated in this research, had potential of falling rock. Although this study focused on excavations from 5 feet to 20 feet, many faces were evaluated of much greater height. As long as the methodology was followed on areas of similar geologic composition, the results were equally as valid as those that fit into the 5 to 20 foot range. This provision allows the user to evaluate any size rock mass. This has potential use in a falling rock mitigation program.

9.8 F6: overhang adjustment

Overhangs cantilever out over the rock face, have higher toppling moments, and are therefore potentially more dangerous, thereby requiring an adjustment to the HRMR. This parameter deals directly with the safety of the excavation and encourages the removal of overhangs to enhance the safety of workers beneath the outcropping.

10 ROCK FACE CLASSIFICATION

Rock faces have five classifications (Connolly and Singh, 1997):

Class I: HRFR = 80 to 100; Fully stable
Class II: HRFR = 60 to < 80; Stable
Class III: HRFR = 40 to < 60; Partially stable; requires partial mitigation

Class IV: HRFR = 20 to < 40; Unstable; mitigation necessary
Class V: HRFR = 00 to < 20; Fully unstable, extensive mitigation required

11 EXAMPLE

In an excavation site in the town of Hilo, Hawaii at the Kuhio Plaza Shopping Mall, the evaluated rock face was 18 feet high and 25 feet wide, of gray/brown color, primarily hard and intact lava rock, massive in size, moderately jointed, having very close aperture with no infilling. The strength of the intact rock was rated at 15 points. The massive, moderately jointed face rated at 17. The wide spacing of the discontinuities, was rated at 15. The condition of the discontinuities was rated at 21. The area was dry, therefore meriting a rating of 10. From the check sheet, HRMR = 92, which at this point falls within a class I slope rating range that is fully stable.

This example then has the adjustments computed and HRFR value recorded. By adding all the various adjustments and applying them to the HRMR value, the holistic HRFR is obtained. HRFR = HRMR + (F1 x F2 x F3) + F4 + F5 + F6. Thus the values of the adjustments are as follows: F1 = 0.15; F2 = 1.0; F3 = 0; F4 = 0; F5 = 0; F6 = 0. The HRFR = 92 - 0 = 92. This is Class I, Fully stable (see Section 10 above). Although the rock face is fully stable, there is still the potential hazard of small rocks falling into the excavation, The recommendation from the worksheet table is to excavate as stable rock with vertical sides, and ensure protection from possible falling rock in the form of canvas or construction fencing (Connolly and Singh, 1997).

12 COMPARISON TO ANALYTICAL EVALUATION

During the study a fellow researcher used an analytical method and evaluated 33 of the 47 slopes that were tested. The results of the empirical HRFR Method were studied and compared to results obtained from analyzing the stability using a 3-D stereographic approach (Gowda, Singh, and Kwong, 1997). The 3D Stereographic Method is the accepted engineering practice in determining stability of slopes either for existing slopes or for the design of slopes that will be excavated. The HRFR results were very consistent with those done analytically. Statistically, each system would evaluate the slope stability significantly the same. When comparing the two methods, the HRFR Method has a definite advantage over the analytical method in ease of use and time saved. The HRFR Method can determine stability in 20 minutes to an hour and is done totally on site. The analytical method requires a couple of hours on site with additional computing in the office. Often, the analytical method could not be totally performed in the field owing to inaccessible regions. Therefore, the empirical HRFR method translates to potential cost savings and incentive for the contractor to use. Moreover, it provides a tool to improve the safety on every excavation site.

13 DISCUSSION

"Are lava rock masses stable? If so, how do you prove it?," was the main purpose of this study. HRFR is a reliable method to determine and satisfy that purpose. The HRFR will determine the true stability of the rock face within the excavation, and by its use, improve site safety.

14 RECOMMENDATIONS

The rock classification system as proposed in this study should be adopted and incorporated into the existing OSHA regulation, Subpart P thus providing a standard methodology to determine the stability of the rock faces. Once the rock and its properties insitu have been clearly identified by the user, then the standardized HRFR methodology can be used to determine the rock mass stability. When the test determines the stability of any face, then there is no real concern as to what excavation method was used. If the face is judged unstable, then contractors will comply with the appropriate type of sloping, benching or shoring. In addition, since blasting is the most common excavation method employed, the contractor may be inclined to improve their blast design techniques (precision blasting) to minimize the fracture of the remaining rock mass. (Singh, Connolly, and Gowda, 1996).

15 ACKNOWLEDGMENTS

Acknowledgements are made to Hawaii Occupational Safety and Health Division, Hawaii Department of Labor and Industrial Relations, for contract Award #00083813.

REFERENCES

Bieniawski, Z. (1989), *Engineering rock mass classifications.* John Wiley & Sons, NY.

Connolly, M. and Singh, A. (1997), *An empirical method to evaluate the construction safety of excavations in Hawaiian rock,* Hawaii Occupational Safety and Health Division (HIOSH), Department of Labor and Industrial Relations (DLIR), Hawaii.

Gowda, R., Singh, A, and Kwong, J. (1997), *A comprehensive analytical strategy for evaluating the safety of rock slopes,* Tech. Report, UHM/CE/97-05, University of Hawaii.

Hazlett, R. and Hyndman, D. (1996), *Roadside geology of Hawaii,* Mountain Press Publishing Co., Montana.

Hoek, E. and Bray, J.W., Rock Slope Engineering, 3rd ed., The Institution of Mining and Technology, 1981.

Romana, M.)1985), *New adjustment ratings for application of Bieniawski application to slopes,* ISMR, Mexico City.

Singh, A. and Connolly, M. (1998), *Judging the safety of construction excavations in volcanic rock,* The 6th East Asia & Pacific-Rim Conference on Structural Engineering & Construction (EASEC-6), Taipei, Taiwan, January 14-16.

Singh, A., Connolly, M. and Gowda, R. (1996), *Blasting safety in volcanic rock,* Proc. 1st Intnl. Conf. on Implementation of Safety and Health on Construction Sites, Lisbon, Portugal, September 4-7.

Stearns, H.)1985), *Geology of the state of Hawaii,* Pacific Books, Palo Alto, CA.

Terzaghi, K. (1946), *Rock defects and loads on tunnel support, in R. Proctor and T. White (eds.), Rock tunneling with steel supports,* Commercial Shearing Co.,Youngstown.

Implementation of Safety and Health on Construction Sites, Singh, Hinze & Coble (eds)
© 1999 Balkema, Rotterdam, ISBN 90 5809 036 1

An evaluation of subsurface utility engineering activities

J.J. Lew
Department of Building Construction Management, Purdue University, W.Lafayette, Ind., USA

ABSTRACT: The Federal Highway Administration (FHWA) commissioned Purdue University to study the effectiveness of subsurface utility engineering (SUE) as a means of reducing costs and delays on highway projects. The concepts and practice of SUE have been developed and refined over many years, but generally were systematically put into professional practice in the 1980s. Virginia and North Carolina were the first states selected to be evaluated.

The reductions in risk for projects utilizing SUE are difficult to quantify. There are many variables and scenarios that may occur. Historical data is difficult to collect. Some savings are easily quantified; others may be qualitative or speculative in nature. The evaluation attempted to categorize savings accordingly.

This paper briefly describes SUE, then summarizes SUE and the evaluation results obtained in Virginia and North Carolina. In Virginia, nine projects were evaluated in-depth, and a cost savings of $4.11 per dollar spent was obtained. In North Carolina, twenty-one projects were studied in detail and a cost savings of $6.63 for each dollar expended on SUE was computed.

1 INTRODUCTION TO SUBSURFACE UTILITY ENGINEERING

Many design and construction projects are taking place in areas where an abundance of underground utilities already exist such as in cities, process plants, airports, highways, and so forth. These existing utilities create risks for the project owner, designer, and constructor. Although there are many reasons for these risks, one of the fundamental reasons is that complete and accurate data on the location (and even sometimes the existence of these *out-of-sight* utilities) is rare. Existing records of underground site conditions are frequently incorrect, incomplete, or otherwise inadequate because:

- The records were not accurate in the first place:
- Seldom were records placed in a single file, and often they were lost;
- References are frequently unusable because they refer to features that are no longer there.
- Lines, pipes, and tanks are abandoned, but aren't removed from the drawings.

1.1 *Liability and utility damage prevention mechanisms*

Engineers recognize this problem of records with incorrect or incomplete information, and attempt to protect themselves through prominently displayed notes on the drawings. Although these notes may vary in wording, a typical example is as follows: *Utilities depicted on these plans are from utility owners' records. The actual locations of utilities may be different. Utilities may exist that are not shown on these plans. It is the responsibility of the contractor at time of construction to identify, verify, and safely expose the utilities on this project.*

1.2 *Subsurface utility engineering*

New technologies such as surface geophysics, surveying techniques, CADD/GIS systems, *etc*., are known as Subsurface Utility Engineering (SUE). Subsurface utility engineers collect utility data and certify the data's quality rather than disclaiming responsibility for it.

In order to understand SUE, it is important to first define the four accepted quality levels of utility information that are available to the design engineer, constructor, and project owner.

- Quality Level D: Information that is derived solely from utility records which are unreliable sources. Unreliable in this case means that there is little, if any, confidence in the accuracy or completeness of the utility information that is supplied. Allocation of risk follows traditional contracting methods which typically treat utility information as an "unknown or differing site condition" requiring location by contractors during construction.
- Quality Level C: Quality level C information is Quality Level *D* information that is augmented by surveying and plotting visible surface utility features. The engineer or surveyor assumes risk for the accuracy of the surveyed visible features. Presumably, correlation of the quality level *D* data yields more accurate design information for construction.
- Quality Level B: These data are obtained through the application of appropriate surface geophysical methods to identify the existence and approximate position of utilities within the project limits. Quality level *B* data is reduced and shown on project documents as horizontal, or x and y designation. Risk for the horizontal designation is assumed by the subsurface utility engineer.
- Quality Level A: These data are highly accurate and are typically obtained by actual exposure and 3 dimensional location of the facility being measured. Accuracy is typically set at 15mm vertical, and to applicable horizontal standards. The accuracy of this data is also typically guaranteed by the subsurface utility engineer. Other data typically characterized include material type, surface elevation, utility size/capacity, outside dimensions, and configurations.

Quality level expectations must be specified by the project owner. If the owner specifies lower-quality information than indicated to the design engineer, the owner must be willing to pay for the associated costs in project delays, bid contingencies, change orders, unnecessary utility relocations, redesign, and utility damages. Most projects currently proceed by owner specification at quality level *C* whether or not they realize it.

2 BACKGROUND OF SUE IN VIRGINIA

The development of Subsurface Utility Engineering (SUE) in Virginia was started in 1984 by the Virginia Department of Transportation (VDOT) as a program by VDOT that gradually evolved into a continuous process. The primary reason for utilizing SUE in Virginia is to reduce the cost of highway construction. Cost reduction can be obtained through the elimination or reduction of claims, change orders, construction delays, disruption of utility services, and damage to utilities and adjacent facilities (accidents).

SUE is an integral part of the complete engineering design process for VDOT transportation improvement projects. An essential part of the damage prevention program involves the obtaining of timely and accurate survey and location information on underground utilities. Since the beginning of the VDOT SUE program in 1984, survey and location information has been obtained through the use of consultant contracts by using a SUE provider.

Before the inception of SUE in Virginia, most highway design was based on assumptions that frequently proved to be incorrect. The result was that during construction, unplanned utility conflicts arose that added costs to projects. The normal policy was to require utility relocation as a matter of policy, even though relocation might not be necessary. Often project design was based on "as-built drawings" that were often inaccurate, and infrequently reflected the field changes made during the project's construction. Virginia now reports a common discovery rate of 25 percent; in other words, employing SUE on a project reveals that 25 percent of the utilities located were not shown on any record drawings.

Virginia annually spends 4 to 5 million dollars on Subsurface Utility Engineering. All highway projects in Virginia are required to use SUE and most projects utilize both location and designation. SUE information has proven useful to utility companies, and is used to reduce utility conflict and redesign.

2.1 *Benefits of employing SUE in Virginia*

There are numerous benefits obtained when using SUE on highway projects in Virginia. By using SUE, significant benefits are derived for VDOT, utility companies, SUE consultants, contractors, and the general public or rate payers. Some of the benefits that have been obtained in Virginia are given as follows:

- Reduction in unforeseen utility conflicts and relocations;
- Reduction in project delays due to utility relocates;
- Reduction in claims and change orders;
- Reduction in delays due to utility cuts;
- Reduction in project contingency fees;
- Lower project bids;
- Reduction in costs caused by conflict redesign;
- Reduced the cost of project design;
- Reduced travel delays during construction to the motoring public;
- Improved contractor productivity and quality;
- Reduced utility companies' cost to repair damaged facilities;
- Minimized damage to existing pavements;
- Minimized traffic disruption, increasing DOT public credibility;
- Minimized the chance of environmental damage;
- Induced savings in risk management and insurance;
- Introduced the concept of a comprehensive SUE process.

3 COST, DATA, AND INFORMATION OF SUE IN VIRGINIA

The level of investment in SUE by the State of Virginia from 1985 to 1995 totaled $28,278,000 representing 2,037 project locations. Virginia presently is spending $4 to $5 million annually for SUE, although exact amounts for the past three years were not available to the evaluation. Under 3.1, *Specific Projects Studied*, a summary of the data and information obtained from the specific projects used in the Virginia evaluation study is presented. The total SUE costs are listed with quality levels along with the total SUE savings obtained in the evaluation study.

3.1 *Specific projects studied*

NOVA, Route 1 Dumfries — Total SUE $37,252.91, Quality Level A

SUE Activities: A sanitary force main was determined not to be in conflict with construction, and a water line was properly located, eliminating relocation costs, delays, and change order claims. *Total SUE Savings: $98,000.00, or $2.63 per $ spent.*

City of Suffolk/Route 629 — Total SUE: $53,168.60, Quality Level A

SUE Activities: Exact conflict sections for a water main located the entire length of the project allowed the relocation to be bid at a unit cost resulting in considerable relocation savings. *Total SUE Savings: $679,720.00, or $12.78 per $ spent.*

Interstate 64, Grove Interchange — Total SUE: $206,249.85, Quality Level A

SUE Activities: The relocation of a 16" gas main with two offsets was eliminated. Drainage structures were designed to avoid conflicts with existing utilities the entire length of the project. *Total SUE Savings: $342,000.00, or $1.66 per $ spent.*

City Co. & York Co. Route 199 — Total SUE: $110,323.2, Quality Level A

SUE Activities: A 16" sanitary sewer for Eastern State Hospital was never designated on the plans. Design of required relocations was minimized saving unnecessary relocation costs.

Total SUE Savings: $100,000.00, or $0.91 per $ spent.

Hopkins Road City of Richmond — Total SUE: $ 93,553, Quality Level A

SUE Activities: 156 potential conflict sites were reduced to only 14 actual conflicts. Design changes eliminated most conflicts without increasing actual construction costs.

Total SUE Savings: $738,425.00, or $7.89 per $ spent.

York County – Waller Mill Road — Total SUE: $26,022.12, Quality Level A

SUE Activities: Locating utility lines eliminated unnecessary relocations. Cost savings were realized in decreased project design and construction costs. A time savings was also realized. *Total SUE Savings: $91,100.00, or $4.22per $ spent*

Iron Bound Road – Williamsburg — Total SUE: $14,492.41, Quality Level A

SUE Activities: A water main was located and found to be in conflict without breaking the line which would have occurred during construction. There were no delays, damage expenses, or redesign expenses. *Total SUE Savings: $100,969.61, or $6.95 savings per $ spent*

City of Emporia – Low Ground Road — Total SUE: $12,697.45, Quality Level A

SUE Activities: SUE located a water main and a sewer line not shown on the project's plans. Project was redesigned and conflict relocations were bid rather than being included in a change order. *Total SUE Savings: $131,638.00, or $10.37 per $ spent*

Bristol District Town of Lebanon — Total SUE: $3,500, Quality Level A

SUE Activities: A force main was located and found to be 10 feet off from what the as-builts indicated. A field change order was eliminated resulting in a $12,000 savings.

Total SUE Savings: $12,000.00, or $3.43 per $ spent.

4 CONCLUSIONS, VIRGINIA

Virginia has been a difficult state for which to analyze SUE cost savings. The reasons for this are varied. Some of the reasons are as follows:

- The program has been in place for a long time. Historical data of "before SUE" projects are hard to find and the data is difficult to analyze due to changing construction practices, utility damage prevention programs, and so forth.
- No projects could be identified where utilities were plotted from record and then designated. A project with these characteristics would allow easy comparison to accuracy, comprehensiveness, and interpretive ability between existing records and quality level *B* data.
- Contractors already are well educated about SUE, making bid savings hard to quantify.
- Project synopses after construction do not easily separate change orders, redesigns, extras, and claims by categories such as would be helpful to identify specific utility problems.
- Restructuring of VDOT with lots of early retirements and downsizing of staff makes contacts with previous project managers difficult or impossible.

Twenty-five or so projects were selected mainly at random for review. From these twenty-five projects, nine were selected for an in-depth cost study. These nine projects had project designers and managers available for interviews. Virginia typically requires a thorough use of SUE on a regular basis on all highway projects.

Overall, the savings of SUE in Virginia appear to be quantifiable at about $4.11 to every $1.00 spent in upgrading to quality levels *B* and *A* data. This does not include qualitative savings, which might make this ratio much higher. Regardless, when extrapolating the savings across their entire highway program, the savings of SUE approximate $20,550,000 per year, minus the $5,000,000 spent in their program, for an effective savings of $15,550,000 per year or more. The current level of investment in SUE for Virginia is approximately $5,000,000 per year.

5 BACKGROUND OF SUE IN NORTH CAROLINA

The development of Subsurface Utility Engineering (SUE) in North Carolina was started in 1991 by the North Carolina Department of Transportation (NCDOT). SUE was started as a program by NCDOT and gradually evolved into a continuous process. The primary reason for utilizing SUE in North Carolina is to reduce the cost of highway construction. Cost reduction can be obtained through the elimination or reduction of claims, change orders, and construction delays, and through the minimization of disruption of utility services.

SUE began as an aid to in-house-designed projects with an initial contract with one provider valued at approximately $300,000. The program was successful and additional SUE consultants were brought under contract. Currently there are four providers; however, the contract values are not equal. For designs performed by outside consultants, i.e., non-state employees, the DOT requires that the outside designers hire one of the four state DOT-approved SUE consultants for their team. Consequently, the two contracting methods, i.e., state contract for in-house design and project contract for outside consultant design, result in a total, state, DOT SUE program valued at approximately $3,000,000 per year. This represents a SUE budget of approximately 2 percent of the total state engineering/construction budget.

When SUE was initially utilized in North Carolina, a formal review procedure was adopted. The procedure was used for one or two periods. The use of the procedure was informally abandoned for no given specific reason. SUE is now employed in North Carolina by an informal procedure based on cooperation between design engineers and area engineers. This informal procedure is accomplished by mutual agreement and judgement between design and area engineers on an *as needed project basis* due to amount of utilities, potential impact, and engineering judgement. Now that many of the design and area engineers have become familiar with the concepts of SUE, the informal process is working well.

5.1 *Benefits of employing SUE in North Carolina*

There are numerous benefits obtained when using SUE on highway projects in North Carolina. By using SUE, significant benefits are derived for NCDOT, utility companies, SUE consultants, contractors, and the general public. Some of the benefits that have been obtained in North Carolina and used in the evaluation are given as follows:

1. Reduced the number of utility relocations, saving time and costs;
2. Reduced project delays due to utility relocates, saving time and costs;
3. Reduced construction delays due to utility cuts, saving time and costs;
4. Reduced contractors' claims and change orders, saving time and costs;
5. Reduced delays caused by redesign due to conflicts, saving time and costs;
6. Reduced accidents and injuries due to utility cuts, saving time and costs;
7. Reduced the loss of service to utility customers, saving costs and user costs;
8. Improved contractor productivity resulting in cost savings;
9. Increased the possibility of reduced bids, resulting in cost savings possibilities;
10. Reduced the cost of project design, saving time and costs
11. Reduced the damage to existing pavements, saving costs;
12. Minimized disruption to traffic and emergency equipment, saving time and costs;
13. Minimized chance of environmental damage, saving time and costs;
14. Induced savings in risk management and insurance, saving costs;
15. Introduced the concept of comprehensive SUE, saving time and costs.

6 COST, DATA, AND INFORMATION OF SUE IN NORTH CAROLINA

SUE Expenditures by Provider and Contract for North Carolina indicates a total expenditure of $8,725,371.64 for SUE on 295 in-house projects in North Carolina from 1991 to the first quarter of 1998. Annual SUE expenditures are currently approximately $2,000,000 per year on the in-house projects. Total SUE savings obtained in the evaluation of the 21 projects was $4,183,432. The total SUE cost was $631,368.42, yielding an overall Savings Analysis of $6.63 savings for each dollar spent on SUE for in-house projects. The information presented under 6.1, *Specific Projects Studied,* briefly summarizes the data and information collected by interviews with key project personnel for each specific project used in the evaluation study. Studied were cost factors of Time, Cost, User, and Risk Management Savings.

6.1 *Specific projects studied*

NC 24 — Total SUE: $80,000.00, Quality Level A

SUE Activities: 40,000 lf of 12" water line was potentially in conflict. Project consisted of symmetrical widening of the existing roadway from two lanes to five lanes. Project located parallel to the beach in a potentially environmentally-sensitive area. Expansion includes two travel lanes in each direction and one turning lane in the center for a total of five lanes. SUE was used on the 40,000 feet of water line and the crossing lines.

Total SUE Savings: $750,000, or $9.38 per $ spent.

Hickory E. Side of Thoroughfare — Total SUE: $10,226,411.00, Quality Level A

SUE Activities: A considerable number of utilities were located by the use of SUE. Utility relocations were reduced, and project delays due to relocation were avoided, along with eliminating the possibility of line cut accidents.

Total SUE Savings: $20,296, or $1.28 savings per $ spent.

NC 168 from US 158 A to VA State Line — Total SUE: $5,005.00, Quality Level A

SUE Activities: Project consisted of 18.5 miles of highway widening from two lanes to five lanes. Using SUE it was determined that a fraction of the water line could remain in place, otherwise it would have all been removed and relocated.

Total SUE Savings: $773,400, or $129.00 per $ spent.

Forsyth Co., bridges over US 421 & SR 1528: — Total SUE: $3,883, Quality Level A

SUE Activities: SUE was used to determine conflicts with a water line. This allowed considerable reduction in unnecessary water line relocation costs and project time savings.

Total SUE Savings: $802,500, or $206.67 per $ spent.

High Point Bridge Replacement, Guilford Co. — Total SUE: $8,789.50, Quality Level A

SUE Activities: The project was a bridge replacement over a RR in the busy downtown area of High Point. Unique scheduling constraints required that the project be completed in a short construction season. Also, the project is in the old part of town where many old utilities are located and records are minimal at best. Use of SUE permitted the utilities to be located with minimal disruption to traffic. *Total SUE Savings: $135,000, or $15.36 per $ spent.*

I-85, Guess Road Interchange — Total SUE: $74,696.21, Quality Level A

SUE Activities: SUE was employed to locate water, sewer, gas, and phone and other cables along the project's length. Utility location was critical to the design and scheduling of the project. There was a time limit on the duration of the project and delays to the project could not be permitted to happen. SUE techniques were able to locate the utilities at all depths and reassure the design of the project. No significant utility conflict delays were encountered.

Total SUE Savings: $51,150, or $0.68 per $ spent.

I-95 @ Halifax & Northhampton — Total SUE: $8,438.00, Quality Level A

SUE Activities: For construction of a bridge with possible utility conflicts consisting of a 6" plastic-coated steel gas line, a phone conduit 6-way-multi-tile duct, and a CATV cable, designers were concerned where to place the pile bents. Accurate location was required to avoid conflict, or complete relocation would be required--the gas line could not be hit. NCDOT determined that the utilities could remain in-place by utilizing the information provided by SUE that the utilities were as submitted. The utilities were test holed for design purposes and flagged during construction. No utility cuts or other problems were encountered during construction. *Total SUE Savings: 24,000, or $2.84 per $ spent.*

I-40 Welcome Center Total SUE: $6917.58, Quality Level B

SUE Activities: The existing sewer line had sagged on each side, but was serviceable between its ends, therefore, SUE revealed that most of the existing sewer encasement could be retained. This meant that DOT could replace the sewer line without reboring under I-40. This saved the installation of a tunnel 150 feet in length. The water lines were found to be in the encasement and could also be replaced without additional boring. Records of the water line, sewer line, and encasement were non-existent.

Total SUE Savings: $100,000, or $14.37 per $ spent.

Raleigh Outer Loop Total SUE: $25,578.00, Quality Level B

SUE Activities: In this new road location around the town, few utility conflicts were found. Since SUE was used for designating, only horizontal locations were found. Normally, depth locations indicate potential problems where they are located on a project. On this project, designating proved to provide sufficient information. No delay-causing utility conflicts were found on the project. *Total SUE Savings: $56,220, or $2.20 per $ spent.*

Raleigh Outer Loop Total SUE: $90,307.00, Quality Level A

SUE Activities: SUE was employed to avoid utility conflicts at crossing locations and to facilitate the relocation of utilities where required. SUE was successfully used to locate utility lines as much as possible before construction, and to assist with utility replacement and upgrading, which helped prevent delays to the project No conflicts with utilities occurred that have resulted in a claim. By working with the phone company to bury a replacement cable for a phone cable found by SUE, possible delays were avoided. If this line had not been located and marked by SUE, it would likely have been cut, resulting in an accident and a delay.

Total SUE Savings: $193,986.00, or $2.15 per $ spent.

Bethel By-Pass Total SUE: $7978.90, Quality Level A

SUE Activities: SUE was used to check the elevations of the existing force main which was in the project's right-of-way for a length of 9000 meters. The normal procedure employed in the past was to have the entire length of the force main relocated. The relocation of the force main was limited to 3000 meters, thus eliminating 6,000 meters of force main from unnecessary relocation. *Total SUE Savings: $243,000, or $30.45 per $ spent.*

Capital Blvd., Raleigh Total SUE: $98,717.46, Quality Level A

SUE Activities: SUE was used to locate the depth of the water line. By checking the actual depth, designers were able to modify the design of the project to avoid unnecessary relocation of a 16 inch water line. It was determined that 5000 feet of the water line would not have to be relocated. *Total SUE Savings: $199,548, or $2.02 per $ spent.*

Fuquay-Varina Loop Total SUE: $22,760.11, Quality Level A

SUE Activities: No design changes related to utility conflicts were encountered during construction. No problems as far as roadway design, hydraulics, and drainage were encountered during construction. Right-of-Way limitations required that the underground utilities had to be located in less Right-of-Way width, which was effectively accomplished with the use of SUE. *Total SUE Savings: $27,000, or $1.19 per $ spent.*

US64 Multi lane Highway Total SUE: $9910.80, Quality Level A

SUE Activities: Using SUE allowed 350' of the 16'' water line to remain in-place and not be relocated as normally would be required. *Total SUE Savings: $45,000, or $4.54 per $ spent.*

US 70, Smithville By-Pass Total SUE: $3401.52, Quality Level B

SUE Activities: Project designers used SUE to locate the existing utilities in the vicinity of the new bridge supports and footings during construction. Originally some of the utilities were not thought to be in conflict with the bridge supports and footings, but indeed were found to be in conflict. SUE supplied the required information to redesign the footings to miss the water line. Though SUE was employed late in the project, savings were still obtained from the use of SUE. There were no claims or change orders as a result of utility conflicts on the project.

Total SUE Savings: $18,000, or $5.29 per $ spent.

NC 105 widening Total SUE: $5778.52, Quality Level B

SUE Activities: A D.I. line replaced an existing 12" AC line. SUE verified that the line had to be replaced. Old surveys and drawings were mostly nonexistent and there were no as-builts for the water line. *Total SUE Savings: $6,000, or $1.04 per $ spent.*

US 311 in High Point, NC — Total SUE: $3918.73, Quality Level A

SUE Activities: There were a lot of conflict locations with an 8 inch gas main along the length of the project. The conflicts were resolved and the gas line was relocated at several locations without project delay. *Total SUE Savings: $11,575, or $2.95 per $ spent.*

Capital Blvd, US 1 — Total SUE: $61,099.76, Quality Level A

SUE Activities: SUE was used to determine that a sewer line and 5,000 feet of water line did not need relocation. *Total SUE Savings: $384,000, or $6.29 per $ spent.*

Duraleigh Road — Total SUE: $56,279.00, Quality Level B

SUE Activities: By obtaining designating information, SUE was used to analyze along the length of the project where utility lines (mostly water) could be in conflict. It was determined and field-verified that only one place required locating. Result was that the water line was not in conflict and line could be left in place. *Total SUE Savings: $61,300, or $1.09 per $ spent.*

Monroe Road, SR 1009 — Total SUE: $27,369.95, Quality Level A

SUE Activities: On this project, it was possible to employ SUE in the initial construction phases to determine actual utility line conflicts. The use of SUE determined that utility lines did not have to be relocated at three different locations. Savings were based on bid unit costs. *Total SUE Savings: $55,753, or $2.04 per $ spent.*

US 17, Onslow County — Total SUE: $9,253.38, Quality Level A

SUE Activities: With the aid of SUE information, it was determined that only 1500 linear feet of the water line needed to be relocated. Without this information, approximately 10,000 linear feet of additional water line would have been relocated.

Total SUE Savings: $220,000, or $23.78 per $ spent.

7 CONCLUSIONS, NORTH CAROLINA

The evaluation study has computed a cost savings of $6.63 for every dollar spent for SUE. The total amount of expenditures for SUE in North Carolina is $8,725,371.97. This represents a projected savings of $57,849,211.39 since the SUE process was started in North Carolina. The SUE savings computed in this study are related to the in-house projects designed and constructed by the NCDOT.

NCDOT appears to have figured out how to use SUE effectively in their state and are doing so for the benefit of the taxpayer and ratepayer. NCDOT has utilized SUE for eight years, with a progressive amount of contract value. They are currently funding SUE at levels in excess of $3 million per year. It is difficult to estimate non-quantifiable savings resulting from decreased utility damages, bid prices, construction delays, and so forth; however, quantifiable savings (after studying about 7 percent of NCDOT's in-house projects on both a cost and project basis, indicating a return in excess of $6.63 for every dollar spent) were obtained. Therefore, a quantifiable savings per year for NCDOT projects is approximately 19.8 million dollars. The majority of projects utilizing SUE showed no delays due to utility conflicts and none with serious accidents, an improvement over past engineering practices.

8 REFERENCES

Lew, J. J. (1998). *An evaluation of subsurface utility engineering of the North Carolina Department of Transportation* (FHWA Preliminary Report Draft DTFH61-96-00090). West Lafayette, Indiana: Purdue University, Building Construction Management.

Lew, J. J. (1998). *An evaluation of subsurface utility engineering of the Virginia Department of Transportation* (FHWA Preliminary Report Draft DTFH61-96-00090). West Lafayette, Indiana: Purdue University, Building Construction Management.

Implementation of Safety and Health on Construction Sites, Singh, Hinze & Coble (eds)
© 1999 Balkema, Rotterdam, ISBN 90 5809 036 1

A heuristic approach for the identification and prevention of trench failure incidents

S.T.Ariaratnam & D.M.Cruden
Department of Civil and Environmental Engineering, University of Alberta, Edmonton, Alb., Canada

R.S.Joseph
Northwestern Utilities Limited, Edmonton, Alb., Canada

ABSTRACT: The Province of Alberta, Canada has experienced its share of fatalities due to trench failure incidents over the past 20 years. From 1976 to 1996, there were 30 recorded fatalities due to trench collapses in the province. This paper describes the development of a heuristic approach to aid in the identification and prevention of potentially dangerous trench conditions. Actual accident investigation reports obtained from Alberta Labour, Occupational Health & Safety Division are used to analyze and determine factors contributing to trench collapses. From these reports, four main categories of variables are identified: 1) Project Parameters; 2) Environment Factors; 3) Ground Vibration; and 4) Trench Parameters. By dividing these factors into four main categories, critical measures are developed to aid in the warning of potentially dangerous scenarios. Through the development of this heuristic approach, it is anticipated that there will be an increased awareness for the safety of crews working in open trenches.

1 INTRODUCTION

Occupational Health & Safety (OH&S) is the enforcement arm of the Alberta government and looks after on the job enforcement of government regulations regarding on the job safety. A major role of OH&S is to investigate all accidents reported on construction sites. Each fatality is closely investigated and a detailed report written documenting all findings that may have contributed to the accident. Alberta Labour has recognized the danger of excavation and trenching related activities and have addressed safety issues in General Safety Regulation AR 448/83, Section 173 (excavation) and 174 (trenching), as part of the Occupational Health and Safety Act.

All fatality reports in the Province of Alberta are held on file by Alberta Labour. These records were obtained through the Freedom of Information and Protection of Privacy (FOIP) Act of 1995. Obtaining these reports proved to be a lengthy process as it was required that all references to names of persons and organizations involved in each fatality incident be removed prior to public release. There were a total of 30 fatalities from trench-related accidents in Alberta between 1976-1996. Of these, 24 fatalities were a direct result of sewer and utility projects, while 6 were from mining operations. This study only examined fatalities resulting from sewer and utility construction projects. Each of the reports was thoroughly examined and information separated into four main categories for analysis: 1) Project Measures; 2) Environment Factors; 3) Ground Vibration; and 4) Trench Parameters. These form the basis for the failure event causation model illustrated in Figure 1.

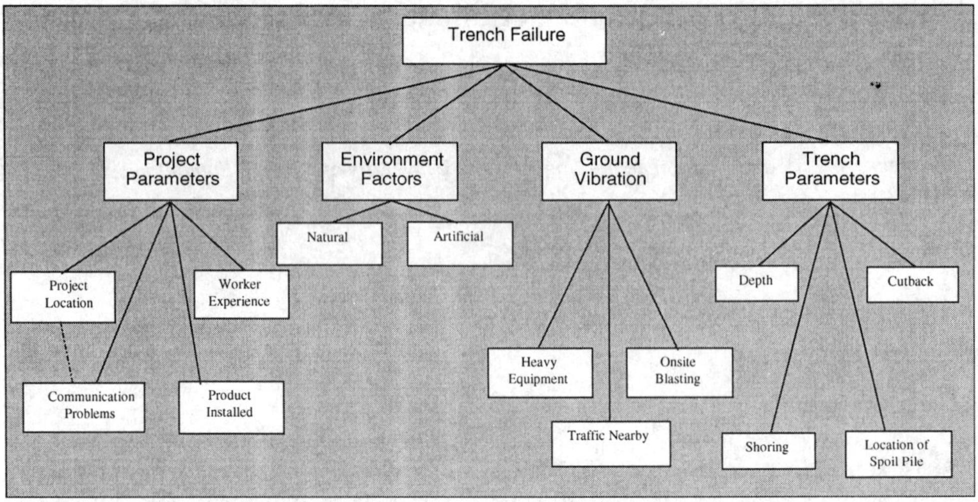

Figure 1. Failure Event Causation Model

Much literature is available documenting research conducted in the area of construction safety. Several authors recognize the substantial financial consequences of negligence while investigating the costs associated with construction injuries (Hinze and Applegate 1991; Everett and Frank 1996). De La Garza et al. (1998) analyzed safety performance of contractors using the results from a previous study that examined the roles of contractors regarding worker's compensation (WCB) issues (Hancher et al. 1997). Research conducted by Lan and Arteau (1997) concluded that safety codes regulating excavation and trench work should be more specific about required cutback angles in relation to the types of soil and the conditions of the water table. The importance of proper soil identification prior to conducting trenching was also identified in a study by Brown et al. (1992). Davies and Tomasin (1996) cited the main cause of accidents in trench work as being collapse of the side burying or partially burying those in the trench. About one in five of this type of accident results in death and collapse can occur quickly and without warning. Often workers consider deep trenches as the most hazardous, however, a survey of 265 trench fatalities in North America between 1974-1998 conducted by Rice (1992) concluded that a fair number of fatalities occurred in trenches 3 m deep or less. Common causes of trench collapse cited by Davies and Tomasin (1996) include: mechanical failure of the soil; vibration of equipment; storage of spoil material near the edge of the trench; and previously excavated ground.

The main objective of this research is to analyze reports from previous trench-related fatalities in the Province of Alberta to develop a heuristic approach for identifying and preventing future incidents.

2 HISTORICAL PERSPECTIVE OF TRENCH FATALITIES IN ALBERTA

This section provides an analysis of trench fatalities in Alberta that occurred between 1976 and 1996. Workers buried at the bottom of trenches when walls collapse result in a minimum of 1.14 deaths per year. Thirteen (or 50%) of these fatalities occurred in the period of the 1970's and early 1980's. This was a time of growth in Alberta's construction industry compared to the slow period experienced during the mid-1980's to early 1990's. According to the Alberta Construction Association, there has been an explosion of construction activities across Alberta since 1995. The markets are experiencing a shift into the heavy industries such as oil and gas, forestry and agriculture and related manufacturing industries such as chemical and food processing. Additionally, activities in commercial, institutional, light, and residential construction continued to increase. Currently, there are between 90,000 to 100,000 people employed in Alberta's construction sector (ACSA Industries 1998). This has lead to a shortage of skilled and competent workers, which may ignite an increase in job-related injuries across the province.

Culver et al. (1990) determined that there were 3,496 construction deaths, 239 of which occurred during work in trenches through a study of the U.S. construction industry between 1985-1989. From 1992 to 1996, there were 54 construction deaths in the Province of Alberta according to statistics compiled by Alberta Construction Safety Association Industries (1998). During the same period, there were 7 trench-related fatalities recorded, or approximately 13% of all construction fatalities.

Figure 2 demonstrates the significant (at the 5% level) concentration of fatalities in the last half of the year when there may be some urgency to complete work before the ground freezes and excavation becomes more difficult as is the case in Alberta. Fatalities recorded during the months of September and October are of particular interest to note since these occur during the end of the construction season when projects may be fast-tracked to meet schedule deadlines. As a result, shortcuts are often taken in the safety area. The financial payback of strictly adhering to safety regulations is not immediate in that it may cause delays in an already tight schedule. However, the prevention of injury and cost of WCB increases should outweigh this cost.

The bimodal frequency distribution with peaks in early summer and in the fall is reminiscent of the annual distribution shown in Figure 3 of landslides in the province (Cruden 1996). The early summer peak in both landslides and trench collapse may be due to ground saturation by snowmelt. Long daylight hours prevalent in mid-summer tend to dry out and stabilize the ground.

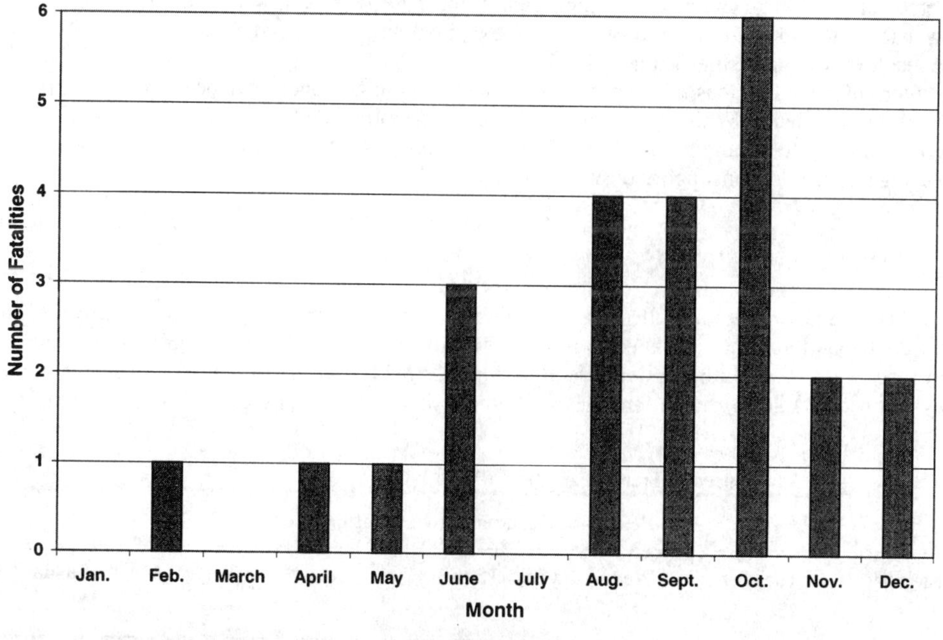

Figure 2. Monthly Occurrences of Trench Fatalities in Alberta, 1976-1996

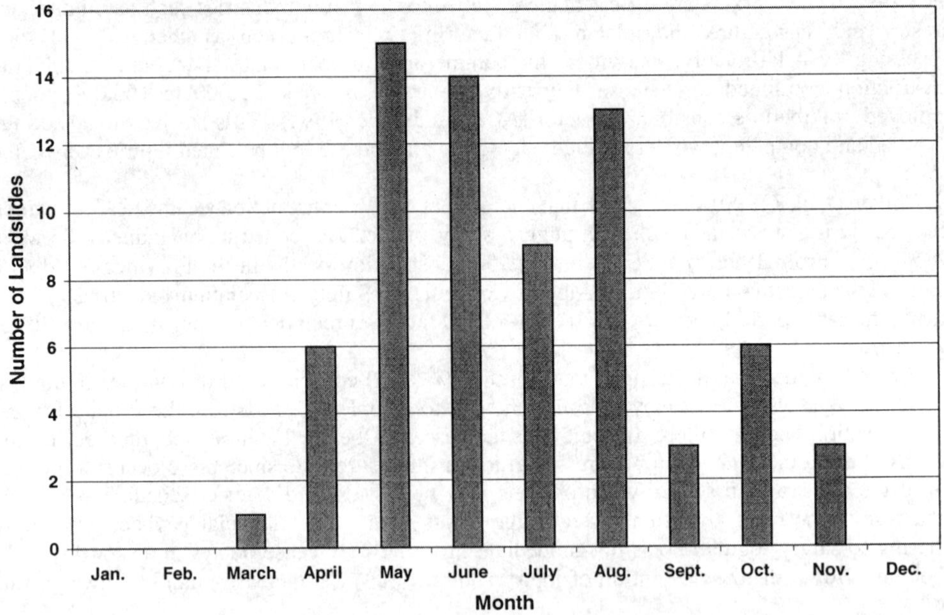

Figure 3. Monthly Occurrences of Landslides in Alberta, 1900-1991 (Cruden 1996)

The hypothesis that fatalities are uniformly distributed throughout the week can be tested by comparing the departures of the daily fatality rates (Table 2) from the calculated mean rate with departures expected from a chi-squared distribution (Benjamin and Cornell 1970) with 5 degrees of freedom. The result is significant at the 10% level, but not at the 5% level. If the fewer fatalities on Saturday and Sunday are excluded and the weekdays tested for uniformity, the result is no longer significant.

Our information does not include for what duration trenches had been left unsupported before they failed. While the mortality rate on Monday and Tuesday is not statistically significantly higher than on other weekdays, the number of fatalities might suggest instances of entering a trench left unsupported over a weekend.

3 PROJECT PARAMETERS

Project parameters were collected to gather general information on the contractor and miscellaneous information related to the specifics of the project. Information was collected on the type of business, either independently owned or public, to investigate if this factor played any significant role in trench fatalities. Although there is no strong correlation in fatality rate

Table 1. Daily Occurrences of Trench Fatalities, 1976-1996

Percent of Occurrence by Day of the Week						
Monday	Tuesday	Wednesday	Thursday	Friday	Saturday	Sunday
21	25	8	25	4	4	13

based on type of business, safety training of all employees involved in trenching-related construction activities is strongly advocated. A majority of the OH&S reports indicated worker experience. Although not empirically categorized, forty-six percent of the cases reported that the fatality occurred on an "experienced" worker who had performed trenching-related construction activities on numerous occasions in the past. According to Alberta Labour (1996), about 25% of all injured workers are those that were hired less than six months prior to being injured. Inexperienced workers comprise about 17% of all persons employed in the construction industry in Alberta. Many of the projects at rural locations conducted in the 1970's revealed lack of adequate communication resulting in wait times of greater than forty-five minutes before medical assistance could be received on site. Therefore, it is imperative that all workers be briefed on safety protocol prior to commencing work. An alarming statistic revealed that, in 96 percent of the cases, other workers entered the trench after collapse to assist their injured colleague. Although none of these cases resulted in fatalities, it is not a recommended practice due to safety concerns.

4 ENVIRONMENT FACTORS

Environment factors, presented in Table 2, were divided into two major categories: 1) natural; and 2) artificial. Natural factors are the presence of freeze/thaw, ingress of water, unfrozen ground, and soil type. Artificial factors include daylight or nightlight conditions during excavation and previously excavated ground.

Soils in Alberta are complex; their strengths may vary across a wide range of values within a site and within a trench. Glacial deposits vary from stiff, cohesive tills to running, loose sands. Both the deposits and the often weak and weathered bedrock they sit on, may be sheared and displaced by glacial tectonics.

The soils excavated are either glacial tills (which are usually cohesive) or post-glacial lake and river deposits, which are often non-cohesive. Tills cover approximately 60% of the province while the lake and river deposits are more common in urban areas close to present day rivers. Therefore, the proportion of trench failures in non-cohesive soils is not surprising.

The till may be stiff, fissured, and contain a substantial proportion of coarse-grained sediment. Previously excavated ground reduces the inherent strength of the adjacent ground as it may have been of variable consistency and not well consolidated. As backfill, it is not easily compacted and may not develop substantial cohesion. Backfilled excavations in till may, therefore, represent non-cohesive soils re-deposited in extensive areas of cohesive soils. Consideration might be given to including some reflective material in the backfill in these areas. This would flag this potentially hazardous situation in any re-excavation of the disturbed soil. While only four fatalities occurred due to backfill collapse in cohesive soils, these represented two-thirds of the fatalities that occurred in soils identified as cohesive.

Soil properties may vary annually. In the winter, freezing penetrates metres into the ground substantially increasing the cohesion of soils with high water contents. Granular soils above the water table may be unaffected. Thawing of ice-rich, poorly drained soils in the spring can produce critical stability conditions on slopes. In the summer, southern Alberta has a moisture deficit and soils there stiffen and may fissure.

Twelve fatalities are reported in ground that was supposedly frozen. This number includes all the cases of failure in backfill in cohesive soils and 6 of 13 cases of failure in non-cohesive soils. While the surface of the ground may be frozen, the freezing front penetrates the ground slowly and soils towards the bottom of the trench may still be unfrozen in the late fall. The development of substantial cohesion in ground below 0°C requires the presence of water in the soil. Coarse, well-drained soils above the water table may flow into a trench even when ground temperatures are below freezing. Indeed, that excavation was possible under frozen conditions might act as a warning that trench failure in this soil is also possible under frozen conditions.

Table 2. Environment Factors

Causation Factor	Yes (%)	No (%)	Insufficient Info. (%)
Was there evidence of freeze/thaw conditions?	37	63	0
Was the ground unfrozen?	50	50	0
Was water present in the trench?	63	4	33
Was the soil non-cohesive?	54	25	21
Did the collapse occur in an area of previously excavated material?	54	21	25
Did the accident occur during night light conditions?	17	83	0

Table 3. Ground Vibrations

Causation Factor	Yes (%)	No (%)	Insufficient Info. (%)
Was there evidence of heavy equipment operating nearby?	92	8	0
Was there evidence of traffic or other vibrations nearby?	17	62	21
Was there any blasting operations performed on site?	0	100	0

Of interest is that fact that water was present in the trenches in 63 percent of the fatalities. This suggests an occurrence of recent precipitation in the site, which may be a factor to observe prior to entering a trench. Ingress of water not only removes particle adhesion and reduces resistance to movement, but also increases the weight on the supported walls of the trench.

5 GROUND VIBRATIONS

Heavy machinery operating near the trench may produce sufficient ground vibrations and heavy surcharge load on the unsupported sides that may cause the trench to collapse. In 92 percent of the fatalities (Table 3), there was evidence of heavy machinery operating nearby. Blasting of rock, though not common practice, may loosen the soil cohesion resulting in unstable effects. The effect of ground vibrations is addressed in General Safety Regulation AR 448/83, Section 174(3)(d) which clearly states that additional protection is required where machinery or a heavy object is placed or is working within a distance from a vertical line drawn from the near edge of the bottom of the trench equal to the depth of the trench.

6 TRENCH PARAMETERS

Trench Parameters, presented in Table 4, are defined as those related to the physical characteristics of the trench. This category was examined to determine what, if any, safety

Table 4. Trench Parameters

Causation Factor	Yes (%)	No (%)	Insufficient Info. (%)
Was the trench depth greater than 1.5 m?	100	0	0
Was the spoil pile found at the edge along the same side as the collapse?	37	46	17
Was the excavation unsupported?	92	8	0
Was there shoring in place, but not sufficient?	8	92	0
Did the contractor attempt to cutback the trench?	25	71	4
Did the degree of cutback meet safety regulations?	0	96	4

precautions were undertaken by the contractor prior to entering the trench. In 100 percent of the fatalities, the trench depth was greater than 1.5 m. Ninety-two percent of the fatalities resulted from unsupported trenches with 8% having shoring, albeit not sufficient. General Safety Regulation AR 448/83, Section 174(1) outlines the protection requirements for trenches greater than 1.5 m in depth.

Excavated spoil located too close to the edge of the trench may produce a surcharge load, which may be too great for the trench walls to withstand. According to Davies and Tomasin (1996), no materials or spoil should be placed within 1 m of the trench and the height of spoil heaps should be limited to between 1 and 2 m. This is further reiterated in General Safety Regulation AR 448/83, Section 173(3)(e). The analysis revealed that in 37% of the fatalities, a spoil pile was found along the same side as the collapse.

Cutback was attempted in 25% of the cases, however, none fulfilled the degree required to meet safety regulations set forth by the General Safety Regulation AR 448/83, Section 173(2). It states that where the cutback method is used, the walls must be cut back: 1) in hard and compact soil, to not less than 30 degrees from the vertical, and 2) in other soils to not less than 45 degrees from the vertical.

Section 169(b) of the General Safety Regulation AR 448/34 defines "hard and compact" as soil that can only be excavated by machinery and shows no sign of cracks after excavation. Such a soil may still move into a trench along pre-existing discontinuities in the mass of soil surrounding the trench. The performance of this mass reflects not only the properties of the soils shown in the bucket of the excavator, but also the properties of discontinuities in the soil mass which may be too widely spaced or poorly sampled to be obvious in the excavated soil. This discontinuity might be the freezing front, a sand or clay lens within the till or gravel, a sheared surface formed in the till as it was deposited under the glacier or even artificial surfaces created by previous construction activity.

There are, then, risks in relying on observations of the excavated soil to predict the performance of a trench. Some judgment is required and inexperienced workers may be optimistic in their evaluations. A more conservative approach might, for instance, require the approval of an engineer for unsupported slopes steeper than 45 degrees. This would be a step towards more elaborate classification schemes.

7 CONCLUSIONS

Following safety guidelines is the first step in reducing injuries and fatalities during trenching construction operations. However, this is not sufficient. Dividing information contained from accident reports written by OH&S into four main categories formed the basis for a heuristic approach for the identification and prevention of trench failure incidents. Workers can reduce the incidents of trenching-related fatalities by paying attention to these categories and adhering to pertinent safety regulations. Additionally, workers must also be particularly cognizant of the condition of the soil they are excavating. Prior knowledge of soil composition and ingress of water may dictate proper shoring requirements and cutback slope angles. Unfortunately, Alberta's General Safety Regulation AR 448/83 does not provide soil classification nor a description of hydraulic conditions, but rather a set of general criteria. Therefore, it is recommended that OH&S revisit General Safety Regulation AR 448/83 and develop a more conservative approach to cutback requiring the approval of an engineer for unsupported slopes steeper than 45 degrees. This would be a step towards a more elaborate classification scheme.

REFERENCES

Alberta Construction Safety Association Industries 1998. *Construction '97: A summary of occupational injury and disease 1993 to 1997.* Edmonton, Alberta.

Alberta Labour 1996. *Occupational injuries and disease in Alberta.* www.gov.ab.ca/lab.

Alberta Labour 1995. *Occupational Health and Safety Act and General Safety Regulations.*

Benjamin, J.R., and D.H. Cornell 1970. *Probability, statistics, and decision for civil engineers.* McGraw-Hill, New York, 684p.

Brown, L.C., K.L. Kramer, T.L. Bean, and T.J. Lawrence 1992. *Trenching and excavation: safety principles.* Report AEX-391, Agricultural Engineering Department, Ohio State University Extension, Columbus, Ohio.

Cruden, D.M. 1996. An inventory of landslides in Alberta, Canada. *Proceedings of the 7th International Symposium on Landslides.* Trondheim, Norway, June 17-21, 1996, 1877-1882.

Culver, C., G. Florczack, R. Castell, and G. Pelton 1990. *Analysis of construction fatalities – the OSHA database 1985-1989,* November.

Davies, V.J., and K. Tomasin 1996. *Construction safety handbook, 2nd Edition,* Thomas Telford Publishing, London, 297p.

de la Garza, J.M., D.E. Hancher, and L. Decker 1998. Analysis of safety indicators in construction. *J. of Constr.Engrg. and Mgmt.*, ASCE, 124(4), 312-314.

Everett, J.G., and P.B. Frank 1996. Costs of accidents and injuries to the construction industry. *J of Constr. Engrg. and Mgmt.,* ASCE, 122(2), 158-164.

Hancher, D.E., J.M. de la Garza, and G.K. Eckert 1997. Improving workers' compensation management in construction. *J. of Constr.Engrg. and Mgmt.*, ASCE, 123(3), 285-291.

Hinze, J. and L.L. Appelgate 1991. Costs of construction injuries. *J. of Constr.Engrg. and Mgmt.*, ASCE, 117(3), 537-550.

Lan, A., and J. Arteau 1997. Limits and difficulties in applying the Quebec safety code for the construction industry – the case of excavations. *Proceedings of the a Session on Construction Safety Affected by Codes and Standards,* ASCE, Minneapolis, MN, Oct. 5-8, 16-34.

Implementation of Safety and Health on Construction Sites, Singh, Hinze & Coble (eds)
© 1999 Balkema, Rotterdam, ISBN 90 5809 036 1

Safety and productivity issues in utility trenching

D.M.Abraham & D.W.Halpin
School of Civil Engineering, Purdue University, W.Lafayette, Ind., USA

ABSTRACT: For the majority of underground utilities placed in the U.S., open trench technology has been the preferred method. Advances in hydraulic excavators and the development of new stabilization technologies for trench walls have contributed greatly to increased productivity of the open-cut method. However, excavation and trenching remain, statistically, 'the most hazardous type of work in the U.S. construction industry.' This paper will explore some of the products used as protective systems in excavations and trenching, and will analyze safety features and productivity aspects of employing such systems. Two methods for ensuring trench safety will be explored. These include sloping and shoring. Selection criteria for appropriate shoring systems will be discussed. Productivity and safety associated with the use of conventional wood shoring systems and with hydraulic shoring will be studied. The paper will also discuss the simulation studies performed to analyze the productivity of trenching operations.

1 INTRODUCTION

A field study conducted by the National Institute of Standards and Technology (formerly the National Bureau of Standards) indicates that contractors cite two factors that contribute to poor safety planning in trenching and excavation processes. First, the soil classification in subpart P of the OSHA regulations for Excavations, Trenching and Shoring (NBS 1980) is unclear. Second, contractors and compliance officers often disagree over identification of soil types. Field studies of practices in excavation (trenching and shoring) also indicated that:

a) 50% of the work utilized sloped sides, and 43% utilized bracing systems.
b) Most trenches deeper than 15 feet were braced.
c) The majority of the contractors left their trenches open less than one day (NBS 1979).

OSHA's 1990 standard, the 29 CFR Part 1926, Subpart P provides clearer guidelines on safety standards in utility trenching and increases the flexibility of contractors in the choice of protective systems (TSSA 1992). Thus, contractors have the responsibility to analyze the production rates and safety features associated with protective systems used in trenching.

2 NATURE OF CAVE-INS IN TRENCHING OPERATIONS

Every type of excavation unbalances the forces in the soil that were in equilibrium prior to the excavation. In the typical trench excavation, the impact of the excavation is to introduce unbalanced horizontal forces to the walls of the trench. Unless the excavation opening is consistent with the internal angle of friction, failure starts immediately after excavation. Depending on the type of soil and its shear strength properties, collapse may be delayed for a certain time. In cohesive soils, the walls of the trench can maintain their verticality for some time, whereas in dry or fully saturated granular soils, the trench walls fail rapidly. Therefore, after a trench is opened, the unbalanced horizontal forces on trench walls must be balanced by some type of support.

3 ASSURING TRENCH SAFETY

There are three ways of ensuring the safety of workers in a trench. These are: sloping, shoring and shielding. The potential of a slope failure in a trench decreases as the sides of trench are cut with flatter slope. The appropriate slope of a given trench must be determined by careful consideration of the characteristics of the soil including the type of soil and the short-term and long-term shear strength, existence of ground water, etc.

In trench shoring, the basic idea is to restore the equilibrium of horizontal forces acting on the trench walls. The function of shoring is to bear against the exposed side walls of the trench so that the horizontal forces acting on the trench walls will be balanced. Shores are under compression when used to support trench walls in a typical trench shoring design problem. Parameters that influence the design of shores include the compressive force on each shore, the distance between two consecutive points of shoring support, shear strength values of the soil, level of the ground water table, etc. Another factor that must be considered in the design and installation of trench shores is the need for adequate working space in the trench. Installation of shoring should be from top to bottom. Likewise, removal of shores must be from bottom to top. The key point is that the worker installing or removing the shoring system is always in the protected zone.

A trench shield is defined as a structure that is able to withstand the forces imposed on it by cave-ins and thereby protects workers within the structure. Shields can be permanent structures or they can be designed to be portable and moved along as the work progresses. Shields used in trenches are usually referred to as "trench shields" or "trench boxes." Trench shields, when appropriately used, permit narrower trenches. This reduces the amount of excavation, bedding, backfill, compaction and restoration work required.

The main difference between shoring systems and shielding systems is that the former is designed to prevent the failure of the trench walls, while the latter is designed to protect the workers within the trench when failure occurs. Thus, the shoring system is an active system while a shielding sytem is a passive one.

4 SELECTION OF APPROPRIATE SHORING SYSTEMS

4.1 Timber shoring

Timber shoring is a traditional type of shoring system in which, structural elements made of timber (Figure 1) are used to prevent failure of trench walls. Depending on the soil type, a maximum horizontal spacing is usually specified. When the soil conditions are poor and the base soil has a potential to run into the trench between consecutive shores, additional vertical timber or plywood has to be installed to provide close sheeting. This may not be necessary in a cohesive soil with high shear strength whereas for poor soil conditions, close sheeting is mandatory.

As the quality of the soil deteriorates (i.e., as the soil type changes from A to C), the required cross sections of vertical upright members and cross braces will be larger, leading to congestion in the trench. This problem can be solved by replacing the timber vertical upright sections with steel sections (typically H-sections).

4.2 Pneumatic and hydraulic cross braces

These devices act as modern substitutes for traditional timber and screwjack struts. They are generally used in conjunction with timber or steel sheeting and wales. Since the braces are each independently installed, they are used with most types of vertial uprights or horizontal

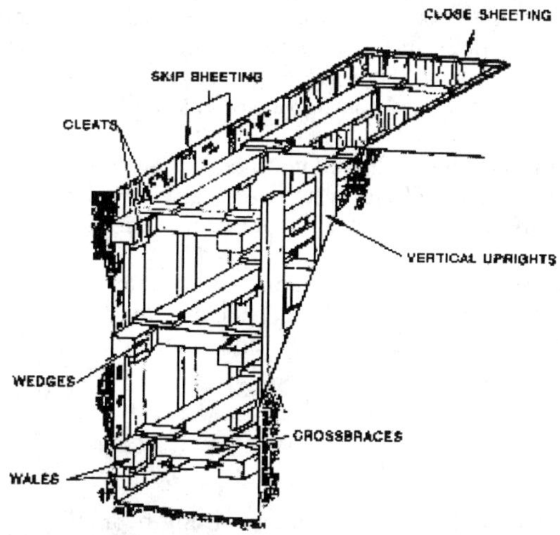

Figure 1. Timber shoring

wales and often have swivel-end adapters for shoring sloped walls. Some air and hydraulic braces may be installed and removed from above the trench, while others require traditional installation from top down and removal from bottom up. When using a pneumatic activated strut, the cylinder must always be mechanically locked to prevent contraction under load, as air is compressible.

4.3 Aluminum hydraulic trench shoring

These pre-engineered systems consist of aluminum hydraulic cylinders (cross braces) used in conjunction with vertical rails (uprights) or horizontal rails (walers). Such a system is designed specifically to support the sidewalls of an excavation and to prevent cave-ins.

These systems are constructed of lightweight, high strength aluminum alloys that permit easy handling. They are installed and removed from above the ground to insure worker safety throughout the shoring process. The equipment operates on a water-based fluid mixture and is pressurized with a gage-regulated pump to insure even pre-load of the trench walls. Various extension systems are available to extend the operating range of the cylinders and they may be field installed to quickly adapt to changing trench conditions. The cylinders also incorporate a safety bleed-off hole to prevent expansion beyond their working width.

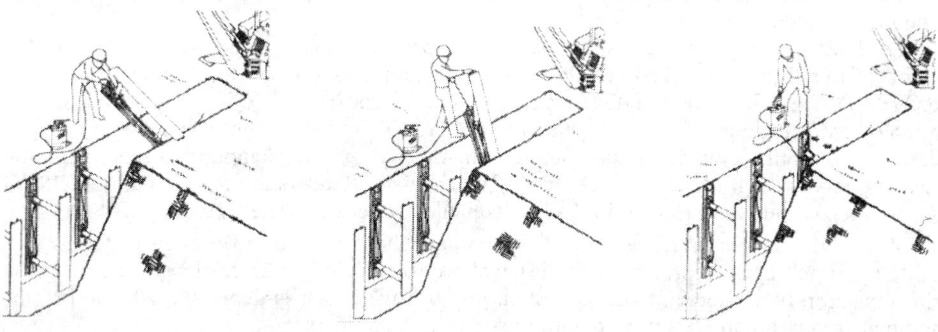

(a) (b) (c)

Figure 2. Installation Procedures for Vertical Shores

Some of the common hydraulic trench shoring systems used today are: vertical shores and waler systems. Vertical shores (Figure 2) are normally installed and removed by one person, and are quick and easy to install from above the trench. They are extremely durable and can be reused many times. Vertical shores are primarily suited for stable ground conditions with good cohesion.

Waler systems are designed for use in unstable soils in conjunction with close or intermittent sheeting along the walls of the trench. These systems are also installed and removed from above the ground but may require the use of a bachhoe or excavator in order to position them properly in the trench.

In addition to vertical shores and walers, there are a number of specialty systems designed for more specific applications. These include lightweight hydraulic shields for municipal maintenance and repair, manhole shoring devices for square and rectangular shaft excavations, and a host of other hydraulic shoring devices.

5 PRODUCTIVITY ANALYSIS OF SHORING METHODS

A study (Halpin, et. al 1995) of the productivity of trenching operations using hydraulic and timber shoring systems, examined two sets of parameters: environmental and shoring system related parameters. The environmental parameters included:

a) Trench properties that were governed by soil conditions, presence/absence of ground water, method of access into the trench, disruptions in trenching that were caused by the presence of obstacles, etc.
b) Weather conditions.
c) Presence of toxic wastes in the vicinity of the trench.

The parameters related to the shoring system included:

a) The time required for installation of the shoring device.
b) Experience of the persons installing the shoring device, operating the backhoe, laying the pipe, etc.

DISCO, a graphical construction simulation system, was employed to determine the productivity of the trenching operations under two cases:

1. in dry and cohesive soil conditions
2. in dry and non-cohesive soil conditions.

5.1 The DISCO Program and CYCLONE models

DISCO (Dynamic Interface for Simulation of Construction Operations) is a prototype program developed for the modeling and simulation of construction operations. The CYCLONE modeling methodology is employed for the abstract model building and its set of standard modeling elements constitute the basic icons in the DISCO graphical menu bar. Readers are referred to Halpin and Riggs (1992) for details regarding the CYCLONE modeling methodology.

The DISCO program serves as both a preprocessor and a post-processor for the MicroCYCLONE program (Halpin 1990), which is a microcomputer-based program that uses the CYCLONE methodology. DISCO generates the MicroCYCLONE input file and takes the chronological list file generated by running the MicroCYCLONE program as its input. It then recreates the entire course of simulation dynamically on the computer screen with the schematic model diagram used as the display mechanism. Readers are referred to the DISCO User's Guide (Huang and Halpin 1993) for detailed description of the features of DISCO.

CYCLONE models were developed to simulate the trenching process in different soil conditions. Each cycle represented the excavation, installation of 20 ft of pipe, and backfill and compaction of a "standard" trench which was 6 ft wide, 8 ft in depth and 20 ft in length. Durations used in the models were deterministic.

5.2 CYCLONE Model 1: Trenching in dry and cohesive soil conditions

In this model (Figure 3), the simulation starts with the excavation (Node 2) of one section (4 lf) of the trench. The excavation is followed by the shoring process. The sequence of tasks in the shoring process is as follows: connect the pump to the shore, position the shore across the trench, unfold the shore, drop the shore into the trench, pressure the pump and finally disconnect the pump. Once the pump is connected to the shore, the other activities can continue without disruption.

Once the excavation and shoring of five sections are complete (as depicted by Node 11), the installation of the pipe begins. After the pipe has been installed, the shores are removed. This is followed by backfilling the trench (Node 13).

5.3 CYCLONE Model 2: Trenching in dry and non-cohesive soil conditions

In granular soils, trenching operations can become problematic in dry conditions. Such soils have a potential to cave in very rapidly during trenching operations because of the lack of cohesion to withstand the tensile stresses that develop in unsupported trenches. Hence, the trenching process in such soil conditions is slightly different than that in cohesive soil conditions. A fin board has to be installed to support the exposed trench walls. The simulation model (Figure 4) is modified accordingly. In this model, a new work task (Node 4), namely, "lay fin board" is inserted after the work task "excavate one section" to model the placement of fin boards after the excavation of one section. The queue nodes for the excavator and shoring labor are also altered slightly in this model, in order to reflect the modified relationships between activities.

5.4 Comparison of productivity results

The two models were simulated using the durations shown in Table 2. These durations were estimated using published data and former MPDM studies done at Purdue University.

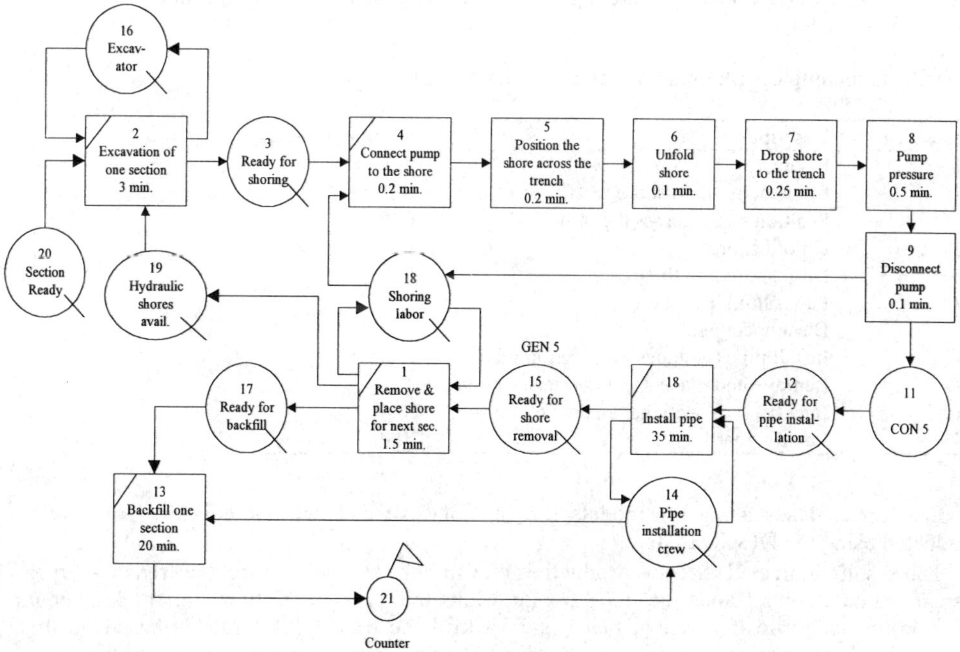

Figure 3. CYCLONE model – Trenching in dry and cohesive soil conditions

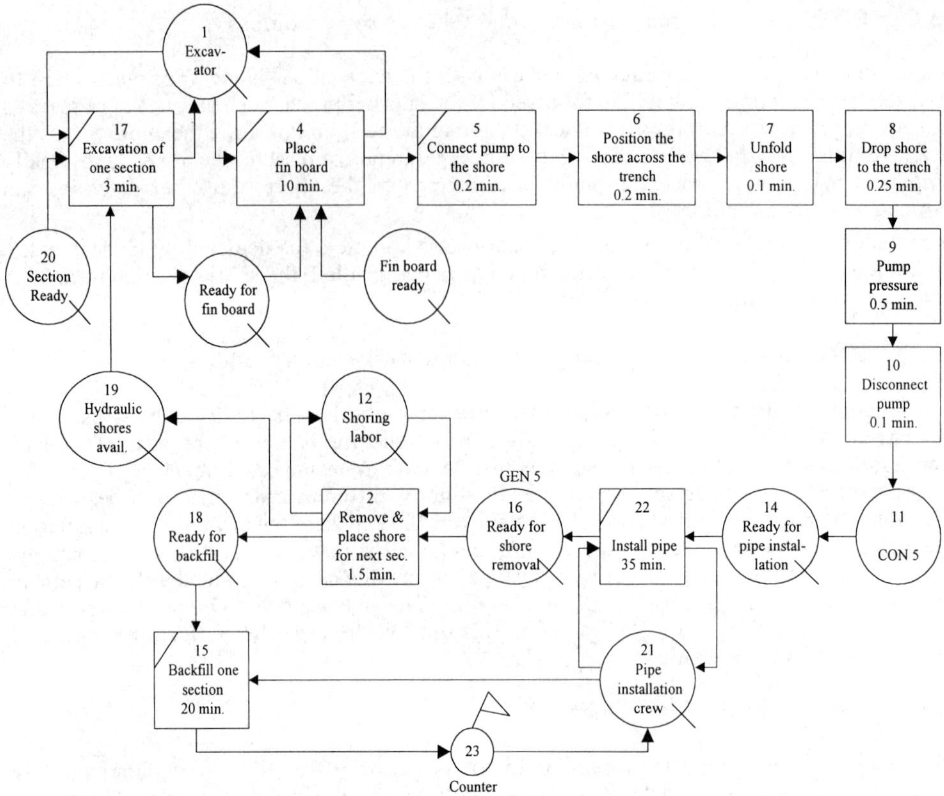

Figure 4. CYCLONE model – Trenching in dry and non-cohesive soil conditions

Table 2. Trenching operation sequence and work task durations.

Work Task	Description	Duration (minutes)
2	Excavate 4 ft	3.0
4	Connect pump to shore	0.2
5	Position shore across the trench	0.2
6	Unfold shore	0.1
7	Drop shore into the trench	0.2
8	Pump fluid into shore	0.5
9	Disconnect pump	0.1
18	Install pipe (includes preparation time	35.0
1	Remove and place shore for next section	1.5
13	Backfill (4 ft sections, assume 6” lifts)	20.0
4	Lay fin board	10.0

Productivity analyses using the models for dry cohesive and dry non-cohesive soils were conducted using the DISCO system.

The results indicated that the production rate in cohesive soils using the trench shoring system and based on a standard 6 ft x 8 ft x 20 ft long trench was 8.53 ft/hr, i.e., it takes about 2.5 hours to excavate the trench, place and backfill the trench. This rate is based on the assumption that 6 sets of shoring are available. The rate for a similarly configured trench in non-cohesive soils (using the fin board inserts) was found to be 7.84 ft/hr. The work associated with the added task of installing the fin board leads to a slight reduction in productivity.

Figure 5 shows the change in production rate as the number of shores is varied in each case. In both cases, five to six shores appear to be the optimum number of shores required to maintain a reasonable level of production. When the number of shores is increased beyond six, the marginal change in production is negligible. Further analysis (see Figure 6) shows that increasing the number of shores increases (at an increasing rate), the percentage of time they are not in use.

Conventional wood shoring that entails the use of vertical elements, lagging, cross bracing, and walers would require (Puerifoy 1975):

a) an additional eight hours for driving the 2-by 12-in. lumber 8 ft long for sheeting with two horizontal rows of 4-x 6-in. wales on each side of the trench.
b) an additional two hours for placing the trench braces 4 ft apart along each row of wales.
c) An additional two and one-half hours for removing the sheeting, wales, and braces.

Given this information, it would take at least 14 hours to install conventional shoring, excavate, install and backfill 20 ft of pipe. Based on simulation analysis, the hydraulic shores yield a production rate increase, which is 5 to 6 times that of conventional labor-intensive wood shoring systems.

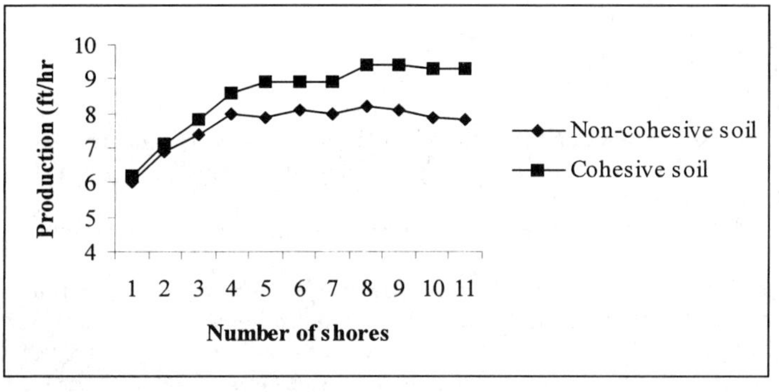

Figure 5. Change in production rate with increase in the number of shores

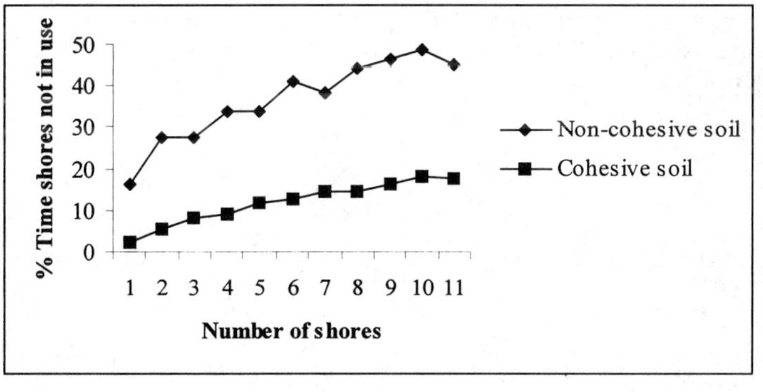

Figure 6. Percentage of time that the shores are not in use.

6 CONCLUSIONS

Two methods for securing trenches, namely sloping shoring, were discussed. Trench shoring devices provide a quick means for maintaining a safe, productive environment, since shoring systems are installed from above ground before construction personnel enter the trench.

Preliminary DISCO simulation models were developed based on literature review, demonstrations and interviews with users of traditional and hydraulic shoring. Productivity analysis using the simulation models indicated that the production rates (including excavation, pipe placement and backfilling the trench) using the hydraulic shores was 7.84 ft/hr in the case of non-cohesive soils (using fin board inserts) and 8.53 ft/hr in the case of cohesive soils. These rates show a five to six-fold increase in production compared to similar trenching operations performed using conventional wood shoring systems.

Apart from the improved production, the more important advantage in using the hydraulic shores in trenching operations is the additional safety offered by the system. At no time, is a worker required to be in an open trench. Accidents involving workers in open, unshored or inadequately shored trenches carry a high cost not only financially, but also legally. Proper planning of trenching operations requires careful attention to detail, including analysis of hidden obstructions, disturbed ground, water conditions, weather conditions, vibrations and equipment usage. Hydraulic shoring systems provide an efficient mechanism by which many of these considerations can be addressed in providing a safe environment for trenching operations.

REFERENCES

Halpin, D. W., D. M. Abraham, R-Y. Huang, and Y. Ersoz 1995. *Speedshore Trench Shoring Vs. Conventional Shoring Methods: A Simulation Based Comparative Study*. Project Report, School of Civil Engineering, Purdue University.

Halpin, D. W. and L. Riggs 1992. *Planning and Analysis of Construction Operations*, John Wiley and Sons, NY.

Huang, R-Y and D. W. Halpin 1994. *Disco User's Guide*. Division of Construction Engineering and Management. Purdue University, West Lafayette, Indiana.

National Bureau of Standards (NBS) 1979. *An Analysis of the Responses from an Associated General Contractors of America Survey of Trenching and Shoring Practices*. Washington, D.C.

National Bureau of Standards (NBS) 1980. *Soil Classification for Construction Practice in Shallow Trenching*, Prepared for the Occupational Safety and Health Administration. Washington, DC.

Puerifoy, R. L. 1975. *Estimating Construction Costs*. McGraw Hill Company, New York.

Trench Shoring and Shielding Association (TSSA) 1992. *An Introduction to Modern Trench Shoring and Shielding*. New York.

Implementation of Safety and Health on Construction Sites, Singh, Hinze & Coble (eds)
© 1999 Balkema, Rotterdam, ISBN 90 5809 036 1

Risk management for trenchless technology projects: Applications to horizontal directional drilling

S.T. Ariaratnam, E.N. Allouche & J.S. Lueke
Department of Civil and Environmental Engineering, University of Alberta, Edmonton, Alb., Canada

ABSTRACT: The management and subsequent mitigation of risk is particularly important in underground construction due to the uncertainty in estimating subsurface ground conditions. This paper focuses on the management of risks associated with a special type of trenchless underground construction, Horizontal Directional Drilling (HDD). Traditionally, horizontal directional drilling has focused on utility and pipeline applications, however, a recent trend has been towards the geo-environmental arena. The first section of this paper details the identification and management of risks associated with utility and pipeline installation, while the second section documents risks on environmental projects. HDD contractors are discovering the need for a comprehensive safety and training program to alleviate potential accidents, especially on remote sites. Through the identification of various types of risks associated with both utility installation and remediation projects, effective methods of risk management can be developed. As more and more HDD contractors enter into the environmental remediation market, new risks must be identified and new safety procedures and manuals developed to minimize loss.

1 INTRODUCTION

It is virtually impossible to discuss construction projects without mentioning the term "risk." Risks are inherent in virtually every type of construction project and have been addressed by previous researchers (Al-Bahar and Crandall 1990, Kangari 1995, Hinze and McClelland 1997). The area of risk, however, has not been formally addressed in the area of underground trenchless technology. This paper focuses on risks associated with a special type of underground construction, Horizontal Directional Drilling (HDD). This trenchless construction operation has many of the same risks as traditional construction; however, there are distinct differences in their magnitude. Information and experiences provided by several contractors and other individuals working in the HDD industry form the basis for this paper.

While written mainly from the contractor's perspective, owners and consultants can greatly benefit from gaining an understanding of the risks being faced by the HDD contractor. By recognizing these risks prior to contract tendering, the owner/consultant may reduce the average bid price submitted by sharing ownership of these risks.

2 AN OVERVIEW OF HDD

Horizontal Directional Drilling (HDD) makes it feasible to drill horizontal boreholes between 50 mm and 1200 mm in diameter beneath surface/subsurface obstacles over extended horizontal distances. Traditionally, HDD has been used in the oil, gas, and utilities industries for the installation of pipelines and conduits, although applications of HDD for the remediation and sampling of contaminated soils and ground water has been gaining momentum in recent years (Allouche et al. 1998, Wilson 1995). Other applications of HDD technology include: de-watering; pipe reaming; cathodic protection; and slope stability.

Typical HDD rigs use a chain drive, rack and pinion drive, or a hydraulic cylinder to push or pull a carriage to advance and retract the drill string. A carriage slides forward on a frame and pushes the pipe into the borehole. The carriage is then retracted and a new pipe segment is attached. HDD rigs are classified as mini, midi, or maxi rigs depending on their thrust/pull-back and torque capacities, as shown in Table 1. Regardless of their classification, typical drilling rigs operate using the method described above.

The bore is launched from the surface and the pilot bore proceeds downward at an angle until the necessary depth is reached. Then the path of the bore is gradually brought to the horizontal, and the bore head is steered to the designated exit point where it is brought to the surface following a curved path. A directional monitoring device, located near the head of the drilling string, is used to track the position of the drilling head. After the pilot string breaks the surface at the exit location, the bit is removed from the drill string and replaced with a back-reamer. The pilot hole is then back-reamed, enlarging the hole to the desired diameter while simultaneously pulling back the pipeline behind the reamer. The above description is typically referred to as a continuous borehole. In some instances it is desired to drill a single-ended borehole, commonly named a blind borehole. In this case construction, reaming and all other activities that take place are initiated at the entry pit. Blind boreholes are used only in environmental applications for either collecting soil samples or installing horizontal wells beneath structures.

During the drilling process, the bore path is traced by interpretation of signals sent by electronic sensors located immediately behind the drill head. Continuously along the drilling path, the operator receives information regarding the position, depth and orientation of the drilling tool, allowing navigation of the drill head towards its intended target. Accuracy of tracking the drilling head varies according to the method and type of equipment used. Typically, depth and position readings may be assumed to be accurate within ±5% in terms of the drilling head true depth, although tracking systems which allow accuracy of ±2% in both plan and profile, regardless of the borehole's depth, are available. Current commercial HDD equipment can operate in a wide range of soil conditions, from extremely soft soils to full face rock formations with unconfined compressive strengths of 210 MPa (30,000 psi).

3 IDENTIFICATION OF RISKS

The term "risk" in this paper refers to the an occurrence or situation which may results in: 1) compromising the health and safety of the crew or the public; 2) the of loss or damage to equipment; 3) damage to surrounding utilities; 4) damage to the product installed or a reduction

Table 1. General Specifications of Directional Drilling Rigs (May 1994)

	MINI RIGS	MIDI RIGS	MAXI RIGS
Thrust/Pullback	< 90 kN	90 kN – 355 kN	> 355 kN
Maximum Torque	< 2.7 kN-m	2.7 kN-m – 27 kN-m	> 27 kN-m
Drilling Speed	> 130 rpm	130 - 200 rpm	< 200 rpm
Carriage Speed	> 30 m/min.	27 – 30 m/min.	< 27 m/min.
Carriage Drive	Cable or Chain	Chain or Rack & Pinion	Rack & Pinion
Drill Pipe Length	1.5 – 3 m	3 – 10 m	10 – 12 m
Drilling Distance	< 210 m	210 – 600 m	> 600 m
Power Source	< 150 hp	150 - 250 hp	>250 hp

in its performance; or, 5) excessive lost of time and unnecessarily incurred costs. The technical and financial success of a project hinges on a contractor's ability to minimize these risks.

3.1 Utility Construction

The installation of utilities is the most common application of horizontal directional drilling. This typically involves the installation of relatively small diameter (up to 150 mm) lines of relatively short distances (up to 300 m) and includes the installation of telecommunication (eg., fiber-optic), electrical conduits, and natural gas distribution systems.

3.1.1 *Subsurface Soil Conditions*

In order to plan a boring project effectively, adequate geotechnical information should be obtained. Depending on the soil structure, specific choices should be made regarding the type of drilling rig, drilling head, and back-reamer to be used on a particular project. Inaccurate and incomplete information on soil condition could results in the loss of down hole equipment and drilling rods. For example, the presence of voids can cause loss of drilling fluid circulation, while a gravel bed or quick sand condition could cause the drill string to jam.

3.1.2 *Existing Underground Obstacles*

An accurate survey of the proposed alignment is necessary. Accidentally damaged utility lines causes inconvenience to the affected consumer, reducing their productivity and level of comfort. Utility line strikes are costly for the utility companies and those who are responsible for the damage, and more seriously can cause injury or even death. Striking a power cable can electrocute workers, while looking at severed fiber optic cables can affect eyesight. The greatest danger comes when live natural gas lines are damaged, potentially causing explosions. The risk of accidental strikes increases with the desire to use directional drilling to place new lines near existing lines. One contractor in a recent project had to maneuver through 43 buried utilities over a distance of 90 m. This risk is of greater concern when the operator and/or drilling head locators lack the necessary experience or training.

3.1.3 *Limitation of Tracking Capabilities*

When using a walk-over tracking system, the ability to trace the path of the drilling head is greatly influenced by the presence of electromagnetic disturbances such as overhead power lines, buried metal objects, or reinforcing steel. Reduced ability to navigate the drilling head increases the risk of drilling outside of the designated "drilling window", and can lead to a collision with surrounding utilities. Another type of navigation problem may occur when the borehole profile at a particular location is at a depth greater than the operating range of the walk-over system.

3.1.4 *Experience of Crew*

A contributing factor influencing the success of a project is the crew running the drilling equipment. Even with all the utilities marked, an inexperienced crew has a higher risk of striking buried utilities. An experienced and well-trained crew has a better chance of navigating the drill around buried obstacles. Another consideration is that an experienced crew more likely would be able to salvage a drilling project when unexpected conditions are encountered.

3.1.5 *Inclement Weather*

Inclement weather conditions can have a great effect on the drilling project. Though a drilling crew and machinery can operate in many conditions, extreme cold can freeze the drilling mud, making it impossible to operate or damage the pumps. If the ground freezes, the drilling operation is slower as productivity drops. The effects of water and cold temperatures has varying affects on the soil depending on the its composition. It is important to identify how weather conditions will affect the ground at a particular site. While beyond the contractor's control, being prepared for extreme whether conditions (i.e. steam trucks; proper protected clothing; enclosed operator shack) can help to reduce the negative impact on productivity and safety (i.e. reduce exposure to cold, frost-bite, etc.).

3.2 Environmental Projects

Aside from the risks identified during utility construction, environmental projects present several additional financial, health, and safety risks. The presence of contaminants places the crew members at immediate risk, while at the same time placing the company in a position of long-term liability. Knowledge of the type of the contaminants suspected to be present on site is essential to ensuring that all proper precautionary measures are taken and the crew is provided with the necessary protective gear. The drilling crew must be appropriately trained to operate the required protective gear (e.g., respirators) and monitoring equipment. Failure to provide a full disclosure could result in harmful effects to the workers. Contractors should insist on full disclosure clauses in their contracts, as these clauses transfer the risk of unknown site contaminants from the contractor to the owner. Special care must also be taken when the contaminants on the site are radioactive or are a by-product of chemical warfare. Exposure to such contaminants can be potentially lethal, and the long-term liability should be considered before entering such a contract. When entering an environmental project, the contractor must be aware of the additional expenses incurred in this type of project including proper protective gear, pre-construction safety training, the need to decontaminate the equipment before leaving the site, and the need to treat/dispose of the returned drilling fluids as a hazardous material.

The technical success of an environmental project relies on the ability to restore the formation's initial permeability (well development) and closely follow the alignment called for by the remediation engineer. Proper well development will maximize the well's zone of influence and thus its productivity. Accuracy in installation alignment is necessary since the target zone (e.g., free pool of hydrocarbons on top of the GWT) may be only a few feet deep. The contract must state the exact drilling, reaming, and installation procedures. Also, type and duration of well development efforts must be established before construction commences. Defining an acceptance criteria for the well is crucial, since the contractor is not responsible for the design of the well, and thus can not guarantee the well performance in term of yield.

4 METHODS TO MINIMIZE RISKS AND LOSSES

4.1 Utility Construction

After the potential hazards are identified, risk management and mitigation begins. Part of risk management is finding ways to minimize, if not to eliminate, the identifiable risks from the job. The following section identifies various risk reduction methods a contractor can adopt in order to reduce the frequency of occurrence and impact of various risks during utility construction.

4.1.1 *Design Considerations*

The contractor must become familiar with the bore design and must inspect the route of the proposed pipeline, and be familiarized with the conditions likely encountered along the route. Vertical boreholes drilled along the proposed route may provide valuable information regarding the type of soil, depth of the water table, and depth of the bedrock, all of which affect the type of equipment required to perform the operation. In some cases, although the bore path may be designed correctly, inaccurate surveys by the design engineer may cause difficulties in implementing the design. There could be locations along the route where the cover depth is too shallow, for example when crossing beneath a river. Another difficulty could occur if the radius of curvature is too sharp for the proposed pipeline product or for the contractor's equipment. With the proper pre-construction survey, inconsistencies in design can be found and corrected.

4.1.2 *Pre-Construction Site Investigations*

Contractors should perform pre-construction site investigations. During this investigation, issues that relate to site access should be identified. Ensuring that the contractor can access the site without disrupting traffic or interfering with adjacent site operations is essential. There must be adequate space on the site for the placement of all necessary equipment. On the rig side of the line, there must be room for the rig, drilling rods, slurry mixing facilities, slurry pump and in some cases, a cuttings settlement pit. On the pipe side of the project, there may also be a cuttings settlement pit along with welding/fusion equipment and the pipeline product. The product pipe can take a considerable amount of space, as it is preferable that the pipe be inserted in a continuous pull. As mentioned previously, knowledge of ground conditions to be encountered is important in ensuring successful execution of the project, as it determines the type of rig, drilling head, and drilling mud which are to be used on a particular project. Traditionally, vertical boreholes are used to characterize the soil conditions on a particular site. The contractor should ensure that the borehole has been drilled and logged down to the bedrock, or beneath the installation target depth. When the size of the project does not justified the cost of vertical drilling, a hand held core barrel may be used to collect soil samples along the route. Visual inspection of an exposed river bank or a near by excavation may also reveal valuable information regarding the type of soil to be encountered.

Traffic control may be required in urban areas or near highways and any site specific hazards should be identified. In environmental projects, some sites may have ground and air borne contamination which require the use of specialized protective equipment. It is also essential to identify the availability of fresh water required for the mixing of the slurry. Through a good site investigation, many uncertainties may be eliminated, in turn reducing the risk during construction.

4.1.3 *Differing Subsurface Site Conditions*

As with any underground construction, the uncertainty of what lies below ground can influence the success of a drilling project. Most ground investigations consist of a sample of the conditions expected, however, that is all they are - a sample. Quite often, the conditions expected a few meters to either side of the sampling location may vary. Subsurface conditions that potentially cause difficulties in a drilling operation include artesian aquifers, quick sand conditions, and gravel/cobble beds. If an artesian aquifer is drilled into, depending on the amount of hydraulic head, there is the possibility of damage to equipment as water could be forced back up the bore. On the other extreme, when drilling through a gravel bed, slurry pumped down the bore will be lost to the formation rather than creating a mud cake. Course granular material is also more difficult to navigate the drill string through.

Contractors should be protected from excessive loss of time and equipment under such circumstances. This may be done by entering a contract type where both the contractor and the owner share the risk, or by incorporating an appropriate "different conditions" clause into a lump sum contract.

4.1.4 *Existing Utility Locations*

In utility installation, it is important when working near existing service lines, that these lines be identified and marked (i.e. first call). This action will reduce the chances of striking an existing utility line. It should be noted that some buried utility detection equipment may not be able to detect the proper depth or location of buried utilities if there are many buried utility lines in the same location, or if there are large metal structures present. In these situations, hydro-excavation may be used to expose the utilities prior to commencement of drilling. The cost of marking the lines and/or performing excavation investigations is small when compared to the cost of striking a buried utility. Not all utilities or buried objects are marked. Therefore, it is recommended to first walk along the drilling alignment and look for depressions in the ground which may indicate the possible location of any buried objects. Walking along the alignment with a "walk-over" locating device can help to identify buried metal objects and power lines.

4.1.5 *Experience of Crew*

A contributing factor influencing the success of a project is the experience and amount of training of the crew running the drilling equipment and locating the drilling head. Even with all the utilities marked, an inexperienced crew has a higher risk of striking buried utilities. Additionally, a more experienced crew would more likely be able to salvage a drilling project when unexpected conditions are encountered. The investment of time and money in training or hiring an experienced crew is minimal compared to repairing damaged utility lines, mechanical damage to the machinery, loss of drilling equipment underground, and safety risks. Training is particularly important in a prosperous market where contractors are expanding rapidly and experienced personnel are in shortage.

4.1.6 *Site Safety Procedures*

As with any operation, it is important to work safely and have contingency plans in place to handle emergencies. Safety procedures should be well established and all personnel on site should be instructed on protocol. The following is a list of several issues that should be addressed when designing a safety procedure:

- Maintaining properly trained persons who could administer first aid on site,
- First aid equipment readily available, and
- Knowledge of the travel route to the nearest hospital or care facility.

When working in a remote location, the following additional issues should be addressed:
- A vehicle should be designated to transport injured personnel to a hospital or care facility,
- Procedures to call for helicopter evacuation on very remote sites,
- Identification of the nearest landing site for a helicopter.

It is imperative to develop site specific safety procedures as every site has unique safety issues that need to be addressed. For example, industrial sites often have a safety procedure in place. Crews should be aware of these safety procedures, and how to notify the proper personnel when an emergency situation arises.

4.1.7 *Site Meetings*

Another important mechanism used to minimize risk is site meetings. The value of site meetings is sometimes overlooked. They provide an opportunity for all parties involved to convene and discuss potential problems, establish lines of communication, identify potential risks, and to work together in completing the project. At pre-construction meetings, safety issues such as site layout and potential underground hazards should be discussed. Ensuring that all personnel are aware of potential ground hazards such buried power, gas, fiber-optic, and chemical lines reduces the risk of exposure to dangerous situations. While sometimes underrated, proper utilization of site meetings serves to reduce risk on a project.

4.2 Environmental Projects

While all of the above risk minimization methods are also applicable to environmental projects, the special nature of these projects requires the following additional issues to be addressed.

4.2.1 *Protective Gear*

After determining the type of contamination found on a site, proper protective gear should be attained. The gear can range from respirators and face shields to fully sealed protective suits with dedicated compressed air supply. The amount of protection required is determined by the type of contaminant(s) suspected to be present at the site as well as the comfort level of the crews working.

4.2.2 *Training on Safety Procedures*

In all environmental remediation projects, personnel should receive sufficient on-site training on emergency and evacuation procedures, handling of protective gear, protective gear exposure limits, location of first aid facilities, and operation of any necessary monitoring devices.

4.2.3 *Contracts*

When planning to undertake a project where there are contaminants present, the incorporation of a full disclosure clause in the contract in essential. A full disclosure clause forces the owner to have a thorough knowledge of the contaminants on his site, and to inform the contractor of these contaminants. This clause transfers the risk from the contractor to the owner. Any costs or damages incurred from the discovery of contaminants not identified by the owner are the responsibility of the owner. This protects the contractor against future health issues, as many side effects resulting from exposure to contaminants are not immediate. Future health problems resulting from personal contamination become the responsibility of the owner, not the contractor if the damage results from undisclosed contamination. The full disclosure clause is an important method of minimizing health risks for a contractor involved in an environmental remediation project.

4.2.4 *Migration of Contaminant*

During environmental remediation projects involving the removal of contaminants, it is important to realize that the horizontal borehole itself can serve as a conduit to enhance contaminant migration. Penetration of an impermeable layer may cause the contaminant to migrate into previously unaffected zones resulting in health and safety issues. These considerations must be evaluated when designing the alignment of the borehole. Additionally, the ends of the borehole should be sealed with grout before leaving the site. The contractor is not legally responsible for any breach of containment due to the drilling operation, provided that the borehole is installed within the tolerances specified by the remediation engineer.

4.2.5 *Drilling Disposal and Equipment Decontamination*

During the drilling process, a large volume of drilling fluid is utilized. While some of the drilling fluid is lost to the surrounding strata, a significant quantity, depending on the soil composition, may end up in collection ponds. This fluid is considered, and should be treated as a hazardous material. Proper removal and disposal procedures for the returned drilling fluids should be determined before the project commences as they may vary with the type of contaminant encountered. Also, the cost of removal and disposal of drilling fluid should be estimated and included in the cost estimate before the project commences.

Before leaving the site, all equipment involved in the remediation project must be decontaminated. The availability of proper decontamination facilities should be determined before the project commences. If a wash down facility is not available, the cost of renting or installing such a facility, including the costs of any special solvents or cleaners, should be determined and incorporated into the cost estimate. This step will greatly reduce the occurrence of future public health issues.

5 SUMMARY AND CONCLUSIONS

While written from the contractor's point of view, this paper is intended to provide owners and consultants with an understanding of how to identify and manage risks associated with horizontal directional drilling projects. This is essential in the prevention of health and safety issues that may be potentially hazardous on the construction site. The next step should be committed to developing a method of quantifying risk using mathematical models to enable both contractors and owners to determine the extent of risk for a given HDD project.

REFERENCES

Al-Bahar, J.F., and K.C. Crandall 1990. Systematic risk management approach for construction projects. *J. Constr. Engrg. and Mgmt.*, ASCE, 116(3), 533-546.

Allouche, E.N., S.T. Ariaratnam, K.W. Biggar 1998. Environmental remediation using horizontal directional drilling: applications and modeling. *Practice Periodical of Hazardous, Toxic, and Radioactive Waste Mgmt.*, ASCE, 2(3), 93-99.

Hinze, J., and T. McClelland 1997. Risk assessment in underground construction: uncertainty is certain. *No-Dig Engineering,* 4(3), 19-22.

Kangari, R. 1995. Risk management perceptions and trends of U.S. contractors. *J. Constr. Engrg. and Mgmt.,* ASCE, 121(4), 522-429.

May, D. 1994. The use of horizontal wells for subsurface soil and aquifer restoration. *Drilling Technology,* Vol. 56, 227-239.

Wilson, D.D. 1995. Introduction to 1995 NGWA environmental horizontal well seminar. *The Green Book,* Charles Machine Works Inc., Perry, Okla., 11-13.

19 Medical and health management – I

Implementation of Safety and Health on Construction Sites, Singh, Hinze & Coble (eds)
© 1999 Balkema, Rotterdam, ISBN 90 5809 036 1

Prevalence of carpal tunnel syndrome among construction apprentices

J.C.Rosecrance, T.M.Cook, C.L.Zimmermann & D.Johnson
Injury Prevention Research Center, Department of Preventive Medicine and Environmental Health, The University of Iowa, Iowa City, Iowa, USA
K.P.Farrell
Department of Physical Therapy, St. Ambrose University, Davenport, Iowa, USA

ABSTRACT: The purpose of this cross-sectional study was to investigate the prevalence of carpal tunnel syndrome (CTS) in construction workers during their apprenticeship. To determine the prevalence of CTS, 207 apprentices from the sheet metal and electrical trades were evaluated. Nerve conduction studies and clinical exams were performed on both hands of each apprentice. A case definition of CTS included symptoms as well as an abnormal nerve study across the carpal tunnel. The data indicated that 10.6% of the apprentices fit the case definition of CTS in at least one hand. The prevalence of CTS increased with the number of years in the apprentice program. Although many apprentices had symptoms, history, and nerve studies suggesting CTS, very few had seen a physician for their condition. Because many of the CTS cases go undetected, ergonomic education and CTS screenings may help identify cases earlier and prevent permanent disability.

1 INTRODUCTION

Within the construction industry, hand and wrist disorders are responsible for many lost time injuries, second only to low back disorders. According to the U.S. Bureau of Labor Statistics (1996), a large percentage of the hand and wrist disorders in the construction industry are carpal tunnel syndrome (CTS) and involve more days away from work than any other work-related disorders. Carpal tunnel syndrome is a neurological disorder that results from a compression of the median nerve at the wrist. Common symptoms include tingling and numbness of the fingers, gradual weakness in the hand, and eventual disability. The disorder has been associated with highly repetitive work and / or forceful exertions of the hands and wrists (NIOSH 1997).

As reported by the U.S. Bureau of Labor Statistics (1996), 1253 construction workers (0.03%) had carpal tunnel syndrome that resulted in lost workdays. In the 1988 National Health Interview Survey of Occupational Health Supplement, 2.12% of males in the construction industry reported that they had CTS (Tanaka et at. 1995). Thus, the rates of CTS reported to the BLS for the construction industry may be an underestimation. The self-reported national average for CTS in male and female workers is 1.55% (Tanaka et al. 1994).

Although data is available on the prevalence of CTS in active construction workers, there has not been specific data reported on the rate of CTS in construction apprentices. As part of an overall prevention strategy for CTS in the construction industry, it may be enlightening to determine when in a construction worker's tenure they begin developing the disorder. As part of a longitudinal study investigating the natural history of musculoskeletal disorders in construction workers, annual screenings for CTS have been performed on apprentices in several construction trades.

Screening construction apprentices for CTS may help identify the disorder earlier in its course. Early recognition and subsequent treatment may lead to faster recovery and prevention of a permanent disability. Additionally, workers with evidence of CTS could be identified have their job tasks modified to reduce exposure to occupational and non-occupational traumatogens. The purpose of this investigation is to determine the prevalence of abnormal median nerve conduction across the carpal tunnel in apprentices of construction trades.

2 METHODS

Although various screening tests for CTS have been proposed, nerve conduction studies provide an objective measure of median nerve function and have been found to be highly sensitive and specific in the diagnosis of CTS (AAEM 1993). Several studies have incorporated both nerve conduction results and the presence of symptoms in the case definition for CTS (Atterbury et al. 1996; Barnhart et al. 1991; Bernard et al. 1993).

Of the 217 eligible apprentices from the sheet metal and electrical trades, 207 participated (95%) and were evaluated for signs of CTS at their respective union hall. Apprentices completed a symptom and job history questionnaire and were interviewed for assessment of work and non-work factors that have been associated with CTS. Standardized nerve conduction studies were performed on both hands of the apprentices. The nerve conduction study included measurements of sensory orthodromic latency from mid-palm to the wrist of the median and ulnar nerves. Sensory palmar latencies were determined over an 8-cm palmar segment with the recording electrode 8-cm proximal to the cathode of a bipolar stimulator. Median nerve latencies were measured with the active recording electrode placed over the median nerve two centimeters proximal to the distal wrist crease. The ulnar latency was determined with the active recording electrode placed 2-cm proximal to the distal wrist crease over theulnar nerve. Sensory latency (peak latency) was determined as the time interval between onset of the stimulation artifact and the peak of the negative response. Standard 8cm orthodromic motor latencies were measured if a sensory block through the carpal tunnel was present.

The primary electrophysiologic measurement used to determine a median mononeuropathy was the difference between the median nerve sensory latency to the ipsilateral ulnar nerve sensory latency. Thus, the median-ulnar latency difference (MULD) in the right and left hands was used to determine the prevalence of median mono-neuropathy. Median mono-neuropathy was defined as a median ulnar latency difference of 0.5 ms or greater. A 0.5 ms difference was established because it represents a very conservative criterion for determining median neuropathy within the carpal tunnel (Redmond & Rivner 1988). For surveillance purposes, a case definition of CTS was defined as an abnormal electrophysiologic study indicating median neuropathy (MULD >0.5 ms) and symptoms consistent with CTS (numbness or tingling, and brachaligia paraesthesia nocturna localized to an anatomical area that includes the median nerve distribution of the hand)

3 RESULTS

First through forth year sheet metal apprentices (n=101) and first through fifth year electrical apprentices (n=106) were tested and completed questionnaires. The apprentices ranged in age from 19 to 41 years with an average of 25 years. Eleven (5%) of the 207 apprentices tested were female. Apprentices did not report any medical conditions that were relevant to CTS (i.e. rheumatoid arthritis, diabetes mellitus, hypothroidism, cervical radiculopathy, or pregnancy).

Preliminary analysis of the data indicated that on average 10.6% of the apprentices fit the case definition of CTS in at least one hand. The prevalence of CTS generally increased as the number of years in the apprentice program increased. The prevalence of CTS by apprentice level was 5%, 9%, 13%, 17%, and 15% in the first through fifth years, respectively. The fifth year apprentice class consisted of electrical workers only. Although approximately 11% of the apprentices had symptoms, history, and nerve studies suggesting CTS, only one of the apprentices had seen a physician for their condition. Apprentices with CTS, including those with the most advanced cases (atrophy of the thenar muscles), were reluctant to file a health care report or workers' compensation claim.

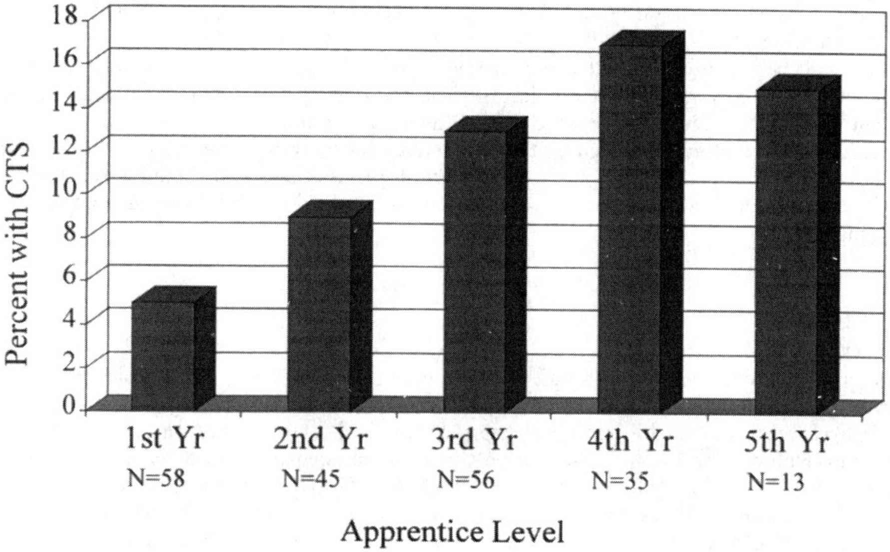

Figure 1. Prevalence of CTS in construction apprentices by year in apprentice program.

4 DISCUSSION

There were several limitations in this cross-sectional study. This cross-sectional design only provided a measure of this disease process at a single point in time. Workers with CTS that have transient symptoms or transient electrophysiologic abnormalities may not have met our case definition of CTS. Some studies reporting the prevalence of CTS only use symptoms as the criterion for a case definition for CTS. Thus, our case definition was considered highly specific and may have underestimated the true prevalence. Additionally, workers who were absent due to illness of left the workforce were not included in this sample. This study used a relatively small and restrictive sample and may not be representative of other apprentices, trades, or construction workers.

The results of this study indicate that more than 10% of the relatively young construction workers tested have CTS. Unfortunately there are no other reports of the prevalence of CTS in construction apprentices or in non-unionized construction workers during their first few years in their trade. As compared to other investigations of CTS in the construction industry (regardless of age), the prevalence of CTS determined in this group of workers is much higher than that previously reported. In a study investigating work-related musculoskeletal disorders (WMD) in 514 carpenters (average age 42 years), approximately 4% met a case definition of a hand/wrist WMD and had nerve studies consistent with CTS (Atterbury et al. 1996). The data from the U.S. Bureau of Labor Statistics (BLS 1996) and the Occupational Health Supplement to the National Health Interview Survey (Tanaka et al. 1995) indicate that the prevalence of CTS in construction workers is even lower.

Although the purpose of this cross-sectional analysis was not to specifically address the association of age with the prevalence of CTS, it is interesting to note the high prevalence of CTS in this group of relatively young workers. Additionally, within this group of apprentices, there is a trend of increasing CTS prevalence from the first to fifth years in the apprentice program. Although several studies have found age to be a factor associated with the prevalence of musculoskeletal disorders (Guo et al. 1995; Biering-Sorensen 1983; English et al. 1995; Ohlsson et al. 1994), others have not (Herberts et al. 1981; Punnet et al. 1985). The age group with the highest rate of compensable low back pain and strains is the 20-24 age group for men (Guo et al. 1995). Wisseman and Badger (1976) reported that the median age of workers with chronic hand and wrist injuries was less than the median age for the unaffected workers. It should be noted that age is a cofounder with number of years working and may explain part of the relationship between work and disease.

The majority of apprentices that we identified with CTS (including the sever cases) had not sought medical attention for their condition. The reasons for not seeking medical attention were numerous and included economic as well as psychosocial factors. Typical reasons for not seeking medical attention given by the apprentices for their symptoms included: assuming the disorder is "just part of the job," fear of employer reprisals, not being hired after the apprentice program, being labeled a complainer, unfamiliarity with filing a compensation claim, and loss of pay.

5 CONCLUSIONS

Construction apprentices have a relatively high prevalence of CTS that appears to increase with the number of years in the trade. Many construction workers may begin developing CTS in their first few years in the trade and not seek medical attention until a severe functional limitation develops. Additional research on this topic is needed to; 1) determine when CTS first develops, 2) investigate the prevalence of the disorder in specific age groups of construction workers, 3) identify coping strategies that help maintain employment in the construction industry for workers with CTS, and 4) develop prevention strategies to limit the functional limitation and disability caused by CTS.

REFERENCES

AAEM Quality Assurance Committee, C.K. Jablecki, M.T. Andary, T.S. Yuen, D.E. Wilkins, & F.H. Williams 1993. Literature review of the usefulness of nerve conduction studies and electromyography for the evaluation of patients with carpal tunnel syndrome. *Muscle Nerve* 16:1392-1414.

Atterbury M.R, J.C. Limke, G.K. Lemasters, Y. Li, C. Forrester, R. Stinson, & H. Applegate 1996. Nested case-control study of hand and wrist work-related musculoskeletal disorders in carpenters. *Am. J. Ind. Med.* 30:695–701.

Barnhart S, P. Demers, M. Miller, W. Longstreth Jr., & L. Rosenstock 1991. Carpal tunnel syndrome among ski manufacturing workers. *Scand. J. Work Environ. Health* 17(1):46–52.

Bernard B, S. Sauter, & L. Fine 1993. Hazard evaluation and technical assistance report: Los Angeles Times, Los Angeles, CA. Cincinnati, OH: U.S. Department of Health and Human Services, Public Health Service, Centers for Disease Control and Prevention, National Institute for Occupational Safety and Health, *NIOSH Report No. HHE 90–013–2277.*

Biering-Sorensen F. 1983. A prospective study of low-back pain in a general population. I. Occurrence, recurrence, and aetilogy. *Scand. J. Rehabil. Med.* 15:81–88.

English C.J., W. Maclaren, C. Court-Brown, S. Hughes, R. Porter, & W. Wallace 1995. Relations between upper limb soft tissue disorders and repetitive movements at work.*Am. J. Ind. Med.* 27(1):75–90.

Guo H.R., S. Tanaka, L. Cameron, P. Seligman, V. Behrens, J. Ger, D. Wild, & V. Putz-Anderson 1995. Back pain among workers in the United States: national estimates and workers at high risk. *Am. J. Ind. Med.* 28(5):591–602.

Herberts P., R. Kadefors, G. Andersson, & I. Petersen 1981. Shoulder pain in industry: an epidemiological study on welders. *Acta. Orthop. Scand.* 52:299–306.

NIOSH 1997. Musculoskeletal disorders and workplace factors: A critical review of epidemiologic evidence for work-related musculoskeletal disorders of the neck, upper extremity, and low back. Cincinnati, OH: U.S. Department of Health and Human Services, Public Health Service, Centers for Disease Control and Prevention, National Institute for Occupational Safety and Health, *DHHS (NIOSH) Publication No. 97–141.*

Ohlsson K., G. Hansson, I. Balogh, U. Stromberg, B. Palsson, C. Nordander, L. Rylander, & S. Skerfving 1994. Disorders of the neck and upper limbs in women in the fish processing industry. *Occup. Environ. Med.* 51(12):826–832.

Punnett L., J. Robins, D. Wegman, & W. Keyserling 1985. Soft-tissue disorders in the upper limbs of female assembly workers: impact of length of employment, work pace, and selection. *Scand. J. Work Environ. Health* 11(6): 417–425.

Redmond M.D. & M. Rivner 1988. False positive electrodiagnostic tests in carpal tunnel syndrome. *Muscle* Nerve 11:511-517.

Tanaka, S., D.K. Wild, P. Seligman, V. Behrens, L. Cameron, V. & Putz-Anderson 1994. The US prevalence of self-reported carpal tunnel syndrome: 1988 national health interview survey data. *Am. J. Public Health* 84:1846-1848.

Tanaka, S., D.K. Wild, P. Seligman, W. Halperin, V. Behrens, & V. Putz-Anderson 1995. Prevalence of work-related carpal tunnel syndrome among U.S. workers: analysis of the occupational health supplement data of 1988 National Health Interview Survey. *Am. J. Ind. Med.* 27(4):451–470.

US Department of Labor, Bureau of Labor Statistics 1997. *Workplace Injuries and Illnesses in 1996.* Washington, DC: US Government Printing Office.

Wisseman, C., & D. Badger 1976. Hazard evaluation and technical assistance report: Eastman Kodak, Windsor, Colorado. Cincinnati, OH: U.S. Department of Health, Education, and Welfare, Public Health Service, Center for Disease Control, National Institute for Occupational Safety and Health, *DHEW (NIOSH) Report No. TA 76–0131.*

Implementation of Safety and Health on Construction Sites, Singh, Hinze & Coble (eds)
 ISBN 90 5809 036 1

Hardhat ergonomists: A regional training strategy for identifying and reducing musculoskeletal hazards in construction

S. Moir & B. Buchholz
Construction Occupational Health Project, Department of Work Environment, University of Massachusetts Lowell, Mass., USA

M. Vogel
Massachusetts Construction Safety Congress, Dorchester, Mass., USA

ABSTRACT: Two levels of ergonomics training were developed and delivered to construction workers and managers and were evaluated by participants and researchers. The objectives of the training and evaluation activities were (1) increase the capacity for the identification and reduction of ergonomic hazards on construction sites and (2) identify of sources of support for and resistance to ergonomic changes in construction. Popular education methods were used in the design and implementation of the activities. Modest successes in creating capacity have resulted. Construction workers were highly supportive of the goals and activities of the training. The primary barriers to change were identified as lack of support from management, resistance to change among union representatives and the acceptance of the precedence of contractor profit over health and safety concerns by all parties. Two initiatives for investigating organizational and economic barriers are reported.

1 INTRODUCTION

Musculoskeletal disorders, including back pain and other strains and sprains, are the most common injuries in the US construction industry (Center to Protect Workers' Rights 1993). Ergonomics is the science of preventing these injuries through workplace redesign. In the absence of a federal standard for ergonomics in the workplace, the work of identifying and reducing hazards that cause these injuries is left to the discretion of management and the power of workers and their representatives to legitimate the issue, identify problems and propose viable solutions. As part of a national effort to increase awareness of ergonomic hazards in construction and to develop and promote solutions that reduce rates of injury, a regional training program was developed. Efforts focused on providing workers, union representatives and frontline managers with the knowledge base necessary to practice ergonomics on construction sites. Each training included participatory activities on barriers to change which provided qualitative data on sources of support for and resistance to ergonomic improvements in construction.

In his proposal to "devolve ergonomics," John Wilson argues "that ergonomics [in the workplace] can be applied successfully through individuals doing something about their own jobs" and that, for these frontline changes to occur, these individuals need "a changed attitude, *some* ergonomics knowledge, tools and techniques of their own and a back-up resource." [emphasis in original]. Wilson calls this strategy "spreading and embedding." (Wilson 1994) The regional training program focused on spreading and embedding through the development of "hardhat ergonomists," that is, workers and frontline managers with the ability to apply ergonomics to hazard reduction on construction sites. Training addressed this goal through two objectives. The first, creating capacity for recognizing hazards and implementing interventions, gave workers knowledge and some of the tools and techniques needed to analyze their own jobs and proposed solutions to musculoskeletal hazards. The second objective, identifying opportunities for and resistance to change within the industry, sought to identify "back-up resources" and understand processes by which those resources would act as allies to workers who were trained to "[do] something about their own jobs." To address "changed attitudes," popular education methods were used in the design and delivery of the training. Popular education methods are *not* used to change individual worker behavior, but to develop institutional responsibility and capacity for identifying problems and developing and implementing solutions. Within the training participants are challenged to address power relationships in the workplace and obstacles to workers affecting change. Pre-training planning activities included focus groups and informal interviews to assess the level of knowledge of ergonomics and attitudes toward change among workers, union representatives, contractors and project owners. A bi-level strategy for training and intervention developed from the analysis of the data gathered through these qualitative research activities.

- Awareness-level: A two-hour awareness-level training program was developed and delivered to journeylevel operating engineers. Using popular education and problem-solving methods, workers were trained in the fundamentals of musculoskeletal hazard identification and reduction. The local leadership of the operating engineers union received the two-hour awareness training and participated in an evaluation of the program.
- Train-the-trainers: A train-the-trainers program for union and management health and safety representatives provided a working knowledge of ergonomics and popular education teaching methods to industry leaders who were capable of conducting the awareness-level training in their areas.

2 MULTI-LEVEL TRAINING ACTIVITIES

In the unionized sector of the construction industry, primary responsibility for training rests with the joint labor-management training centers. The training staff of these centers is usually composed of former or current construction workers. Learning takes place in several ways:

- Initially workers learn their trade through apprenticeship programs which include classroom and hands-on training as well as work experience.
- As journey-level workers, they often take upgrade and refresher training.
- Increasingly, construction workers are taking specialized training to qualify them to do specific types of work, such as lead abatement, asbestos abatement and hazardous waste removal.

In addition to the training delivered at the training centers, contractors are responsible for on-site, job-specific training. This is usually delivered by the contractor's safety officer through site-wide safety meetings and "toolbox meetings," short pre-shift meetings which emphasize specific hazards.

2.1 Methods: Popular education

In recent years increasing use has been made of "popular education" training methods for teaching workers about health and safety (Wallerstein 1992). Popular education de-emphasizes didactic learning methods and is delivered through participatory learning activities where students utilize their own experience as a catalyst for learning new material. Learning occurs through the use of open-ended questions and group activities which make the learning experience immediate and relevant. These methods draw primarily on two pedagogic theories: adult education and liberatory education. Adult education focuses on respect for the learner and the previous knowledge that s/he brings to the teaching situation. Liberatory education builds on that theory by addressing the social context of learning and the effect of power relationships on the learners' ability to use new knowledge outside the classroom.

An important component to popular education is the willingness of the participants to engage in the process. The construction industry is dominated by forces resistant to change. In attempting to effect change through training, researchers designing the regional training activities have described their sampling strategy as "low hanging fruit," or the LHF factor. Training has occurred where the participants have been "ripe" for the experience. Following this purposeful sampling method, participants included members and leaders of a progressive regional union which actively supports outside research into changes in the construction industry and volunteers from the safety departments of supportive construction contractors. The LHF factor, combined with the experience of construction industry members with experiential participatory training in the hands-on acquisition of the skills involved in the trades, created opportunities to research the usefulness of popular education methods in a "best possible case scenario." Results have shown that, even where there is willingness to engage in innovative and more interactive teaching, barriers to effective implementation and diffusion of new ideas are many and complex. Among the barriers are: 1) instructors with little training in participatory methods who are fearful of trying new techniques, 2) a culture that changes slowly and values doing things "the way they have always been done," 3) severe time limitations imposed by production pressures and 4) day-to-day uncertainty in a high hazard industry ruled by external factors such as weather, supply deliveries and the vicissitudes of the schedule of the "guy in front of you."

2.2 Awareness-level training program: developing "hardhat ergonomists"

Construction is an industry where workers retain a high level of autonomy and are expected to address problems as they arise through their own expertise and by engaging in cooperative problem solving. A strategy for educating "hardhat ergonomists" draws on these industry traditions.

A two-hour awareness-level training program was developed and delivered to 400 journey-level operating engineers over a period of two years. Using popular education and problem-solving methods, workers were trained in four fundamentals of ergonomics: work-related musculoskeletal disorders, risk factors, job analysis/redesign and barriers to change. To begin discussion of work-related musculoskeletal injuries, participants were asked, "Where does your job hurt you?" Learning activities included "mapping" the students' own injuries on wall sized body outlines and grouping injuries common to construction workers into back, and upper and lower extremity categories. To learn "risk factors," participants answered the question, "How does your job injure you?" After identifying the causes of their injuries, an exercise in matching knowledge aided the participants in associating the factors generated by the class with the scientific risk factors of force, awkward posture, repetition, psychosocial and environmental stressors. As a process of learning job/analysis and redesign, participants drew risk maps of their jobs and answered the question "What could be changed?" Small groups used their risk maps to proposed the elimination of risk factors in two categories: short-term/inexpensive and long-term/more expensive. Each small group presented its analysis to the class. Finally, participants discussed, "What prevents these types of changes?" and explored power relationships and economic and political barriers to change on the construction worksite. Classes ranged from 15 to 40 participants and were included as a two-hour segment of an annual upgrade training which the participants attended voluntarily on their own time. A key finding from this activity is that recent changes in technology have led to rapid changes in the nature and intensity of the ergonomic risks involved in operating heavy construction equipment. While force and heavy exertions have often been reduced in new equipment, repetition and awkward postures have sometimes been increased.

2.3 Evaluation of the awareness level training program

The leadership of the operating engineers union (business managers and stewards) participated in the two-hour awareness training and conducted an one-hour evaluation. This included critiquing the curriculum and analyzing the barriers to ergonomic changes in construction that had been identified by their members in previous classes. Researchers have provided the leadership with formative evaluations of all the training activities conducted with the union over the past three years. A participatory evaluation is planned for the next year. This will involve focus groups and interviews to elicit specific feedback and strategic suggestions from the local leadership and the production of a joint union/research summative report.

An early finding from the evaluation of the awareness-level training was that the job conditions differ greatly between experienced journeylevel workers and apprentices and that training activities require designs that meet the separate needs of each group. Apprentices do much more varied tasks and often more physically demanding work than more experienced workers. In training sessions and qualitative data gathering activities, such as interviews and focus groups, journeylevel workers are consistently much more open and forthcoming about their work-related injuries. Apprentices are more likely to not be injured (yet) and to believe that injury is a matter of luck or individual behavioral choices. It was decided to develop a separate apprenticeship training program in ergonomics.

The leadership evaluation has validated a key finding from the awareness level trainings that the economic barriers to change which are primary to all parties in the construction industry are particularly important among operating engineers and owners of heavy construction machinery due to the large capital expenses involved in improving and purchasing heavy equipment.

2.4 Train-the-Trainers: Master Ergonomics Trainers

A regional joint labor-management advisory board, established to participate in the development and design of a range of research activities in the construction industry, initiated a train-the-trainers program to develop Master Ergonomics Trainers (MET). The goal of this activity is to train a group of union and management trainers in both ergonomics and popular education methods. The METs will use popular education methods to teach the two-hour ergonomics-awareness curriculum in training centers and on construction sites throughout the region. The plan calls for training 25 METs. The first training was held in 1998 and was attended by 11 participants, including instructors representing Sheet Metal Workers, Laborers, Electrical Workers, Operating Engineers, and contractors. Participants received a 13-hour course using popular education methods. Along with learning the curriculum and popular education teaching methods, participants discussed obstacles to implementing training programs that are more participatory and that address power and control in the workplace. The Master Ergonomics Trainers have begun delivering the awareness-level training at training centers. A second MET course is planned.

Key findings of the first MET course include: (1) the identification by participants of the on-site steward/foremen relationship as a critical point where negotiations for change often happen and where ergonomic changes may be successful and (2) the limitations on dealing with workers' daily safety and health issues within training courses given at the joint labor/management training centers. Specific on-site conflicts, including health and safety concerns, are viewed as being within the exclusive sphere of the union and not the business of trainers or the training centers.

3 EVALUATION

In addition to the leadership evaluation of the program, participatory formative evaluation was conducted at the conclusion of each training session. First, the participants were asked, "What did you like about this training?" Participants consistently liked the level of participation and the opportunity to talk about their own experience. Next they were asked, "What suggestions do you have for change?" They often wanted the classes to be longer (regardless of the course length) and asked for more technical information. In evaluating the 2-hour awareness-level training, participants wanted to spend more time on how to make changes at work.

A random phone survey of 33 operating engineers who had taken the awareness-level training over a period of a year showed a very favorable response to the course. On a scale of 5, the average overall rating of the course was 4.2 and the participatory methods were rated at 4.4. While specific retention of facts was, as expected, not high (about 45% could approximate a definition of ergonomics and 33% could name risk factors) participants were enthusiastic about taking another course and recommending it to others.

The participants in the MET train-the-trainers course completed written evaluations wherein they responded very favorably to both the content and presentation of the course.

4 SUMMARY AND CONCLUSIONS

In all other aspects of public health, except in the workplace, two assumptions prevail: that the public health is best served when the public at risk is involved in solutions and that health is in part a function of power relationships in society (Minkler 1997). The regional ergonomics training program attempts to apply these fundamentals of public health policy to the occupational health problem of reducing musculoskeletal disorders through improving ergonomics in the construction workplace. The program differs from many workplace training programs in that the goal is not to change individual worker behavior, but to develop industry-wide responsibility and capacity for identifying ergonomic problems and implementing solutions. The strategy behind this intervention posits that the education of the workers at risk is an effective and efficient means to reduce risk and injury. However power and control are potential workplace hazards where workers have the knowledge and capacity to reduce their risk, but not the power to act. Ergonomics is a particularly fertile field for these hypotheses because workers embody all the necessary expertise to reduce hazards.

The goal of industry penetration is difficult to quantify, but modest successes have been demonstrated. Significant increases in awareness of ergonomics and the relationship of ergonomic risk factors to musculoskeletal injury have resulted in a "buzz" in the regional industry. A multimillion-dollar tunnel project was redesigned to reduce extensive overhead drilling in part because researchers and frontline managers demonstrated the potential ergonomic risks in the operation. Workers continue, as they did before research, to redesign jobs which hurt them and some of these redesigns have been disseminated across sites due to the impetus provided by education and research. An example is the widespread use of a modular guardrail system which eliminates the need to tear apart traditional wood systems. Operating engineers report some changes such as adding mirrors to reduce awkward neck postures and improving handholds for mounting and dismounting equipment. There appears to be a greater awareness of seat maintenance since the early years of the research project when it was common to see operators sitting on milk crates.

However, there is much to be done. The primary barriers to change have been identified as lack of support from management, resistance to change among union representatives and acceptance of the precedence of contractor profits over health and safety concerns by all parties. Two initiatives will address these barriers in the next phase of the research.

To address the organizational and cultural barriers to change in construction, an extensive needs assessment and empowerment evaluation (Fawcett 1996) is being undertaken with operating engineer apprentices. This process will produce an apprentice-level ergonomics training program for the local union and will explore the potential for impacting the industry through educating its newest members. The question that will be asked is whether a class of apprentices, who will both study their trade together and begin to experience the injuries of their trade over their first four years in the union, can function as a cohort for change if they have shared educational experiences.

To address the economic barriers to change in construction, the development of a practical "cost effectiveness" instrument that can be used by construction industry personnel is being investigated. Although questions of contractor profitability and job creation are raised by all parties in construction as the single most important obstacle to reducing ergonomic and other health hazards, no workable model for calculating the cost of intervention presently exists. Recent research into prevention effectiveness (Gobsky1996) has opened up the possibility that an accessible and practical model may make it possible for contractors and/or union leadership to cost out the payback period for interventions.

REFERENCES

Center to Protect Workers' Rights 1993. *An agenda for change: Report of the national conference on ergonomics, safety and health.* Report No. G2-94. Washington, DC.

Gobsky R, A. Haddix & P. Shaffer 1996. Cost of an intervention In A. Haddix, & S. Teutsch (eds.), *Prevention Effectiveness.* New York: Oxford University Press.

Fawcett S & A Paine-Andrews 1996. Empowering community health initiatives through evaluation. In Fetterman D & S. Kafterian (eds.), *Empowerment evaluation: knowledge and tools for self-assessment and accountability*, Thousand Oaks: Sage.

Minkler, M. (ed.) 1997. *Community organizing and community building for health.* New Brunswick: Rutgers University Press.

Wallerstein N. & M. Weinger, 1992. Health and safety education for worker empowerment. *American Journal of Industrial Medicine* 22:637-649.

Wilson J., 1994. Devolving ergonomics: the key to ergonomics management programmes. *Ergonomics 37(4):579-594.*

Implementation of Safety and Health on Construction Sites, Singh, Hinze & Coble (eds)
© 1999 Balkema, Rotterdam, ISBN 90 5809 036 1

The work related health atlas for the construction industry

L.Akkers
Arbouw, Amsterdam, Netherlands

ABSTRACT: The new Work related Health Atlas for the Construction Industry describes in detail how construction workers in over 50 different building professions experience their health and work situation. The Atlas is a health and safety thermometer. On the basis of this Atlas the construction industry, construction companies and occupational health and safety centres can carry out an even more effective health and safety policy.

The Atlas registers for each profession exactly how often complaints occur. This facilitates a comparison between industrial workers and between various building professions, which is necessary for establishing priorities.

A figure is depicted for each profession, indicating which body parts are burdened more than on average.

The Atlas is a valuable instrument for construction companies in determining an effective policy for health and safety care and a usefull tool for experts from occupational health and safety centres.

1 CHARACTERISTICS OF THE DUTCH CONSTRUCTION INDUSTRY

In the construction industry in the Netherlands about 275,000 people work in about 20,000 companies. There are a very large number of small or very small companies. Only 400 companies employ more than a hundred people. There is a great degree of mobility within the employees' group, because for the most part employees have no fixed contracts and do not feel closely connected to the company. They do, however, feel closely connected to their profession and to the branch of industry as a whole. The construction work site can be regarded as a work place without a roof, where all weather conditions have free access. The location of the work site varies continuously, according to each building project. As you can imagine these factors do not make it easy to carry out health and safety care at company level. There are simply too many logistic, organisational and financial problems involved. Working conditions in the construction industry are hard: in general, the physical workload is heavy and there are many ergonomic problems, exposure to noise and vibrations and exposure to chemical substances. Moreover, employees alway work outdoors. The psychosocial work stress is high and there is a high risk of accidents. These working conditions resulted in the previous years in high sick leave and disability rates of more than 10 %, compared to 2,8 % in all other branches of industry.

Nowadays, the average sick leave percentage of 4.8 % is relatively low, compared to other branches of industry with an average of 5.3 %. At the end of my presentation I will come back on these promising trends for the construction industry.

2 THE ARBOUW FOUNDATION

The Arbouw Foundation was established in 1986 by employers' and employees' organisations in the Construction and Allied Trades on the basis of parity. The new foundation was commissioned to encourage safety and health in the construction and allied trades and to reduce sick leave.

The Arbouw Foundation provides services for companies in the civil and utilities construction sector, the ground works, the road-building and waterway construction sector and finishing works.

The Arbouw Foundation is the national institute for Health and Safety in the Construction Industry in the Netherlands. Its main activities are research and development, information and documentation and the organisation of Occupational Health Care for 275,000 construction workers. With regard to the latter activity the Arbouw Foundation has a contract with about 50 certified occupational health and safety centres (OHSC). Every construction worker is entitled to occupational health care. This right is set out in the various Collective Labour Agreements for the construction industry and is collectively financed. Since 1993 the Arbouw Foundation has been collecting all the occupational health data concerning health, safety and work environment relating to all construction workers in the so-called VISA-system. This is an Occupational Health and Safety Contact Registration System or Monitoring System.

3 MONITORING SYSTEM

VISA contains information on pre-employment examinations, periodical medical check-ups, specific periodical monitoring and screening programmes and medical guidance on work-related complaints and diseases.

Per year about 30,000 construction workers have a voluntary periodical medical check-up. They fill in questionnaires concerning their perceived health status and their perceived working conditions (subjective data). Also, the occupational health physician records his/her medical opinion (objective data) on a contact registration form. I shall restrict myself in this presentation to the subjective data.

The subjective data, collected over a two-year period, are used for the so-called 'Atlas' and can be compared with the data of a representative reference population of all sectors of industry in the Netherlands. 55 Occupations are included in the Atlas. The minimum size for an occupational group has been set at 100 people. All the complaint percentages are adjusted for age by direct standardisation. Differences between occupations or sectors are therefore not due to differences in age.

4 ATLAS

The objective of the Atlas is to present information about the perceived health and safety of the workers and their working conditions in the construction industry.

Together with other data such as sick leave and disability percentages, this information can be used for trend analysis, policy-making and the evaluations of sector policy and priority setting with regard to health protection and promotion, improving working conditions and reducing sick leave and disability. Another important goal of the Atlas is to demonstrate improvements as a result of preventive measures relating to certain health and safety aspects in a specific branch or occupation.

It is Arbouw's intention to update the Atlas of Work and Health in the Construction Industry every second year. For each professional group this will enable progress in the field of working conditions to be measured and new bottlenecks to be spotted in time and dealt with. The Atlas is therefore a valuable instrument for construction companies in determining an effective policy for health and care.

Also, the Atlas is a useful tool for experts from occupational health and safety centres. It provides them with an up-to-date insight into the specific risk factors relating to a particular profession. Should an occupational physician be confronted with a steel bender with back, wrist and hip complaints, the Atlas will instantly tell him/her that these are typical problems among steel benders. The physician then knows than that a so-called profession-specific complaint pattern is likely. So he/she has to put most of his/her efforts into modifying the steel bender's work site or tasks.

5 RESULTS

In the Atlas the results of the analyses are discussed per occupation, per age group and per sector and presented in the form of profiles. These profiles show the mean sum scores on the various clusters of complaints about the working conditions as a percentage of the maximum score. The health complaints concerns single items and therefore represent the real prevalence of these complaints.

Complaints which are reported more often than the reference population are presented in red bars, and complaints which are reported less than the reference are presented in green bars.

For the blue-collar and white-collar population of the construction industry the reference groups are the blue-collar and the white-collar populations of all sectors of industry in the Netherlands. I will show you now some results.

6 COMPARISON WITH INDUSTRIAL WORKERS

Some findings in the Atlas are indeed striking. It is especially remarkable that employees in the construction industry are in many ways more positive about their work than industrial workers, with whom they were compared.

In particular, construction workers are more positive about the appreciation, payment and interesting aspects of their work. The Atlas explains this phenomenon as follows: 'A possible cause of this favourable result is that in the construction industry a greater appeal is made to the workers' craftmanship and that there is a wider variation in tasks than in industry'. It is also conspicuous that in comparison with industrial workers building site workers complain less about noise. This could be connected to a less permanent noise exposure in the construction industry. Furthermore, construction workers are less frequently absent from work as a result of illness or accidents than industrial workers. On the other hand construction workers complain a lot more about the strenuous aspect of their work and about dust exposure. They also experience more complaints connected with the awkward and static postures during work, vibrations and climate. The complaint percentages are highest among older employees. The Atlas also discusses the health situation of staff and administrative employees. Generally speaking, more or less the same picture arises among this group. However, the number of complaints among staff and administrative workers concerning the required mental effort and time pressure is significantly higher than in industry. But in total the opinions of both building site and staff and administrative employees regarding health and work situation are more favourable than the opinions of their colleagues in industry.

Now I will show you some profession profiles, the paviour, the steel bender, the painter in maintenance and the foreman.

7 PAVIOUR

Generally speaking, paviours judge their work to be physically demanding. They often associate health complaints with their work. Factors they reported often are working in the same posture and repetitive bending. Lower back complaints and hand, hip, and knee complaints occur considerably more often among paviours than in other building professions. Also, paviours are treated more often for muscular skeletal problems. Complaints about noise and vibrations also occur frequently. Yet complaints about bad hearing are not mentioned more than by other construction workers. The paviour profession involves a lot of complaints concerning the weather. Paviours are particularly bothered by the heat and cold and temperature fluctuations. Bronchial tubes complaints are not more frequent than among other building professions, nor are paviours been treated more often for these complaints. Paviours are not very satisfied with the appreciation of and payment for their work.

All in all, they assesse their work situation in a more negative way than their colleagues in other building professions do.

8 STEEL BENDER

The steel bender's job is very physically demanding, as a result of standing for too long and working in the same position for too long, and frequent bending down. Back pain or stiffness occurs relatively often, as well as hip and wrist complaints. Steel benders are treated more often for muscle and joint problems and are also been forced to stay home more as a result of accidents or illness. Steel benders say they have to work under time pressure a lot. They experience their job as being extremely strenuous. More often, steel benders work in very unfavourable working postures. The large number of fatigue complaints is probably related to this factor. Among this professional group there are relatively many complaints about noise and climatic factors such as heat and cold. In comparison with other building professions the steel benders do not find their work very interesting and do not enjoy their work very much. They do not have a lot of complaints about the work organisation and daily management, but are often not pleased with the teamwork. Moreover, they are of the opinion that the work does not offer enough security and rewards. Therefore, steel benders' opinions of their work situation are less positive than the overall opinion of the average construction worker. Steel benders report slightly more complaints about their own health and attribute these to their work considerably more often.

9 PAINTER IN MAINTENANCE

These painters do not often characterise their work as physically strenuous. However, they are troubled by having to stand for too long. Neck-, shoulder- and knee complaints are mentioned more than on average. The main complaints are exposure to dust and toxic substances. The relatively large number of psychological complaints such as nervousness is striking. Painters in maintenance are troubled a lot by dust and climatic factors, such as heat, cold, temperature fluctuations and draughts. They also complain about bad lighting and lack of safety. The lack of everyday management and insufficient say in their work are irritating factors for this group. The teamwork leaves much to be desired. The overall opinion of the work is less favourable than that of other construction workers. Health complaints are made more than on average, but are less often attributed to the job.

10 FOREMEN

The foremen suffer more complaints, relatively speaking, of the back, arms and legs. They are also frequently treated for these complaints. Furthermore hearing problems and work related-complaints are reported. Also significantly more complaints are reported relating to static postures, heavy work, noise, vibrations, toxic substances, climate and lighting. Note the high percentages of complaints about mental effort and work under time pressure. You also can see more complaints relating to co-operation, appreciation and payment.

Significantly fewer complaints are reported about skin problems and the interesting aspects of the job. In spite of all these complaints only 4 % of foremen have a negative opinion about their work situation.

11 TRENDS 1989-1996

Now I will present some of the first results of our trend analysis for the time period 1989-1996. At the start of my presentation I mentioned the relatively low percentage of sick leave compared to the other branches of industry. The results of our analysis show the same trend, namely that the number of reported complaints in the Dutch construction workers has decreased substantially.

The first example shows the back and neck complaints reported by workers at construction sites. The mean percentage of the complaints is in the period 1989/1990 27 %, in 1993/1994 24 % and in 1995/1996 22 %. A reduction of 5 %! The columns in the figure represent the different age categories. Of course the younger construction workers reported fewer complaints than the older ones. But in the older age categories there is the same reduction in back and neck complaints.

In this next figure you can see the same trend for staff and administrative workers in the construction industry. A slight reduction in complaints is reported, but obviously not such a decline as with the work-site population.

The last figure shows the psychological work demands reported such as mental effort, concentration, accuracy and time pressure. Of all the Atlas aspects this is the only factor that increased. For the work-site population it increased from 41 % (1989/1990) to 45 % (1993/1994) to 46 % in 1995/1996. This increase is also reported in other sectors of industry in the Netherlands and other countries in Europe. For staff and administrative workers the absolute percentage starts at 70 % in 1989 and increases to 74 % in 1996. Psychological demands are nowadays the highest risk factors for staff and administrative workers.

12 DISCUSSION

The reliability of the subjective date from the questionnaires is very high because of the large number of people involved from each occupational group and the fact that the data were collected on a national scale. The date were adjusted for age by direct standardisation.
Therefore a comparison between occupations, sectors etc. is possible. Former research did not show a significant difference between the results in the response and non-response population. We can therefore assume that the complaints percentages concerning health and working conditons are representative for the whole construction industry.

There are some limitations, related to the cross-sectional design of this research and to the registration quality, which are relevant for the results of the Atlas. First of all there is the influence of the healthy worker effect which is more predominant in the construction industry than in many other industries. In the older age categories especially, there is a relatively healthy working population. Therefore the percentages of the health complaints and the working conditions complaints will most likely be an underestimation of the real health effects of the working condition. This also counts for the medical diagnoses of the occupational health physicians.

The use of the whole population of blue and white collar construction workers as reference groups for the 55 different occupations and 16 sectors of the construction industry has disadvantages as well as advantages. In general the physical workload of blue collar construction workers is relatively high. So an occupation with the same high percentage of health complaints as the reference group could in fact have a serious work-related problem. On the other hand, the advantages are the limited number of significant health problems monitored which represent the really big problems in the construction industry.

13 CONCLUSIONS

The Atlas gives a detailed overview of working conditions and health problems for the whole construction industry, the various occupations, the different sectors and of all age categories. The information is very useful for trend analysis, priority setting and evaluation studies. For policy-making a combination of other data sources such as sick leave and disability rates and more detailed objective risk inventories and evaluations are often necessary. The reliability of the subjective data registered by the workers is high. Also, the face validity seems good when you compare the results of the profession profiles with your own expert opinion.

For the coming years we plan to do more secondary analysis e.g. on ageing and on sick leave and disability rates.

REFERENCES

Arbouw 1997. *Bedrijfstakatlas, Arbeid en gezondheid in de bouwnijverheid*. Amsterdam.

Implementation of Safety and Health on Construction Sites, Singh, Hinze & Coble (eds)
© 1999 Balkema, Rotterdam, ISBN 90 5809 036 1

Skin cancer and construction industry workers: Is it time to cover up?

C. Pritchard
Department of Building and Environmental Health, The Nottingham Trent University, UK

ABSTRACT: The high risk nature of the construction industry in relation to health and safety has long been recognised. Research has predominantly focused on accidents in the industry with less consideration on the long-term health effects in particular those arising from exposure to adverse weather conditions.

Excessive exposure of the skin to ultra-violet radiation from sunlight is the most important risk factor in the development of skin cancer. It is estimated that the incidence of skin cancer is doubling every decade and this factor, coupled with high mortality rates, has made skin cancers an important health priority. In particular, outdoor workers are being recognised as potential victims of non-melanoma skin cancers and the workplace a valuable setting in which to carry out skin cancer interventions.

The paper highlights the risks to outdoor workers in the construction industry from excessive exposure to sunlight and considers the variety of prevention methods available. Comparisons are made to other sectors of outdoor workers. The conclusion will inform the debate that the construction industry has a valuable role to play in the protection of the workforce against this disease.

1 INTRODUCTION

Cancers are the uncontrolled proliferation of cells that may prevent an organ from functioning. In particular circumstances metastases arise that affect other parts of the body. On a worldwide basis the prevention and treatment of cancers is a public health priority.

Exposure to sunlight is known to be beneficial. However, excessive exposure can result in harmful health effects. Increasingly, evidence indicates that long-term damage may occur to the eyes and skin (MacKie 1987) and has also been linked to Non-Hodgkinson Lymphoma (Adami 1995).

The incidence of skin cancer has increased worldwide. This increase is inspite of evidence of the causal link between exposure to sunlight and skin cancer, the various risk factors known and the prevention strategies available.

As a major employer of outdoor workers, the construction industry possesses the potential to improve the health of its workers by reducing exposure to the risk of skin cancer. The paper outlines the various types of skin cancers and the international incidence of the disease. The risk factors associated with skin cancer will be identified together with the implications of these for the construction industry. The challenge to health promoters is how best to ensure that this knowledge is translated into practise in the construction industry that isnot based on static sites.

2 CONTEXT

Initial exposure of the skin to ultra-violet radiation results in the increased production of melanin that acts to protect cells from the harmful effects of sun rays. Under specific circumstances, and dependant upon an individuals risk factors, the skin may become reddened and result in sun burn in the short term, but in the longer term skin cancer may occur. There are three forms of skin cancer: cutaneous malignant melanoma and the non-melanoma skin cancers in the form of basal cell carcinoma and squamous cell carcinoma.

Malignant melanomas, whilst only accounting for 10% of all skin cancers (Ross 1988), result in the highest mortality rates (MacKie 1987) as it is able to metastasis. It has been found that early detection is of the utmost importance as prognosis is directly dependent on the thickness of the lesion (Campbell & Birdsell 1994). It is believed that individuals at greatest risk are those not necessarily exposed to sunlight on a daily basis, therefore seasonal outdoor workers (Campbell & Birdsell 1994) as well as office workers may be at increased risk. The age of onset has been found in people as young as 15 (Skinner 1997), but more commonly it is in the 45-50 age range (MacKie et al 1991 & 1987).

The non-melanoma skin cancers are more common in the population and in particular in outdoor workers (Marks 1991) as typically found in the construction industry. The development of non-melanoma skin cancers is associated with longer-term exposure to sunlight (MacKie 1987) and the lesions are on exposed parts of the body, commonly the face, hands, legs and torso. An older age of onset is usually linked with these cancers being found in people over 60 years old (Campbell & Birdsell 1994). Of the two forms of non-melanoma skin cancers, basal cell carcinomas account for 70% of all lesions (Ross 1988). However, the prognosis is excellent following treatment of this form of cancer (MacKie 1987). Squamous cell carcinomas account for 20% of the cancers (Ross 1988) and the prognosis is dependant upon early detection and treatment as they are able to metastases and therefore have a higher mortality rate (Skinner 1997).

Statistics about cancer are always problematic however, they are more comprehensive than other disease incidence data. It has been noted that incidence of skin cancers are increasing in the population on a worldwide basis. Estimates vary for the rate of increase of skin cancer with an estimated doubling of incidence of malignant melanoma every 10 years, with most populations showing an average rate of increase of between 3-6% per annum (Skinner 1997). Pale-skinned individuals in Australia have the highest incidence of malignant melanoma. Rates are also high in the United States of America and northern Europe. The countries with the lowest reported rates are India and Japan (CRC 1995). In Britain, skin cancers account for 12% of all reported cancer cases; second only to lung cancer (CRC 1995). Mortality rates have increased for 60 years, however, with improved detection, treatment and effective public health campaigns, many countries are now noting a levelling of these rates (Giles 1996).

Much of the research related to incidence and mortality have focused on outdoor workers as a generic group in preference to considering specific trades or industries. This provides a valuable insight into the potential effect of this disease in the construction industry. Outdoor workers are more likely to suffer from the non-melanotic skin cancers (Marks 1983) with a two-fold increase in incidence over those who work indoors (Gallagher 1989). In addition, outdoor workers have higher rates of skin damage than the general population (Webb 1992). Indeed, it is considered that exposure to sunlight over a long time period may provide a protective effect against malignant melanoma (Gallagher 1989). However, a study that focused on mortality in the construction industry in the USA found elevated mortality rates for brickmasons caused by melanomas (Robinson et al 1995).

Male mortality from skin cancer is twice that of women (Marks 1991). This may well be linked to the fact that only 35% of men believe that sun exposure increases their risk of contracting skin cancer (Campbell & Birdsell 1994) and the resultant impact that this has on prevention practices. The construction industry predominantly employs men and as they are more likely to be outdoor workers (Campbell & Birdsell 1994), they are at increased risk of non-melanotic skin cancers.

3 RISK FACTORS ASSOCIATED WITH SKIN CANCER

Exposure to sunlight is the most important risk factor for the development of skin cancer (CRC 1995). In addition, extensive research has identified the key individual risk factors for skin cancer (Elwood 1990, Mackie 1987). Skin types have been classified by dermatologists on a scale of 1-6. Type 1 is fair skin type that never tans yet always burns, whereas type 6 is genetically black skin. It is estimated that those with skin type 1 and 2 will burn readily after 20-40 minutes in the midday sun (MacKie 1987). Associated with skin type are those individuals with a large number of naevi and freckles who are at increased risk, and those with blue, grey or green eyes and red or fair hair (Elwood 1990). History of sunburn in childhood is also believed to be a risk factor (Elwood 1990). However, the findings are not consistent in the research on this subject (Skinner 1997).

Incidence of skin cancer are highest for fair-skinned people living near to the equator. Therefore, the length of day light, altitude, latitude, weather conditions and time in the sun will all have implications. People with type 1 and 2 skin who have migrated to Australia have a ten-fold increased risk of contracting skin cancer in comparison to their Northern Europe counterparts (Skinner 1997). Increasingly, it is also being suggested that the reduction in the ozone layer may well have a detrimental effect on skin cancer and it is estimated that it may result in an increased incidence of between 30-50% (Jones 1992).

In addition to direct exposure to the sun, construction industry workers may also be exposed to a synergistic effect from the use of materials on site such as polycyclic hydrocarbons found in a range of products including cutting oils and tars (Ross 1988). It is believed that these products may not only act as carcinogens themselves, but also increase photosensitisation (Webb 1992). The construction industry therefore needs to consider more fully the health of its workforce, by carrying out comprehensive risk assessments that focus on the individuals risk factors as well as the work undertaken.

4 PREVENTION METHODS AVAILABLE

Health promotion, prevention and education each have an important role in the long-term control of skin cancer in the construction industry. The primary prevention method on site should be the completion of a comprehensive risk assessment. Assessing the risks of skin cancer on site are of paramount importance in striving towards a safe and healthy work place. In order to ensure that all the key issues are addressed, the risk assessment should include the following:

* Identification of hazards - exposure to sunlight and to polycyclic hydrocarbons
* Assessment of risks - likelihood that harm will be realised therefore individual risk factors need to be taken into consideration
* Identification and implementation of suitable control measures - measures to control or eliminate the risk. A range of options are available.
* Monitoring of control measures. Are these methods being utilised by staff?
* Review of the control measures.

Whilst it is recognised that outdoor work is inextricably linked to the construction industry, effective job planning can help. The avoidance of operations during the mid-day sun may be the only control method available in certain countries together with the adoption of flexible working hours. Greater use of shaded areas, so that workers can remove themselves from the worst excesses of the mid-day sun, is also a valuable method of control. These methods could ensure the minimisation of risks by planning a suitable system of work. The use of personal protective clothing may provide the only defence in the initial stages of a construction project where most of the work is executed in the open. The use of tightly woven clothes covering the torso and arms and the use of hats that cover the neck are methods of protection. Sun-screen, with a protection factor of 15+, could also be used. However sun-screens should only be used as the final line of defence when the other options above have been adopted.

Health education interventions carried out in Australia on outdoor workers (telecommunication and electricity supply workers) have yielded some interesting initiatives and results provide a useful starting point for the construction industry (Borland et al 1991)(Girgis 1994)(Godkin 1991). A range of education health intervention methods were used in order to encourage outdoor workers to cover up. Use was made of informative posters, videos, individual resource material including protective equipment and dermatological screening of the outdoor workers. Support was obtained from both staff and management in order to ensure the maximum rates of participation in these studies. In all of the interventions there was a significant improvement in the outdoor workers' knowledge about skin cancer and their behaviour in the sun in comparison to a control group, thus demonstrating the benefit of an educative approach.

5 DISCUSSION AND CONCLUSION

Construction workers predominantly work outside and therefore are at increased risk of non-melanoma skin cancers (Marks 1991). Some other skilled groups, such as brickmasons, may also be at increased risk of malignant melanomas. Borland (1991) found that only 27% of outdoor workers wore hats and 13% covered their arms. Therefore, outdoor workers are an important risk group. In addition to concerns about exposure to sunlight, the use of carcinogenic products on construction sites may increase skin cancer risks.

The construction industry is male dominated and in relation to their risk of contracting skin cancer there are two important factors to consider. Firstly men are less likely to consider exposure to sunlight to increase their risk of skin cancer thus limiting the use of prevention methods, this is compounded by the second fact that men are less likely to be recipients of health information than women (Hill 1992).

Risk assessments are an important starting point to ensure that all factors related to skin cancer have been taken into consideration and a variety of intervention options are utilised. Risk assessments must take into consideration the range of risk factors associated with skin cancer including individual hazards. Thoroughly executed risk assessments may reveal the need for change in working practices. These changes should be underpinned by education initiatives that may well influence behaviour of workers both on and off the site. The success of studies carried out in Australia (Borland et al 1991)(Girgis 1994)(Godkin 1991), demonstrate the value of working together with management and staff in order to promote skin cancer awareness and behaviour change in workers. This is of value in the construction industry because the attitudes and behaviours of managers on site are extremely influential on the workforce (Leather 1988) (Marsh 1995), therefore, the role of managers in promoting sun-safe behaviour is crucial.

One of the difficulties with the development of health promotion strategies in the construction industry, is the largely transient nature of the workforce. Also, a large proportion of hourly paid and blue collar workers are unlikely to participate in such initiatives (Gebhant

& Crump 1990)(Stonecipher & Hyner 1993). However, the detrimental health effects of skin cancer place a moral if not a legal obligation on the construction industry to attempt to address this important issue. Currently evidence suggests that in Britain only 32% of home builders provided any form of advise to their employees on the risks of skin cancer. Of these respondents, only 28% stated that they followed an active role of requiring their staff to cover up (Pritchard & McCarthy 1997), therefore the opportunity to effectively reach the workforce is currently being missed.

As the construction industry are a large employer of outdoor workers they have a valuable role to play in promoting sun safe behaviour on site. This is of importance because increasingly evidence indicates that there may be a range of diseases associated with exposure to sunlight and skin cancers may prove to be the tip of the iceberg . As risk factors are well documented and prevention methods well known and relatively easy to implement surely to help prevent skin cancer, it is time for workers in the construction industry to be encouraged to cover up.

REFERENCES

Adami, J. et al. 1995. Evidence of the association between Non-Hodgkin's lymphoma and skin cancer. *British Medical Journal.* 310:1491-5.

Borland, R.M. et al. 1991. The impact of skin cancer control education package for outdoor workers. *The Medical Journal of Australia.* 154 May 20.

Campbell, H.S. & Birdsell J.M. 1994. Knowledge, beliefs, and sun protection behaviours of Alberta adults. *Preventative Medicine.* 23: 160-166.

Cancer Research Campaign. 1995. *Fact sheet.*

Elwood, J.M. 1990. Malignant melanoma in England: risks associated with Naevi, freckles, social class, hair colour and sunburn. *International Journal of Epidemiology.* 19 (4):801-810.

Gallagher, R.P. et al. 1989. Is chronic sunlight exposure important in accounting for increases in melanoma incidence? *International Journal of Cancer.* 44:813-15.

Gebhandt, D.L. & Crump C.E. 1990. Employee fitness and wellness programmes in the workplace. *American Journal of Psychology.* 45:262-72.

Giles, G.G. et al. 1996. Has mortality from melanoma stopped rising in Australia?Analysis of trends between 1931-1994. *British Medical Journal.* 4th May. 312:1121-1125.

Girgis, A. et al. 1994. A workplace intervention for increasing outdoor workers' use of solar protection. *American Journal of Public Health.* 84 (1) 77-81.

Godkin, G.A. 1991. Changing workplace behaviour: Skin cancer protection. *Journal of Occupational Health and Safety-Aust NZ.* 7(6):477-482.

Hill,I.M. 1992. Health promotion in the British workplace: a suitable case for treatment? *Occupational Medicine.* 42:175-178.

Jones, R.R. 1992. Ozone depletion and its effects on human populations. *British Journal of Dermatology.* 127(41):2-6.

Leather, P.J. 1988. Attitudes towards safety performance on construction work: an investigation of public and private sector differences. *Work & Stress.* 2: 155-167.

MacKie, R.M. et al. 1987. Links between exposure to ultra-violet radiation and skin cancer. *Journal of the Royal College of Physicians of London.* 21(2): 1-6.

Marks, R et al. 1983. Non-melanotic skin cancer and solar keratoses in Victoria. *Medical Journal of Australia.* 2:433-34.

Marks, R. 1991. Public health evaluation of skin cancer campaigns. *Journal of Occupational Health and Safety-Aust NZ.* 7(5): 425-430.

Marsh, T. 1995. Behavioural goal-setting and feedback on construction sites. *The Safety and Health Practitioner.* January: 14-17.

Naidoo, J. & Wills, J. 1994. *Health promotion: foundation for practice.* Balliere Tindall.

Pritchard, C. & McCarthy, A. 1997. Promoting health in the construction industry? *Journal of Occupational & Environmental Medicine.* In print.

Robinson, C. et al. 1995. Assessment of mortality in the construction industry in the US 1984-1986. *American Journal of Industrial Medicine.* July 28(1):49-70.

Ross, D. 1988. A look at skin cancer. *Occupational Health.* January 420-422.

Skinner, E. 1997. *Skin cancer. A brief review of incidence and causes.* pp. 1-14 London:Health Education Authority.

Stonecipher, L.J. & Hyner, G.C. (1993). Health practices before and after worksite health screening. *Journal of Occupational Medicine.* 35(3):297-306.

Webb, G. 1992. Solar radiation protection for outdoor workers. *Journal of Occupational Health and Safety-Aust NZ.* 8 (6): 470-485.

Implementation of Safety and Health on Construction Sites, Singh, Hinze & Coble (eds)
© 1999 Balkema, Rotterdam, ISBN 90 5809 036 1

Pre-planning of health and safety

John Smallwood
Department of Construction Management, University of Port Elizabeth, South Africa
Claire Wheeler
OCCUMED cc, Cape Town, South Africa

ABSTRACT: Pre-planning for health and safety realises a structured approach to health and safety [H&S] related issues by both designers and constructors.

The system to achieve H&S is prevention, which in turn implies planning. Accidents and disease, although caused, are unplanned events which are not project requirements. By considering H&S throughout all phases of a project, designers can contribute to pre-planning health and safety. Optimum interaction with clients will minimise variations during the contract. Completeness of design enhances contractor pre-planning such as: selection of major plant and equipment; site layout; general H&S provisions, and budgeting.

This paper tables the findings of a pilot study conducted among contracts managers in the employ of general contractor members of the Master Builders & Allied Trades Association [Cape Peninsula]. Findings include, inter alia: legislation predominates among the reasons for including H&S at the pre-planning stage. However, aspects such as previous experience, productivity and constructability are most frequently referred to in determining approach.

1 INTRODUCTION

According to Oluwoye & MacLennon [1994], pre-planning health and safety can be defined as the determination of the most appropriate construction method and construction programme that will protect workers from any dangers or risks. The planning of H&S is performed in conjunction with the planning and construction work during design and construction. Liska [1994] maintains there are two parts to pre-planning: pre-project and pre-task, and that pre-planning provides the foundation for the project H&S programmes. Pre-planning identifies all the ingredients of and resources required for the H&S programme to be effective and efficient.

According to the Commission of the European Communities [1993] 36% of construction fatalities are traceable back to design decisions and 27% are attributable to lack of planning.

2 ASPECTS RELATED TO PRE-PLANNING

Pre-qualification in the health and safety sense provides a standardised method for selecting contractors on the basis of demonstrated safe work records, H&S commitment and knowledge and the ability to work in a healthy and safe manner [Levitt & Samelson, 1993]. Pre-qualification engenders pre-planning through the requiring of, inter alia, policy, health and safety plans, and work procedures [SWP's].

A project health and safety plan is comprehensive and addresses all aspects of on-site construction operations and activities associated with a particular contract [Hinze, 1997]. This plan could contain some job-specific items such as: competent safety personnel who have training and experience in recognising H&S hazards, legal requirements, job hazard analysis, training and safety orientation to be given. The health and safety plan alerts contractors to H&S issues that will need to be taken into account in tenders or similar proposals [Munro, 1996].

Research conducted in Sydney, Australia, among project and site managers indicates a high level of reference [Oluwoye & MacLennon, 1994] to drawings, site inspections and legislation with respect to gathering information for the planning of H&S. Although specifications and geotechnical reports were referred to less frequently, they should be focused upon, due to their being more detailed than drawings. Finishes specified could be hazardous, or their position on the structure may require specific plant and, or equipment. Models, being three-dimensional, can facilitate identification and appreciation of hazards by both management and workers and the planning of H&S [Smallwood, 1996].

Medical surveillance entails the examination of individuals expressly for the purpose of detecting health effects at pre-employment and periodic monitoring [Welch and Roto, 1995]. Design and implementation of a medical surveillance programme should follow a structured process, inter alia, risk assessment, identification of the potential organs which could be damaged, scheduling tests, actions and ethics; employee fitness, evaluation of control and record keeping.

3 MULTI-DISCIPLINARY APPROACH

Liska [1994] maintains it is critical that all project stakeholders be involved in pre-planning. Hinze [1997] suggests that all parties to a construction project must be included in some way in the health and safety programme.

The Construction [Design and Management] [CDM] Regulations 1994 promulgated in the United Kingdom require the appointment of a planning supervisor whose function it is to integrate design and construction in terms of H&S and the preparation of a health and safety plan which evolves at the inception of a project through commission and maintenance [Jeffrey & Douglas, 1994].

Partnering brings the various stakeholders involved in a project: client; designers; general contractor; subcontractors, and suppliers together and entails inter alia, developing mutual goals and mechanisms for solving problems, which effectively complements health and safety [Levitt & Samelson, 1993]. Partnering engenders pre-planning by facilitating multi-disciplinary contributions.

The research conducted by Oluwoye & MacLennon [1994] indicates that project and site managers are highly involved in planning H&S, while health and safety officers are only fairly involved. Estimators, planning engineers and sub-contractors are involved to a lesser extent, and suppliers and union representatives having little or no involvement. Every person should contribute to H&S. Project and site managers develop the method of construction, estimators should allow for healthy and safe methods of construction in their estimates. Planning engineers will sequence construction work to avoid clashes of trades and H&S problems. Health and safety officers should advise project and site managers regarding healthy and safe methods of construction that are practical, legal and cost efficient.

Designers influence construction H&S directly and indirectly. Directly through the selection of frame type, design, detail and specification of materials. Indirectly through selection of procurement system, preparation of contract documentation and decision regarding contract duration [Smallwood, 1996]. Given the reference to drawings, details and specifications for the pre-planning of health and safety, then designers can contribute substantially to pre-planning.

Clients influence H&S both positively and negatively, either directly or indirectly. Indirectly through: project documentation; optimum project schedule; requiring of quality management systems [QMS's], and pre-qualifying contractors on H&S. Directly through: imposition of permit systems; conducting health and safety audits; educating and training the personnel of contractors, and referring to H&S throughout the construction process [Smallwood, [1997]. Optimum interaction with designers averts variation orders which maginalise planning, result in 'out-of-sequence' work and which can be the catalyst for a series of events which result in an accident on site [Jeffrey & Douglas, 1994].

4 STRATEGIES AND PROCESSES

Constructability is a system for achieving optimum integration of construction knowledge and experience in planning, engineering, procurement and site operations, and the balancing of various project environmental constraints to achieve overall project objectives [McGeorge & Palmer, 1997].

Total Quality Management [TQM] is a strategy towards realising a quality product i.e. an error-free one by preventing errors in the construction process by integrating health and safety, productivity and quality [Levitt & Samelson, 1993]. Dreger [1996] maintains preference should be given to designers and contractors who practise TQM.

Problems related to H&S, productivity and quality can frequently be traced to substandard, inconsistently applied or non-existent safe work procedures [SWP's], operating procedures and practices. Standard operating practices and procedures are the core component of quality management and health and safety management systems as they guarantee uniformity of operation throughout an organisation. SWP's effectively ensure that each time task is performed, it is done consistently, correctly and safely [Hood, 1994].

5 BENEFITS OF PRE-PLANNING

Reduced accidents, followed by increased worker health and safety awareness and increased manager health and safety awareness were the top three objectives of planning quoted by the project and site managers surveyed by Oluwoye & MacLennon [1994]. Reduced accident induced time loss and reduced accident induced costs were also identified by the majority of project and site managers.

Hinze [1997] maintains that employees work more safely when they like the firm and have developed a loyalty to it. He further suggests that if project activities are carefully planned uncertainty and risks are minimised.

Further benefits include: less injuries; less property damage; less down time; improvement in morale; enhanced industrial relations; increased productivity; reduced cost, and less compensation insurance [Pomfret, 1997].

6 RESEARCH

Thirty two [32] construction managers in the employ of general contractor members of the Master Builders & Allied Trades Association [Cape Peninsula] in the Western Cape were surveyed to determine the nature and extent of pre-planning for health and safety. Eighteen [18] questionnaires were returned, which constituted a response rate of 56.25%.

The contractors employing the respondents undertook more commercial [77.8%] and industrial [77.8%], as opposed to other types of construction work: domestic [50%] and infrastructure [16.7%].

88.9% of contractors undertook work in excess of R25m [US$3.88m] and 11.1% between R10m and R25m [US$1.55m and US$3.88m] in 1997.

Based on value of construction, the mean level of labour only subcontracting was 11.6% and full [labour, material, etc.] was 50.8%

The greater percentage of contractors [44.4%] worked mostly on structures between 10 and 20m in height, approximately 3 to 6 storeys.

The contractors employed on average 409 workers, 236 on a permanent and 203 on a temporary basis.

27.9% of respondents were contracts directors, 38.7% were contracts managers, 27.9% were contract managers and 5.5% were site agents.

The respondents had worked 20.3 years on average in construction and 12.2 years on average for their current employer.

Before possession of site [61.1%] is the occasion respondents refer most frequently to H&S. Only 50% do so at pre-tender/bid and 33.3% at pre-activity.

H&S should be considered on all occasions.

The majority of contractors had the possible guides that complement H&S: a written H&S programme [66.7%]; written SWP's [77.8%]; a documented QMS [66.7%], and 'toolbox talks' [72.2%].

A written health and safety programme is the first step a contractor can take to demonstrate reasonable efforts to provide for H&S [Dunn & Lehrer, 1987]. SWP's ensure that each time an activity is undertaken work is done safely and consistently [Hood, 1994]. Documented QMS's complement both H&S and quality as they identify the systems, procedures, checklists, resources, activities and responsibilities necessary to assure work [Smallwood, 1997]. 'Toolbox talks' are an effective training method [The Associated General Contractors of America, 1990] and enable the review of projected activities on hazard recognition.

Table 1 indicates the sources/opportunities referred to/used for H&S information on various occasions. The ranking of each source based on frequency of reference/use is noted in parenthesis. Legislation predominates as the source referred to for H&S information on virtually all occasions. Legislation [72.2%], pre-tender/bid site visit [66.7%], geotechnical reports and contractors [61.1%] and client's specific requirements [55.6%], predominate among sources referred to during the pre-tender/bid period. Legislation [61.1%], drawings [55.6%], contracts [50%] and specifications and client specific requirements [44.4%] predominate among sources referred to before possession of site. Weather reports [66.7%], legislation [61.1%], 'toolbox talks' [55.6%] and drawings and project health and safety plans [50%] predominated among the sources referred to during the pre-activity period.

61.1% of respondents stated that their organisations prepared project specific health and safety plans.

The various parameters, agencies, aspects and contributors in tables 2-5 have been ranked according to firstly, median, then mean and then lastly, standard deviation.

Productivity followed closely by constructability are clearly the underlying basis of deciding on a construction method, H&S is only ranked fifth [Table 2].

Table 1 Sources/opportunities referred to/used for health and safety information on various occasions.

Source/opportunity	Occasion [%] Pre-tender/bid	Before possession of site	Pre-activity
Pre-tender bid site visit	66.7 [2]	38.9 [6]	27.8 [11]
Geotechnical reports	61.1 [3=]	27.8 [9=]	38.9 [7=]
Model [[3D]	16.7 [9=]	27.8 [9=]	5.6 [12]
Drawings	44.4 [7]	55.6 [2]	50.0 [4=]
Specifications	50.0 [6]	44.4 [4=]	44.4 [6]
Schedules	38.9 [8]	33.3 [7=]	38.9 [7=]
Contract	61.1 [3=]	50.0 [3]	33.3 [10]
Client, specific requirements	55.6 [5]	44.4 [4=]	38.9 [7=]
Legislation	72.2 [1]	61.1 [1]	61.1 [2]
Toolbox talks	16.7 [9=]	22.2 [12]	55.6 [3]
Weather reports	22.2 [11=]	27.8 [9=]	66.7 [1]
Project health and safety plan	22.2 [11=]	33.3 [11=]	50.0 [4=]

Table 2. Basis of deciding on a construction method.

Basis	Rank	SD	Median	Mean	Frequency [%] Always	Sometimes	Rarely	Never
Cost	4	12.7	100.0	94.5	83.3	16.7	0.0	0.0
Health and safety	5	12.7	100.0	94.5	83.3	16.7	0.0	0.0
Environment	6	30.1	67.0	70.4	38.9	38.9	16.7	5.6
Productivity	1	0.0	100.0	100.0	100.0	0.0	0.0	0.0
Quality	3	10.7	100.0	96.3	88.9	11.1	0.0	0.0
Constructability	2	0.0	100.0	100.0	100.0	0.0	0.0	0.0

Table 3. Agencies contributing to determining the danger related to a construction activity/process.

Agency	Rank	SD	Median	Mean	Frequency [%] Often	Sometimes	Rarely	Never
H&S consultant	3	33.3	67.0	68.6	38.9	38.9	11.1	11.1
Legislation	2	26.1	100.0	85.3	66.7	27.8	0.0	5.6
Client advice/ requirements	7	28.2	33.0	44.4	5.6	38.9	38.9	16.7
Practical experience	1	0.0	100.0	100.0	100.0	0.0	0.0	0.0
Publications	6	26.1	33.0	46.3	5.6	38.9	44.4	11.1
Records	4	34.4	50.0	55.5	27.8	22.2	38.9	11.1
Suppliers [MSDSs]	5	29.5	50.0	40.8	0.0	50.0	22.2	27.8

Table 4. Extent to which health and safety influences the planning of various aspects.

Aspect	Rank	SD	Median	Mean	Frequency [%] Often	Sometimes	Rarely	Never
Site accommodation	6	16.0	100.0	89.0	66.7	33.3	0.0	0.0
Ablutions	2	10.7	100.0	96.3	88.9	11.1	0.0	0.0
First aid	3	10.7	100.0	96.3	88.9	11.1	0.0	0.0
Mess facilities	12	37.0	67.0	64.9	38.9	33.3	11.1	16.7
Storage	8	23.3	100.0	87.1	72.2	16.7	11.1	0.0
Access ways	1	7.8	100.0	98.2	94.4	5.6	0.0	0.0
Materials and handling	4	18.3	100.0	92.6	83.3	11.1	5.6	0.0
Personnel movement	7	16.0	100.0	89.0	66.7	33.3	0.0	0.0
Plant and equipment	5	18.3	100.0	92.6	83.3	11.1	5.6	0.0
Temporary services	11	20.2	67.0	79.8	44.4	50.0	5.6	0.0
Temporary structures	9	16.9	100.0	85.3	55.6	44.4	0.0	0.0
Waste/rubbish removal	12	20.6	100.0	83.4	55.6	38.9	5.6	0.0

Practical experience, followed by legislation, predominated among agencies contributing to determining the danger related to a construction activity/process [Table 3].

Access ways, followed closely by ablutions and first aid predominate in terms of the extent to which H&S influences the planning of various aspects. The identification of access ways is probably related to the identification of productivity as the primary basis for adopting a construction method. The identification of ablutions and first aid is probably attributable to the identification of legislation as the second most important agency contributing to determining the danger related to a construction activity/process. Materials and handling, plant and equipment and personnel movement are also related to productivity [Table 4].

Previous experience predominated among sources referred to when planning various aspects of H&S. Thereafter, legislation was referred to most frequently, followed by client requirements, union awareness and to the least extent, past projects [Table 5].

Table 5. Frequency of reference to various sources when planning various aspects of health and safety.

Aspect	Sources [%]				
	Legislation	Client requirements	Union awareness	Past projects	Previous experience
Site accommodation	33.3	38.9	11.1	22.2	100.0
Ablutions	50.0	27.8	27.8	16.7	77.8
First Aid	77.8	11.1	5.6	11.1	72.2
Mess facilities	33.3	5.6	38.9	11.1	83.3
Storage	44.4	22.2	0.0	11.1	100.0
Access ways	55.6	38.9	5.6	11.1	94.4
Materials & handling	55.6	16.7	11.1	11.1	88.9
Personnel movement	44.4	22.2	16.7	11.1	100.0
Plant & equipment	61.1	22.2	11.1	11.1	100.0
Temporary services	61.1	27.8	11.1	11.1	100.0
Temporary structures	38.9	22.2	11.1	11.1	94.4
Waste/rubbish removal	38.9	27.8	11.1	11.1	94.4

Table 6. Frequency of contribution by various stakeholders to the pre-planning of health and safety.

Aspect	Rank	SD	Median	Frequency [%]					
				Mean	Often	Some times	Rarely	Never	No answer
Architects	7	29.3	33.0	25.8	5.6	11.1	38.9	44.4	0.0
Clients	6	25.7	33.0	33.3	0.0	27.8	44.4	27.8	0.0
Engineers	5	34.2	33.0	37.0	11.1	22.2	33.3	33.3	0.0
Health & safety consultant	2	31.5	67.0	59.3	22.2	44.4	22.2	5.6	5.6
Sub-contractors	3	22.5	33.0	42.6	0.0	38.9	50.0	11.1	0.0
Suppliers	8	20.5	16.5	18.4	0.0	5.6	44.4	50.0	0.0
Unions	4	32.2	33.0	37.1	5.6	33.3	27.8	27.8	5.6
Workers	1	25.4	67.0	70.4	33.3	44.4	22.2	0.0	0.0

Table 7. Benefits of planning of health and safety.

Benefit	Response [%]
Assure health and safety	38.9
Reduce accidents	100.0
Reduce illness/disease	44.4
Improve productivity	83.3
Improve quality	61.1
Reduce cost	77.8
Other	5.6

Based on the mean frequency of response, workers, followed by health and safety consultants, predominated among the various stakeholders contributing to the pre-planning of H&S [Table 6].

Reduction of accidents [100%], improvement of productivity [83.3%] and reduction of cost [77.8%] predominated among the benefits of planning of H&S. Improvement of quality was also identified by the greater percentage of respondents. The level of response relative to assurance of health and safety and the reduction of illness/disease indicates a lack of appreciation of the role of pre-planning in the case of the former and the impact illness/disease in the case of the latter [Table 7].

72.2% of respondents stated that H&S is very important and 22.2% that H&S is important.

61.1% of respondents had more than one comment in general regarding the planning of H&S. 38.9% had no comments.

7 CONCLUSIONS

Literature relates the benefits of pre-planning of H&S at the initiation of a project, namely the optimum interaction of all the major role players which will reduce costs, reduce hazards and injuries, increase productivity and achievement of quality.

Given that 72.2% of respondents stated that H&S is very important and 22.2% that H&S is important, then the respondents are likely to constitute the more committed group of construction management.

Construction managers generally consider H&S at pre-tender/bid stage and before taking possession of site. The construction managers have a number of formal means available to enhance health and safety: programme; SWP's; QMS, and 'toolbox talks'.

Legislation predominates among the sources referred to and opportunities used for H&S information. Other frequent referred to sources and opportunities used are: drawings; specifications; contract, and client specific requirements.

Although productivity and constructability are the underlying reasons for deciding on a construction method, quality, H&S, and cost are close seconds.

Practical experience and legislation are the agencies that contribute most to determining the danger related to a construction activity or process. Similarly, previous experience, legislation and client requirements are the sources most frequently referred to. Productivity and resource related aspects predominate in terms of extent to which H&S influences planning: workers and H&S consultants are the stakeholders that contribute most frequently to the pre-planning of H&S.

Pre-planning H&S realises synergistic benefits which result in an overall improvement in financial performance.

The implications of the findings are that: design and details need to be complete prior to the commencement of a project; all stakeholders need to contribute to pre-planning and given the benefits arising from pre-planning, it needs to feature in all H&S programmes.

ACKNOWLEDGEMENTS

The authors are grateful to the construction managers who responded to the written survey and to Danie Venter, Director, Institute for Statistical Consultation and Methodology, University of Port Elizabeth, and Wilma Slater of OCCUMED cc for the typing.

REFERENCES

Commission of the European Communities 1993. *Safety and health in the construction sector.* Luxembourg: Office for Official Publications of the European Communities.

Dreger, G.T 1996. Sustainable development in construction: Management strategies for success. *Proceedings of the 1996 CIB W89 Beijing International Conference – Construction Modernisation and Education.* Beijing, October 1996. CD-file:///D1/papers/160-169/163/p.163.htm

Dunn, T.R. & Lehrer, G.H 1987. Construction Safety: The duty of contractors. *American Professional Constructor*, July: 2-5.

Hinze, J 1997. *Construction Safety.* Prentice Hall, New Jersey.

Hood, S 1994. Developing operations procedures: 9 steps to success. *Accident Prevention*, May/June: 18-21.

Jeffrey, J & Douglas, I 1994. Safety performance of the UK construction industry: *Proceedings of the 5th Annual Rinker International Conference focusing on Construction Safety and Loss Control,* Florida, October 1994: 233-253.

Levitt, R.E. & Samelson, N.M 1993. *Construction Safety Management.* 2nd Edition. New

York: John Wiley & Sons Inc.
Liska, R.W 1994. Zero injury techniques. *Proceedings of the Fifth Annual Rinker International Conference Focusing on Construction Safety and Loss Control*, Gainesville, Florida, 12-14 October: 293-303. Gainesville: M.E. Rinker Sr. School of Building Construction, University of Florida.
McGeorge, D. & Palmer, A 1997. *Construction Management New Directions*, Oxford: Blackwell Science Ltd.
Munro, W.D 1996. The implementation of the Construction [Design and Management] Regulations 1994 on UK construction sites. *Proceedings of the First International Conference of CIB Working Commission W99 – Implementation of Safety and Health on Construction Sites*, Lisbon, 4-7 September: 53-60. Rotterdam: A.A. Balkema.
Oluwoye, J. & MacLennon, H 1994. Designing for safety and the environment. *Proceedings of the Fifth Annual Rinker International Conference Focusing on Construction Safety and Loss Control*, Gainesville, Florida, 12-14 October: 175-185. Gainesville: M.E. Rinker Sr. School of Building Construction, University of Florida.
Pomfret, B 1997. Psychology of Safety. *National Safety*, May/June: 14-15.
Smallwood, J.J 1996. The influence of designers on occupational health and safety. *Proceedings of the 1st International Conference of CIB W99 – Implementation of Safety and Health on Construction Sites*, Lisbon, September: 203-213.
Smallwood, J.J 1997. Total Quality Management in Construction. *ProjectPro*, November: 17-19.
The Associated General Contractors of America [AGC] 1990. *AGC Guide for a Basic Company Safety Program*. Washington: AGC.
Welch, L. & Roto, P 1995. Medical Surveillance Programmes for Construction Workers. *Occupational Medicine: State of the Art Reviews*. 10[2] April: 421-432.

20 Medical and health management – II

Implementation of Safety and Health on Construction Sites, Singh, Hinze & Coble (eds)
© 1999 Balkema, Rotterdam, ISBN 90 5809 036 1

Differences in musculoskeletal symptoms and job factor perceptions between Slovak and US construction workers

T.M.Cook, C.L.Zimmermann & J.C.Rosecrance
Injury Prevention Research Center, Department of Preventive Medicine and Environmental Health, The University of Iowa, Iowa City, Iowa, USA
K.Hatiar
Ergonomics Department, Institute of Clinical and Preventive Medicine, Bratislava, Slovakia

ABSTRACT: The purpose of this project was to generate descriptive data regarding musculoskeletal complaints and problematic job factors among construction workers in the United States and the Slovak Republic with the overall goal of reducing occupational injuries and illnesses within the construction trades in both countries. Data using a two-page self-administered symptom and job factors questionnaire were collected from unionized construction workers in the U.S. (2,516) and the Slovak Republic (1,096). Analysis of the data indicates many similarities in the prevalence of musculoskeletal complaints between the two groups of workers but also interesting differences, particularly in the wrist/hand and upper back. Workers in the Slovak Republic reported a five-fold increase in the portion of workers who reported missing work to workers who reported symptoms, compared to the U.S. Overall, reports of problematic job factors were very similar between the two countries, although there were some factors that were different. This type of data is being used as the basis for participatory ergonomic problem solving in both countries.

1 INTRODUCTION

Work-related musculoskeletal disorders are becoming increasingly recognized as widespread and costly problems in construction, as well as in other industries (Schneider, et al.1995). In the last several years it has become clear that both profitability and worker safety and health can be maximized through the use of a participatory process of problem solving (Cohen, 1997). This process consists of involving all relevant stakeholders, including managers and workers, in a cyclical process of problem identification, problem analysis, solution development, implementation, and evaluation (Imada, 1990). Evidence is rapidly mounting that a participatory approach to workplace safety and health has many benefits including: decreases in injuries, in health care costs, in insurance costs, in absenteeism, in re-training costs and in worker turnover, and better product quality, better customer service, and increased productivity (Urlings et al. 1990). Enterprises that are proving to be competitive in the global marketplace are those that have high levels of involvement by a stable workforce and that devote a substantial portion of their resources to employee education and to health and safety measures (Rosecrance and Cook, in press). These successful companies maximize their profitability by aggressively preventing workplace injuries, illnesses, and deaths, rather than expending their resources, and the resources of their governments, to deal with the consequences of these problems.

As the initial phase of implementing a participatory ergonomics process, the authors have used a symptom/job factor questionnaire as a tool for problem identification and analysis. We have collected data on over 10,000 workers in the United States in a variety of industries, including manufacturing (Rosecrance, et al. 1997), health care (Bork, et al. 1996), newspaper production (Cook, et al. 1994) and construction (Rosecrance, et al. 1996; Cook, et al. 1996;

Zimmermann, et al. 1997). This report describes the results of data collected using this questionnaire in a sample of unionized construction workers from the Slovak Republic and contrasts those results with data obtained from a comparable sample of construction workers from the mid-Western United States. The primary purpose of this project was to generate descriptive data regarding musculoskeletal complaints and problematic job factors among construction workers in the Slovak Republic with the overall goal of reducing occupational injuries and illnesses within the construction trades in both the U.S. and the Slovak Republic. A secondary purpose was to begin to identify tools, materials, work methods, or work practices which might account for any observed differences in symptoms and job factors between the two groups of construction workers, especially within specific trades.

This project utilized the close working relationship developed between investigators at The University of Iowa Injury Prevention Research Center and The Institute of Preventive and Clinical Medicine in Bratislava, Slovakia. At the time of the study, investigators from both institutions were participating in a five-year NIH-sponsored Fogarty International Center research and training project that includes five countries in Central Europe. It was hoped that this project would serve as a model for similar projects to be conducted in other Central European countries.

2 METHODS

The primary data collection method was a two-page self-administered questionnaire that contained questions about personal characteristics, work history, recent history of musculoskeletal complaints in nine body regions, and perceptions of the relative importance of fifteen different job factors in causing musculoskeletal complaints. The questionnaire has two major sections and can be completed in ten minutes. The symptom survey section of the questionnaire is a modification of the standardized Nordic Questionnaire (Kuorinka, et al 1987) and consists of questions referring to nine body areas. A simple body figure with body areas highlighted (3 upper limb, 3 lower limb, 3 trunk) was used to help the respondents answer "yes or no" to the question, "During the last 12 months have you had a job-related ache, pain, discomfort, etc." in each of nine different body segments. If the respondents indicated that a work-related musculoskeletal symptom had occurred, they were then asked to answer "yes or no" to two additional questions. Those questions were, "During the last 12 months have you been prevented from doing your day's work due to this condition?" and, "During the last 12 months have you seen a physician (M.D., Chiropractor, Osteopath) for this condition?" If the respondent indicated yes to a "job-related ache, pain, discomfort, etc." they fit our case definition of work-related musculoskeletal symptom. This case definition was used for surveillance purposes only and was not considered to represent any specific pathology.

Workers' perceptions of the physically stressful elements in their jobs were assessed using the "job factors" portion of the questionnaire. This page of the questionnaire contained a listing of fifteen conditions and tasks with the following instructions, "This list describes things at work that could contribute to job-related pain and injury. Please indicate, on a scale of 0 to 10, how much of a problem (if any) each item is for you by circling the appropriate number." Zero was equivalent to a job factor being "No Problem" while ten was used to indicate that a job factor was considered a "Major Problem." The job factor descriptions are contained in Figure 2. Questionnaires were distributed and collected using a combination of mailing, personal distribution, and distribution through union safety representatives. In all cases, the identity of the respondents was kept confidential and employers did not have access to individual data. Participation was strongly supported by the union leadership in both countries.

3 RESULTS

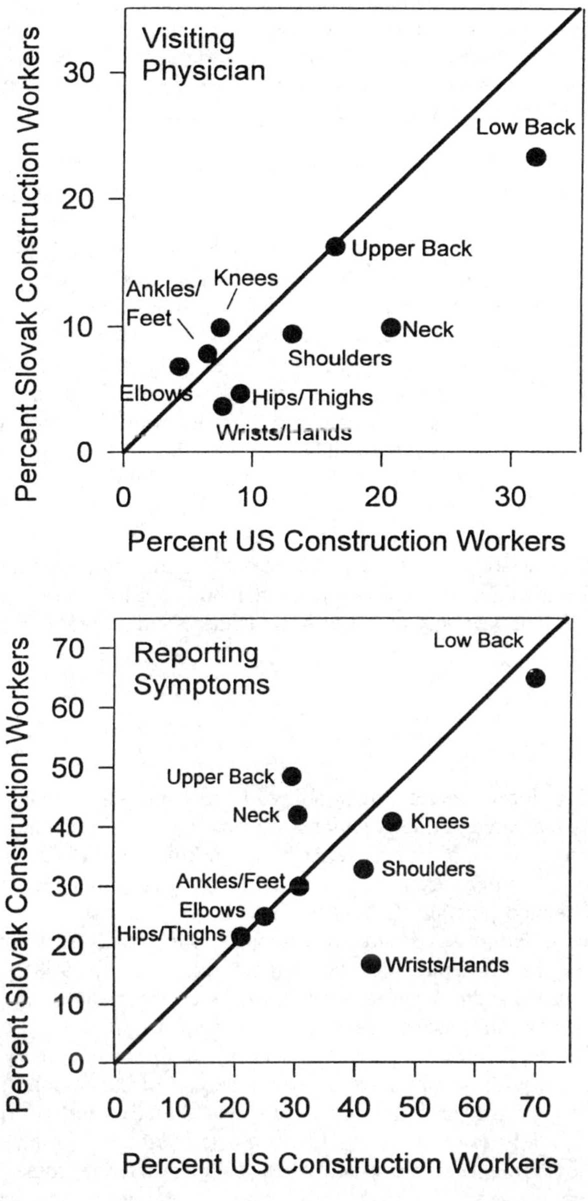

Figure 1. U.S. and Slovak construction workers reporting symptoms and visiting a physician.

In the U.S., 2,516 (35%) unionized construction workers mailed back completed questionnaires for analysis. The respondents had an average age of 42.5 years, an average of 19.2 years of experience in their respective trade, and 69.4% had received apprenticeship training. At the time of the survey 89.9% of the respondents reported that they were currently working and the average number of weeks worked in the previous year was reported to be just over 45. In the Slovak Republic, 1,096 (16.3%) workers returned completed questionnaires for analysis. The respondents had an average age of 40.2 years, an average of 15.2 years of construction work experience, and 67.9% had received apprenticeship training. At the time of the survey 94.7% of the respondents reported that they were currently working and the average number of weeks worked in the previous year was reported to also be just over 45. There were no statistically significant differences between the two groups in terms of age, years of experience, weeks worked per year, or percent of workers having apprenticeship training.

The percent of respondents from the United States and Slovakia who reported work-related musculoskeletal symptoms (Figure 1) was very similar for the hips/thighs (21%), elbows (24%), knees (46%, 41%), and low back (70%, 65%), respectively. U.S. workers reported more symptoms in the shoulders (43% Vs 33%) and the wrists/hands (43% Vs 17%) but fewer symptoms in the neck (31% Vs 42%) and upper back (30% Vs 48%), respectively.

Among U.S. construction workers, the proportion of workers reporting missing work to workers reporting symptoms ranged from 0.07 (elbows) to 0.16 (low back). Among Slovakian workers, this proportion ranged from 0.37 (neck) to 0.60 (low back), indicating approximately an overall five-fold difference between the two groups. The proportion of reported physician visits to reported symptoms was highly dependent on the anatomical area; compared to Slovak workers, the proportion was higher among US construction workers for the neck (0.68 Vs 0.23), the upper back (0.56 Vs 0.35), and the hips/thighs (0.45 Vs 0.21).

The profile of mean scores of problematic job factors was remarkably similar between the two groups of workers with a few exceptions (Figure 2). U.S. workers expressed a greater concern about: (1) bending and twisting awkwardly; (2) working in the same position for long time periods; and (3) reaching overhead or away from the body. Slovakian construction workers reported a mean score that was only one-fourth the value of U.S. workers for working at or near their physical limits.

4 DISCUSSION

The validity of self-reporting for collecting data on working conditions is an important question. A recent report has indicated good agreement between workers describing their own working conditions and occupational physicians describing working conditions in the same factories (Hays, et al. 1996). Another recent study compared worker self-reports to expert-observed estimates of work-related physical factors related to carpal tunnel syndrome. This study found that, "although imperfect, exposure information collected from workers' self-reports is useful for many ergonomic epidemiology studies" (Nordstrom, 1998). There are important advantages in using self-administered questionnaires as the basis of participatory problem solving related to ergonomic issues. First of all, data can be collected on a large number of workers in a rather short period of time. Secondly, the questionnaire is relatively inexpensive to administer and requires minimal expertise on the part of the person distributing it and explaining its purpose. Thirdly, information is obtained directly (and confidentially) from the workers who are actually doing the jobs that may be leading to problems. Both company officials and government bureaucracies are not involved, thus, common impediments to accurate reporting are avoided. And lastly, the information obtained is immediately useful for prioritizing and developing detailed solutions to problems.

Low response rates and response bias are important considerations in a study such as this one. According to union officials in the U.S., the union locals involved in this study routinely have a response rate of 20 to 25 percent on the questionnaires and other information they solicit from their membership. Using this criterion, the response rate for this survey can be considered typical, but lower than desirable. Union officials confirmed to the investigators that the average age, experience level, and rate of apprenticeship training were in close agreement with their records for the overall union membership. Since union officials in the

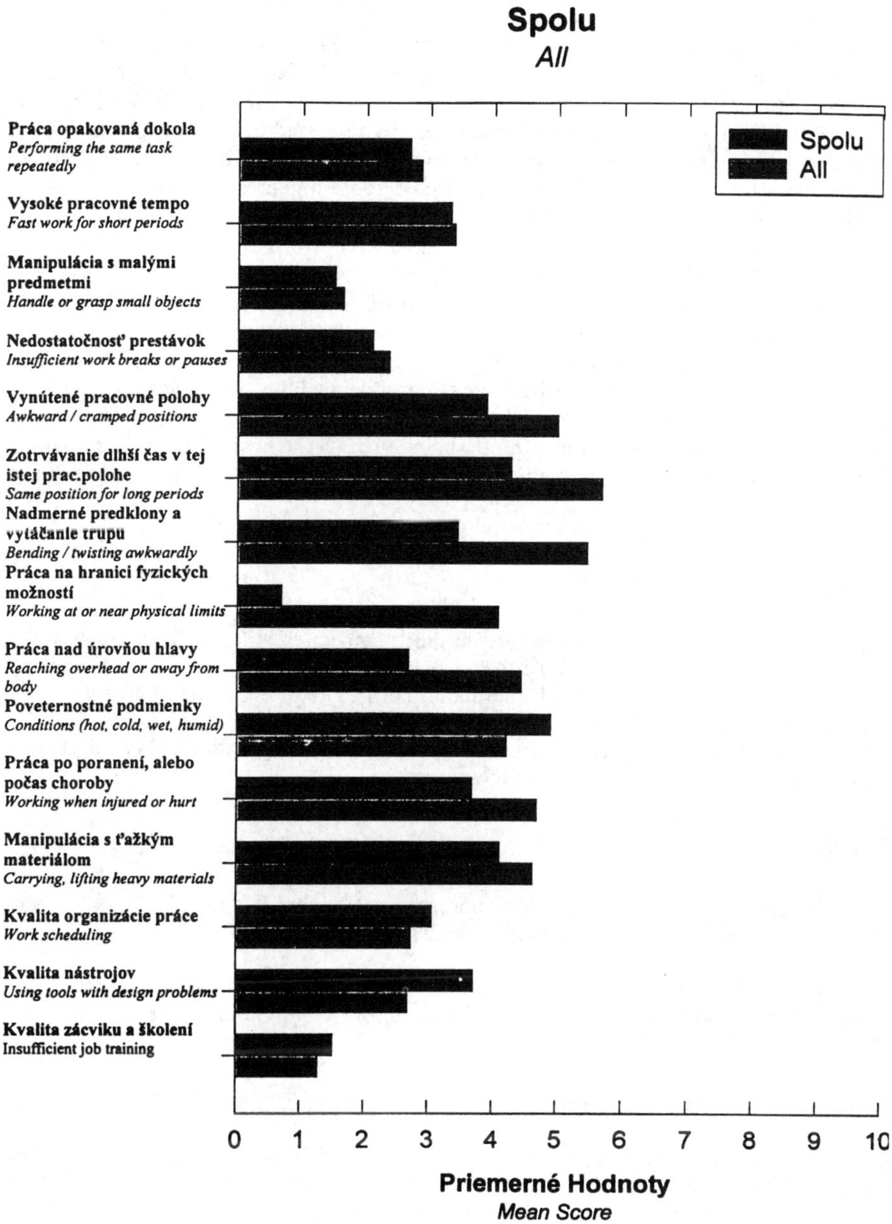

Figure 2. Percent of Slovak (Spolu, upper bars) and U.S. (all, lower bars) construction workers reporting problematic job factors.

Slovak Republic have limited experience with soliciting questionnaire information from their membership, it is difficult to evaluate the low response rate for the survey in that country. From a purely epidemiological perspective, the possibility of response bias clearly exists in this study. It is quite possible that construction workers who have had work-related musculoskeletal problems were more likely to complete and return the questionnaire than those who had not experienced such problems. However, the main purpose of this project was to gather information regarding the relative anatomical distribution of musculoskeletal complaints, rather than their absolute prevalence.

Despite the relatively low response rates for the two groups of construction workers and possible response bias, there are some very interesting comparisons and contrasts in the data. In four of the nine anatomical regions, the percent of workers who report work-related complaints is virtually identical in the two countries. These findings would seem to lend credibility to the data from other anatomical regions where there are differences in the percent of complaints, some higher in the U.S. (wrist/hand) and some higher in the Slovak Republic (upper back). Possible explanations for the differences reported between these two groups of construction workers include differences in: work methods, work organization, materials, tools, psycho-social factors, economic factors, health care delivery systems, and others. For example, the authors have observed a much greater use of powered hand tools in the Slovak Republic. This difference may account for the fact that the percent of reported wrist and hand complaints was approximately one-third that reported in the U.S. More detailed analysis of data from this and other comparative studies may provide insights into the factors which contribute to occupational musculoskeletal injuries and illnesses in the construction trades worldwide. The results of this study are currently being used as the basis for joint discussions by union officials and contractors in both countries on how to develop safer construction worksites.

5 CONCLUSIONS

The results of this study indicate both marked similarity and interesting differences between the U.S. and the Slovak Republic in regard to the percent of unionized construction workers who report work-related musculoskeletal complaints and in regard to their perceptions of problematic job factors which may be contributing to their problems. This data seems useful as the basis for a participatory process of ergonomic problem solving with the overall goal of better understanding and improving construction worksites.

REFERENCES

Cohen, A., et al. 1997. Elements of ergonomics programs. National Institute for Occupational Safety and Health, Publication 97-117, Cincinnati, OH.

Hays, M., et al. 1996. Conformity of workers' and occupational health physicians' descriptions of working conditions. *International Journal of Occupational and Environmental Health*, 2(1): 10-17.

Imada, A. 1990. Ergonomics: influencing management behavior, *Ergonomics*, 33(5):621-628.

Kuorinka, I., B Jonsson, A. Kilbom, et al. 1987. Standardized Nordic questionnaires for the analysis of musculoskeletal symptoms. *Applied Ergonomics,* 18:233-237, 1987.

Nordstrom, D., et al. 1998. Comparison of self-reported and expert-observed physical activities at work in a general population. *American Journal of Industrial Medicine*, 34:29-35.

Rosecrance, J., T. Cook In press. Upper extremity musculoskeletal disorders: occupational association and a model for prevention, *Central European Journal of Occupational and Environmental Medicine*.

Schneider, S., L. Punnett, T.Cook 1995. Ergonomics: applying what we know. *Occupational Medicine: State of the Art Reviews* 10(2): 385-394.

Urlings, I., D. Nijboer, J. Dul 1990. A method for changing the attitudes and behavior of management and employees to stimulate the implementation of ergonomic improvements, *Ergonomics*, 33(5):629-637.

Implementation of Safety and Health on Construction Sites, Singh, Hinze & Coble (eds)
© 1999 Balkema, Rotterdam, ISBN 90 5809 036 1

Health Trak: A participatory approach for ergonomic improvement in construction

Bryan Buchholz, Susan Moir & Joel Garrett
Construction Occupational Health Project, Department of Work Environment, University of Massachusetts Lowell, Mass., USA

ABSTRACT: Health Trak is a participatory approach that involves construction workers and management, as well as occupational health researchers, in identification, evaluation and control of health hazards. Each eight-week Health Trak cycle has three key components: 1) participatory committee meetings with worker and contractor representatives, 2) toolbox meetings with workers on-site, and 3) participant observation. Ergonomic hazards, such as those that may lead to musculoskeletal disorders, are the current focus of Health Trak. Two Health Trak cycles have been completed to date. In the first cycle, the participants identified the lack of sanitary facilities as the health hazard for study. Though an intervention was not implemented, the efforts of this committee have had a profound impact on the local industry. In the second cycle, the committee focused on ergonomics and was successful in implementing interventions to control manual material handling hazards.

1 INTRODUCTION

Construction workers are at great risk to injury, disability and early death as a result of exposure to health and safety hazards at work (Center to Protect Workers Rights 1993). The broad goal of this research project is to reduce the incidence and prevalence of injuries and illnesses to construction workers. The barriers to this objective are not primarily technological, but cultural and organizational. A wide range of possible interventions is available to be introduced at the industry, contractor and individual levels (Schneider 1996). These interventions include: technological changes, improved personal protective equipment, and administrative reforms. The imminent question is not so much what to change, but how to change and who will affect that change.

Ringen *et al.* (1995) cite several characteristics which make construction unique, including multiple employers, short-term employment and the ever-changing nature of the construction workplace and its hazards. In this study, rapid appraisal and participatory methods are combined in the design of Health Trak to address the uniqueness of construction. A compelling body of literature contends that successful introduction of ergonomics into the workplace is preconditioned on participation by the affected workers (e.g. Gjessing *et al.* 1994; Schurman *et al.* 1994). Rapid appraisal uses popular education and participatory techniques in an intensive, systematic and semi-structured method carried out by a team composed of outside researchers and a representative cross-section of the affected population to enter a research setting, do assessment, and introduce and evaluate interventions in the shortest feasible time (e.g. Chambers 1993; Murray 1994).

2 PARTICIPATORY METHODOLOGY

Each Health Trak cycle is conducted over eight weeks and utilizes three key components: 1) Participatory meetings of Health Trak committees composed of 3-4 workers, the contractor safety representative, and other on-site personnel responsible for production; 2) Toolbox meetings involving larger groups of workers; and 3) Participant observation, focus group meetings and interviews with workers and contractor representatives. These three components are integrated and function concurrently during the cycle (Figure 1).

1. Health Trak committee meetings. Participation on the committee involves attendance at a weekly one-hour meeting, which consists of participatory learner-centered activities, designed to identify and/or solve problems through eliciting the experience of the participants in a group process. These activities include risk mapping of hazards in the workplace, team building exercises, communications skill development, storytelling, role plays and other psychomotor methods for learning and sharing knowledge. As part of each committee meeting, participants design a data collection instrument with which to survey co-workers in the week following the meeting. For example, when participants conduct risk mapping in the Health Trak committee meeting to identify hazards in the workplace, they then take the risk maps with them at the end of the meeting and survey their co-workers. The feedback from this survey is reviewed and evaluated at the next Health Trak meeting. The committee meetings are divided into four steps during the eight-week cycle:

a. Problem identification and definition (weeks 1-2). The objective of these sessions is to identify problems with the health component of the current health and safety program. The product of these two weeks is a written definition of the identified health problem, which serves as the "research question" for the participants and will be the focus for the remainder of the Health Trak cycle.

b. Analysis of hazards and strategies for intervention (weeks 3-5). The objectives of these weeks are to analyze the written research problem and to develop strategies for

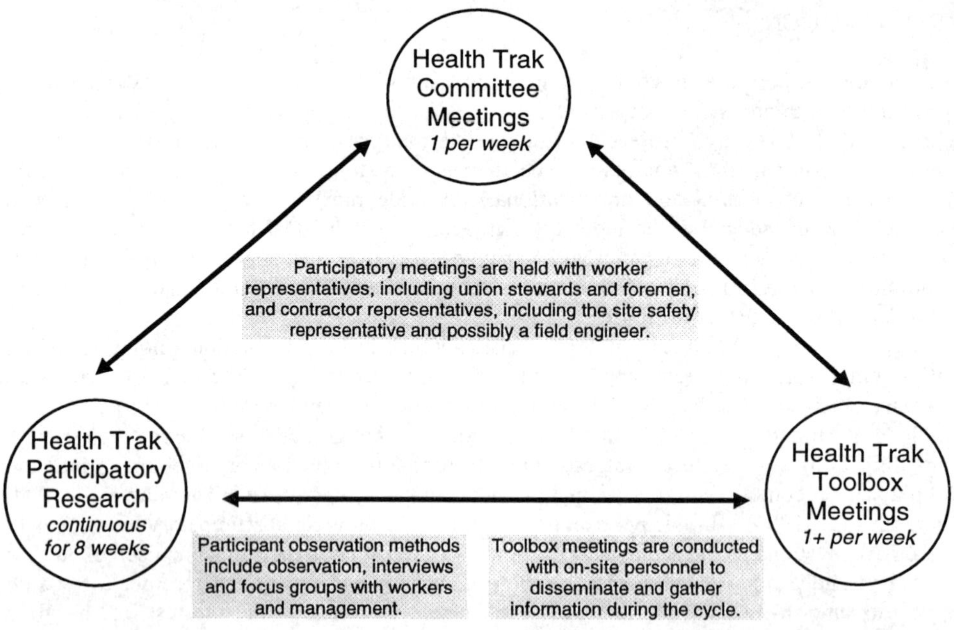

Figure 1. Interaction of the three components of a Health Trak cycle.

intervention. Participants conduct a needs assessment, which is used to assess their current health and safety working conditions relevant to the defined research question, envision possible solutions, and assess potential strategies for change, in order to move from the present conditions to their desired conditions. The product of these sessions is possible interventions that address the research question.

c. Implementation of proposed intervention (weeks 6-7). Participants will prioritize interventions and introduce them at the worksite. The product of these sessions will be the intervention and its evaluation.

d. Evaluation by participants (week 8). Impact and outcome measures for evaluation include self-assessment by participants of the Health Trak process and their participation in it, evaluation of changes in the health and safety program, a review of injury data for the eight-week period, and evaluation of the effectiveness of the intervention.

2. Health Trak toolbox meetings. At least once a week during each eight-week cycle, a Health Trak researcher conducts a Health Trak participatory "toolbox meeting" for a group of workers not participating in the Health Trak Committee. Toolbox meetings are 10-15 minute lectures (often having a safety component) held on many construction sites before work in the morning. The focus of these toolbox meetings is the research question defined in the committee meeting. A checklist is used by a researcher to identify immediate problems and track the perceptions of changes and response from this larger group of workers. Results of the toolbox meetings are reported to the Health Trak committee for its review and action.

3. Health Trak participant observation. Researchers also spend 1-2 days per week on the site using participant observation, interviews and informal focus groups, to collect data about conditions and perceived changes by labor and management during the Health Trak cycle.

3 EVALUATION OF PARTICIPATORY METHODOLOGY

Qualitative and quantitative evaluation occurs at multiple levels:

1. Surveys of Health Trak committee members. A survey is administered to each participant before and after each Health Trak cycle. Participants are asked to quantitatively and qualitatively evaluate their workplace health and safety program and the perceived changes that occur as a result of Health Trak.

2. Quantitative assessment of hazards pre- and post-intervention. Interventions are evaluated using appropriate ergonomic risk assessment methods. These include structured ergonomic job analyses (Keyserling *et al.* 1991), ergonomic checklists, and PATH, a work-sampling based ergonomic job analysis method (Buchholz *et al.* 1996).

3. Qualitative analysis of data collected during each Health Trak cycle. Qualitative evaluation is conducted at the end of each cycle using data collected from committee meetings, toolbox meetings and participant observation. Data include: data from the committee meetings, in the form of transcripts, notes and artifacts from participatory activities; the results of participatory field work conducted by participants in the Health Trak committee; checklists from toolbox meetings; trend in injury data, including OSHA 200 logs, workers' compensation data, and Health Trak injury logs; and data collected by researchers in meetings or on-site through participant observation. Outcomes include changes in the perceptions of Health Trak participants and other employees regarding the effectiveness and responsiveness of the health and safety program and changes in exposure to hazards.

4. Site walk-throughs by external occupational health and safety experts. Experts in health and safety (e.g. construction safety consultants, union safety staff, state and federal health and safety professionals) conduct site walk-throughs and give written evaluations of the health and safety programs on these sites.

5. Evaluation by an advisory board. An advisory board is convened twice a year to review and evaluate Health Trak. This board consists of representatives from the participating contractors and unions and experts in occupational health and safety, participatory research, rapid appraisal methods, and evaluation research.

4 RESULTS TO-DATE

The Boston Central Artery/Third Harbor Tunnel (CA/T) construction project has been the site for these research activities. The CA/T is currently the largest on-going public works project in the United States. The construction project is comprised of numerous individual sites, each with its own general contractor, as well as many smaller sub-contractors. The Massachusetts Highway Department is the project owner and they have hired a firm to manage the project, which oversees all aspects of the construction.

4.1 *Results of Cycle 1*

The first Health Trak cycle was held on a site constructing an elevated portion of the Central Artery construction project. A planning meeting was held with the site safety representative, the contractor's corporate safety representative and the assistant resident engineer from the Central Artery project construction manager and the first cycle began within a month. The Health Trak Committee was composed of a carpenter, an operating engineer, two laborers, the site safety representative and a field engineer.

The weekly Health Trak committee meetings went as planned, with the curriculum specifics unfolding week-to-week. The committee decided during week three that the lack of sanitary facilities on site would be the hazard to prioritize for intervention. The site in question had only portable toilets and, more importantly, no facilities to wash and change before lunch or leaving the site. The next four weeks were spent attempting to develop and implement an intervention for this hazard. Two separate proposals for intervention were developed: one involving connection to the city sewer (~$150,000) and one using a holding tank (~$50,000). Two obstacles to implementation were identified: the contractor's concern for cost and acceptable area practice (i.e. sanitary facilities are not normally provided on construction sites). Despite meetings with the contractor, construction manager and owner, the intervention was never implemented. We have continued to work on solutions to this hazard, with the hope that, at a minimum, the owner will include the need to address the hazard in the bid specifications for new sites on the Central Artery.

In this cycle, the weekly toolbox meetings were used primarily as a communication tool to provide information on the status of Health Trak. Future cycles will attempt to use toolbox meetings as training and education sessions. Both production constraints and requirements for other safety training limit time for this purpose. In addition, more on-site worker involvement will be solicited.

The committee communicated well and labor/management/researcher barriers were not apparent. The committee members were performing research early in the cycle, which underscores this productive relationship. Because an intervention was not successfully implemented and because of the nature of the health hazard, quantitative assessments were not performed. The surveys of workers on the site indicated an increase in awareness of health hazards. The comparison of the pre- and post-cycle evaluations by the independent health and safety expert showed a similar result indicating that this Health Trak cycle had a positive influence on site.

4.2 *Results of Cycle 2*

The second Health Trak cycle was held on a utility relocation site on the Central Artery construction project. An orientation meeting was held with the site safety manager and the cycle began shortly thereafter. The Health Trak Committee was selected by the site safety manager and was composed of a laborer steward, a laborer, an operating engineer, the site safety manager and the owner's environmental health and safety representative. Representatives from the owner's occupational health clinic observed at five meetings on a rotating basis.

The researchers made a conscious effort to steer the committee towards an ergonomic intervention in this cycle. This was explained to the committee in the first meeting, along with the broader goal of creating a general process for reducing health-threatening exposures. The weekly Health Trak committee meetings went as planned, with the curriculum specifics unfolding week-to-week. In the second week, the committee identified lifting oxygen-acetylene canisters, repetitive shoveling, pressure from rushing their work, and general exposures to dust, noise, fumes, and hot temperatures as important exposures occurring on their site. During the next two meetings, they decided to implement interventions to reduce exposure during manual materials handling tasks. The controls they decided to use were to increase the number of manual materials handling aids on site, to improve housekeeping and to improve work surfaces. Through the participatory process, hand trucks, work surface improvements, and wheel wells were identified as feasible controls. Two hand trucks were purchased at once and the work surface improvements also began immediately. Wheel wells were designed for use during lifting and lowering materials from/to the jacking and receiving pits. However, the redesigned wheel wells were never completed.

Although we focused on ergonomics during this cycle, sanitary facilities were also an issue, as they were in the first cycle. The committee actually recommended that a hand washing station be the first control implemented and though the committee did not spend a lot of time on this issue, the laborer steward worked with the safety manager to create a prototype. Simultaneously, the project owner mandated washing facilities for all Central Artery sites, and lavatories for permanent work areas.

In this cycle, researchers communicated with the 70 other workers on site through the weekly toolbox meetings. These meetings were used to provide limited ergonomic training and to provide information on the status of Health Trak. The researchers also conducted participant observations, including interviews and focus group meetings, with workers and contractor representatives.

The evaluation by the committee members indicated that the committee and toolbox meetings were successful. The members thought Health Trak was a method that they could adapt and continue to use on their site. The surveys of workers on the site did not show any significant changes in workers perceptions of health hazards. The comparison of the pre- and post-cycle evaluations by the independent health and safety expert showed an increased awareness of hazards from manual materials handling.

5 SUMMARY AND CONCLUSIONS

Health Trak, a participatory model for intervention in construction that involves construction workers and management, as well as occupational health and safety researchers, is being used to reduce the risk of injuries and illnesses for construction workers. It has three key components: 1) participatory meetings with worker and contractor representatives, 2) toolbox meetings with workers on-site using a checklist which is designed to collect data on worker-identified hazards, and 3) participant observation methods including observation, interviews and focus groups with workers and management.

The first two Health Trak cycles were different from each other. In the first cycle, the committee was ambitious, but had no real decision-making authority. They wanted to tackle big issues, but were not successful in generating real change. In the second cycle, the committee was more passive and was only willing to take on small issues. They were moderately successful in developing interventions around these issues, but because the issues were small, they had only a minor impact on exposure to potential health risks.

Formative analysis of qualitative data from these two cycles has provided two significant findings:

1) A strong committee is needed to generate real lasting change. Stewards and foreman play key leadership roles on site and are responsible for the day-to-day negotiations that resolve the constant contradictions arising between production pressure and safety and health. The participation of stewards and foremen, as well as involvement of a management representative with decision-making authority, will produce a strong Health Trak Committee capable of improving health conditions on site.

2) The scope of the goals for each cycle has to be laid out in advance and in proper perspective. If the goals are too large, the committee will be unsuccessful; if the goals are too small, they will have little impact.

Our efforts to-date have given us an early indication that this participatory methodology will work in construction. Future Health Trak cycles will be modeled using these two findings.

REFERENCES

Buchholz, B., V. Paquet, L. Punnett, D. Lee & S. Moir 1996. PATH: a work sampling-based approach to ergonomic job analysis for construction and other non-repetitive work. *Applied Ergonomics* 27: 177-187.

Center to Protect Workers Rights 1993. *An Agenda for Change: Report of the National Conference on Ergonomics, Safety and Health in Construction*. Report No. G2-94. Washington, DC.

Chambers R. 1993. *Relaxed and Participatory Rural Appraisal:* Notes on practical approaches and methods, unpublished document for participants in workshops at the Harvard Center for Population and Development Studies. Washington, DC.

Gjessing. C., T. Schoenborn & A. Cohen (eds.) 1994. *Participatory Ergonomics in Meatpacking Plants*. Publication No. 94-124. Cincinnati: NIOSH.

Keyserling, W.M., T.J. Armstrong & L. Punnett 1991. Ergonomic job analysis: a structured approach for identifying risk factors associated with overexertion injuries and disorders. *Applied Occupational and Environmental Hygiene* 6: 352-363.

Murray, S.J., J. Tapson, L. Turnbull, J. McCallum & A. Little 1994. Listening to local voices: adapting rapid appraisal to assess health and social needs in general practice. *British Journal of Medicine* 308: 698-700.

Ringen, K., A. Englund, L. Welch, J. Weeks & J. Seegal 1995. Why construction is different. *Occupational Medicine: State of the Art Reviews* 10: 255-259.

Schneider, S. 1996. Making ergonomic changes in construction. Proceedings of *CybErg 96*.

Schurman, S., B. Silverstein & S. Richards 1994. Designing a curriculum for healthy work: Reflections on the United Automobile, Aerospace and Agricultural Implement Workers-General Motors Pilot Project. *Occupational Medicine State of the Art Reviews* 9: 283-304.

Implementation of Safety and Health on Construction Sites, Singh, Hinze & Coble (eds)
© 1999 Balkema, Rotterdam, ISBN 90 5809 036 1

Prevention of silicosis in concrete and masonry workers

K.D.Linch
National Institute for Occupational Safety and Health, Cincinnati, Ohio, USA

ABSTRACT: Respirable crystalline silica dust is a respiratory hazard commonly found in the construction industry. When concrete or other masonry is disturbed by either sawing, grinding or hammering, dust controls such as water application or local exhaust should be used to reduce or eliminate the high concentrations of respirable crystalline silica that are generated. Since power equipment produces high concentrations of dust in a short time, if not properly controlled, working in enclosed or partially enclosed areas which are not ventilated increases exposures. A number of National Institute for Occupational Safety and Health (NIOSH) field studies have found a need for improved control of respirable crystalline silica during sawing and drilling of masonry material. For example, NIOSH obtained a measurement of 14.2 mg/m^3 of respirable crystalline silica dust, while a plumber used a handheld masonry saw on a concrete floor, without dust control, inside an unventilated office building that was being renovated.

1. INTRODUCTION

Respirable crystalline silica dust is a respiratory hazard commonly found in the construction industry. Crystalline silica is found in abundance in the earth's crust. Silicosis is permanent lung damage caused by breathing dust containing extremely fine particles of crystalline silica. The three major forms of crystalline silica are quartz, cristobalite and trydimite.

Quartz is by far the most common form of crystalline silica found at construction sites and is the focus of this discussion. Quartz is a constituent of several common building materials including concrete, concrete blocks, mortar, and bricks. The source of quartz in these materials is the sand and/or rock that are used to make them. Many types of power tools are used to grind, saw, plane, chip, hammer, core and drill concrete and other masonry. Some of these tools are available with dust collection or for use with water, but some types of tools or models may not be designed to help control respirable dust. Even tools with vacuum or fan type dust collection systems may not be designed to effectively eliminate the respirable quartz dust hazard. Concrete finishers and masons use many of these tools. For example a concrete finisher may use an angle grinder to smooth poured concrete walls and ceilings as well as a floor grinder. They may also use a chipping hammer to remove excess concrete after concrete forms have been removed. During highway construction they may use a concrete saw. Masons may use either a hand-held or table masonry saw. Masons also use special grinders or joint cutters to remove mortar from between bricks during repointing of masonry. Masons or their helpers mix silica sand with Portland cement and water to make mortar. They may also be exposed to abrasive blasting of masonry walls during cleaning or refurbishment. If silica sand is used as the abrasive, quartz dust will be generated from both the abrasive and the masonry.

The key to silicosis prevention is to prevent dust from being in the air through the use of proper dust control. Crystalline silica overexposures occur frequently in some parts of the construction industry as evidenced by the Occupational Safety and Health Administration (OSHA) sampling data. Air sample data in the OSHA Integrated Management Information System (IMIS) indicate that during the years 1980 to 1995, for Standard Industrial Classification (SIC) 1741- Masonry, Stone Setting, and Other Stone Work, 39% of the OSHA silica samples were over the Permissible Exposure Limit (PEL) (Linch 1997). A recent NIOSH examination of the OSHA compliance data indicates that the more general SIC 174 - Masonry, Stonework, Tile Setting, and Plastering- may have more workers exposed to crystalline silica dust than any other 3-digit SIC (excluding mining & agriculture). This study estimates that 13,800 (1.8%) workers employed in SIC 174 in 1993 were exposed at levels of 10 times or more the NIOSH Recommended Exposure Limit (REL) of 0.05 mg/m^3 (Linch 1998). This data agrees with the NIOSH finding that, with the exclusion of the coal mining industry, the construction industry is indicated on death certificates with mention of silicosis more often than any other industry (NIOSH 1996a).

Since building materials containing crystalline silica are so common, there are numerous ways in which construction workers can become exposed to respirable crystalline silica. However, construction companies can protect workers through: judicious specifications of masonry materials; implementation of a strong hazard communication program; use of all feasible engineering controls and respiratory protection (NIOSH 1996b).

In order to better understand the crystalline silica exposure potential for construction workers who use concrete and masonry saws, air samples of respirable crystalline silica dust were obtained from construction sites where this type of work was being performed.

In addition to the dust hazard discussed here, contractors must be concerned with the carbon monoxide poisoning risk associated with the use of gasoline powered tools such as masonry saws in situations where adequate ventilation is not present (Earnest 1997).

2. SAMPLING AND ANALYTICAL METHODS

NIOSH obtained personal breathing zone and general area air samples during various concrete sawing operations to determine the potential exposure concentrations of respirable crystalline silica dust. The personal respirable dust samples were taken using 10-mm nylon cyclone pre-separators, 37-mm PVC filters, and constant-flow pumps calibrated at 1.7 lpm. Area samples were collected using tripods and baskets to hold the sampling equipment or by mounting the basket of equipment on the saw housing. The area respirable dust samples were obtained by using the same sampling method as the personal samples with the exception that sometimes a high-volume sample was also obtained. The high volume samples were collected on 37-mm PVC filters using a ½ inch metal cyclone at its designed flow rate of 9.0 lpm. At some of the work sites samples of settled dust (bulk dust) was collected and analyzed.

The air samples were analyzed at the NIOSH contracted laboratory in Salt Lake City, Utah. The analysis included total weight gain by gravimetric analysis according to NIOSH Method 0500. The limit of detection for this procedure is 0.02 mg. The air samples were then analyzed for quartz and cristobalite using X-ray diffraction as per NIOSH Analytical Method 7500. The limit of detection for quartz on the filters is 0.01 mg and the limit of quantification is 0.03 mg. The limit of detection for cristobalite on the filters is 0.015 mg and the limit of quantification is 0.03 mg. The bulk samples of settled dust were analyzed also by the use of X-ray diffraction. A two-milligram portion of the bulk sample was weighed onto a PVC filter prior to analysis per NIOSH method 7500.

Samples 1-4 in Table 1 were taken during the renovation of a multi-story office building. Each floor of the building had been gutted. A plumber was using a hand-held gasoline powered circular masonry saw to saw the concrete floor around floor drains. An indentation in the concrete floor around each floor drain was desired. The masonry saw was used to make several cuts into the concrete around the drain holes so that a jackhammer could then be used to gouge out the indentation. This procedure needed to be performed where new rest rooms were to be installed on each of the 16 floors of the building. Since the pedestrian sidewalk at ground level was open for use, the general contractor did not want any windows to be open for fear of falling objects. Special permission was given to have one window open on the floor that the concrete sawing was being done. A floor stand fan was situated in an attempt to blow the dust out of the building through the open window. No other means of controlling the dust was used. The worker wore a single use particulate filter respirator during the task. Visible clouds of dust were generated while he sawed. The resulting respirable quartz concentration was measured at 14.15 mg/m^3. If no exposure is assumed for the unsampled portion of the work shift the 8 hr time weighted average (TWA) for this sample is 10.3 mg/m^3 respirable quartz. This concentration is over 200 times the NIOSH REL. Analysis of three samples of bulk settled dust indicated a mean quartz content of 43%. The area samples were obtained at a distance of approximately 25 feet from the saw. The area samples indicate that workers not involved with this particular task may be temporarily exposed to significant levels of crystalline silica dust. Since it is common for many sub-contractors to be engaged in work at a large work site it is important that each contractor control dust at the point of generation so that others are not exposed.

Hand-held gasoline powered masonry saws of this type can be purchased that are equipped to provide water to the diamond impregnated saw blade and should be used whenever feasible. If water is not available at the site, drums of water can be used by either gravity feeding or pumping the water to the saw.

As seen in Table 1 when concrete is sawed with water being applied to the blade to cool and prevent wear of the blade there is the added benefit of reduced respirable dust being released into the air. Samples 5-11 in the table were obtained during the sawing of expansion joints in fresh concrete during interstate highway construction. The area sample (sample 8) was obtained from a basket attached to a tripod that was moved as the work progressed. A large walk behind saw is used for this type of work. Since the saw blades are expensive most contractors that do this type of work take care to make sure that they only saw with water. Concrete which is curing still contains water in it and so may not produce as much dust as old concrete. Analysis of six samples of bulk settled dust indicated a mean quartz content of 20%. For expansion joint sawing the entire thickness of the concrete pavement is not sawed; therefore, the dust generation rate for this work is probably lower than when entire blocks of pavement are sawed and removed during repair work.

Samples 12-23 were taken during the use of a walk behind concrete saw in which old concrete pavement was sawed in order to remove blocks of it from the roadway. The area samples were obtained from baskets attached to the saw housing.

3. SUMMARY & CONCLUSIONS

The provision of water to diamond impregnated masonry saw blades is intended to protect the blade; however, it also has the beneficial effect of reducing respirable dust containing quartz. Some saws, for example electric saws or saws with composite blades, are not suitable to be used with water. When concrete or other masonry is sawed dry and without effective respirable dust collection, very high concentrations of respirable dust containing crystalline silica in the form of quartz must be expected. If the dust from these operations is not controlled at the source, workers not involved with the task may also be exposed at unacceptable levels.

The scientific literature does not offer guidance on how much water should be supplied in order to assure that respirable quartz dust is not released into the working environment while concrete is being sawed; therefore, care should be taken to assure that plenty of water is used. Since a large amount of respirable dust can be generated very quickly with these saws, diligence in the use of water on the blades is recommended. It should also be noted that, since the amount of quartz in concrete varies from application to application, the results of sawing other concrete may be significantly different. It is typical for contractors to allow the slurry of water and

Table 1. Dust containing crystalline silica generated from concrete sawing.

Sample number	Type p/a[a]	Flow rate (lpm)	Sample duration (min.)	Respirable dust (mg/m^3)	Respirable quartz (mg/m^3)	Dust control
1	p	1.7	350	28.89	14.15	none
2	a	9.0	350	9.13	4.11	none
3	a	1.7	350	7.68	3.36	none
4	a	1.7	350	8.07	3.19	none
5	p	1.7	240	0.02	nd[b]	water
6	p	1.7	242	0.05	nd[b]	water
7	p	1.7	238	0.07	nd[b]	water
8	a	1.7	160	nd[2]	nd[b]	water
9	p	1.7	545	0.05	nd[b]	water
10	p	1.7	545	0.05	nd[b]	water
11	p	1.7	535	0.03	nd[b]	water
12	a	1.7	176	0.33	nd[b]	water
13	a	1.7	174	0.71	nd[b]	water
14	a	9.0	170	0.31	0.02	water
15	p	1.7	271	0.43	nd[b]	water
16	a	9.0	275	0.30	0.02	water
17	p	1.7	231	0.89	(0.05)	water
18	a	9.0	230	0.93	0.10	water
19	p	1.7	369	0.72	(0.03)	water
20	a	9.0	261	0.49	0.05	water
21	a	9.0	112	1.40	(0.10)	water
22	p	1.7	383	0.32	nd[b]	water
23	a	9.0	383	0.27	0.01	water

a. p = personal, a = area. b. Denotes "not detected".
( ) Denotes that the value is between the limits of detection and quantification.

concrete dust to dry on the concrete surface. In the case of highway pavement work, this dried slurry is often reintroduced into the air by mobile equipment and construction vehicles. Ideally, the slurry should be removed before it is allowed to dry. This can be done with a vacuum made to pick up water.

Respirators should not be used as the primary means of preventing or minimizing exposures to airborne contaminants. Effective source controls, such as substitution, automation, enclosed systems, local exhaust ventilation, wet methods, and good work practices, should be implemented to minimize worker exposure to silica dust. Such measures should be the primary means of protecting workers. However, when dust source controls cannot keep exposures below the NIOSH REL, controls should be supplemented with the use of respirators.

When respirators are used, the employer must establish a comprehensive respiratory protection program as outlined in the NIOSH Guide to Industrial Respiratory Protection (NIOSH 1987) and as required in the OSHA respiratory protection standard (29 CFR 1910.134), (29 CFR 1926.103). Some important elements of this standard are:

- periodic environmental monitoring,
- regular training of personnel,
- selection of the proper NIOSH-approved respirators,
- an evaluation of the workers' ability to perform the work while wearing the respirator,
- respirator fit testing,
- maintenance, inspection, cleaning, and storage of respiratory protective equipment.

REFERENCES

Earnest, GS, Mickelsen RL, McCammon JB, O'Brien DM 1997. Carbon monoxide poisonings from small, gasoline-powered, internal combustion engines: Just what is a "well-ventilated area"? *AIHA Jour* (58) 787-791.

Linch KD 1997. Crystalline silica dust exposure: Compliance in the construction industry. *Appl. Occup. Environ. Hyg.* 12(9) 577-580.

Linch KD, Miller WE, Althouse RB, Groce DW, Hale JM 1998. Surveillance of respirable crystalline silica dust using OSHA compliance data (1979-1995). Accepted for publication, *Amer. J. Ind. Med.*

NIOSH 1987. *Guide to industrial respiratory protection.* DHHS (NIOSH) Pub. No. 87-116.

NIOSH 1996a. *Work-related lung disease surveillance report.* DHHS (NIOSH) Pub. No. 96-134.

NIOSH 1996b. *ALERT Request for assistance in preventing silicosis and deaths in construction workers.* DHHS (NIOSH) Pub. No. 96-112.

Implementation of Safety and Health on Construction Sites, Singh, Hinze & Coble (eds)
© 1999 Balkema, Rotterdam, ISBN 90 5809 036 1

Mortality patterns among construction site electrical workers

C.F. Robinson, M. Petersen & S. Palu
National Institute for Occupational Safety and Health, Cincinnati, Ohio, USA

ABSTRACT: This study evaluated the mortality of 31,068 U.S. members of the International Brotherhood of Electrical Workers who primarily worked in the construction industry and died 1982-1987. Comparison to the U.S. population showed significantly elevated proportionate mortality for the diseases caused by asbestos: lung cancer (PMR=117), asbestosis (PMR=247), and malignant mesothelioma (PMR=356) and from fatal injuries, particularly electrocutions (PMR=1180). The findings of excess deaths for prostate cancer (PMR=107), musculoskeletal disease (PMR=130), suicide (PMR=113), and disorders of the blood-forming organs (PMR=141) were unexpected. Elevated mortality for leukemia (PMR=115) and brain tumors (PMR=136) is not inconsistent with previous reports of electricians with occupational exposure to electric and magnetic fields. Excess deaths due to melanoma skin cancer (PMR=123) are consistent with other findings of PCB-exposed workers. Results suggest that more detailed investigations of occupational risk factors are needed to prevent excess mortality for this occupation.

1 INTRODUCTION

In 1992, the Bureau of Census estimated that there were 548,000 electricians currently employed and this occupation was projected to increase 29% between 1990-2005 (Census U.S. Bureau of Statistical Abstract of the United States 1992, 1992). The leading U.S. labor representative for electrical workers is the International Brotherhood of Electrical Workers (IBEW). Founded in 1981, IBEW has 900,000 US members that belong to about 1300 local unions. Of these, 257,000 or 28% are active members in locals that are dominated by construction industry work. (Personal communication, Mark Evert, IBEW)

This is the first study of electrical workers who work on residential or industrial construction sites. There are two subgroups of electrical workers in the construction industry. Outside work involves high voltage wiring and takes place in between a community substation and the consumer. Inside work involves wiring of interior circuitry. According to union officials, members tend to stay with either indoor or outdoor work most of their working life, because of the degree of skill required. The IBEW has 643,000 members in other industries including utility, manufacturing, radio, television, railroad, and others. Previous studies of electrical workers in the utility and manufacturing industries focused on the effects of occupational exposures to electric and magnetic fields. They have reported findings of excess brain tumors, leukemia, melanoma skin cancer, breast cancer, musculoskeletal disorders, depression, suicide, electrocution, and

Alzheimer's disease among electrical workers in the utility or manufacturing industries (Bowman 1992, Casini 1993, Floderus 1993, London 1994, Savitz 1995, Theriault 1994). A previous study of U.S. construction workers found excess risks for asbestos-related disease, including mesothelioma, lung cancer, and asbestosis, as well as other cancer, digestive diseases, poisonings, homicide, and fatal injuries and reported exposures to a wide variety of work site hazards (Robinson 1995).

The construction industry presents unique challenges for occupational health research and prevention because it involves large numbers of relatively small employers, multi-employer work sites, many hazardous exposures, and a highly mobile workforce. In 1990, the Congress of the United States allocated funds for the National Institute for Occupational Safety and Health (NIOSH) to develop a comprehensive prevention program directed at health and safety problems affecting construction workers. One facet of the program is to conduct union-based studies that characterize the cause-specific mortality experience of the construction trades. The purpose of the current study was to identify any excess mortality and potentially hazardous exposures to electrical workers on construction sites as a starting point for prevention efforts.

2 MATERIALS AND METHODS

2.1 Study Population

The present study was based on the death benefit files maintained by the IBEW. Under the IBEW bylaws, the family of the deceased electrical worker was eligible for the fraternal benefit if he had belonged to the IBEW for at least 6 months and had died while an active member or after retirement. Only *electrical workers in construction* are eligible for the fraternal death benefit. The national union administers the death benefit and requires that the local union submit a death certificate as proof of death. The national union also tries to ensure that all deceased members are identified and scans Social Security tapes every few years for its members. In addition to the fraternal benefit, the IBEW has a Pension plan, started in 1928.

Our study was limited to the deaths of 31,068 members whose computerized records were obtainable from the IBEW national office and whose death certificates could be located. Only deaths that occurred from January 1, 1982 through December 31, 1987 and deemed eligible by the union for survivor benefits were included. The records of members who were in arrears on dues and subsequently terminated are not maintained by the union. Thus, terminated electrical workers could not be included in the study. Any electrical worker in arrears on dues is dropped from membership after six months of nonpayment. At that time, the record is purged from the computer file. All death certificates were coded by an experienced nosologist according to the Ninth Revision of the International Revision of the International Classification of Diseases (WHO 1977). Members who died outside the United States were excluded from the study, due to the unavailability of appropriate death rates for analysis and small numbers of deaths.

2.2 Statistical Analysis

Mortality analyses consisted of standard life-table analyses (Steenland 1990, Waxweiler 1983) comparing the mortality of the electrical workers to the population of the United States, after stratification of deaths and person-years at risk by potential confounders (age, calendar time). Rates were available for 99 different categories of death. The measure of effect was the proportionate mortality ratio (PMR) (Hernberg 1992).

3 RESULTS

3.1 *Number in the study*

Included in the study were 30,652 white males, 114 white females, 300 non white males, and 2 non white females whose families were eligible to receive a union death benefit. These electrical workers were either working or retired at death. Cause of death data could not be obtained for less than 0.1% of members known to be deceased.

3.2 *Proportionate Mortality Ratios*

Figure 1 shows the distribution of the number of years in the union for deceased electrical workers. We found that based on union records, over 80% of deceased electrical workers had over 30 years of employment (membership) as electrical workers.

Table 1 shows selected PMRs for all electrical workers combined. The results in Table 1 indicate that electrical workers have significant proportionate excesses of mortality due to mesothelioma or cancer of the pleura, as well as leukemia and cancer of the lung, skin (melanoma), and prostate. In addition, significant excess of deaths due to benign or unspecified tumors of the eye, brain, and central nervous system was observed. An evaluation of death certificates revealed these to be mostly brain tumors. Results for non cancer outcomes indicated excesses for diseases of the blood forming organs, cirrhosis of the liver, asbestosis, and musculoskeletal disease. Results for fatal injuries indicated considerable excess mortality.

Transportation injuries involving railway and motor vehicles (traffic and nontraffic); water, air, and space transport fatalities; and other injuries were significantly elevated. This latter category (other injuries) includes 137 deaths due to worksite machinery or equipment, with circumstances of being struck by, striking against, or caught between. There were 127 fatal injuries that were caused by electric current.

Based on information on the death certificate, we found that 115 or 90% of electrical

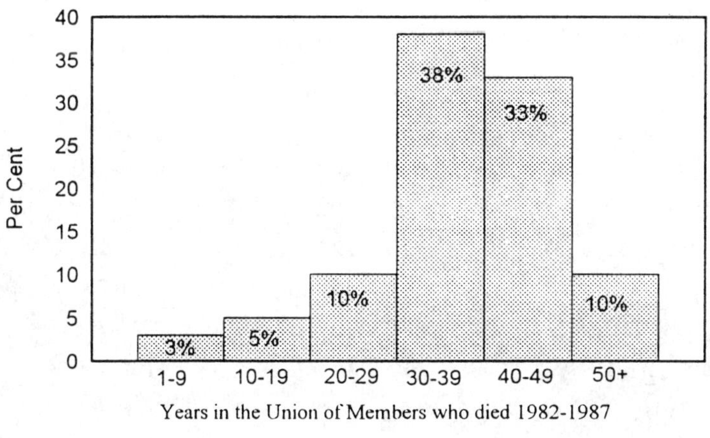

Figure 1. Electrical workers by number of years in the union.

Table 1. Deaths, PMRs, and confidence intervals for all electrical workers, 1982-87

Cause of Death (ICD-9)	Observed No. Of Deaths	Proportionate Mortality Ratio(PMR)	95% Confidence Interval
All Electrical Workers			
All Causes (000-999)	31,068	100	99-101
All Malignant Neoplasms (140-208)	8,154	110**	107-112
Cancer of the Lung (162)	2,977	117**	112-121
Cancer of Pleura (Mesothelioma) (163)	19	356**	214-555
Cancer of Skin (Melanoma) (172)	119	123*	102-147
Cancer of Prostate (185)	870	107*	100-114
Cancer of Other Unspecified Sites, Minor (190-199)	638	120*	111-130
Leukemia (204-208)	312	115*	102-128
Benign or Unspecified Neoplasms of Eye, Brain/Central Nervous System (210-239)	118	136**	112-163
Other Diseases of the Blood Forming Organs (288-289, 17-21)	51	141*	105-186
Ischemic Heart Disease (410-414)	9,993	97**	95-99
Asbestosis (501)	13	247**	131-422
Emphysema (492)	350	108	97-120
Other Resp. Diseases (470-478,494-519)	1,370	98	93-103
Musculosketal System Diseases (710-739)	70	130*	101-165
Cirrhosis of the Liver (571)	512	111*	102-121
Arthritis and Spondelitis	34	140	97-196
All Fatal Injuries (E800-949)	1,192	117*	110-124
Rail, Water, Air Transportation (800-807, 830-845)	60	138*	105-177
Machine Injuries, struck, caught, explosion, cut, radiation, heat (914-921, 923-924, 926-928)	138	166*	139-196
Electrocutions	127	1,180**	984-1,404
Suicide (E959)	505	113*	104-124
Females (White)			
All Causes (000-999)	114	100	82-120
All Malignant Neoplasms (140-208)	27	109	72-158
Intestine Cancer except Rectum (152, 153)	7	223	89-458
Cancer of Breast (174)	6	195	24-704
Leukemia (204-208)	2	195	24-704
Mental Disorders, Not Alcoholism (290-302)	5	601**	195-140
Ischemic Heart Disease (410-414)	19	62**	37-96
All Fatal Injuries (E800-949)	6	129	47-280

*There were only 2 deaths among black female electrical workers.

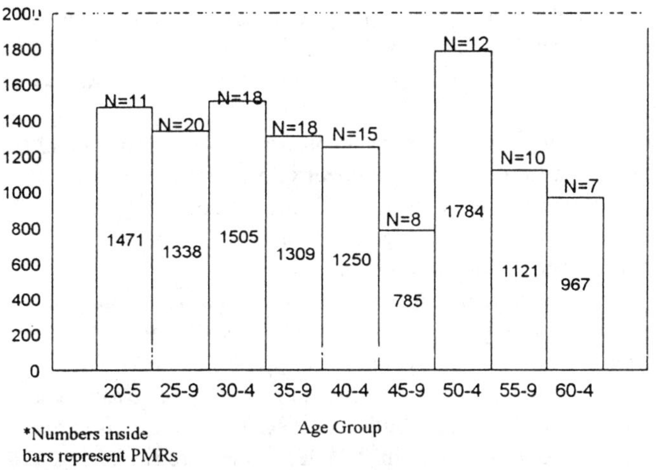

Figure 2. PMRs by age groups for electrocutions before age 65.

fatalities among electrical workers occurred at the construction worksite. One hundred nineteen of the 127 electrical fatalities were under the age of 65. PMRs were significantly elevated for all five year age groups between age 20 and 64. Forty-nine of 119 victims (41%) were less than age 35 when they died (Figure 2).

4 DISCUSSION

Our results showed a small but statistically significant excess of brain tumor deaths, based on 118 benign or unspecified tumors of the eye, brain and central nervous system. This is consistent with earlier studies of electricians and with a meta-analysis of brain cancer and occupational exposure to EMF that reported an overall small increase in risk of 10-20% (Kheifets 1995). Published epidemiologic studies of electricians have reported elevated risk for brain cancer and tumors of the central nervous system (Theriault 1994, Milham 1997Thomas 1986, Lin 1985, Preston-Martin 1989).

Our data indicated a significant excess of leukemia among electrical workers in construction. This estimate is consistent with previous studies of workers exposed to electric and magnetic fields that are based on job title or occupation (London 1994, Milham 1997, Loomis 1990, Loomis 1991). Although excess leukemia mortality has been a persistent finding of studies of electricians and electrical workers, a recent meta-analysis (Kheifets 1997) indicated a weaker association with leukemia than that observed for brain tumors (Kheifets 1995).

Electrical workers on construction sites had excess deaths due to fatal injuries, particularly electrocutions. The excess occurred in young and old workers. Forty-one percent of electrical workers who died by electrocution were less than age 35 when they died (Figure 2). Other studies have reported a higher percent electrocutions in general for younger workers who died at work compared with older workers (Casini 1993, Jenkins 1993), but only one study has reported on the electrical fatalities among construction electrical workers (Ore 1996).

In addition to observing excess proportionate risk within all age strata (Figure 2), our study observed significantly elevated PMRs for all 5 year membership strata between 5 and 35. Ore and Casini reported that electrical fatalities in general have been decreasing over the last decade, but are still elevated four fold (Ore 1996). It is important to find out if the rate of electrical fatality among the electrical workers on construction sites also has been decreasing.

The most striking excess, after electrocutions, was for respiratory diseases (including cancers) related to asbestos exposure, a well-documented hazard on construction sites (Sullivan 1995). The excess risk for mesothelioma clearly indicates exposures to asbestos. When examined by years of exposure, the PMR was significant only after 30 or more years, a typical latency period for asbestos-related mesothelioma.

Construction site electrical workers experienced a significant excess of skin cancer that was due entirely to malignant melanoma (PMR=123, based on 119 observed deaths) (Table 1). When PMRs were evaluated by years of membership, the less than 20 year stratum exhibited the highest significant risk, PMR=153. Previous studies have associated melanoma with potential exposure to polychlorinated biphenyls (PCBs), that are used in electrical transformers (Rockey 1994, Loomis 1997, Sinks 1992). Other studies have reported excess melanoma associated with exposures to hydrocarbon solvents and formaldehyde, used in insulating materials, wiring, and soldering.

Our data showed a significant excess of prostate cancer. As in the U.S. population (Anon 1997), risk was higher for black male electrical workers (PMR=132). Our results confirm earlier reports where electrical power workers were found to be at moderate risk for prostate cancer in a population-based case control study of cancer and occupation in Montreal (Aronson 1996), and by a study of electrical utility workers in Canada and France (Therault 1994). In other studies, cadmium exposure was associated with increased prostate cancer incidence (Checkoway 1987). Electrical workers on construction sites may be exposed to lead and cadmium in welding fumes.

Our data showed a modest excess of lung cancer (Table 1) consistent with the moderately elevated risk found for all electrical workers in published population-based studies (COMS 1987, Zahm 1989). The lung cancer PMR was significantly elevated for the less than 20, 20-30, and 30+ years of union membership, typical latency periods for asbestos. Recent data obtained from a large national survey (Nelson 1994) showed that cigarette smoking prevalence among electricians and electrical power installers and repairers was 29% during 1987-1990. This was considerably lower than other workers in the category of Craftsman/Kindred whose prevalence was estimated at 39% by the same survey so it is unlikely that confounding by cigarette smoking explains any part of the lung cancer excess. Although, most previous studies of electrical workers have not reported excesses, one study showed that pulsed electromagnetic fields were strongly associated with lung cancer in Quebec (Armstrong 1994), and the effect could not be explained with confounding by smoking or other exposures. Our findings taken with the findings of others suggest the need for intervention activities directed at the prevention of lung cancer and other respiratory disease.

Mortality due to diseases of the blood forming organs was significantly elevated (PMR=141, 51 deaths observed, $p<.05$) (Table 1). Most of the deaths were due to white blood cell disorders (ICD 288-289). A recent study reported an exposure-response relationship for benzene and decreased white blood cell count (Ward 1996).

The dangers of work in the construction industry are well known for catastrophic outcomes. A death certificate review showed that 90% of electrocutions had occurred at work. High rates for traumatic fatalities that occurred at work in the construction industry have been previously documented (Jenkins 1993, Bell 1990, Stout-Wiegand 1988). Limitations of PMR studies have been widely discussed. They include potential inaccuracies in cause-of-death on death certificates, lack of data for occupational exposures, and the loss from the study of short-term members. These are common limitations of most PMR studies.

5 CONCLUSION

In summary, our study of construction site electrical workers found striking excess risk for electrocution at work and modest excesses for brain tumors, leukemia, melanoma skin cancer, prostate cancer, and asbestos-related illness- lung cancer, mesothelioma, and asbestosis. The excesses are consistent with an occupational etiology and are consistent with others' findings. Further investigations of these excesses, via case-control studies may be warranted.

REFERENCES

Anon. Cancer Facts and Figures--1997. American Cancer Society, Atlanta, Georgia.

Armstrong B, Theriault G, Guenel P, Deadman J, Goldberg M, and Heroux P (1994). Association between exposure to Pulsed Electromagnetic fields and cancer in electric utility workers in Quebec, Canada, and France. Amer J Epid 140(9):805-820.

Aronson KJ, Siemiatycki J, Dewar R, and Gerin M (1996). Occupational risk factors for prostate cancer: results from a case-control study in Montreal, Quebec, Canada. Amer J Epid 143(4):363-373.

Bell CA, Stout NA, Bender TR, Conroy C, Crouse W, and Myers JR (1990). Fatal Occupational Injuries in the U.S. 1980 through 1985. JAMA. 263:3047-3050.

Bowman JD, Sobel E, London SJ, Thomas DC, Garabrant DH, Pearce N, and Peters JM. (1992). Electric and Magnetic Field Exposure, Chemical Exposure and Leukemia Risk in "Electrical" Occupations. The EPRI Report, TR-101723, University of Southern California, Los Angeles, California.

Casini VJ (1993). Occupational Electrocutions. Professional Safety: 34-39, 1993.

Census U.S. Bureau of Statistical Abstract of the United States, 1992. Washington, DC, 1992.

Checkoway H, DiFerdinando G, and Hulka B (1987). Medical, lifestyle, and occupational risk factors for prostate cancer. Prostate 10:79-88.

COMS--California Occupational Mortality 1979-1981. Sacramento, 1987.

Floderus B, Persson T, Stenlund C, Wennberg A, Ost A, and Knave B (1993). Occupational Exposure to electromagnetic Fields in Relation to Leukemia and Brain Tumors: a case control Study in Sweeden. Cancer Causes Control 4:465-476.

Hernberg, S (1992). Introduction to Occupational Epidemiology: Lewis Publishers; Chelsea, Michigan.

Jenkins EL, Kisner SM, Fosbroke DE, Layne L, Stout NA, Castillo DN, Cutlip PM, and Cianfrocco R. (1993). Fatal Injuries to Workers in the United States, 1980-1989: A Decade of Surveillance: National Profile. Washington, D.C.: US Government Printing Office. DHHS (NIOSH) Publication no. 93-108.

Kheifets L, Afifi A, Buffler P, and Zhang ZE (1995). Occupational electric and magnetic field exposure: A Metanalysis. J Occ Env Med 37:1327-134.

Kheifets LI, Afifi A, Buffler PA, Zhang ZW, Matkin CC (1997). Occupational Electric and Magnetic Field Exposure and Leukemia, a Meta-Analysis. J Occ Environ Med 39:1074-1091.

Lin RS, Dischinger PC, Conde J, and Farrell KP (1985). Occupational Exposure to Electromagnetic Fields and the Occurrence of Brain Tumors. J Occ Med 27(6):413-419.

London SJ, Bowman JD, Sobel E, Thomas DC, Garabrant DH, Pearce N, Bernstein L, and Peters JM (1994). Exposure to magnetic fields among electrical workers in relation to leukemia risk in Los Angeles County. Amer J Ind Med 26:47-60.

Loomis D, Browning SR, Schneck AP, Gregory E, and Savitz D (1997). Cancer Mortality Among Electric Utility Workers Exposed to Polychlorinated Biphenyls. Occ Env Med 54:720-728.

Loomis D and Savitz DA (1990). Mortality from Brain Cancer and Leukemia Among Electrical Workers. Brit J Ind Med 47:633-638.

Loomis D, and Savitz DA (1991). Occupation and Leukemia Mortality Among Men in 16 States: 1985-1987. Am J Ind Med 19:509-521.

Milham SJ. Occupational Mortality in Washington State 1950-1989. NIOSH Pub No. 96-113, 1997.

Nelson DE, Emont Sl, Brackbill RM, Cameron LL, Peddicord J, and Fiore MC (1994). Cigarette Smoking Prevalence by Occupation in the United States: A Comparison Between 1978-1980 and 1987-1990. JOM 36(5):516-525.

Ore T and Casini V (1996). Electrical fatalities among U.S. construction workers. J Occ Env Med 38(6):587-592.

Preston-Martin S, Mack W, and Henderson BE (1989). Risk Factors for Gliomas and Meningiomas in Males in Los Angeles County. Cancer Research 49:6137-9143.

Robinson CF, Stern F, Venable H, Petersen M, Frazier T, Burnett CA, Lalich N, Sestito J, Fingerhut M, and Halperin W (1995). The Assessment of Mortality in the Construction Industry in the United States 1984-86. Amer J Indust Med 28:49-70.

Rockey PF, Trieff N, Wagner RF, and Tyring SK (1994). Nonsunlight risk factors for Malignant Melanoma Part I: Chemical Agents, Physical Conditions, and Occupation. International Journal of Dermatology 33(6):398-406.

Savitz DA, Boyle CA, and Holmgreen P (1994). Prevalence of Depression in Electrical Workers. Amer J Indust Med 25165-176.

Savitz D and Loomis DP (1995). Magnetic field exposure in relation to leukemia and brain cancer mortality among electric utility workers. Am J Epid 141:123-134.

Sinks T, Steele G, Smith AB, Watkins K, and Shulte RA (1992). Mortality among workers exposed to polychlorinated biphenyls. Am J Epid 136:389-398.

Steenland K, Beaumont J, Spaeth S, Brown D, Okun A, Jurcenko L, Ryan B, Phillips S, Roscoe R, Stayner L, and Morris J (1990). New Developments in the Life Table Analysis System of the National Institute for Occupational Safety and Health. JOM 32(11):1091-1098.

Stout-Wiegand N (1988). Fatal Occupational Injuries in the U.S. in 1980-84: Results of the First National Census of Traumatic Occupational Fatalities. Scand J Work Environ and Health 14:Suppl 1:90-92.

Sullivan PA, Bang KM, Hearl FJ, and Wagner GR (1995). Respiratory Disease Risks in the Construction Industry. Occ Med: State of the Art Reviews 10(2):313-334.

Theriault G, Goldberg M, Miller AB, Armstrong B, Guenel P, Deadman J. Imbernon E, To T, Chevalier A, Cyr D, and Wall C (1994). Cancer risks associated with occupational exposure to magnetic fields among electric utility workers in Ontario and Quebec, Canada, and France: 1970-1989, Amer J Epid 139(6):550-572.

Thomas T and Waxweiler RJ (1986): Brain tumors and occupational risk factor. A Review. Scand J Work Environ Health 12:1-15.

WHO International Classification of Diseases. Ninth Revision. Geneva. World Health

Organization, 1977.
Ward E, Hornung R, Morris J, Rinsky R, Wild D, Halperin W, and Gutherie W (1996). Risk of low red or white blood cell count related to estimated benzene exposure in a rubber worker cohort (1940-1975). Amer J Ind Med 29:247-257.
Waxweiler RJ, Beaumont JJ, Henry JA, Brown DP, Robinson CF, Ness GO, Wagoner JK and Lemen RA, (1983). A Modified Life-Table Analysis System for Cohort Studies. J Occup Med 25(2):115-124.
Zahm SH, Brownson R, Chang J, and Davis J (1989). Study of Lung Cancer Histologic Types, Occupation and Smoking in Missouri. Amer J Indust Med 15:565-578.

Implementation of Safety and Health on Construction Sites, Singh, Hinze & Coble (eds)
© 1999 Balkema, Rotterdam, ISBN 90 5809 036 1

Occupational health in construction

J.J. Smallwood
Department of Construction Management, University of Port Elizabeth, South Africa

R. Ehrlich
Department of Community Health, Medical School, University of Cape Town, South Africa

ABSTRACT: Health hazards in construction include ergonomic stresses such as bending, lifting and repetitive movement and vibration; environmental stresses such as heat, sun, noise, poor illumination and wet or damp work; skin and respiratory exposure to chemicals and dust, as well as mental stress among managers and supervisors. In South Africa, these may add to the health problems already suffered by construction workers because of poor community health, substance abuse and inadequate health services.

Almost no research has been done on the health problems in construction in South Africa and the means to control them. As a start to addressing this lack, a questionnaire was sent to Health & Safety Officers of general contractors who are members of BIFSA and SAFCEC, non-student members of the SAIB, and interviews were also conducted with construction workers. The results of the survey indicate that: certain construction materials and processes are hazardous; musculoskeletal injuries predominate among health problems; designers influence occupational health; alcohol and drug/other substance abuse constitutes a problem; the provision of personal protective equipment predominates as a preventative measure; the construction process is stressful; AID's and tuberculosis are encountered, and contractors are aware of the importance and need for occupational health.

1 INTRODUCTION

Health and safety in construction has traditionally been concentrated at the safety end of things. Occupational health services, which in any case exist in only about 11 to 18% of South African workplaces (Department of Health, 1996), are more difficult to organise on temporary construction sites. Construction management, primarily concerned with costs (Levitt & Samelson, 1993), has also been reluctant to add what are considered 'unprofitable overheads' by spending on occupational health services. The strenuous nature of construction work has probably also produced a 'healthy worker effect', by which construction workers with health problems, whether occupationally induced or not, tend to leave the industry.

2 STATISTICS

Construction is an important sector in South Africa employing about 352 919 workers in the formal sector in 1991 (Central Statistical Service, 1996).

In 1991, only 104 occupational diseases were reported for all industries, excluding the mines, to the Compensation Commissioner. This low assessment can be attributed to the drastic under-ascertainment of occupational disease.

In the United Kingdom, a national survey of employee-reported work-related ill health found approximately a million prevalent cases 'caused by work' and a half a million cases 'made worse' by work (Health & Safety Executive, 1990). Musculoskeletal disorders dominated, followed by hearing loss, respiratory disease, stress/depression/headaches and skin disease. Construction poses a risk of all of these disorders.

In the USA, estimates based on employer reports amounted to half a million occupational illnesses in 1992, in a workforce of around a 100 million. The largest single group was associated with repetitive trauma, followed by skin diseases, respiratory diseases and disorders due to physical agencies such as noise (Markowitz, 1994).

If, in South Africa, construction employs about 6.5% of the formal workforce of around 8 million (Central Statistical Services, 1995 & 1996), the order of magnitude of the annual number of occupational illnesses in construction is likely to be around 2 600. Seldom are these ever reported.

3 LEGISLATION

The Occupational Health and Safety Act, 1993, requires contractors to have a written health and safety policy, to train and inform workers of occupational health hazards and to carry out risk assessments of prescribed work. This is carried further in the Regulations for Hazardous Chemical Substances, 1995, which specifies 750 substances requiring special actions. Where necessary the employer is required to obtain environmental measurements and arrange for medical surveillance.

In parallel with this Act, the Compensation for Occupational Injuries and Diseases Act, 1993, greatly expands the list of compensable occupational diseases. Of importance to construction work is that noise induced hearing loss is now formally listed, as is musculoskeletal overexertion, strain due to repetitive work, tuberculosis related to silica inhalation and any pathological manifestation due to chemical substances such as hydrocarbon compounds including organic solvents.

4 HEALTH HAZARDS IN CONSTRUCTION

As the potential health hazards of construction work have been well reviewed elsewhere, an overview is provided in Table 1. The incidence or prevalence of these conditions in South Africa are unknown.

Table 1. Health hazards related to construction work including stresses/agencies and disease.

System	Stress/Agency	Illness/Disease
Musculoskeletal	Lifting/loads Repetitive strain Abnormal postures Whole body vibration	Muscular pain syndromes Tenosynovitis Bursitis Osteoarthrosis
Sensory	Noise	Hearing loss
Skin	Cement (chromates), rubber Thinners, epoxies Tar, pitch Solar radiation	Allergic contact dermatitis Irritant contact dermatitis Acne, Skin cancer Keratoses, Cancer
Respiratory	Silica Asbestos Spraypaints, woods, epoxies Irritant dusts, welding fumes	Silicosis, TB, Asbestosis, Cancer Asthma Bronchitis
Psychosomatic	Physical stress Psychosocial stress	Headaches Depression Fatigue Substance abuse
Nervous System	Lead Organic solvents	Peripheral and central neuropathy Headaches, Dizziness, Mood disorder, Dementia

5 PREVENTATIVE MEASURES

Preventative measures include the traditional hierarchy of control approaches (La Dou, 1994): substitution for the particular material or agent under consideration; engineering controls such as enclosure, isolation and ventilation; controls involving human behaviour such as control of behaviour patterns and work practices, and last in the hierarchy, the use of personal protective equipment. With respect to construction the influence of design, partnering and enabling legislation should be noted.

Designers influence the health and safety performance of a project Tenah (1994), and the International Labour Office (ILO) (1992) recommends that those involved with the design and planning of projects should: not include anything in the design and planning which would necessitate the use of dangerous structural or other procedures and, or hazardous materials which could be avoided by design modifications or by substitute materials.

Levitt and Samelson (1993) advocate partnering as it brings together the various groups involved in a construction project: the client; the general contractor; the subcontractors; the designers and the suppliers – to develop, inter alia, mutually acceptable goals for running a construction project. One such goal could well be 'zero injuries and disease'.

The promulgation of the Construction (Design & Management) Regulations in the United Kingdom resulted in the introduction of a planning supervisor with the responsibility for co-ordinating the health and safety aspects of the design and planning phases of projects. The health and safety plan and health and safety file are the primary documents which, inter alia, collate health and safety information and contain information on toxic materials respectively (Jeffrey & Douglas, 1994).

6 RESEARCH

Recently completed research by the authors includes two descriptive surveys and consultations of construction workers.

1111 metropolitan based general contractors (GCs) who were either members of the Building Industries Federation of South Africa (BIFSA) or the South African Federation of Civil Engineering Contractors (SAFCEC) were surveyed using a written questionnaire. 48 questionnaires were returned which constituted a response rate of 4.3%. 790 non-student members of the South African Institute of Building (SAIB) were surveyed using an identical written questionnaire. 27 questionnaires were returned which constituted a response rate of 3.4%. The low level of response is in itself a finding in that normally approximately a 10 to 15% level of response to a comprehensive postal survey is received. The response is likely attributable to the populations not being familiar with the subject. Given that the respondents are likely to constitute the more knowledgeable and committed, then the resultant bias could mean that the situation with respect to certain responses is far worse. 145 Construction workers were consulted using a structured questionnaire.

Concreting (45.5%) predominated in terms of the work workers were mainly involved with, the next common trades being excavating (31.7%), formwork (31%) and bricklaying (29.7%). 30.3% were general workers, 13.8% semi-skilled and 28.3% artisans (craft workers).

Given that many questions required a response in terms of frequency, extent or degree, it was necessary to compute an importance index. Although the number of variables and definitions may differ from question to question, the following formula constitutes the basis for computing the importance index:

$$\text{Importance Index} = \frac{5n_1 + 4n_2 + 3n_3 + 2n_4 + 1n_5}{5(n_1 + n_2 + n_3 + n_4 + n_5)}$$

Where n_1 = frequency of 'Daily'; n_2 = frequency of ' Weekly'; n_3 = frequency of 'Fortnightly'; n_4 = frequency of 'Monthly', and n_5 = frequency of 'Rarely'

This enabled the problems to be ranked in terms of severity based on a value out of 5 derived from weighting the range of responses.

Table 2 indicates the extent to which materials, activities or processes constitute a health problem based on the daily, weekly, fortnightly, monthly or non-exposure thereto.

When the top five problems for each population are consolidated, it is apparent that 10 of the 15 problems are ergonomic related. The top five 'worker' problems are all ergonomic related, one of which was identified by both GCs and SAIB members and another by GCs. 'Handling heavy loads' was ranked in the top five by GCs and SAIB members. However, cement dust is ranked first by both GCs and SAIB members, concrete dust fifth by GCs and block/brick dust third, by SAIB members. Thereafter sun exposure was ranked sixth, seventh and sixth by GCs, SAIB members and workers. The ranking of 10 for cement dusts according to the worker responses is probably due to the generally low level of exposure of workers to trades entailing the use of cement: concreting (45.5%); bricklaying (29.7%); plastering (23.4%); screeding (20.7%), and blocklaying (13.1%).

Table 3 indicates that GCs and SAIB members rarely or never conducted occupational health interventions. The low level of affirmative response by workers reinforces this finding. The GCs' responses resulted in AIDS education being ranked first. AIDS education was also the occupational health intervention workers were most frequently exposed to. The responses by GCs and SAIB members resulted in in-employment medicals being ranked second and first respectively and pre-employment second by SAIB members. The other occupational health interventions workers were exposed to most frequently were fluid samples (blood and urine) and drug and other substance abuse education.

Table 2. Occupational health problems in construction as perceived by GCs, SAIB members and workers.

Problem	Importance Index			Ranking		
	GCs	SAIB	Worker	GCs	SAIB	Worker
Chemicals:						
Acids/Alkalis	0.75	2.00	0.63	29	21	25
Bitumen/Pitch/Tar	1.42	1.22	0.30	22	29	28
Epoxy-resins	1.29	1.44	0.56	23	26	26
Fumes:						
Metal cutting	2.31	2.70	1.23	14	11	20
Soldering/ Welding	1.92	2.52	0.63	19	14	24
Waterproofing	1.23	1.56	0.63	24	24	23
Mineral wools	0.46	1.15	+	30	30	+
Oils/Petrol	2.23	2.59	1.72	15	13	14
Vapours (adhesives/paints/solvents)	1.94	2.37	1.61	17	17	16
Cold	1.56	1.44	1.36	20	25	19
Dusts:						
Asbestos	0.98	2.07	0.39	26	20	27
Block/Brick	2.73	3.30	1.95	11	3	12
Cement	3.44	3.44	2.46	1	1	10
Concrete	3.23	2.93	2.28	5	9	11
Quartz	1.19	1.44	1.71	25	27	15
Silica	0.94	1.41	*	27	28	*
Ergonomic:						
Bending/ twisting the back	3.23	3.00	3.34	4	8	1
Climbing & descending	3.42	3.37	3.16	2	2	3
Handling heavy loads	3.33	3.15	2.87	3	4	8
Reaching away from the body	2.81	2.85	3.08	9	10	4
Reaching overhead	2.83	3.11	2.98	8	6	7
Repetitive movements	2.89	2.41	3.22	7	16	2
Use of body force	2.79	2.44	3.01	10	15	5
Vibration	2.56	2.11	1.57	13	19	17
Heat	1.94	2,59	1.88	18	12	13
Illumination (poor)	1.46	1.81	1.21	21	23	21
Noise	2.71	3.15	2.66	12	5	9
Sun exposure	3.06	3.04	2.99	6	7	6
Ventilation (poor)	0.88	1.93	0.90	28	22	22
Wet or damp work	2.04	2.30	1.44	16	18	18

* Silica combined with quartz + Not asked

Table 3. Current contractor occupational health interventions according to GCs and SAIB members.

Problem	Importance Index		Ranking		Workers	
	GCs	SAIB	GCs	SAIB	Yes response (%)	Ranking
Biological monitoring:						
Hearing tests	0.70	0.32	6	9	23.4	5
Fluid samples	0.40	0.41	9	8	42.4	2
Environmental measurement:						
Air quality/Breathing tests	0.64	0.57	7	6	20.0	6
Lighting	0.87	0.87	5	3	N/A	N/A
Noise	0.87	0.83	3	4	-	-
Health education:						
AIDS	1.04	0.50	1	7	50.3	1
Drug/Other substance abuse	0.87	0.64	4	5	37.2	3
Medicals:						
Pre-employment	0.64	0.96	8	2	19.3	7
In-employment	0.87	1.00	2	1	24.8	4

Table 4. Contractor actions taken to reduce health hazards encountered on site according to GCs and SAIB members.

Action	Yes response %	
	GCs	SAIB
Adapted/changed tools & equipment	73.2	84.2
Changed work practices	70.7	84.2
Provided personal protective equipment:		
Earplugs	68.3	76.5
Earmuffs	80.5	76.5
Gloves	95.2	94.1
Goggles	95.2	94.1
Hard-hats	92.7	95.0
Rainsuits	82.9	82.4
Respirators	78.0	70.6
Safety boots/shoes	87.8	72.2
Substituted hazardous materials	15.0	5.9
Training: Supervisors	82.9	83.3
Training: Workers	75.6	77.8

Table 5. Use of drugs or illegal substances by workers as perceived by GCs and SAIB members.

Occasion	Yes response (%)		Mean % of workers using	
	GCs	SAIB	GCs	SAIB
Before work	43.7	50.0	13.3	7.3
Working hours	35.5	34.6	5.9	8.0
Tea time	27.1	34.6	9.1	17.6
Lunch time	29.2	34.6	8.7	17.6

Most GCs (83.3%) and SAIB members (77.8%) responded that they encountered health hazards on site of which 97.5% and 100% respectively tried to reduce the hazards. Table 4 indicates that the provision of PPE as well as training was readily resorted to, by both GCs and SAIB members. SAIB members identified, 'adapted/changed tools and equipment' and 'changed work practices' more frequently than GCs.

Table 5 indicates 'before work' to be the occasion workers used drugs or illegal substances most frequently. The mean percentage of workers doing so was estimated to be 13.3% and 7.3%.

Both the GCs and SAIB members identified 'before work' to be the occasion workers used alcohol most frequently. However, percentage wise SAIB members identified all occasions more frequently than GCs. The highest estimated mean percentage of workers using alcohol was before work by the GCs and at lunch time by the SAIB members (Table 6).

GCs and SAIB members identified Contracts Management as the most stressful occupation, the degree of stress increasing commensurately with the level of responsibility in the organisation. The worker consultations determined that 35.9% of workers experienced stress daily, 20.7% weekly, 11% fortnightly, 11.7% monthly and 10.3% never.

Table 6. Use of alcohol by workers as perceived by GCs and SAIB members.

Occasion	Yes response (%)		Mean % of workers using	
	GCs	SAIB	GCs	SAIB
Before work	43.7	50.0	12.1	5.0
Working hours	27.1	42.3	5.0	4.0
Tea time	16.7	26.9	6.0	5.0
Lunch time	27.1	42.3	5.2	13.1

Table 7. Top six causes of stress as perceived by GCs and SAIB members.

Cause	Importance Index		Ranking	
	GCs	SAIB	GCs	SAIB
Crisis management	1.46	1.85	6	3
Inadequate staff	1.54	1.77	2	6
Incompetent staff	1.54	1.81	4	5
Job complexity	1.54	1.92	3	2
Overburdening	1.49	1.85	5	4
Project schedule	1.88	2.04	1	1

Table 8. Reported extent of incorporation of legislation into management practices.

Act/Regulation	Importance Index		Ranking	
	GCs	SAIB	GCs	SAIB
Compensation for Occupational Injuries & Diseases Act	1.29	1.15	3	2
Occupational Health & Safety Act	1.54	1.52	1	1
Environmental Regulations	0.98	0.93	4	4
General Administrative Regulations	1.29	1.11	2	3
Hazardous Chemical Regulations	0.87	0.74	5	5
Lead Regulations	0.48	0.48	6	6

Table 7 indicates that with respect to the causes of stress constituting a problem to construction management/supervision, production workers and others, the top 6 of 27 causes are common to both GCs and SAIB members. Project schedule is clearly the biggest cause of stress followed by job complexity.

Based on the level of responses to: extensively; to some extent; not at all; never heard of it, and don't know, the Occupational Health and Safety Act attained a ranking of first for both GCs and SAIB members with respect to the extent. The General Administrative Regulations and the Compensation for Occupational Injuries and Diseases Act both attained rankings in the top three (Table 8).

A similar percentage of GCs (50%) and SAIB members (48.1%) always/often made reference to the health aspect of a job when an instruction to execute a task is given. The workers responded to a lesser extent (41.9%).

The majority of GCs (70.8%)and SAIB members (81.5%) responded that occupational health is either very important / important.

Table 9 indicates that according to both GCs and SAIB members flu and back ailments predominated among ailments suffered by workers. Given the recognition of ergonomic related problems as the primary 'group' of health problems then the incidence of back ailments and muscle and joint ailments is not surprising. Similarly, given the recognition of dusts as the second most important group of health problems then the incidence of chest illness and eye problems is not surprising. However, the incidence of chest illness and tuberculosis could also be attributable to socio-economic conditions referred to in the literature.

The importance indices resulting from the range of responses: more; adequate; less, and don't know, indicate worker training to be the aspect related to occupational health requiring the most universal attention. Other aspects appearing in the top five of all populations include: education (management); awareness; information; training of management, and worker participation (Table 10).

Table 9. Perceived incidence of ailments suffered by workers.

Affliction	Response (%)						Ranking	
	Yes		No		Don't know			
	GCs	SAIB	GCs	SAIB	GCs	SAIB	GCs	SAIB
AIDS	27.1	25.0	41.7	45.8	31.2	29.2	13	13
Back ailments	72.9	70.8	6.2	20.8	20.8	8.3	2	2
Chest illness	58.3	58.3	16.7	33.3	25.0	8.3	5	4
Colds/Flu	97.9	83.3	2.1	12.5	0.0	4.2	1	1
Epilepsy	45.8	29.2	35.5	54.2	18.7	16.7	8	12
Eye problems	60.4	54.2	25.0	33.3	14.6	12.5	4	5
Hearing problems	45.8	45.8	35.4	45.8	18.7	8.3	9	9
Muscle & joint ailments	62.6	50.0	18.7	41.7	18.7	8.3	3	7
Respiratory problems	39.6	37.5	39.6	54.2	20.8	8.3	11	11
Sexually Transmitted Diseases	47.9	54.2	14.6	29.2	37.5	16.7	7	6
Skin problems e.g. Dermatitis	43.7	50.0	33.4	37.5	22.9	12.5	10	8
Substance abuse	39.6	45.8	27.1	41.7	33.3	12.5	12	10
Sunburn	18.7	20.8	66.7	66.7	14.6	12.5	14	14
Tuberculosis	56.3	58.3	22.9	16.7	20.8	25.0	6	3

Table 10. Perceived degree of attention required with respect to aspects pertaining to occupational health.

Aspect	Importance Index			Ranking		
	GCs	SAIB	Workers	GCs	SAIB	Workers
Awareness: Industry	1.52	1.85	-	3	1	-
Awareness: Workers	-	-	1.38	-	-	5
Education (Management)	1.52	1.81	1.05	2	2	6
Focus	1.23	1.42	1.01	6	6	8
Information:	1.48	1.58	-	4	5	-
Disease	-	-	1.55	-	-	1
Hazardous materials	-	-	1.53	-	-	2
Legislation	0.88	1.08	-	8	8	-
Research	1.19	1.38	-	7	7	-
Inspections	-	-	1.00	-	-	9
Training of Management	1.42	1.69	1.02	5	4	7
Training of Workers	1.56	1.77	1.52	1	3	3
Worker participation	-	-	1.44	-	-	4
Union involvement	0.83	0.88	-	9	9	-

7 CONCLUSIONS

With the exception of cement, concrete and block/brick dusts, ergonomic issues predominated among the problems experienced. The extent to which contractors had encountered certain ailments among workers: colds/flu; back ailments; muscle and joint problems, and Tuberculosis indicates that occupational health has both a work and an off-the-job component. This is further substantiated by the extent to which alcohol and drug or illegal substance abuse is acknowledged to be a problem as reflected in the estimated mean percentage of and perceived occasions workers use alcohol, drugs or illegal substances. Contractors have taken action as evidenced by the ranking of AIDS and drug/other substance abuse education among occupational health interventions.

However, the priority of actions taken to reduce health hazards encountered on site is the reverse of the hierarchy of actions recommended in literature. The fact that contractors focused firstly on the provision of personal protective equipment followed by training of supervisors and workers, then adaptation/changing of tools and equipment and lastly substitution of hazardous materials, is likely an indicator that the processes and materials used are largely prescribed by designers and not by contractors. The problem is exacerbated by the separation of design and construction.

The research findings clearly indicate the construction environment and process to be a stressful environment, project schedule being ranked first among causes. These findings generally concur with those tabled in literature.

Generally, GCs and SAIB members have not incorporated occupational health related Acts and Regulations into their management and work practices as evidenced by the low importance indexes. However, the realisation of the status quo and extent of the occupational health problem in construction by GCs and SAIB members is reflected in: the importance attached to occupational health; the frequency of reference to the occupational health aspects of a job when instruction to execute a task is given; and the perceived need for more attention to be given to industry awareness, training and educating of management and workers, and the provision of information.

8 RECOMMENDATIONS

The conclusions clearly indicate a need to raise the level of awareness for occupational health in construction. Therefore the curricula of all: construction related disciplines presented at tertiary educational institutions; construction management, management development, and foremen training programmes; and trades courses, should include occupational health. Contractors should also address occupational health during induction and should focus on AIDS, tuberculosis, alcohol and drug/other substance abuse education.

Constructability reviews should address occupational health and contractors should oppose the specification of hazardous materials and processes arising from design and details.

Ultimately. the process of construction should be reviewed and re-engineered to improve the general environment.

REFERENCES

Central Statistical Service (CSS). March. 1995. *October Household Survey - 1994.* CSS report P0317. Pretoria: CSS.

Central Statistical Service (CSS). 1996. *Statistics in Brief 1995.* Pretoria: CSS.

Department of Health. January. 1996. *Report of the Committee on Occupational Health.* Pretoria.

Grandjean P. *Occupational Health aspects of construction work.* World Health Organisation. Regional Office for Europe. EURO Reports and Studies 86.

Health and Safety Executive. 1990. *Occupational Health Review.* London.

International Labour Office (ILO). 1992. *Safety and health in construction.* Geneva: ILO.

Jeffrey, J. & Douglas,I. 1994. Safety performance of the UK construction industry. *Proceedings of the 5th Annual Rinker International Conference focusing on construction safety and loss control,* University of Florida, Gainesville.

La Dou. J. 1994. *Occupational Health and Safety.* 2nd Edition. Itasca. Illinois: National Safety Council.

Levitt. R.E. & Samelson. N.M. 1993. *Construction Safety Management.* 2nd Edition. New York:.John Wiley & Sons Inc.

Markowitz. S. 1994. *Counting Occupational illness. Draft Report.* Division of Occupational and Environmental Medicine. Mount Sinai Medical Center. New York.

Pritchard. C. 1996. Where is the health in construction health and safety management? *Proceedings of the 1996 CIB Beiring International Conference Construction Modernization and Education.* Beijing. PRC. CD Rom:\ papers\ 180-189\189\p 189.htm.

Tenah. K.A. 1994. Incorporating Safety Mechanisms into Engineering Design. . *Proceedings of the 5th Annual Rinker International Conference focusing on Construction Safety and Loss Control.* 211-221. Gainesville. Florida.

21 Medical and health management – III

Implementation of Safety and Health on Construction Sites, Singh, Hinze & Coble (eds)
© 1999 Balkema, Rotterdam, ISBN 90 5809 036 1

The NIOSH respirator decision logic – Interactive edition

J.S. Nelson & K. Manchego Crase
Conceptual Arts Incorporated, Gainesville, Fla., USA

R.W. Niemeier
National Institute for Occupational Safety and Health, Cincinnati, Ohio, USA

ABSTRACT: The "NIOSH Respirator Decision Logic" was published as a guide for respirator selection. This paper describes the adaptation of the decision logic contained in the publication for interactive delivery. The interactive version is a computer-aided respirator selection tool that queries users on-screen and guides them through the logic. As a final product the program displays the result and additional information (e.g. respirator limitations, use of applicable exposure limits, warning properties, protection factors, oxygen limitations, and medical evaluation of suitability to wear respirators) when appropriate to the situation described by the user. Test subjects have found the interactive version very approachable and easier to use than the print version.

1 INTRODUCTION

In May 1987, the National Institute for Occupational Safety and Health (NIOSH) published the "NIOSH Respirator Decision Logic" (NIOSH, 1987). The intent of the Respirator Decision Logic was to provide industrial hygienists and other professionals knowledgeable in respirator selection with a procedure for selecting suitable classes of respirators for particular concentrations of specific contaminants. The decision logic publication includes a flowchart comprised of a series of questions about conditions related to proper respirator selection.

2 THE ORIGINAL RESPIRATOR DECISION LOGIC

The Respirator Decision Logic Sequence is represented in the NIOSH publication both as a series of detailed questions (Figure 1) and as a flowchart of abbreviated questions. After determining a defined set of relevant criteria, the user answers the questions and follows the accompanying instructions to work through the logic sequence.

C. **Respirator Decision Logic Sequence**

After all criteria have been identified and evaluated and after the requirements and restrictions of the respiratory protection program have been met, the following sequence of questions can be used to identify the class of respirators that should provide adequate respiratory protection:

1. Is the respirator intended for use during fire fighting?

   a. If yes, only a self-contained breathing apparatus (SCBA) with a full facepiece operated in pressure demand or other positive pressure mode is recommended.

   b. If no, proceed to Step 2.

2. Is the respirator intended for use in an oxygen-deficient atmosphere, i.e., less than 19.5% oxygen at sea level? (Refer to Subparagraph 1, page 21, for a discussion of oxygen deficiency.)

   a. If yes, any type of SCBA or supplied-air respirator (SAR) with an auxiliary SCBA is recommended. Auxiliary SCBA must be of sufficient duration to permit escape to safety if the air supply is interrupted. If additional contaminants are present, proceed to Step 3.

   b. If no, proceed to Step 3.

Figure 1. Two Questions from the NIOSH Respirator Decision Logic Sequence

As an example to demonstrate how the publication is used, consider the following respirator selection criteria:

- Oxygen deficient atmosphere
- No other contaminants are present.

We begin with question 1 on page 8 (Figure 2).

1. Is the respirator intended for use during fire fighting?

   a. If yes, only a self-contained breathing apparatus (SCBA) with a full facepiece operated in pressure demand or other positive pressure mode is recommended.

   b. If no, proceed to Step 2.

Figure 2 - Question 1 of the Respirator Decision Logic Sequence

The answer here is "No" so we continue on to Question 2 (Figure 3).

2. Is the respirator intended for use in an oxygen-deficient atmosphere, i.e., less than 19.5% oxygen at sea level? (Refer to Subparagraph 1, page 21, for a discussion of oxygen deficiency.)

   a. If yes, any type of SCBA or supplied-air respirator (SAR) with an auxiliary SCBA is recommended. Auxiliary SCBA must be of sufficient duration to permit escape to safety if the air supply is interrupted. If additional contaminants are present, proceed to Step 3.

   b. If no, proceed to Step 3.

Figure 3 - Question 2 of the Respirator Decision Logic Sequence

Before answering this question, we are directed to page 21 for a discussion of oxygen deficiency (Figure 4).

Subparagraph 1: Oxygen-Deficient Atmosphere

The National Institute for Occupational Safety and Health (NIOSH) defines an oxygen-deficient atmosphere as any atmosphere containing oxygen at a concentration below 19.5% at sea level [1]. NIOSH certification of air-line or air-purifying respirators is limited to those respirators used in atmospheres containing at least 19.5% oxygen, except for those air-line respirators equipped with auxiliary self-contained breathing apparatus (SCBA).

The minimum requirement of 19.5% oxygen at sea level provides an adequate amount of oxygen for most work assignments and includes a safety factor. The safety factor is needed because oxygen-deficient atmospheres offer little warning of the danger, and the continuous measurement of an oxygen-deficient atmosphere is difficult.

At oxygen concentrations below 16% at sea level, decreased mental effectiveness, visual acuity, and muscular coordination occur. At oxygen concentrations below 10%, loss of consciousness may occur, and below 6% oxygen, death will result. Often only mild subjective changes are noted by individuals exposed to low concentrations of oxygen, and collapse can occur without warning [2,3,4].

Since oxygen-deficient atmospheres are life-threatening, only the most reliable respirators are recommended; the most reliable respirators are the self-contained breathing apparatus or the supplied-air respirators with auxiliary self-contained units. Because a high protection factor is not necessary to ensure an adequate supply of oxygen even in an atmosphere containing no oxygen, any certified self-contained unit is adequate. All aspects of a respiratory protection program must be instituted for these recommendations to be valid.

Figure 4 - Subparagraph 1 of the Respirator Decision Logic Sequence

Returning from page 21, we answer the question with "Yes." We get the answer listed at 2a and are asked one more question (Are additional contaminants present?) to which we answer "No" and are finished.

For criteria that yield an answer quickly such as the example above, use of the decision logic sequence as presented in the publication is not difficult. For criteria that require many questions to be answered, use of the publication is awkward and cumbersome. Users must follow the sequence as it jumps from the questions in the middle of the publication, to the subparagraphs in the back of the publications, back to the questions in the middle, and again to the subparagraphs in the back, and so on.

With certain criteria, the user is provided partial answers at several points in the sequence making it necessary to remember them all and to assemble them at the end of the sequence for the complete answer. For example, consider question 2 (Figure 3). If the answer is "Yes" and there are additional contaminants present, the user would receive the partial answer at 2a and then continue through the questions to receive additional information.

3 ADAPTING THE GUIDE FOR INTERACTIVE MEDIA: DESIGN ISSUES

It was precisely the problems of non-sequential presentation (i.e. having to jump around) and record-keeping (i.e. remembering a series of partial answers) that made this decision logic sequence an excellent candidate for adaptation to interactive media. In designing an interactive version, we would let the computer do the jumping around and the record-keeping. Additionally, we wanted to the interactive version to keep a history that would allow the user to back-up and review or change any answer in the sequence of responses previously given.

The interactive edition is to be distributed in two forms – one as part of a CD-ROM-based CBT module and the other as a part of the NIOSH website. There were NIOSH-imposed design restrictions with regard to the website version: in order to reach a wide audience, we would be restricted to developing in version 2.0 of HTML. Additionally, it is always important to consider download-times when designing for the web (Gillespie, 1998).

4 THE INTERACTIVE EDITION

A traditional "left hand side" navigational bar (Figure 5) was implemented to provide the user with the ability to quickly start over, get help, find out more about the Respirator Decision Logic, go to the home page of the Respirator Decision Logic, and go to the NIOSH homepage. The information for the decision logic includes Questions, Answers, and More Information and is laid out in the middle of the page. A footer bar is also included to enclose the information. The footer bars for the Questions include "yes" and "no" buttons.

The history section (on the right-side of the page) enables the user to see how they answered each question. The user can click on any of these previously answered questions to be sent back to that question and continue from that point.

To illustrate how the system is used, consider again the respirator selection criteria presented earlier in this paper. We start the Respirator Decision Logic with Question 1 (Figure 6).

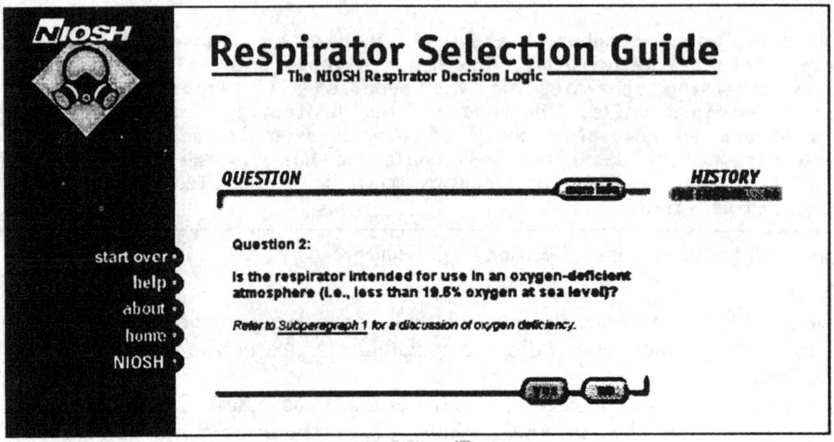

Figure 5 - Question 2 of the Respirator Decision Logic Sequence

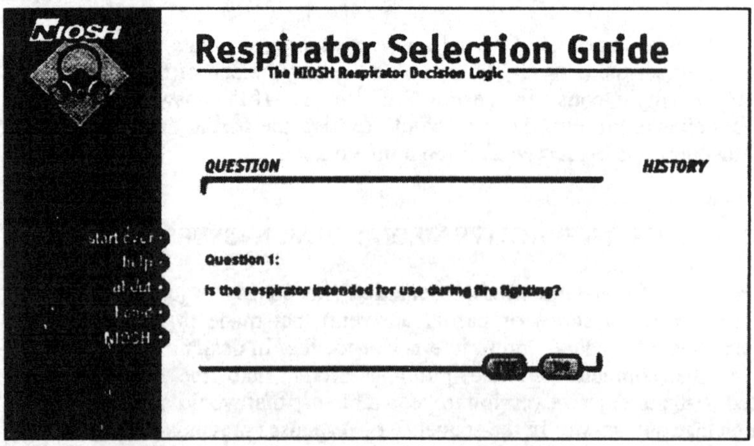

Figure 6 - Question 1 of the Respirator Decision Logic Sequence

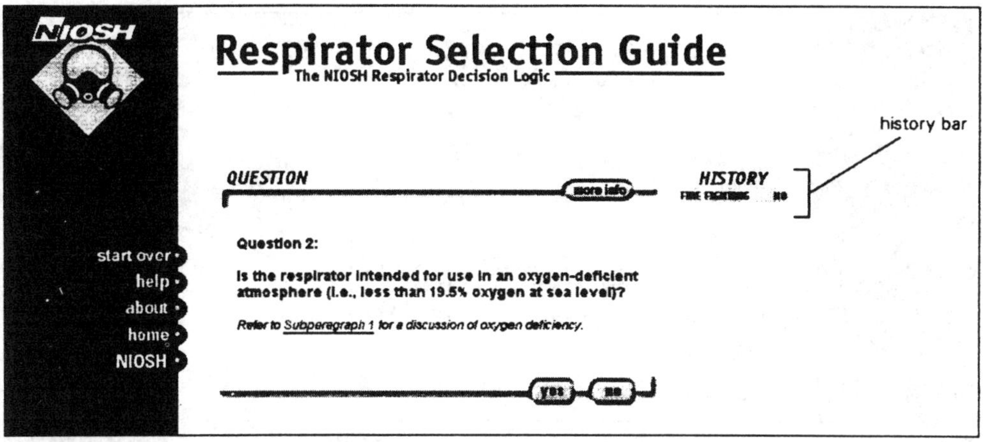

Figure 7 - Question 2 of the Respirator Decision Logic

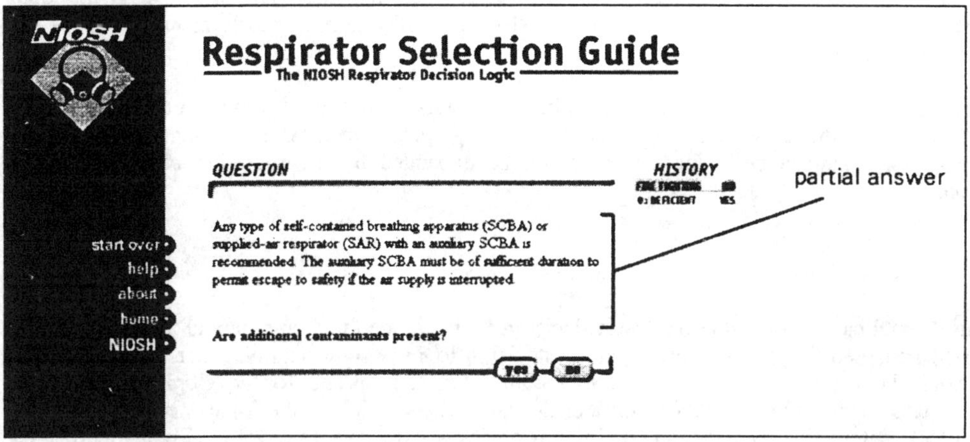

Figure 8 - Partial answer to the criteria chosen

We answer "No" by clicking on the "No" button at the lower right hand corner. Then we get the next screen shown in Figure 7. Note the history bar on the right -- it indicates that user answered "No" to the question about fire fighting.

Now in order to get to Subparagraph 1, we simply click on the underlined link below the question to see that information on another page. From there, we use the "Back" button on the browser to get back to the question we were on. Answering "Yes" to Question 2 brings us to the following screen (Figure 8.)

Here is an example of getting a partial answer and then going on to another question. The user's problem of having to remember each partial answer is solved by leaving the answer at the top of the information section while going on to the other questions. We answer "No" to this question since there are no additional contaminants in our criteria. Now we have a complete answer, shown in Figure 9:

5 FUTURE DEVELOPMENT

This interactive module is to be incorporated into a computer-based training (CBT) product dealing with ventilation and respiratory protection. Plans are also underway for installing a

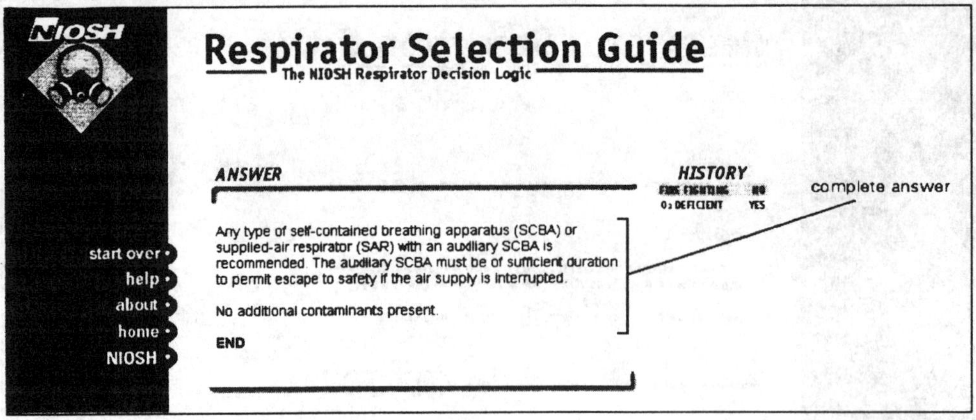

Figure 9 - Answer of the Respirator Decision Logic

world-wide web version of this interactive respirator selection tool. This would make the logic available to a larger audience and enhance NIOSH's ability to more easily manage revisions to the respirator decision logic.

The selection of an appropriate breathing apparatus is but one component of respiratory protection. Other aspects are important as well, e.g., proper medical evaluation, fit-testing, and respirator maintenance. The system could be expanded by incorporating training in these topics.

6 CONCLUSION

In its publication form, the respirator decision logic can be cumbersome and awkward to use. When adapted for interactive media, the decision logic is significantly easier to use. Efforts should be made to utilize interactive media when appropriate for developing information products such as the "NIOSH Respirator Decision Logic". When the logic was first published in 1987, interactive media was in its infancy and very expensive to develop. Today, there are a wide variety of interactive media development and delivery tools available. The costs of interactive media development continue to decrease. Interactive media delivery via CD-ROM and the World Wide Web can actually be less expensive than delivery via traditional printed publications.

7 REFERENCES

Gillespie, J. 1998. *Web Design for Designers*. URL http://www.wpdfd.com

NIOSH (National Institute for Occupational Safety & Health). 1987. *NIOSH Respirator Decision Logic*. Publication No. 87-108, U.S. Department of Health and Human Services. Washington, D.C.

Implementation of Safety and Health on Construction Sites, Singh, Hinze & Coble (eds)
© 1999 Balkema, Rotterdam, ISBN 90 5809 036 1

What's killing carpenters? Injury and illness prevention for the future

T. Alterman, C.F. Robinson & K. Sieber
National Institute for Occupational Safety and Health, Cincinnati, Ohio, USA
J. Durst
United Brotherhood of Carpenters and Joiners, Washington, D.C., USA

ABSTRACT: A pilot intervention was conducted with 79 attendees at a national conference held by the United Brotherhood of Carpenters and Joiners of America (UBCJA) in 1997. NIOSH education documents and mortality study results were provided, as well as an opportunity for participants to share their experiences. Participants were carpenters, business agents, health and safety personnel, trainers, and apprenticeship coordinators. NIOSH investigators, along with a UBCJA member led two workshops. Focus groups were used to discuss injuries, illnesses, and stresses that carpenters experience. Training needs were also discussed and results from a recently completed NIOSH mortality study of 27,362 UBCJA members was presented. Although mortality studies have been instrumental in identifying areas for further research and examination of risks, results have typically not been disseminated to workers. Suggestions for improved injury and illness prevention were made, and demonstrated the need to take into account social, economic, situational, behavioral, and environmental factors.

1 INTRODUCTION

Data from the Bureau of Labor Statistics indicate that Carpenters, the largest group of building trades workers, held about 992,000 jobs in 1994 and are employed throughout the country in almost every community. The majority worked for contractors who build, remodel or repair buildings and other structures. Most of the remainder worked for manufacturing firms, government agencies, wholesale and retail establishments, and schools. Carpenters learn their trade through on-the-job training and through formal training programs. It is expected that more than 100,000 jobs will become available each year as carpenters transfer to other occupations or leave the labor force (U.S. Dept. Of Labor, 1997).

Occupational injury rates in the construction trades, including carpentry are high compared to the general workforce in the United States. The Bureau of Labor Statistics (BLS) (U.S. Department of Labor, 1997) reported a rate for lost time due to medical injuries or illnesses of 9.9 per 100 full-time workers in the construction trades, compared with an overall rate of 7.4 per 100 full-time workers in 1996.

The United Brotherhood of Carpenters and Joiners of America (UBCJA), formed in 1881, is the leading labor representative for carpenters, joiners, and related trades in the United States. It is the largest trade union in the Building and Construction Trades Department of the AFL-CIO. The membership is composed of approximately 600,000

U.S workers. Of these, about 475,000 members are involved in construction carpentry, including carpenters, millwrights, pile drivers, dry wall workers, and lathers. About 125,000 are involved in industrial wood processing products carpentry including wood furniture, cabinetmaking, plywood, and lumber or sawmill work. Carpenters may be employed in various industries such as building and construction, shipbuilding, manufacturing industries and mining, where they may also be exposed to hazardous materials from the work of other crafters. Carpenters encounter many hazardous exposures including asbestos, wood dust, noise, formaldehyde, elevated working areas, hyper baric pressure, wood preservatives, solvents and power tools (Robinson et al., 1996).

NIOSH evaluated the mortality of 27,362 members of the UBCJA who died 1987-1990. Elevated mortality was found for white males in the construction trades for lung cancer, bone cancer, asbestosis, emphysema, transportation injuries, and falls. Leading causes of death in the industrial wood products locals were due to stomach cancer, male breast cancer, and transportation injuries (Robinson et al., 1996).

Carpenters's work can also result in increased back, neck and shoulder problems and traumatic knee problems. Work-related musculoskeletal disorders were associated with the use of certain tools, such as overhead bolt pistols, carpet knee-kickers, and straight handled claw hammers (Albers et al. 1997). After conducting and evaluating an ergonomic education program for apprentice carpenters, Albers et al. (1997) recommended that construction safety and health education and training should be developed to improve simultaneously the capability of all stakeholders to recognize potential hazards and act to prevent their occurrence. They also recommend improving the traditional construction "tailgate" or "toolbox" safety meeting, such that presentations engage participants, provide information relevant to the job site and encourage discussion about potential hazards and the means to prevent injuries and illnesses.

This paper presents the results of two workshops held in Palm Springs, California in March 1997 using a focus group format to elicit the input of carpenters and related workers in identifying risks and suggesting preventive strategies for the improvement of occupational safety and health among carpenters.

2 MATERIALS AND METHODS

2.1 Workshop Participants

Participants were 79 volunteer attendees at the Annual UBCJA Apprenticeship Training and and Health and Safety Conference. Participants attended one of two sessions titled "What's Killing Carpenters? Injury and Illness Prevention for the Future". Because of space limitations, some potential participants were turned away once the room was full.

2.2 Methods

NIOSH investigators led two 1 ¼ hour long workshops at the UBCJA Health and Safety Conference. Each workshop had approximately 30-40 participants. Educational materials from the UBCJA and NIOSH were distributed to participants at the beginning of the workshop. During each workshop, participants divided into three focus groups to identify and discuss strategies and prevention methods for addressing the injuries, illnesses, and stresses that carpenters experience on the job. Participants provided demographic and occupational information. The workshops ended with a presentation of the NIOSH mortality study of carpenters (Robinson, 1996).

The moderators were two epidemiologists from NIOSH, and the Director of Occupational Safety and Health from the UBCJA who recorded responses on a flip chart. Questions to the participants included: What type of injuries can occur on the job? What can you do to prevent each of these injuries from occurring? What type of diseases can be caused or made worse by the work that you do? What might cause you to lose a day of work or go home feeling ill? What can you do to prevent each of these diseases or to keep

Table 1. Examples of Injuries and Preventive Action Identified by Workshop Participants

Injuries		Preventive Actions
Falls	➤	Guard Rails
		Appropriate housekeeping
		Use of anti-skid mats
		100% tie-off
Back strain	➤	Ergonomic training
		Proper use of mechanical equipment
		Increased manpower
		Observance of lifting standards
Blows to head	➤	Use of personal protective equipment
		Use of toe boards
		Proper rigging of loads
		Guard nets and tie offs
		Overhead protection
		Restricted access areas
Chemical burns	➤	MSDS* & identification of chemicals
		Use of cut-off valves
		Chemical awareness
		Use of personal protective equipment
Lacerations	➤	Use of gloves and protective equipment
		Operative equipment guards
		Training to use equipment
Electrocutions	➤	Observance of lockout/tagout procedures
		Inspections & proper maintenance
		Training
Hearing loss	➤	Implementation of engineering controls
		Use of personal protective equipment
		Implementation of administrative controls
Vehicle & lift injuries	➤	Eye to eye contact
		Use of back-up alarms
		Use of seat belts
		Training certified drivers
		Adherence to strict substance abuse policy
Heat & welding burns	➤	Use of personal protective equipment
		Use of welding shields
		Safety watch
		Warning signs
		Barricades
		Red flag tape

*MSDS refers to material safety data sheets.

from getting sick? What type of stresses are you exposed to that might put you or your co-workers at risk on the job? What can you do to prevent or deal with theses stresses? What kind of training do you need to stay healthy and prevent injuries? The goals of the workshops were to involve workers' in identifying and discussing strategies and prevention methods for addressing the risks faced by carpenters.

3 RESULTS

Participants were predominantly white men, along with 2 non-white men, and 3 women drawn from 26 states across the United States. They were involved in both wood processing and construction, and included administrators, apprenticeship coordinators and directors, business representatives or agents, carpenters, health and safety trainers, and millwrights. Most had more than 10 years experience, worked more than 40 hours a week, both indoors and outdoors. The majority of participants were 40-49 years of age, followed by those 50-59 years, and more than half had some college education.

Participants provided the moderators with illnesses, injuries, prevention strategies, and training suggestions relevant to work as a carpenter. These responses have been aggregated over the six groups and presented in the tables below. Because of time and space limitations, not all responses are included in this presentation.

3.1 Training

Participants felt that both classroom and hands on training was needed. They stressed the need for certification, qualified people, and current updates and information. Joint labor-

Table 2. Examples of Diseases & Preventive Action Suggested by Workshop Participants

Diseases		Preventive Actions
Lung Disease	➤	Use of personal protective equipment
		Implementation of engineering controls
		Protection against asbestos exposure
		Education regarding carcinogens
		Smoking cessation programs
Skin Disease	➤	Use of personal protective equipment
		Chemical substitutions
		Availability and use of MSDS*
		Labeling & reading directions
Frostbite & hyperthermia	➤	Awareness of temperature
		Proper attire (e.g. hard hat liners)
		Training and recognition
Lead poisoning	➤	Engineering controls
		Monitoring of workers
		Training
Liver damage	➤	Use of personal protective equipment
		Avoidance of chemical exposures
		Chemical substitutions
		Alcohol awareness
		Substance abuse treatment
Mental problems	➤	Identification & proper labeling of solvents
		Use of personal protective equipment

*MSDS refers to material safety data sheets

Table 3. Examples of Stresses Identified by Workshop Participants

Communication problems
Interpersonal conflicts
Family responsibilities and problems
Confined spaces and extreme heights
Sexual harassment
Under bidding jobs and scheduling difficulties
Labor conflicts
Disrupted sleep schedules
Excessive travel and transportation time
Economic issues - layoffs and firing

Table 4. Example of Ways to Reduce Stress Suggested by Workshop Participants

✔	Training in interpersonal problem solving
✔	Training in gender issues
✔	Counseling
✔	Availability of Employee Assistance Programs
✔	Mediation
✔	Established grievance procedures
✔	Training in healthy lifestyle factors
✔	Health check-ups
✔	Availability of health insurance

management classes were recommended which should include problem solving and resolution of conflicts. Other areas emphasized were first aid and cardiopulmonary resuscitation (CPR).

4 DISCUSSION

Despite their widespread use, focus groups have been the object of little systematic research in recent years. Focus groups may be used an any point in a research program, but are particularly useful for exploratory research. They are used widely because they provide useful information and offer the researcher a number of advantages (Stewart and Shamdasani, 1990). Advantages of focus groups include: (1) provision of data from a group of people more quickly and at less cost than individual interviews; (2) allowing a researcher to interact directly with respondents, including opportunities for clarification and follow-up questions; (3) providing an opportunity to obtain large and rich amounts of data in the respondents own words; (4) allowing respondents to react to and build upon the responses of other group members which may result in ideas that might not be uncovered in individual interviews and (5) results of focus groups are easy to understand. Limitations include: (1) the convenience nature of most recruiting practices which may limit generalization to a larger population; (2) responses are not independent and results may be biased by a very dominant or opinionated member; (3) open-ended nature of responses obtained may make summarization and interpretation of results difficult and (4) the moderator may bias results by providing cues about what types of responses are desirable (Stewart and Shamdasani, 1990).

Other issues particular to this study is that of generalizability from unionized to non-unionized workers. Union workers may have safer work practices which are learned over time or in training (Lipscomb et al., 1997). Concerns raised by other focus groups include possible effects of peer pressure for women to perform comparably to men, lack of fit of basic tools for women due to smaller body, and reduced upper body strength (Lipscomb, et al., 1997).

Participants in this project were volunteers who may have had more awareness of safety and health issues or training than non-unionized workers, or non-participants. Information on the health hazards of women carpenters is badly needed. Our findings are limited by the participation of only 3 women in the groups. Minorities were also under-represented in the workshops. Moderators attempted to control more dominant participants and encourage responses from others.

Since the workshops were held, the UBCJA has obtained a grant (New England Project) from NIOSH to assist small employers by offering to do workplace inspections to identify workplace hazards, and promote accurate record keeping. The UBCJA has also been involved in several noise and hearing loss intervention projects. Investigators at NIOSH plan to develop a fact sheet to disseminate study results from the PMR study to workers and a NIOSH document for dissemination to the carpenter and construction trade community that summarizes what was learned during the focus groups.

Responses by workshop participants suggest that workers recognize the multi-faceted nature of injury and illness prevention. Many of the suggestions involved increased use of personal protective equipment, implementation of administrative controls, increased training, proper housekeeping, and taking appropriate safety precautions. Future efforts toward reducing illness and injury should take into account cognitive, behavioral, social, economic, situational, and environmental factors.

ACKNOWLEDGMENTS

Appreciation is extended to the Health and Safety Fund of the United Brotherhood of Carpenters and Joiners of America, and to those carpenters and attendees at the National Apprenticeship and Health and Safety Conference who participated in the focus groups.

REFERENCES

Albers JT, Yuhua L, Lemasters G, Sprague S, Stinson R, Bhattacharya A. 1997. An ergonomic education and evaluation program for apprentice carpenters. *Am J Ind Med* 32:641-646.

Lipscomb HJ, Dement JM, Loomis DP, Silverstein B, Kalat J. 1997. Surveillance of work-related musculoskeletal injuries among union carpenters. *Am J Ind Med* 32:629-640.

Robinson CF, Petersen M, Sieber WK, Palu S, Halperin WE. 1996. Mortality of carpenters' union members employed in the U.S. construction or wood products industries, 1987-1990. *Am J Ind Med* 30:674-694.

Stewart DW, Shamdasani PN. *Focus groups: Theory and practice*. 1990. Newbury Park, California: Sage.

U.S. Department of Labor, Bureau of Labor Statistics 1997. "Workplace Injuries and Illnesses in 1996". Publication No. USDL-97-453.

Implementation of Safety and Health on Construction Sites, Singh, Hinze & Coble (eds)
© 1999 Balkema, Rotterdam, ISBN 90 5809 036 1

Implementation of safety and health on construction sites

Ian M. Eilenberg
Construction Management, RMIT University, Melbourne, Vic., Australia

ABSTRACT: The Australian Construction Industry is aware of drug taking by its employees. Limited education is undertaken by the employers and unions with the main message sent out by these bodies being 'Drugs Should Not be Used On Site'. In the second part of the ongoing research, 90% of the construction employees interviewed had not received any drug education, or details of their own company drug policies. Of this 90%, only 48% felt that they wanted any education on this issue. The problem is not so much covered up, as ignored by most people within the industry. By comparison, in secondary and technical education areas, the drug problem is widely recognised, but is largely concealed. Further, the two groups considered drug use differently. The construction industry considered the problem was directly related to the industry, whilst the teaching fraternity considered drug use was widespread and transcended any industry or group. Even the conception of the problem was different. The teaching fraternity identified the use of prescription drugs as the major problem within its ranks. The construction industry identified that after alcohol, marijuana is the drug most commonly used.

1 INTRODUCTION

The Australian Drug Foundation defines a drug as any substance that 'alters the operation of the mind, emotions or the central nervous system'. In a report by the Victorian Building Industry Association, and in the first part of this ongoing research project being undertaken by RMIT University, it was clearly demonstrated that there was a problem with drug (other than alcohol) usage amongst the younger employees of the construction industry. This part of the research was aimed at investigating what education was available on drug use/abuse for the members of the construction industry and how successful any such education was in respect to the younger employees.

The other aim was to confirm that drug usage and its related problems was not just confined to the construction industry. Whilst any number of industries could have been examined (and others may be in later research) it was decided to consider the education sector (teachers), especially the secondary school (high) and technical and further education (TAFE) sectors. The reasons were two fold. First because the education industry was also involved with younger and older members of the public and secondly as it was a relatively simple group to contact for interview purposes being fixed in the school.

All questionnaires were administered verbally in face-to-face interviews with the answers recorded during the interview.

2 DRUG LEGISLATION

There are three government acts that cover occupational health and safety. These include the Occupational Health and Safety Act 1985, which makes no reference directly to drugs, but includes the requirement for the government to support of education and training of employees in matters of health and safety, which amongst others areas includes drugs. Under this Act there is a requirement for the formation of Occupational Health and Safety Committees in all work places, including schools and construction sites.

The Drugs, Poisons and Controlled Substances Act 1987 plays a role in the distribution, containing and labelling of drugs. Part of this Act established the Drug Rehabilitation and Research Fund, which was funded from the collection of fines, government grants and donations. Any money so raised was to be used specifically for research, education, distribution of information about drugs and the prevention of drug abuse.

The other related Act is the Drug, Poisons and Controlled Substances Regulations 1987. This Act describes who are entitled to have permits for the holding, prescribing and distributing drugs.

3 DRUG POLICIES

There are a number of policies within the construction industry covering drug related matters.

3.1 *Victorian building industry alcohol and other drugs*

Whilst there is no specific equivalent for the education sector, the major policy in the construction arena is the Victorian Building Industry Alcohol and other Drugs project, commenced in 1992. This is controlled by an Advisory Committee consisting of union and employer representatives. This project was funded by the Victorian Building Industry Redundancy Payment Central Fund (Incolink). The aim is to involve the people working on the construction site in a program based on peer intervention, education and, where appropriate, assistance to employees and families.

The Advisory Committee believed the policy provides a practical way for employees to take responsibility and assist each other with issues which could affect building workers. They also believe that many companies have formulated policies which restrict and ban the consumption of alcohol and specific drugs in the workplace, but not many of them contain provisions for rehabilitation or education for prevention and harm minimisation strategies. The policy was so formulated that it can be effective on both large and small work sites. It can be included in consultative agreements or implemented independently. As will be discussed later the problem with this policy is the word CAN – but apparently seldom IS.

The implementation and the management of the policy and program was designed to be through the Occupational Health and Safety Committees – established under the OH&S Act referred to above. The Occupational Health and Safety Representatives are not expected to become alcohol and drug counsellors, but to maintain site safety and to provide information and assistance to colleagues. The role of the employer is to assist in the smooth and confidential application of the policy. This is largely carried out at inductions of new employees.

3.2 *Union policy*

The Victorian Trade Hall Council adopted its current revised policy, applicable to all associated unions including the education and construction sectors, on drug and alcohol use in 1992. The trade union movement decided to focus its attention on those aspects of the workplace which influence worker's decisions in relation to consumption and alcohol. Broadly speaking the unions do not support random blood or alcohol testing, considering them an infringement on the rights and privacy of the individual. Generally it only when drugs and alcohol are misused to the extent that the user cannot perform properly and safely their regular duties that there is the need for control and preventative measure – on the worker – not on the general situation that led to that use or abuse.

3.3 *Construction company policies.*

The policy of a number of construction companies were investigated. One large company included the policy within the Enterprise Bargaining Agreement between the company and the employees. However if any person is found on a construction site to be under the influence of drugs, they will be told to stop work and arrangements made for them to go home. Only after a worker has been so approached a number of times will the OH&S representative suggest counselling and rehabilitation. A person that has a drug dependence problem will be given a verbal warning, followed by a written warning and only then after further misdemeanours, dismissed.

Another company was more pro-active. The Drugs and Alcohol Policy states that its aim is to 'eliminate safety hazards created by drug and alcohol abuse'. The main responsibility lies with the site manager and the project manager, who are required to be pro-active in detecting drug users and to notify the health and safety committee. The first action is a verbal warning followed by a written warning. The role of the site manager is not only to supervise, but to actively assist the health and safety committee in educating and providing other avenues for the treatment of people with drug problems on site.

One company forbids the use of alcohol and drugs by its employees, and specifically mentions prescribed medicines. There is no provision for post discovery education or rehabilitation once an employee is found to under the influence of drugs.

One of the best and most thorough policies is that of a major civil contractor. This company policy document included a number of headings, with considerable details provided in each. These include: Workplace requirements; Education and Prevention; Rehabilitation/Counselling; Medication through prescribed Drugs; Alcohol in the workplace; and disciplinary action upon breaches of any of the other. The company will offer rehabilitation and counselling to its workers before resorting to disciplinary action – from censure to sacking depending upon severity and history of the worker and the offence.

Most of the companies, from quite small to major contractors had some reference to drug use and ways that it would be managed, although some were very brief and with no specific management action plans. Removal from the site was a common inclusion, but without the back up of rehabilitation or counselling.

3.4 School Policies

The Victorian Education Department has an overall drug policy of 'no drugs in schools', but it is up to the individual schools to implement the policy. This usually consists of education sessions for the students by specialists. There is little or no education for the teachers themselves. Within the school many have introduced television monitoring of toilet blocks and associated buildings to cut down on usage and selling/buying transactions. There are few, if any school policies for the employees, other than they should not use them.

4 INDUSTRY EXPERIENCE

4.1 Health and safety managers

Interviews were carried out with a number of construction Health and Safety Managers from a range of different sized construction companies, and Occupational Health and Safety Representatives from various schools. Within the construction industry only one in three of them admitted to having encountered any of their employing drugs, other than alcohol. Those that had other experienced had encountered both heroin and marijuana. Whilst none had actually seen anyone taking the drugs, needles and spoons for heroin as well as the smell of marijuana had been noted on more than one occasion. Marijuana was the most commonly reported drug, with one Health and Safety Manager admitting that it was a regular occurrence on a particular project (being more than once a week). The age of the people believed by the Health and Safety Managers to be using these drugs were the younger employees in the 18-25 year age group. The most common group noted as probably users were the labourers and then trades people. One manager stated that in his opinion: 'It (usage) seems to be the most common in the least skilled.' None of the Health and Safety Managers interviews could state for sure that any accidents or injuries could be directly related to drug (other than alcohol) usage.

One outcome of the interviews was that most of the Health and Safety Managers had not received specific training in drug related matters. Only a few had attended any lectures and these had largely been run on a voluntary basis. Very little, if any, training on how to detect drug users, and even less on how to treat the drug user when they were s suspected of using drugs. Even for the Health and Safety Representative, their training on drug related matters was only part of a larger First Aid and safety courses.

Within the teaching profession, Occupational Health and Safety Representatives said that they had never encountered illicit drugs amongst the teachers. Further they had minimal information on drugs and did not know what signs to look out for. One Representative went so far as to say that there was no need for education programs as no problem exists.

4.2 *Employees*

Individual employees were interviewed in a face-to-face situation. In the construction industry, the age of the interviewee ranged from 17 to 58 and covered a cross section of trades including carpenter, plumbers, labourers, riggers and technicians. The results from the survey can be summarised as follows:

- Majority of non-alcohol drug usage was amongst the under 24 year olds
- 86% of respondents believed that the non-alcohol drug of preference was marijuana
- approximately 33% of respondents regularly used prescribed drugs
- main reasons given for the use of non-alcohol drugs was either peer pressure, or no particular conscience reason at all
- only 10% had not seen any attempt on the construction site to reduce drug use
- 76% of respondents believed that if an employer would do 'something about it' if they knew there was a drug problem.

The construction related interviewees who admitted to having used drugs, readily acknowledged that they knew of accidents on construction sites which were related to the use of drugs by the people involved in the event. None knew of any serious accidents requiring hospitalisation or of a death as a result of drug taking. The majority believed that it was not serious accidents that were the main problem but rather that drug users were responsible for faulty or below standard final products. Being younger, they did not have the developed skills and required more concentration to produce the high quality product so that using drugs reduced the quality and often led to the need to rework the item. The respondents did not restrict the usage to any trade but between them knew people right across the trade spectrum.

With respect to education and prevention, none of the respondents admitted to having attended a formal training session on drugs. They had seen the pamphlets and posters in site sheds. Further 91% strongly believed that the use of blood tests was an invasion of privacy and should not be used. With respect to urine sampling, the response was less definite with approximately 50% being for and 50% against urine sampling. Yet in the USA in a Gallop poll performed in 1990 found that 81% of construction workers supported testing for drugs. (Hinze 1997)

Within the education sector, the interviewees ranged in age from 25 to 54 – rather older than in the construction sector. The answers were somewhat different. Only 18% of teachers used marijuana, with speed and LSD being 6% & 12% respectively. The main drugs were either pain killers – 19% or prescription drugs – 61%. The frequency of usage for the latter two categories was close to 100% - but was not perceived as a problem.

The statistics published following the Survey on Drug Abuse in Australia in 1992 showed that in the four days preceding the survey on drug usage, 78% of all respondents had used a drug of some kind:

Table 1. DRUG USAGE – past 4 days

Medical Prescription Drugs	37%
Over the Counter – esp. pain killers	44%
Legal Social Drugs – include. Alcohol and cigarettes	33%
Illicit Drugs	11%
Any Substance	78%

4.3 1998 SURVEY

A survey undertaken in April 1998 amongst 5000 teenagers, showed that the drugs of preference amongst school and tertiary students were generally in line with the earlier RMIT survey.

DRUGS preferred by Teenagers (%)

DRUGS	Total	Males	Females
Marijuana	32%	35%	28%
Ecstasy	6	8	3
LSD	6	9	3
Speed	6	9	3
Heroin	4	6	2
Cocaine	4	6	1

***Survey undertaken by Herald Sun and tabulated by NCS Australia – 3325 responses analysed**

The other data to come out of the survey was that drug usage was more popular at TAFE and universities than at either state of private schools. Of those who used drugs, 88 per cent said it was their own choice (but admitted a lot of peer pressure) whilst others said that they turned to drugs to escape their problems. The main drug was still alcohol which was recorded at – 75 per cent of 17 year olds drank regularly, and dropped to 30 per cent of 13 year olds.

4.4 Drug recognition

One fact that came out of the survey was the lack of awareness by both those in construction and in the education sector of the effect of various drugs on human performance. In the older age group, mainly teachers, this was often drugs taken for blood pressure. In the younger age group, drugs taken for sports related injuries. The use of antihistamines is common across both groups. The following simplified table shows a sample range of some of the more commonly taken drugs and their possible side effects.

Table 3. DRUG DATABASE

DRUG Category Common Name	Habit Forming	Main Usage	Adverse Reaction	Availability	Effects of combining
Alpha Blockers Pressin	NO	Relieves High Blood Pressure	Lose Consciousness	1	+Alcohol Excessive low blood pressure
Amiloride Amizide	No	Ditto	Faintness	1	+Alcohol Over sedation
Antihistamines Many including Palaramine etc	No	Hayfever relief	Dizziness	2	+Alcohol Over sedation
Benzodiapines Lorazepan	Yes	Relieves Tension	Dizziness	1	+Alcohol Over sedation
Cough Suppressants Pentoxyerine	Yes	Suppress coughs	Drowsiness	2	+Alcohol Over sedation
Dantrolene Dantrium	No	Muscular Pain relief	Fatigue	1	+Alcohol Over sedation
Narcotic Analgesic Pethidine	Yes	Pain Relief	Nausea	1	+Alcohol Loss of coordination
Orphenadrine Disipal	No	Relieves muscular pain	Drowsiness	1	+Alcohol Over sedation
Tetrabenazine Nitoman	No	Treat muscle disorders	Confusion	1	+Alcohol Over sedation

1 = prescription only
2 = over the counter

Chart 1. Category of Drugs Used by Positive Responses

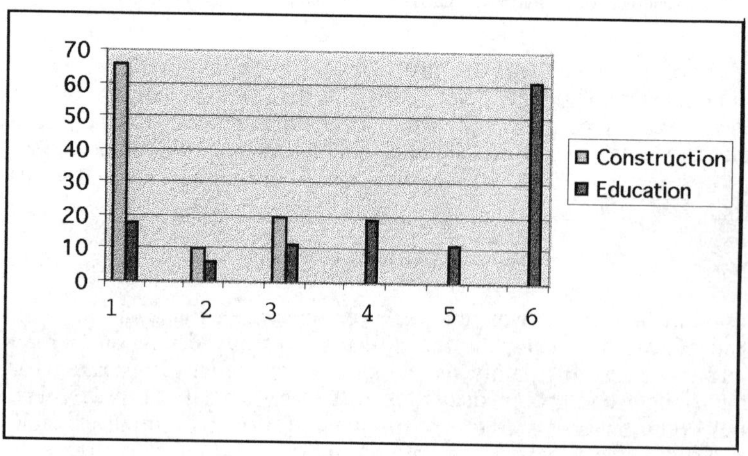

1 Marijuana
2 Speed
3 LSD
4 Pain Killers
5 Ecstasy
6 Prescription

Whilst not as severe a problem in the confines of a class room, the very nature of the construction industry and its propensity for alcohol usage (esp. in Australia as reported in previous research) makes this a major issue when considering youth and drugs in relation to onsite construction health and safety.

5 INDUSTRY COMPARISONS

The following graphs compare the responses from the construction and the education sector. The first graph provides the specific types of drugs used by those who responded 'yes' to the question as to whether they had used drugs.

In Chart 1, it can be observed that the drug most commonly used in the Construction Industry is Marijuana, whilst amongst teachers it is the prescription and over the counter drugs. The older age of drug users in the teaching profession as shown below may account for this difference.

Chart 2 Perceived Drugs Used In the Workplace by Industry

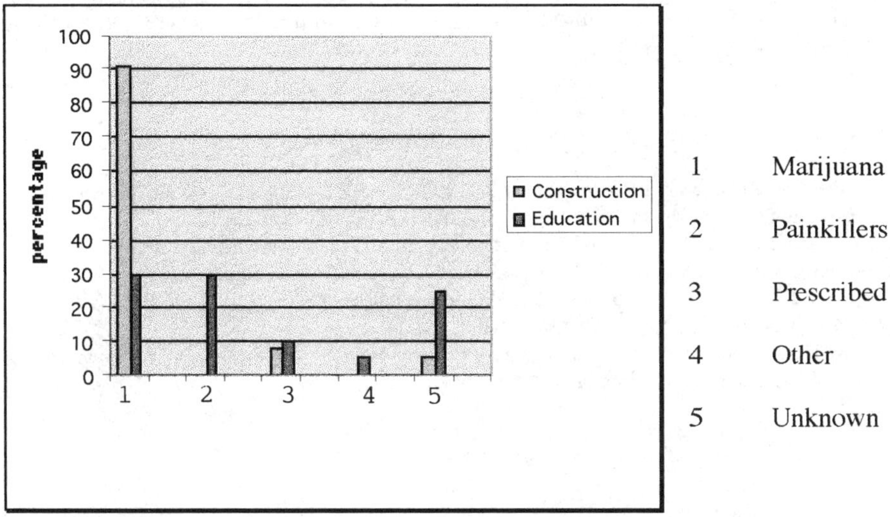

Chart 3. Percentage of Employees to have Received Drug Education

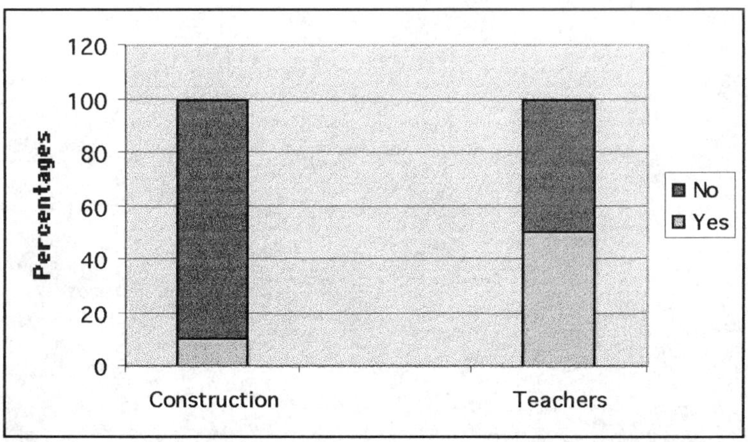

6 DRUG USAGE.

In a comparison between the construction and the education industry, the type of drug used by those who admitted to usage was quite different in each case. In the construction industry, marijuana was clearly the winner, whilst with teachers it was the prescription/over the counter, type drugs. The age difference may be one factor in this difference.

CONCLUSION

The independent survey on teenage drug usage, and the specific construction industry survey produced similar results. Drug usage of the younger construction workers mirrored that of their peers in society. That is up to one third of them have or are using a drug which will have an affect on their ability to carry out their work in a safe and workmanlike manner. Yet there is little training for their employers, at any level, to recognise those using these drugs or how to deal with the problem when it is recognised. Whilst the education sector employees do not seem to have the same illicit drug usage, they do have the problem of prescription drugs which still have an affect on those who take them. There is even less education on drug user recognition in this sector than in the construction sector.

REFERENCES

Drug Use on Victorian Construction Sites – Part 2, 1997, – RMIT University, Dept. of Building & Construction Economics, Ian Eilenberg Research Coordinator, Australia

Research into Teenage Drugs, 1998, Herald Sun with NCS Australasia, Melbourne Australia

Victorian Building Industry Alcohol and other Drugs Steering Committee – stage 2, 1993, Australia

Alcohol and Other Drugs in the Workplace, 1991, Victorian Trades Hall Council, Australia

Hinze J.W., 1997, *Construction Safety*, Prentice-Hall, USA

Hall W, Solowij N, Lemon J, 1994, *The Health and Physiological Consequences of Cannabis Use*, National Drug and Alcohol Research Centre, Australia

Implementation of Safety and Health on Construction Sites, Singh, Hinze & Coble (eds)
© 1999 Balkema, Rotterdam, ISBN 90 5809 036 1

Buildings for healthy living: Building related illness

Ingrid Senitkova
Department of Civil Environmental Engineering, Technical University of Kosice, Slovakia

ABSTRACT: The technical and socio-economic changes have profoundly influenced the methods now employed to plan, design, construct and operate buildings. As a consequence, a dilemma has developed for those responsible for the performance of buildings during the four main phase of a building's life as design, construction, operation and demolition. The indoor environment became one of the main determinants of a man's life. Therefore it is obvious that the living and especially the working environment becomes at present the matter of broad interest. Numerous scientific work related to health problems among employers in buildings have been described. The design of healthy buildings is a complex task due to the large number of factors that must be addressed and the lack of knowledge about the effects of interactions among the various environmental factors and the human and environmental responses to them. Guidelines to address interactions are lacking. Designers must review the research literature, establish priorities, and design accordingly. Indoor environmental factors of importance are identified by examining the literature on sick building syndrome and building related illness. This allows prioritization of design attention and rationalization of design decisions. Many risk factors for indoor environments can be easily addressed by design. Others require operation and maintenance solutions, although these can be facilitated by good design. Overall, some design solutions will benefit both the quality of the indoor environment and the impact of the building on the larger environment. Some of the most significant solutions are identified and recommended.

1 INTRODUCTION

The design of intelligent buildings is a complex task due to the large number of factors that must be addressed and the lack of knowledge about the effects of interactions among the various environmental factors and the human and environmental responses to them. Guidelines to address interactions are lacking. Designers must review the research literature, establish priorities, and design accordingly. Indoor environmental factors of importance are identified by examining the literature on Sick Building Syndrome and Building Related Illness. This allows prioritization of design attention and rationalization of design decisions. Many risk factors for indoor environments can be easily addressed by design. Others require operation and maintenance solutions, although these can be facilitated by good design. Overall, some design solutions will benefit both the quality of the indoor environment and the impact of the building on the larger environment Some of the most significant solutions are identified and recommended.

2 INTELLIGENT BUILDINGS

If we want to have intelligent buildings, we have to make them intelligent. The intelligence of a building is manifesting itself during the construction, realization and exploitation.

The main characteristics of intelligent building are the following:

- ecological building - is a healthy building for all its occupants
- solar or low-energy building - the exploitation of the building is highly economically effective
- building with automatic service control - optimalization is based upon both ecological and energetic criteria for creation of artificial environment.

Architecture of the building is characterized by many factors interacting as: its function, aesthetics, economy and engineering. Building is strongly by the structures and materials applied. Intelligent buildings have to employ ecologically clean materials and composites. On the other hand intelligent building is characterized by indoor climate with its components: thermal comfort, acoustical comfort, optical comfort, aerodynamical comfort, cleanliness of the air.

Energy consumption is determined by numerous factors building orientation, accumulation within thermal mass of walls and roof, total area of transparent constructions, physical properties of envelope structures, air change, heating temperature, furniture and carpets, window curtains and shutters, heat transmission loss, air change loss, solar radiation gain and free heat gains.

Required characteristics of intelligent building are expressed by interaction of following tools:

- interaction of annual energy consumption, ecologically optimal indoor climate and thermo-optical properties of transparent constructions of intelligent building,
- interaction of annual energy consumption, ecologically optimal indoor climate and intelligent building orientation toward cardinal points,
- interaction of annual energy consumption, ecologically optimal indoor climate and both regulated and non-regulated air change in intelligent building,
- interaction of solar system and energetic regime of intelligent building.

The coin intelligent building in the present, is not only the concept principle of the future development of theory and creation of buildings but it is requiring team work of architects and technical and environmental engineers from the very beginning. The significant factor of intelligent building development is computer and graphic equipment.

3 HEALTHY BUILDINGS

In order to design building for healthy living designers must have criteria. Some factors for which criteria are requited are listed in Table 1. It can be seen that this list, although it encompasses only fairly broad categories of indoor environmental factors, is rather lengthy. When the details of each of these factors is articulated, it is clear that the design of healthy buildings is a complex and challenging task. While data and standards exist to guide design relative to the separate factors there is little guidance for designers that allows understanding of the complex interactions among these factors and the combined effects of multiple environmental exposures on building occupants (Levin 1996).

Humans respond to all facets of the indoor environment The human body integrates its responses to environmental factors as mediated through various receptors: visual, tactile, aural, thermal, and posture. Most research separately addresses these various receptors and the related human responses. Standards and criteria generally available for design are derived from this focused research. Very little research has addressed the interactions among the multitude of factors. Thus, the designer is challenged to discover the inter-relationships among various factors and ascertain the effects of interactions among environmental factors and the human responses to them. This is not a trivial task, and little guidance exists to assist designers effectively to address the inherent problems

Table 1. Important factors for which "Healthy Building" criteria should be established.

Environmental focus	Criteria focus
Indoor environmental quality	Thermal environmental quality
	Indoor air quality
	Illumination
	Acoustics
	Functional support
	Security
	Privacy
	Way-finding
General environmental quality	Mineral resource consumption
	Energy consumption
	Natural resource consumption
	Habitat destruction, Biodiversity loss
	Land use
	Atmospheric pollution
	Water pollution
	Soil pollution

4 BUILDING RELATED ILLNESS

Buildings may cause illness to their occupants, producing symptoms like irritation in the nose, throat, eyes, and skin, as well as shortness of breath, nausea, dizziness, and fatigue. These symptoms are commonly referred to as the sick building syndrome. Such illness is usually minor and disappears when the sufferer leaves the building.

Low rates of ventilation are often mentioned as a contributory factor. Allergy to indoor air contaminants is the main factor in sick building syndrome These contaminants can be in the form of fine dust, fumes, fibers, odours or gases which are not removed by air conditioning filters or cleaning.

4.1. *Sick building syndrome*

The World Health Organisation has defined the sick building syndrome as an excess of work related irritations of the skin and mucous membranes and other symptoms, including headache, fatigue, and difficulty concentrating, reported by workers in modern office buildings (World Health Organisation 1983).

The term Sick Building Syndrome describes a collection of non-specific symptoms. These symptoms have been grouped into five categories (Table 2):

- sensory irritation of the eyes, nose, and throat, including dryness, pain, stinging sensations, hoarseness, and changes in the voice and in sounds from the respiratory system
- skin irritation, including blushing, pain, stinging or itching sensations, and dry skin
- neurotoxic symptoms, such as headache, nausea, drowsiness, tiredness, lethargy, reduced mental capacities, and fatigue
- unspecific hyperactivity reactions, such as runny eyes, runny nose, and asthmalike symptoms among nonasthmatic

Table 2. Sick Building Syndrome

SBS Symptoms	Explanation
Sensory irritation	including dryness, pain, stinging sensations, hoarseness, and changes in the voice and in sounds from the respiratory system
Skin problems	including blushing, pain, stinging or itching sensations, and dry skin
Neurotoxic problems	such as headache, nausea, drowsiness, tiredness, lethargy, reduced mental capacities, and fatigue
Unspecified hyperactivity	such as runny eyes and nose, and asthma-like symptom among non-asthmatics
Odour or taste complaints	including changes in odour or taste and unpleasant odours or tastes

- odour or taste complies, including changes in odour or taste and unpleasant odours or tastes (Molhave 1989).

On the whole, research has been focused on the workplace where media attention to sick building syndrome originated, but sick building syndrome is not the exclusive domain of modern air-conditioned offices. Inadequate fresh air supply and uncomfortable conditions can also be found in naturally ventilated buildings which often receive less fresh air, and certainly less filtered air than mechanically ventilated structures. Maintaining comfortable thermal conditions can also present a substantial challenge in naturally ventilated buildings.

Sick building research has identified a number of physical and chemical factors that may combine to create an uncomfortable indoor environment. Without some idea of background pollution levels, however, anomalous conditions are impossible to ascertain. The plethora of building styles and conditions, coupled with the vast range of indoor pollution sources, the number of reactive surfaces and the impact of penetrating outdoor pollutants provide indoor air researchers with enormous challenges. Over-simplification can lead to misdiagnosis, which may prove costly in time and money and produce future design problems. In particular, the removal of a single pollutant or pollutant source will do little to reduce the incidence of sick building syndrome if, as it is so often the case, the causes of sick building syndrome are multifactorial and probably synergistic (Senitkova 1996).

4.2 *Causes of health problems indoors*

Despite the fact that indoor environments vary from one country (or even region) to the other, due to differences in construction habits, climatic conditions and life-style, the industrialised societies tend to face a growing number of similar indoor air quality problems. Indoor environmental problems have five common causes, and more than one may be active at any time: an inadequately cleaned or maintained environment, insufficient ventilation, pollutants emitted from sources and activities inside the building, contamination from outside sources, and biological contamination due to a lack of moisture control. These causes can often intensity or add to the stress that occupants suffer from inadequate temperature, humidity, or lighting, or by excess noise. Exposure to pollution indoors also adds to stress of occupant density, job dissatisfaction, lack of personal privacy and control over the environment. The inadequately maintained environment is the consequence of inattention to the different emissions and by-products of activities indoors and the need for constant ordering. Poor ventilation sometimes is the result of dirty air filters that need periodic cleaning or replacement. Emission from cooking or tobacco products always need to be removed. Left alone, they accumulate damaging materials, causing odours, and in some cases increasing cancer risks to humans. Particles from the outside are constantly being tracked or blown inside. They need to be removed through cleaning and maintenance. Before energy crisis, most buildings were designed to provide maximum comfort to inhabitants. Heating, ventilating, and air conditioning systems were designed to provide fresh outside air for each building occupant. Later, buildings have been designed to save energy. The size of spaces for heating and cooling has been reduced, and outdoor air ventilation lowered. Moreover, many ventilation systems do not effectively distribute air to people in buildings Inadequate air diffusion combined with reduced ventilation causes pollution levels to grow and, as pollutant concentrations rise, so do health complaints. As already discussed, many indoor pollutants come from sources inside the building, while others may enter the building with outdoor air or from the soil (Maroni 1997).

5 RISK ASSESSEMENT - RISK MANAGEMENT

Since indoor air qualities affects directly people's health, it should be taken as a risky factor. Missing its importance can cause hazardous effects from headache to death of a person. To identify and control the risky conditions, indoor air qualities can be covered by Risk Assessment and Risk Management work (Öztürk 1997).

5.1 *Risk Assessment*

Risk Identification: Adverse health effects of monitored or suspected indoor air pollutants, groups of occupants who are affected, sources that are likely to release air pollutants, and possible spaces that indoor air pollution occurs are examined in detail and determined by risk identification.

Risk Estimation: Emission rates, concentrations and combinations of indoor air pollutants, occurrence of exposure in long or short term, and effects of results of indoor air pollution on people's health and productivity, non-human based occupant's health (animals or plants) or contents of the indoor spaces are stated or estimated during risk estimation.

Risk Evaluation: Indoor air qualities related actions, outcomes, exposures and results of all actions are judged by giving ranks, which are based on various criteria, to them on some scale during risk evaluation.

5.2 *Risk Management*

Risk Acceptability: The acceptable levels of emissions, concentrations and combinations of indoor air pollutants are determined by indoor air qualities related standards, regulations, legislation, codes of practices, etc..

Risk Control: The stages of risk control are the prevention, monitoring, control and treatment which are operated either before indoor air pollution incident happens or after the occurrence of indoor air pollution.

6 DESIGN STRATEGIES

The overal design should be based on a determination of what is important and established priorities design efforts to create healthy buildings:

- this should include an assessment of the amounts of surface area and mass of major building materials and products that will be used
- it should include an overview of use of the building and the appropriate ventilation strategy
- it should include a detailed plan for the operation and maintenance of the building over its entire life cycle.

Material selection should include evaluation of the following

- identification of quantities of major materials and products being used

- emissions of new materials and decay profiles
- maintenance and cleaning requirements
- surface renewal requirements
- durability - expected useful life.

Develop guidelines to protect air quality during construction and renovation

- isolate construction areas
- provide abundant air exchange during installation of "wet" products
- protect building surfaces (sinks) from adsorption of emissions
- exhaust construction fumes to outdoors
- provide extra ventilation during and after installation of new materials.

Develop guidelines to protect against important airborne pathogens, allergens, and asthmagens

- evaluate effectiveness of building design and construction methods for moisture control
- evaluate effectiveness of ventilation and filtration to prevent infection from specific important inhaled pathogens such as rhinovirus, Legionella pneumophila, B tuberculosis
- evaluate effectiveness of housekeeping in protecting against dust and allergens.

REFERENCES

Levin, H. 1996. Designing healthy buildings strategies and issues. Proc.: Int. Conf. *Building for Healthy Living*, Praque 21-24 April 1996: 1-19.

Maroni, M. 1997. Management of Indoor Air Quality in Buildings. Proc.: Int. Conf. *Environmental Engineering*, Kosice 6-8 May 1999: 5-15.

World Health Organisation, 1983. Indoor Air Pollutants: *Exposure and Health Effects*, EURO Reports and Studies 78, World Health Organisation

Molhave, L. 1989. The Sick Building and Other Buildings With Indoor Climate Problems, *Environment International*, No. 15: 65-74.

Senitkova, I. 1996. Health issues of indoor building construction. Proc.: Int. Conf. *Implementation of Safety and construction Sites*, Lisbon 4-7 September 1996: 431-443.

Öztürk, A. 1997. Indoor Air Quality Process. Proc.: Int. Conf. *Healthy Buildings'97*, Washington 27 September - 2 October 1997: 535-539.

Implementation of Safety and Health on Construction Sites, Singh, Hinze & Coble (eds)
 ISBN 90 5809 036 1

Occupational health management in construction – Problems and solutions

A.G.F.Gibb, D.E.Gyi & R.A.Haslam
HASEC Loughborough University, European Construction Institute, UK

ABSTRACT: Historically, occupational health has been neglected in favour of an emphasis on safety. The health of the workforce is difficult to assess and even more difficult to manage. However, recent increases in litigation on health grounds; realisation of the affect on productivity and image of organisations; and, continued moral pressure has caused some construction organisations to review their approach towards health management. This has led the European Construction Institute (ECI) to produce a management guidance manual on occupational health in construction. The background work for the ECI manual is reviewed in this paper.

1 INTRODUCTION

1.1 *What is occupational health?*

Back in 1946 the World Health Organisation (WHO) defined health somewhat idealistically as "a state of physical, mental and social well being, not merely the absence of disease or infirmity" (Pike & Forster 1995). By 1950 WHO had developed this definition to 'the maintenance of the highest degree of physical, mental and social well being of workers in all occupations. In 1997 the UK Health and Safety Executive (HSE) stated that 'ill health includes acute and chronic ill health caused by physical, chemical or biological agents as well as adverse effects on mental health' (Ahmed 1998).

1.2 *H&S = safety without health*

For as long as the terms health and safety (H&S) have been linked together they have been used as a euphemism for safety alone. H&S officers have been trained in safety matters with little or no regard to health matters. Even the UK's enforcement body, the Health and Safety Executive, has traditionally concentrated on safety to the detriment of health issues. The authors have found that construction managers understand their role in safety and the prevention of injury, but are less clear regarding the prevention of ill-health (Gyi 1998). Some would argue that this emphasis on safety, especially in construction, has been necessary due to the significant and major risks of accidents leading to serious injury or death. However, the authors contend that this attitude can no longer be tolerated.

This paper discusses why occupational health is neglected, demonstrates that this poses a very real problem for the construction industry and identifies key implications for the industry if the current situation is not improved. Finally the paper explains how attitudes are changing, albeit slowly, and how a health management manual from the European Construction Institute can help improve health management in construction.

2 WHY IS OCCUPATIONAL HEALTH NEGLECTED?

2.1 *Health issues are difficult to identify at an early stage*

Pritchard (1996) found that 'health problems are far more complex to try and tackle than safety because: the cause and effect of exposure to certain substances is unclear; certain diseases or conditions may have a long latency period; and, the solutions may require the involvement of a number of different people'.

The fact that the results of ill-health may not appear for several years after exposure to the causal factors is one of the biggest problems facing health managers in any industry.

Death certificates record up to three causes of death and only the last occupation of the individual. The eventual cause of death registered at the time of death may never be related back to the initial circumstances that triggered the problems in the first place.

The industry also has a major impact on the rates of morbidity that may result in people leaving the industry to find other forms of employment. The true reason for the person leaving is rarely stated and certainly not recorded in a manner that could inform industry health statistics.

2.2 *Health risks may occur in materials and activities that are part of 'everyday working life'*

Asbestosis is a prime example, in that asbestos was an accepted part of modern life until relatively recently. In the early days nobody imagined that there could be a risk. Notwithstanding, there was then a considerable period after the warning bells started to ring before effective action was taken. It is cases such as these that are now a regular occurrence in litigious court actions all over the world. Without the benefit of hind-sight one can understand the reluctance of some to accept initial scientific concerns, particularly when the first indications are rarely conclusive and are often disputed by equally well-meaning experts. Currently in the UK more than 3500 people die each year as a result of exposure to asbestos and most of these deaths are in the construction industry (Gordon 1998).

Recently we have seen concern expressed over materials that are just as common as asbestos was before the 1980s. In 1997 in the US, the Occupational Safety & Health Administration (OSHA) moved to place hazard warnings on materials containing more than 0.1% silica. Free silica has been shown to cause silicosis, which, in medical terms, is similar to asbestosis. Silica occurs in so many construction products such as gravels, aggregate, clays, crushed stone and gypsum as well as sand. There is currently a reaction against the blanket labelling of these materials with many calling for a genuine risk assessment approach rather than simple hazard identification (Russell 1997).

2.3 *The industry has an itinerant workforce*

Workers rarely stay with one company throughout their working life. Furthermore, in the UK since the early 1970's much of construction work has been done by labour-only subcontractors (although there is a recent return to more direct employment). These factors, combined with the difficulty in making early diagnoses of health problems has resulted in the industry claiming that it cannot respond effectively to the occupational health challenge. It can be difficult for a person to demonstrate that one particular previous employer was responsible for the acts or omissions that lead ultimately to a health problem.

2.4 *Life-style issues complicate occupational health assessment*

An individual's lifestyle choices will have a major affect on their health. This complicates the occupational health factors and leads some to play down, or even dismiss, the significance of health risks at work. Pritchard (1996) found that lung cancer in the construction industry population is higher than the levels found in the average population and generally those affected are younger. She commented that this may be due, in part, to the high use of tobacco (41%) and alcohol (34%).which are obviously lifestyle factors (Robinson & Burnett 1994). However, even when these figures were adjusted for smoking habits, the risk of lung cancer was found to be highest in unskilled workers in road construction (Vaughan 1989). Pritchard claims that this may be due to exposure to Polynuclear Aromatic Hydrocarbons emitted from coal tar and its products. There are clear risks to occupational health in construction that can not be completely dismissed due to 'statistical noise' from life-style issues.

2.5 There is an inertia in the industry preventing effective action

The construction industry is very conservative and traditionally has been slow to adopt new techniques or approaches. This is true in health as well as other matters. For example, skin diseases such as dermatitis or eczema are among the most widespread occupational health conditions (Rosen & Freeman 1992). The situation is exacerbated in construction due to many factors including:

- Variety of products used on site and a lack of thorough training in their use
- Rapidly changing and often extreme weather conditions with little adequate protection
- Lack of adequate welfare facilities on many sites

Exposure to cement is believed to be the cause of between 20 and 40% of all cases of occupational dermatitis (Beck, 1990). In the UK and the US, it is estimated that the construction industry workers have the third highest rate of skin disorders (Burkhart 1993). This costs between US$222 million - US$1 billion per year (Rosen & Freeman 1992). The risk of occupational dermatitis due to cement can be controlled by reducing the water-soluble chromate from 10 ppm to 2 ppm (Avnstorpe 1991). However, it appears that despite further research this practice is rarely used (Berger 1998).

3 IS OCCUPATIONAL HEALTH A PROBLEM FOR THE CONSTRUCTION INDUSTRY?

3.1 *What the papers say*

The UK HSE state that 'the problem of occupational health is enormous, unacceptable and unnecessary. The 1995 Reported Work Related Illness Survey indicates that 2 million people in Great Britain suffer from ill health caused by work. Muscularskeletal disorders were the most commonly reported work related illness (1.2 million people), followed by stress (500K), noise (170K), skin diseases (66K) and headaches or eyestrain (50K). An estimated 19.5 million working days are lost as people have time off because work made them ill. All this involves unnecessary suffering for individuals (545K in this sample) and costs UK society at least £4.5 billion.' (HSE 1998)

In a paper called 'Where is the health in construction health and safety management?', Pritchard (1996) identified a number of health conditions where construction workers were at greater risk than the average person. Construction industry trade groups had an increased risk (2.1) of contracting Non-Hodgkins Lymphoma. This increased to 6.0 for painters and plasterers 12.0 for carpenters, brick and stone masons, plumbers and roofers. Within these occupational groups, exposure to particles, (eg dust, sawdust, fibres) resulted in an increased risk, along with exposure to chlorophenols or phenoxyacids used as wood preservatives (Scherr 1992). White unskilled labourers were also at an increased risk of cancers of the lip and oral cavity, oesophageal cancer and leukaemia. Road construction workers had high Standardised Mortality Ratios (SMR) for leukaemia (SMR = 425) and urological cancer, (SMR = 292) (Burkhart 1993) when compared to the average SMR of 100 within the population.

Carpenters and painters with at least ten years experience were almost five times more likely to suffer from squamous cell cancers of the pharynx and sinonasal cavity (Jockel 1992).

The standard mortality ratios (SMR) for construction labourers due to mental disorders, alcohol abuse, suicides and ischemic heart disease are twice those of other occupational groups (Burkhart 1993).

3.2 *Survey of construction projects*

The authors completed case studies on seven UK construction sites representing a cross section of projects in the Engineering Construction sector. The full results have been published elsewhere (Gyi et al 1998). A summary of the central questions is given in Table 1 and the health and welfare provision on each of the sites in Table 2. All sites also had first-aid facilities, trained first-aiders and emergency procedures. The key findings are summarised after the tables.

Table 1. Key issues raised during case study interviews (Adapted from Gyi et al 1998).

Topic	*Central questions and issues*
Health and welfare	Facilities? (hand-washing, toilets, clothing storage, treatment room etc) Number of health professionals? (doctor, nurse, physiotherapist) trained first-aiders?
Health assessments	Pre-employment? Includes subcontractors? Health questionnaires?
Occupational health data	Sickness absence (not accident related)? Fitness tests? Hearing tests? Central collation? Analysis of trends? Insurance claims?
Health management of subcontracted workers	Mobility of workers? Proportion of small firms/self employment? Nature/levels of exposure to risk? Worker attitude?
Health surveillance	Medicals? Specific health checks? (e.g. hearing, lung function, eye tests etc.) Includes subcontractors? Other ways of ensuring fitness?
Alcohol and substance abuse screening	Post-accident? Random? Pre-hire? Includes subcontractors? Policy?
Other ways of safeguarding health	Risk assessments? Health education? (e.g. videos, tool box talks) Human factors? PPE choice and replacement? Validation of interventions?

Table 2. Health and welfare provision on case study sites (Adapted from Gyi et al 1998)

Site	*Site health facilities*	*Site welfare arrangements*
Power Station	Large, fully equipped treatment room set up by the principal contractor with full-time nurse. Retained OH physician (0.5 days / 2 weeks)	In line with and above UK regulations (WCs, sinks, canteen, drying rooms)
Sewage Plant	No medical centre or medical/health professionals	In line with and above UK regulations (WCs, sinks, canteen, drying rooms)
Process Plant	Client-run medical centre with staff (nurse, physio and OH physician)*	In line with and above UK regulations (WCs, sinks, canteen, drying rooms)
Pipe-laying	No medical centre or medical/health professionals	Mobile facilities (WCs and sinks) Mobile drying rooms/showers were difficult to arrange
Power Station Maint-enance	Client-run medical centre with full-time OH nurse*. Physio and OH physician visited regularly	Excellent facilities including hand-wash areas, eye wash areas, showers, canteen and industrial chaplain. Contract workers had own building with canteen, offices and car park
Tunnel	Client-run medical centre*. Variability in provision for individual contractors suspected.	In line with and above UK regulations (WCs, sinks, canteen, drying rooms)
Power Station	Fully equipped clinic set up by the principal contractor with a full-time nurse.	In line with and above UK regulations (WCs, sinks, canteen, drying rooms)

* *This facility was not available to contract personnel except for serious injuries*

The key conclusions from the case-study research included:

- Data and records on health-related absence were limited and inconsistent, particularly for subcontractors
- Pre-employment questionnaires, medicals, health assessments, surveillance were rarely used.
- Many managers were concerned about the potential large costs of facilities and difficulty of demonstrating cost benefits to senior management; the sizable, temporary and mobile workforce; and the lack of interest of workers.
- Managers have limited health expertise coupled with a lack of understanding of the potential role of health professionals.
- There are very few health professionals involved at any level in the industry.

4 WHAT ARE THE IMPLICATIONS IF CONSTRUCTION CONTINUES TO IGNORE HEALTH ISSUES?

4.1 *Cost implications*

Litigious court actions on occupational health grounds are now commonplace in countries such as the US and are starting to be seen in Europe as well. The UK HSE estimates that injury and ill health costs British business £5–9 billion every year in lost production, insurance, compensation, training and replacement costs. This inevitably leads to higher prices for goods and services. The UK's Trades Union Congress claims that when added to the costs from the National Health Service (NHS) and the Benefits Agencies, which are met from taxes, this results in a real cost penalty on the average British Family of £21 a week (Anon 1997a).

Work-related muscular skeletal disorders, often related to work posture or work-place design, are becoming significant causes of large compensation payments. For example a UK health authority has recently settled out of court by paying £70 000 compensation for an office worker who became unable to work due to upper limb disorder caused by her work inputting computer data (Anon KHa 1998). Muscoskeletal disorders due to poor training and ergonomic design are a major cause of morbidity accounting for 27% of all reported accidents in Britain (Oliphant 1995) and are the leading cause of compensation claims in the USA (Hunting 1994).

Introducing the UK HSE's health campaign in May 1998, Health Directorate Head Peter Graham stated that 'in the past year there have been some substantial compensation awards for vibration white finger to employees in various industries. The HSE has estimated that around 36000 people in the UK suffer from this disease to some extent, but indications from new HSE research suggest that over a million people in the UK may be at risk.' (HSE 1998)

A record payment in the UK of £76 000 for the industrial disease of hand-arm vibration syndrome was won by a damp-proofing operative in Middlesborough (Anon 1998). The condition developed through prolonged use of heavy hammer action drills between 1972 and 1994 when his employment was terminated through ill health. The court found against the employer because no advice or warnings were given despite the fact that the dangers were known about since 1976. In another case in Manchester operatives won awards of between £6500 and £20 000 each for pain, suffering and loss of amenity due to vibration white finger. The judge found that at least by 1975 there were numerous precautions that should have been taken ranging from changes to the work and equipment to training, maintenance and monitoring (MCC 1998).

In January 1998 the UK's High Court found British Coal liable for causing respiratory diseases in six former miners. The crucial aspects were that British Coal knew that exposure to mine dust was injurious to health, but made no committed efforts to control dust until 1970 at the earliest. The judge considered that such action should have occurred 20 years earlier (Anon KHb 1998). Estimates of the liability in related cases fall between £50 million and £1 billion.

In 1997, damages totalling £185 000 were awarded to a widow of a 54 year old man who died of mesothelioma following workplace exposure to asbestos (Anon 1997b). Asbestos-related diseases are continually resulting in significant claims, to the extent that they are rarely considered news-worthy enough to be reported in the press.

4.2 *Productivity implications*

No studies exist to establish the extent of lost productivity due to poor occupational health. However, clearly workers who are ill will not work as effectively as those who are fit. Furthermore, it is clear from even a cursory glance at the incidence of ill health at work that the number of people affected is very great. In addition, workers who are ill but still at work, especially when their illness has been caused by work, are very likely to produce a negative effect on the morale of other workers and hence reduce their productivity.

4.3 *Moral implications*

In developed societies it is inconceivable for an employer to disregard the moral implications of work-related occupational health problems. Nevertheless, it occurs and some employers use many of the difficulties covered earlier in this paper (Section 2) as excuses to take no action.

4.4 *Image implications*

The affect of occupational ill health on the image of a business or industry sector can best be seen by considering an example such as the coal industry in the UK. The public perception was that work 'down the mine' was dirty, unsafe and unhealthy. Miners were expected to have chronic health problems that would reduce their life-expectancy. Many of the sons of mining families would do anything to avoid a career as a miner – The image of the coal industry was tarnished forever. To a large extent this tarnished image has been also earned by the construction industry.

5 HOW IS THE ATTITUDE TO HEALTH MANAGEMENT CHANGING?

5.1 *UK's HSE Health campaign*

In May 1998, the UK's Health and Safety Executive (HSE) launched a substantial package of measures to help prevent work-related ill health. At the launch of a new phase of it's biggest ever health campaign, 'Good Health is Good Business', HSC Chair Frank Davies explained that the particular focus areas for this phase were hand-arm vibration and industrial solvents. (HSE 1998) This initiative demonstrates a change of emphasis towards health in health and safety management and the attitude of the regulatory and enforcement bodies in the UK.

5.2 *Key industry players are now concentrating on health*

Whilst accepting that the battle against accidents and injuries is never completely won, the larger, more forward-thinking organisations in construction are seeing occupational health as the area where most improvement can be made (Gyi et al 1998). Several years ago this would have been extraordinary, but now it shows an increased awareness of health issues and increased efforts to change the current culture. However, it should be said that in many organisations the emphasis on health still falls well short of the ideal situation.

6 HOW CAN THE ECI MANUAL HELP TO IMPROVE CONSTRUCTION HEALTH?

6.1 *What is the European Construction Institute?*

The European Construction Institute (ECI) is a client/contractor organisation committed to the continual improvement of the construction industry in Europe. Members come from many EU states with a bias towards the Engineering Construction Sector. The ECI's Safety, Health and Environment (SHE) Task Force has been in place for more than eight years and has produced a number of deliverables (e.g. ECI 1995). The Task Force decided that there was a real need for a practical management guide to occupational health in construction.

6.2 *What does the ECI health manual cover?*

The ECI manual provides information and best practice solutions for the major health problems in construction. Key actions are provided for each stage in the design construction continuum. For each major health risk advice is given on the following:

- Risk indicators
- Pre-employment assessment
- Practical advice for managers
- Signs and symptoms
- Preventative or protective measures for workers
- Health surveillance

The context of the manual is the European Union and appendices cover the main distinctives of health management for each of the EU member states.

6.3 *How should the ECI health manual be used?*

The manual should be used as part of an integrated health management strategy which includes:

- Setting targets for health improvement
- Implementing risk assessments and control measures
- Monitoring performance
- Informing, instructing and training workforce
- Surveillance (Team work, OH physicians, nurses, hygenists)
- Improving communications
- Audit, Audit, Audit (Ahmed 1998)

CONCLUSION

'Health matters – Good health is good business'. Despite the difficulties in managing improvement in this area, financial and moral considerations require the construction industry to take action. The European Construction Institute's Health Management Manual, implemented as part of an integrated health management strategy, is a good start in an area where so little has so far been done.

ACKNOWLEDGEMENTS

The ECI Health Manual featured in this paper is the result of task force collaboration by:

Mahmood Ahmed	BG plc	Peter Brown	Bechtel
Ian Burgess	GEC Alstom Engineering Systems	Ray Canning	Taylor Woodrow
Alan Costley	Stone and Webster	Jean Luc Dumas	Technip
Susan Fink	Masons	Roy Greenslade	AMEC Process & Energy
Dirk Hessing	ABB Lummus Global	Carsten Mink	Lurgi Ol
Terry Pierce	*Previously* M W Kellogg	Terry Skinner	National Power
David Stewart	Foster Wheeler Energy	Trevor Thompson	Kvaerner Process & Energy
Mike Totterdell	*Previously* Fluor Daniel		

The technical editing team comprised Alistair Gibb, Trevor Thompson and Diane Gyi with secretarial support from Debbie Weston and Vera Cole of ECI.

The background research survey was completed by Diane Gyi on behalf of HASEC (Health And Safety and Ergonomics in Construction) at Loughborough University. HASEC is a multi-disciplinary collaboration between Roger Haslam (Human Sciences); Diane Gyi (Industrial Design); and, Alistair Gibb (Civil & Building Engineering).

REFERENCES

Ahmed, M. 1998. *Occupational health management – ECI Initaitive.* Presentation to BGplc (unpublished).

Anon, 1997a. Injuries at work cost British families £21 a week. *Safety and Health Practitioner.* May: 2.

Anon, 1997b. Widow of asbestos victim wins substantial damages. *Safety and Health Practitioner.* May: 5.

Anon, KHa. 1998. Workstation and job design failures cost employer. *Health & Safety Bulletin.* IRS, Industrial Relations Service, Eclipse Group Ltd, 18-20 Highbury Place, London N5 1QP, January, 265: 3.

Anon, KHb. 1998. British Coal loses landmark compensation case. *Health & Safety Bulletin.* IRS, Industrial Relations Service, London. April, 268: 22-23 (The full High Court judgement from 23/1/98 can be down-loaded at http://www.open.gov.uk/courts/court/qbdind.htm).

Anon, 1998. Record payment for hand vibration disease. *Health & Safety at Work.* 147, May: 6.

Avnstorpe, C. 1991. Risk Factors for cement eczema.. *Contact Dermatitis.* 25, pp. 81-88.

Beck, M H. 1990. Letter to editor. *British Medical Journal.* 301: 932.

Berger, J. 1998. Cement Dermatitis – an International Problem, Proceedings of CIB W99 Conference, *Environment, Quality and Safety in Construction*, Lisbon, June: 1-8.

Burkhart, G. et al. 1993. Job Tasks, potential exposures and health risks of labourers employed in the construction industry. *American Journal of Industrial Medicine.* 24: 413-425.

Dedobbeleer, N. 1990. Safety performance among union and non-union workers in the construction industry. *Journal of Occupational Medicine.* 32 (11): 1099-1103.

ECI, 1995. *Total Project Management of Construction Safety, Health and Environment.* 2nd Edition, Thomas Telford, London: 206, ed. Gibb, Tubb, Thompson. ISBN 0-7277-2082-1.

Gordon M. 1998. Industry unites behind plan to end 3500 deaths every year. *Contract Journal.* 25 March: 14-15.

Gyi, D.E., Gibb, A.G.F. & Haslam, R.A. 1998. Case studies of Occupational Health Management in the Engineering Construction Industry. *Occupational Medicine.* 48 (4):263-271, ISSN 0962-7480/98.

HSE, 1998. Ignore workers health and profits will suffer warns chairman at launch of new HSE health campaign. *Health and Safety Executive Press Release.* 22 May, E114: 98.

Hunting, K L. 1994. Surveillance of Construction Worker Injuries through an Urban Emergency Department. *Journal of Occupational Medicine.* 36 (3): 356-364.

Jockel, K H. et al. 1992. Occupational and Environmental Hazards associated with lung cancer. *International Journal of Epidemiology.* 21 (2): 202-213.

MCC, 1998. Lead vibration white finger action against British Gas successful. *Manchester County Court judgement records.* HH Judge Ensor 7/4/98, C7000034, LTL 1/5/98.

Oliphant, J. 1995. Sore Sites. *Health and Safety at Work,* March: 36-37.

Pike S & Forster D. 1995. *Health Promotion for All.* Churchill Livingstone, Edinburgh.

Pritchard, C. 1996. Where is the health in construction health and safety management? *Proceedings CIB W89 Beijing International Conference*, 21-24 October, China Architecture Press, ISBN 7 112 02264 9.

Robinson, C F. & Burnett, C A. 1994. Mortality patterns of US female construction workers by race. *Journal of Occupational Medicine.* 36 (11): 1228-1233.

Rosen, R H & Freeman, S. 1992. Occupational contact dermatitis in New South Wales. *Australian Journal of Dermatology.* 1-10.

Russell, H. 1997. Health clamp down on silica. *NCE New Civil Engineer.* Telford, London, 20 November: 3.

Scherr, P A. et al. 1992. Non Hodgkins Lymphoma and Occupational Exposure. *Cancer Research.* (Supp) 52: 5503s-5509s.

Vaughan T.L. 1989. Occupational and Squamous cell cancers of the pharynx and sinonasal cavity. *American Journal of Industrial Medicine.* 16(5): 493-510.

22 Accident statistics and risk analysis

Implementation of Safety and Health on Construction Sites, Singh, Hinze & Coble (eds)
© 1999 Balkema, Rotterdam, ISBN 90 5809 036 1

Analysis of accidents in bridge construction

Fernando A. Branco & Luís M. Alves Dias
Instituto Superior Técnico, Technical University of Lisbon, Portugal

ABSTRACT: Bridges are special structures where, besides the use of current building techniques, particular construction methods are applied from which safety problems frequently arise. In this paper, examples of accidents related to these situations are analysed, and methods for their prevention are referred. The role of the designer, contractor and quality and safety inspector in the construction site is also discussed, aiming to a better prevention of these situations.

Keywords: Construction Safety; Occupational Accidents.

1 INTRODUCTION

Construction safety relies on the knowledge of technologies that were developed considering mainly construction problems. Those technologies are also currently applied in bridges, but due to their special construction characteristics, problems frequently arise.

Typically, in bridge construction there are two types of occupational accidents: those that arise from deficient personal or collective safety measures and those that are associated with equipment causing structure collapse during construction phase.

The first type includes all the current construction occupational accidents, perhaps with more emphasis in fall accidents, and can be covered with appropriate safety and health plans and its effective implementation by actions of the safety inspectors.

The second type includes the bridge accidents that arise due to the method of construction, namely from inadequate use of construction equipment causing structure collapse, situations that frequently lead to a big number of fatal accidents and to important economical losses.

These last situations that will be analyzed in this paper, belong to an area that is not usually well covered by the safety and health plans, as they have specific characteristics and frequently need a real and effective interaction between the designer, the contractor and the quality and safety inspector.

2 BRIDGE CONSTRUCTION METHODS

Besides the application of the classical methods, bridge construction had a significant development in the last decades, leading to special equipment that allow cantilever construction, large precast segmental construction, etc.. These equipment are different for each bridge, leading to working adaptation problems of the workers and safety inspectors.

2.1 *Small and medium span structures*

Most of the roadway bridges have short to medium spans (up to 30 meters) and their decks are close to the soil (less than 10m). The deck of these viaducts is typically built using either precast elements or in place cast concrete with formwork and a shoring system allowing the passage of vehicles.

Here the safety procedures are similar to the general construction ones, except for the interaction with traffic that usually flows under the bridge deck formwork. The use of precast beams has the advantage of reduction the period of this interaction as only during the positioning of the beams special safety measures must be taken.

2.2 *Large Span Structures*

Large span bridges are built following special techniques that have been developed in the recent years. These can be summarized in the following types:

- precast of large spans;
- span after span movable formwork;
- incremental launching method;
- in situ cantilever construction.

The "precast of large spans" is a technique that uses concrete precast segments, built in a precast plant. They are then joined in situ to build the span. Alternatively, the segments can be joined in the precast plant and the whole span is placed in situ. Both solutions need special safety measures associated with the transportation and in in situ placing of heavy loads.

The "span after span movable formwork" is a method that uses movable big steel trusses, as supporting beam for the formwork of each span. The deck is cast in situ span after span. Safety problems are usually associated with deficient procedures during the movement of the steel truss, where high jacking forces are applied.

The "incremental launching method" is a technique where the bridge is built by in situ casting, close to an abutment, and afterwards is sequentially pushed over the columns, with the help of a provisional front beck. Safety problems may also arise during the pushing operations where high forces are involved.

The "in situ cantilever construction" is a method nowadays frequently used in frame type structures or cable-stayed bridges. Beginning from the columns, the span is built with a movable formwork on the top of the cantilevers, allowing the placing of concrete of the several elements of the deck. The movable structure has frequently only a few fixing points to the cantilever, what led to several accidents associated with deficient procedures during the movement of the formwork.

The accidents that have been occurred with these methods, clearly point out the need of skilled workers and engineers, who must know how to use these construction equipment and the operation instructions including the quality and safety measures.

3 TYPICAL ACCIDENTS ARISING FROM BRIDGE CONSTRUCTION

3.1 *Interaction with traffic*

Most of the urban bridges (viaducts) are built over roads with important traffic conditions. This leads to special safety problems arising from the interaction between the construction procedures and the existing traffic.

Clearances - The construction of a viaduct frequently implies a periodic reduction of the normal horizontal and vertical clearances, due to the formwork and shoring. This reduction leads frequently to traffic accidents associated with the shock with the shoring or formwork.

These situations must be clearly posted through speed reduction and adequate signally. The single adoption of signs indicating new clearances is usually not sufficiently effective, as most of the truck drivers don't know the effective height of their trucks. The best way is the placement of a frame device with the exact clearances provoking a light shock for bigger vehicles. Traffic interruption should also be definitively envisaged during more critical operations like concreting or dismounting of the formwork.

Fall of Objects - During construction, elements or light equipment may fall from the deck and damage traffic passing under. The construction small holes existing in the deck are special dangerous points, as frequently steel and other elements fall through them.

It is recommended that safety nets are used during all construction period to prevent the falling of these elements. These safety nets should not be removed before formwork is taken off and the holes in the deck are closed.

Electrical Railways - The construction of a viaduct over an electrical railway line is always a delicate situation, requiring special measures to be taken. Accidents have occurred with the approach of a steel element to the electrical cables or even with the drop of water.

In these situations work must always be performed at a minimum distance from the electrical cables, typically, 3 or 5 meters as the cable tension is less or more than 50000 Volts. Signally must be quite effective, alerting to the dangerous and the use of water without protection forbidden. Safety nets should also be applied. For delicate operations, it is also recommended to cut the electricity during some periods.

3.2 *Fire accidents*

The existence of several inflammable materials during the construction of the bridge (for example, the wood formwork), have given rise to several fire accidents, frequently leading to the partial collapse of the bridge and consequently to occupational accidents.

Formwork fire - The burning of wood formwork is usually associated with misuse of torches used in construction for the operations of welding or cutting steel bars. Also some fires developed at the soil may lead to accidents, specially, if the shoring has wood elements.

To prevent these situations safety measures should be implemented associated with the use of torches or any fire during the construction of the bridge.

Traffic - The interaction of the construction site and traffic also can give rise to fires, namely from cigarettes threw from vehicles. In the construction site there are several elements easily inflammable, as for paper, wood, paints, etc.. A simple burning cigarette may be enough to start a fire.

Adequate signally is fundamental, alerting to possible fire situations. Besides that, cleaning and protection of zones close to the road with traffic should be implemented.

3.3 *Environmental actions*

Storm wind - The construction of important bridges with significant height (50 meters) or in open valleys, usually exposes the workers and equipment to direct wind action. This is specially important during storm periods, as most of the construction equipment used in the formwork is not always prepared to support these actions, leading to collapses and fall of these equipment.

The movable and some fixed equipment should have a manual with instructions associated with their use in these situations, namely as for the fixing systems.

Flood - Sudden river flood is usually associated with economical and human losses, related to construction sites placed in the river embankments.

These accidents should be prevented with monitoring of the water level during risk periods and preventive measures should be taken.

Temperature effects – Differential environmental temperature lead to important movements, specially in steel bridges. Unstable situations may arise, leading to the fall of equipment or even of bridge decks.

These rare effects should be analyzed during construction preparation and shadow preventive measures should be defined for abnormal temperature situations.

3.4 *Construction Equipment*

Movable cranes - Accidents with cranes frequently arise from deficient weight transportation (not well fixed, balanced, excessive, etc.) leading to drop of the weight or to the collapse of the crane. Particular accidents may also arise from the use of cranes close to electrical lines or even, radio emission antennas where the crane may receive an electrical shock eventually leading to the death of the operator.

To prevent these risks, a careful design of the operation manoeuvres must be taken and, if electrical problems may arise, it is useful to make the earth protection of the equipment.

Hydraulic jacks - This equipment is associated with the application of important forces as for the prestress, formwork equipment movements, or even bridge deck movements. Accidents may arise from unexpected braking of material (for example, the prestress wires) or from lack of equilibrium during movement.

These situations should be prevented with safety instructions associated with the maneuvers during force application.

3.5 *Dismounting procedures*

Accidents during dismounting of supporting structures (for example, shoring) are frequently associated to situations with lack of equilibrium for some of the dismounting phases. In fact, the dismounting must follow specific rules and sequential programs and can not be stooped in any phase. Shoring structures can not be left partial dismounted while they are exposed to accidental loads.

For safety reasons, dismounting programs must include the sequential phases, the location of the dismounted equipment, the positioning of the workers during these operations, etc..

3.6 *Construction procedures*

Safety Plans are usually developed based on the construction procedures and phases and based on the experience of past activities. Nevertheless accidents still occur associated with the construction procedures in bridges and this arises mainly due to the fact that the formwork equipment are different for each bridge and due to some innovative procedures adopted in the site, out of the safety plan.

The use of the movable formwork equipment needs a definition of sequences of procedures, without which collapse may arise. This must be understood by the workers, and deeply controlled by the safety and quality inspector.

Frequently, accidents don't occur in the first movements of the formwork, but after a few ones, when the workers become confident and neglect all the sequential operations. Besides, the structural behaviour of the movable formwork usually is not well understood by the workers or even the control engineers. So, any changes in the Plan should always be approved by the designer, as the risk of an accident is always very high.

4 SAFETY ORGANIZATION IN THE CONSTRUCTION SITE

The organization of the work in a bridge construction site is typically based in the following procedures.

The bridge design is developed by a bridge engineer, that frequently indicates only general procedures for the construction. For important bridges this design is followed by a revision carried on by an independent consultant, to assure to the bridge owner that it is in good conditions.

After the tender, the contractor develops the design of the construction equipment, usually by a specialized equipment company. This design is frequently analyzed by the bridge designer in terms of feasibility, and a revision by an independent consultant is seldom performed. The operation of this equipment needs also, at design stage, the development of a manual, indicating all the sequential phases, and the associated safety procedures. Besides, the equipment design is frequently an adaptation of an old existing equipment, where some elements are not new and the operation procedures are not always the best.

In the bridge construction site, the contractor must implement the construction procedures defined by the equipment designer. As specific problems may arise, slight adaptations are sometimes performed. The quality and safety engineers should also verify in the bridge construction site, all the procedures defined in the design and they may improve them. Anyway, changes should always be confirmed by the equipment designer, that should also follow the first operations of the equipment.

After all the procedures and adaptations are approved it is also fundamental that during all the main operations (not only the first ones) the procedures are checked by the control engineers.

5 CONCLUSIONS

Prevention methods to avoid typical accidents during the construction of bridges, namely with construction equipment, were suggested in this paper. They are based in the analysis of most recent construction methods, which are summarized, and on some real accidents known by the authors.

The discussion presented suggest that more attention must be paid to a better interaction between the designers, contractors and the safety and quality inspectors during the construction process, aiming to a work environment with no accidents.

REFERENCES

Dias, L. Alves and Fonseca, M. 1996. *The Safety and Health Plan in the Construction Industry* (Portuguese version). IST/IDICT. Lisbon. Portugal.

Neves, I.; Branco, F.; Valente, J. 1997 - Effects of Formwork Fires during Bridge Construction. *Concrete International - Design & Construction, American Concrete Institute*, vol.19, nº3, pp. 41-46, Detroit.

Machado, L.; Branco, F. A. 1997 - The Vasco da Gama Bridge Project. IABSE - FIP Int. Conference on *New Technologies in Structural Engineering*. Lisbon, Portugal.

Implementation of Safety and Health on Construction Sites, Singh, Hinze & Coble (eds)
© 1999 Balkema, Rotterdam, ISBN 90 5809 036 1

Application of risk mapping to construction project jobsites

J.A.Gambatese & P.J.Stewart
University of Nevada, Las Vegas, Nev., USA

ABSTRACT: The identification and elimination of jobsite safety hazards is an important part of providing a safe worksite. One method for assessing and improving workplace safety is titled "Risk Mapping". The Risk Mapping method incorporates all workplace personnel in a participatory process of addressing safety hazards using a map of the workplace. The Risk Mapping method has been used successfully in a variety of workplace settings, but its applicability to and success on construction jobsites is not well known. A study involving the application of the Risk Mapping process on a construction project revealed that it is well suited to construction jobsites. Modifications to the present Risk Mapping model may be required to achieve optimal benefit on construction projects. It is anticipated that Risk Mapping can ultimately be used on construction jobsites to effectively address and mitigate safety hazards.

1 INTRODUCTION

Jobsite hazard analysis is an important part of providing and maintaining a safe work environment. The analysis of hazards begins with the determination of the work practices and conditions that create an unsafe workplace and put workers at risk of injury. Once safety hazards have been noted and their causes understood, effort can then be expended to minimize or eliminate the hazards by revising the work practices or altering the worksite conditions.

For the construction industry, workplace safety hazard analysis is both an important and often difficult task. When compared with other work industries, the construction industry has historically experienced a disproportionately high rate of disabling injuries and fatalities for its size (Hinze 1997). Improvements in construction worker safety have been achieved, but the industry continues to lag behind most other industries with regards to safety (NSC 1952-1997). Construction's poor safety record can be attributed in part to the unique aspects of construction jobsites. The construction workplace encompasses many different activities, materials, and equipment, varying jobsite personnel, and a constantly changing worksite. The complex, changing nature of a construction jobsite can make it difficult to determine the safety hazards and maintain a safe workplace. The process of identifying, assessing, and mitigating safety hazards, whether performed prior to or during construction, must take into account construction's complexity and changing work conditions to be successful.

Construction personnel perform jobsite hazard analyses by using various methods and tools. Regularly scheduled jobsite safety inspections provide one means of assessing safety during construction. Safety inspections are typically performed at various stages of construction by a designated safety inspector, job superintendent, project manager, or safety committee. These inspections often consist of walking the site to ensure that safe work practices are followed, search for unsafe conditions, and plan safety measures for upcoming work stages. To facilitate the inspections, some firms use safety checklists. Computer software has been developed that can be used to create project-specific checklists for use during site inspections (Gambatese & Hinze 1996).

Another method of addressing workplace safety is titled "Risk Mapping". The Risk Mapping method incorporates all workplace personnel in a participatory process of locating and assessing safety hazards through the use of a map of the workplace. The map constitutes a visual tool that all personnel can readily comprehend and utilize to address safety hazards in an efficient manner. Personnel use the map to conduct open-ended, group discussions to brainstorm ideas on how to eliminate hazards. Safety improvement measures are sketched on the map and then implemented through a plan of action.

The Risk Mapping process has been used successfully in a variety of workplace settings, but information is not readily available on its applicability to and success in construction. It would appear that the visual aspect of Risk Mapping would be well suited to construction projects. The design plans and elevations used during construction are an ideal tool for mapping safety hazards. Construction personnel work closely with project drawings on an everyday basis and, as a result, might feel comfortable with the graphic nature of the map. In addition, the method's participatory process might help foster communication between the labor force and management regarding safety issues, an interaction that is often strained on construction sites.

Though the Risk Mapping method appears on the surface to be quite applicable to construction jobsites, the complexity and variability of the construction workplace may hamper application of the process. The constantly changing workforce, multitude of work activities, and varying work conditions may limit application of the process as it is currently structured. Modifying or augmenting the process may be needed to achieve effective implementation on construction projects.

This paper describes a study to research the Risk Mapping method in its application to construction projects. The objectives of the study are to reveal obstacles to its implementation, determine how Risk Mapping affects safety, and learn how to best apply Risk Mapping on construction projects. The study is intended to be a preliminary evaluation and lead to further informed research of Risk Mapping as an effective method for reducing or eliminating safety hazards on construction jobsites.

2 THE RISK MAPPING METHOD

The original development and implementation of Risk Mapping occurred in an Italian automobile plant in the early 1960's (LOSH 1996). Use of the method spread to Mexico in the 1970's and to various institutions in the United States and other countries in the 1980's. The participants in the process include primarily the employees themselves. Jobsite personnel implement the process and benefit from its implementation. The basic process is executed according to the following seven steps (LOSH 1996).

1. Form a planning committee, or use an existing committee, to implement the method. It is suggested that the committee be composed of employees who work together and represent others who are likely to have similar safety concerns.
2. Select or develop a questionnaire to solicit information about the safety hazards present on the worksite including a description of the hazards and their location. The questions should apply to the particular conditions of the targeted worksite and contain language with which the employees are familiar.
3. Distribute the questionnaire among the employees in the workplace. The group should be homogeneous in some manner - such as all forklift operators, all work in a self-

contained work area, or all part of the same work process - and share most of the significant workplace hazards.

4. Transfer the findings from the questionnaire onto a Risk Map. The map can be a large scale floor plan or map of the worksite, and should indicate the locations of all machinery, storage areas, work stations, exits, and other pertinent information. Use different colored markers and different symbols to indicate each type of hazard, such as a red diamond (◇) for physical hazards, a blue circle (O) for chemical hazards, and a green square (□) for ergonomic hazards. Identify the intensity of each hazard (how much of the hazard the employees are exposed to) by varying the size of the symbol, i.e., a larger symbol means high intensity and a smaller symbol indicates low intensity. Indicate the frequency of the hazard by writing inside the symbol the number of employees exposed to the hazard. Provide additional information about the hazards or related safety regulations in the margin of the map as needed.
5. Bring together the employees to review the Risk Map. Encourage additions or modifications to the map as necessary.
6. Ask the employees to propose specific solutions to minimize or eliminate the hazards and come to an agreement on the priority areas that need correction.
7. Take action, or have the employees take action, to improve the worksite conditions. Revise the Risk Map to show where the improvements have occurred.

The Risk Mapping process is repeated as often as necessary to address the safety hazards present in the workplace. The process should be modified to reflect workforce characteristics (size, education level, native language, etc.), worksite organization, and other project- or company-specific limitations. For example, a national telecommunications company used Risk Mapping for safety awareness among employees in an office building (LOSH 1996). The company had floor plan posters placed in common areas on each floor level with the question "Where are the health and safety hazards here?" written above each poster. The employees were given instructions on how to enter hazards on the plans using symbols. Several employees noted on the map the location where they banged their heads on a wall-mounted air-conditioning unit as they walked around a corner of the building. In response to this hazard, a sign was posted on the wall warning of the hazard around the corner and arrows were painted on the pavement to warn anyone turning the corner not to walk close to the building. The date on which action was taken to correct the hazards was then written on each floor plan poster.

In another case, Risk Mapping was implemented during an employee workshop on lead hazards at a company that performs lead shielding (LOSH 1996). During the workshop, a large floor plan of the facility was posted on the wall and small-sized plans were handed out to each employee at the workshop. Colored symbols were developed for marking specific hazards on the maps. Using colored markers, the employees noted the hazardous areas on the maps. Air sampling information generated from company medical monitoring data and a previous Occupational Safety and Health Administration (OSHA) inspection was also noted on the maps. Lastly, the employees drew on the maps the locations of "clean" areas and equipment, and analyzed how each area could become contaminated with lead and how to prevent such contamination.

3 METHODOLOGY

The study involves implementation of the Risk Mapping method on a project under construction. The project selected for the study is the construction of the Paris Resort Hotel and Casino in Las Vegas, Nevada. The project sits on a 10.5 ha. (26 acre) site and consists of a 3000 room, 31-story hotel tower plus a low-rise structure that includes gaming, entertainment, and convention areas (Figure 1). The tower is of post-tensioned reinforced concrete, flat-slab construction with an elevator core at its center. The low-rise portion is a three-story, steel-framed structure with 10.6 meter (35 ft.) story heights. The project is expected to be completed in 1999 at an estimated construction cost of $750 million.

The project's general contractor, Perini Building Company, implemented the Risk Mapping method using its employees. As the general contractor on the project, Perini's scope of work includes self-performing the concrete construction, clean up, operation of forklifts and cranes, and overseeing the work of more than thirty subcontractors. Over the duration of the study period, Perini employed approximately 150 workers. This workforce included a project manager, general superintendent, several area superintendents, a dozen construction engineers, a safety engineer, and a safety intern. The remainder of the workforce was comprised of carpenters, operating engineers, and laborers.

At the time of the Risk Mapping implementation, work was beginning on the upper floors of the hotel tower and on miscellaneous sections of the low-rise portion. Construction of the tower superstructure progressed at a rate of approximately one floor every seven to nine days.

4 RESULTS

The Risk Mapping method was applied following the basic steps described previously. The authors, along with the safety engineer on the project, acted as the planning committee to organize and implement the process.

4.1 *Development of the questionnaire*

The study began with the development of an employee questionnaire. The questionnaire initially asked for personal information regarding occupation (trade), level of experience, length of time on the project, type of facilities previously constructed, and typical workday characteristics (repetition, heavy lifting, work hours, crowding). Types of safety hazards typically found on construction sites - such as falling objects, slipping, and tripping - were then listed. Space was provided for the employees to note specific hazards and their location on the project site. A cover letter was attached that provided directions for completing the questionnaire and explained the confidentiality of the information given.

4.2 *Distribution of the questionnaire*

The contractor's safety program on the project includes safety toolbox meetings conducted by each foreman during the first fifteen minutes of every Monday morning. The meetings cover topics related to general safety procedures, safety requirements for upcoming work, and accident and injury causes. The Risk Mapping questionnaires were distributed by the foremen during a regularly scheduled toolbox meeting. The foremen were given an extra copy of the cover letter to read to the employees before filling out the questionnaires.

4.3 *Development of the Risk Map*

At the end of the toolbox meetings, the questionnaires were collected by the foremen and returned to the safety engineer. Of the 152 employees that attended the toolbox meetings on the day the questionnaires were distributed, a total of 144 questionnaires were completed and returned. A list of the jobsite hazards most frequently cited in the questionnaire responses (a total of approximately 20 hazards) was created (Table 1). Each hazard was numbered and categorized as either a hazardous worksite condition or a hazardous work practice. Hazardous worksite conditions were noted on the list with a red diamond and hazardous work practices were noted with a blue circle.

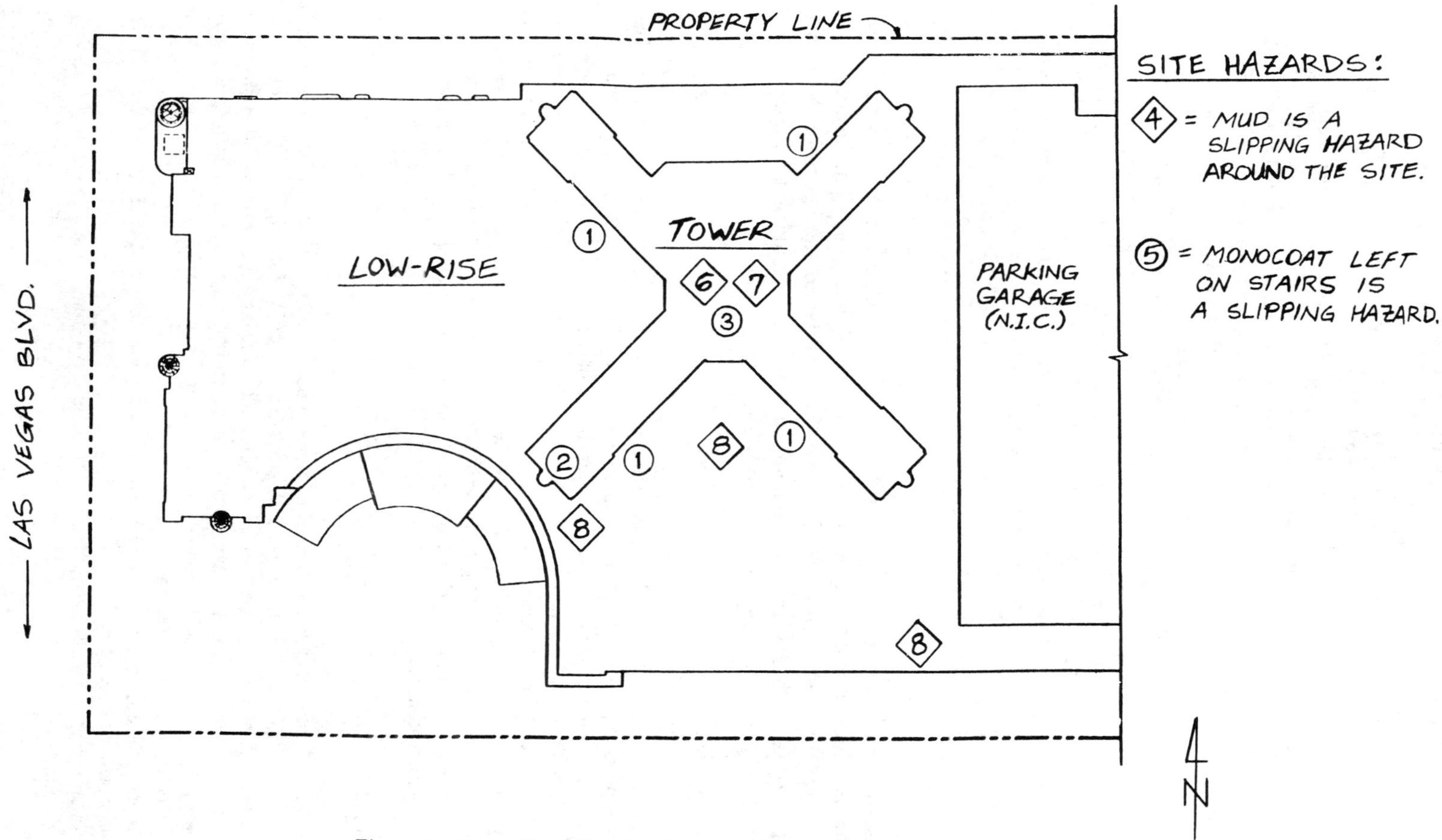

Figure 1. Plan of Paris Hotel and Casino Project and Risk Map

Table 1. Sample of Safety Hazards Most Frequently Cited

No.	Hazard Type	Hazard Description
1.	Falling object	Cranes are possibly causing a greater danger than necessary by hauling material over work areas.
2.	Fire	Slag from welders on the metal stairs around the perimeter of the building is posing a fire hazard.
3.	Slipping	Excessive form oil on the decks is creating a slipping hazard.
4.	Slipping	Mud is creating a slipping hazard around the site.
5.	Slipping	Monocoat left on the stairs is a serious slipping hazard, especially if the stairs remain open for passage.
6.	Tripping	A tripping hazard exists when walking on rebar and post-tensioning cables on the decks. (Most frequently cited hazard.)
7.	Tripping	No temporary lighting is being provided on the top four floors of the tower for work during night shifts.
8.	Obstructions	It is hazardous maneuvering between material, equipment, and other workers on the site. The corners of the building and material that is stacked overly high are creating blind corners. There is no designated path for workers around the site.

The list of hazards was used to develop two Risk Maps. One map was a typical floor plan of the hotel tower and the other map was an overall site plan. The hazards were represented on the maps using the diamonds and circles at the specified locations (Figure 1). Hazards that were more general in nature that could not be specifically located on the site were listed in the margins of the maps.

4.4 *Review of the Risk Maps*

Review and discussion of the Risk Maps occurred at a weekly meeting of Perini foremen. The regular meeting is set up for the project superintendent to discuss project progress and other issues with the foremen. Twenty-two foremen participated in the meeting along with the project superintendent, safety engineer, and authors. The two Risk Maps were posted on a wall for all to view. Each foreman was given a copy of the list of hazards with the related colored symbols that matched those shown on the Risk Maps. The authors led the meeting discussion which included reviewing and discussing each hazard on the list and determining means for mitigation. The Risk Maps were used to illustrate the location of the hazards and possible mitigation strategies.

4.5 *Hazard mitigation*

Following the Risk Map review meeting, the safety engineer and intern organized the proposed methods for mitigating the hazards and assigned specific foremen to oversee their implementation. A memo was written to each foreman describing the hazards and appropriate actions to take. A summary of the meeting was written and also distributed to each foreman to read at the next weekly toolbox meeting. The following is a sample of the actions taken in response to the Risk Mapping session.

- Large-sized aggregate was laid down on the main foot traffic areas of the site to minimize the mud hazard.
- Concrete barriers were situated along the main travel road on site to create a walkway for pedestrians.

- Materials were arranged on site to prevent blind corners and minimize material storage in high travel areas.
- Crane lifting zones were surrounded by caution tape and bright orange signs warning of the falling object hazard in the zones were posted.
- Lights were provided on the tower floors that were being worked on at night.
- A process was created to more carefully check crane riggers for the necessary experience and knowledge of their important job. Training and/or other measures were provided when necessary.

From the start of the project up to the point of Risk Mapping implementation - approximately nine months - Perini employees reported almost 50 injuries. A wide variety of injuries were reported including concrete burns, sprained ankles and wrists, lacerations and abrasions, and back strains. The primary causes of the injuries were slipping, tripping, heavy lifting, roughly handling material, and improper use of equipment and tools. The majority of the injuries occurred at locations within the hotel tower, the area primarily under construction during the initial construction period.

At this time, information revealing the injury incidence rate and severity following implementation of Risk Mapping is not yet available for review. In addition, an additional cycle of the Risk Mapping method has not been implemented because of minimal change in the worksite characteristics.

5 CONCLUSIONS

Implementation of Risk Mapping on the project provided a formal method of jobsite hazard analysis. Its implementation led to the exposure and elimination of safety hazards and ultimately a safer worksite. It is expected that a reduction in worker injuries will be realized as a result of the elimination of safety hazards.

Safety improvements on many jobsites are designed and implemented by a safety engineer with little input by those at risk. One of the favorable aspects of the Risk Mapping method is the involvement of the employees in the effort to improve safety. Risk Mapping promotes the participation of all employees in the safety effort. Requesting employee participation can increase safety hazard awareness among all personnel, promote open and positive communication between management and labor forces, and demonstrate a proactive stand on safety by a company. Brainstorming with the employees for possible means of hazard mitigation can lead to innovative solutions that improve safety yet do not disrupt the work environment or limit productivity.

The successes resulting from implementation of the Risk Mapping method in this study were limited by several factors. The foreman's meeting during which the Risk Maps were discussed included more than 25 participants. The large-sized group prohibited efficient discussion of the hazards and full participation by all present at the meeting. Reviewing the Risk Maps in several smaller groups of approximately 5 to 7 participants would have led to more efficient discussion of the hazards and greater participation of all employees involved. The results of each smaller group meeting could then be accumulated and given to a selected group of several employees to review and implement.

A lack of labor force representation at the foreman's meeting allowed for one-sided discussions of the Risk Maps. This led to discussions that placed blame for the hazards primarily on the workers and prohibited objective discussion of mitigation strategies. In addition, the Risk Maps used at the foreman's meeting were displayed far away from some of the participants. This lessened their effectiveness as visual tools for use in discussing the hazards and possible remedies. Better use of the maps would have been realized by displaying them on the center of the table amidst the foremen or including them in the handout along with the list of hazards.

6 RECOMMENDATIONS

Based on the information obtained from this initial study, the following recommendations are offered for implementation of Risk Mapping on construction sites.

- Have those most experienced in addressing safety issues develop the questionnaire.
- Organize the questionnaire for ease of completion and understanding by all employees.
- Attach a cover letter to the questionnaire explaining the intent of the study, employees' role, and participant anonymity. Stress the importance of the questionnaire to the project safety effort so that workers give serious consideration in completing it.
- Provide adequate time to complete the questionnaire.
- When reviewing the completed questionnaires, serious hazards revealed should be corrected immediately and not run through the rest of the process.
- Organize the Risk Map discussion meetings to minimize overcrowding, include all levels of employees, and promote participation by all attendees.
- Make clear at the beginning of the meetings that the object of the discussions is to find ways to create a safer work place, rather than to place blame for particular hazards.
- Locate the Risk Maps at the meetings so that they can be used as visual tools to generate discussion and are easily readable by all meeting participants.
- Inform the workers of the Risk Mapping results, i.e., the hazards recorded and how the hazards were mitigated.

Additional studies of the application of Risk Mapping on construction sites are needed to further refine the process for the construction industry. These studies might include an evaluation of when Risk Mapping should be initially implemented during the project, how often it should be implemented, and who should be involved. It is expected that modifications to the process will vary depending on the size and type of project and how often the jobsite conditions change significantly. Lastly, further effort is needed to benchmark the effect Risk Mapping has on eliminating construction site injuries and fatalities.

7 ACKNOWLEDGMENT

The authors would like to thank the Perini Building Company and its employees on the Paris Resort Hotel and Casino project for providing the opportunity to perform the study and for the time committed by its employees to participate in the study.

8 REFERENCES

Gambatese, J.A. & Hinze, J.W. 1996. A Guide for Safety Inspections. Proceedings of the First International Conference of CIB Working Commission W99: 553-563. Rotterdam: Balkema.

Hinze, J.W. 1997. Construction Safety. New Jersey: Prentice-Hall.

Labor Occupational Safety and Health Program 1996. Risk Mapping. Center for Occupational and Environmental Health, University of California, Los Angeles, CA.

National Safety Council 1952-1997. Accident Facts. National Safety Council, Itasca, IL.

Implementation of Safety and Health on Construction Sites, Singh, Hinze & Coble (eds)

Identification of work organization risk factors in construction work

P.L.Grubb & N.G.Swanson
National Institute for Occupational Safety and Health, Cincinnati, Ohio, USA

ABSTRACT: The term "work organization" refers to the way work is structured and managed, including topics such as workload, work schedule, training, interpersonal relationships, and organizational factors such as safety climate and company culture. In the present study, qualitative techniques were used to explore work organization risk factors among construction workers employed by small, non-unionized firms engaged in residential construction. Focus groups, comprised of construction workers, contractors, and company owners drawn from a broad range of occupations, were asked about work organization issues. Results indicated that safety climate was generally poor, and that there was very little training in this regard. Job demand/workload was a major stressor, particularly for company owners. The findings suggest that interventions targeting safety climate, workload, and training may be particularly effective in preventing injuries and stress among construction workers.

1 INTRODUCTION

Long-term surveilance data indicate that, across industries, the largest number of fatalities on the job are associated with the construction trade (U.S. Department of Health and Human Services, 1993). Construction workers also experience the highest rates of illness and injury leading to lost work time, and it has been estimated that 15% of all workers compensation expenditures are for construction-related injuries.

Due to the risk posed by construction work, Congress allocated funds specifically for health and safety research aimed at reducing and preventing deaths, injuries, and illnesses among construction workers (U.S. Department of Health and Human Services, 1997). In the newly-developed National Occupational Research Agenda (NORA; U.S. Department of Health and Human Services, 1996), 18 of the 21 research priority areas outlined by the National Institute for Occupational Safety and Health (NIOSH) and outside partners bear directly on the construction industry. Thus far, NIOSH has developed recommendations and interventions for physical factors in construction work such as the prevention of fatal falls, electrocutions, and exposures to toxic substances.

There has been limited research, however, on non-physical aspects of construction work which may affect safety and health. In studies of Swedish construction workers, psychosocial factors such as lack of job control, low levels of job satisfaction, and high job demands have been associated with increased risk of musculoskeletal disorders (Engholm, 1993), and time pressure and lack of social support have also been reported as a significant stressors in the construction industry (Eriksson and Lindell, 1991).

The changing nature of the workforce as well as the work environment itself in the construction industry provides a unique challenge to health and safety researchers. For example, there may be many different laborers and companies interacting on a given construction project. Worksites and co-workers may vary from day to day, and, as a consequence, each project provides a diverse set of potential hazards. The construction industry relies heavily upon part-time and temporary workers as well as varied work schedules, and there is often a seasonal aspect to the work. In addition, frequent layoffs may be common due to economic and other factors. Recent studies suggest that work organization factors such as these may broadly impact worker safety and health (U.S. Department of Health and Human Services, 1996).

The term "work organization" refers to the way work is structured and managed, and encompasses topics such as work schedules, shift work, job design and control, workload, interpersonal relationships on the job, management style, and organizational characteristics such as safety climate and company culture (U.S. Department of Health and Human Services, 1996). Recent research indicates that work organization factors including high job demands and low levels of worker control may contribute to musculoskeletal disorders and cardiovascular disease (U.S. Department of Health and Human Services, 1996). It has also been shown that work climate factors, particularly company safety climate, can influence worker compliance with safety procedures such as the use of universal precautions in health care settings (DeJoy, Murphy, and Gershon, 1995; Murphy, Grosch, Gershon, and DeJoy, 1997).

To date, there has been virtually no research examining work organization factors such as job demands, work scheduling, and safety climate in construction work. Consequently, very little is known about how these organizational aspects of construction work may affect worker health and productivity. The goal of the current exploratory study was to determine the major work organization stressors that may influence health and safety among construction workers. The findings have important implications for the design of effective safety and health interventions for construction work and for directing the focus of ongoing research.

2 METHOD

Qualitative research methods were chosen for this study since it was exploratory in nature (Hamilton, 1994; Krueger, 1988; Miles and Huberman, 1994). Focus groups were drawn from small local companies and contractors (20 or fewer employees) involved in residential construction, across a broad range of occupations (i.e., drywallers, carpenters, plumbers, etc). Three separate focus groups were formed for company owners, contractors, and workers, with each group consisting of 8 to 10 men (N=26). All participants were caucasian males. An experienced moderator led the focus group discussions, asking participants open-ended questions regarding the following work organization variables: safety climate (e.g., What about your job keeps you from following safety guidelines?), hours of work (e.g., Do you work overtime/weekends?), job demands (e.g., Is your workload heavy?), interpersonal relationships (e.g., Do you work alone or with others?), and training (e.g., What kinds of training are offered to you?).

Written transcripts, video tapes, and audio tapes of the sessions were obtained. Written transcripts were checked for accuracy using the video and audio tapes. After verifying the content, electronic copies of the transcripts were analyzed using QSR-NUD*IST, a software package for organizing qualitative or ethnographic data (Qualitative Solutions and Research Ltd., 1997). First, responses to questions were sorted by topic into the appropriate work organization category (i.e., safety climate, hours of work, job demands, interpersonal relationships, and training). Then, within each of the work organization categories, the content was analyzed and major themes emerging from the discussions of a given topic were indexed separately for owners, contractors, and workers. Consistent themes occurring throughout the data on a given work organization topic for all three groups were then identified and summarized.

3 RESULTS

3.1 *Safety climate*

Several themes emerged from the discussions of safety policies and practices. In general, there was little in the way of formal safety training or written policies, and safety issues were typically communicated on the job according to perceived need. Attitudes that surfaced consistently across the three groups were the notions that safety is mostly a matter of "common sense" and that an individual's experience and personal comfort level dictate whether an action is safe (e.g., not using a saw guard or a fall protection harness). However, a clear distinction was drawn between unsafe behaviors that might result in injury to the person engaging in the action and those that might lead to others being injured. Workers were more willing to confront someone whose behavior was posing a threat to co-workers' safety (e.g., bypassing or disconnecting electrical safety devices onsite; excessive clutter). As far as conforming to safety regulations, the primary reasons stated for non-compliance by owners and contractors were cost and time pressure, whereas inconvenience, time pressure, and interference with the task at hand played a stronger role in employees' willingness to adhere to guidelines. Examples of remarks concerning safety climate include:

"It would have to be something that you were either going to hurt yourself bad or someone else. If you weren't wearing your safety goggles, I wouldn't say a word, but if you were holding a 2x4 in your hand and started running a power saw through it, I would say 'What are you? Some kind of nut?'"
"Most of the guys I work with have 20-30 years experience...they don't need somebody coming up to them giving them a hard time. If a guy picks up a piece of wood and saws it, it is his common sense, whatever he wants to do."
"Ninety-nine percent of what we do is common sense and you cut your thumb off because you're in a hurry and you know you can do it, not because you don't know how to run a saw."
"I haven't been to any safety training class and I don't know too many guys really in the trade that have. I think you learn a lot of it out in the field. You learn from another man's mistake."
"The problem is it's just cost prohibitive. You cannot deal with what OSHA wants you to do. You couldn't stay in business, period."

3.2 *Hours of work*

Workers and contractors tended to work more than 40-hour weeks, with some side jobs on the weekend, but felt their schedules were satisfactory. Some owners, on the other hand, typically worked 12 or more hour days, and reported that work interfered with their personal lives. Examples of comments are:

Owner: *"We swing the hammer during the day and at night we're at home doing the books, paper work, and contracts and estimates."*
Worker: *"Sometimes we might get done at 2:00, but the next day we might work until 6:00 p.m."*
Worker: *"I work about 40 hours a week, average, and it doesn't really cut in on my personal life."*

3.3 *Job demands*

Workload was heaviest for company owners, and they reported high levels of stress. Workload became a problem for employees and contractors when environmental demands (e.g., inclement weather), physical demands, and time pressure were increased. Examples of statements are:

Owner: *"Sometimes I'll work all day painting. But sometimes I have to stop and either go look at something, meet somebody somewhere, come home and do some kind of crap in the office and go back out and get back home from work, eat supper, and typically go back into the office for another hour after the kids go to bed and then fall into bed dead tired."*
Worker: *"My wife works by the airport and I just mentioned to one of my bosses how much longer we're going to be on 10 hour days because it's tough trying to get the kids from the babysitter by 5:30."*

3.4 *Interpersonal relationships*

Workers and contractors typically worked in small crews (e.g., 2-3 people) or independently. Crews tended to get along well interpersonally and work as teams, and job satifaction in all three groups was generally quite high. Examples of comments are:

"My boss leaves me alone, that is why I stay there. He doesn't bother me at all."
"Where I work, everyone seems to be on par with each other. And that is the key, I think, in small companies. The guys with the lousy personalities tend to get weeded out soon enough."
"I'm pretty laid back. I just do my job and let them do what they want and work at whatever rate they want."

3.5 *Training*

Most training was on-the-job. Owners expressed reluctance to pay for various types of formal job training because then the employee would become too knowledgeable about the business and then become a competitor by starting their own company. Examples of remarks are:

Worker: *"You can't learn anything sitting in a classroom reading a code book. You have to be on the job training and asking the old timer to explain it to you right in front of you."*
Owner: *"It's terrible to say, but you're training these people to be YOUR competitors. So, I hate to say this, it sounds terrible, but, the less school training for your employees, the better it is for you personally."*

4 DISCUSSION AND CONCLUSIONS

The data collected in this exploratory study provided some important insights into organizational risk factors in construction work in small, non-unionized firms. Hours of work and job demands were major work organization risk factors for company owners, but not for front-line workers or contractors. Interpersonal relationships in these small construction companies were quite good, and this did not seem to be a significant stressor for any of the three groups. Most training was provided on-the-job, and all three groups seemed to see very little benefit in more formalized training.

The most striking work organization risk factor was safety climate. Little formal safety training is provided. The consensus seemed to be that safety regulations were only there to get in the way of your job and "cause you grief," and that the monetary cost of adhering to all of these guidelines is exhorbitant. For front-line employees, the attitude expressed was that safety is a matter of "common sense", and that each worker determines what is safe based on their experience and personal comfort level with a given risk, and that one basically learns what is not safe from another person's mistakes.

These findings suggest that potential interventions might initially target company owners' views regarding the importance of safety practices and more formalized safety training for employees. In addition, interventions need to address employees' attitudes concerning personal risk and responsibility for maintaining a safe work environment.

REFERENCES

DeJoy, D.M., Murphy, L.R., & Gershon, R.M. 1995. Safety climate in health care settings. In A.C. Bittner and P.C. Champney (ed.), *Advances in industrial ergonomics and safety VII*: 923-929. Bristol, PA: Taylor and Francis.

Engholm, G. 1993. Physical loads and constraints in the construction industry: Problems and solutions. A paper presented at the *International Congress on Occupational Health, Nice, France, September 26-October 1 1993.*

Eriksson, B., and Lindell, J. 1991. Psychosocial work environment profiles. *Bygghälsan*

Bulletine, No. 91-09-16, 48-49.
Hamilton, D. 1994. Traditions, preferences, and postures in applied qualitative research. In N.K. Denzin and Y.S. Lincoln (ed.), *Handbook of qualitative research*: 60-69. Thousand Oaks, CA: Sage Publications.
Krueger, R.A. 1988. *Focus groups: A practical guide for applied research*. Newberry Park, CA: Sage Publications.
Miles, M.B., and Huberman, M.A. 1994. *Qualitative data analysis*. Thousand Oaks, CA: Sage Publications.
Murphy, L., Grosch, J., Gershon, R., and DeJoy, D. 1997. Safety climate and injuries: The case of occupational exposure to HIV. In P. Seppälä, T. Luoparjärvi, C. Nygard, and M. Mattila (ed.), *Proceedings of the 13th Triennial Congress of the International Ergonomics Association, Tampere, Finland, June 29-July 4 1997*: 666-668. Helsinki, Finland: Finnish Institute of Occupational Health.
Qualitative Solutions and Research, Ltd. 1997. *QSR-NUD*IST 4 User's guide (2nd ed.)*. Thousand Oaks, CA: Sage Publications Software.
U.S. Department of Health and Human Services. 1993. Fatal injuries to workers in the United States, 1980-1989: A decade of surveillance. Public Health Services, Centers for Disease Control, National Institute for Occupational Safety and Health, DHHS (NIOSH) Publication Number 93-108. Washington D.C.: U.S. Department of Health and Human Services.
U.S. Department of Health and Human Services. 1996. National occupational research agenda. Public Health Services, Centers for Disease Control, National Institute for Occupational Safety and Health, DHHS (NIOSH) Publication Number 96-115. Washington D.C.: U.S. Department of Health and Human Services.
U.S. Department of Health and Human Services. 1997. NIOSH reseach projects August 1997: Construction. Public Health Services, Centers for Disease Control, National Institute for Occupational Safety and Health, DHHS (NIOSH) Publication Number 97-152. Washington D.C.: U.S. Department of Health and Human Services.

Implementation of Safety and Health on Construction Sites, Singh, Hinze & Coble (eds)
© 1999 Balkema, Rotterdam, ISBN 90 5809 036 1

Work accident evaluation: The basis for safe and occupational health programs planning

S.M.S.da Cruz
Departamento de Química e Física, Universidade de Santa Cruz do Sul, Brazil

V.L.D.V.Pereira
Departamento de Engenharia de Produção, Universidade Federal de Santa Catarina, Florianópolis, Brazil

ABSTRACT: Several authors have elected the planning as the main factor for coordination among the several activities of an undertaking success. In this sense, the planing of an occupational health and safety program is primordial for employee's life and working conditions and improvement, besides company quality and productivity improvement.

At civil construction companies, this process requires an over attention, since a series of intervening agents has been involved, with cultural, position, and special differences, manipulating daily everyday a great amount of information required for these activities development. These information usually are spread around and do not follow any sort of well structured or defined order at the company, making difficult any management task that requires reliable enough information, for a decision making.

This survey demonstrates that a detailed work accident control has played an important role on safety and occupational health on civil construction programs planning. Through such a control important information are found for these programs planning process at a private management system as well and public.

1. INTRODUCTION

Quality on Brazilian market nowadays is vital for company's survival. According to Fruet (1993) is then indispensable that civil construction performs an improvement of the employee's quality of life and productivity. The acquisition of a quality status is closely linked to safe, hygiene and health conditions at work improvement. Lima (1995) maintains that is also improbable a company get to reach an excellency level neglecting the quality of life of those who produces its goods. In this sense, "safety and hygiene at work get a more generic aspect than questions concerning the humanity, economical and the company image," (Miranda Jr., 1995).

The planning has been placed for several authors as a preponderant factor for the success in the coordination is reached between some activities of an enterprise; it is considered as basic a management function. This is essential for problems solving, mainly for those ones related to human resources. In this sense, the planning of a security and occupational health program is primordial for employee's life and working conditions improvement, besides the quality and productivity of the company improvement.

According to Bernardes (1996) at civil construction companies this process requires doubled attention therefore, are involved a series of intervening agents, with different cultures, functions and specialties, manipulating, daily, a great amount of necessary information need for its activities development. These information usually are dispersed, with no formal structure at the company, making it difficult, the management task, that needs enough trustworthy information for decision.

This article has as an objective to demonstrate that the data of work accidents in civil construction has a lot of information that is supposed to be used for an occupational safety program planning in these companies.

2. SAFETY MANAGEMENT

The main goal of a security management program at work, in small civil construction companies, is to assure the physical integrity of the construction industry worker. So, the program can be divided into three main concern areas as follow.

a. Assure worker a safe workstation - this guarantee will be provided by the following stages:
- risks recognition;
- security prevention routines implantation;
- effective evaluation of the security routines.

b. Educate and acquire knowledge for the security - providing employees information, training and stimulating the management awareness.

c. Provide safe machines and equipments - executing the following step:
- equipment need evaluation;
- equipment use schedule;
- equipment availability in the market analyzes;
- design new equipment.

To guarantee the effectiveness of the management program, the analysis of the initial situation, i.e., what really occurs at the workstation daily it is primordial for the posterior steps to be followed so that implantation of the improvements is really effective. Thus, for a correct existing risk recognition at the workstation it is necessary to carry out an exhausting analysis of work methods and processes besides the design to be executed. In this analysis is searched:
- to identify the risks;
- to determine and place the respective sources;
- identify paths and means for propagation;
- determine functions and numbers of employees exposed;
- characterize activities and types of exposure;
- identify any possible health damages;
- describe existing control actions.

This analysis, however, does not have to be restricted only to the initial phase of the management planning. For a continuous improvement of the security and hygiene system of the work it is necessary a periodic analyses in order to verify the effectiveness of the procedures adopted, besides working as a kind of internal editorship. This evaluation could be carried out through the collection and analyzes of data concerning the results of the security, or either, data on accidents and illnesses at work.

3. BRAZILIAN LEGISLATION

Work hygiene and security management in the civil construction is foreseen on the specific legislation through the requirement of a PCMAT (Program of Conditions and Environment of Work in the Industry of the Construction) elaboration. This program must be constituted of the following documents:
- A description concerning environment and conditions of working activities and operations, taking into consideration the risks of accidents and illnesses at work and its respective prevention procedures;
- collective protections execution design according with the execution stage;

- collective and individual protections devices to be used technical specification ;
- schedule defined for PCMAT implementing;
- workstations initial layout of the contemplating, also, forecasting the size of the experience areas;
- educative program contemplating the prevention of accidents and work illnesses.

The concerning rule also foresees as a first step the analyses of risky situations to which employees are exposed at work environment. A tool that must be used is the statistics evaluation of accident related matters. Through it, occupational accidents are analyzed, regionally registered by the federal agency responsible for the Social insurance.

4. METHODOLOGY

First the survey on occupational accident number has been performed at the National Institute of the Social Security, referring to accidents occurred in the period between July of 1994 to July of 1996, in the city of Saint Maria, in Rio Grande do Sul State, in Brazil.

The following data had been then gathered:

- day and hour of the accident;
- occupation of the victim;
- victim day of work return;
- victim wage;
- description of the accident.

Based upon these data, the accidents has been classified according to the part of the body harmed, the kind of the accident, the agent of injury and the kind of the injury. For this sorting it has been used a file for each employ-related accident according to the NR-18 (Brazilian Federal Bafety Norms) that deals with the conditions and work environment of the construction industry.

It has also been noticed, that in many accidents description it was not possible to get these information which are of vital importance for a minimum recognition of its occurrence. For a more deep analysis the information had been used as a data base tool and its results presented in a table showing the percentages on the collected accidents.

5. RESULTS DISCUSSIOUS

Through the sorting and analysis of the accidents descriptions collected has been found results that point out failures on human resources management and employees security at civil construction companies. The results found have shown undoubtedly that the main problems concerning safety to be faced in order to met better results for productivity and quality on civil construction.

5.1 *Information gathered directly from accidents descriptions*

Concerning the part of the body injured, according to table 1, 42% had upper members injuries, hands and arms, and 35%, the lower members, feet and legs. The fact of only 3% of the accidents had multiple parts demonstrates the gravity of these occurrences was not, in most of times, very severe.

A characterization of the nature of the occurred accident had great variability of causes, shown on table 2. The falls from difference level and extreme or improper physical effort had presented some meaning, both at 13.6% rate. The accidents due the impact, over 20%, object falls on the worker are also included. They are sufficiently significant and demonstrate to the lack of joined organization a majority of the workstations of the sector. The item referring to strange body focus nails and sticks that generally characterize neglects of EIP use (Equipment of Individual Protection).

Table 1. Part of the body injured

PART OF THE BODY INJURED	ACCIDENT NR.	ACCIDENT (%)
Upper Members	237	42,25
Lower Members	198	35,27
Thorax	61	10,90
Not specified	28	5,00
Head	18	3,20
Multi-parts	16	2,85
Eyes and Organs	3	0,53

Table 2. Accident kinds

ACCIDENT KIND	ACCIDENTS NR	ACCIDENT (%)S
Impact	115	20,50
Fall from a different level	76	13,55
Improper physical effort	76	13,55
Impact toward	69	12,30
Fall on the same level	53	9,45
Strange body	49	8,73
Not defined	42	7,49
Pressings	27	4,81
Others	22	3,92
Body reactions and its movements	13	2,32
Burring	6	1,07
Electrical Shock	4	0,71
Harmful Substances Contact	4	0,71
Attrition or abrasion	3	0,53
High Temperature Exposition	1	0,18
Living beings Attack	1	0,18

Table 3. Injuries agents

INJURY AGENT	ACCIDENTS NR	ACCIDENTS (%)
Others	78	13,90
Not Specified	76	13,55
Wood (lose particles)	62	11,05
Tool with no motive power	53	9,45
Machinery or equipment	50	8,91
Floor and wall	43	7,66
Nails	41	7,31
Scaffolding	34	6,06
Metallic parts	33	5,88
Containers or disposal recipients	33	5,88
Portable devices	28	4,99
Trash, or discharges	14	2,50
Chemical	7	1,25
Ceramics	5	0,89
Doors gates, windows, etc.	2	0,36
Particles in the air	2	0,36

To identify injury agent, Table 3, it has been found a lack on definition of the accidents descriptions, probably due to a lack of a correct information to fill in the file necessary to inform

Table 4. Accident lesion kind

LESION KIND	ACCIDENT NR	ACCIDENTS (%)
Not defined	294	52,41
Sprains	66	11,76
Puncture	53	9,45
Muscles pains	52	9,27
Cut	38	6,77
Smashing	14	2,50
Multiple injuries	14	2,50
Excoriation	8	1,43
Breaking	8	1,43
Electric Shock	4	0,70
Burning	3	0,53
Multiple Injuries	3	0,53
Eyes Irritation	2	0,36
Bruise	1	0,18
Amputation	1	0,18

Table 5. Injured position

OCCUPATION	ACCIDENTS NR	ACCIDENT (%)
Servant	355	59,66
Mason	97	16,30
Carpenter	91	15,29
Workmanship	16	2,69
Ironworker	15	2,52
Painter	5	0,84
Foreman	4	0,67
Electrician	3	0,50
Tiler	2	0,34
Scream operator	2	0,34
Plumber	2	0,34
Topography auxiliary	1	0,17
Pile Driver	1	0,17
Driver	1	0,17

the accident to the specific commission. It has also been found a high index (14%) of total absence of data on accidents description, according to NR-18. This would be a good opportunity to perform a normative program for procedure standardization.

When characterization is carried out taking in consideration the injury kind, see table 4, it is noticed once again failure on form fulfilling on description of the accident. The fact of 52.4% of the told accidents not to have clearly defined the nature of the injury disables any relative action toward these data, leaving clearly, one more time, the necessity of a prompt answer of this communication form.

Table 5 shows that approximately 60% of the victims had maid position, what points clearly to a poor educational background. The lack of training, previous qualification and of even experience in the position even makes this professional a more vulnerable to accidents inside of the at work.

5.2 *Crossing data generating information*

By using data bank tools is possible also, cross the different collected data categories. The information obtained from these manipulation will give to the planner a detailed view of haw the accidents occurs.

From these crossing process some of them show more significant results, they are without doubt reference points to priority the safety and health plan in the civil construction work place installation.

When crossing data related to injured occupation, with those about accident classification, the more significant values showed that:

- 20,6% of injured servant (355 accidents), and 26,4% injured carpenters (91 accidents) are injured by object impact in the accident occurrence. The majority of these impacts are caused by loose material around the work place. Generally this kind of material, such as scraps fall from superior levels, showing that there are a lack of cleanliness and organization in the site;
- 30% of accidental mason (97 accidents) fall due to level difference, and the majority, from stairs and scaffold without protection and improvised under unsafe circumstances;
- more than 17% of masons (97 accidents) suffered accidents due to physical inadequate effort, showing lack of preparation for this activity.

Crossing the information related to the body region injured and the factor causing these injuries led to the following results:

- from the accidents occurred over the upper body member (237 accidents) 16% are due to hand tools and 16% by moving machines and equipment, showing that there are a need for better training;
- from the accidents occurred over the lower body member (198 accidents) 18% are caused by nails, showing low level of cleanliness and low individual protection equipment usage.

Now, regardily to the correlation among occupation and body regions injured the crossing shows:

- the servant (355 accidents) as carpenters (91 accidents) shown higher accidents frequency injury in the upper body pants, 40% and 44% respectively;
- the mansons7 (97 accidents) are more frequently injured in the lower body parts 35%.

In the crossing, relating accident nature with the accident factor causing the injury, the more relevant results are:

- from occurred impacts (115 accidents) more than 35% are due to loose wood piece;
- from falling of level difference (76 accidents) more than 30% are caused by scaffolds;
- from inadequate effort (76 accidents) 27,6% are due to moving containers or disposal recipients,
- from work impact against objects (69 accidents) 43,5% are caused by inadequate use of hand tools.

These results shown the basic deficiencies of civil construction work place, lack of cleanliness and organization, equipment and machine improvisation and lack of training. The analysis and observation show clearly the inefficient use of individual protection equipment, which an example are the use of helmets and boots due to great number of workers injured by nails and loose material.

6. CONCLUSIOUS

An amount of other information are available through a detailed work accident through the detailed control of the occurred occupational accidents, evidencing the important function of this in the planning of security programs and occupational health in the civil construction that take care of the real necessities of the sub-sector of constructions in the civil construction, of this region.

This control also assists the process of planning of programs of the public management. The information are important in such a way for the aiming of campaigns of education, perfectioning and awareness as for the execution of the monitoring carried through for the agencies Ministry of the Work.

7. RECOMENDATIONS

Fill the CAT correctly is supposes to the main concern of the company management as a government institution. For accomplish this is necessary the training. Also the correct cost information are fundamentals for the management. The statistical data is suppose to be regularly collected for more effective measures.

BIBLIOGRAPHY

Bernardes, M. Junho 1996. Método de Análise do Processo de Planejamento da Produção de Empresas Construtoras Através do Estudo de seu Fluxo de Informação: Proposta Baseada em Estudo de Caso. *Dissertação de Mestrado em Engenharia Civil - Universidade Federal do Rio Grande do Sul.* Porto Alegre. 126p.

Fruet, G.M. & Formoso, C.T. 1993. Diagnóstico das Dificuldades Enfrentadas por Gerentes Técnicos de Empresas de Construção Civil de Pequeno Porte - *II Seminário Qualidade Na Construção Civil - Gestão e Tecnologia – Anais.* Porto Alegre: NOIRE, UFRGS. pp.1-52.

Gyi, D.E., Haslam[a], R.A. & Gibb, A.G.F.1997. Interviews to explore the quality of accident and health data in the construction industry. *Proceedings of the 13th Trienal Congress of the International Ergonomics Association.* Tampere, Finland . pp. 109-111

Hinze, J.W. & Gambatese, J.A.. 1996. Using injury statistics to develop accident prevention programs. *Implementation of Safety and Health on Construction Sites.* CIB W99 Conference Lisbon, September. pp. 117-127.

Lima, I.S. & Heineck, L.F. 1995. Uma Metodologia para a Avaliação da Qualidade de Vida no Trabalho Operário da Construção Civil *Gestão da Qualidade na Construção Civil: Uma Abordagem para empresas de pequeno porte, 2ª ed.* Porto Alegre: NOIRE. pp. 169-198.

Miranda Jr., L.C. 1995. Prevenção, o novo enfoque. *Rev Proteção.* Novo Hamburgo – RS. pp. 26-28.

Norma Regulamentadora No. 18. 1996. *Manuais de Legislação Atlas. Segurança e Medicina do Trabalho, 32ª ed.* São Paulo: Atlas S.A. pp. 216-255.

Implementation of Safety and Health on Construction Sites, Singh, Hinze & Coble (eds)
© 1999 Balkema, Rotterdam, ISBN 90 5809 036 1

A Pareto analysis of construction and maintenance operations accidents

J.A. Kuprenas
Vanir Construction Management, Los Angeles, Calif., USA, and University of Southern California, Los Angeles, Calif., USA

M.D. Kenney, Jr
United States Navy, San Diego, Calif., USA

E.B. Nasr
California Department of Transportation, Los Angeles, Calif., USA

ABSTRACT: This work describes an application of the Pareto principle to accident data and suggests that in most cases, accidents are distributed in a way such that a vital few types of accidents and types of injuries make up the largest portion of overall safety losses either in terms of frequency of accidents or lost work hours. This study uses actual data from United States Navy weapons facility construction and maintenance operations to create six diagrams which examine the relationship between body part injured, types of injuries, severity of injuries (measured through lost work hours), and frequency of occurrence. Conclusions detail shortcomings and successes of the Pareto Analysis approach, as well as summarize the safety improvement recommendations for the naval facility.

1 INTRODUCTION AND BACKGROUND

The Naval Weapons Station, Seal Beach is located on 5500 acres in southern California. The station employs 1000 civilian staff and 200 military staff and is responsible for receiving, storing, segregating, and issuing military ordnance. The station consists of buildings, yards, roads, and a wharf with a physical plant value of $400,000,000. The station is under the authority of a base commander. Reporting directly to the base commander is a safety manager. The safety manager maintains records of all construction., maintenance, and operations accidents that occurred at the base. This study concentrates on five years of safety records of accidents recorded from July 1, 1993, to June 30, 1997, consisting of 369 separate incidents.

This work describes an application of the Pareto principle to the accident data and suggests that in most cases, accidents are distributed in a way such that a vital few types of accidents and types of injuries make up the largest portion of overall safety losses either in terms of frequency of accidents or lost work hours. This study uses presents six diagrams which examine the relationship between body part injured, types of injuries, severity of injuries (measured through lost work hours), and frequency of occurrence.

Accident records include data related to the individual involved in the accident as well as the injury itself. (Figure 1). This study concentrates on three elements of the injury records – type of injury, body area injured, and medical attention required. The first element considered in the study was the part of the body injured. Body parts injured in this study include:

- Abdomen
- Ankle
- Arm
- Back
- Chest
- Ear
- Elbow
- Eye
- Face
- Finger
- Foot
- Hand
- Head
- Hips
- Knee
- Leg
- Mental
- Mouth
- Neck
- Respiratory system
- Rib
- Shoulder
- Skin
- Stomach
- Toe
- Wrist

#	File	Name	SSN	Age	Day	Type of Injury	Body Area	Lost Hrs	Medic	Location
31	1090	XXXX, A	####	34	Friday	Laceration	Hand	24.0	LT	BLDG 416

Description
Employee was feeding 1" wooden material thru the table saw when a piece of wood kicked back and hit him between the fore finger and index finger on the left hand causing a laceration that required five stiches. Kickback guard was installed during the incident.

Figure 1. Typical Safety Record

In some cases, more than one body part was listed as injured. In these situations (two occurrences), the first body part listed was used in the study because this was where the primary injury was located.

The second element analyzed is the type of injury. Injury types are based on the physical mechanism of the injury and vary greatly. Type listed include:

- Bite
- Burn
- Contusion
- Crushing
- Cut
- Dislocation
- Fracture
- Hernia
- Irritation
- Puncture
- Splinter
- Sprain/ Strain
- Trauma

Again, in some cases, more than one element was listed. In these situations (21 occurrences), the first type of injury listed was used in the study because this was the primary injury.

The last element of the study is medical attention required. The type of medical attention required is a measure of severity of injury and is labeled as one of the following:

- Not reported (NR) – no first aid required, no medical bill incurred
- No lost time (NL) – medical bill incurred, but no lost time
- First Aid (FA) – a least two visits to the doctor beyond the day of the accident
- Lost Time (LT) – a lost time from work beyond the accident date

Accidents which have been classified as lost time accidents also include the number of work hours lost due to the accident.

2 PARETO ANALYSIS

A Pareto analysis was conducted based on the above three elements of the 369 injury records – type of injury, injury type, and body part injured. The Pareto Principal is based upon the observation of Vilfredo Pareto in nineteenth century Italy that 20% of the population controlled about 80% of the wealth. Researchers have applied Pareto's concept to many other topics other than wealth distribution and have found that in most cases, occurrences are distributed in a way such that a vital few make up the largest portion of the population of outcomes -- but not always strictly in a 20-80 relationship (Juran 1989).

The Pareto diagram is a graphic representation of this concept. The Pareto diagram itself is a histogram with the categories of data arranged in order from the largest the smallest and a cumulative curve for all outcomes. Used in research applications, Pareto diagrams graphically allow the separation of the vital few items from which the majority of occurrences are generated from the trivial many. Resources are then directed to the vital few, thus maximizing the effective use of available resources. Used in such a fashion, Pareto diagrams have been used or proposed in a number of applications such as quality control (Kuprenas 1998) and engineering management (Graves 1993).

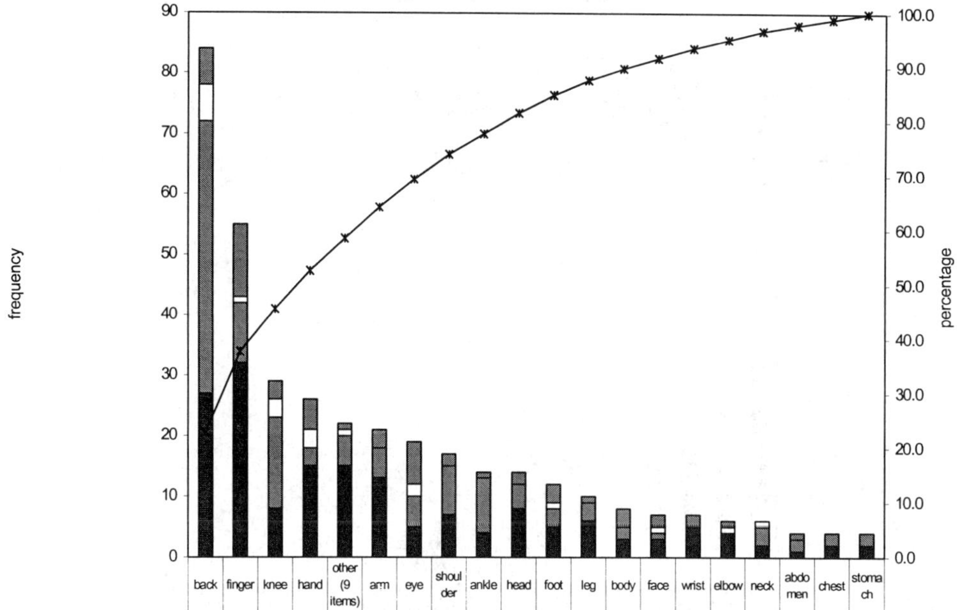

	back	finger	knee	hand	other (9 items)	arm	eye	shoulder	ankle	head	foot	leg	body	face	wrist	elbow	neck	abdomen	chest	stomach
Not Reported	6	12	3	5	1	3	7	2	1	2	3	1	3	2	0	1	0	1	0	0
No Lost Time	6	1	3	3	1	0	2	0	0	0	1	0	0	1	0	1	1	0	0	0
Lost Time	45	10	15	3	5	5	5	8	9	4	3	3	2	1	2	0	3	2	2	2
First Aid	27	32	8	15	15	13	5	7	4	8	5	6	3	3	5	4	2	1	2	2
Cumulative Percentage	22.8	37.7	45.5	52.6	58.5	64.2	69.4	74.0	77.8	81.6	84.8	87.5	89.7	91.6	93.5	95.1	96.7	97.8	98.9	100.0

Figure 2. Frequency analysis by body part

2.1 *Analysis by Body Part*

The initial analysis of the data from the Naval Weapons Station, Seal Beach focused on which part of the body was most frequently injured and how sever were the injuries. A Pareto chart was created for frequency of injuries by body part (Figure 2). The chart shows the body part injured on the x-axis and frequency of occurrence and cumulative percentage of occurrence on the two y-axes. Review of the figure shows that the most commonly injured body part is the back, the second most common is the finger, and the third most common is the knee. The stacking of bars based on the medical attention required does show, however, that the most common injuries were back injuries requiring lost time (45 cases) and a finger injuries requiring first aid (32 cases). The trivial few (although no safety issue can truly be categorized as trivial) body parts injured ranged from hand to stomach but represented over 50% of the total accidents. The Pareto principal was somewhat visible for this date. In order to most dramatically reduce accident occurrences at the base, safety management efforts should focus on protection of the back, fingers, and knees.

The research team then considered the amount of time lost due to accidents for each body part. A Pareto chart for lost hours by body part injured was created (Figure 3). The chart again shows the body part injured on the x-axis, but the left y-axis shows the total time lost due to accidents for each body part. The second, right, y-axis shows a percentage of cumulative time lost. Review of the figure shows that the injured body part requiring the most time lost from work is the back, the second time loss required comes from the finger, and the third most time loss comes from knee injuries. These are the same three body parts identified through the frequency of injury study (Figure 2), but the cumulative percentage attributed to back injuries is doubled from 20% to 40%. The "trivial" few body parts injured (with respect to lost hours) now represent only 30% of the total accidents. The Pareto principal was more valid in this case. Again, in order to most dramatically reduce lost work hours at the base, safety management efforts should focus on protection of the back, fingers, and knees.

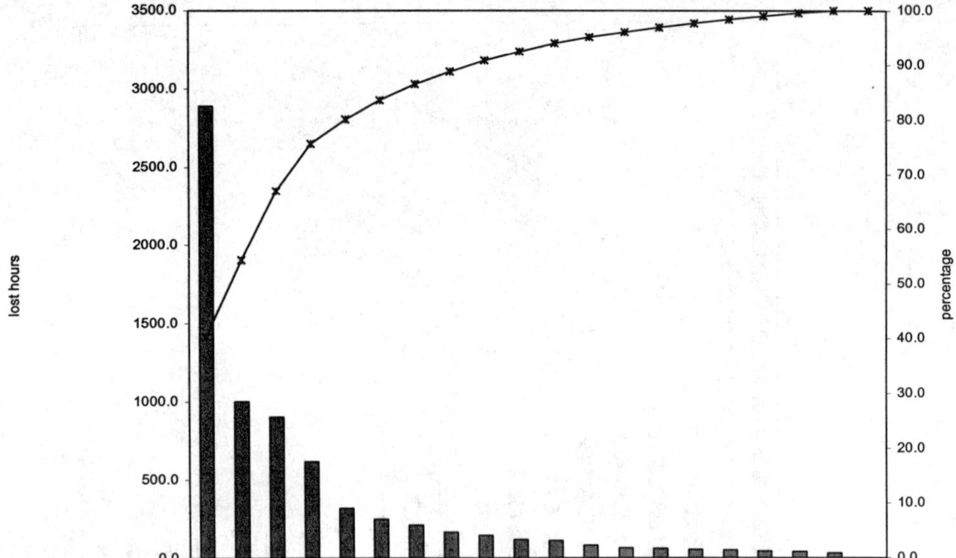

	back	finger	knee	hand	other	arm	eye	shoulder	ankle	head	foot	leg	body	face	wrist	elbow	neck	abdomen	chest	stomach
Lost Time	2889.5	1003.0	904.8	620.0	318.0	248.0	209.5	164.5	144.0	117.5	110.5	82.5	64.0	60.0	54.3	51.0	44.0	40.0	32.0	0.0
Cumulative Percentage	40.4	54.4	67.0	75.7	80.1	83.6	86.5	88.8	90.8	92.5	94.0	95.2	96.1	96.9	97.7	98.4	99.0	99.6	100.0	100.0

Figure 3. Lost hours analysis by body part

2.2 Analysis by Injury

The second type of analysis of the data from the Naval Weapons Station, Seal Beach was construction of a Pareto chart for the body part injured. A Pareto chart was constructed for frequency of injury based on type of injury (Figure 4). The chart shows the type of injury on the x-axis and frequency of occurrence and cumulative percentage of occurrence on the two y-axes. Review of the figure shows that the most common type of injury is strain/sprain, the second most common is cut, and the third most common is contusion. The most common injury type – medical attention combination was lost time – strain/sprain (73 occurrences). The "trivial" few injuries ranged from trauma to dislocations and account for only 30% of all types of injuries. The Pareto principal was well demonstrated by this data.

The research team then considered the amount of time lost due to accidents for each body part (Figure 5). The chart shows the type of injury on the x-axis, but the left y-axis shows the total time lost due to each type of injury. The right y-axis shows a percentage of cumulative time lost. Review of the figure shows the best, most dramatic illustration of the Pareto Principal. Two types of injuries accounted for 80% of the total lost hours. These two types of injuries were sprain / strain and cuts. Hence, in order to most dramatically reduce lost work hours at the base, safety education and protection efforts should focus on how to avoid strains/sprains, cuts, and contusions.

2.3 Pairwise Analysis

Analysis showed that the back is the most critical body part to protect, while sprain/strain is the most critical injury to be prevented against. Analysis then questioned whether most common combination is backs that were sprained/strained backs. A pairwise analysis Pareto chart for frequency of injuries (body part to injury type) was conducted to check this hypothesis (Figure 6).

The Pareto chart shows the body part injured on the x-axis and frequency of occurrence and cumulative percentage of occurrence on the two y-axes. The chart was created for all data pairs which had over five occurrences. Several dozen pair combinations occurred only one or two

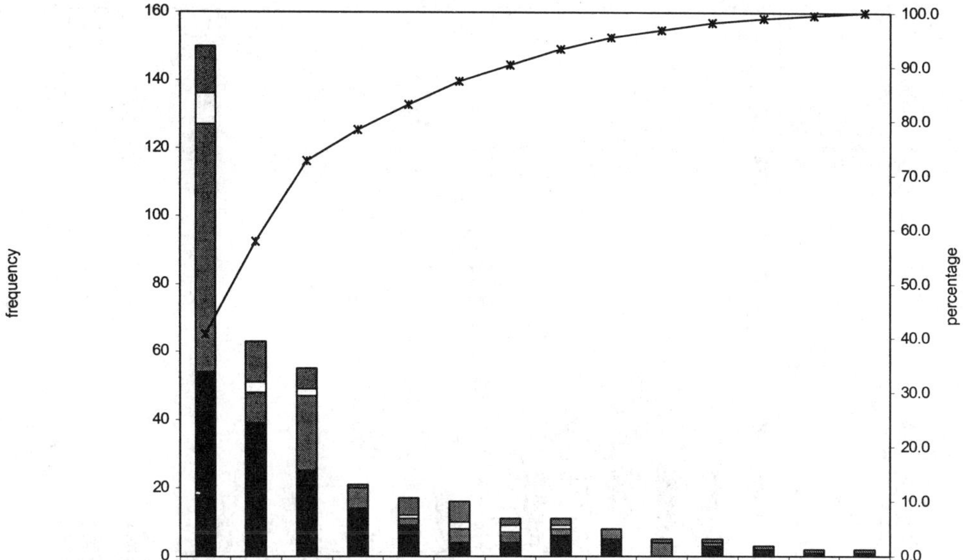

	strain / sprain	cut	contusion	trauma	puncture	splinter	bites	other	crushing	fracture	burns	irritation	hernia	dislocation
Not Reported	14	12	6	1	5	6	2	2	3	1	1	0	0	0
No Lost Time	9	3	2	0	1	2	2	1	0	0	0	0	0	0
Lost Time	73	9	22	6	2	4	3	2	0	4	1	1	1	1
First Aid	54	39	25	14	9	4	4	6	5	0	3	2	1	1
Cumulative Percentage	40.7	57.7	72.6	78.3	82.9	87.3	90.2	93.2	95.4	96.7	98.1	98.9	99.5	100.0

Figure 4. Frequency analysis by injury type

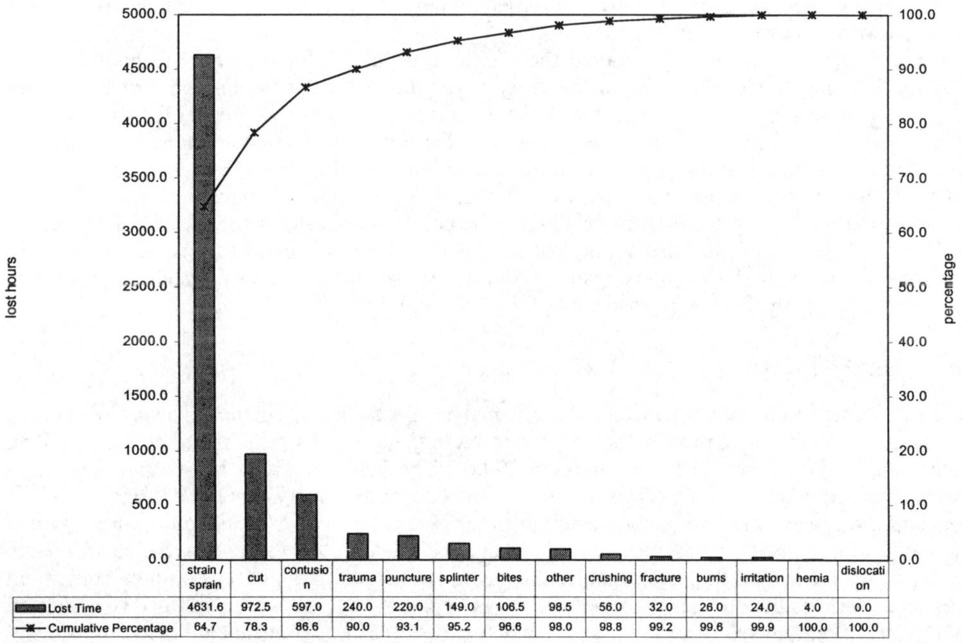

	strain / sprain	cut	contusion	trauma	puncture	splinter	bites	other	crushing	fracture	burns	irritation	hernia	dislocation
Lost Time	4631.6	972.5	597.0	240.0	220.0	149.0	106.5	98.5	56.0	32.0	26.0	24.0	4.0	0.0
Cumulative Percentage	64.7	78.3	86.6	90.0	93.1	95.2	96.6	98.0	98.8	99.2	99.6	99.9	100.0	100.0

Figure 5. Lost hours analysis by injury type

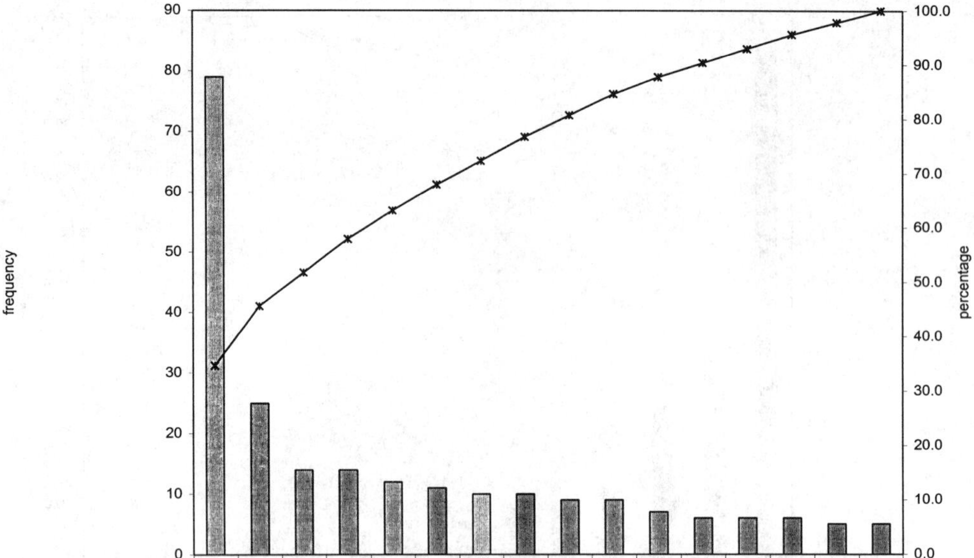

	back - sprain / strain	finger - cut	knee - sprain / strain	eye - splinter	knee - contusion	shoulder - sprain /	ankle - sprain / strain	finger - contusion	hand - cut	head - cut	hand - contusion	neck - sprain / strain	finger - puncture	finger - crushing	wrist - sprain / strain	arm - sprain / strain
No. of Accidents	79	25	14	14	12	11	10	10	9	9	7	6	6	6	5	5
Cumulative Percentage	34.6	45.6	51.8	57.9	63.2	68.0	72.4	76.8	80.7	84.6	87.7	90.4	93.0	95.6	97.8	100.0

Figure 6. Frequency analysis by pairwise body part - injury type

times and were not shown in the analysis. Review of the figure shows that the most common combination is in fact the back – sprain/strain pair, the second most common is the finger -- cut combination, and the third most common is a tie between the knee – sprain / strain and the eye -- splinter pairs. The remaining few pairs represented only 40% of the total accidents. The Pareto principal was somewhat valid in this case.

The research team then considered the lost time accidents for the pairwise combinations (Figure 7). The Pareto chart shows the body part injured on the x-axis and lost hours and cumulative percentage of occurrence on the two y-axes. The chart was created for all data pairs which had over 100 hours of lost time. Review of the figure shows that the most hours are lost from the back – sprain/strain pair. The second most hours are lost from the knee – sprain/strain pair, and the third most hours are lost from the knee – contusion combination. The finger-cut combination is not even inclused in the chart. The back – sprain/strain combination represented 50% of the lost hours -- the Pareto principal was valid in this case. In order to most dramatically improve safety at the base, safety management efforts should focus on methods to protection against sprains of the back and protection of the knee and finger.

3 SAFETY MANAGEMENT ACTIONS

The researchers recommend an accident chain analysis for the Naval Station. The accident chain accident management approach focuses on correcting the events which lead to an accident, rather than the accident itself (Loosemore 1998). The Pareto analysis has shown that backs being sprained or strained is a significant problem. Rather than focusing on the lifting itself (the reason for the accident) and recommending lifting belts and proper lifting procedures (which probably already occurs), the accident chain analysis focuses on the events that lead to one trying to lift more than they are capable. Management actions should focus on how equipment is stored (perhaps the bundles/packages are too heavy), how the loads are packaged (the pick-up points may be in awkward locations), and whether proper lifting assistance devices are being used (hoists, pulleys, or lifts). An accident chain analysis, however, is never simple. A study of the problem of knee injuries at the facility would likely show hundred of possible events which led to the injury. Other issues beyond the event chain may also be of value to

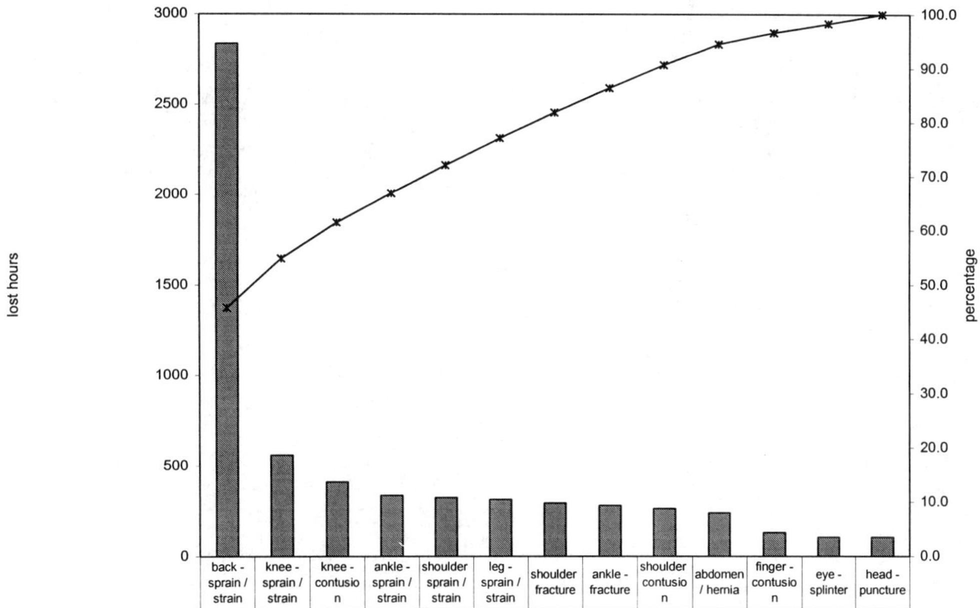

	back - sprain / strain	knee - sprain / strain	knee - contusion	ankle - sprain / strain	shoulder sprain / strain	leg - sprain / strain	shoulder fracture	ankle - fracture	shoulder contusion	abdomen / hernia	finger - contusion	eye - splinter	head - puncture
Lost Time	2838.5	560	411	336	323.8	313	293	280	264	240	130.5	106.5	106
Cumulative Percentage	45.8	54.8	61.4	66.8	72.1	77.1	81.8	86.3	90.6	94.5	96.6	98.3	100.0

Figure 7. Lost hours analysis by pairwise body part – injury type

management plans. Perhaps worker age is an impact, particularly at the Naval station where many injuries occurred to staff over 50 years old. Additional research is needed in these areas.

4 CONCLUSIONS

This work has described an application of the Pareto principle to accident data and has shown that, accident frequency and losses are distributed in a way such that a vital few types of accidents and types of injuries make up the largest portion of overall safety losses either in terms of frequency of accidents or lost work hours. This study has examined data from United States Navy weapons facility construction and maintenance operations and found sprains and strains of backs to be the most common injury, as well as the injury pair that results in the most time lost from work. A accident chain analysis is recommended to identify events which lead to the accidents. Additional research in the station to develop safety management plans and to further study the safety data is recommended.

REFERENCES

Graves, R. 1993. Total Quality -- Does it work in engineering management? *J. of Mgmt. in Engrg.* ASCE. 9(4):444-455.

Juran, J. M. 1989. *Juran on leadership for quality.* New York:Free Press.

Kuprenas, J. A. 1998. A Pareto analysis of spacecraft component manufacturing defects. *Project Management Journal.* Project Management Institute, accepted for publication.

Loosemore, M. 1998. Psychology of accident prevention in the construction industry. *Journal of Management in Engineering.* ASCE. 14(3):50-56.

23 Program implementation and audits

Implementation of Safety and Health on Construction Sites, Singh, Hinze & Coble (eds)
© 1999 Balkema, Rotterdam, ISBN 90 5809 036 1

Safety and health design and management on the construction site: Guidelines for roles and application tools

Arie Gottfried & Marco L.Trani
Building Engineering and Territorial Systems Department, Polytechnic of Milan, Italy

ABSTRACT: This paper provides some guidelines which enable a safety-oriented organization of the building process during the design and execution phases. The guidelines have been structured in order to comply with the Italian legislative Decree No. 494/1996, which implements the European Directive 92/57/EEC concerning workers safety and health on temporary or mobile construction sites. The aim is to organize and support relationships between safety managers, building-site technicians and work-site foremen in order to maximize the positive effects of their co-operation.

1 INTRODUCTION

Recent years have witnessed profound innovations in measures in defence of the safety and health of workers in Italy, with the adoption in the *corpus* of national standards of a series of European directives, namely the EEC framework directive 89/391/EEC and the nine applicatory directives linked to it. These directives have been adopted in Italy through a number of legislative acts: the l.Dec. No. 626/1994, No. 493/1996 and No. 494/1996. The basic aim of the first is the introduction of the new approach to safety in work places in general, through the institution of "risk assessment". The second regards specifically signs and notices that are to be displayed in work places. The specific purpose of the third decree regards temporary or mobile construction sites for works of building or civil engineering. The l.Dec. No. 626/1994 and the l.Dec. No. 494/1996 thus render the new European approach to the safeguarding of workers operative in Italy by introducing new professional figures and by instituting the obligation to draw up a series of safety-oriented documents for the planning of working activities. The l.Dec. No. 494/96, for the first time in Italy, makes the Client responsible for the safety of building workers. This responsibility may be, at the Client's own discretion, shared with a Work Safety & Health Supervisor, and must be delegated, in certain codified cases, to the Planning-phase and to the Execution-phase Coordinators. In addition, the decree rounds out the design documentation with the introduction of the Safety and Coordination Plan, the General Safety Plan, and the Building Maintenance Safety File. It therefore emerges clearly how such an abundance of new professional figures and documents entails a set of practical tools for the synergistic and effective management of worker safety and health on construction sites.

2 SAFETY DESIGN TOOLS

2.1 *The Operating Design*

The need for methods and tools for managing safety in the context of a building process has aroused interest, in Italy, for a specific design discipline: Operating Design.

The operating design of a building process is a set of documents aimed at defining the specifications and indications for the procedures of execution of the various parts of the technological system of a building. By its very nature, this *corpus* of documents must be prepared jointly by the professional responsible for the executive design and the Planning-phase Coordinator, each of these within his own specific sphere of competence. The Operating Design documents are aimed at the choice of construction methods – and methods of building production organization – capable of meeting the qualitative expectations expressed in all the previous design phases. It is therefore a matter of optimizing the material, human and economic resources that are to be made available for use in the subsequent executive phases of the building process.

On the basis of the foregoing considerations, it is therefore possible to characterize a building intervention also from the point of view of the operating quality that this is able to present. This kind of response in terms of performance regards the capacity of the technical solutions and the products chosen to satisfy the expected levels of a performance programme regarding the reduction in the resources and times necessary for their handling. In particular, the operating quality assumes as indispensable requisites all the aspects that concern safeguarding of workers on the construction site, which may be expressed in terms of working safety, health and hygiene. The operating quality of a design thus becomes the terrain for a comparison of proposals and synergistic action between the Designer and the Planning-phase Coordinator.

Operating design is to be understood in a dynamic sense; it must in fact be organized in two phases distinct in time but linked together. Both these phases characterize the aspects of executive safety of the works. The first phase is connected to the decisions prior to a contract, and hence involves the making of choices that render it possible to do without the actual organization of building production, since the firm that is to carry out the building is unknown *a priori*. The second phase regards the final operating choices, and hence the ones that concern the actual conditions of the production context of an intervention. The choices are tied both by the prescriptions represented by safety priorities of the various processes (or of the individual activities that make up these processes, as expressed in the operating schedules of the Safety and Coordination Plan), and by the conditions of acceptance regarding the structure to be built, which are specified in the appropriate operating plans (Gottfried, 1997a).

2.2 *Quality and operating safety requisites*

In order to manage the executive safety of a building process it is therefore necessary for the operating design to assume a construction-site system as a set of spatial elements equipped for production and/or at the service of production (work places), linked to one another by conditions of relationship, with each element being aimed at a given physical transformation of the environment in which it is located. In other words, by availing oneself of the instrumentation proper to the operating design of an intervention, it is possible to investigate the conditions of relationship of a Worker/Machine/Environment (W/M/E) system, assigning minimum requisites of interface between the elements of the system itself. The spatial and temporal compatibility of the processes, as also the behavioural models of the workers and the suitability of the materials, means and equipment, represent in fact the key safety aspects of a construction site (Trani, 1996).

Taking as an example the construction of the complex front of a building, the Planning-phase Coordinator is called upon to identify the corresponding W/M/E system, breaking it down into parts reflecting the individual working places for construction/maintenance operations (Environment), to be equipped with machinery and systems suitable for the processes envisaged (Machine), which are to be carried out by workers appropriately trained for the task and adequately informed about surrounding conditions, in terms of time and space, that characterize their activity (Worker).

In particular, the study of the Man-Machine relationships presupposes a joint evaluation made between the Designer and the Coordinator regarding the technological choices adopted by the Designer as the activity of executive design progresses. For this purpose, it is possible to refer, from the instrumental point of view, to two lists of requisites associated to the needs of

Table 1. Requisites of operating quality

1.	*Dimensional flexibility*: capacity of the product for adapting to the design dimensions without adjustments on the construction site.
2.	*Dimensional adaptability*: capacity of the product for being adapted to the design dimensions via processes on the construction site.
3.	*Limitation in the use of equipment*: capacity of the product for being transferred and worked on the construction site with a minimal use of equipment.
4.	*Limitation in the use of materials and tools*: capacity of the product for being transferred and worked on the construction site with a minimal use of accessory materials and tools.
5.	*Ease of use of the product*: capacity of the operating characteristics of the product for enabling a good laying or assembly precision with few transformations.
6.	*Limitation of overall dimensions*: capacity of the product for being handled and moved on the construction site with limited space for movement.
7.	*Production of toxic residue:* capacity for the reduction of hazards for the operator deriving from the toxicity of the materials used during work processes.
8.	*Resistance to agents*: capacity of the product for resisting environmental agents or agents deriving from laying operations.
9.	*Limitation of extent of checks*: capacity of the product for being treated properly, with easy and fast checks, in the various operating phases.
10.	*Limitation of size of labour force*: capacity of the product for being worked on the construction site with the use of a limited number of operators.
11.	*Non-tiring of operator*: capacity of the product for not leading to operator tiredness during the operating phases.
12.	*Safety of use*: capacity of the product for being transferred and worked on the construction site with the highest degree of safety as far as machinery is concerned.

rationalization and optimization of the constructional/maintenance processes. The first list (Table 1) concerns the requisites of operating quality of products, components, technical elements and production technologies (Trani, 1997a); the second (Table 2) concerns the operating safety requisites in the use and management of the equipment or of the potentially harmful agents (Trani, 1997b). These lists enable the specification of the expected performance levels as regards health and safety in the course of the work. The list of requisites of operating quality, in particular, proves very useful for structuring the relationships between Designer and Planning-phase Coordinator.

For example, we shall consider the first four requisites of Table 1. The reduction, in general, of the adjustment operations, as well as the limitation of the necessary equipment or of the auxiliary products, or else the real possibility of adapting a technical element on the work site without having to resort to operations that are unsuitable for the situation of the operating context, in actual fact diminishes the number of possible material agents that might be a cause of injury or illness with which the operator could come into contact in a given "working place" of the construction site. A choice in this direction, then, has a positive effect on the overall danger potential inherent in a given process that characterizes a work place.

The ease of use of a product (requisite No. 5), on the other hand, by reducing the number of "constructive acts" necessary for laying the product in place, may have a positive bearing, for example, on the duration of exposure to a possible hazard linked to it that cannot be eliminated.

The limitation of overall dimensions (requisite No. 6), if respected, may in turn favour the exploitability of the operating areas of a construction site.

The production of toxic residue (req. No. 7), whether this be in the form of a solid, liquid or gas, must be taken into account also in assessing the environmental impact of the building site.

Table 2. Requisites of operating safety

Requisites for hazardous situations with user active	
• Preventing contact with ...	• Preventing piercing oneself with …
• Preventing treading on …	• Preventing bumping against …
• Preventing hitting oneself with …	• Preventing the possibility of tripping …
• Preventing ingestion of …	• Control of uncoordinated movements …
• Effort control involved in handling of heavy objects...	• Preventing getting caught up in or hooked onto …
• Effort involved in handling of light objects …	
Requisites for hazardous situations with user passive	
• Preventing getting grabbed by …	• Preventing getting pierced/stung by …
• Preventing getting hit by …	• Preventing getting crushed by …
• Preventing getting banged by …	• Preventing getting submerged by …
• Preventing getting run over by …	• Preventing getting buried/swept away by …
• Preventing getting bitten by …	• Preventing getting jammed between …
Requisites for specific situations of environmental risks	
• Preventing inhalation of …	• Preventing exposure to …
Requisites for specific situations involving risk of falling	
• Preventing falling from a height …	• Preventing falling down into …
• Preventing falling on the flat …	
Requisites regarding the use of means of transport	
• Preventing accident on board …	• Preventing accident at the wheel of …

The resistance to environmental agents, whether physical or chemical, during the storage period or in the phase of laying in place (requisite No. 8) must be traced by the Coordinator via reference to the product technical sheets. It may in fact occur, for instance, that packaging that has been subjected to extreme weather conditions is no longer able to guarantee adequate levels of safety during handling.

The limitation of the labour force and of the extent of the checks (requisites Nos. 9 and 10), in addition to the evident implications of an economic nature, contributes to the reduction of the subjects exposed to a given risk, thus affecting any possible interference between activities that are contiguous in space.

The requisite of avoiding undue tiring of the worker (requisite No. 11) refers the Coordinator to a verification regarding compliance with the provision of the Legislative Decree No. 626/1994 concerning manual handling of loads.

Finally, if on the one hand safety in use (requisite No. 12) might represent a requisite that embraces the previous requisites, on the other hand it may assume a specific place of its own alongside the other requisites. In certain conditions, for example, it may be useful to opt for a supply of ready-mixed concrete, thus eliminating the hazards associated to its production and handling on the construction site.

The requisites of operating safety, on the other hand, provide an aid to the Coordinator in foreseeing the possible harmful events that may be associated to the envisaged processes as regards the material agents of injury or illness which are necessary for carrying out construction-site operations. The list presented in Table 2 has been drawn up on the basis of the mechanism for recording accidents adopted by INAIL. This list enables a more reliable assessment of risk, also owing to the availability of the Institute data, which are divided into temporary injuries, permanent injuries, and fatal injuries.

2.3 *Safety Operating Schedules*

The heart of the safety operating design is represented by the safety operating schedules and the risk assessment schedules contained in the Safety and Coordination Plan. In general, these sheets refer to the work items and to the processes identified in the Standard Work Plan drawn up by the Work Safety and Health Supervisor (e.g. work item: vertical structure; work process: north side columns). Here we shall concentrate our attention on safety operating schedules.

Drawing-up of the safety operating schedules involves the analysis of the working places associated with the spatial elements of the construction-site system. In this phase, the Coordinator must thus pay particular attention to the following points: (a) access (preparation of horizontal and vertical pathways for workers and machinery: tracks, ramps, gangways, catwalks, ladders, etc.); (b) boundaries (vicinity of empty spaces or spatial elements equipped for processes or activities that may interfere); (c) exploitability (availability of spaces suitable for carrying out the activities that make up a particular process).

Figure 1 provides an example of a safety operating schedule regarding a process to be carried out on a construction site for the restructuring of a building. Note, first of all, the contextualization of the schedule as regards design documentation via a set of alphanumeric codes which can enable loop retrieval of work order data.

In particular, the empty spaces in the schedule, which are available for the identification of the subcontractor (Company title), employer and site foreman, indicate that these data must be entered when the contract is assigned, in order to facilitate the activity of the Coordinator and of the main Contractor during execution of the work. The purpose is to facilitate coordination between the various subjects present on the construction site during building processes and exchange of information between employers.

The second section of the schedule envisages the description of the process, inferred from the special tender specifications with the addition of operating details referring to particular measures aimed at worker safety and health required for carrying out the process (e.g., protective sheets and scaffolding, as illustrated in Fig. 1). This description may be useful also for separating the amount for accident-prevention and protection devices off from the tender base price for the work process, so as to prevent the discount offered by the building firm from adversely affecting levels of safety at the construction site. In fact, it is not possible to allow tender rebates in the sphere of worker safety.

SAFETY OPERATING SCHEDULE No. 07	WBS code: 3.1 Dwg. Reference: D-04
Work item	DEMOLITION
Process	Ground-floor foundation slab demolition for accommodating elevator platform for disabled persons
Encharged Firm	(Company title)
Employer	(Name)
Site Foreman	(Name)
Description	Demolition of cement-brick slab, using electromechanical and manual means, for creating rectangular space for accommodating elevator platform for disabled persons to provide connection with basement. Includes preparation of complete boarding on trestle at intrados of slab for accumulation of rubble to be removed. Includes preparation of polyethylene sheets in basement for collecting dust. Manual loading of rubble for subsequent removal.
Activity Priorities	A.1 Demolition ⇒ Check lay-out and need for shoring with Works Manager; ⇒ Check stability of provisional structures; ⇒ Check that the boards of the floor system are set close together for subsequent accumulation of rubble; ⇒ Check conditions of ventilation; ⇒ Check that no workers are present in the basement.
Activity Priorities	A.2 Laying-out of sheets ⇒ Check suitability of the provisional structures necessary for their positioning (double ladder).
Activity Priorities	A.3 Removal of rubble ⇒ Check suitability of area of temporary accumulation and of dump-bodies/containers; ⇒ Prepare a gangway suitable for unloading into dump-body of lorry

Figure 1. Example of Safety Operating Schedule

The Design Coordinator must then split the building process into a series of elementary activities that enable identification of the possible sources of danger and the consequent operating priorities for ensuring safety. These priorities constitute internal safety controls regarding the work place. The procedures for proper use of the machinery and equipment should, in fact, be contained in the Risk Assessment Document of the Contractor and individual Sub-contractors. In the event of the Design Coordinator deeming that given safety measures contribute to the reduction in residual risk characterizing the process in question, the firm must adapt its own production procedures to these specifications, unless it is able to show that it can achieve the same level of safety, at the same cost, using alternative technologies with which it is more fully acquainted. It remains to be noted that, should the binding choices made by the Client in favour of safety lead to an increase in production costs, the firm must be acknowledged, at the level of estimatory calculation of safety operation costs, the heavier economic burden it has to face. This acknowledgement, which is compulsory by law, is another of the novelties introduced in Italy by the Legislative Decree No. 494/1996.

The priorities and safety measures indicated in the Safety Operating Schedules must necessarily take into consideration the Standard Work Plan prepared ahead of the Operating Design, without the actual production reality of the firm carrying out the work being known. Were there to arise, in the course of the work, significant modifications to the work schedules envisaged, such as to reveal situations of interference between the various processes, the Execution-phase Coordinator will find himself faced with two alternatives. On the one hand, he can act on the work schedule slippage allowed by the programme, in such a way as not to jeopardize the critical path of the P.E.R.T. grid. Or else, he can take into account the interference, adding specific operating prescriptions to the Safety Operating Schedules in order to eliminate or reduce the risk due to the interference noted. The safety operating design must therefore be strictly correlated to the preventive operating programming activity and to its subsequent updates in the course of execution (Gottfried, 1997b).

2.4 *General Safety and Health Plan*

The safety plan for the construction site must be filled in with a set of information of a general nature on the characteristics of the building context and with a list of minimum safety and health requisites that the Contractor must meet at the moment of organizing the work. In this connection, the Legislative Decree No. 626/1994 sanctions the obligation, on the part of the employer, to develop a thorough-going design of the construction site. Table 3 provides an example of the contents regarding the general part of the safety operating design to be integrated with the information of the Safety and Coordination Plan.

Table 3. Contents of the general part of the Operating Design

Section A - Context and environmental conditions in surrounding area
A.1 - <u>Intrinsic characteristics of area</u>: Geomorphological characteristics of soil; Underground structures and system subservices; Non-removable ties above ground; Overhead electricity lines; Emission of polluting agents into the environment in the direction of the construction site; Projection and falling of objects into construction site.
A.2 - <u>Environmental impact of construction site</u>: Emission of polluting agents; Interference with neighbouring sites; Interference with neighbouring buildings and conditions of road/track; network in surrounding area; Projection and falling of objects outside construction site.
Section B - Requisites of construction site
B.1 - <u>Boundaries and conditions of road/track network</u>: Fencing; Access to site; Road/track network conditions.
B.2 - <u>Services provided on construction site</u>: Logistics services; Hygiene-health services.
B.3 - <u>Storage of materials and semi-finished products</u>: Procurement and storage; Stores/depots.
B.4 - <u>Stable working sites</u>: Working of steel; Working of wood; Mixing of mortar; Mixing of plaster; etc.
B.5 - <u>Construction-site systems</u>: Electric system; Water system; Hoisting equipment.
B.6 - <u>Waste disposal</u>: Solid waste; Liquid waste.
B.7 - <u>Assistance in emergencies</u>: Health and first-aid facilities available on site; Health services available locally; Fire Brigade; Municipal Police; Local Health Welfare Corporation
B.8 - <u>Fire prevention</u>: Escape routes; Fire-extinguishing means; Worker information.

3 SAFETY MANAGEMENT

During the design phase, the management aspects of the safety documentation in practice amount to a good synergy and exchange of information between the Designer and the Coordinator. This synergy may be implemented through a constant exchange of opinions between the two professional figures on the technical solutions adopted, starting from the list of requisites previously illustrated. The Client (or the Work Safety and Health Supervisor), in this phase, must in any case guarantee the observance of the general principles of safeguarding of workers. For this purpose, he may avail himself of the checklist presented in Figure 2, applying it to the work processes necessary for carrying out the work.

Before starting construction-site operations, it is advisable to call a meeting of coordination at which the following persons must be present: the Execution-phase Coordinator, Works Manager, Site Technical Manager, and Site Foreman, as well as the persons responsible for any subcontractors. The aim is to enable the firm carrying out the work to present a request for variations in certain items of the Safety Plan, based upon the firm's own experience and production capacity. The variations that are approved, however, cannot entail modifications of the contract price set.

SAFETY-PLAN CHECKLIST			
Work item:			
Process:		Y	N
1)	Have the risks for the health and safety of workers been assessed?	☐	☐
2)	Have the risks been eliminated taking into account the know-how acquired on the basis of technological progress?	☐	☐
3)	Have the non-eliminable risks been reduced to the minimum, preferably at the source?	☐	☐
4)	Has prevention been programmed, integrating it in the technical, production, organizational and environmental conditions characterizing the process?	☐	☐
5)	Have the sources of danger been eliminated?	☐	☐
6)	Have the non-eliminable sources of danger been replaced with less dangerous ones?	☐	☐
7)	Have the principles of ergonomics been observed in the design of the work places?	☐	☐
8)	Have the principles of ergonomics been observed in the choice of equipment?	☐	☐
9)	Have the principles of ergonomics been observed in the definition of work methods?	☐	☐
10)	Have monotonous or repetitive work methods been avoided or attenuated?	☐	☐
11)	Has priority been given to collective protection measures as against individual ones?	☐	☐
12)	Has the number of workers who are or may be exposed to a risk been limited?	☐	☐
13)	Has the use of chemical, physical or biological agents been limited in the work places?	☐	☐
14)	Has an (anonymous) verification been made on the results of the health checks on workers?	☐	☐
15)	Has a procedure been envisaged for removing workers from the work site for individual health reasons?	☐	☐
16)	Have the necessary measures of work hygiene been provided?	☐	☐
17)	Have individual protective devices been provided?	☐	☐
18)	Have emergency measures to be adopted for first-aid been envisaged?	☐	☐
19)	Have emergency measures to be adopted for fire-fighting been envisaged?	☐	☐
20)	Have emergency measures to be adopted for evacuation of workers been envisaged??	☐	☐
21)	Has the use of warning and safety signals been envisaged?	☐	☐
22)	Has regular maintenance of work places been envisaged?	☐	☐
23)	Has regular maintenance of systems, machines and equipment been envisaged?	☐	☐
24)	Has a check on the degree of worker training been envisaged?	☐	☐
25)	Has a procedure for proper information of workers regarding the environmental hazards to which they could be exposed been envisaged?	☐	☐
26)	Has a procedure for proper training of workers as regards work methods been envisaged?	☐	☐

Figure 2 – Example of safety-plan checklist

The activity of coordination during the execution phase is characterized by three distinct events: coordination meetings, on-the-spot inspections (safety audit), and inspections on paper. Approximately one week before the start of a new work process, the Coordinator calls a meeting with the aim of getting a clear picture of the state of progress of the work, in connection, above all, with the simultaneous presence of a more than one firm on the construction site. In practice, the aim is to define the situation in order to check whether the measures contained in the Safety Plan remain valid or whether it is necessary to make any modifications. In the latter case, on the ensuing days the Coordinator shall see to formalizing the variations agreed upon, re-drafting the safety operating schedules and updating the work programme.

In the course of the operations, the Coordinator must make a series of on-the-spot inspections with the aim of verifying whether the systems of prevention, protection and coordination contemplated in the Safety Plan have been implemented. For each visit, a record is drawn up aimed at providing useful indications for the progress of the work process and for checking the frequency of any non-compliance that has been encountered. In the event of serious and imminent danger, the Coordinator has the power to stop operations immediately. In less serious cases (e.g., absence of compulsory documentation regarding safety to be kept on the construction site), some gradual mechanisms of sanctions must be envisaged for the firm, which are to be applied at the discretion of the Works Safety and Health Supervisor. For this reason, the latter must in all cases receive a copy of the records of the coordination meetings and on-the-spot inspections in order to be able to verify that the requirements of the checklist shown in Fig. 2 have been met also during the execution phase.

4 CONCLUSIONS

Apart from their objective complexity in terms of application, the new laws regarding worker health and safety will certainly have, in the medium and long term, a positive effect as regards management of the building process in a broad sense. This effect will be all the more evident and immediate, the better equipped the market will be for selecting building firms, also as regards the latter's effective capacity for operating safely (and at a high qualitative level), and not merely on the basis of the most "advantageous" offer. A cultural growth of the building market is thus called for, so that the sector may be able to understand that a price that is more "advantageous" at the expense of quality and safety, in fact is not truly advantageous. The work of the new professional figures represented by the Works Safety and Health Manager and the Safety Coordinators introduced into the building process ought to be of considerable help in this direction.

On their own part, the building firms will have to devote increasing attention to the training of workers, equipping themselves with a true Safety System, to be integrated, as the needs arise, with the Safety Operating Design produced by the Client and contextualized for the individual construction site.

REFERENCES

Gottfried, A. 1997a. *Basis of Operating Design.* In A. Gottfried & M.L. Trani (eds), The Construction Planning and Execution phase Coordinator: 459-475. Rimini: Maggioli.

Gottfried, A. 1997b. *Basis of Operating Program.* In A. Gottfried & M.L. Trani (eds), The Construction Planning and Execution phase Coordinator: 477-501. Rimini: Maggioli.

Trani, M.L. 1996. *Workers' Safety and Health management in construction industry.* Ph.D. Thesis, ISET Dept. Polytechnic of Milan.

Trani, M.L. 1997a. *Safety Design and Management.* In A. Gottfried & M.L. Trani (eds), The Construction Planning and Execution phase Coordinator: 503-539. Rimini: Maggioli.

Trani, M.L. 1997b. *Risk Assessment.* In A. Gottfried & M.L. Trani (eds), The Construction Planning and Execution phase Coordinator: 657-677. Rimini: Maggioli.

Implementation of Safety and Health on Construction Sites, Singh, Hinze & Coble (eds)
© 1999 Balkema, Rotterdam, ISBN 90 5809 036 1

Impact of safety programs on project performance

A.P.C.Chan, F.K.W.Wong & K.K.Lo
Department of Building and Real Estate, The Hong Kong Polytechnic University, Kowloon, People's Republic of China

ABSTRACT: Hong Kong has a bad reputation for construction safety. In 1996-97, construction site accidents accounted for more than one-third of all industrial accidents. Safety management system was introduced in government contracts in 1992 (for Airport Core Program projects) and in 1993 (for non-ACP projects). Two major safety initiatives, the Pay for Safety Scheme and the Independent Safety Audit Scheme, were introduced in 1996 to encourage the setting up and implementation of efficient safety management systems and to enhance the standard of safety performance. The study of the impact of these safety measures on project performance deserves paramount attention.

This paper reports the findings of a research study, which was undertaken to examine the impact of safety programs on project performance. The effectiveness of the safety programs, client's satisfaction on cost, time, quality, functionality and overall satisfaction; as well as design team leader's and construction team leader's satisfaction of 110 recently completed projects were measured and analysed. Using factor and stepwise multiple regression analysis impact of safety programs on project performance was identified. It was shown that projects where effective safety programs were implemented could lead to better quality, higher designer's overall satisfaction, higher client's satisfaction on cost, improved functionality, higher client's overall satisfaction and higher contractor's overall satisfaction.

1 INTRODUCTION

The number of industrial accidents in Hong Kong remains at a very high level. Despite a downward trend in the total of industrial accidents from the peak year of 1988, the construction industry continues to account for over one third of all industrial accidents and the bulk of the fatalities. The average probability of a construction worker getting involved in industrial accidents between 1985 and 1994 is 5.8 times higher than the overall average for all industrial workers. The construction industry has an average fatality rate of 0.868 per thousand workers between 1985 and 1994 that is 10.5 times higher than the overall average for all industrial workers (Education and Manpower Branch, 1995). The unsatisfactory construction safety record of Hong Kong is a matter of concern to the Government and the public alike. Table 1 shows the number of construction accidents by causes in selected items from 1994 to 1996.

This paper reports the findings of a research study, which was undertaken to examine the impact of safety programs on project performance. The effectiveness of the safety programs, client's satisfaction on cost, time, quality, functionality and overall satisfaction; as well as design team leader's and construction team leader's satisfaction of 110 recently completed projects were measured and analysed. Using factor and stepwise multiple regression analysis impact of safety programs on project performance was identified.

Table 1: Number of construction accidents by causes in selected items from 1994 to 1996
Data Source: Report of the Commissioner for Labour, 1995-97

Causes		1994	1995	1996
Machinery	a) Prime mover & transmission	13	12	18
	b) Working and other parts	682	707	800
Transport	a) Power driven	54	56	54
	b) Non power driven	3	16	34
Explosion of fire		100	97	85
Hot or corrosive substances		235	149	206
Gassing, poisoning and other toxic substances		8	13	10
Electricity		33	32	38
Fall of people		2598	2412	2983
Stepping on, striking against or struck by object		6898	6905	8000
Falling object		1125	910	967
Fall of ground		8	2	11
Handling with machinery		3703	3204	2532
Hand tool		539	561	613
Miscellaneous		423	192	118
Total		16422	15268	16469

2 LEGISLATIVE FRAMEWORK FOR SAFETY AND HEALTH AT WORK

The principal legislation for safety and health at work in Hong Kong is the Factories and Industrial Undertakings Ordinance (FIUO), Cap. 59. Under the Ordinance, 27 Regulations have been made to prescribe detailed safety and health standards on work situations, plant and machinery, processes, and substances. The FIUO is enforced by the Labour Department. The focus of the FIUO was originally standards setting and enforcement. An amendment in 1989 introduced the general duties provisions in line with an international trend towards self-regulation. Under the provisions, the employers are responsible for taking all reasonably practicable steps to ensure the health and safety of all persons employed at the workplace. The workers also have the duty to exercise reasonable care at work and cooperate with the employers on safety measures.

In 1986 new regulations were introduced under the FIUO to require the employment of safety officers and safety supervisors at construction sites. In 1988 the Occupational Safety and Health Council (OSHC) with tripartite membership from the Government, employers and employees was established by statute to promote occupational safety and health. New regulations under the FIUO were introduced in 1993 and 1994 to implement a certification system in the training and testing of operators of cranes and suspended working platforms respectively.

The building laws are kept under review to ensure that there is adequate safety assurance on construction sites in the private sector for the general public and site personnel. Amendments to the Building Ordinance are being proposed to provide more detailed control of contractors by reference to their qualifications to undertake works and to provide for more controls over dangerous and potentially dangerous building sites. The Buildings Department also issues practice notes to building professional and contractors to brief them regularly on safety requirements.

3 SAFETY INITIATIVES FROM THE PUBLIC AND PRIVATE SECTORS

3.1 Public sector

The onset in 1991 of the Airport Core Program (ACP) brought about a situation under which civil engineering and construction works of unprecedented scale were completed within demanding program deadlines. The Government, as employer, introduced contract conditions and various special initiatives to enhance the culture of safety at ACP sites. It also set safety standards and strictly monitored compliance. Similar action is being taken for Public Works Program (PWP) and other public works sites. ACP and PWP safety manuals were introduced by the Government in 1992 and 1995 respectively. It is already a contractual requirement for all ACP sites that all workers engaged must be given at least induction safety training. A new accident reporting system was introduced in late 1994 and any public works contractor with more than 5 convictions on breaches of safety legislation over a rolling period of 6 months may be suspended from tendering for Government works.

The Housing Authority has also taken a number of initiatives to improve safety at its construction sites in recent years. It has made access to tendering opportunities conditional on the safety performance of the contractors. The Authority also sponsors research on construction safety.

Both the Government and the Housing Authority require their contractors to employ sufficient safety officers and to implement safety measures such as safety plans, site safety committees, safety audits, and tool box meetings with workers. There are also standard terms in their contracts on safety procedures and annual safety campaigns are held to promote safety awareness on sites.

In 1993 the Construction Advisory Board was formed. Members of suitable background and experience were drawn from Legislative Councillors, trade and professional associations and institutions in the construction industry, vocational and skill training institutions, tertiary institutions and relevant Government departments. It is chaired by the Secretary for Works and responsible for advising on issues relevant to the construction industry such as education, training, safety and health. A Safety Committee has been set up under the Board.

3.2 Private sector

The private sector has also contributed to the improvement in construction safety in Hong Kong. Most notable examples are the provision of safety training to their workers by some large contractors. Some leading contractors also follow the Labour Department's initiatives to implement the concepts of "Safety Charter". The Safety Charter calls for collaboration from the contractor and the key subcontractors. A typical Safety Charter will normally be agreed and signed by all key construction team members at the outset of the construction project. It covers issues on safety policy, safety programs, safety training, safety procedures, promotion, safety audit, and penalties for non-compliance. The Hong Kong Construction Association also takes the initiative to introduce on-site safety training and issue a green card to those workers who have completed the course. The Construction Industry Training Authority (CITA) is now fully responsible for the issue of green card to construction workers, which is recognised by the Labour Department. So far 24,000 construction workers are being certified, and it is expected that another 20,000 workers will be certified in 1998.

4 DEVELOPING A RESEARCH MODEL

To study the impact of these safety programs on project performance, a research (Chan, 1996) was undertaken to reveal the underlying factors leading to the success of a construction project. A research model was developed which sets out to depict the relationships between the variables in the building process. The aim of the model is to facilitate investigation on which factors are important for achieving an outstanding project result and how these factors relate to project success. The model can also assist in determining how important are these factors when they are used to predict project success. There are six main groups of independent variables, namely, client variables, project variables, project environment variables, project team leader variables, project procedure variables and project management variables. The impact and interaction of these independent variables will determine the dependent variable, which is, in this study, project success.

5 RESEARCH APPROACH

The research approach adopted in this study comprised of a preliminary pilot study and a series of detailed case studies. The pilot study, conducted following a review of the literature, took the form of structured interviews with the industry's participants. The pilot study was conducted to gain an understanding of the construction practice in Hong Kong as well as to provide information sought in the development of the questionnaire. The main aim was to ensure that the information sought in the questionnaire would be relevant to the current practice as well as relatively convenient for the respondents to answer or "respondent friendly". The questionnaire was refined a number of times based on feed back from these discussions before it was used as a base for the structured interviews carried out at the data collection stage. The case study sample was obtained by approaching clients, professional practices and contractors in the Hong Kong industry. Those who expressed a willingness to cooperate were sent the questionnaires, followed by an interview. During the interview, the authors had the opportunity to clarify any ambiguities that the respondents might have encountered. Each of the respondents was asked to complete the questionnaire for two recently completed projects with which they were associated. The respondents were senior personnel who were able to provide informed and competent information. A total of fifty five (55) respondents agreed to contribute giving a total of one hundred and ten (110) cases.

With these 110 cases, the majority are residential and commercial projects (31.5% and 22.2% respectively). Education and Industrial developments represent the next categories on the list (14.8% and 7.4% respectively). Hotel and Health developments only represent a small portion of the projects (5.6% and 4.6% respectively). Figure 1 shows the percentage of cases by project type.

The contract values range from HK$1m to HK$2750m, with more than 65% of the projects falling in the range of HK$10m to HK$300m. Figure 2 shows the number of projects by contract value. The total number of projects recorded in the figure is 98 because there are some missing values from the research samples.

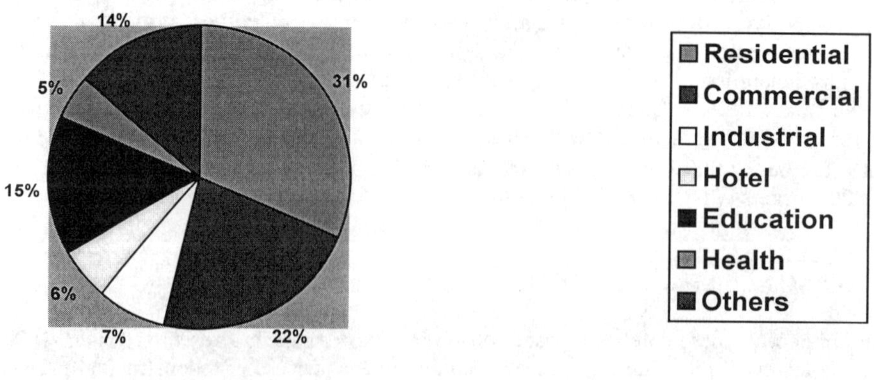

Figure1: Percentage of cases by nature of project

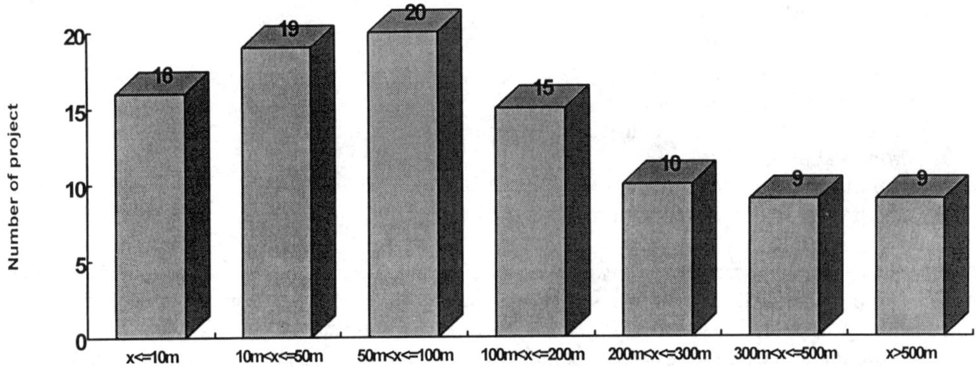

Figure 2: Number of projects by contract value HK$million

6 REVISED RESEARCH MODEL

The original research design solicited 77 independent variables, some of which may be interrelated to each other. Since the presence of large intercorrelations between the independent variables could affect the results of multiple regression analysis, factor analysis[1] is performed to group these interrelated variables to a smaller number of underlying factors prior to the multiple regression analysis. The underlying factors are each computed with a factor score, which are then fed into the multiple regression model[2] as independent variables. The multiple regression will, in turn, identify the important factors determining project success.

Principal component analysis was used to identify factors affecting project success. A principal component analysis with oblimin rotation conducted on the 77 variables produced sixteen-factor solution. The factors affecting project success are: effectiveness of design team leader; effectiveness of client's project manager; complexity; effectiveness of construction team leader; project scope; support from parent company; nature of client; environment; competency of client; client's emphasis on quality; project management actions; procurement method; client's emphasis on cost; client's emphasis on time; nature of project; and client size (Chan, 1998).

A revised research model, comprising of these 16 factors is produced to replace the original model. Each of these factors will be considered as an independent variable and fed into a multiple regression model to determine its relationship to project success.

7 RESULT OF THE ANALYSIS

Sixteen factors were measured for 110 cases and stepwise multiple regression analysis was applied to determine the relationships of these underlying factors with the performance variables. The performance variables tested were speed of construction; percentage net variation; unit cost; client's satisfaction on time, cost, quality and functionality; client's overall satisfaction; designer's overall satisfaction and contractor's overall satisfaction (Chan, 1997).

Project management action of the project team is associated with six of the ten performance measures. It has two highest beta coefficients[3] amongst other determining factors. Therefore project management action of the project team has a significant impact on project success. Examining the components of the factor analysis, project management action of the project team is measured and represented predominantly by the following eleven variables:

Implementing an effective safety program
Development of a good reporting system
Monitoring and updating plans

Development of standard procedures
Holding of regular meetings
Communication system for the project
Ability to organise an appropriate structure
Implementing an effective quality assurance program
Up-front planning efforts
Control of subcontractor's work
Overall project management action

Most of these variables are geared towards measuring the project team's ability to plan, develop, implement, monitor and control the safety programs. The two performance measures where project management actions of the project team have the highest beta coefficients are quality of the final product, and designer's overall satisfaction. The other four performance measures which were found to be significantly affected by this factor are client's satisfaction on cost, functionality of the final product, client's overall satisfaction, and contractor's overall satisfaction. It was shown that effective safety programs will result in better quality, higher designer's overall satisfaction, higher client's satisfaction on cost, improved functionality, higher client's overall satisfaction and higher contractor's overall satisfaction. Figure 3 shows the impact of safety programs on the various performance variables.

↑ Safety programs	☞	↑ Client's satisfaction on quality ↑ Designer's overall satisfaction ↑ Client's satisfaction on cost ↑ Functionality of the final product ↑ Client's overall satisfaction ↑ Contractor's overall satisfaction

Figure 3: Impact of safety programs on various performance variables

8 CONCLUSION

The number of construction accidents in Hong Kong remains at a very high level despite a downward trend in recent years. The unsatisfactory safety record in the construction industry of Hong Kong is a matter of concern to the Government and the general public. The successful development of a safety program in Hong Kong will depend on the inculcation of a safety culture among all parties concerned through an on-going and enhanced program of education and training (Wong, 1996).

This paper examined the impact of safety programs on project performance. The effectiveness of the safety programs, client's satisfaction on cost, time, quality, functionality and overall satisfaction; as well as design team leader's and construction team leader's satisfaction of 110 recently completed projects were measured and analysed. Using factor and stepwise multiple regression analysis impact of safety programs on project performance was identified. It was

---

[1] Factor analysis is a statistical technique used to identify a relatively small number of factors that can be used to represent relationships among sets of many interrelated variables.

[2] A regression model is a mathematical model, which can relate a number of independent variables to a dependent variable.

[3] Comparing the beta weights (coefficients) of each variable can reveal the relative importance of variables. Beta coefficients are the coefficients of the independent variables when all variables are expressed in standardised (Z score) form.

shown that projects where effective safety programs were implemented could lead to better quality, higher designer's overall satisfaction, higher client's satisfaction on cost, improved functionality, higher client's overall satisfaction and higher contractor's overall satisfaction.

The safety program is still an unfamiliar concept to the local industry. The employers and workers must therefore be introduced to this new concept and convinced to support its implementation. They will also need to be trained on their new roles under the safety program. Even the safety professionals should receive training to update and sharpen their skills on new safety management measures such as safety audits.

9 REFERENCE

Chan, APC (1996) Determinants of project success in the construction industry of Hong Kong. Unpublished PhD Thesis, University of South Australia.

Chan, APC (1997) Measuring success for a construction project, The Australian Institute of Quantity Surveyors – Refereed Journal, Volume 1, Issue 2, 1997, 55-59.

Chan, APC (1998) A predictive model for project success, Second International Conference on Construction Project Management, 19-20 February 1998, Singapore, 311-359.

Education and Manpower Branch (1995) Consultation paper on the review on the industrial safety in Hong Kong, Hong Kong Government.

Wong, FKW (1996) Construction safety management in Hong Kong. Proceedings of the CIB W99, 1st International Conference – Implementation of Safety and Health on Construction Sites, Technical University of Lisbon, Portugal, 243-249.

Implementation of Safety and Health on Construction Sites, Singh, Hinze & Coble (eds)
© 1999 Balkema, Rotterdam, ISBN 90 5809 036 1

Auditing for excellence – The value of safety program audits

W.G.Thompson IV
Parsons Brinckerhoff, Herndon, Va., USA

ABSTRACT: Auditing is best described as the system used to assemble pertinent information in order to substantiate compliance with established criteria. However, a successful auditing program is contingent upon development of applicable criteria, standardization of technique and assessment of information.

1. INTRODUCTION

The Occupational Safety and Health Administration (OSHA) has concluded that effective management of worker safety and health protection is a decisive factor in reducing the extent and severity of work-related injuries and illnesses (OSHA 1996).

If this statement is true, than to properly assess the "effective management" of any safety and health program, a company must audit on a regular basis.

2. BASIC VALUE OF AUDITS

In today's climate, audits are used extensively to measure everything from how well cars are constructed to whether enough widgets are being successfully turned out per hour of production time. Successful auditing provides a methodical and comprehensive approach to safety and health program analysis.

2.1 *Assisting program evolution*

Auditing will assist the evolution of a safety and health program by facilitating independent goal setting. Various work departments or job sites may establish different goals based upon types of hazard exposures or past audit performance.

A good auditing program will also identify new areas of concern as a program evolves. In addition, it may be necessary to adapt a flexible auditing design to account for changes in the overall program, or changing location conditions.

2.2 *Independent evaluation*

Effective auditing also provides an independent evaluation of the safety and health program to a variety of interested parties. For example, insurance loss control representatives, a client or owner of a particular project, or members of Management not directly involved with the administration of the safety and health program.

However, to accomplish this independent analysis, auditors must be trained to be objective and impartial and reported results reviewed in a purely constructive fashion.

2.3 *Corrective action system*

Use of well-structured audits, on a regular basis, helps to establish a corrective action system by highlighting which areas need improvement and the level of progress made since the last examination. This type of system will help ensure that corrective actions are cumulative, rather than repetitive.

3. WHEN AND WHERE SHOULD AUDITS BE CONDUCTED?

Audits must be conducted anytime a company has concerns about the effectiveness or direction of their safety and health program. Any program that appears to be in trouble from outside indicators such as high insurance rates, regulatory agency citations, or outside independent evaluation, such as from an owner or client, should be examined.

Audits should also be conducted continually to maintain the integrity of existing safety and health programs. This will prevent programs from becoming weak and unfocused over a period of time.

3.1 *Annual audits*

Although time periods are not firmly established, annual audits should be performed at a minimum. Other factors which may influence frequency are: major changes in program policies or procedures, excessive losses through accidents, or poor performance on previous audits.

3.2 *Locations*

All company areas should be audited regardless of whether the areas are divided by department, work site location, or other factors. This enables the largest and most complete assembly of applicable information and, ultimately, the best basis for analysis.

4. WHAT MATERIALS, METHODS AND PROCEDURES SHOULD BE USED?

The materials, methods and procedures used must be developed and implemented by persons knowledgeable in the specific type of work, or exposures, being audited. The most important aspect of this concept is in only allowing these type of qualified individuals to execute the actual audit. However, it should not be assumed that only safety and health professionals are capable of performing these tasks. Also, consideration may need to be given to the development of specialized audit teams to cover larger locations or specific operations.

4.1 *Background information*

The importance of background information cannot be underestimated and may take several forms. Previous accident/incident reports, insurance company loss information, as well as, prior audits may all provide a good preliminary location picture prior to the actual audit.
Other information, such as project type or scope of work services may also be helpful to the auditing process.

4.2 *Use of standard format and design*

The use of a standard format is paramount. All parties must agree that the designed format is easy enough for auditors to use, yet will provide all the necessary information to properly assess the safety and health effort. In addition, the design must be exclusive for the particular application that it is intended to address. A generic audit may be used as an example, or template, but trying to use this type of tool in its raw form can cause several problems.

4.3 *Documentation and follow-up*

Documentation must be thorough and a follow-up system developed as part of the process. It is preferred, for example, that if numerical values are used to denote a level of compliance, that corresponding documentation regarding the basis for numerical assignment be included within the audit.

Follow-up should be on a regular basis and should consist of repeating the entire audit or, at a minimum, examining the areas of substandard compliance. This follow-up should be conducted unannounced and provide a detailed report separate from the original audit.

5. MAJOR ELEMENTS THAT SHOULD BE COVERED BY AUDITS

Although specific elements included in audits may vary substantially from location to location, or business to business there are at least five major elements that should be covered.

5.1 *Management commitment*

The level of management commitment must be thoroughly addressed within the audit. This will include not only policy statements and the level of safety authority given to various supervisory levels, but also to Management's demonstrated example of safe and healthful behavior and the adequacy of company resources allocated to the safety and health program.

5.2 *Employee involvement*

Employee involvement must be documented and include issues such as procedures for reporting injuries and illnesses, and procedures for reporting potentially hazardous situations. Details concerning employee involvement could be determined by conducting employee interviews using established questionnaires.

5.3 *Work site hazard assessment*

Hazard assessment focuses mainly on the physical or potential hazards present in a particular area to be audited. Typical components of this element would include: information on periodic site self-inspections conducted by a qualified person, review of accidents or near-miss accidents, review of overall injury and illness experience, procedures to identify accident causes and needed corrective actions and the in-house documentation related to these components.

5.4 *Hazard prevention and control*

This element is composed of items such as: proper engineering and personal protective equipment controls, procedures for correction/control of newly-identified hazards, administrative controls and safe-work procedures, system for rule enforcement and discipline, ongoing monitoring and maintenance of equipment, and emergency planning and procedures.

5.5 *Safety and health training*

Safety and health training includes the following components: how supervisors explain, teach and enforce rules, procedures and work practices for hazard control, whether employees can explain how and why they perform their jobs safely, how training is reinforced or repeated when needed, whether employees know when, where and why particular personal protective equipment is required, as well as, its limitations and proper maintenance.

6. CONCLUSION

As the needs of a safety and health program change so will the format and procedures of the auditing program. This is a natural progression and should be viewed as a continuing challenge to constantly improve the overall safety and health effort. In addition, it is proven that in order to establish and maintain a proactive and progressive safety program it is essential to develop and implement a strong and vibrant auditing program.

REFERENCES

Occupational Safety and Health Administration (OSHA) 1996. *Safety and Health Program Management Guidelines.*

Implementation of Safety and Health on Construction Sites, Singh, Hinze & Coble (eds)
© 1999 Balkema, Rotterdam, ISBN 90 5809 036 1

A proactive approach to monitor construction safety using indicators of safety and quality audits

A.W.K.Kwok & S.L.Tang
Department of Civil and Structural Engineering, The Hong Kong Polytechnic University, Kowloon, People's Republic of China

ABSTRACT: In the past decades, poor construction quality and safety records in the construction industry in Hong Kong are being criticised. Both problems inherently reflect the insufficient effort input by the management of construction contractors. Auditing systems both in quality and safety are developed to monitor the management performance. In this regard, the authors aim to work out a method which will help improve construction safety proactively by correlating the two auditing scores (quality and safety) with safety performance. The principle objective of this work is to develop two indicators. The first indicator is the relationship between accident rates and quality auditing scores and the second one is the relationship between accident rates and safety auditing scores. The indicators would be in the form of simple graphs relating accident rates with the scores achieved in quality audit and safety audit exercises being used by local construction contractors.

1 CONSTRUCTION SAFETY IN HONG KONG

Though having tuned down a bit after the financial storm in the Asia Pacific Rim in late 1997, the construction industry has been long playing a very important role in Hong Kong's economy. The property and construction sectors contributed about 25% of Hong Kong's GDP (Rowlinson and Walker 1995). The earned per capita GDP for Hong Kong was about US$ 24,500, and some skilled workers in the construction industry earned as much as US$ 135 a day in 1996. However, there had been a high price to pay for occupational safety and health.

The reportable accident rate for the industrial sector was 52.4 per thousand workers in 1996, compared to 53.2 in 1986. For the non-industrial sector, the rate was 10.9 in 1996, compared to 26.7 in 1986. In absolute terms, there were 59,000 work-related injuries in 1996, of which 40,200 arose from industrial undertakings which used 32.6% of the workforce. The safety record in the construction industry has been the worst for many years. In 1996, the construction industry which employed about 75,000 workers contributed about 40% of the total industrial accidents while the manufacturing industries which employed 347,000 workers contributed only 20% (Lee 1997). The accident rate per thousand workers per year in the construction industry was 220 (including 51 fatal cases) which was about ten times as that of the developed countries (Rowlinson 1997).

2 THE GOVERNMENT'S POLICY ON INDUSTRIAL SAFETY

Over the past three decades, the Hong Kong Government adpoted an "prescription approach" to enforce industrial safety regulations in Hong Kong. It has been considered to be reactive and non-incentive. In this approach, actions would be taken only after accidents occurred. In response to the very bad situation in workplace safety, in particular the construction industry, the Government disseminated a "Consultation Paper On The Review Of Industrial Safety In Hong Kong" in July 1995 proposing new strategies in industrial safety of transforming the latter from an enforcement approach to a safety management approach. A new set of regulations will be enforced which will encourage employers and employees to manage safety on a self-regulatory manner through education, training, promotion of safety awareness and a better understanding of the costs of accidents (Hong Kong Government 1995). Then, in 1997, the Governement enacted the Occupational Safety & Health Ordinance offering a wider coverage of safety in the workplace by the inclusion of workers from the non-industrial settings. In 1998, the self-regulatory safety management regulations which include the implementation of Safety Auditing System became legislated.

3 SAFETY AUDITING IN PUBLIC WORKS

The Works Bureau is a government department responsible for the construction of public civil engineering works in Hong Kong. Before the commencement of the Safety Auditing Systems required by law, it initiated the Pay For Safety Scheme (PFSS) (Hong Kong Government 1996a) to encourage safety promotion among its contractors. About 2% of the contract sum is allowed in the bill of quantities for provision of site safety organisation, safety inspection, safety training and safety auditing. The audit system being adopted in this scheme is designated as an Independent Safety Audit Scheme (ISAS) (Hong Kong Government 1996b) which was developed by the Occupational Safety & Health Council by modifying an UK proprietary safety auditing system named "CHASE". At the same time, the Housing Authority of the Government supported the move and imposed the ISAS in housing contracts as well. The audit system delineated fourteen key process elements for monitoring the safety performance of the contractors' management (Hong Kong OSHC 1996).

4 QUALITY PERFORMANCE IN CONSTRUCTION INDUSTRY IN HONG KONG

Apart from safety performance, poor quality of public housing blocks built in the past decades were also much criticised. The Hong Kong Housing Authority initiated to implement a quality assessment system named "Performance Assessment Scoring Scheme (PASS)" for the new housing work since 1991 and a similar system named "Maintenance Assessment Scoring Scheme (MASS)" for maintenance work since 1994. The schemes are used to measure and review the contractor's quality performance against defined standards as well as their management capability. There are three main components in the "PASS" which includes outputs, inputs and maintenance period assessment. The output component includes structural work, architectural works, external works and other obligations, while the input component includes management of works, resources, co-ordination and control, documentation and programming aspects. The third component is a quarterly assessment during the maintenance period that divides into three parts: (1) outstanding works, (2) execution of works of repairs, and (3) management, response and documentation (Kam & Tang 1997).

5 OBJECTIVES OF THE RESEARCH

It is a universal trend that the mechanism of control of management systems is shifted from a reactive approach to a proactive approach. Auditings in quality and safety are developed to monitor the management efforts of construction contractors. Both safety and quality problems inherently reflect the insufficient effort input by the management. It is logical to consider that high scores in safety and quality audit exercises indicate companies' commitment to sustaining good management performance. Now, the two auditing mechanisms are carried out independently. It is likely that a correlation between quality audit scores and safety performance will exist and that a correlation between safety audit scores and safety performance will also exist.

The principal objective of this work is to develop two indicators to monitor proactively safety management of construction companies in Hong Kong. One indicator is to establish a relation of accident rates against quality audit scores and the other is to establish a relation of accident rates against safety audit scores. The required level of safety performance can be predicted by using the results of the quality and safety audits well before the occurrence of accidents. The scores will serve as signals to contractors and thus the method is proactive.

6 THE TWO INDICATORS

The two indicators may be illustrated by two simplified graphs as shown below.

6.1 *Indicator I*

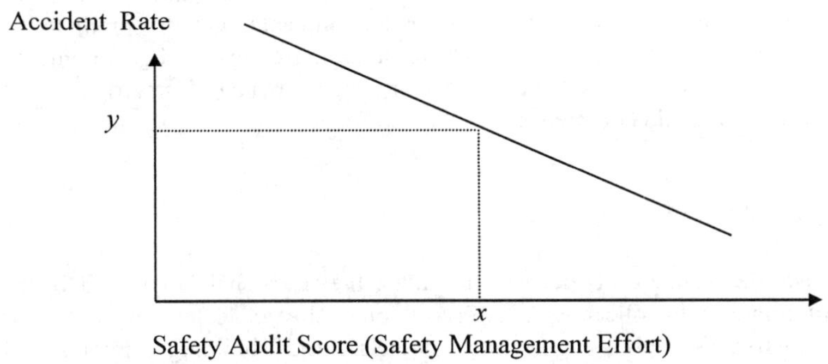

6.2 *Indicator II*

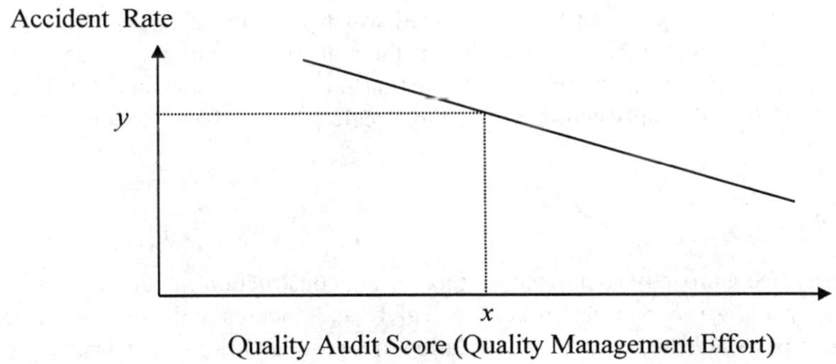

6.3 *Accident Rate*

One of the key parameters in formulating the performance indicators is the accident rate. As mentioned above the accident rate per thousand workers in the construction industry in Hong Kong is about 220 per year. A reportable accident is defined as one that requires more than three days absence from work resulting from injury (Rowlinson 1997). Currently there are a number of different methods used to define accident rate by different occupational safety institutions in different countries. The most common way to define accident rate is usually the incident rate (per thousand workers per year) (*IR*) or the frequency rates (number of accidents per 100,000 man-hours worked) (*FR*).

$$IR = \frac{\text{No. of reportable accidents}}{\text{Working mandays for the months}} x1{,}000x12x30$$

$$FR = \frac{\text{No. of reportable accidents}}{\text{No. of working}} x100{,}000$$

6.4 *Application*

If there is a requirement in the contract documents of a civil engineering/building contract that the accidents rate should not be greater than a certain figure, say, y per thousand workers per year, then one can find out the value x which is the corresponding safety audit score (or quality audit score) of the accident rate y from the correlation graphs (indicators I or II). Before the commencement of any construction processes, a safety audit (or quality audit) will be conducted. An audit score which is greater than x will indicate that the predicted future accident rate of the project will be lower than y (i.e. satisfying contractual requirement), and vice versa. If the audit score is less than x, the contractor's management effort should be improved before any work could commence.

7 METHODOLOGY AND STUDY DESIGN

In order to increase the accuracy of the two indicators, the research is supported by a large amount of data information. In collecting data of scores on quality and safety audits, a survey on 100 construction contracts, 50 civil engineering contracts and 50 building contracts, is being conducted. The target groups of contractors are those in the List I Groups B & C and List II contractors on the Government's list for Works Bureau projects and the contractors in the approved list of Building Contractors for Housing Authority projects.

In the collection of data, a questionnaire is designed which basically consists of three parts. The first part collects the demographic information of the construction companies and the type and stage of work. The second part collects information on accident rates and the final part details of scores of safety and quality audits. Such a large scale survey is being conducted.

8 CONCLUSION

This paper has described a proactive approach for monitoring construction safety using indicators of safety and quality audits. A correlation between safety audit scores and safety performance (accident rates) is to be established. Another correlation of quality audit scores and accident rates is also to be established. The level of safety performance in a construction contract can be predicted by using the correlation graphs if the score of a safety audit (or a quality audit) is known. The safety audit (or quality audit) can be carried out before a contract commences. The score will serve as a signal to the contractor and thus the method is proactive. The correlation graphs are being established. It is necessary to have a big scale survey to have accurate graphs and such a survey is being conducted. It is hoped that some figures will be available at the time of presentation of this paper at the conference.

REFERENCES

Hong Kong Government 1995. Consultation Paper On The Review Of Industrial Safety in Hong Kong, *Education and Manpower Branch, Hong Kong Governement, July 1995.*

Hong Kong Government 1996a. Works Branch Technical Circular No. 4/96: Pay for safety scheme (PFSS). *Works Branch, Hong Kong Government Secretariat, 4 March 1996.*

Hong Kong Government 1996b. Works Branch Technical Circular No. 5/96: The independent safety audit scheme. *Works Branch, Hong Kong Government Secretariat, 5 March 1996.*

Hong Kong Occupational Safety & Health Council 1996. Works Branch Safety Auditing Systems (WBSAS) Version 1.0. *Hong Kong Occupational Safety & Health Council.*

Kam, C.W. & S.L Tang 1997. Development and implementation of quality assurance in public construction works in Singapore and Hong Kong. *International Journal of Quality & Reliability Management.* 14(9): 909-928.

Lee, K.F. 1997. Managing a shared responsibility - From safety charter to new legislation. *APOSHO 13 Conference Proceedings, Hong Kong, 21-22 April 1997:* 7-13.

Rowlinson, S. & A. Walker 1995. *The Construction Industry in Hong Kong*. Longman Asia.

Rowlinson, S. 1997. *Hong Kong Construction - Site Safety Management*. Sweet & Maxwell.

Implementation of Safety and Health on Construction Sites, Singh, Hinze & Coble (eds)
© 1999 Balkema, Rotterdam, ISBN 90 5809 036 1

The implementation of a safety management behavioural audit: Field study results

I.Cameron
Department of Building and Surveying, Caledonian University, Glasgow, UK
A.R.Duff
Department of Building Engineering, UMIST, Manchester, UK

ABSTRACT: In 1997 Cameron and Duff reported the design of a Safety Management Behavioural Audit to use with participative goal setting and public/ graphical performance feedback to improve performance. These techniques have since been used to motivate managers to improve their existing safety performance across the following seven areas of management performance on two case study sites:

Induction Training for Contractors Operatives
Tool-Box Training for Site Operatives
Safety Committee Meetings
Sub-contractor Safety Meetings (Pre Starts and Induction's)
Safety Records and Documents
Safety Adviser Inspections and Actions
Safety Considerations (i.e. operative evaluations)

These dependent variables were measured on a monthly basis within a longitudinal experimental design. The baseline comprised a three month period of measurement coupled with no goal setting and no feedback, whereas, the six month intervention period comprised a period of specific and difficult goal setting coupled with public performance feedback in relation to these goals. Results from a £12 million sterling project and a £24 million sterling project within the United Kingdom construction industry are discussed and analysed.

1 INTRODUCTION

Peterson (1988) charts the evolution of safety management through an engineering control era and a legislative era. Peterson suggests that in reality an effective safety strategy represents the integration of a number of different perspectives. Notwithstanding this, a human factors era has emerged in recent years and it is this era that would appear to offer the greatest utility for safety practitioners (e.g. McAfee and Winn, 1989). These views appear to be shared by the UK's Health and Safety Executive (HSE). A recent Director (Rimmington, 1993) stated:

> "we seem to have passed the era where the need was for further engineering safeguards .... what we need to do now is capture the human factor ....".

The construction sector of the Executive therefore attempted to capture this human factor through two different, but related, contract research studies. Whittington *et al.* (1992) adopting an essentially top-down approach identified management or latent factors that materially influence the accident causation sequence in the construction industry (i.e.

management system failures e.g. poor risk assessments etc). However, these practitioners stopped short of implementing a safety strategy on a real project site to better the status quo. Duff *et al.* (1993), adopting an essentially bottom-up approach, identified operative or proximal factors that influence the accident causation chain in the construction industry (i.e. unsafe acts or conditions e.g. poor edge protection etc). These practitioners went on to design and implement a safety strategy on a number of real project sites to better the status quo.

It could be argued that it was this later study that offered the potential to reduce the number of accidents on construction sites and help fulfil the reduction in accidents suggested by the UK's Construction Task Force (Egan, 1998). These researchers demonstrated that behavioural techniques may be appropriate for the construction process e.g. by improving housekeeping standards for instance. However, sometimes the results were mixed (e.g. Marsh *et al.,* 1994) as management factors could undermine the measure. These management safety factors identified previously (e.g. Whittington *et al.*) were not specifically addressed. Therefore the authors decided to apply these behavioural techniques to management safety factors.

The authors' believe it is reasonable to expect safety interventions to operate more effectively through management. In other words operative behavioural safety (e.g. Duff *et al.*) essentially targets proximal factors (i.e. slipping, tripping etc) at the expense of distal factors i.e. management failures (e.g. design of safe systems of work etc). These relationships may be shown in Bird and Loftus' (1976) "Updated Domino Sequence of Accident Causation: A Behavioural Perspective" (Figure 1).

Figure 1 demonstrates that classifications such as falls from heights represent the "final common pathway" only (Krause *et al.,* 1980). Management failures or "resident pathogens" (Reason, 1987) such as poor communication or poor risk assessments etc represent the underlying cause of accidents in industry. Therefore behavioural safety interventions (using self-set improvement goals and public performance feedback) should more adequately reflect

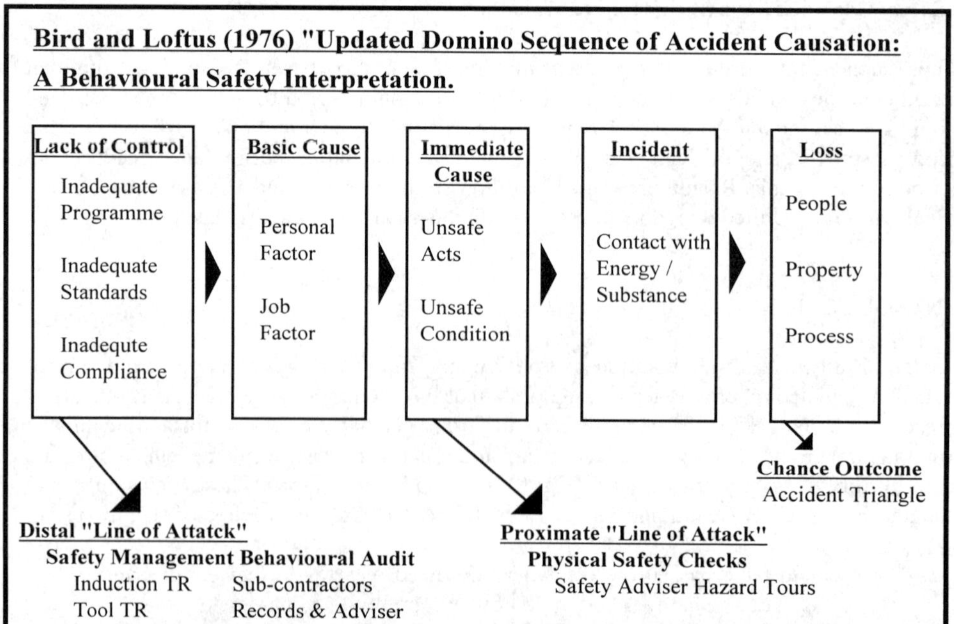

Figure 1. Bird and Loftus (1976) "Updated Domino Sequence" of Accidents: Behavioural Perspective

and evaluate these management factors in their programmes. This lead to the development of the research problem / proposal.

2 PROBLEM STATEMENT, AIMS, OBJECTIVES

The project represents a partnership between industry and academia with two problems.
(1) How can construction organisations reliably measure the safety performance of their managers ?
(2) How can construction organisations motivate their managers to improve their safety performance ?

The project represents the following overall aims.
(1) To develop and implement a reliable measure of safety management performance for the construction industry.
(2) To improve management's safety performance by the application of a behavioural safety audit utilising goal setting and feedback .

These aims are achieved through the following objectives.
(1) To measure management safety performance before the implementation of a behavioural safety audit on a large construction project (pre measure).
(2) To measure management safety performance after the implementation of a behavioural safety audit on a large construction project (post measure).
(3) To quantitatively evaluate the impact of a behavioural safety audit on management's safety performance (pre measure and post measure comparison and analysis).

3 THE DEVELOPMENT OF A SAFETY MANAGEMENT BEHAVIOURAL AUDIT

This evidence suggests that a valid strategy would be to design a behavioural safety audit to reduce distal failures e.g. ineffective methods i.e. poor communication and safety motivation. Few safety audits however are designed for construction. Indeed, audits have been criticised for using superficial questions which fail to identify underlying issues (HSC, 1992). Therefore, the authors designed a "Behavioural Safety Management Audit". This audit concentrates on critical safety actions that must be performed by management during any programme month.

For instance, a question like: “Is your company safety policy signed by the director ?”, is unlikely to change from one month to the next nor does the question demonstrate behavioural commitment to the safety role. Whereas a question like: “How many safety awareness training talks were conducted this month ?”, is likely to change from one month to the next plus the question demonstrates behavioural commitment (i.e. time / effort etc) to the managers safety role. This later question style is critical and forms the core of the behavioural safety audit.

The audit therefore integrates the philosophy of measurement enshrined in behavioural safety initiatives with the sharp management focus associated with proprietary auditing. The integrated perspective may be viewed in figure 2.

This approach combines the strengths of the behavioural approach with the strengths of the auditing approach thus reducing the inherent weaknesses of both.

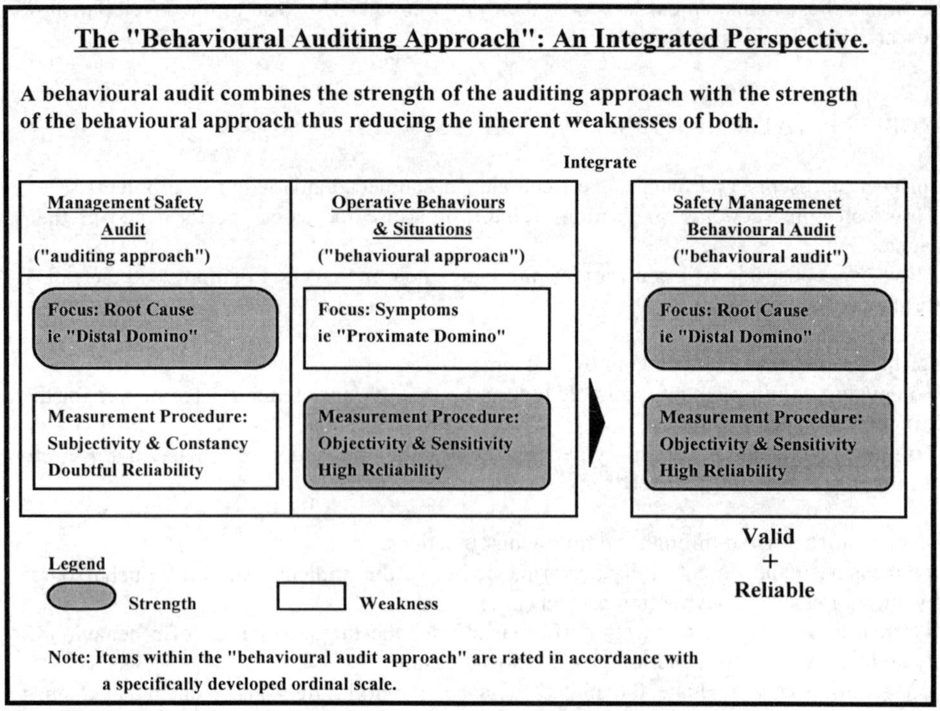

Figure 2. The Safety Management Behavioural Audit: An Integrated Perspective.

4 METHOD EMPLOYED

The behavioural audit contains seven individual performance items extracted from a literature review (Cameron and Duff, 1997): Induction Training; Tool-Box Talk Training; Safety Committees; Sub-contractor Pre-Starts; Maintenance of Safety Records; Safety Adviser Inspections; Safety Consideration. Five of the items represent essentially management recalled objective data from a management interview; one of the items represents data extracted from organisational records from a review of safety records; and the remaining item represents essentially operative recalled objective data extracted from a series of operative interviews. Measurements are conducted on a monthly basis throughout the study/audit period

4.1 *Scales and scoring*

Question conformance is measured by either percentage bands or absolute descriptive labels. These guides ensure the scale exhibits high discriminatory meaning. Measurements therefore represent an objective, and not subjective, rating of performance. In other words, ratings are not dependent on the judgement of the auditor. Scale bands are incontrovertible which ensures that different auditors arrive at the same rating. The data produced by the scale are therefore believed consistent and sensitive for each of the seven individual items within the audit.

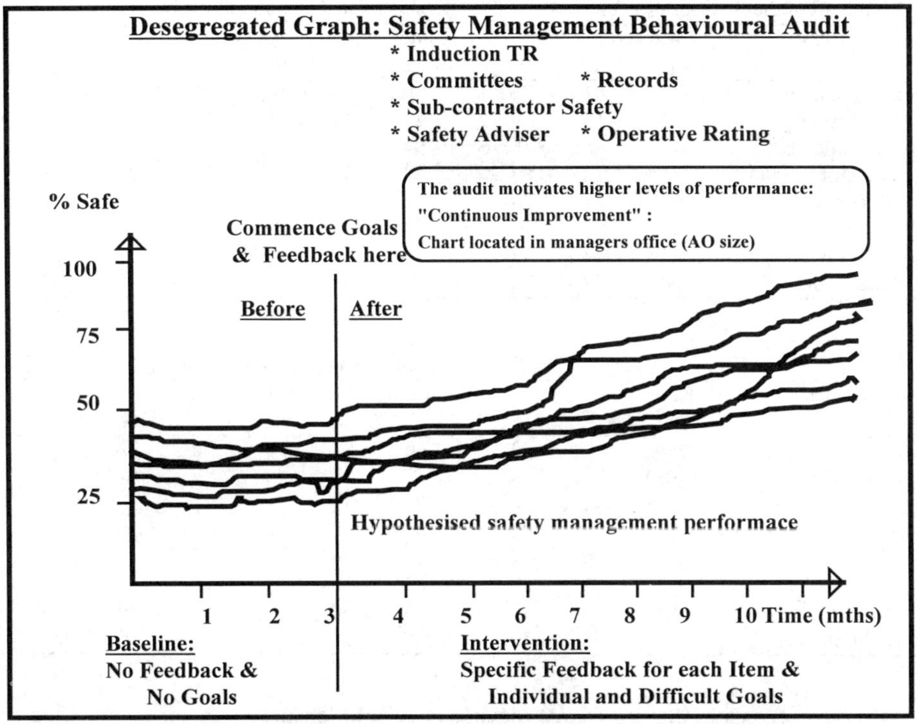

Figure 3. Method Employed on Site to Implement the Behavioural Safety Audit.

4.2 *Motive force*

The audit utilises goals and feedback as a motivational tool to improve safety performance. These variables are a proven management technique (e.g. Locke and Latham, 1990). They have previously improved safety performance in manufacturing industries (e.g. McAfee and Winn, 1989). The utility of these variables has recently been extended to construction safety research (e.g. Duff *et al,* 1993). These variables therefore represent a valid proposition for use in construction management safety. Also, the audit represents a low-cost method of improving safety. Expensive incentives or gimmicks (Peterson, 1988) are not required.

Baseline measurements were first taken for a period of three months to establish "normal" performance. No feedback or goal setting occurred during this period. Thereafter a six month period of specific continuous improvement safety goal setting commenced. Performance results (after the elapse of the three month baseline) were illustrated to site staff by graphical means via an AO wall chart located in the project manager's office. Figure 3 refers.
The procedure outlined in figure 3 forms the basis of the method used to:
(1) Demonstrate the utility of a behavioural safety audit; and,
(2) Motivate management to improve their current safety performance:

4.3 *Instruments and materials*

Performance Standards: This document was issued to site staff at the outset of the project. The document contained the audit parameter i.e. measurement scope of the audit. These performance dimensions were extracted from the contractors existing Safety Management

Table 1. Results from the Scottish Project for The Safety Management Behavioural Audit (%'s).

Scotland: Stirling	Base			Inter	ven	tion			
	1	2	3	4	5	6	7	8	9
Induction Training	0	81	75	89	95	95	95	95	94
Tool-Box Training	0	46	50	86	61	80	90	0	66
Safety Committees	0	0	0	94	92	100	92	0	84
Sub-contractor Pre Start	76	72	80	98	98	97	84	97	97
Safety Records	63	61	79	96	93	92	90	79	90
Safety Adviser	83	67	67	83	75	88	83	92	83
Safety Consideration	43	44	42	59	66	65	60	61	55
SSM Audit Score	38	52	56	87	87	89	85	63	83

Table 2. Results from the English Project for The Safety Management Behavioural Audit (%'s).

England: Manchester	Base			Inter	ven	tion			
	1	2	3	4	5	6	7	8	9
Induction Training	69	71	80	90	87	90	90	92	93
Tool-Box Training	0	0	0	57	43	44	0	61	0
Safety Committees	0	64	0	71	0	0	73	0	77
Sub-contractor Pre Start	76	69	82	87	89	90	85	90	91
Safety Records	67	56	67	85	76	77	76	75	75
Safety Adviser	67	67	83	83	50	75	75	75	75
Safety Consideration	40	48	40	63	50	55	56	60	55
SSM Audit Score	48	58	50	78	58	61	69	66	72

System. Compliance with this standards document would ensure a good audit score throughout the contracts life. Essentially the document represents audit rules.

Feedback Graph: Performance results were shown graphically after the completion of the baseline period. This method of presentation is recognised as a powerful motivational medium. The graph shows the seven items expressed as a group of percentage scores forming a time-series. The graph also shows next months percentage safety goals (which the team set and agreed to try and attain next over the next month).

Action-Plans: These represent qualitative feedback which explains how to attain next months new safety goals. This qualitative information supplements the quantitative feedback shown on the chart i.e. the action-plans may be construed as sub-goals (information) which clear a path to the macro-goals (percentage targets).

5 RESULTS

Results from a £24 million shopping precinct in Scotland and a £12 million (sterling) Superstore in England are shown in tables 1 & 2. The reliability of the data shown in tables 1 & 2 was confirmed by independent secondary rating.

5.1 Scottish Case

Figure 3 represents the Scottish project which in this study appears to represent the best-practice case. The project performed in a typical fashion during baseline with activities such as Induction Training, Tool-Box Training, and Safety Committees being performed in an ad-hoc manner. Likewise, Safety Records such as scaffold inspections sometimes failed to occur. However, during the intervention the site management team consciously tried to attain their own difficult performance goals. In practice, problem solving behaviours were in evidence - an indirect goal mechanism in complex tasks - to achieve better performance. For instance: a programme of relevant Tool-Box Training was devised and added to the contract programme; an appropriate agenda and subcontractor representatives were appointed for the Safety Committee; a site safety supervisor was appointed to oversee goal directed effort and report ongoing progress/delay back to the Project Manager. These strategies caused a consistent continuous improvement trend across global scores and individual items such as Induction Training and Safety Committees. Further, by the end of the project there was strong evidence to suggest that this site consistently practiced the procedures detailed in the organisations Safety Management System Procedures Manual and the Site Health and Safety Plan. The only negative aspect of this site was the detrimental effect twenty-four hour working and the impending handover had on safety towards the end of the contract.

The results on this site seemed to be directly related to management behavioural commitment and it seems reasonable to conclude that the Safety Management Behavioural Audit helped to promote this enhanced commitment by virtue of the audits focus on the Project Manager who became the Champion of the Intervention.

5.2 *English Case*

Figure 4 represents the English project which in this study appears to represent a typical-practice case. Although scoring did improve globally and some items became more consistent (e.g. Induction Training and Tool-Box Training) in general the site seemed to lack goal effort and seemed incapable of initiating strategies to attain their goals. Indeed, goal focus seemed reactive and related to the demands of the Area Safety Manager. There was no strong evidence to suggest the behavioural audit guided safety management on a daily basis.

Despite this, site staff said that they liked the concept but simply did not have the time to consistently perform all of their safety duties outlined in the organisations Safety Management System. This issue of resources is a real one and worthy of further research. The site did seem at least one manager short for a project of this size (£12 million with 3 production staff).

5.3 *Commentary: underlying factors*

Notwithstanding the debatable issue of resources, management commitment appeared to be a key difference between the two sites. The Scottish site Project Manager reported regularly to an Area Director who was passionate about safety and acted accordingly on site. The English site Project Manager reported periodically to an Area Director who was indifferent towards safety and likewise acted accordingly on site. Therefore, perhaps the behaviours of these respective Project Managers simply reflected differences in organisational values as evinced through the very director persons who symbolise what the organisation really values.

Also, there was some evidence to suggest that the management team at the Scottish project appeared to be better organised, better educated and trained, more strategic and pro-active, younger and generally more professional than their counterparts at the English site. This finding tends to support a long held belief of safety practitioners that in order to manage safety better, it is first necessary to become a better manager i.e. safety management and general management are inextricably linked.

6 CONCLUSIONS

Results to date tentatively suggest the measures and methods are both valid and reliable and can lead to improvements in safety which helps to promote management accountability thus addressing the findings of Blackspot Construction (1988): "70 % of accidents could be prevented by better management of the safety function". It would appear however that this relationship is moderated by the behavioural commitment of executive head office managers who, although not based on site, appear to influence safety much more than they possibly realise. Generally a case of actions (or omissions) speaking louder than words.

REFERENCES

Bird, F.E. and Loftus, R.G. (1976). *Loss Control Management.* Institute Press. Georgie

Cameron, I and Duff, A.R. (1987) The Development of a behavioural safety audit for the construction industry. *Paper presented at the CIB/W99 First South African Health and Safety Conference.* Cape Town. October 7-10, 1997.

Duff, A.R., Robertson, I.T, Cooper, M.D., Phillips, R.A. (1993). Improving safety on construction sites by changing personnel behaviour. *HMSO Report Series CRR 51/93*: HMSO. London

Egan, J (1988) *Rethinking Construction.* The report of the Construction Task Force. HMSO

HSC (1993). Health and Safety Commission. *Third Report: Organising for safety.* ACSNI. (Advisory Committee on the Safety of Nuclear Installations) study group on human factors.

HSE (1988). *Blackspot Construction.* The Health and Safety Executive. HMSO London

HSE (1992). *Successful health and safety management.* HSE. HS (G) 65. HMSO

Komaki, J.L., and Jensen, M. (1986). *Within group designs: an alternative to traditional control groups.* In : M.F. Cataldo and T.J. Coates (Eds), Health and Industry, A Behavioural Medicine Perspective. Wiley, New York, NY, pp.

Krause, T.R., Hidley, J.R., Hodson, S.J. (1990). *The behaviour based safety process.* Van Nostrand Reinhold. NY.

Locke, E.A. and Latham, G.P. (1990). *A theory of goal setting and task performance.* Prentice-Hall International (UK) Ltd, London

Marsh, T.W., Duff, A.R., Phillips, R.A., Robertson, I.T., Cooper, M.D., and Weyman, A (1994). Practical problems encountered attempting to improve behavioural interventions. *Fifth Annual Rinker Conference on Building Construction Focusing on Construction Safety and Loss Control*, Florida, Oct. pp65-77

McAfee, R.B. and Winn, A.R. (1989) The use of incentives/feedback to enhance workplace safety a critique of the literature. *Journal of Safety Research,* 20, 7-19.

Peterson, D. (1988). *Safety Management: A Human Approach.* Aloray Inc., New York

Reason, J.T. (1987). *A framework for classifying errors.* In Rasmussen, Duncan, and Leplat's (eds) New Technology and Human Error. John Wiley and Sons.

Rimmington , J. (1993). Does health and safety at work pay ? *Journal of Safety Management*, 9, pp. 59-62

Whittington, C., Livingston, A., Lucas, D.A. (1992). Research into management, organisational and human factors in the construction industry. *HMSO Report Series CRR45/92:* HMSO. London

24 Standards/fall protection

Implementation of Safety and Health on Construction Sites, Singh, Hinze & Coble (eds)
© 1999 Balkema, Rotterdam, ISBN 90 5809 036 1

Safety-campaign 'Fall risk in construction work'

U. Berg
Bau-BG Frankfurt am Main, Germany

ABSTRACT: Fall prevention is key to protecting construction workers from work-related injuries and deaths. Worldwide, falls are the leading cause of deaths in construction. Therefore the BBG Frankfurt launched a safety campaign, "Fall Prevention in Construction Work," comprising information, supervision, and advice. The campaign has been conducted by the Technical Inspectorate of the BBG Frankfurt, Department for Occupational Safety and Health. The objective has been to inform member firms, especially contractors and site managers, about the hazards and ways to prevent them; to assist the firms in injury prevention using all suitable methods; and to ensure enforcement of injury-prevention regulations.

1 STARTING POINT

Our safety-campaign was envisioned after our construction statistics revealed an alarming trend. When we considered the types of incidents leading to serious injuries and deaths, we realized that most were caused by falls. This is not only a trend in Germany. Falls are the leading cause of fatal injuries in the U.S. construction industry, also (fig. 1).

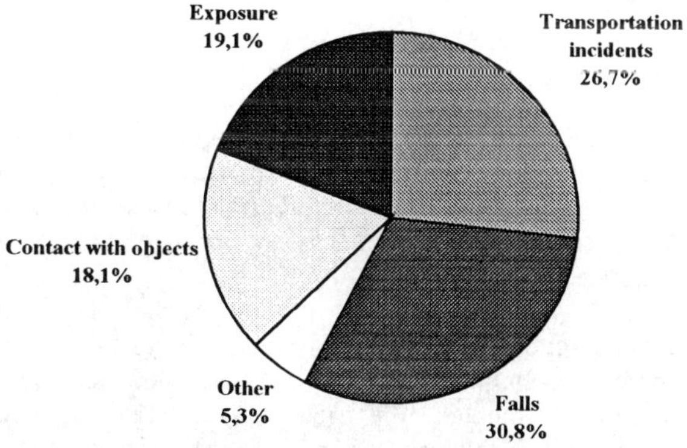

Figure 1 Distribution of leading causes of deaths from injuries in the construction industry, 1995

Source: THE CENTER TO PROTECT WORKERS' RIGHTS, Washington D.C. (1998)

The data on deaths are derived from the 1995 Census of Fatal Occupational Injuries conducted by the Bureau of Labor Statistics and show that 30.8% of all fatal injuries in the construction industry of the U.S. are caused by falls. In Germany the percentage is about 50%.

2 BASIC IDEAS

We wanted to improve our efforts in injury prevention by using new methods and more effective strategies for supervision and advice. We realized that there are too many construction sites and activities for our personnel - especially our technical inspectors - to monitor regularly. This is particularly true because we must, at times, repeatedly check some sites where problems are found. Therefore we decided to try a new approach in small regions that have a high concentration of construction activities.

The "Fall Prevention in Construction Work" began in the summer of 1995. This campaign has 3 main elements:

- We launched a comprehensive information campaign in the entire area of Germany we are responsible for: These are the Federal States of Hesse, Thuringia and parts of Rhineland-Palatinate.
- Then, we began a comprehensive program of supervision in those regions where the most construction work is under way.
- Last, we intensified our advice activities by using a standardized checklist for the target groups (mainly contractors and site managers) concerning responsibility, organization, and management for safety and health at work. The checklist did not focus only on falls.

In addition to these 3 main elements, we focused our regular safety-training activities on fall prevention. We invited the contractors, site managers and foremen of these branches of the industry

- high-rise construction
- scaffolding
- roofing
- carpentry.

3 INFORMATION CAMPAIGN

The BBG Frankfurt sends a quarterly magazine to all member firms and interested parties. It contains information on the BBG, the Statutory Industrial Accident Injuries Insurance, and safety-and-health issues, such as changes in the injury-prevention regulations. In the magazine, we published articles describing the planned safety campaign and safety provisions that were already in place to prevent falls and to minimize fall risks.

In addition to this general information, we wanted to establish contact with construction firms that were in or near regions where the supervision campaign was going to take place. Therefore we sent a mailing two weeks before starting the supervision activities in each region and described the main features of the safety campaign. We tried to summarize the most important provisions covering safety measures for fall prevention. In addition to text, this summary tried to appeal to the contractors and construction workers with illustrations showing typical examples.

An order form was attached to the mailing. The form enabled anyone to obtain more-detailed information from our institution - for instance, injury-prevention regulations, safety guidelines, and brochures. These publications are available for everyone free of charge. During the

Figure 2 “Prevent Falls“

supervision activities, we wanted mainly the support of contractors and site managers, but also of all construction workers. Therefore we offered special posters, stickers, and a leaflet about our safety campaign "Prevent Falls" (fig. 2) to the target groups free of charge, also.

These publications have been available by mail. They have been included with every letter that the construction firms have received containing a report on a safety check of one of their construction sites. Another efficient way to disseminate our posters and stickers has been for the technical inspectors to hand out the posters for the safety campaign during site visits or meetings with contractors or site managers. Our technical inspectors have tried to motivate construction-firm staff to use these stickers and display the posters. With this publicity, we have intended to involve everybody working on site in this campaign as well as inducing the construction workers themselves to confirm their commitment to fall prevention.

The leaflet shows a hierarchy of measures to prevent fall injuries. The best way is through engineering by covering or closing off openings and by using guard rails or barriers. These means sharply reduce the chance of anyone falling from one level to another. This is important for scaffolds, edges of roofs, or elevated building surfaces.

If these barriers are impractical for technical reasons, safety measures must be focused on minimizing the effects of falls. This is done by providing ways to catch anyone who falls, with arrestor bars or roof arrestor bars, protective walls, or safety nets.

Only if use of these two general solutions to prevent falls is not feasible do we suggest using personal protective equipment (PPE). A safety harness cannot prevent a fall but can reduce the height of a fall and reduce any impact that might cause severe or, even, fatal injuries. The effectiveness of safety harnesses and belts always depends on the proper use and behavior of the workers using them. Human beings' willingness and behavior combined with sufficient knowledge to protect themselves involve a large number of imponderables. Therefore, PPE is better than no protection but not the best way to ensure safety. We don’t favor the use of PPE in case of fall protection and recommend to choose PPE only when there are no other alternatives on site.

4 SUPERVISION CAMPAIGN

For the supervision campaign, each region having a large proportion of construction activities is subdivided into 4-to-6 sectors. Each sector is generally small enough that two technical inspectors are able to visit all the construction sites and also get to know every construction activity in the sector. Over the 2-to-4 weeks the program is in effect in a region, the technical

Checklist for Contractors and Site Managers of Construction Firms
Safety and Health at Work
Responsibility – Organization – Management

General questions

- Do you think that safety and prevention are as important as other company activities, e. g. marketing, profit, quality?
- Do you know that safety and health – like other objectives of your company – depend on planning as well as on organizing and carrying out the measures?
- Are you aware the fact that company's basic preconditions are indispensable to success in the field of safety and health in construction?

Planning management

- Did you plan safety and health measures even when you were planning operations on building site and thinking about the costs?
- When do you think about fall risks on the construction site and about measures to prevent falls?

Personnel management in matters of safety and health at work

- How do you let workers know that safety and health at work are important for you and your company?
- In particular, how do you communicate the importance of safety and health to your managers and foremen?
- Do you appreciate a positive attitude of your staff toward safety and health?
- What are your means of praising employees about attitudes and results in matters of safety and health at work?
- Are you aware the fact that your behaviour and personality are the best example even in matters of safety for all the members of your staff?
- Do you properly use personal protective equipment, e. g. helmet, safety shoes, ear protection?

Enforcement of safety and health on construction sites

- Is there any written document in which the responsibilities for safety and health in your company are explained and assigned in detail?
- Do you provide sufficient devices and material on the job site to meet the safety requirements and to carry out the necessary measures to protect falls?
- Do you provide written instructions concerning how to safely carry out special tasks on the job site, e. g. scaffolding, demolition work? Do you provide an owners' manual for special devices in use on the job site?
- Do you arrange any regular meetings or safety training events especially for the site managers, foremen and workers?

Monitoring of safety measures

- Does your company regularly check compliance with the stated requirements for safety and health on the construction site?
- Do you have regular meetings or discussions with your staff to exchange experiences (problems or improvements) in enforcement of safety and health on construction sites?
- In which situations do you obtain advice of safety and health experts, e. g. manufacturer of safety devices (scaffold or protective equipment), Technical Inspector of the BBG or industrial physician (company doctor) of the occupational health service of the BBG?

Figure 3 Checklist for the conversations with contractors and site managers inspectors regularly check the construction sites in every sector. The inspectors monitor the construction sites weekly and, if necessary, nearly every day.

Of course, our member firms are not used to having inspectors visit so often and most find the frequent visits to be inconvenient. We think, however, that this intensive supervision is the main strength of the safety campaign. And we think that this strategy improves safety in the long term in the companies being supervised.

After a workplace is checked, the technical inspectors give company officials a form listing the hazards found on site. At the same time, a notice is mailed to officers of each company by mail, with more-detailed descriptions of the main safety deficiencies creating fall hazards on site. When basic safety provisions have been neglected, the responsible representative on the construction site (site manager or foreman) or the contractor is fined by the committee for offences of the BBG Frankfurt.

5 ADVICE ACTIVITIES

The objective of our advice activities has been to give not only advice on specific safety deficiencies on the inspected construction sites, but in all aspects of responsibility, organization, and management of safety and health. We have wanted to know more about the structural problems pointed up by the enforcement of safety measures on the construction sites and to suggest ways to solve the problems and improve the awareness of the target groups for our fall-prevention campaign. Therefore we use a checklist as a guideline for all the conversations that our technical inspectors have with contractors and site managers. This checklist, shown in figure 3, has about 18 questions in the following 5 categories:

- general questions
- planning management
- personnel management in matters of safety and health at work
- enforcement of safety and health on construction sites
- monitoring of safety measures.

6 SUMMARY

The safety-campaign has been carried out in the following 9 regions during 1995 and 1996: Frankfurt, Gießen/Wetzlar, Mainz/Wiesbaden, Darmstadt, Kassel, Eisenach, Erfurt, Jena and Gera. The districts with frequent supervision activities lasting up to 4 weeks have been mainly located in or near these cities. All the staff of our Technical Inspectorate including nearly 50 technical inspectors have been involved in this campaign and these are the main points - and results - of our safety-campaign "Fall Prevention in Construction Work":

- We are convinced that this safety-campaign has contributed substantially to increase safety and health on inspected construction sites and in the construction firms involved in the several regions.
- We think that our activities could help contractors and construction workers be more aware of the fact that falls can have the most severe outcomes and thus should be the main focus of prevention. Risk always depends on probability and potential severity. In terms of both of these factors, falls have very high values compared with other risks in construction.
- We hope that our information campaign and advice activities have been able to demonstrate that falls can be prevented as well as can other types of injuries and that a high risk of falls in construction work is not acceptable until at least everything has been done to meet the requirements of the accident prevention regulations.
- We are convinced that our safety-campaign, "Fall Prevention in Construction Work," helped to point out a range of hazards on the construction site and to address structural problems in

each company's safety and health program. This campaign has been an occasion for the BBG Frankfurt and all the staff of our Technical Inspectorate to offer to help find solutions for safety-and-health problems and thus confirm our partnership with our member firms.

REFERENCES

BAU-BERUFSGENOSSENSCHAFT FRANKFURT AM MAIN- BBG - 1995 / 1996. *Mitteilungsblatt der BBG Frankfurt*: Safety-Campaign „STOP dem ABSTURZ - WIR SIND DABEI“. Frankfurt am Main

THE CENTER TO PROTECT WORKERS‘ RIGHTS - CPWR -; April 1998 Second Edition. *The Construction Chart Book*: The U.S. Construction Industry and Its Workers. Washington D.C.

Implementation of Safety and Health on Construction Sites, Singh, Hinze & Coble (eds)
© 1999 Balkema, Rotterdam, ISBN 90 5809 036 1

An evaluation of safety programmes in construction companies

Franco Geminiani
Faculty of Civil Engineering, Building, Architecture and Agriculture, Port Elizabeth Technikon, South Africa

ABSTRACT:The safety of the human being at work or on a construction site is a complex problem. The construction industry in South Africa is known to be one of the most hazardous with an unacceptably high level of injuries and fatalities resulting in considerable human suffering.

This paper focuses on the recommendations of a completed Master's study evaluating safety programmes in selected construction companies against a normative safety programme model.

Information was sought and obtained from respondents of construction companies by means of a questionnaire survey and personal interviews conducted. The results of the survey indicated that the safety programmes of the majority of construction companies did not compare favourably with the normative model. The results also led to the deduction that management in general is not fully committed to safety.

The findings of the study with recommendations are also included. The recommendations of the study are put forward to indicate to management which elements should be included in a safety programme. This normative safety programme model will guide them in safety related matters as well as instil safety awareness in workers.

Keywords: construction safety, elements, safety, safety programme, safety awareness.

1. INTRODUCTION

It is generally known that the construction industry is hazardous in terms of safety. From statistics released throughout the world and also from past historical perspectives it is clear that safety in construction has always been a matter of concern.

Although literature is available on the topic of construction health and safety, very little specific information can be found with regards to safetyprogrammes being used by construction companies in South Africa.

This study is based on the findings of a study done on the evaluation of safety programmes in selected construction companies. The purpose of this study was to: (i) determine which elements should be included in a safety programme (ii) to develop a normative model of safety programme elements and (iii) to evaluate the elements included in the safety programmes of construction companies against the proposed normative model.

Out of a total of 102 (small, medium and large) construction companies, 21 respondents returned their questionnaires to the researcher. This poor response rate of 20.58% could be attributed to the fact that some companies had closed down their operations in the Port Elizabeth metropolitan area due to a decline in the building industry.

It could also indicate that the respondents who are still in operation are not interested in completing questionnaires on safety but rather in attaining production results. The results obtained from the respondents were analysed and compared to the normative safety programme model as developed from the literature.

2. LITERATURE SURVEY

The purpose of the literature study was to identify the elements that should be contained in a safety programme. Table 1.1 shows the safety programme elements as identified in literature by various authors namely, Blake (1964), Boylston (1990), De Reamer (1958) and Mansdorf (1993). In reviewing their opinions, it is noted that most authors regard the element of management, training and the identification of hazards as essential elements of a safety programme.

It was therefore assumed that these three elements mentioned above are the most essential parts of a safety programme. However, having reviewed the literature, it was concluded that management is not directly an element of a safety programme, although so emphasised by the authors, but rather a prerequisite for efficiently and effectively implementing and maintaining a safety programme.

As is noted in table 1.1 various authors have different views of what should be contained in a construction health and safety programme. Therefore, no generally accepted formal safety programme model seems to exist in literature. This justifies the need for the research.

Table 1.1 Elements of safety programmes as identified by four authors

Blake	Boylston	Mansdorf	Do Reamer
Executive leadership	Management leadership	Management commitment	
		Hazard prevention	
	Identification of hazards	Analysis of hazards	Job hazard analysis
	Employee & Sup training	Safety training	Safety training
	Safety & health recordkeeping		

	First Aid & Medical		
	Employee awareness		
Safe plant & equipment			Plant safety
	Assignment of responsibility		
			Communication
			General administration
			Safety programme
			Safety rules
Competent supervision			
Employee co-operation		Employee involvement	

Sources: Blake (1964, p. 77), Boylston (1990, p. 16), Mansdorf (1993, p. 15), De Raemer (1958, pp. 333-340).

3. INFORMATION THAT PROVES THE NEED TO IMPROVE SAFETY PRACTICES

Some progress has been made in understanding the dynamics of the unexpected loss phenomenon which is more carefully referred to as the accident process (Petersen and Goodale, 1980, p. 3). The progress which has been made over the years in the transportation field has been phenomenal (Davis, 1988, p.9). Progress ranges from the steam age to the mechanized age of underground tubes to the most modern forms of transport such as the Concorde and the most recent achievements in space.

However, the development in safety management has not shown similar progress and therefore the approach to industrial safety must change. It is regularly reported in the media that industrial accidents take place where workers are injured, disabled or killed at the workplace, in factories, construction sites, public places, mines as well as in general road accidents (Kruger, 1996, p. 9).

According to Mr. Kruger, the spokesman for the National Occupational Safety Association, five people are killed and 640 injured every day in South Africa in industry-related accidents throughout the country. He stated that employees are not being properly informed about the hazards of their work and are not being provided with proper training as required by the Occupational Health and Safety Act Number 85 of 1993.

From statistics obtained from the Workman's Compensation

Commissioner's Annual Report (1990, unpaginated) there were a total of 322 252 persons employed in the Building and Construction Industry in South Africa in 1990.

There were 15600 reported accident-related cases in the Building and Construction Industry in 1990 (Workman's Compensation Commissioner's Annual Report 1990, unpaginated).

The report shows that:
(i) 53.9% were medical aid cases
(ii) 18.4% were temporary disabled cases
(iii) 3.0% were permanent disabled cases
(iv) 0.2% were fatal cases.

Therefore 4.8% of the total workforce in the Building and Construction Industry were involved in accident-related cases in 1990. The Department of Manpower Annual Report (1993, unpaginated) shows a comparative survey of fatalities by all industries as a whole in South Africa. It indicates that the Building and Construction Industry has amongst the highest fatality ratings in the country. The statistics and fatalities in the Building and Construction Industry between 1986 and 1993 show a decrease in the number of fatalities. However this figure is still unacceptably high (see table 1.2).

From the statistics provided by the Workman's Compensation Commissioner's Annual Report (1994, unpaginated) for every R100m of construction completed: (i) 0,8 workers were killed (ii) 4,0 workers were permanently disabled (iii) 32,1 workers were temporarily disabled and (iv) 42,5 workers received medical aid. These statistics illustrate the urgency for the need to improve safety practices in the building and construction industry.

4. ELEMENTS OF A SAFETY PROGRAMME

The remainder of the paper is devoted to the elements contained in a safety programme illustrating a model outlining the elements forming part of a safety programme.

TABLE 1.2 Number of incidents in the building and construction industry

YEAR	INCIDENTS	FATALITIES
1986	: 967	: 967
1987	: 924	: 102
1988	: 902	: 76
1989	: 1041	: 112
1990	: 877	: 117
1991	: 864	: 123
1992	: 741	: 92
1993	: 894	: 97

Source: Department of Manpower, Annual Report : 1993.

4.1 *Proposed normative model of elements of a safety programme linked to managerial levels*

Introduction
The proposed integrated model (see figure 1.2) of an effective safety and health programme is introduced, based on guidelines as drawn from the literature reviewed previously.

4.2 *Elements making up a safety programme model*

The following elements are grouped under the following management levels.

Top management
Compilation of a safety policy statement

Middle management
Education and training
Emergency management procedures
Safety manual
Safety promotion on site
Accident prevention and investigation

Operational management
Inspections on site
Record keeping of information
Safety competitions
Medical and first aid
Personal protective equipment

It can be concluded that management provides the central leadership role within the organisation. An additional requirement for achieving safety excellence is that the leader must recognise leadership as a responsibility rather than a particular rank or privilege. This normative model is proposed as a guideline for construction companies. An exposition of the empirical study undertaken among construction companies in the Port Elizabeth metropolitan area is illustrated next, in order to analyse what elements make up construction companies' safety programmes.

5. CONCLUSIONS DRAWN FROM THE EMPIRICAL STUDY

- Thereis a possibility that accidents do occur on site because management is not adequately trained in safety management.
- It appears that management is mostly concerned with increasing production to supplement fast track construction, while at the same time neglecting to practice safety management for the benefit of it's workers.
- The majority of construction companies do not employ safety officers because of financial reasons.
- A large percentage (47.61%) of construction companies do not implement a structured safety programme that compares favourably with the developed normative safety model.

- Most construction companies do not have a formally structured safety policy statement.
- Not enough emphasis is placed on safety education and training and very little is done in respect of safety orientation of new workers entering the industry for the first time.
- Most construction companies do not have a formally structured safety programme, but the ones that have, have their own made up safety programme.
- Most construction companies do not have a safety policy statement. Where there is one it is more in the form of a mission statement.
- Most construction companies do not carry out structured investigation procedures according to legislation.
- Most construction companies do not employ a full-time safety officer because of financial reasons.
- The majority of elements contained in the safety policy statement as well as the safety manuals of construction companies do not compare favourably with the normative model developed.
- The majority of operational workers have a "don't care" attitude.
- Management of construction companies are generally aware that an unsafe site reduces productivity.
- The management of construction companies are not fully committed to safety and tend to neglect safety.
- Minimum safety qualification requirements are not adequate.
- Workers are still being employed at the gate without any proper education and training given.
- Illiteracy is a problem with regards to safety in the construction industry.
- The Department of Labour's inspectors do not carry out frequent surprise inspections.
- Workers lack concentration due to emotional stress caused by insufficient satisfaction of basic human needs.

6. RECOMMENDATIONS FOR SAFETY PRACTICES

The following recommendations are put forward to assist companies to minimize accidents, injuries and deaths.

- Construction companies should employ a full-time safety officer to implement and maintain an effective safety programme.
- Companies should implement a safety programme comparable to the normative safety programme model.
- Management should endeavour to become fully committed to safety, acknowledging that a safety programme does cost a company money, but realising that in the long term it is financially beneficial. A structured safety policy should be developed to guide both employer and employees towards a common safety objective. Safety should be a priority along with production.
- Management should introduce formal education and training programmes suitable for all employees at all levels.
- Minimum qualifications should be raised relevant to the positions attained. Employees should not be taken on without the minimum required safety

qualifications and should not be allowed on to a site without having being inducted, orientated, trained and made aware of the dangers existing on site.

- Management should rather educate and train employees for future projects during periods of decline in operations instead of laying them off.
- Workers should change their attitude towards safety by identifying the dangers that exist on a site and taking remedial action.
- The Department of Labour should appoint specialists who could offer specialist consultation to companies who are not aware of the requirements of the law.
- The inspectors of the Department of Labour must show improved activity in carrying out their duties by initiating more frequently surprise inspections.
- Safety committees and safety representatives should be encouraged to perform their duties with more enthusiasm and diligence. An incentive scheme should be introduced for safety representatives.

7. FINAL CONCLUSION

It can therefore be concluded that the elements identified in the normative model of a safety programme can be utilised to improve the safety programmes of construction companies. In conclusion, it is evident from the research findings that management must become fully committed to safety in order to preserve it's most valuable resource, the human being.

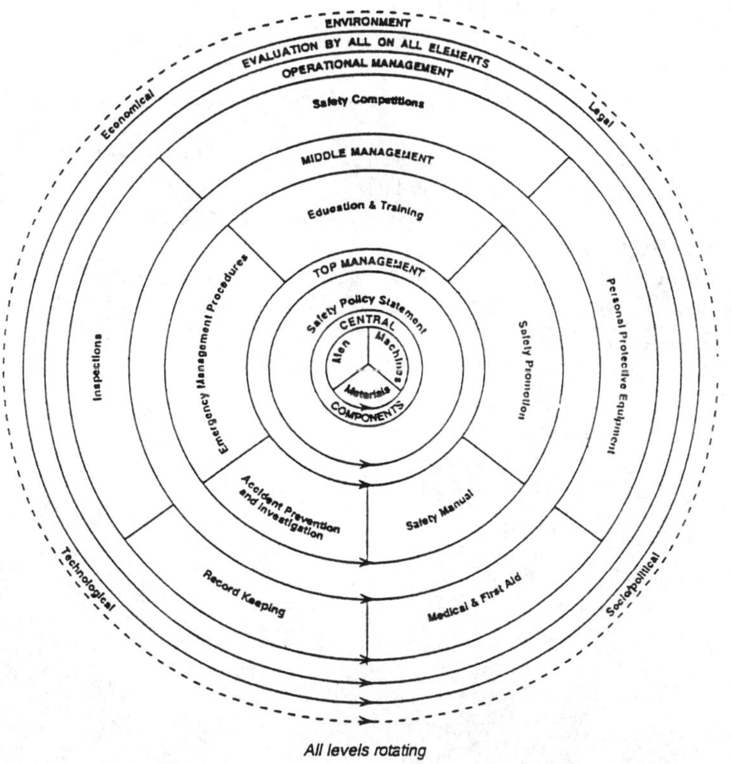

FIGURE 1.2 Model of Elements of A Safety Programme Linked to Managerial Levels

Construction companies need to promote and develop a culture of safety thinking, ensuring that the employees identify the constant dangers which exist on a construction site and take remedial action immediately.

Management of construction companies must realise that money spent on safety programmes initially is not wasted but is a benefit to the company in the long term. Working safely means increasing production which increases profits.

REFERENCES

Blake, R.P. 1964. *Industrial safety.* 3rd Edition. New Jersey: Prentice Hall.

Boylston, R. 1990. *Managing safety and health programmes:* New York. Van Nostrand Reinhold.

Davis, F.E. 1988. *Effective Safety Management New horizons in safety and risk management for the manager.* Burwood: The Craftsman Press.

Department of Manpower. 1990. *Annual report.* Cape Town: Cape & Transvaal printers.

De Reamer, R. 1958. *Modern safety practices.* New York: John Wiley & Sons.

Geminiani, F. 1998. Dissertation. *An evaluation of safety programmes in selected construction companies.* Port Elizabeth Technikon: P.E. Bookbinders.

Hodgetts, R.M. 1977. *Introduction to business.* Reading Massachusetts: Addison Wesley.

Kruger, P. 1996, 28 June, p9. *Industry related accidents.* Eastern Province Herald. Port Elizabeth South Africa.

Mansdorf, S.Z. 1993. *Complete manual of industrial safety.* New Jersey: Prentice hall.

National Occupational Health and Safety Act No.85 of 1993. Pretoria: Government Printers.

Petersen, D. and Goodale, J. 1980. *Readings in industrial accident prevention.* New York: McGraw Hill.

Working hazards to health, 1996, June 28. Eastern Province Herald. p.9. Port Elizabeth South Africa.

Workmen's Compensation Commissioner, 1990. (Report on the 1990 statistics). Cape Town: Cape and Transvaal printers.

Workmen's Compensation Commissioner, 1994. (Report on the 1994 statistics). Cape Town: Cape and Transvaal printers.

Implementation of Safety and Health on Construction Sites, Singh, Hinze & Coble (eds)
© 1999 Balkema, Rotterdam, ISBN 90 5809 036 1

Elements of a fall safety through design program

F.E.Weisgerber
Department of Civil and Environmental Engineering, University of Cincinnati, Ohio, USA
M.C.Wright
LJB Group, Inc., Dayton, Ohio, USA

ABSTRACT: Falling from elevated work positions is the second leading cause of accidental deaths in the construction industry. The large number of deaths and injuries due to falling is especially tragic because the majority of these casualties could be prevented. There are established programs, practices, systems and technologies which have been shown to drastically reduce the hazards and risks of falling. This paper discusses the safety through design approach which is particularly appropriate for construction. The outline of a comprehensive program is provided and a general review of the methods of controlling fall hazards is included.

1 INTRODUCTION

In each of the past five years, approximately 600 workers fell to their deaths at construction sites in the United States (BLS 1997). Falling from elevated work positions, such as roofs, ladders, platforms and scaffolds, is the second leading cause of accidental deaths in the construction industry. Furthermore, in each of those years, about 25,000 construction workers missed at least one day of work because they were seriously injured due to falling to a lower level. Most importantly, these statistics represent a multitude of personal tragedies and overwhelming human suffering. Also, these statistics represent a tremendous expense to contractors, to owners, to the families of the fall victims, and through this, a staggering loss to society as a whole.

The present rates of casualty and economic loss are particularly tragic when viewed from the position of knowing that the majority of these deaths and injuries could be prevented. There are recommended practices, training programs, systems and technologies which have been shown to greatly reduce casualty rates due to falls when they are adopted as components of comprehensive safety programs. While these measures are being utilized with increasing frequency, the statistics indicated above suggest they are still under-utilized.

Many employers accepted the responsibility to provide a fall safe work environment a decade or so ago. Some companies have been active in this regard for much longer. Nevertheless, it appears difficult to motivate a great number of employers to commit to a strong program of fall safety: failure to have fall protection is presently one of the most frequently cited OSHA violations. Employers become committed to fall safety for several reasons. Some employers become pro-active and develop strong safety programs on the basis of ethical responsibility. Others have had commitment thrust upon them as they react to OSHA citations. It is indeed unfortunate that commitment to a fall safety program is often embraced only after disaster strikes the work site.

Employers which resist fall safety programs frequently cite economics as the primary obstacle to utilization (Ellis 1993). However, it has been shown that it is directly economically advantageous for employers to utilize these safety measures. Providing for fall safety is sound business practice when accomplished in a pro-active and informed manner.

2 FALL SAFETY THROUGH DESIGN

In the history of attention to fall safety, the earliest approach was typically to identify an existing fall hazard and devise a method of control which usually involved an addition to the physical system, such as guardrails or personal fall protection gear worn by the worker-at-risk. Fall hazards were addressed one by one until an acceptable level of safety was achieved. More recently, a program approach has gained favor. In this approach, all identifiable hazards are analyzed and prioritized before devising methods of control. Priorities may be assigned on the basis of the degrees of risk associated with the various hazards. This approach is significantly more effective and efficient than the one-by-one approach, but is still is largely reactive rather than pro-active. The reactive program approach is often appropriate for existing operations which are found to possess fall hazards. However, for future operations, a pro-active program may result in means for controlling fall hazards which are much closer to the optimum compromise between safety and production. A pro-active program considers fall safety in the earliest stages of planning the operation or designing its associated equipment or physical systems. Thus the fall safety enhancing measures are not "add-ons" but integral features which will tend to be much more effective, efficient and economical to incorporate in the operation.

This pro-active program approach, because it is applied during initial planning and design stages of an operation has been dubbed "Safety Through Design", whether considering fall safety, chemical safety, electrical safety or any other safety issue. The National Safety Council strongly endorses this approach and has formed the Institute for Safety Through Design to promote, guide and aid the adoption of this practice (ISTD 1996).

Practically any construction project can be viewed as a future operation rather than an ongoing operation and consequently construction operations are likely candidates for using the safety though design approach to controlling fall hazards. Using this approach will require additional time for planning and this may involve additional start-up expenses, especially the first time such a program is followed. However, there are strong expectations that the early planning for fall safety will lead to more economical methods of enhancing fall safety. Moreover, the planning for subsequent operations will benefit from the first few attempts and the cost of this approach will therefore decrease with time.

3 REVIEW OF FALL HAZARD CONTROL METHODS

There are several methods used to control fall hazards (OSHA 1998). The established hierarchy of effectiveness in enhancing fall safety is indicated by the order of the following methods, going from most effective to least.

3.1 *Elimination or Substitution*

There are several methods providing increased fall safety is to remove the hazard of falling. This may be accomplished by eliminating an operation or by substituting an operation which has no fall hazard for the existing one which possesses the fall hazard. It is understood that this approach is not applicable to all, or even to many, operations in construction, but it is also true that many traditional construction operations are assumed to be immutable when indeed they are not. Foresight, creativity and strong motivation to enhance safety have together lead contractors to find alternatives to traditional operations or construction sequences which are much safer but are not significantly more costly. Occasionally, these revised operations or sequences are intrinsically more economical. They can frequently be justified as a least cost approach to establishing the required level of fall safety. That is, given that something must be done to rectify an unsafe operation, elimination or substitution may be less expensive than the costs associated with adding engineering controls or personal safety equipment to the operation. Employing the method of elimination or substitution may have a relatively high initial cost, relative to the other approaches, but generally this is a one time cost rather than a continuing one. A second redeeming feature of this approach is that it involves the exercise of considering the construction operations and sequences in depth, considering their inter-relationships and their fundamental objectives. This exercise very often leads to modifications which yield increased efficiency, economy or quality of final product which would not be achieved without completion of the exercise. A simple example of the elimination or substitution method would be, for construction of concrete/steel composite beams, to weld shear studs to the steel beams in the fabricator's shop instead of after the steel beams are erected. This eliminates the welding as an operation at an elevated work position and thereby eliminates a fall hazard. Generally, the shop welding will be less expensive than field welding. Certainly, this elimination must be considered in its relationship to other operations which may require walking the steel beams. The complex inter-relationships between various operations must be foreseen, well-understood and fully considered.

3.2 *Engineering Controls*

The approach of using engineering controls is often associated with industrial operations involving the use of machines where, for example, a machine won't operate unless the operator places both hands on buttons which are located in safe locations. The general idea is to provide mechanisms or guards so that the user would need to actively and consciously overcome them to place himself in jeopardy. At the construction site, engineering controls refers to guardrails, handrails and the like which are passively engaged to control the fall hazards.. Workers-at-risk do not need to act in order to employ engineering controls (once they are constructed, of course).

3.3 *Warnings*

Warning mechanisms inform and remind workers-at-risk to avoid circumstances and areas which constitute fall hazards. This approach often employs means to identify areas such as holes in the floor or work platform, roof or floor edges or areas with insufficient strength. Signs which instruct personnel to keep away from specific areas may be used. Areas may be more distinctly identified by roping them off or painting. When the work area changes rapidly or frequently with the construction operation, an individual may be assigned to issue verbal warnings to those conducting the operation. Applying composition roofing is one operation which may exemplify the circumstances which warrant the use of an oral warning mechanism. This approach is considered relatively effective because it prevents the worker from approaching the fall hazard. However, if the employer is not perceived as serious in enforcing the warnings, this approach will fail.

3.4 *Administrative Controls*

Administrative controls include the policies and procedures which serve to enhance fall safety and are adopted, promulgated and enforced by the employer . These frequently include company policies that require fall safety training for all workers and supervisory personnel. The administrative controls may include simple, precise and specific rules for operations, such as: "personnel shall not walk on metal decking until it has been secured to the structure with adequate screws or welds." This method may also include more general rules, such as: "all work crews shall include at least one member, preferably the foreman, who has completed sufficient training in fall safety so as to be a "Competent Person" as defined by OSHA."

3.5 *Personal Protective Equipment*

Although the use of personal fall protection equipment is the lowest item in the hierarchy of methods used to enhance fall safety, this method is very frequently utilized. Personal fall protection equipment includes harnesses worn by the workers-at-risk, retraining ropes, fall arresting lanyards, self-retracting lanyards, vertical lifelines, horizontal lifelines, ladder climbing devices, etc. and all of their appurtenances. These items are needed by personnel who are face to face with the hazard of falling. This approach has the lowest position in the hierarchy of fall safety approaches because in most applications the hazard has not been removed, nor the worker removed from the hazard. Additionally, this method always requires the worker to erect assemblages, complete attachments, or perform other actions each time it is employed. Personal fall protection equipment is frequently utilized because it can be retroactively applied to circumstances and procedures which are left practically unmodified and because this approach may be the most economical in the short run or appears to be the most expedient. However, this method is justified only when the other methods have been considered but deemed not feasible and it is often relatively expensive in the long run.

4 THE FALL SAFETY PROGRAM

The nature of the program which is best for a specific industry, company or department to effect improved fall safety depends on the specific circumstances of the unit. What needs to be done depends upon what has already been accomplished, the level of satisfaction with existing circumstances and the degree to which current operations satisfy company goals and prevailing regulations. Therefore, the following paragraphs present the outline of a general fall safety program which may be used as a model for the majority of cases. The program may be modified, extended or abbreviated as required by specific circumstances. However, it may be noted that companies with a history of concern for fall safety generally do not abbreviate this program. Rather, the more conscientious employers periodically rework all of the following steps. The following steps are presented in an approximate order of completion but several steps may progress concurrently. The nature of the last three step forces the program to include an iterative aspect. That is, the consideration of rescue methods and maintenance, along with worker feedback, generally leads to a revisit to several or all of the first five steps.

4.1 *Establish Policy*

The first step in a comprehensive fall safety program is to establish a company policy concerning fall safety. The policy statement should be brief: it should clearly and precisely indicate the desire of the employer to achieve a safe environment and assert the commitment of the employer to this goal. It may directly establish specific goals and may define the measures to be used to assess the degree to which those goals are met. It may define a time frame or plan to achieve the goals. Establishing the policy comes first because this will drive several of the subsequent steps in the program. The policy statement comes from the highest level of management but the active involvement of all tiers in its development contributes to the development of a sense of ownership and aids cohesiveness among management. The fall safety policy statement may stand alone or be combined in a more general document addressing several safety issues.

4.2 *Develop Fall Protection and Prevention Manuals*

The fall protection and fall prevention manuals include: directions, consistent with the company policies, for determining which of the various fall safety methods approaches are to be considered under various circumstances or or in response to certain fall hazards; details of the procedures and physical systems to be employed; instructions to managers, forepersons and workers for recognizing fall hazards; the assignment of responsibility for recognizing fall hazards, ensuring an appropriate approach is chosen and ensuring that workers-at-risk in fact employ the chosen method; and other any other details which are needed to carry the concept from the drawing board to actual practice. Over time, as more experience is gained, both by the employer and by the fall safety community at large, these manuals would be updated to include modifications of systems and procedures which better suit the field conditions and more effectively provide fall safety.

The manuals cannot be completed until subsequent steps have been addressed, but their development should be initiated early in the program since they will be useful to subsequent steps. Much of the information contained in these manuals may be viewed as an extension of the policy statement which itself should be concise.

Safety through design suggests that the fall safety procedures and systems become integral with the task, operation or equipment to be used. In the manufacturing industries, a design may be completed once and applied to many subsequent similar operations. Then these manuals may take on a formulaic format, stating that for such-and-such a case, a certain procedure or physical system is to be used. However, in the construction industry, it may be more difficult to develop a formulaic type of manual. Although some operations are similar in many construction projects, each project may present a fairly unique set of circumstances. Thus, in construction, the manuals may emphasize principles and fundamentals rather than plug-in solutions. It is obvious that the manuals must provide a sound basis for designing fall safety into the construction operation but still permit, even encourage, creative solutions which effectively rectify fall hazardous conditions.

4.3 *Conduct Fall Hazard Analysis*

The fall hazard analysis typically consists largely of identifying and enumerating the hazards which exist and, whenever possible, quantifying the level of risk associated with each identified fall hazard so that priorities may be assigned for rectification. Conducting a comprehensive fall hazard analysis is an extremely critical step in any attempt to enhance fall safety at the workplace. When fall protection is to be retroactively applied to an ongoing operation, this analysis involves observation of the operation. When the goal is to effect safety through design, the fall hazard analysis is conducted relative to future operations rather than existing ones which cannot be physically observed, measured or otherwise studied. Thus this step requires much experience both in fall protection and in the operation itself. It is essential that the analysts have the ability to clearly visualize the entire operation sequence and mechanics so that their associated fall hazards may be foreseen and evaluated. While this may appear to be a Hurculean task, it will be recognized that the successful construction manager generally already possesses a well-developed ability to plan and clearly visualize the entire construction sequence and the inter-relationship between the multitude of operations. Safety through design therefore additionally requires expertise in fall hazard recognition. The construction planner may gain personal expertise in fall safety through training and experience or may team with existing experts.

Generally, it is strongly recommended that the fall hazard analysis be conducted separately from and completed prior to the subsequent step, determining appropriate hard control methods, even when working in the safety through design mode and therefore dealing with future operations. Iterating cyclically through these two steps is expected to some degree, but combining the analysis with the design of rectifying actions as each hazard is identified generally is inefficient for several reasons, not the least of which is that combining the steps does not permit reasonable prioritization of corrective activities. Attempting to rectify each hazard as it is identified also obscures the inter-relationships between the hazards associated with the many different operations at the work site.

4.4 *Determine Appropriate Hazard Control Methods*

After the fall hazard analysis is completed, the appropriate hazard control methods are determined. A method is selected, based on the hierarchy of the methods reviewed above, for each fall hazard, beginning with the hazards which have been assigned the highest priority. This step may be regarded as the meat of the fall safety through design process. It will employ the manuals, which ideally guide this process, the creativity and the experience of the designer or design team. It results in feasible plans, procedures and physical systems which are expected to control specific fall hazards. Several alternate designs may be initially identified for certain fall hazards and these considered and compared to determine which offers an acceptable compromise of effectiveness, cost and practicability.

4.5 *Orientation and Training Program*

The policy statement in step one will generally declare that all workers-at-risk, as well as managers and other personnel will be trained, to various levels according to needs and responsibilities, in fall protection concepts and methods. Personnel in administrative positions will require, at a minimum, orientation to the precepts and regulations associated with fall safety. Foremen typically need training to certify them as "Competent Persons". Some of the project managers, job superintendents and especially engineers should be trained to be certified as "Qualified Persons" since OSHA regulations require that only "qualified persons" may design and supervise the erection and use of certain types of fall protection systems.

4.6 *Rescue and Evacuation*

No design or method of fall protection is complete unless plans are drawn for timely rescue of persons who may become fall victims. The method of controlling fall hazards referred to as fall arrest systems, at the lowest point in the hierarchy of preferred methods but frequently employed in construction, controls the hazard by reducing consequences of the fall, yet permits the fall to initiate. This method generally leaves the fall victim suspended somewhere between the elevated work position and the floor or obstacles below. Certainly, this method should never be used unless a reliable plan and means for immediate rescue are in place.

4.7 *Inspection and Maintenance*

When the method of controlling a fall hazard involves physical systems, inspection and maintenance must be performed periodically to ensure all components are sound and functioning as planned. The frequency of inspection should be preplanned and stated in the fall protection manual. Operations should be observed periodically, if not constantly, to ascertain that safe practices are being followed and that workers-at-risk neither circumvent the physical systems, nor neglect to don appropriate gear. Foremen and other supervisory personnel should be frequently reminded to enforce the procedures and practices assumed in the design for fall safety.

4.8 *Worker Feedback*

The opinions and observations of the workers-at-risk are important for several reasons. A surprisingly high percentage of workers initially resist fall safety programs, claiming that it hinders their productivity or that harnesses, lanyards and such decrease their safety. It is uneconomical in the long run to let these individuals fall. Consideration of their concerns and actions based on their suggestions give the everyone a sense of ownership and helps to ensure compliance with the required procedures. It is important that the workers sense a true regard for their safety. The worker-at-risk has a view of the fall hazard and its control which may be considerably different from the view of those who direct the operation or only observe it. The view of the worker-at-risk will often be very realistic and practical, and based on direct experience which can strongly contribute to the fall safety program. Care must be taken to differentiate between worker input based on knowledge and facts from that which is founded on intuition and emotion.

5 PROGRAM REQUIREMENTS

Developing an effective program of fall safety, especially one based on safety through design, requires commitment, resources and action. Obviously, there will be some expenses associated with starting this program and it may be necessary to justify the costs. Neglecting altruistic rationale, justification for incurring the expense may be based on the following points:

1) Something must be done to reduce the casualty rate due to falls at the workplace. It is, in the least, necessary to understand and comply with government regulations related to fall safety.

2) Of the possible actions to enhance fall safety, the most effective is to develop a comprehensive program.

3) The safety through design program is well-suited to the construction industry and offers significant advantages over the reactive program.

4) The safety through design program in particular offers potential for saving more than it costs, especially in comparison to the one-by-one reactive approach. These savings result from least-cost control of fall hazards, reduced insurance rates, reduced risk of citations for infractions of regulations, not to mention the potential benefits from rethinking the basic operations.

The comprehensive fall safety program may require the development of a modified attitude towards safety on the job site. The process of initiating the program may help to generate a spirit of a teamwork, especially if, to some degree, workers-at-risk attend fall safety training activities along with supervisory and management personnel. By forming a committee of personnel, all tiers may be involved in the inventory of fall hazards, and in other steps in the program.

It is essential that appropriate individuals be identified and invited to assume leadership roles in the process. The leaders should have demonstrated interest and have a strong desire to contribute to the program. The individuals should have solid experience in the operations so that they can readily foresee the progress of the construction project. Training in fall safety must be obtained by the leaders early in the process so that they become intimately familiar with the applicable regulations and develop the ability to recognize fall hazards in the future operations. These individuals need to have authority to require actions, or to have direct access to those who have such authority. Energetic persons with a pro-active attitude and demonstrated creativity in devising solutions to complex problems should be encouraged to assume positions of leadership.

There are a variety of useful tools and resources which help the safety through design fall safety program to be successful. Training programs for safety leadership, workers-at-risk, support personnel such as engineers and supervisory personnel are available from a variety of sources including consultants specializing in fall safety and manufacturers of fall safety gear. OSHA can help to identify training programs as well as provide literature on many topics, from general discussion of programs to particular details regarding its regulations. It is much better to develop a close relationship with OSHA as a resource in preventing falls than as an adversary after a fall or citation. The National Safety Council and the Institute for Safety Through Design have a strong interest in providing information to those who wish to start or improve a safety program. Regretfully, very little can be found on fall safety in published literature base and college curricula pertaining to construction include next to nothing on fall safety. Even the handful of colleges with safety engineering programs primarily relate to manufacturing rather than construction.

Many companies have found that it is cost-effective to engage a consultant in the specialty area of fall safety, especially in the initial stages of forming a comprehensive program, rather than to attempt to perform all activities in-house. In addition to immediately providing expertise and leadership at the start, where it is needed most, the consultant can remain available on an as-needed basis for reference on less routine issues after the program is firmly rooted.

6 CONCLUSION

Based on the statistics on casualties due to falls at construction sites, employers must give strong attention to this issue. The comprehensive, pro-active, safety through design approach appears to be the most effective and economical approach to controlling fall hazards, but this requires a strong commitment and progressive attitude on the part of the employer. This design program requires anticipation of the hazards associated with future operations and as such requires training and talents somewhat beyond that required by the more common reactive approaches. However, a variety of tools are available to the employer to create an effective program and there is a strong potential for the benefits of the safety through design program to exceed those of the alternatives.

REFERENCES

Bureau of Labor Statistics (BLS), U.S. Department of Labor, in cooperation with State and Federal agencies 1997. *Census of Fatal Occupational Injuries, 1992-1996*.

Ellis, N. 1993. *Introduction to Fall Protection*, Des Plaines, IL: American Society of Safety Engineers.

Institute for Safety Through Design (ISTD) 1996. *Compendium of 1996 Integrating Safety Through Design Symposium*, Crystal Lakes, IL: ISTD.

Occupational Safety and Health Administration (OSHA) 1998. *Fall Protection in Construction, OSHA Publication 3146*: OSHA.

Implementation of Safety and Health on Construction Sites, Singh, Hinze & Coble (eds)
© 1999 Balkema, Rotterdam, ISBN 90 5809 036 1

Findings of fall protection on Maui, Hawaii

Elaine Baker
County of Maui, Department of Public Works and Waste Management, Hawaii, USA
Amarjit Singh
University of Hawaii at Manoa, Honolulu, Hawaii, USA

ABSTRACT: A mail survey of roofing contractors and building contractors was conducted on Maui in September 1997. The survey asked contractors for their roof safety practices and their opinions and perceptions of factors which influence those practices such as cost and standards used in local industry. Roof slope does not control practice, with contractors practicing a range of methods from monitoring to Personal Fall Arrest Systems (PFAS) on roof slopes ranging from 1:12 to 6:12. Over half, 57%, work on roofs with slopes 5:12 and less. There is a substantial positive association between practice and perception of the industry standard. Contractors perceiving the degree of usage to be 'minor fall protection usage' practice monitoring. Those who practice monitoring do not believe costs outweigh benefits; those who use equipment believe costs outweigh benefits. A surprising finding was that, when asked for causes of Hawaii's high accident rate, 60% replied workers' compensation fraud.

1 BACKGROUND OF STUDY

This study of fall protection practices on Maui (Baker & Singh, 1998) follows a study of fall protection practices on Oahu, by Johnson & Singh, (1997). This research was sponsored by the Hawaii Department of Labor & Industrial Relations, Hawaii Occupational Health and Safety Division (HIOSH).

2 SCOPE OF STUDY

Study objectives were twofold:

- To find out the standard of practice for fall protection on Maui,
- To find out the status of compliance with fall protection regulations on Maui.

3 SIGNIFICANCE OF FALL PROTECTION STUDY

Three fatalities due to falls occurred in Maui since 1990. The object of fall protection is to avoid death and minimize injury from falls, but there are other considerations such as cost and perceptions of the standard practiced by the competition. This study explores those issues to determine their effect on non-compliance with the regulations. Efforts to contend with the factors that contribute to non-compliance can benefit the industry and the people who make their living from it.

4 QUESTIONNAIRE MAIL SURVEY

A mail survey was the most cost effective, least intrusive way of finding out what the standard of practice for fall protection is on Maui. It promised anonymity and occurred in a limited time frame, September 1997.

The mailing list for roofing contractors and building contractors was obtained from the yellow pages of the GTE phone book for Maui, Molokai, and Lanai (May 1997). A self-addressed stamped envelope was included in the mailing packet with a cover letter. Seventeen (17) roofing contractor surveys and one hundred ten (110) building contractor surveys were mailed.

The surveys were mailed out between September 24-26, 1997. All responses were received within a few weeks, well before follow-up post cards were mailed on October 29, 1997. These post cards were an effort to acknowledge the responses received as well as to stimulate additional responses and further comment.

5 SURVEY RESPONSE RATES

Thirty-three out of one hundred ten (33/110) building contractors who received surveys responded, a response rate of 30%. Seven out of seventeen (7/17) roofing contractors who received surveys responded, a response rate of 41%.

Overall the response rate for contractors is 40/127 or 32%. This study assumes that the most vocal and interested responded for the others who preferred to remain silent. This assumption is based on the fact that many respondents made comments, although unsolicited. Their responses are considered representative.

The Roofing Contractor Survey is given in Appendix 1. Unanimous response was received to five questions. Five more received a response rate of 90% or greater. The ten closed-end questions had a higher response rate than the two open-end questions. The single open-end question with the lowest overall response rate was one about common roof safety practices, 80%. The second part to the other open-end question about the state construction accident rate, asking for causes, received a 60% response rate.

6 SURVEY VARIABLES

Variables are defined as classifications which sort information into inclusive categories which are mutually exclusive from each other. Variables have four components: names, definitions, sets of categories, and a classification procedure for sorting information into the categories.

In this project, the variables are *fall protection importance, fall protection standard of practice or degree of usage, cost-benefit ratio, fall protection impact on productivity, accident investigation, fall protection regulations description, roof safety practices, injuries impact on competitiveness, accident rate cause, roof slope, and work pressure impact on safety.*

Two-thirds of the questions investigating these variables are opinion based. Two questions ask about behavior: accident investigation and common roof safety practices. One each asks for knowledge, i.e., roof slope, and perception - the degree of usage of fall protection.

7 ROOFING PRACTICES ON MAUI

The typical residential roof on Maui is a gable roof with a 4:12 slope. Common roofing materials are asphalt composition shingles, diamond rib metal sheeting, and cedar shakes. The selection of roofing materials controls the framing appropriate to its installation.

For asphalt composition shingles, 5/8" to 3/4" thick, 4' x 8' sheets of plywood sheathing are nailed directly to the prefabricated trusses. Roofing paper is tacked to the plywood sheathing prior to stapling the shingles.

For diamond rib metal roofs, 2"x 4" purlins are nailed 12" on center to the trusses; the 4'x 8' metal sheets are nailed or screwed to the purlins.

For shake roofs, 1"x 4" purlins are nailed 12" on center directly to the trusses; the typically two foot long shakes are then nailed to the purlins.

In response to the question of whether there are more single story residential buildings than two story, Tavares (1997) says the County reviews just as many single story plans as two story plans.

8 ROOF SLOPE

The most common roof slope is 4:12 for 39% of the contractors--36% for roofing contractors and 40% for building contractors. Second most common slope is 5:12 for 18% of all contractors--14% for roofing contractors and 19% for building contractors. No roofing contractors reported working on roofs with slopes >5:12.

Majorities of roofing contractors (57%) and building contractors (73%) responded with a single slope to the question "What is the most common roof slope you work on?" A range of slopes were given by 29% of the roofing contractors and 18% of the building contractors.

9 ROOF SLOPE AND FALL PROTECTION

Does roof slope control roof safety practices? All four contractors reporting "none" for the standard of practice or degree of usage of fall protection work on 4:12 and 5:12 roofs. Safety practices are: "keep your eyes open, be careful, don't hurry, don't fall, never work when uncomfortable, have journeymen with helper as apprentice."

Of the seventeen contractors reporting "some" as the degree of fall protection usage six mention tying-off. Two of the six also mention monitoring activities in addition to eight who cite monitoring alone. Scaffolding was included by one for steep slopes at the overhang. "Safety line and harness" and "tag lines anchored to ridge with safety harness" are cited by only two (2) contractors, who cite it exclusively. They work on slopes ranging from 1:12 to 12:12.

Of the eleven contractors reporting "a lot" as the degree of usage of fall protection, two mention harnesses, two mention tying and jacks, two cite monitoring, one states "standard" coded as tying and monitoring, one cites scaffolds, and three don't respond. They work on slopes ranging from 1:12 to 6:12.

Of all seven contractors reporting "complete" as the degree of usage of fall protection, two cite the usage of monitoring, two tying, one cites jacks, one cites rails as common practices, with one non response. They work on slopes ranging from 1:12 to 5:12.

Roof slope does not control practices with contractors practicing a range of methods from monitoring to PFAS on slopes ranging from 1:12 to 6:12.

10 FALL PROTECTION METHODS AND PERCEPTION OF THEIR USAGE

Common roof safety practices are categorized as follows: monitoring, tying, jacks, scaffolds, rails. Are these methods related to the perception of the degree of usage for fall protection or opinions about costs? Do contractors use methods which they see other contractors using?

Table 1 examines the relationship between two practices, monitoring and tying, and perceived degree of fall protection usage as "minor" or "major" usage.

Table 1. Fall protection methods and perception of standard.

	Monitoring	Tying
Minor Usage	12	6
Major Usage	4	7

Yule's Q, a simple statistical test to measure significance in relationships, measures here the degree of association between the two methods of monitoring and tying and degree of usage.

Q = (12*7)-(6*4)/(12*7)+(6*4) = 84-24/84+24 = 60/108 = 0.56, a substantial positive association. Contractors perceiving the standard to be minor fall protection usage practice monitoring. Perceptions are consistent with methods of practice adopted.

The association is stronger when jacks, rails, and scaffolds are included with tying (Table 2). Q = (12*10)-(7*4)/(12*10)+(7*4) = 120-28/120+28 = 92/148 = 0.62, a higher value for the association of perception of the standard and fall protection method.

Table 2. Categories of fall protection methods and perception of standard.

	Monitoring	Tie, Jacks, Rails, Scaffold
Minor Usage	12	7
Major Usage	4	10

11 FALL PROTECTION METHODS AND OPINIONS OF THEIR COSTS

Table 3 shows, according to method used, the "yes" and "no" responses to "Do you think fall protection costs outweigh the benefits?

Table 3. Categories of fall protection and opinions of their costs.

	Monitoring	Tie, Jacks, Rails, Scaffold
No	10	8
Yes	6	9

Q = (10*9)-(8*6)/(10*9)+(8*6) = 90-48/90+48 = 42/138 = 0.30, a moderate positive association. Those who practice monitoring do not believe costs outweigh benefits; those who use equipment believe costs outweigh benefits. Opinions of costs follow from method used, whether or not it involves equipment.

It was discovered that roof slope does not control degree of usage. Over half of the contractors work on slopes 5:12 or less, low sloped roofs. The type of fall protection system used, and its cost, controls the perception of costs and benefits. The costs of more equipment intensive systems outweigh the benefits of using them.

12 CAUSES OF ACCIDENT RATE

One-third of the contractors (8/24) responding "yes" to the question "Is there something specific in Hawaii that contributes to the high accident rate in construction?" place responsibility for accidents on workers who either overwork, lack skill, abuse substances, need training, are complacent, careless, or inattentive.

Only one respondent placed the responsibility for accidents on management with the statement that "Many companies do not comply with HIOSH regulations." This reflected the findings of Johnson and Singh (1997) that "the construction managers are saying that the workers are a source of the problem."

One contractor noted the effects of wind as an accident cause. Fifty eight per cent (14/24) responded that the cause of Hawaii's high construction accident rate is fraudulent workers' compensation claims.

A surprisingly large majority of those responses, 86% (12/14), target false injury fraud on the part of the worker such as fabricating accidents and lying about injuries.

13 ACCIDENT INVESTIGATION

On what basis do 60% of the contractors attribute the state's high construction accident rate to workers' compensation fraud? All contractors answered Question #5, "Do you investigate accidents?" with a response rate of 100%. Ninety per cent investigate accidents, thereby implying that their investigations--both of the accident and post-accident administration--reveal unfair practices

14 LAID UP OR LAID OFF

One contractor interviewed under conditions of anonymity noted that if there are rumors of pending layoffs, accidents increase. He cited two examples. He told one employee one morning he was going to lay him off that afternoon. Immediately following the conversation, the man left the office, picked up a beam, and returned to the office with an injured back. Laid up is preferable to laid off. Another man facing layoff fell down a flight of stairs injuring his shoulder.

Another anonymous contractor, himself injured, overheard others in physical therapy discussing ways to prolong being laid up. One scenario is to close your garage door to ostensibly move some boxes; when you reopen the door, you may be bent over from having injured or re-injured your back.

Previous injuries are a recurring problem for contractors. Most likely, a worker who has been injured will experience re-injury. The problem for any subsequent contractor is that the re-injury is a new claim to be covered by that contractor's insurance carrier. Frequently, several insurance carriers cover a recurring injury, at different times and for different contractors. Shoulders, knees, and backs are common sites of re-injury.

Both anonymous contractors had observed that when a project is near completion or when the economy is bad and there are no jobs with new projects, accidents increase. Both contractors suggested looking at the weekly benefits which are tax-free.

They also noted the ease with which benefits may be extended. The doctor sends a letter to the worker instructing him to return for a visit or the case will be closed. The worker invariably makes the appointment.

In 1993 Hawaii was second in the nation for per capita workers' compensation payments "Gale" (1996). Contractors on Maui target fraud as the leading cause of the state's high construction accident claims, artificially inflating the rate.

15 SUMMARY AND CONCLUSIONS

The following findings emerged from the Maui survey of contractors' fall protection opinions, perceptions, and practices.

- 32% (40/127) of roofing contractors and building contractors responded to a mail survey sent in September 1997.

- Roof slope does not control practices with contractors using a range of methods from monitoring to PFAS on slopes ranging from 1:12 to 6:12.
- There is a substantial positive association, as measured by Yule's Q, between fall protection method and the perceived degree of usage. Contractors perceiving the standard to be minor fall protection usage practice monitoring.
- This relationship is stronger when jacks, rails, and scaffolds are included with tying rather comparing only tying and monitoring (Q = 0.56 vs. 0.62).
- There is a moderate positive association between fall protection method and opinion that costs outweigh benefits. Contractors who practice monitoring do not believe costs outweigh benefits; those who use equipment believe costs outweigh benefits.
- Two thirds (24/36) of the contractors responded that there is something specific in Hawaii that contributes to the high accident rate in construction; one third (12/36) said "no."
- Fifty eight per cent (14/24) of the contractors responded that the cause of Hawaii's high construction accident rate is fraudulent workers' compensation claims.
- Most of those target false injury fraud on the part of the worker such as fabricating accidents and lying about injuries, 86% (12/14).
- Ninety per cent of the contractors investigated accidents.

REFERENCES

Gale State Rankings, 1996. Gale Publications.

Baker, E. & Singh, A. 1998. "*A Survey of Fall Protection on Maui for Roof Construction Works*", Research report, Hawaii Occupational Safety & Health Division, DLIR, Honolulu, HI.

Jessop, K., HIOSH Safety Inspector, 1997. Wailuku, HI, Personal communication.

Johnson, H. M. & Singh, A. 1997. *"Residential Roof Construction Fall Protection Analysis and Recommendations*", Research Report UHM/CD/97-06 University of Hawaii, Manoa, HI.

Tavares, W., 1997. Building Inspector, County of Maui, Wailuku, HI, Personal communication.

APPENDIX 1: ROOFING CONTRACTOR SURVEY

1. Do you think fall protection is becoming more important or less important or staying the same for roofing jobs? Check one:
   ______ more important
   ______ less important
   ______ the same

2. What is the standard of practice* for fall protection? Check one.
   _____ none
   _____ some
   _____ a lot
   _____ complete

3. Do you think fall protection costs outweigh the benefits?
   Check one: ______ yes _____ no

4. Do you know think fall protection slows down the work? _______ yes ______ no

5. Do you investigate accidents? ______ yes ______ no

6. Fall protection regulations: (Check whatever apply)

_______ are too complicated
_______ are too expensive
_______ slow down the work
_______ are irrelevant
_______ are unnecessary
_______ are okay
_______ are overkill
_______ conflict with the degree of usage
_______ are compatible with the degree of usage

7. Common roof safety practices are: ____
____

8. Do injuries reduce a company's competitiveness? ____ yes ____ no

9. Is there something specific in Hawaii that contributes to the high accident rate in construction?
____ yes ____ no

10. If yes, what is it? ____

11. What is the most common roof slope you work on? ___________________________

12. Does work pressure reduce safety? _______ yes _______ no

---

* Interpreted as Degree of Usage

Implementation of Safety and Health on Construction Sites, Singh, Hinze & Coble (eds)
© 1999 Balkema, Rotterdam, ISBN 90 5809 036 1

Benchmarking safety management systems in contracting organizations against best practice in other industries

M. Loosemore
School of Building, University of New South Wales, Sydney, N.S.W., Australia
H. Lingard & D. H.T. Walker
Department of Building and Construction Economics, Royal Melbourne Institute of Technology, Vic., Australia
J. Mackenzie
School of the Built Environment, University of Glamorgan, Pontypridd, UK

ABSTRACT: In issues of health and safety, the construction industry can learn from other industries that have higher performance levels. The benefits of improved health and safety performance are; increased efficiency, competitiveness and profitability; reduced delays, disputes and conflict; positive publicity; and more highly satisfied and motivated staff. This paper describes a research project that is investigating the hard and soft factors that determine safety performance in construction contracting organisations. The "critical incident technique" and "fault-tree analysis" are used to compare construction industry performance with best practice in other industries. The paper focuses upon the justification for such research and upon the methodology to be used. This methodology is designed to both increase understanding of health and safety influences and to bring about a change in performance levels within the construction industry. The latter is achieved through the incorporation of a period of reflection, implementation and feedback, where the researcher can assist participating contractors to bring about recommended changes.

1 INTRODUCTION

The construction industry has a poor health and safety record in comparison to other industries (HSE *et al* 1992, Jaselskis *et al* 1996). Dangers to health and safety exist in the construction industry because of its fragmented nature, the uncertain and technically complex nature of construction work, the uncontrollable environment in which production takes place, employment practices and financial and time pressures imposed upon project participants (Eves 1989, Helander 1991, King and Hudson 1985).

Despite its poor health and safety record, the construction industry has demonstrated a reluctance to change (HSC 1994, Hunter 1997). A sense of collective responsibility for health and safety is lacking. For example, fault for accidents is often automatically attributed to contractors but construction clients rarely indicate health and safety issues as a priority. Client requirements typically focus upon cost, time, quality, functional and aesthetic criteria. (NEDO 1975, Bennett and Flanagan 1983, Hewitt 1985, Masterman 1994) and health and safety are considered a relatively unimportant issue in the selection of consultants and contractors for construction projects (Russell *et al* 1992, Bishop 1993, Preece and Male 1997). Thus, the construction industry's poor safety record has been, in part, a natural response to an environment that has created little incentive to consider health and safety issues in the development of project management strategies. While there is evidence to suggest that clients are taking the lead in elevating the importance of health and safety issues (Whitington *et al* 1992, Ayoade and Gibb 1996), it still remains the case that contractors gain little competitive advantage from a good safety record (Preece and Male 1997). The construction industry has a long way to go in changing attitudes towards health and safety issues and ultimately the lead must come from clients. However, there are likely to be difficulties in responding to such a lead because of a lack of experience in developing efficient health and safety management control systems. The aim of this paper is to describe a research project that is seeking to address this deficiency in contracting organisations. The proposed research is based upon an "action research" methodological strategy that is designed to actively involve the research team with participating organisations to bring about an improvement in health and safety performance. In essence, the research comprises three main phases:

1) Investigating the hard (system, organisational) and soft (motivational, cultural) influences upon health and safety performance in contracting organisations;

2) Investigating the transferability of best practice in the areas identified in stage 1, from other industries to the construction industry; and

3) Implementing transferable best practice from other industries (such as nuclear, aerospace, telecommunications and chemical) to the construction industry to gauge their potential to improve performance levels.

2 THE AUSTRALIAN OPPORTUNITY

The Australian construction industry has so far avoided the punitive health and safety legislation that has been introduced in the UK (Construction Design and Management Regulations 1994). This legislation was a coercive means of controlling a construction industry that had repeatedly demonstrated an inability to improve its safety performance levels (Lander 1994, HSE 1994, Bishop 1994). However, the Australian legislative framework for health and safety has followed that of the UK (Brooks 1988, Forlin 1995) and there is no reason to suppose that it will not do so again. The Australian construction industry should pre-empt such action by pro-actively and voluntarily improving its health and safety record, without the need for restrictive legislation and taxpayer expense.

3 THE RESEARCH FOCUS

3.1 Managers

Poor management is widely accepted as a fundamental cause of many accidents within and outside the construction industry (Bird and Loftus 1976, Lingard and Rowlinson 1995) therefore, the focus of the research will be at managerial level. In particular, it will focus upon the development of improved control systems relating to health and safety issues. A model of control developed by Loosemore (1997) will be used as a basis for this research since it identifies a series of cyclical managerial activities which have to be performed efficiently if a control system is to successfully detect and mitigate the impact of potentially dangerous events. These are monitoring (the process of detecting potentially dangerous events), comparing (recognising their danger), effecting (deciding whether and how to react), implementation (implementing the final decision) and feedback (supplying information about the success of the control process for further action if necessary). Loosemore (1997a) identified a range of psychological and organisational reasons for poor performance and thus the research will focus upon hard and soft variables that may influence the performance of such systems.

3.2 Design and build projects

Procurement issues are also important in health and safety research because of power balances between consultants and contractors who are collectively responsible for health and safety performance. Traditional procurement systems reduce the ability of contractors to influence other project participants. Therefore to study the influence of contractors' managerial strategies in such an environment would be inappropriate. In contrast, a design and build procurement approach, places the contractor in a position of relative power and authority relating to project performance. This does not absolve consultants of their responsibilities but, in such an environment, it is possible to more rigorously study the influence of contractor management upon health and safety performance. For this reason the focus of this research will be upon construction managers, at a range of levels, within design and build projects. It is important to point out that while the research will be conducted within specific design and build projects as individual case studies, its focus will be at company level. Projects will be the location for data collection because they represent the arena within which company policies are implemented and where their effect can be best gauged. Furthermore, it is important to point out that while the research focuses upon control systems within contracting organisations, the high interdependency of companies contributing to a construction project, means that the influence of other participants, upon contractor performance, will have to be considered. This will make the research results of wider value within the construction industry and help to emphasise the collective responsibility which exists for improving the industry's health and safety record. Finally, the focus upon design and build contractors reduces the need to distinguish between the design and construction phases of construction projects for research purposes. The reason is, that compared to traditionally procured projects, the distinction between design and construction phases is far less obvious in design and build projects. The research will therefore take place throughout the life of a construction project.

4 THE BENEFITS OF THE RESEARCH

The results of the research will be of potential value to the construction industry as a whole but in particular to construction companies. The benefits of improved health and safety performance would include:

- Increased efficiency, competitiveness and profitability from reduced costs associated with accidents;
- Reduced delays/disputes and conflicts associated with accidents;
- Positive publicity for the construction industry and individual companies; and
- More highly satisfied and motivated staff.

5 BENCHMARKING AS A TOOL OF CONTINUOUS IMPROVEMENT IN HEALTH AND SAFETY PERFORMANCE

Benchmarking is a technique that has been developed in the manufacturing sector to bring about continuous, long-term improvement in business performance (Camp 1989, Zaire 1994). In essence, it is a management technique that provides organisations with a competitive edge by encouraging constant comparisons with best practice in key business processes (Rolstadas 1995). Comparisons can be made internally (within a single organisation) or externally (with direct competitors). Through comparison, deviations from best practice can be investigated to provide explanations of success and failure and lessons that provide the stimulus for learning, innovation and continuous improvement. In this sense, the benchmarking approach has great potential to provide the improvements in health and safety performance which are needed in the construction industry. The development of a successful benchmarking system would be a great step forward, since the benefits of benchmarking have yet to be realised in the construction industry because of misunderstanding, scepticism and resistance to change (Whitehead 1995).

6 METHODOLOGY

Stage one: Identify managerial activities in relation to organisational systems and human resource management strategies that influence health and safety performance.

The "critical incident" technique developed by Tarrants (1980) will be used to identify the conditions which lead to accidents on site. This technique involves a trained interviewer encouraging people to incidents through a special interviewing process. This will involve managers from all levels of a contractor's organisation, being asked to recall workplace incidents which they either experienced or witnessed within the month preceding the interview. Incidents are defined as unplanned occurrences with the potential to cause injury or ill-health. The methodology is essentially cross-sectional since the interviews will take place at standard predetermined periods throughout the life of each case study project. During the interviews, respondents will be guided to focus upon hard and soft variables. That is, a whole range of possible influences ranging from those which are procedural and organisational in nature to those which are cultural, attitudinal and behavioural. For each incident, subjects will be asked to complete an incident card, using a checklist format, which asks the subject to identify immediate unsafe acts and conditions that preceded the incident and more deep-rooted systems or human issues that may have contributed to an incident. By collecting this data from a variety of managers in relation to any one incident, it will be possible to cross-reference accounts and produce a reliable portrait of reality.

From the variety of accounts that result from the above process, it will be possible to construct a "fault tree". A fault tree is a logic diagram that traces all branches of events that could have contributed to an incident or accident. It is a technique similar to those used in TQM (total quality management) which are commonly used in combination with quality circles as a means of continuous improvement (Saski and Hutchins 1984, Rapid Results College 1994). A fault tree diagram depicts causses and effects and may be used to identify interdependencies between causal factors that can be critically analysed to identify their probable contribution to the incident to develop holistic managerial strategies for improved performance. A typical fault tree is depicted in Figure 1.

Once produced, the fault tree will be presented to the original respondents within a group context and their task will be to consider and refine it through open discussion, managed by an experienced interviewer. This reiterative and exhaustive process will result in a refinement of the original fault tree and a robust model of causal influences relating to a particular incident or accident.

It is anticipated that, to produce representative results, at least thirty five fault trees will be have to be produced relating to a whole range of incident types. To minimise inconvenience to participating organisations, it is anticipated that no more than two incidents should be investigated in any one case study. This means that a minimum of eighteen case studies will be used. Overlaps between the thirty-five fault trees will indicate areas of management activity and systems which are important influences upon health and safety performance. It is important to point out that those factors that will emerge from this overlapping process will be general in nature and not specific to any particular type of incident. Further research, in the future, will need to be done on identifying a typology of incidents and specific causes associated with each type. Nevertheless, the general factors identified will form the basis of the next stage of the methodology.

Stage two: Developing performance measures for modelled factors.
Having identified a range of interdependent factors that are related to safety performance, a method of measuring performance in these areas will be developed. Until the actual factors are identified it is impossible to elaborate upon the details of these measurement systems. However it is likely that the measurement will involve the development and use of a series of positive performance indicators which represent more pro-active and meaningful measures of the OH&S management effort than traditional outcome-based measures such as injury/incident rates (Worksafe Australia 1994).

Once developed, actual measurement will involve further retrospective data collection from the case study projects. This will involve liaising with participating contractors in order to more deeply investigate their existing systems relating to the processes identified as particularly important in stage one of this methodology. This will enable the production of performance indices and descriptive accounts of these aspects of each organisation.

Stage three - Identify best practice.
Research will be conducted to identify best practice in the areas identified as being of influence upon health and safety performance. This will be focused upon organisations outside the construction industry that are recognised as being leaders in each particular area. This research will first be primarily literature based but then may involve some interaction with and research within identified organisations. The aim of this research will be to use the methods of measurement developed in stage two, to develop comparative measures from best practice.

Stage four - Comparison of performance measures and descriptions from the six construction based case studies with those measured in other best practice industries.
The measures of best practice will represent the base-line against which the construction industry results will be compared. This process of comparison will enable areas of good and poor performance to be identified.

Stage five - Production of best-practice guide and implementation report to participating contractors.
This report will identify areas of strength and weakness within the sample provided by pointing to areas of divergence and concurrence between the construction industry and best practice industries. Its aim will be to provide managers with structural and behavioural recommendations that will enable them to maintain high levels of performance in successful areas and to improve performance in areas of low performance. It will also provide a basis for discussion and will serve the important function of focusing managers' minds upon the need and potential to improve health and safety performance by adopting or adapting the practices of other industries.

The report will be organised into sections according to the managerial levels typically found within construction companies and there will be a series of interdependent recommendations for each level. To facilitate easier implementation they will be presented in the form of checklists and flow charts. There will also be policy recommendations that will be presented in a more descriptive format and advice on implementation strategies that will help employees of participating contractors to internalise the proposed changes with maximum enthusiasm and minimum resistance.

Stage six - Monitoring implementation.
This stage will involve monitoring the actions of participating contractors in implementing the recommendations of the report identified in stage five of this methodology. This process will involve repeating stage four, thus enabling the identification of how the participating organisations have improved their performance in relation to best practice by adapting their managerial strategies in line with the research recommendations.

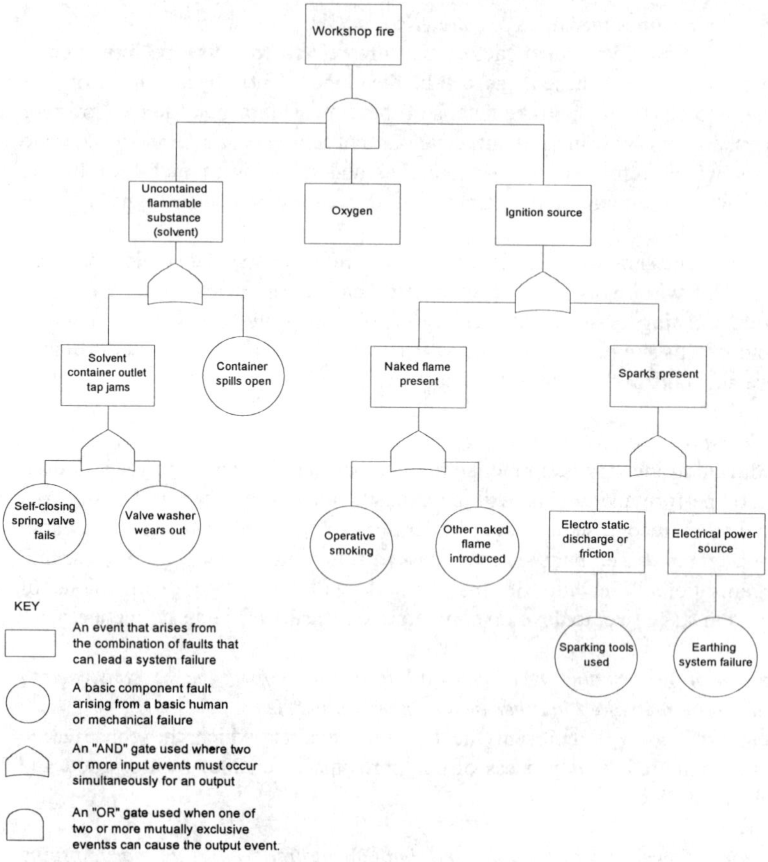

Figure 1: A typical fault tree diagram

Stage seven - Production of final research report to participating contractors.
This report will be based upon the results of stage six. That is, it will address the success of the research in assisting organisations in bringing about both cultural and organisational changes to improve the health and safety performance levels. This report will complement the intermediate report prepared in stage five by identifying areas of relative success and failure in bringing about improved performance. By doing so it will also identify where further attention, by the participating contractors, is necessary in the future.

7 CONCLUSION

This paper has demonstrated the potential value of benchmarking as a means of bringing about improved health and safety performance in contracting organisations. It has proposed a seven-stage methodology that is able to identify specific areas, both hard and soft, from which we could benefit from the experiences of other industries. This methodology is illustrated in Figure 2.

One unique characteristic of this methodology is that it is designed to both increase understanding of health and safety influences and to bring about a real change in performance levels. The latter aim is achieved through the incorporation of a period of reflection, implementation and feedback, where the researcher can assist participating contractors to bring

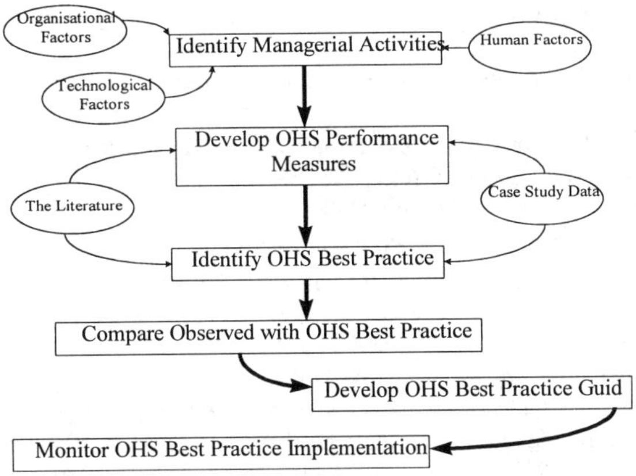

Figure 2: Illustration of overall methodology

about the recommended changes. This is important to ensure that the research has its desired impact rather than becoming a valueless symbol of good intentions which is merely used to pacify management's conscience.

REFERENCES

Ayoade, A I and Gibb A G F 1996. Integration of quality, safety and environmental systems, In *Implementation of safety and health on construction sites*, Alves-Dias, L M and Coble, R J (eds) Balkema, Rotterdam,pp 11-20.

Bennett, J. and Flanagan, R. 1983. For the good of the client, *Building*, 1 April, 26-27.

Bishop, D. 1994. The professional's view of the health and safety commission's draft construction (Design and management) Regulations, *Construction management and Economics*, 12(4), 365-72.

Camp, R.C. 1989. The search for industry best practices that lead to superior performance, *American Society for Quality Control,* Quality Press, Milwaukee.

Construction (Design and Management) Regulations 1994. HMSO, London.

Hewitt, R.A. 1985. *The procurement of buildings: Proposals to improve the performance of the industry*, Unpublished report to the college of estate management for the RICS Diploma in Project Management, College of Estate Management, Reading.

Health and Safety Commission 1992. *Conference of the professions,* HMSO, London.

Health and Safety Commission 1994. *Review of health and safety regulations - main report,* HMSO, London.

Health and Safety Executive 1988. *Black-spot construction*, HMSO, London.

Hunter, W. 1997. *Safety needs to contribute to productivity*, Safety Management, 13(4), 6.

Jaselskis, E.J., Anderson, S.D. and Russell, J.S. 1996. Strategies for achieving excellence in construction safety performance, *Journal of Construction Engineering and Management,* 122 (1), 61-70.

Lander, V. 1994. Safe as houses, *Chartered Builder*, Feb, 7.

Lenard, D., Raftery, J and McGeorge, D 1997. Designing a research methodology, *Journal of Construction Procurement*, 3 (2), 39- 51.

Loosemore, M 1997. The psychology of accident prevention, *ASCE Journal of Management in Engineering*, in press.

Masterman, J.W.E. 1994. *A study of the basis upon which clients of the construction industry chose their procurement systems,* Unpublished PhD Thesis, UMIST.

Murdoch, J. and W. P. Hughes 1996. *Construction contracts - Law and management*, E and FN Spon, London.
National Economic Development Office 1975. *The public client and the construction industries*, HMSO, London.
Preece, C. and Male, S. 1997. Promotional literature for competitive advantage in UK construction firms, *Construction Management and Economics*, 15 (1), 59-69.
Rolstadas, A 1995. *Benchmarking theory and practice*, Chapman and Hall, London.
Saski , N. and Hutchins, D. 1984. *The Japanese approach to product quality*, Pergamon.
Sawacha, E.O. 1993. *An investigation into safety attitudes and safety performance in the construction industry*, Unpublished PhD thesis, Brunel University.
Tarrants W. E. 1980. *The Measurement of Safety Performance*, Garland Press, New York.
Turner, J R 1993. *The handbook of project based management*, McGraw-Hill, London.
Walker, D H T 1995. *An investigation into factors that determine construction time performance*, Unpublished PhD Thesis, Royal Melbourne Institute of Technology, Melbourne, Australia.
Whittington, C., Livingston, A. and Lucas, D.A. 1992. *Research into management, organisational and management factors in the construction industry*, HSE Contract Research Report No. 45/1992, HMSO, London.
Whitehead, T 1995. *Benchmarking - A window on better performance*, Contract Journal, July, 18-20.
Worksafe Australia 1994. *Positive Performance Indicators – Beyond Lost Time Injuries* (Part 2), Canberra.
Yin, R K 1994. *Case study research design and methods*, Second Edition, Sage, California.
Zaire, M 1994. *Practical benchmarking: The complete guide*, Chapman and Hall, London.

25 Safety management

Implementation of Safety and Health on Construction Sites, Singh, Hinze & Coble (eds)

Safety on the construction site

H. Hessing
STV Incorporated, Manhattan, N.Y., USA

ABSTRACT: Recently, several requests for proposals have come into our office which include wording such as "the Resident Engineer will insure (or ensure) the safety of all personnel on the construction site." It seems that the client who asks for this insurance from the firm providing Resident Engineering and Construction Inspection services in effect is shifting responsibility from themselves as Owner unto others.

Indemnification provisions are common place; however, it seems every party in the construction process probably feels they are asked to shoulder a disproportionate share of the safety risk on construction sites. How can we address safety risks?

1 INTRODUCTION

The Resident Engineer on a construction site is there not just to ensure that the design is implemented but implemented in a safe manner. Good Contractors are safety conscious and develop safety programs and procedures as part of their initial construction process. The Contractor must both control his personnel and operations and perform his work in a safe manner, not just for his work force but also for the public welfare.

2 SAFETY THE NUMBER ONE PRIORITY

The emphasis on safety as the number one priority in both the design and construction phase of a project is illustrated by two examples: rehabilitation of the Long Island Expressway and demolition of an FAA tower.

Mr. Gary L. Brinderson of Brinderson Companies, General Contractor was quoted in Engineering News Record, p 93, November 17, 1997 in an article entitled "Don't Punish Safety Violators." He suggested that the construction industry "emphasize safety over expediency..." Safety was a priority in both the design and construction process.

2.1 *Safety over expediency*

The Long Island Expressway is a major east - west corridor on Long Island, New York. The average daily traffic (ADT) is more than 100,000 vehicles per day. In the mid eighties, there was a rehabilitation project on the Long Island Expressway. The project required the Contractor to do concrete roadway repairs at night. To accomplish this the Expressway was closed at 9:00 P.M.

and reopened at 5:00 A.M.. Truck traffic was diverted to Jericho Turnpike while vehicular traffic was diverted to Northern State Parkway.

Three hundred and fifty (350) maintenance of traffic signs were installed. Approximately, one hundred and fifty (150) were lit. These were spaced over twenty miles. In addition, the overhead motorist information system was constantly in use providing information to the traveling public. Two hundred signs were covered or uncovered on a nightly basis.

Lit barricades were used to cordon off the four mile stretch of highway. Behind them were placed inertial barrier systems at both ends of the expressway and each ramp. Behind the inertial barriers additional lit barricades were placed. The Contractor also left a ten yard dump truck transversely behind the barricades. Contractor and inspection personnel used blinking safety lights on the tops of their vehicles. H.T. Schneider, General Contractor exercised safety over expediency in establishing a safe work site.

In addition, inspector and two men from the Contractor's work force verified that each sign was lit, covered or uncovered as appropriate for each night's work at the beginning and end of each work period. Also, three Nassau County police cars helped open and close the Expressway daily.

All these activities were noted in the design drawings and specifications. It was clear that the design intent emphasized safety.

This work took time and patience. Another contractor might have tried for expediency. The Contractor on this project took a professional attitude towards safety in fulfilling his contractual responsibilities.

2.2 *Three unexpected and uncontrollable incidents*

In spite of these precautions, the west bound inertial barrier was hit on consecutive Friday nights by the same driver. He and his partner had stolen cars and were deliberately chased into the barrier by the police (who knew the inertial barrier was in place).

In another incident, one driver went around the barricades and inertial barrier on a ramp, drove down the embankment and onto the roadway work area because he "saw there was no traffic on the Expressway." The final incident involved a drunk driver who drove around all the barriers and was arrested by the police.

2.3 *Emphasis*

The emphasis on safety as indicated by the design drawings, as carried out in the field by the Contractor, as reviewed on a daily and nightly basis by the resident engineering staff, and the Contractor's personnel prevented these three incidents from being fatal accidents.

2.4 *Demolition of an air traffic control tower*

The Port Authority of New York and New Jersey (PANYNJ) removed an existing 150 foot tall air traffic control tower (ATCT), situated north of the International Arrivals Building (IAB) at John F. Kennedy International Airport (JFK), and constructed a new air traffic control building. To allow for future expansion of the IAB the existing tower, constructed in the late forties and reconstructed in 1953, had to be demolished. Demolition was completed in June, 1996.

2.5 *Concerns*

There were several safety concerns involved with demolishing an existing structure in an active airport. The ATCT was situated between a parking lot, several roadways, and the IAB. Pedestrians crossing between the parking lot and the IAB, and vehicular traffic on the roadway system had to be separated and protected from the demolition.

2.6 *Safety over expediency*

From the onset the PANYNJ was concerned with coordinated site safety. This was expressed in the design documents, emphasized during revue stages, pre-bid, bid and award phases. Environmental plans and permits were obtained by PANYNJ. This work preceded demolition and was closely monitored.

Leon D. De Matteis Construction Corp. of Elmont, New York was the Contractor. He was authorized to use part of the adjacent parking lot for his staging area. This area was enclosed with a painted, plywood fence and protected with white concrete barriers. Painting the plywood fence was an esthetic requirement.

Scaffolding was required to facilitate glass removal from windows and to try to ensure that glass would not blow off the building. Approval was obtained from the PANYNJ Risk Management Office; the Fire Department of New York; and the Bureau of Fire Prevention. Scaffolding could not interfere with access to the stand pipe (Siamese connection). A blue light had to be installed to illuminate it. Scaffolding was installed around the tower in December, 1995. The work area was protected by painted plywood fence and white concrete barriers. Gates were padlocked. Regional, the subcontractor, installed a wire mesh debris net as described in the Contract documents. A sidewalk shed with a parapet, and lighting, was installed to protect pedestrians. Scaffolding remained in place during all phases of the tower demolition as demolition progressed from roof to slab. This work was done in compliance with the New York City Building Code, Safety of Public Property During Construction Operations, Article 19.

The monthly meetings between the various managers emphasized work zone and environmental safety. Safety tool box meetings were held weekly. Copies of minutes were submitted to the construction manager. Typical discussions included use and storage of acetylene tanks; wearing hard hats, vests, eye protection, and keeping the site clean and dust free. The construction manager had fire extinguishers placed on every floor. With the cooperation and efforts of the Contractor, this was the safest and cleanest job seen in recent memory.

3 RESPONSIBILITY FOR SAFETY

If the goal is Safety on the Construction Site as the number one priority in both design and construction, then responsibility for safety belongs to everyone who is involved in the constructed project . Safety starts with the design engineers who prescribe safe operations and procedures in their drawings and specifications and continues with safety conscious personnel in the field.

3.1 *Safety plan*

To create a safe working environment, to comply with OSHA, and to encourage others to do so as well we should carefully plan each project and enforce safety . We can create a safety plan or management system to accomplish this task.

We should audit our safety plans, procedures and management system on each project. We should be flexible in making adjustments and course corrections to promptly comply with the most logical and safe construction conditions.

3.2 *Safety training*

We can increase safety on the construction site by implementing a safety training program (as noted in STV's construction management manual) and instruct individual employees in the use and care of protective equipment. This has been part of our company process since 1987. We can hold regularly scheduled safety meetings. We can prepare check lists to be used at these meetings and in the field.

3.3 *Emergency response*

We can minimize emergency response time for police, fire and ambulance personnel by meeting with them and establishing open communication and practical site specific procedures. This should be done immediately after the "get start meeting" in conjunction with the Contractor's Site Safety Plan.

3.4 *Minimize hazards*

We can minimize hazards to the public by access control and protection plans to secure the construction site. Each project has its specific safety hazards that need to be addressed. Implementing a site safety program is a starting point.

4 SUMMARY AND CONCLUSION

The two projects cited had safety as the priority in both design and construction:

(a) The Long Island Expressway project design emphasized safety. The contractor practiced safety over expediency. The Resident engineering staff monitored his efforts.

(B) Demolition of the air traffic control tower at JFK airport required safety be emphasized in the design phase, and during demolition at weekly tool box meetings, and monthly progress meetings.

Everyone involved in the constructed project is responsible for safety. We can supply the materials and methods to support the construction team's efforts to maintain a safe job site. We can prepare a safety plan or management system. We can make it site specific.

The more often it is stated that safety is our first concern eventually it will become habit both in design and in construction.

REFERENCES

New York City Building Code, 1989, Article 19, *Safety of Public and Property During Construction Operations,* Building Trades Employers Association

Seelye Stevenson Value & Knecht, 1987, *Construction Management Manual - Safety Program*

Implementation of Safety and Health on Construction Sites, Singh, Hinze & Coble (eds)
© 1999 Balkema, Rotterdam, ISBN 90 5809 036 1

Evaluation of safety content of registrar of contractors' examinations

K.D.Walsh & H.H.Bashford
Arizona State University, Tempe, Ariz., USA
S.A.Rashid
BKS (Pty) Limited, Engineering and Marketing, Johannesburg, South Africa

ABSTRACT: Registrars of Contractors are tasked with licensing contractors, which is usually done (at least in part) through the administration of examinations, usually conducted in at least two phases. The first consists of a business management examination. The second consists of a trade-specific examination(s), which relate to the practice and important codes for the trade. In this study, the relative weight given to worker safety on the examinations was investigated. Relatively little weight was given to worker safety on the trade examinations. Only 15-20% of the trade examinations contained any worker safety questions at all. Some 6-8% of the questions on these examinations concerned worker safety. The business management examinations contained about 10% worker safety related questions. Based on the needs of each trade area, recommendations for appropriate safety content for each trade were made. A summary of examination safety content and recommended additions are presented.

1 INTRODUCTION AND BACKGROUND

In most jurisdictions in the United States, a person seeking to enter business as a contractor must obtain some form of license in order to enter contracts and receive payment. These licenses are typically provided by a state agency or quasi-state agency (usually called a Registrar of Contractors, a Contractors Licensing Board, or similar), although several large municipalities maintain their own licensing organization, and several states have no licensing procedure at all. Clearly, there are no national standards for this process, and no apparent effort to develop such standards, which is part of the motivation for a national certification program being advanced by the American Institute of Constructors.

Because of the lack of standards, the process varies from place to place. However, generally the prospective licensee must demonstrate applicable experience and submit to one or more examinations. Some degree of standardization exists in the examination process, because a single company, the National Assessment Institute (NAI), provides and conducts examinations for over 30 state licensing agencies. This consistency is less than one might expect, because the registration agency in each state is advised or directed by a board of local contractors, who have some say in the type of questions and the manner of testing in each jurisdiction. Therefore, the laws and examinations may show significant differences in different locations, depending on the makeup and philosophy of the advisory board and the amount of influence accorded to contractor associations, unions, or individual contractor representatives. Still, the NAI does maintain a national question bank, and there is enough uniformity to allow national publication of examination review manuals and seminars (Gladstone 1989).

A large electrical utility in Phoenix, Arizona, recently undertook a safety improvement program based largely on the Construction Industry Institute Zero Injury Techniques (Zero Accidents Task Force 1993). This program led the utility to investigate the safety programs of their contractors, and they began to receive documentation of contractor's licensing as evidence of acceptable safety performance. Contractors argued that the Arizona Registrar of Contractors was charged with assuring the degree to which the contractor adhered to standard business norms, and licensing was therefore evidence of industry-standard safety programs. The utility was generally unimpressed with this argument, primarily because they were looking for safety performance above industry standards. However, the discussion did awaken curiosity about the licensing process and the degree to which worker safety was a consideration in the licensing process. As a result, the utility asked the Alliance for Construction Excellence (ACE) at Arizona State University to obtain data about the examination in Arizona. An academic team consisting of a graduate student and ASU faculty was formed, with a research task force constituted by ACE to serve as a sounding board for the research. The task force consisted of practicing contractor and owner safety personnel and a representative from the Registrar of Contractors. This paper presents the results of this investigation, and while the reader must be mindful of the variability between jurisdictions mentioned above, the authors believe that the evaluation process developed for this investigation provides a useful model for other jurisdictions. Further, as Arizona is an NAI state, the general content issues are believed to be reasonably widespread.

The heavy use of qualifying language in the preceding statement is necessitated, in large part, by the secrecy in which the examination and its development are shrouded. The reason for this secrecy is understandable, and the authors were never permitted access to the actual test bank, nor do they have any intention (or ability) to reveal information which is not potentially available to prospective licensees.

NAI develops questions for their test bank following a standardized process. Subject areas are recommended by the licensing agency which holds NAI under contract. NAI maintains a technical staff and a number of consultants who develop multiple choice questions. The questions are then vetted in a test board process, typically conducted at the NAI main office in Utah with some frequency, and less often at dispersed locations among the states where NAI provides the test. In the board process, licensed contractors and members of the licensing agency evaluate the proposed questions and answers for clarity, accuracy, and applicability. Questions which are deemed acceptable in the test boards become part of the test bank, which is available directly on a computer for instant scoring in many locations. Clearly, there are several opportunities for the local industry to influence content, and one could argue that the examination is thereby an expression of what is important to the building community.

2 DESCRIPTION OF CONTENT

To become licensed in Arizona, the licensee must certify at least four years of related experience and demonstrate qualification for the type of work the potential contractor proposes to perform. The licensee must pass a business management examination in one of two categories (general or specialty). The business management examination primarily tests the applicant's general management and legal acumen, and relates to procedures for maintaining and running a business and to understanding of contracts. Table 1 shows a summary of the content areas for the business management examinations as of the start of this investigation.

Table 1. Content Summary for Arizona Business Management Examinations (source: National Assessment Institute)

	Coverage (% of total)	
Subject Area	General	Specialty
Project Management	19	19
Contract Management	19	18
Safety Regulations (OSHA)	10	10
License Law	10	12
Financial Management	10	10
Labor Law	9	10
Tax Law	7	5
Risk Management	6	6
Business Planning and Organization	5	5
Lien Law	5	5

In addition, the licensee must specify one or more specific trade area(s) for licensing, in one of two categories. The Commercial category includes 62 trades, of which 33 require the completion of an examination. The Residential category includes 75 trades, of which 31 require completion of an examination. Trade examinations focus on the skills required for that specific type of construction. Because there are so many examinations a complete summary of all content is impractical in this space, but as an example, the content areas for two of the commercial classification examinations are summarized in Table 2. The examinations vary by trade, but average about 80 multiple choice questions and require completion of 70% correctly for a passing score. The average candidate usually spends about 3 hours on an exam. Those trades which do not require a trade specific examination require documentation of related experience.

A complete breakdown such as the one shown above was provided by NAI for every trade examination, and was compiled by the task force. Areas in which worker safety was tested were noted. It should be pointed out that the materials provided by the Arizona Registrar of Contractors and the NAI listed problems specific to applicable codes (such as UBC provisions) as "safety," indicating the importance of good understanding of codes to the safety of the public after completion of the work. However, the emphasis for this study is on worker safety during construction, and so only questions which relate to worker safety and OSHA were considered. A summary of the extent of worker safety testing for Arizona contractor licensing is provided in Table 3.

Table 2. Content Summary for Two Arizona Commercial Trade Examinations (source: National Assessment Institute)

Classification	Subject Area	Coverage (% of total)
Blasting (A-3)	Blasting Techniques	15
	Explosives	15
	Safety Regulations (OSHA)	15
	Layout	10
	Electrical Circuits	10
	Non-Electric Circuits	10
	Priming	9
	Blasting Equipment	8
	Storage	8
Piers and Foundations (A-7)	Piles and Caissons	20
	Footings and Piers	15
	Formwork	12
	Concrete	12
	Reinforcement	10
	Excavation	10
	Shoring	8
	Layout	5
	Public Safety (Codes)	5
	Plan Reading	5

Table 3. Summary of Safety Content of Arizona Registrar of Contractors Examinations

	Business Management	Commercial	Residential
Number of Classifications	2	62	75
Number Based on Experience Only	0	29	44
Number Based on Experience and Exam	2	33	31
Number of Exams with Safety Content	2	5	6
Average Safety Coverage on Exams with Safety Content	10%	8%	6%

3 ANALYSIS OF CONTENT

It is clear from Table 3 that the coverage of worker safety on the trade examinations is fairly low. The Business Management examinations include worker safety questions, but as the examination is required for all contractors, the questions relate to home office practices (e.g. record-keeping requirements, first aid kit requirements, OSHA poster requirements). Specific questions regarding the field application of worker safety requirements are relegated to the trade examinations, and it is obvious that there are relatively few trade examinations in which safety is covered.

A further breakdown is provided in Table 4, which provides the specific commercial trade classifications for which an examination is required, with examinations covering safety separated from examinations without safety coverage. A similar table could be provided for the Residential area, but is withheld in the interests of space.

The underlying rationale governing which trades are listed in which column of Table 4 is not obvious. If one wanted to make the argument that worker safety was not an appropriate topic for the ROC examination, that instead the OSHA certification, insurance, and qualified/competent person standards were adequate for this issue, then there should be no trades listed in the first column. If instead one wanted to argue that worker safety is an appropriate topic, then the classifications selected are curious. The trade classifications for which worker safety coverage is present on the examinations include obvious dangers (blasting) and less obvious (steel reinforcing). The roofing trade classification includes written examination coverage of worker safety, but other trades with similar worker safety concerns, fall protection for instance, have no worker safety content (steel and aluminum erection, electrical and transmission lines, carpentry, fire protection, or masonry, among others).

As a result, the authors attempted to develop a basis by which the decision could be made to determine firstly, whether a given trade should include safety coverage, and secondly, what specific topics should be included. As can be seen in Table 2, the examination content summaries available from NAI already included a breakdown of the topic areas considered on each examination. It would be possible to identify a number of OSHA regulations which apply to all of the topic areas. However, the tradition of the ROC examinations is to focus on the areas of importance to successful performance as a contractor. Accordingly, lists of key regulatory issues were obtained.

Two specific lists were identified. First, a listing of the 30 most frequently-cited OSHA standards in Arizona (JN/OTI AZ 1993) provided a means to focus examination topics to regulations of particular financial and regulatory difficulty for Arizona contractors. Another proposed list which could in principle be used would be a listing of the most frequent violations leading to fatalities or injuries. Lengthy and emotional discussions were held by the task force regarding which resource would be the best, with the emphasis on the message sent by the choice of list. Ultimately, the list of frequent citations was used, under the reasoning that this list would include citations arising from investigations of serious incidents or fatalities, but would still maintain the importance of the financial ramifications of violating standards.

This list was then categorized into general topic areas. For example, the availability of a

Table 4: Commercial Trade Classification Listing by Coverage of Safety

Trade Exams with Safety	Trade Exams without Safety
Blasting; Waterworks; Roofing; Steel Reinforcing Bar and Wire Mesh; Plumbing (including Solar)	General Engineering; Piers and Foundations; Swimming Pools; Steel and Aluminum Erection; Electrical and Transmission Lines; Swimming Pools (including Solar); Landscaping and Irrigation Systems; General Commercial; General Small Commercial; Boilers, Steamfitting, and Process Piping; Carpentry; Concrete; Electrical; Elevators; Fire Protection Systems; Masonry; Plumbing; Air Conditioning and Refrigeration; Septic Tanks and Systems; Irrigation Systems; Commercial, Industrial Refrigeration; Water Conditioning Equipment; Comfort Heating, Ventilating, and Evaporative Cooling; Glazing; Low Voltage Communication Systems; Boilers, Steamfitting, and Process Piping (including solar); Air Conditioning and Refrigeration (including Solar); Factory Fabricated Pools and Accessories

hazard communication plan and the needs for employee training might both be described as management issues, and might also be important to trade classifications which involve the use of hazardous materials. This process was conducted for each of the 30 standards. Then, the topic area list for each classification was cross-referenced with the list of citation categories, to develop a recommendation for the most appropriate safety question topic areas for each trade classification. For example, questions related to frequently cited standards regarding use of hardhats, scaffolding, and fall protection were referenced for the masonry trade classification.

The precise level of coverage which should be provided to each topic area was not addressed in the recommendations of the task force. As has been noted, the NAI conducts an extensive vetting process for the test bank, and maintains a dedicated staff for question development. The task force mission was primarily to understand the content of the examinations, and then to make recommendations for suitable content. It is clear that each examination must cover a number of important topics, and as a result the percentage coverage given to any specific topic is a matter of some political importance to the advisory boards and the agency. The task force did, however, make recommendations that the coverage of worker safety be increased, along with specific question subject areas for each examination.

4 SUMMARY AND CONCLUSIONS

A study was reported of the import of safety regulations on the contractors licensing process in Arizona, USA. The contractor licensing process is not standardized across the country, and as a result the procedures and testing requirements vary from jurisdiction to jurisdiction. However, the National Assessment Institute provides contract testing services for a majority of the jurisdictions, which provides some degree of standardization. Even so, the specific findings and observations of the study are most directly applicable to Arizona; the process used to evaluate the examination content is more generally applicable.

NAI provided the authors with a summary of the content areas on all of the business management (2 examinations), commercial (62 classifications, 33 by examination), and residential (75 classifications, 31 by examination). Both of the business management examinations included worker safety coverage (10% of the examination), mostly focused on the corporate management implications of worker safety regulations. Very few (about 17%) of the trade examinations covered worker safety at all, although public safety and code compliance were important issues. There was no obvious rationale for the selection of trade classifications which included worker safety and those which did not.

The trade classification topic areas were then compared with the content of a list of frequently cited standards in OSHA violations. This list was used to identify standards with financial and operational implications for Arizona contractors. Cross-referencing was then performed, matching the trade classification topic areas to the general subject of each frequently cited standard. This cross-referencing led to a list of recommended question topics related to worker safety for each trade classification. Additional lists could be used in other geographic regions to perform a similar analysis.

The recommendations were transmitted to the NAI and the Arizona Registrar of Contractors in 1995. NAI undertook question development over the ensuing 18 months, and an examination board was convened to discuss these (and a number of other questions in unrelated topics) in 1997. Following this process, the examination now includes significantly more coverage of worker safety issues. This is true not only in Arizona, where the study was conducted, but also in other states within the NAI system.

REFERENCES

Gladstone, J., Perpich, G. & Perpich, S., 1989. *Contractor's Exam Book.* Coral Gables, FL, USA: Engineer's Press.

JN/OTI AZ. 1993. *The Most Frequently Cited Violations (Serious), January 1990-December 1992.*

Zero Accidents Task Force. 1993. *Zero Injury Techniques*. Publication 32. Austin, TX, USA: Construction Industry Institute.

ACKNOWLEDGEMENTS

The authors gratefully acknowledge the contributions of Mr. Gary Barras of the Salt River Project, Phoenix, Arizona, who served as an industry advisor to this research and chaired the task force. Appreciation is also due Ms. Allison Staker and Mr. David Buhler of the National Assessment Institute, Mr. Michael Goldwater, the Arizona Registrar of Contractors, and Mr. John Moll, Chief of Examinations at the Arizona Registrar of Contractors. This study could not have been conducted without their assistance and support.

Implementation of Safety and Health on Construction Sites, Singh, Hinze & Coble (eds)
© 1999 Balkema, Rotterdam, ISBN 90 5809 036 1

Construction safety in developing countries

R.J.Coble
M.E.Rinker, Sr. School of Building Construction, University of Florida, Gainesville, Fla., USA

T.C.Haupt
Department of Construction Management and Quantity Surveying, University of Florida, Gainesville, Fla., USA

ABSTRACT: The importance and role of the construction sector in the economy of any country has been confirmed in several studies. In fact the findings of these studies have confirmed that the construction industry reflected the level of economic development within a country. Furthermore, these studies imply that the construction industry must be developed to create and allow accelerated economic development. For this reason, inter alia, the construction industry is more important in developing or emerging countries than it is in developed countries. Characteristically, the construction industries of emerging countries suffer from poor productivity and increasing construction costs. Additionally, poor performance in respect of health and safety in the construction industries of these countries contributes significantly to poor productivity, increased costs, unsatisfactory quality performance, time overruns and undesirable impacts on the environment.

This paper argues that governments can play a proactive role in promoting that safety and health management systems be integrated into the entire construction process. Further it suggests that this commitment from government must be accompanied by commitment from all construction project clients, all levels of management of construction firms regardless of size or capacity, and a reciprocal commitment by construction workers to implement safety and health plans.

Keywords : health and safety, developing countries, costs, productivity

1. INTRODUCTION

Good conditions of health and safety constitute good business practice. Therefore the integration of safety and health measures into a total management system in the construction sector can contribute significantly to cost efficiency, quality assurance and environmental protection. However for this to be achieved there needs to be commitment from all stakeholders in the industry to the process of improving construction safety from government through to the ultimate end users. Ironically, in spite of their threadbare budgets, many governments in developing countries attempt to offer comprehensive care for all – a strategy which results in very inefficient spending (Saxenian 1994). It would be more effective if governments would target some of these funds at efforts to improve safety in an industry which has consistently been responsible world wide for more fatalities than any other. In fact in the industrialised countries, accidents now cause more deaths than all infectious diseases and more than any single illness except those related to heart disease and cancer (Britannica Online 1998a) Governments have the authority to direct public spending to support any of their nationally defined commitments, one of which should be safety and health in construction.

In developing countries it is evident that this commitment is lacking resulting in a safety and health culture which is low. This is not only true of governments but of all participants in construction. It is increasingly acknowledged that only the larger construction organisations use appropriate management systems to control safety, quality and environmental impact, and then only on an *ad hoc* and almost voluntary basis. This is unacceptable if the optimum benefits of improved safety practices in construction are to be derived universally. This paper argues that if added pressure were brought to bear on construction practitioners at all levels by governments in developing countries for these systems to be integrated into the entire construction process from project inception through to final project completion and beyond, there would be significant improvement in productivity, quality, reduced costs and environment protection. This commitment from government must be accompanied by commitment from all construction project clients, all levels of management of construction firms regardless of size or capacity, and a reciprocal commitment by construction workers to implement safety and health plans.

The importance of health and safety as it relates to developing countries is an issue that is essential to the economy of these countries. The pivotal role and the importance of the construction sector in the economy of any country has been confirmed in multiple studies (Strassman, 1975; Turin, 1973; Wells, 1985; Drewer, 1980; Ofori, 1980). The findings of these studies have validated the position that the state of the construction industry reflects the level of economic development within a country. Furthermore these studies imply that the construction industry must be developed to create and allow accelerated economic development. With the increase in construction activity the increase in safety efforts in construction should also increase.

2 THE STATE OF THE CONSTRUCTION INDUSTRY IN DEVELOPING COUNTRIES

The construction industry in developing countries is typically underdeveloped, dysfunctional or non-existent. Where it exists it is dominated by foreign companies with characteristically very little technology transfer to the indigenous construction community. Developing countries, especially in Africa, are littered with 'white elephant' projects that will never be completed due to foreign exchange difficulties and lack of domestic capacity. Additionally, huge uncertainties have been introduced into construction activities as a result. The variability in construction costs, construction material supply constraints, investments in construction technology and personnel have become high risk ventures. Construction entrepreneurs have often found these risks too great to undertake. As a result, the construction capacity of most developing economies has been undermined in various ways (ILO, 1987).

Several of the developing countries, especially those in Africa, do not have uniform national building codes and regulations. Where they do exist they are inappropriate, ineffective, out-moded, out-of-date and based on conditions which prevailed while they were still colonised.

Inadequate capacity has created difficult and onerous market conditions, introducing seasonality to construction demand because of the fact that only the government has been able to operate under these conditions. The fortune of the construction industry in developing countries is largely determined by government budgetary circumstances. During economic recessions, the first area to suffer 'cut-backs' is capital expenditure which underpins most construction projects. While this trend is universal, the precarious financial situation of most developing countries makes the construction sector particularly vulnerable. This stop-go-policy has severe implications for skills training and development. Training in any of the construction related disciplines or professions has become a high risk venture when inadequate capacity makes it difficult to utilise the acquired skills domestically. Under this scenario, the industry remains under-capitalised in terms of human capital formation. Typically, the client base of the construction industry is not well diversified between public authorities, public corporations and private enterprises with the government remaining by far the major client.

The consequence of this scenario has been that safety and health in construction has not received close to the attention that it has in the developed countries, with resultant high accident, injury and fatality rates.

3 CULTURAL ISSUES

Culture has been defined as the integrated pattern of human knowledge, belief, and behaviour (Britannica Online 1998b). It consists of language, ideas, beliefs, customs. taboos, codes, institutions, tools, and techniques. Every human society has its own particular culture. Variation is attributable usually :

- to physical habitats and resources; and
- to the range of possibilities inherent in various areas of activity such as language, customs, the use of tools, and the degree of social development.

The attitudes, values, ideals, and beliefs of individuals are greatly influenced by the culture in which they live. Changes take place within and among cultures :

- by means of ecological and environmental changes;
- by diffusion of advantageous cultural traits;
- by acculturation which refers to the process of change which results from contact of societies with different cultural traditions; or
- by the evolution of cultural elements over a period of time (Britannica Online 1998b).

In most developing countries cultural issues have impact on virtually most aspects of daily living. The construction industry in these countries has certainly not escaped these impacts. In some primitive elements of the societies of developing countries there exists the belief that illness, injuries and even deaths come from the displeasure of ancestral gods and evil spirits, from the malign influence of evilly disposed persons, or from natural phenomena that can neither be forecast nor controlled (Britannica Online 1998c). To therefore demand the blanket introduction, implementation and enforcement of traditional prescriptive safety measures on construction sites while ignoring the cultural impacts of these measures, is unrealistic and doomed to failure. This is exacerbated by an environment typically devoid of uniform building codes. Workers need to be evaluated as to their customs, practices and value systems and a performance based safety and health based programme adapted to these characteristics.

Many issues that would never be acceptable safety and health practices in developed and industrialised countries, may in fact be the best practice in developing countries. Although there are many practices which follow this truth, a good example is the use of bamboo scaffolding which allows for flexibility in use. While frowned upon in developed and industrialised countries it is very acceptable in developing countries because of the culture of the people – by definition an example of an integrated pattern of behaviour. Barefooted workers would never be tolerated on construction sites in developed and industrialised countries. In several developing countries however this is allowed and often considered good practice. These workers have feet which have become conditioned and catalysed to "unsanitary" conditions and reasonably resistant to sharp objects, allowing them to be able to use their feet better to feel unsafe footing and footholds.

It would be expedient therefore, in the context of construction sites in developing countries, to bring about changes in the culture by diffusing only those cultural traits or behaviours which are advantageous and which are directly resultant from environmental and ecological changes. This would imply the introduction and development of an appropriate safety culture.

4 TOTAL SAFETY CULTURE (TSC)

A safety culture is something which develops over time. In developing countries the safety culture is more effective than safety devices and may be the best proactive measure available to construction managers. To this end a total safety culture can, if not already in place, be developed by examining the pre-existing cultural setting and building on the prevailing attributes. For many poor countries the inculcation of a safety-and-quality culture may be one of the best (and cheapest) ways to improve the environment (Cairncross 1994).

This suggestion is supported by the findings of a recent study done in a rural context in Cape Town, South Africa, which suggested that at first glance the presence of any safety culture on the construction site was not apparent. While it seemed that every safety practice was being violated in the process of construction based on what one would expect in a developed and industrialised country, the accident and injury incidence rate on this particular project was extremely low. The workers had adapted to the prevailing circumstances and developed their own successful safety culture.

The question which begs to be answered is whether if one introduced safety and health standards requiring compliance with copious regulations and requirements such as the wearing of safety devices this would significantly improve the already existing safety culture and safety performance.

From the Total Safety Culture (TSC) conceptual perspective, every person involved in the construction process has a personal, individual responsibility on a daily basis in maintaining worker safety and health on construction sites regardless of size, form or location. The TSC perspective according to Blair (1996) requires movement toward a philosophy that embraces total involvement, true empowerment, viewing safety as an intrinsic value rather than merely a priority, and improved and recurrent training methods. This approach promotes safe work practices which are supported via people who actively care about safety on a continuous basis (Geller, 1994). If only management were trying to bring this about, only 3% of the available workforce would be actively engaged while the remaining 97% would be uninvolved (Muller 1997).

Two studies were undertaken in the United States to determine the main causes of accidents. The National Safety Council study produced the following results with respect to the root causes of industrial accidents:

- 10% due to unsafe conditions
- 88% due to unsafe behaviours
- 2% due to unknown causes

The DuPont Company study produced similar results :

- 96% due to unsafe behaviours
- 4% due to unsafe conditions

If one accepts the results of the studies to imply that between 98% and 100% of industrial accidents are caused as a result of a combination of unsafe behaviours and unsafe conditions it is plausible to deduce that both can be addressed resulting in the avoidance of most accidents. Current sophisticated regulatory measures such as occupational safety and health laws, rules and legislation have not been able to address the high proportionate incidence of accidents, injuries and fatalities in the construction industries of the developed nations. They would be even less effective in the context of developing countries. What is needed is an approach which identifies the desired outcomes or results without prescribing definitively how these are to be achieved and establishes a total safety culture.

A performance based approach meets this criteria. There is increasing support for a move away from the traditional focus on strict compliance with procedural requirements and heavy fines for non-compliance in favour of a system based on results while at the same time offering compliance assistance when the requirements are not met.

5 PERFORMANCE BASED SAFETY

More recently the performance based management model has become popular because of its focus on the improvement of worker performance. Perhaps more importantly, it concentrates on outcomes and results rather than actions. Better processes generally lead to better outcomes. It provides an ideal basis for identifying worker behaviour which result in accidents, addressing all the major factors which influence these behaviours, and providing continuous improvement (HPT 1998). Some of the characteristics of this approach include its cost effectiveness, allowance for innovation, the building of both individual and organisational competence, quicker and more sustainable results and more effective training methods (HPT 1998). It has been suggested that performance based safety management systems are easily implemented in a rather short period of time, are inexpensive and lend themselves to "train the trainer" applications and are very flexible. They enable several different solutions to be implemented to satisfy the performance standards, with the best one for the situation able to be chosen (Bowman 1995).

It is recognised that consensus on what would constitute minimum requirements within a performance based safety and health standard could be difficult to achieve without the widest possible consultation and participation in the formulation process. It is in this particular area that governments can play an active role in getting all stakeholders and participants in the construction industry involved. However genuine international approaches are more likely to succeed in the functional area of health which is intrinsically non-conflictive (Alleyne 1997). The standards should be written using simple language which is unambiguous with respect to their meaning regardless of the national context in which they are applied. This could prove to be problematic when it is considered that current construction terminology varies considerably from country to country. The work currently being done by International Alliance for Interoperability (IAI) (Sharpe 1998) could provide new impetus to the development of a common construction language and system of standards. However it is important that these standards, if they are to succeed, promote consistency, thereby eliminating confusion caused by differences in interpretation, perception and application. There is a case for an approach similar to that of international road signs which are universally understood and complied with. For example, everyone knows that when they see a stop sign they should stop regardless of whether they are in Taiwan or Turkey. The stop sign brings about the desired outcome – that of stopping.

The experience of a major contractor in South Africa has demonstrated that legislated (prescriptive) responses with respect to safety and health resulted in an extended time scale with results being delayed and achieved on a drawn out "chain" basis. Improvements in communication were followed by improvements in education, followed by improvements in systems, followed by improvements in quality and only then followed by improved productivity (Muller 1997). The alternative approach resulted in early results in the same areas at the same time.

6 TRAINING AND EDUCATION

The levels of adequate and appropriate construction-industry-related training and education in developing countries where it exists is relatively poor. Education is a cornerstone of development and its influence on and contribution to quality of life cannot be overstated (Hunter 1996). It is contended that before safety and health management measures such as those contained in the EEC Directive 92/57 in Europe, could become effective in developing countries, the level of education and training of workers in construction generally would have to be improved and become a greater priority than at present (Berger 1998). A major consideration would be the literacy levels within developing countries which are usually very low. Some of the measures which can be implemented could include the following :

- implementation of special selection and recruitment procedures;
- improvement of relationships with temporary workers;
- extensive induction training;
- maximum exposure to and participation in regular skills training; and
- availability of literacy training to all employees at receptive times which could at the same time also include life skills training and environmental awareness programmes (Muller 1997).

The introduction of such a programme by a major construction firm in South Africa has resulted in only 5% of their entire workforce remaining unskilled and/or illiterate. Additional tangible benefits of the programme included increased output per employee of approximately 45%, winning of several major safety awards and the development of a relationship of trust and understanding between managers and the workforce (Muller 1997). Other benefits would typically include reduced construction unit costs, increased profitability and competitive edge.

7 CONCLUSION

It has been suggested that by the year 2000 80% of the world's population will live in developing countries (Hunter 1996). With an increasing understanding that the world has become a global village, it is critical that it is appreciated that quality of life, economic development, environmental security, safety and health are all closely linked to what happens throughout the world. When developed countries support developing countries a brighter future for all is being fashioned.

The best way to fight poor safety and health performance on construction sites requires co-operation at all levels and in all forms. It requires the combining of limited resources with a shared resolve and sense of purpose. To this end governments in developing countries can take the lead by putting pressure on construction practitioners to integrate safety and health management systems into the entire construction process from project inception through to final project completion and beyond. This commitment from government must be accompanied by commitment from all construction project clients, all levels of management of construction firms regardless of size or capacity, and a reciprocal commitment by construction workers to implement safety and health plans. The real recipe for development success requires the formation of partnerships with all participants in construction without which all efforts will only be partial, all investments just temporary and all results precarious (Hunter 1996). Furthermore, aid agencies have the clout and indeed a key role to play in reinforcing government policies to ultimately improve resource allocation (Saxenian 1994). This role should be extended to the area of safety and health on aid- funded construction projects.

The development of international minimum safety and health standards in construction which are performance based should be supported.

REFERENCE

Alleyn, George A.O. 1997. Global Health – the Paradigm, Policy and Program implications, *National Council for International Health Monthly Seminar Series* Washington, D.C., Pan American Health Organisation.

Berger, Joachim 1998. The Health and Safety Protection Plan and the File containing Features of the Building according to EEC Directive (92/57), *ME Rinker Lecture Series on Safety and Health in Construction,* University of Florida, Gainesville, USA, 19 October 1998.

Blair, Earl H. 1996. Achieving a total safety paradigm through authentic caring and quality, *Professional Safety, Journal of American Society of Safety Engineers,* vol 41.

Bowman, Richard 1995. The Impact of new standards on the specification of tiling systems and products, http://www......com.au/techpapers/95tilex.html.

Britannica Online 1998a. Safety, *Britannica Online*, http://www.eb.com:180/cgi-bin/g?DocF=micro/516/60.html.

Britannica Online 1998b. Culture, *Britannica Online*, http://www.cb.com:180/cgi-bin/g?DocF=micro/153/45.html.

Britannica Online 1998c. Medicine : The practice of modern medicine : health care and its delivery : medical practice in developing countries ; other developing countries, *Britannica Online*, http://www.cb.com:180/cgi-bin/g?DocF=micro/153/45.html.

Cairncross, Frances 1994. Environmental pragmatism (environmental damage in underdeveloped countries), *Foreign Policy* no 95, pgs 35 – 53.

Drewer, S. 1980. Construction and development: A new perspective, *Habitat International*, vol 5, No. 3/4., pgs 395-428.

Geller, Scott E. 1994. Ten principles for achieving a Total safety Culture, *Professional Safety, Journal of American Society of Safety Engineers,* vol 39.

International Labour Office 1987. *Guidelines for the development of small-scale construction enterprises,* ILO, Geneva.

Maclean, Hunter 1996. Their future is also our future, *Maclean's* vol 109, no 7, pgs 82 – 88.

Muller, Neil 1997. Opening Address, *1st South African Construction Health and Safety Conference,* Cape Town, 8 October 1997.

Ofori, G. 1980. *The construction industries of developing countries, The applicability of existing theories and strategies for their improvement and lessons for the future: the case of Ghana,* Unpublished PhD theisis, University of London.

Saxenian, Helen 1994. Optimizing health care in developing countries, *Issues in Science and Technology* vol 11, no 2, pgs 42 – 49.

Sharpe, Ron 1998. Use of IT to help regulate public health and safety in construction, *ME Rinker Lecture Series on Safety and Health in Construction,* University of Florida, Gainesville, USA, 12 October 1998.

Strassman, W.P. 1975. *Building technology and employment in the housing sector of developing countries,* Michigan State University, East Lansing.

Turin, D.A. 1973. *The construction industry: Its economic significance and its role in development,* Building Economic Research Unit, University College, London.

Wells, J. 1985. The role of construction in economic growth and development, *Habitat International*, vol 9., pgs 55-70.

http://www.qhpt.com/behavioral_safety.html 1998. Start a behavioral safety program – Performance based safety programs, Human Performance Technologies.

Implementation of Safety and Health on Construction Sites, Singh, Hinze & Coble (eds)
© 1999 Balkema, Rotterdam, ISBN 90 5809 036 1

Techniques for risk assessment based on research in the demolition industry

R.Trethewy, J.Cross & M.Marosszeky
University of New South Wales, Sydney, Australia

ABSTRACT: This paper reports on research which has developed a method of risk assessment and management (risk control plans) for potential hazards to health and safety in the demolition industry. The project involved the survey of 60 demolition workers to identify accidents and incidents. An expert group comprising experienced demolishers, a representative from WorkCover, the state government safety agency and the research team used a two stage Delphi approach to identify the hazards in the trade and to develop strategies to manage those hazards. A training course was then developed and delivered as a part of a Technical and Further Education training course designed as a pre-qualification for supervisors working in the industry.

The method of risk assessment that was developed in the project builds on, and adapts, previous work by Fine (1971). The modified graphical technique has been shown to be easy to use in a pilot stage evaluation.

Keywords: safety risk management, risk assessment, management tools, risk rating

1 INTRODUCTION

Building and construction is an inherently hazardous industry with a significant number of injuries recorded annually. In 1995/96 the construction industry in New South Wales alone recorded 53.9 injuries per 1000 workers in regard to number, incidence, time lost and cost. This figure was exceeded only by the mining industry which recorded 76.2 injuries per 1000 workers. The average across all industries in New South Wales, for this period, is 27.4 injuries per 1000 workers (WorkCover 1997, 22).

Despite its poor record in health and safety, formal risk assessment and management techniques are not widely used in the building and construction industry re. Contractor safety systems differ widely – some have formal systems, many don't. The development of such systems are hampered by a lack of hazard identification by workers and industry. Many contractors do not record accidents or incidents, nor do they plot hazard trends in a proactive attempt to increase safety - an incident (near miss) may only be mentioned to the foreman over a beer after work. Another problem is that injury statistics, gathered by bodies such as WorkCover, are usually categorized under the broad umbrella of building and construction, with no specific trade reference; thus it is impossible to identify specific hazards associated with a particular trade.

2 RESEARCH METHOD

The project team, interviewed 64 employees of demolition companies, face to face at work, in order to build a profile of typical industry hazards and associated risks. The duration of the survey was approximately fifteen minutes and asked respondents if they has ever been involved in, or witnessed, an accident/s where an injury resulted in the injured party having to stop work for the rest of the day, or longer. Details of the accident/s were gathered along with any details of incidents (near misses) which the respondent may have been involved in, or witnessed.

A total of ninety eight cases of injury were recorded with the predominant injuries being:

- Burns – 20% mainly caused by working with oxy cutting equipment;
- Bruising – 11% caused by a variety of accidents involving lifting, tools (mainly jackhammers), struck by objects (debris or other);
- Cut stitches – 10% and stressed ligaments – 8 % mainly caused by manual handling of debris; and
- Multiple fractures – 8% mainly due to falls or struck by heavy equipment.
  (Trethewy 1996, p9)

On completion of the survey the project research team developed a two stage Delphi approach to define hazards and strategies to avoid them. The initial document was the report that described the findings of the survey. Four half-day workshops were convened for this purpose. They were attended by industry experts representing five major demolition companies, a senior government inspector for the demolition industry and the authors. The first task of the group was to draw upon the experience and knowledge in the demolition industry to identify, and profile, the major hazards and associated risks in the industry. The hazards identified by the committee expanded on the survey results to ensure that as many hazards as possible were captured and recorded.

The resulting risk profile listed approximately 111 hazards and associated risks. Given budgetary constraints for the project, certain risks were identified for the development of detailed control measures. The risks were selected to give an indication of the way in which risks of varying severity might be treated in the development of control measures (safe work methods) to negate, or reduce risk in a work situation. Similar financial restraint would no doubt reflect the situation in industry where there is a need to prioritize risk to maximize the benefits of budget expenditure on safety.

3 RATING RISK

Once the hazards had been identified, each committee member was asked to individually provide a risk rating for each hazard and associated risk that had been identified in the profile. The aim of this approach was to allow committee members to develop a risk rating in isolation without the influence of the opinions of other members of the group. Ratings were then compared and discussed to derive a consensus on the ratings.

The next stage was to identify high and very high-risk hazards so that safe work methods (control measures) could be developed for those risks.

A number of techniques to derive risk ratings were evaluated and the method outlined by . Fine (1971) in his paper "Mathematical Evaluation for Controlling Hazards" was adopted. Fine provides a mathematical formula for calculating the risk associated with a hazard by considering the following factors:

- the potential consequences of an accident (e.g. serious injury);
- the exposure;
- and probability (likelihood).

He uses the formula "$R = C \times E \times P$" where "R" represents the risk rating, "C" the consequence, "E" the exposure factor and "P" the probability factor (Fine 1971, 158).

Consequence refers to the most likely results of the potential hazard and associated accident, including damage to property and the environment. A numerical value is derived for this aspect of the rating by referring to Fine's table 1 (see attachment 1). For example, an extremely serious injury such as a permanent disability would incur a value of 15.

Exposure refers to the number of times the hazard-event occurs. (a combination of the hazard with the activity being undertake that may start the accident sequence). For example, if you are working at height every day removing roofing materials your exposure to the event of a fall would be 10 (attachment 1) the numerical value for continuously, or many times daily.

Probability refers to the likelihood that should the hazard-event occur, the nominated consequence (injury or damage to property or the environment) will result. For example, "is quite possible, not unusual, has an even 50/50 chance" would give a numerical value of 6 (attachment 1).

Based on Fine's formula $R = C \times E \times P$ the above example would equate to $15 \times 10 \times 6 = 900$. This final figure would represent a very high risk score based on Fine's interpretation which is

- greater than 350 = very high (stop activity until risk reduced);
- 180 – 350 = high (deal with immediately);
- 70-180 = substantial (correction required);
- 20-70 = moderate (attention indicated);
- less than 20 = acceptable (make as low as reasonably practical).

Based on the "R" results provided by each industry committee member, the results obtained from Fine's method of rating were found to vary widely. For example, one of the hazards, and associated risks, for which the committee members were asked to provide an "R" rating for was the possibility of an electric shock (hazard) caused by temporary (intentional and illegal) electrical connections in a building (risk to be managed). Four of the five industry committee members who provided an R figure, calculated the rating outlined in Table 1 below.

Table 1

Committee Member	Rating	Risk Translation
1	150	Substantial
2	100	Substantial
3	450	Very High
4	6000	Very High

The committee member, who provided the "R" figure of 6,000 misunderstood the descriptions, outlined in Attachment 1, and therefore this value was not considered for the final rating result. Based on the calculated ratings in Figure 1 above, the risk rating for this one hazard varied from substantial to very high.

Committee members were asked how they derived their final R rating result. For ***consequence*** each of the three results listed above included a rating of 50, a disaster (fatality, permanent local damage to environment, loss $500,000 - $2,000,000). The project team considered this rating as reasonable as a fatality may be the most likely consequence of an electric shock.

For ***exposure,*** two committee members selected 1; rare (has been known to occur). When questioned they explained that in their experience, which is considerable in the demolition industry, temporary connections have been known to occur on very few occasions. The third committee member selected 3, occasionally (once a week to once a month). When questioned he stated that he had detected such connections on a number of occasions but perhaps not a frequently as he had selected.

For ***probability,*** two committee members selected 3, unusual but possible. In other words they thought that it would be unusual but possible for a fatality (consequence) to occur from an electric shock caused by an undetected temporary connection. The third committee member selected 1, remotely possible (a possible coincidence). When questioned, and the probability of a fatality from an electric shock was discussed, he agreed that this rating was too conservative and that he had misunderstood the definition of ***probability***.

The above example illustrates the large variation that was obtained from a small group of senior industry experts in their assessment of the R rating for this single risk. This sample was typical of a considerable number (approximately 20%) of other ratings derived by the committee for the 111 hazards and associated risks in the profile. Where a significant discrepancy existed, one that would alter the final rating, each risk was discussed and a re-interpretation of the C, E and P figures was undertaken on a consensus basis. In many cases it was apparent to the research team that differing industrial experience in the demolition industry influenced each committee member's selection of the outcomes for each hazard and associated risk. In addition, some committee members misinterpreted the statements in attachment 1 with the result that these were not used to derive an overall risk rating for a selected hazard. In can be concluded that a comprehensive understanding of the terminology developed by Fine (attachment 1) is required before the rating process can yield reliable results. The research team had only provided a brief outline of Fine's method to committee members in an attempt to gauge how readily the method could be understood and implemented.

In order to further evaluate the variation in results obtained by the expert group a further two groups were tested. The first group - thirty-seven students at the University of New South Wales Department of Safety Science were given a detailed briefing of Fine's method of risk assessment. The students were then asked to identify a potential hazard, and provide a risk rating, for a truck driver alighting from a truck parked on the side of a main road.

Figure 1 below represents the results of thirty-five responses by the students. One student failed to provide an exposure rating and another didn't provide any ratings at all. These two results are excluded from the results.

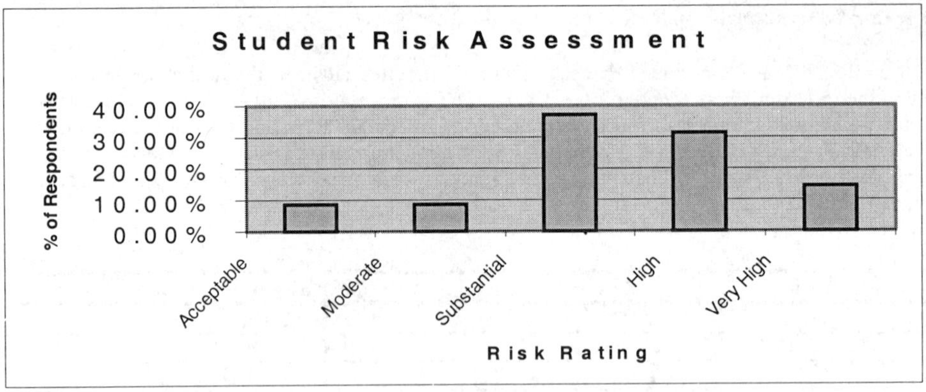

Figure 1

From the above figure it is clear that once again, substantial variations exist in the perception of potential hazards and the level of risk associated with what is by any measure, a common hazard-event. The consequences which the students identified for this scenario ranged from various minor foot, leg and ankle injuries to very serious injuries such as a fractured skull, spinal injury or struck by a passing vehicle.

The second test group was six building subcontractors who were posed the same truck driver scenario. The subcontractors were briefed on the Fine method of risk assessment during a lecture on safety for their trade. Table 2 below lists the results of the six male individuals

Table 2

Building Subcontractor	Rating	Risk Translation
1	150	Substantial
2	100	Substantial
3	150	Substantial
4	350	High
5	450	Very High
6	450	Very High

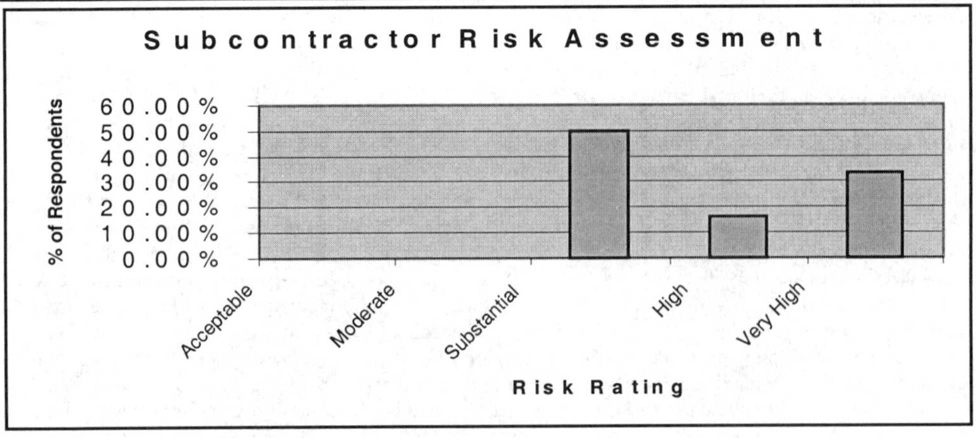

Figure 2

Once again substantial variations resulted in the perception of potential hazards and the level of risk associated with alighting from a truck. The consequences identified by this group for the same hazard-event were similar to those identified by the student group.

The wide spread of ratings and the task of backtracking and identifying each C, E and P ratings proved a very laborious task for the research team. A number of inefficiencies in the risk method became through the rating exercise:

- a final R numerical figure, which equates to a risk rating, gives no indication to the decision making process as to how this figure was achieved, i.e what was the value for C, E and P;
- the misunderstandings regarding definitions and their application clearly indicated that this method should not be simply selected and used by anyone. A very thorough and comprehensive introduction and training in the use of Fine's method is necessary as it can be seen from the examples illustrated in Figures2 and 3 that even when users are briefed, substantial variations still arise;
- the risk rating "R" value is very sensitive to an individual's own experience as this will alter the results considerably from person to person. Further, unless the assessor of a risk using this method provides a detailed explanation of each result (C, E and P ratings) the overall-rating outcome cannot be properly interpreted by others.

4 ALTERNATIVE METHOD

As a result of the above difficulties, Fine's method of risk rating in this form was not considered practical for the research project. A more graphical approach to risk rating was developed in 1976 by Kinney and Wiruth who further enhanced Fine's method. The graphical system (risk rating scale) clearly identifies the decisions made at each step of the rating process – consequence, likelihood and exposure, without the need for numerical figures, and derives a risk rating by drawing straight lines through selected points (see attachment 2).

The risk scale approach provided in attachment 2 is an adaptation of the method developed by Kinney and Wiruth. Their original graphical approach does not provide any explanation, or interpretation, of the terminology of the rating scale on the actual diagram itself. Given that the scale is aimed to be a "stand alone" document, where a user can rate the risk of a particular hazard, annotation was provided by the research team so that users can understand each concept.

In addition, further clarification and refinement of specific terminology on the scale is provided. The original method does not provide a "low" risk category of rating but uses the terminology "risk perhaps acceptable". Discussion with the industry committee for the project highlighted that the term "acceptable" could be interpreted to mean that the potential hazard may be ignored and therefore form part of the accepted consequences of the work in question. This was not agreeable to the committee and the research team decided to change the term to read "low" risk score.

5 RISK RATING SCALE

To derive a risk rating using the scale in Attachment 2, the user must first decide on the potential hazard for the job task. Attachment 2 lists these in the top left corner of the page. The job task is cutting and dropping an overhead steel beam whilst the potential hazard identified is the weight of the debris (cut pieces of steel section) and the associated hazard-event of physical lifting to remove the debris. The potential consequence of this hazard is nominated as a serious non-permanent injury to the back caused by lifting. A user of the scale must decide on the consequence associated with the hazard, or alternately, they can calculate multiple risk ratings for different consequences. For example, the user may decide that a potential back injury from this type of lifting will be very serious (permanent disability), or alternately decide on a serious (non-permanent) injury and so on.

As outlined in Attachment 2 the user of the scale selects a consequence first – indicated by the prompt "start here" labelled on the sheet. The user must then decide on the probability, or likelihood, of the consequence occurring should the hazard-event occur. Here the term "probability" was discarded in favour of the term "likelihood" – the industry committee and the research team agreed that the term probability was more associated with mathematical calculation and may prove difficult to understand.

The likelihood factor for the example in Attachment 2 is "very likely". It was thought that if manual labour was used to clear cut steel sections then there was at least a 50/50 chance of a serious back injury occurring.

The exposure for the example is nominated as "occasional" (physical lifting is carried out once a week to once a month). This selection is relies heavily on detailed knowledge of the work practices associated with this job task. The research team noted that exposure was the most difficult term for the participants to comprehend. A number in each of the three groups thought that because they had never seen or experienced the nominated consequence that the exposure was rare (not known to have occurred). When the research team explained that exposure is the number of times a worker carries out the task for which the potential hazard may occur a clearer understanding of the concept was developed by each group.

Once all three categories are selected i.e. consequence, likelihood and exposure draw a straight line starting at likelihood (very likely) through the exposure location (occasional) to the tie line. From the point reached on the tie line another straight line is drawn through the selected consequence (serious) to eventually derive the risk rating.

In Attachment 2, this rating is substantial. Therefore the work task of removal of cut steel sections using manual labour is a substantial risk to health and safety in the demolition industry.

The above procedure for risk rating has been incorporated into a risk assessment training package, for subcontractors, in a number of trades in the building industry. Initially the research team felt that the concepts of the risk rating scale may have been too detailed for workers to grasp. However two pilot training sessions consisting of approximately 80 workers has indicated that once explained, the method is easily understood and was enthusiastically regarded by the pilot group.

6 CONCLUSIONS

The procedure for calculating risk developed in this research is one of several available to the safety planner. Like the others, the outcome is largely dependent on the individual decision-maker who rates consequence, likelihood and exposure to derive a risk rating. While as with the other methods, there is a possibility that different people may derive different ratings the graphical nature of the risk scale makes these differences easily identifiable in terms of the decisions. In this regard the scale is superior to those other methods of risk scoring which allocate numerical values to the variables involved in such an assessment.

In addition, pilot training in the use of the above risk scale, highlighted that the breakdown and analysis of hazards, and their associated events, into consequence, likelihood and exposure gave workers a much clearer understanding of the analytical processes that can be useful in assessing their own specific work safety risks.

REFERENCES

Fine, W.T, 1971. Mathematical Evaluation for Controlling Hazards, *Journal of Safety Research*, Dec, Vol. 3, No. 4, pp. 157-166.

Kinney, G., and Wiruth, A. 1976. *Practical Risk Analysis for Safety Management (NWC TP 5865)*, China Lake, California: Naval Weapons Centre, June.

Trethewy, R.W. 1996. *Safe Work on Demolition Survey Results Report*, Sydney, Building Research Center and Department of Safety Science, University of New South Wales.

WorkCover New South Wales, 1997. *New South Wales Workers Compensation Statistical Bulletin,1995/96,* Sydney, WorkCover Bookshop.

Attachment 1 (Fine 1971, 159)

RATINGS FOR RISK CALCULATION

Factor	**Classification**	**Rating**
1. Consequences (most probable result of the potential hazard)	• Catastrophe; numerous fatalities; damage over $1,000,000; major disruption of activities • Multiple fatalities; damage $400,000 to $1,000,000; • Fatality, damage $100,000 to $500,000; • Extremely serious injury (amputation, permanent disability); damage $1,000 to $100,000; • Disabling injury; damage up to $1,000; • Minor cuts, bruises, bumps; minor damage.	100 50 25 15 5 1
2. Exposure (the frequency of the occurrence of the hazard event).	Hazard-event occurs: • Continuously (or many times daily); • Frequently (approximately once daily); • Occasionally (from once per week to once per month); • Unusually (from once per month to once per year); • Rarely (it has been known to occur); • Remotely possible (not known to have occurred).	 10 6 3 2 1 0.5
3. Probability (Likelihood that accident sequence will follow completion).	Complete accident sequence: • Is the most likely and expected result if the hazard-event takes place; • Is quite possible, not unusual, has an even 50/50 chance; • Would be an unusual sequence or coincidence; • Would be a remotely possible coincidence; • Has never happened after many years of exposure, but is conceivably possible; • Practically impossible sequence (has never happened).	10 6 3 1 0.5 0.1

Attachment 2. Safe Work on Demolition - Risk Rating Scale

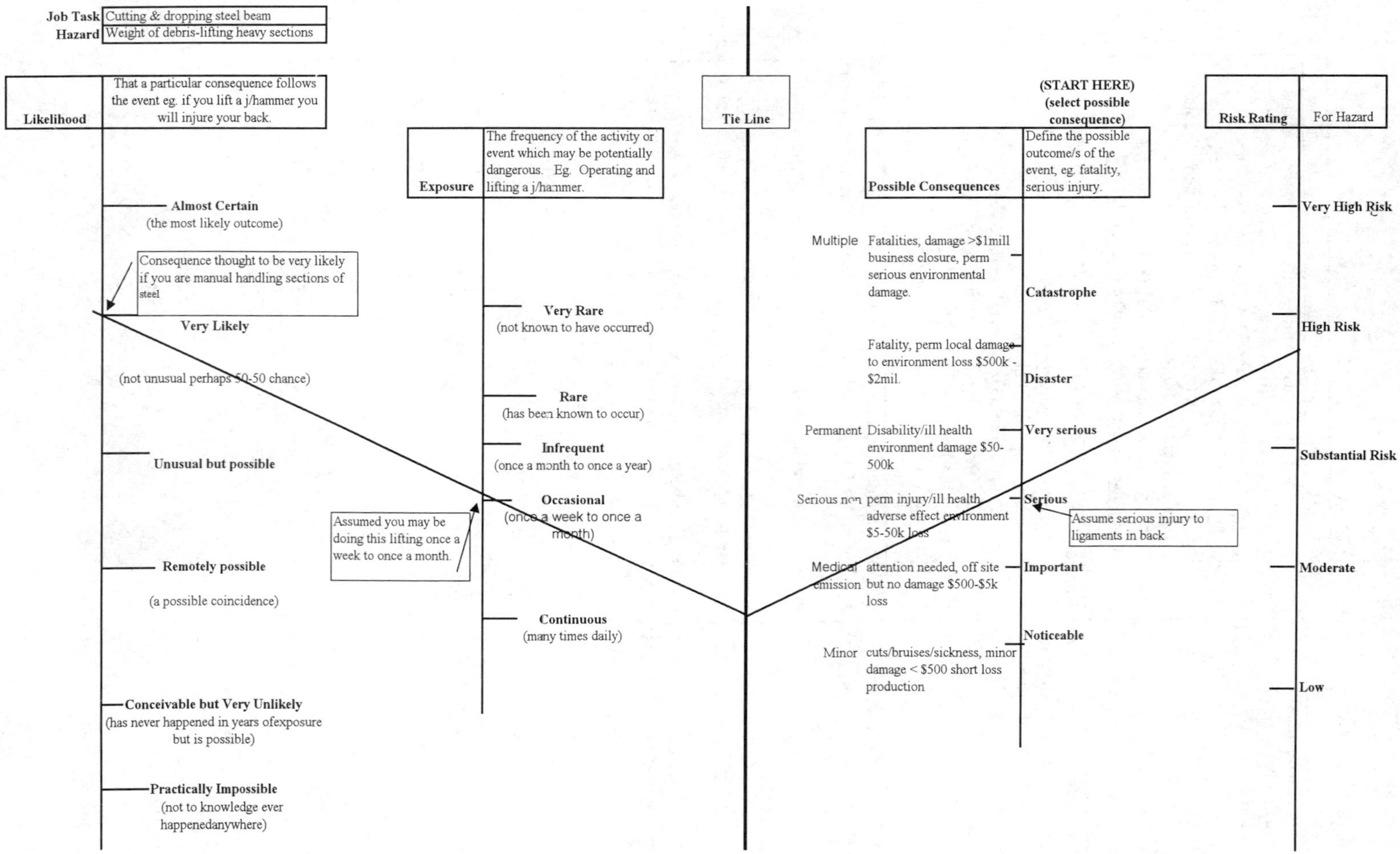

Implementation of Safety and Health on Construction Sites, Singh, Hinze & Coble (eds)
 ISBN 90 5809 036 1

Fall prevention in construction by organizational intervention

P. Becker, M. Akladios & M. Fullen
Safety and Health Extension, West Virginia University, Morgantown, W.Va., USA
G. Hobbs
Statistics, West Virginia University, Morgantown, W.Va., USA

ABSTRACT: Falls are the leading cause of construction industry injuries in the US. The construction safety profession recognizes available engineering controls, work practices, and personal protection which are effective in preventing construction falls. However the equipment and practices are not consistently used in the industry. This project implements a university/construction contractor partnership program called Fall-Safe to improve the use of existing fall prevention methods. This partnership and management program seeks to develop a job site accountability system which can minimize fall hazard exposures and maximize the consistent use of appropriate controls in the dynamic construction setting. The Fall-Safe program incorporates a rigorous evaluation component comparing changes occurring within intervention and control groups of workers and contractors to determine if such changes can be associated with the intervention. Changes are measured using baseline and quarterly measurements of knowledge, attitude, construction site management policies, and construction site fall hazard control.

1. INTRODUCTION

Falls are the leading cause of death in the construction industry (Kisner & Fosbroke, 1994). While the construction workforce represents 5% of the nation's workforce, it accounts for 49.6% of fall fatalities (Cattledge, 1994). In West Virginia, the construction industry accounts for more than a third of all (fatal and nonfatal) occupational falls (Cattledge, Schneiderman, Stanevich, Hendricks, & Greenwood, 1996). A study of 182 claims reported to the West Virginia Bureau of Workers' Compensation reveals that the leading categories for type of work surfaces from which a fall occurred were ladders (33.5%), and scaffolds (13.7%) (Cattledge et al., 1996). This same study reports that the five leading reported causes of falls were slippery substance on surface (17%) slips or trip (13.9%) loss of balance (7.7%), unsafe equipment (6.6%), and a ladder slipping/skidding (7.7%). In this study, 70% of the claims for compensation from falls reported falling 10 feet or less.

Techniques for preventing construction falls are well known. For example, simple techniques such as minimizing work at heights, proper maintenance, placement and tying off of ladders, and appropriate climbing technique will go far to prevent many of the ladder falls identified in the study cited above (Cattledge et al., 1996). However, the dynamic nature of the work at a construction jobsite and the transient nature of the workforce make control of simple hazards relatively difficult (Ringen & Stafford, 1995). If the technical means to prevent many construction falls exists, an effective preventive intervention may be one maximizes the use of the well know fall prevention technologies.

This intervention research project begins with the hypothesis that the public service arm of a land grant university (Safety and Health Extension) can successfully intervene on construction sites to bring about a decrease in injuries and deaths due to falls. Beyond carrying out a successful intervention, this project recognizes the need to state a hypothesis about an intervention program and evaluate the effectiveness of that intervention. A serious research component is needed to give decision and policy makers timely and valid information about proposed preventive intervention (Schulte, Goldenhar, & Connally, 1996).

2 FALL-SAFE CONSTRUCTION FALL PREVENTION PROGRAM

Funded by the Center to Protect Workers' Rights and the National Institute for Occupational Safety and Health, West Virginia University' Safety and Health Extension has devised a partnership program between university and individual contractors as a vehicle to prevent construction falls in West Virginia. The thrust of the partnership is to create an organizational intervention that will provide construction organizations (contractors) with the knowledge, accountability, and incentives to improve their management of construction site fall hazards. Contractors and West Virginia University (WVU) sign a formal contract agreeing that the contractors will conduct fall prevention programs specified by WVU. WVU provides contractors with training, consultation, and marketing and public relations concerning contractors' successful efforts to prevent falls. The contractor must demonstrate that his/her company is carrying out the required fall prevention programs in order to obtain and maintain the public status of a "Fall-Safe Contractor."

2.1 Contractor Responsibilities

Contractors are responsible for creating a series of ongoing programs within their company and at their construction sites intended to improve their control of potential fall hazards. These measures include:

Establish a company written policy
Designate a Fall-Safe Coordinator
Identify the company's typical fall hazards
Train workers and supervisors
Create site specific fall prevention plans for each company construction site
Designate a site competent person
Conduct site specific fall prevention training
Create and maintain a process for employee participation in fall prevention
Conduct weekly and daily fall hazard inspections

2.2 West Virginia University Responsibilities

West Virginia University's Safety and Health Extension provides the Fall-Safe program (Lundstrom, 1997), training and consultation, accountability, and marketing for Fall-Safe contractors. Specifically WVU provides:

Consultation on creating a company policy
Training of supervisors and workers
Quarterly management and site audits
Media coverage of Fall-Safe contractors

Because WVU services are supported by a grant for the Center to Protect Workers' Rights and the National Institute for Occupational Safety and Health, these services are provided without charge during the initial five years of the grant period.

2.3 Fall Safe Training

The Fall-Safe intervention requires that substantial training be conducted to equip managers and workers to prevent construction falls. Construction company upper management is trained by WVU staff in the Fall-Safe management system. Safety personnel and supervisors receive 8 hours of fall prevention training from WVU and a fall prevention manual produced by the Construction Safety Council (Council, 1995a). Workers receive two hours of initial fall prevention training from WVU and a fall prevention manual (Council, 1995b). Additional worker and follow up training is provided by the contractor's fall prevention coordinator.

2.4 Accountability

The Fall-Safe program emphasizes accountability as a means of maintaining consistent control of fall hazards in the face of the ever changing construction site environment. Accountability is understood to begin with accountability within a contractor's organization. Fall-Safe contractors must designate responsible and competent persons, create joint labor and management committees on construction sites, and carry out daily and weekly site inspections.

WVU contributes to accountability by conducting unexpected quarterly audits of a contractor's Fall-Safe sites. These site audits generate a numeric score for the contractors's performance in creating and maintaining an appropriate site specific fall hazard management plan, and a separate score for the contractor's ability to control fall hazards on site. These two audits are conducted on Fall-Safe construction sites using a dedicated software audit tool developed by WVU (Carr, Lundstrom, Becker, Akladios, & Fullen, 1998) and run on a hand held touch screen PC.

The management plan audit involves scoring a site plan for required site management elements, as well as documentation of activities such as inspections and committee meetings. Each management audit results in a score which indicates the percentage of required elements present and documented.

The site audit involves a visual inspection of the site listing all fall exposures, control measures in place to prevent falls, and the condition and appropriate use of the fall prevention measures. The audit elements are based on OSHA's 1926 Subpart M.(29 CFR 1926). Each site audit results in a site audit score which indicates a weighted percentage of compliance with OSHA's Subpart M. Score weight factors emphasize engineering control over individual work practices or personal protection.

Contractors must maintain minimum scores of 70% on both management and site condition audits. If a contractor's scores go below 70% on any audit for two successive quarters, they can no longer be designated a Fall-Safe contractor.

2.4.1 Site Audit Instrument

The Fall-Safe site hazard control audit consists of a list possible fall exposures (i.e sloped roof, pit, scaffold, ladder, etc.) followed by a list of possible control measures for the identified fall exposure (e.g. edge guarding, ladder tie-off, warning lines, harness and lanyard, etc.) Each exposure is then graded on a percentage basis using a checklist of criteria for compliance with proper deployment of the controls. The score is weighted by an algorithm that favors engineering control over personal protection. It also rewards redundancy of control.

Table 1 shows the weighting system used by the Fall-Safe fall hazard control algorithm. Score additions due to redundancy of control can not bring an exposure score over 1.0.

Table 1. Weighting factors for hazard control audit

Control	Primary Control	Redundant Control
Engineering control	0.9	
PPE	0.8	0.4
Nets	0.7	0.2

Example Score: If a contractor was working on a flat roof site, and erected a guardrail which was graded at 0.9 and also used personal fall protection which was graded as 0.8, the score would be computed as follows:

Table 2. Sample calculation for redundant control measures

Control	Score		Weight		Weighted Score
Guardrail	0.6	x	0.9	=	0.54
PPE	0.8	x	0.4	=	0.32
Total Score					0.88

Total score for a contractor on a site is the mean of each of the calculated exposure scores. Reports are generated for each contractor for their management score, hazard control score, and hazard control score broken out by exposure and control. As noted above, a contractor must maintain scores of 70% on both management and site audits to maintain status as a Fall-Safe contractor.

2.5 Incentives for participation

Construction contractors do not normally welcome third parties to their construction sites to evaluate their safety performance. The Fall-Safe program incorporates a set of incentives to persuade potential contractors that they will benefit by carrying out the Fall-Safe program in the context of a partnership with West Virginia University. Central to the Fall-Safe intervention are the incentives built into the program. These incentives and the marketing plan were developed out of information obtained through two focus groups facilitated by WVU staff with construction contractors.

2.5.1 University/contractor partnership

Those contractors who successfully maintain their status as Fall-Safe participants receive plaques, signs, and stickers recognizing their participation. WVU has sought and obtained endorsement of the Fall-Safe program by important construction user groups in the region. West Virginia University is a respected institution in the state. A partnership between a private entity and the state university carries significant status. Contractors have told WVU staff that they believe that attaining Fall-Safe status will improve their ability to market their firm's services.

2.5.2 Marketing to the construction community

Fall-Safe has developed a series of marketing vehicles intended to create a sense of presence and value for Fall-Safe participants on a construction site. These include fall-safe signs for the contractor, fall safe stickers for all Fall-Safe trained workers, Fall-Safe carpenter pencils and cups for all fall-Safe trained workers, Fall-Safe clipboards for Fall-Safe coordinators and competent persons, and Fall-Safe cards for Fall-Safe trained workers. Cards and clipboards all contain lists of responsibilities for the appropriate party. These marketing items have been professionally designed to present an energetic and professional image of the Fall-Safe program.

2.5.3 Marketing to the public

WVU professional marketing staff have devised and are executing a public marketing plan to yield positive public relations for Fall-Safe contractors. This campaign began with a kick-off ceremony hosted by WVU President David Hardesty with press packets sent to local media honoring the accomplishment of each of the initial Fall-Safe contractors. Subsequent press events and press releases will work to continue to generate positive public relations for Fall-Safe contractors in the state and in their local communities.

3. EVALUATION/RESEARCH

Intervention research in occupational safety and health often lacks important elements present in other areas of research. These research projects are often lacking in theoretical basis, intensity to cause a desired (and measurable) change, and are frequently non-experimental or quasi-experimental.(Goldenhar & Schulte, 1994) Difficulty of obtaining access to workplace populations, limited resources, and ethical issues contribute to the difficulty of achieving more rigorous investigations in this field. Regardless, this study accepts its real world limitations and attempts to provide some information on the effectiveness of the Fall-Safe intervention.

3.1 Working Hypothesis

This study suggests that a university extension service can decrease construction falls by intervening in the organizational behavior of construction contracting companies through a contractor/university partnership. That partnership is based on the following three essential elements:

Application of appropriate fall management techniques by contractors
An accountability system to ensure systematic and continuing use of the fall prevention system
A series of incentives to provide benefits for participating contractors

This working hypothesis was developed out of the experience of the WVU Center for Excellence in Construction Safety and through two separate focus groups facilitated by WVU staff. The first focus group consisted of nationally recognized construction safety professionals and managers. The second consisted of contractors and construction safety personnel from the state of West Virginia.

3.2 Study Design

This study employs a quasi-experimental study design. WVU has recruited 10 intervention and six control contractors for participation in this study. It is interesting to note that project staff experienced serious difficulty in recruiting control contractors when discussions used the terminology "control". Potential control contractors consistently asked to become intervention participants, and would not participate as controls. Ultimately WVU devised a strategy of recruiting Phase I and Phase II contractors. Phase I contractors begin the Fall-Safe program immediately, while Phase II contractors will become active participants after providing comparison data for 1.5 years.

Measures of intervention impact include measures of employee knowledge and attitude, employer knowledge and attitude, site management plan audit, and site fall hazard control audit. Attitude and knowledge questionnaires will be administered prior to Fall-Safe enrollment and after 1.5 years to both intervention and control contractors and their employees. Management and hazard control audits will be conducted initially and quarterly for 1.5 years. The program will be deemed successful if the intervention contractor group shows statistically more improvement in knowledge, attitudes and audit scores than the control group.

3.3 Subject Selection and Number

Because of the newness of the Fall-Safe program, and the difficulty of recruiting contractors for this experimental project, subject selection is acknowledged to be non-random. It is anticipated that contractors may choose to participate who already believe their fall prevention management is excellent. Conversely it is deemed doubtful that contractors with poor safety practices will initially enroll. WVU has enrolled 10 intervention and 6 control contractors. In general the first ten contractors entering our system became intervention contractors, and the next six became our controls.

While the number of measurement subjects is small (10 intervention and 6 control contractors), power will be enhanced by making paired comparisons of improvement of scores over time.

3.4 Intensity of Intervention

Compared to other occupational safety and health interventions, Fall-Safe is quite intensive. The Fall-Safe program is comprehensive and demanding. The accountability requirements are strenuous and continuous. The investigators believe the intervention is sufficiently intense to bring about measurable changes in the identified outcome variables. Table 3 (below) indicates that the baseline scores of Fall-Safe contractors do not meet the Fall-Safe criteria and have significant headroom for improvement.

3.5 Measurement Instruments

Three similar questionnaires were created for construction company management, supervisors and owners. These questionnaires seek information regarding knowledge of fall hazards and their control, and attitudes concerning individual and organizational ability to correct these hazards. These questionnaires were developed with the assistance of staff of the Division of Safety Research, NIOSH and were adapted from existing similar questionnaires. Questions addressing specific Fall-Safe issues were added to the generic knowledge and attitude questions.

The Fall-Safe management audit instrument is based directly on the " Contractor's Guide to Managing Fall Hazards"(Lundstrom, 1997). The Fall-Safe fall hazard control audit is adapted from OSHA's construction fall hazard requirements as set forth in 20 CFR 1926 Subpart M (29 CFR 1926).

The audit tools were initially bench tested for usability and validity by WVU staff. After revisions, the tools were field tested twice at construction sites. Initial mean contractor scores of 0.60 for fall hazard control and 0.59 for fall management, support the appropriateness of a cutoff of 0.7 for Fall-Safe participants.

The authors considered an examination of fall injury data for WV contractors available through the WV Bureau of Workers' Compensation. In the current time frame this study would not have yielded sufficient power to show an effect by this method. Such a comparison may be possible in the future if the Fall-Safe program continues to operate past the initial five years.

3.6 *Analysis*

The measures of interest are changes of score or answers for the five measurement instruments. If paired analysis of scores for intervention contractors improves significantly more than paired analysis of scores for control contractors, this study will deem the intervention successful. Likewise if knowledge and attitudes about fall prevention increase more for the intervention group than the control group, the intervention will be deemed positive.

3.7 *Results*

Baseline data is reported for the management and hazard control audit in Table 3. "Locations" signifies the number of work areas examined on each construction site. "Exposures" indicates the number of potential fall hazards on the sites, and "interventions" indicates number of control measures observed (i.e. use of an engineering control and PPE would count as two interventions).

Table 3 - Means of key baseline measurements for fall safe contractors

	N	Locations	Exposures	Interventions	Mgmt Score%	Control Score%
Phase 1	10	3.3	6.8	7.1	60.2	64.5
Phase 2	6	5.3	6.8	6.8	60.3	49.6
All Contractors	16	4.0	6.8	7.0	60.8	59.0

The means are in relatively good agreement between the two phases except for the hazard control score. One of the phase 2 values is zero skewing that mean considerably. Even with this zero included, a comparison of phases 1 and 2 hazard control scores yields a probability $t > |t| = 0.17$, indicating that the groups are not significantly different.

Initial tests were conducted on baseline measures to look for relationships within variables:

Table 4 - Tests for association between variables on baseline data

Variables compared	Association	F Ratio	Probability $F > \|F\|$
Control Score by Management Score	0.39	8.80	0.01
Control Score by Experimental Status (Phase 1 or 2)	0.13	2.10	0.17
Control Score by Contractor Type	0.36	1.56	0.25
Control Score by Number of Interventions	0.00	0.01	0.91

Table 4 indicates a high score on a contractors's hazard control audit can be associated with a high score on the management plan for that site. This result is statistically significant. Hazard control score can also be associated to some extent with contractor type, though this result is not significant. Hazard control score does not appear to be associated with the experimental status of the contractor or the number of interventions audited.

Follow up data is not available for report and analysis at this time.

4. DISCUSSION/CONCLUSIONS

WVU has developed a fall prevention intervention for the construction industry based on a partnership between a land grant university's extension service and construction contractors. The project has developed a fall hazard management program (Lundstrom, 1997), that can be used by other similar institutions. WVU has also developed and executed a marketing plan based on contractors' perceptions and needs. The need for rigorous evaluation of this intervention has considerably slowed the pace of bringing the intervention to the field. In several instances, the marketing plan and intervention strategy had to be modified to make them susceptible to evaluation.

Given the ease WVU has experienced in recruiting 16 contractors to sign Fall-Safe contracts with West Virginia University, this study has established the efficacy of its intervention marketing strategy. More complete conclusions can be drawn concerning the marketing as the study continues, and as contractors are asked to bear the cost of the Fall-Safe partnership.

Preliminary data indicates that the non-random recruitment of intervention and control (Phase 1 and Phase 2) contractors will not bias study results. The preliminary data also supports the importance of a site management plan to controlling site fall hazards. As expected, the type of of contracting work carried out may influence fall hazard control performance.

WVU has succeeded in executing a quasi-experimental research project in the highly dynamic construction industry. WVU has gathered baseline data that indicate the study will generate meaningful results determining the efficacy of the intervention. Additional papers will analyze the continuing results of this study.

REFERENCES

Carr, M., Lundstrom, W., Becker, P., Akladios, M., & Fullen, M. (1998). Fall-Safe Audit Tool (Version 2) [DOS]. Morgantown, WV: West Virginia University.

Cattledge, G. H. (1994). Fatal Occupational Falls in the U.S. Construction Industry, 1980-1989. *Journal of Occupational Medicine, 28*(5), 647-654.

Cattledge, G. H., Schneiderman, A., Stanevich, R., Hendricks, S., & Greenwood, J. (1996). Nonfatal Occupational Fall Injuries in the West Virginia Construction Industry. *Accident Analysis and Prevention, 28*(5), 655-663.

Council, C. S. (1995a). *Competent Person Manual for Fall Prevention in Construction.* Morgantown, WV: West Virginia University.

Council, C. S. (1995b). *Worker Manual for Fall Prevention in Construction.* Morgantown, WV: West Virginia University.

Goldenhar, L. M., & Schulte, P. A. (1994). Intervention Research in Occupational Health and Safety. *Journal of Occupational Medicine, 36*(7), 763-775.

Kisner, S. M., & Fosbroke, D. E. (1994). Injury Hazards in the Construction Industry. *Journal of Occupational Medicine, 36*(2), 137-143.

Lundstrom, W. (1997). *A Contractor's Guide to Fall Hazard Management.* Morgantown, WV: West Virginia University.

OSHA. Fall Protection, *29 CFR 1926 Subpart M* . Washington, D.C.: USGPO.

Ringen, K., & Stafford, R. J. (1995). Intervention Research in Occupational Safety and Health: Examples from Construction. *American Journal of Industrial Medicine, 29*(4), 314-320.

Schulte, P. A., Goldenhar, L. M., & Connally, L. b. (1996). Intervention Research: Science, Skills, and Strategies. *American Journal of Industrial Medicine, 29*(4), 285-288.

Implementation of Safety and Health on Construction Sites, Singh, Hinze & Coble (eds)
© 1999 Balkema, Rotterdam, ISBN 90 5809 036 1

Author index